T0311941

DISTRIBUTED ENERGY RESOURCES IN MICROGRIDS

DISTRIBUTED ENERGY RESOURCES IN MICROGRIDS

Integration, Challenges and Optimization

Edited by

RAJEEV KUMAR CHAUHAN
Department of Electronics and Instrumentation Engineering, Galgotias College of Engineering and Technology, Greater Noida, India

KALPANA CHAUHAN
Department of Electrical and Electronics Engineering, Galgotias College of Engineering and Technology, Greater Noida, India

Academic Press is an imprint of Elsevier
125 London Wall, London EC2Y 5AS, United Kingdom
525 B Street, Suite 1650, San Diego, CA 92101, United States
50 Hampshire Street, 5th Floor, Cambridge, MA 02139, United States
The Boulevard, Langford Lane, Kidlington, Oxford OX5 1GB, United Kingdom

British Library Cataloguing-in-Publication Data
A catalogue record for this book is available from the British Library

Library of Congress Cataloging-in-Publication Data
A catalog record for this book is available from the Library of Congress

ISBN: 978-0-12-817774-7

For Information on all Academic Press publications
visit our website at https://www.elsevier.com/books-and-journals

Publisher: Brian Romer
Acquisition Editor: Lisa Reading
Editorial Project Manager: Aleksandra Packowska
Production Project Manager: Sruthi Satheesh
Cover Designer: Matthew Limbert

Typeset by MPS Limited, Chennai, India

Contents

11. Transmission system-friendly microgrids: an option to provide ancillary services

THOMAS KRECHEL, FRANCISCO SANCHEZ, FRANCISCO GONZALEZ-LONGATT, HAROLD CHAMORRO AND JOSE LUIS RUEDA

12. Energy management of various microgrid test systems using swarm evolutionary algorithms

BISHWAJIT DEY, KUMAR SHIVAM AND BIPLAB BHATTACHARYYA

13. Development of the synchronverter for green energy integration

S. KUMARAVEL, VINU THOMAS, TUMATI VIJAY KUMAR AND S. ASHOK

14. Power converter solutions and controls for green energy

VIJAY K. SOOD AND HAYTHAM ABDELGAWAD

15. Safety and reliability evaluation for electric vehicles in modern power system networks

FOAD H. GANDOMAN, JOERI VAN MIERLO, ABDOLLAH AHMADI, SHADY H.E. ABDEL ALEEM AND KALPANA CHAUHAN

16. Load forecasting using multiple linear regression with different calendars

HANNAH NANO, SAVANNA NEW, ACHRAF COHEN AND BHUVANESWARI RAMACHANDRAN

17. Unintentional islanding detection

M. SUMAN AND M. VENKATA KIRTHIGA

18. An analysis of the current- and voltage current–based characteristics' impact on relay coordination for an inverter-faced distributed generation connected network

KRUTIKA R. SOLANKI, VIPUL N. RAJPUT, KARTIK S. PANDYA AND RAJEEV KUMAR CHAUHAN

19. On the topology for a smart direct current microgrid for a cluster of zero-net energy buildings

FRANCISCO GONZALEZ-LONGATT, FRANCISCO SANCHEZ AND SRI NIWAS SINGH

20. Energy-management solutions for microgrids

MOSTAFA H. MOSTAFA, SHADY H.E. ABDEL ALEEM, SAMIA GHARIB ALI AND ALMOATAZ Y. ABDELAZIZ

List of contributors

Shady H.E. Abdel Aleem Department of Mathematical, Physical and Engineering Sciences, 15th of May Higher Institute of Engineering, Helwan, Cairo, Egypt

Almoataz Y. Abdelaziz Faculty of Engineering & Technology, Future University in Egypt, Cairo, Egypt

Haytham Abdelgawad University of Ontario Institute of Technology (UOIT), Oshawa, ON, Canada

Abdollah Ahmadi The Australian Energy Research Institute and the School of Electrical Engineering and Telecommunications, the University of New South Wales, Sydney, NSW, Australia

Samia Gharib Ali Department of Electrical Power and Machines, Kafrelsheikh University, Cairo, Egypt

S. Ashok Department of Electrical Engineering, National Institute of Calicut, Kattangal, Kerala, India

Alberto Berrueta Department of Electrical, Electronic and Communication Engineering, Institute of Smart Cities, Public University of Navarre, Pamplona, Spain

Biplab Bhattacharyya Department of Electrical Engineering, Indian Institute of Technology (Indian School of Mines), Dhanbad, India

Harold Chamorro KTH Royal Institute of Technology, Stockholm, Sweden

Kalpana Chauhan Department of Electrical and Electronics Engineering, Galgotias College of Engineering and Technology, Greater Noida, India

Rajeev Kumar Chauhan Department of Electronics and Instrumentation Engineering, Galgotias Educational Institutions, Greater Noida, India

T. Chinnadurai Department of ICE, Sri Krishna College of Technology, Coimbatore, India

B. Chitti Babu Department of Electronics Engineering, Indian Institute of Information Technology, Design and Manufacturing, Chennai, India

Achraf Cohen Department of Mathematics and Statistics, Hal Marcus College of Science and Engineering, University of West Florida, Pensacola, FL, United States

Arturo Conde Enríquez Department of Electrical Engineering, Autonomous University of Nuevo Leon, San Nicolás de los Garza, Nuevo Leon, México

Bishwajit Dey Department of Electrical Engineering, Indian Institute of Technology (Indian School of Mines), Dhanbad, India

Gianluca Fadda Department of Electrical and Electronics Engineering, University of Cagliari, Cagliari, Italy

Mauro Fadda Department of Electrical and Electronics Engineering, University of Cagliari, Cagliari, Italy

Foad H. Gandoman Research Group MOBI—Mobility, Logistics, and Automotive Technology Research Centre, Vrije Universiteit Brussel, Brussels, Belgium; Flanders Make, Heverlee, Belgium

Emilio Ghiani Department of Electrical and Electronics Engineering, University of Cagliari, Cagliari, Italy

Francisco Gonzalez-Longatt Wolfson School of Mechanical, Electrical and Manufacturing Engineering, Centre for Renewable Energy Systems Technology – CREST. Loughborough University, Loughborough, United Kingdom

M.C. Gustavo Pérez Hernández Department of Electrical Engineering, Autonomous University of Nuevo Leon, San Nicolás de los Garza, Nuevo Leon, México

Trapti Jain Discipline of Electrical Engineering, Indian Institute of Technology Indore, Indore, India

Thomas Krechel Centre for Renewable Energy Systems Technology – CREST. Loughborough University, Loughborough, United Kingdom

Deepak Kumar Electrical and Electronics Engineering Department, School of Engineering, University of Petroleum and Energy Studies, Dehradun, India

S. Kumaravel Department of Electrical Engineering, National Institute of Calicut, Kattangal, Kerala, India

Mostafa H. Mostafa Department of Electrical Power and Machines, International Academy of Engineering and Media Science, Cairo, Egypt

Hannah Nano Department of Electrical and Computer Engineering, Hal Marcus College of Science and Engineering, University of West Florida, Pensacola, FL, United States

Savanna New Department of Electrical and Computer Engineering, Hal Marcus College of Science and Engineering, University of West Florida, Pensacola, FL, United States

Rupendra Pachauri Electrical and Electronics Engineering Department, School of Engineering, University of Petroleum and Energy Studies, Dehradun, India

Alejandra Pérez Pacheco Energy Control National Center, Peninsular Control Area, Mérida, Mexico

Kartik S. Pandya Electrical Engineering Department, Charotar University of Science and Technology, Changa, India

Virginia Pilloni Department of Electrical and Electronics Engineering, University of Cagliari, Cagliari, Italy

N. Prabaharan School of Electrical and Electronics Engineering, SASTRA Deemed University, Thanjavur, India

A. Prakash Department of EEE, Sri Krishna College of Technology, Coimbatore, India

Vipul N. Rajput Electrical Engineering Department, Dr. Jivraj Mehta Institute of Technology, Mogar, India

E.S.N. Raju P Discipline of Electrical Engineering, Indian Institute of Technology Indore, Indore, India

Bhuvaneswari Ramachandran Department of Electrical and Computer Engineering, Hal Marcus College of Science and Engineering, University of West Florida, Pensacola, FL, United States

Juan M. Ramirez Research Center and Advanced Studies, National Polytechnic Institute, Zapopan, Mexico

Ajit Renjit Electric Power Research Institute (EPRI), Palo Alto, CA, United States

Jose Luis Rueda Section Intelligent Electrical Power Grids, Department of Electrical Sustainable Energy, Delft University of Technology, Stockholm, Sweden

Idoia San Martín Department of Electrical, Electronic and Communication Engineering, Institute of Smart Cities, Public University of Navarre, Pamplona, Spain

Francisco Sanchez Wolfson School of Mechanical, Electrical and Manufacturing Engineering, Centre for Renewable Energy Systems Technology – CREST. Loughborough University, Loughborough, United Kingdom

Pablo Sanchis Department of Electrical, Electronic and Communication Engineering, Institute of Smart Cities, Public University of Navarre, Pamplona, Spain

S. Saravanan Department of EEE, Sri Krishna College of Technology, Coimbatore, India

R. Senthil Kumar Department of EEE, Sri Krishna College of Technology, Coimbatore, India

Kumar Shivam Department of Electrical Engineering, Indian Institute of Technology (Indian School of Mines), Dhanbad, India

Sri Niwas Singh Department of Electrical Engineering, Indian Institute of Technology Kanpur, Kanpur, India

Krutika R. Solanki Electrical Engineering Department, Dr. Jivraj Mehta Institute of Technology, Mogar, India

Vijay K. Sood University of Ontario Institute of Technology (UOIT), Oshawa, ON, Canada

M. Suman National Institute of Technology, Tiruchirappalli, India

Vinu Thomas Department of Electrical Engineering, National Institute of Calicut, Kattangal, Kerala, India

Ramji Tiwari Department of EEE, Bharat Institute of Engineering and Technology, Hyderabad, India

Alfredo Ursúa Department of Electrical, Electronic and Communication Engineering, Institute of Smart Cities, Public University of Navarre, Pamplona, Spain

Joeri Van Mierlo Research Group MOBI—Mobility, Logistics, and Automotive Technology Research Centre, Vrije Universiteit Brussel, Brussels, Belgium; Flanders Make, Heverlee, Belgium

M. Venkata Kirthiga Department of Electrical and Electronics Engineering, National Institute of Technology, Tiruchirappalli, India

Tumati Vijay Kumar Department of Electrical Engineering, National Institute of Calicut, Kattangal, Kerala, India

Dhanashree Vyawahare Electrical Design Group, Nuclear Power Corporation of India Limited, Mumbai, India

Preface

The use of renewable energy resources requires integration of the available energy resources. There is a need to mention and consider the challenges that arise during their integration. The idea of this book originated while discussing the utilization of renewable energy resources. The book gives a technical and mathematical view of the integration of distributed energy resources (DERs) to microgrid (MG) and optimization approaches to obtain the most economical energy for customers and energy sectors. As the market demand increases in the sector of DERs, there has been an increment in the challenges of MG reliability, efficiency, stability, etc. The use of intelligent energy management systems, control strategies, power electronics devices, optimization techniques, and protection algorithms will be helpful in meeting this objective. The book provides solutions to the above problems and describes the advanced MG techniques.

Chapter 1, Microgrids architectures, related to MG architectures, giving a brief outline of common MG architectures based on AC, DC, and hybrid AC/DC buses. Furthermore, comparisons are made between different MG architectures. Positive and negative features of different architectures are given as a guide for further MG system studies. Chapter 2, Distributed energy resources and control, presents various types of DER technologies, including both distributed generation units as well as distributed energy storages and their control in three levels. In Chapter 3, Use of agents for isolated microgrids with frequency regulation, a multiagent system (MAS) is proposed for managing energy on a MG. Buses in the MG constitute different electrical zones, which are coordinated by an agent that shares information amongst the other agents. In Chapter 4, Su-Do-Ku and symmetric matrix puzzles-based optimal connections of photovoltaic modules in partially shaded total cross-tied array configuration for efficient performance, a comprehensive performance assay of the power–voltage (P–V) and current–voltage (I–V) curves of existing total cross-tied (TCT), Su-Do-Ku, and symmetric matrix (SM) puzzle-based configurations is carried out. The Su-Do-Ku–total cross-tied (Su-Do-Ku-TCT) and proposed symmetric matrix-total cross-tied (SM-TCT) configurations are considered for extensive investigation. Chapter 5, Dynamics of power flow in a stand-alone microgrid using four-leg inverters for nonlinear and unbalanced loads, gives a new idea on MGs, using four-leg inverters to handle nonlinear and unbalanced loads. Chapter 6, Lithium-ion batteries as distributed energy storage systems for microgrids, presents a comprehensive analysis of Li-ion batteries. The properties of lithium as a charge carrier in batteries are analyzed and the aging mechanisms of Li-ion batteries are discussed. In Chapter 7, Impact of dynamic performance of batteries in microgrids, the MG operation under isolated or high-impedance connection has been presented. This chapter presents the impact of battery operation dynamics in the frequency control of microgrids. Initial conditions are presented as a predictive energy dispatch to ascertain the requirements of the energy capacity in the batteries and then the location of the battery energy storage system is made in the weakest nodes to later perform charging and discharging of energy as required at each node. Chapter 8, Photovoltaic array reconfiguration to extract maximum power under partial shaded conditions, is mainly focused on various methods such as bypass diode configuration, tracking the maximum power, array configuration, and different reconfiguration strategies. These methods are used to reduce the effect of partial shaded conditions and improve the output of the PV system. In Chapter 9, Communications and internet of things for microgrids, smart buildings, and homes, approaches referring to the integration among power, control, and communication systems, as well as the related enabling models and technologies able to manage heterogeneous DERs for the purposes of enhancing the integration and coordination of small-scale power generation for self-consumption, as well as for trading energy or selling ancillary services on the open market and to distribution and transmission system operators, have been presented. Chapter 10, Communications, cybersecurity, and the internet of things for microgrids, gives the protocol briefs for communication protocols that are common between a MG controller and DERs. Communication protocols that enable utility integration of MGs are discussed. Chapter 11, Transmission system-friendly microgrids: an option to provide ancillary services, presents the concept of grid-friendly or smart-converters and their main capabilities. The functionalities are enhanced by a wide area control and the concept of a transmission system-friendly microgrid. Chapter 12, Energy management of various microgrid test systems using swarm evolutionary algorithms, deals with energy management for three MG test systems consisting of a fuel cell, microturbine, storage devices, and renewable energy sources. A proposed whale optimization algorithm is implemented as the optimal tool to minimize the MG operating cost abiding by various equality and inequality constraints and considering load uncertainties and market bids. In Chapter 13, Development of the synchronverter for green energy integration, a mathematical model of the synchronverter is given. Chapter 14, Power-converter solutions and controls for green energy, presents a comprehensive overview of power converter topologies and control structures for grid-connected PV systems. Chapter 15, Safety and reliability evaluation for electric vehicles in modern power system networks, presents the role of reliability and safety in EVs' electric components from different points of view. Furthermore, the challenges and future perspective of EVs relating to their safety and reliability have been investigated. In Chapter 16, Short-term load forecasting in a smart grid using multiple linear regression, the benchmark vanilla model based on the multiple linear regression analysis is considered. The results are compared with the commonly used Gregorian calendar. Cross-validation is used to find the mean absolute percentage error (MAPE). Chapter 17, Unintentional islanding detection, discusses the necessity of detecting the island formation, types of islanding detection techniques, merits and demerits of various islanding detection techniques, and concludes with the gist of the future scope of islanding detection. In Chapter 18, An analysis of the current- and voltage-current-based characteristics' impact on relay coordination for an inverter-faced distributed generation connected network, voltage-based modified overcurrent characteristics of the relay are utilized for the protection of inverter-faced distributed generations (IDGs) connected networks. The proposed characteristic of relay measures both voltage and current for calculating the operational time of relays. In addition, both modified and conventional overcurrent characteristics of the relay are implemented on IDG-connected 19 bus systems using the firefly algorithm. In Chapter 19, On the topology for a smart direct current microgrid for a cluster of zero-net energy buildings, intelligent DC systems for a group of zero net energy buildings and the strategy necessities to permit a negligible energy target across the stipulated time horizon are discussed, with their optimization. Chapter 20, Energy-management solutions for microgrids, gives an overall description and a comprehensive study about energy management strategies of MGs, including the objective function of MGs, all of the MG constraints, and the mode operation of MGs, in addition to showing the impact of optimal allocation of ESS on the MG operation.

Acknowledgments

We want to give huge thanks to our supporter during the editing of this book. Editing a book is harder than we thought and more rewarding than we could have ever imagined. This would have not been possible without the adjustment made by our son Shaurya Chauhan. He has cooperated a lot and gives his continuous emotional support during this journey.

We are eternally grateful to our parents, who taught us discipline, love, manners, respect, and so much more that have helped us succeed in life. They ever encouraged us to do hard work. We would like to thank to all our family members and friends to their direct and indirect support.

"Thanks, to everyone in our publishing team."

CHAPTER

1

Microgrids architectures

Vijay K. Sood and Haytham Abdelgawad

University of Ontario Institute of Technology (UOIT), Oshawa, ON, Canada

1.1 Introduction

Recently, it has been noted that the world's electricity systems are starting to decentralize, decarbonize, and democratize in many cases [1]. These features, known as the "three Ds," are driven by the need to reduce greenhouse gas (GHG) emissions to alleviate climate change, provide higher reliability and resilience for critical loads, reduce electricity costs, substitute aging grid infrastructure, and supply electricity in areas not served by existing grid infrastructure. While the compromise between the technical driving factors and the details of specific solution may differ from one place to another, microgrids came to the picture as a flexible solution for managing distributed energy resources (DERs) that can meet the varying requirements of different communities.

In recent decades, the research, development, and implementation of renewable energy sources (RESs) have been strongly propelled, mainly as distributed generation (DG). Due to variability and intermittence of RESs, their large penetration over traditional energy systems (especially wind and solar) involves operational difficulties that limit their implementation, such as variations of supplied voltage magnitude or imbalances between active and reactive power among generators. Moreover, new flow patterns may require changes to the distribution grid infrastructure with the application of enhanced distribution automation, adapted protection and control strategies, and improved voltage management techniques [2]. A possible way to conduct the emerging potential of DG is to take a systematic approach which views generation and associated loads as a subsystem, or a "microgrid" [3]. The MG can operate either connected to or separated from the distribution system, thereby maintaining a high level of service and reliability. In this sense, distributed energy storage systems (ESSs) become necessary to improve the reliability of the overall system, supporting the distributed generators' power capability in cases when they cannot supply the full power required by the consumers. MGs can also operate isolated in those areas with no access to the utility.

Distributed Energy Resources in Microgrids
DOI: https://doi.org/10.1016/B978-0-12-817774-7.00001-6

The modularity of emerging generation technologies permits generators to be placed and sized optimally to maximize the reliability, security, and economic benefits of DG deployment. For example, installing microgeneration on the customer's side provides an opportunity for the utilization, locally, of the waste heat from the conversion of primary fuel to electricity (combined production of heat and power—CHP), thus accomplishing a better overall efficiency. In recent years, there has been significant progress in developing small, kW-scale CHP applications, which are expected to play a very significant role in upcoming microgrid implementations [4].

A key feature that distinguishes microgrids from active distribution networks with DG is the implementation of the control system. MGs are considered to be the building blocks of the "smart grids," thus integrating the actions of all the DERs including distributed generators and storage devices, plus local loads and the main distribution grid. The target is to deliver sustainable, economic, and secure electricity supply through intelligent monitoring, control, communication, and self-healing technologies, with cost-competitive information and communication technologies (ICT) playing a fundamental role.

Microgrids can be considered as vital components in the smart grid environment, which is being developed to improve reliability and power quality, and to facilitate the integration of DERs. These are defined as medium or small power systems comprising DERs and controllable and uncontrollable loads; either isolated and meeting their own demand needs or connected to the external grid to supplement their supply requirements. Microgrids are connected to a distribution grid at a single point known as the point of common coupling (PCC). Fig. 1.1 shows a conceptual topology of a future smart grid connecting to a small microgrid [5].

Microgrids are considered to be efficient and resilient since they not only allow for a high penetration level of RESs, but also because of their ability to operate in isolated mode when faults occur in the main grid [6]. Consequently, microgrids will offer greater reliability, especially when integrated with smart grids, because, during outages, they continue to

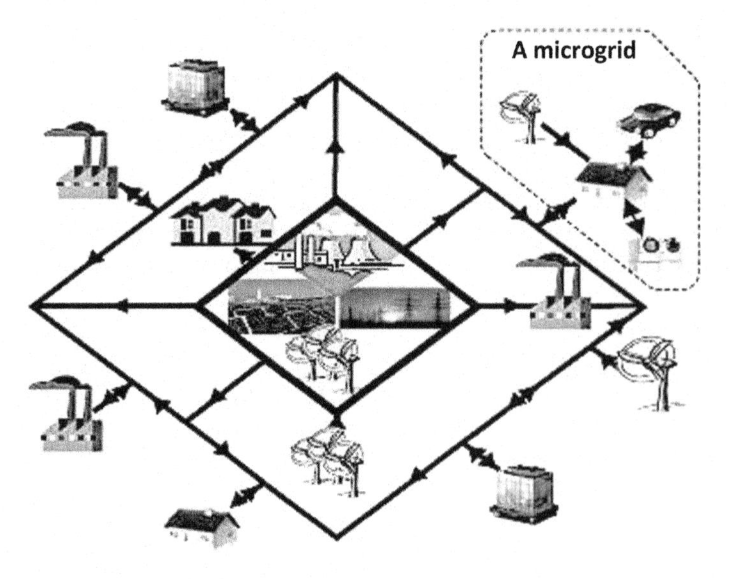

FIGURE 1.1 A conceptual topology of future smart grid [5].

operate in islanded mode or even put power back into the wider grid. Finally, microgrids can offer flexibility, because a variety of resources, including CHP and diesel back-ups can be integrated into the grid, providing reliability, using waste heat used for other purposes, and smoothing out supply/demand spikes. However, the implementation of microgrids encounters a number of challenges. For instance, maintaining demand-supply balance in the presence of RESs is a complex task because of generation intermittencies, load mismatches, and voltage instabilities [7,8].

The microgrid concept is gaining rapid acceptance because of its environmentally friendly energy provision, cost effectiveness, improvement in power quality and reliability, reduction in line congestion and losses, and reduction in infrastructure investment needs. From the customer's point of view, the microgrid is designed to meet their electrical and heat energy demands and avoid loadshedding [9].

In order to integrate DERs into a MG effectively, proper architectures should be performed, based on alternating current (AC), direct current (DC) and hybrid AC/DC systems, seeking the highest reliability and efficiency [10]. The existing grid infrastructure, the DERs to be integrated, as well as specific customer-oriented requirements will determine the most suitable electrical architecture of the MG. Since the late 19th century, AC has been the standard choice for commercial energy systems, based mainly on the ease of transforming AC voltage into different levels, the capability of transmitting power over long distances. Therefore AC distribution is the most popular and commonly used structure for MG studies and implementations. By utilizing the existing AC network infrastructure (distribution, transformers, protections, etc.), AC microgrids are easier to design and implement, and are built on proven and thus reliable technology. However, DC distribution has shown a resurgence in recent years due to the development and deployment of RES based on DC power sources, and the rapid growth of DC loads which today constitute the vast majority of loads in most power systems. DC distribution presents several advantages, such as reduction of the power losses and voltage drops, and an increase of capacity of power lines, mainly due to the lack of reactive power flows, the absence of voltage drops in lines reactance, and the nonexistence of skin and proximity effects which reduce the ohmic resistance of lines. As such, the associated planning, implementation, and operation is simpler and less expensive [11].

This chapter depicts different architectures of microgrids, such as AC, DC, and hybrid AC/DC microgrids, including a general definition of the electrical microgrid, and comparisons are made between different microgrid architectures. The pros and cons of different architectures are given to guide further microgrid system studies.

1.2 Literature review of microgrid studies

This paper references [12] a three-phase power-flow algorithm in the sequence-component frame for the microgrid and active distribution system (ADS) applications. In addition, it presents steady-state sequence-component frame models of DER units for the developed power-flow approach under balanced/unbalanced conditions and develops sequence-component models of directly coupled synchronous machine-based and electronically coupled DER units. The validity and accuracy of the power-flow algorithm of developed models are verified by comparing the power flows of two test systems with those obtained from a time-domain simulation result in the PSCAD/EMTDC.

This paper also references [13] and compares a cost-optimization scheme for a microgrid and also several schemes for sharing power between two generators. The microgrid considered in this paper consists of two reciprocating gas engines, a combined heat and power plant, a photovoltaic (PV) array, and a wind generator. A penalty is applied for any heat produced in excess of demand.

Reference [14] discussed the different interconnections of microgrid and interconnection stability. The evaluation models in the microgrid stability are tie line power flow and frequency deviation. Two research topics discussed here are:

1. Control methods in microgrid for stability and efficiency.
2. Methods of interconnections with microgrid for stability and efficiency. This paper also discusses the benefit of different forms regarding mainly frequency quality under limited tie-line power-flow capacities. Several operational forms of micro grids are expected to be developed in the next decade. For a severe fault, the operational mode can be changed in order to maintain system stability and hence improve the reliability of supply.

Reference [15] proposed the interaction model that has been established between power system and voltage source inverter based series dynamic voltage restorer. On the basis of that, the flowing of energy exchanging between the DVR (dynamic voltage restorer) and the system is discussed in depth and calculated in detail when adopting different compensation strategies. This provides a strong theoretical basis for the design of the DVR system and implementing the control unit as a compensation strategy.

Reference [16] addressed the timely issues of synchronization and application of three-phase power converters connected in parallel utilizing the Power-Hardware-in-the-Loop concept. Without proper synchronization, it is difficult to distinguish the currents circulating between the converters. The paper centers on control methodology for achieving precise phase synchronization for equal load sharing, with minimum current circulation between the paralleled power converter modules, and robust dynamic system control under different transient conditions. One of the possible applications for the configuration presented in this paper is the conceptual virtual microgrid, which utilizes the reactive power compensation ability of the Static Synchronous Compensator (STATCOM).

Reference [17] addressed the hierarchical approach to dealing with the problem of frequency control in a medium-voltage (MV) network comprising several microgrids and DG sources operated in islanded mode. Tasks related to coordinated frequency control were successfully fulfilled, either after islanding or for load-following purposes.

Reference [18] proposed an interaction problem that might be introduced as a result of various types of DGs installed in a microgrid. One of the most important features of MG is the islanding operation. MG does not include large central generators, and all distributed small generators have to share all loads existing in the MG. There is no large inertia enough to compensate the quick load changes in the MG and all distributed generators must compensate them accordingly.

Reference [19] proposed a new approach to managing distribution systems based on adaptive agents that are placed at different locations on the grid. For this purpose a decentralized power-flow method is introduced. Depending on available local information, each agent is able to calculate a system state. Local utilities now face new challenges as they have to apply decentralized EMSs that are based on a high degree of automation. Agent-based distributed power-flow calculation is applicable in distribution grids, and it meets the demands of decentralized EMSs perfectly and leads to correct results.

In Reference [20], a microgrid protection scheme that relies on optimally sizing fault current limiters (FCLs) and optimally setting directional over current relays is proposed. This approach is tested on two medium-voltage networks: a typical radial distribution system and on the institute of electrical and electronic engineers (IEEE) 30 bus looped power distribution system equipped with directly connected conventional synchronous generators (CSGs). Fault current limiters of the inductive type are located at the main interconnection point of the microgrid to the main grid. Without the fault current limiters, it was found that it is difficult to set the relays optimally to satisfy both microgrid modes of operation.

Reference [21] presented a structure of modified DC bus interconnected single-phase microgrid, as well as a topology of a power converter building block (PCBB) based on the DC bus to perform various functions in a distributed energy network. The main functions of the PCBB are: compensating for the current harmonics produced by a nonlinear load in the microgrid, mitigating the voltage sag or swell of the electrical power system (EPS) at the PCC, and facilitating the islanding and re-closure of a microgrid when a severe fault happens to the EPS. A cascaded PI controller is used to regulate the voltage sag compensation.

Reference [22] proposed a unified control strategy that enables islanded and grid-connected operations of three-phase electronically interfaced DERs, with no need for knowing the prevailing mode of operation or switching between two corresponding control architectures. This paper presents the mathematical model on which the proposed strategy is based. Furthermore, the effectiveness of the proposed strategy is demonstrated through the time-domain simulation of a two-unit test microgrid in the PSCAD/EMTDC software environment.

Reference [23] showed that the constant-power band offers a solution for this if a large band is included in renewable energy sources. More controllable units, on the other hand, have a smaller constant-power band to exploit their power flexibility. This paper proposed that the microgrid power is balanced by using a control strategy which modifies the set value of the rms microgrid voltage at the inverter ac side as a function of the dc link voltage. In case a certain voltage, which is determined by a constant-power band, is surpassed, this control strategy is combined with P/V-droop control. This droop controller changes the output power of the DG unit and its possible storage devices as a function of the grid voltage.

Reference [24] described a major research initiative by the British Columbia Institute of Technology for the construction of an intelligent microgrid at its Burnaby campus. Utility companies and researchers are working together to develop architected protocols and models of the evolving intelligent grid.

In Ref. [25], real and reactive power management strategies of electronically interfaced DG units in the context of a multiple-DG microgrid system are addressed in this paper. Based on the reactive power controls adopted, three power management strategies (PMSs) are identified and investigated. These strategies are based on voltage-droop characteristic, voltage regulation, and load reactive power compensation. A systematic approach to develop a small-signal dynamic model of a multiple-DG microgrid, including real and reactive PMSs, is also presented. This paper introduces three PMSs for an autonomous microgrid system.

Reference [26] proposed a small grid with high proportions of nonlinear and unbalanced loads. It is important to control the waveform quality actively in terms of harmonics, transient disturbances, and balance.

Reference [27] introduced the MG planning structure in Shandong Electric Power Research Institute (SEPRI) and discussed the various feasible control approaches used for MG. This was based on modeling different types of DGs and energy storage equipment. MG can maintain stable voltage and frequency stability if different microsources are chosen reasonably and through the design of the controllers. Storage devices which use power droop control play an important role in MG control.

Reference [28] presented the intermittent nature of RESs like PV demands usage of storage with high-energy density. This paper proposed a composite energy storage system (CESS) containing both a high-energy density storage battery and a high-power density storage ultra-capacitor to meet the aforementioned requirements. The proposed power converter configuration and the energy management scheme can actively distribute the power demand among the different energy storages.

Reference [29] provided background on various AC motor types, and the control and protection practices. This paper also discussed various electric motors types and different types of motor starting methods. They are across the line, reduced voltage auto transformer, reduced voltage resistor or reactor, why-delta, part-winding, solid state soft starter, rotor resistance, and adjustable speed drive. It proposed motor protection in the case of abnormal conditions. These protection methods included nameplate values, low voltage motor protection, and mMV motor protection.

Reference [30] proposed a system to limit the flow of large line currents and, hence, protect the microgrids. This paper also proposed two current-limiting algorithms, namely, the RL feed forward and flux-charge-model feedback for controlling a series inverter connected between the microgrids and the utility during utility voltage sags. This paper presented the RL algorithm functions by controlling a series inverter, connected between the micro and utility grids, to insert large virtual RL impedance along the distribution feeder to limit the line currents and damp transient oscillation with a finite amount of active power circulating through the series inverter.

Reference [31] presented advanced sensors and measurement which will be used to achieve higher degrees of network automation and better system control, while pervasive communications will allow networks to be reconfigured by intelligent systems. This report has sought to identify the major research challenges facing electrical power networks and the strategic urgency with which the redesign of key infrastructure must be undertaken. It also serve to create a new electricity system for the next century and generate major economic advantages in providing technical solutions and designs for an international market that is currently facing a similar electrical networks crossroads.

Reference [32] proposed a novel integration of Doubly Fed Induction Generator (DFIG)-based wind farms within microgrids. The voltage and frequency of the microgrid are controlled by the wind generators through droop characteristics. This paper focuses on microgrid application, however, the method is also applicable for AC grid and High Voltage Direct Current (HVDC) link connections. In a power system with a high wind energy penetration, not tracking maximum wind power is not necessarily a disadvantage.

Reference [33] proposed the concept of microgrid village design with DERs. The economic feasibility evaluation of the engineering flow is comprised of four parts: natural environment and demand analysis, selection of applicable DERs and electrical network design, power network analysis, and its economic evaluation. The effectiveness of this suggested scheme has been demonstrated by case studies on realistic demonstration sites.

Reference [34] proved that storage systems used in electrical power microgrids can perform, simultaneously, the task of active power balancing and voltage regulation. In a grid-connected mode of operation, it may ensure load leveling and reducing the power exchange with the network which makes the system operation more efficient and flexible.

Reference [35] presented the results of a research project which is intended to extend the obtained capabilities of a resilient microgrid to a conventional distribution network. The presented work is an advanced interface control system for grid connection of this cluster in order to provide the maximum power in compliance with the distribution network. As a result, the distribution system operator (DSO) considers this locally controlled cluster as a single producer. In addition, an advanced interface control system for a microgrid is presented in this paper. The droop controller for the primary frequency control and power exchange with the distribution network is detailed.

Reference [36] addressed the problem of voltage regulation in MGs that include DFIG-based wind generation. This paper proposed a voltage sensitivity analysis-based scheme to achieve voltage regulation at a target bus in such MGs. The proposed method is local and obviates the need for remote measurements. The proposed methodology can be potentially useful to reactive power management and voltage regulation in MGs.

1.3 Microgrid motivation

When Hurricane Sandy cut off power to millions of homes and businesses in the Northeast, a few areas, mostly parts of universities, kept the lights on using their own power generation systems. This ability to sustain electricity service during widespread natural disasters is one reason for the growing interest in MGs. But they offer other important benefits as well. By increasing efficiency, integrating renewables, and helping manage energy supply and demand, MGs can reduce GHG emissions and save energy. For utilities, MGs can ensure power resiliency and reliability for critical loads such as health care, food and drinking water facilities, transportation, and communications facilities. MGs can also provide electricity in remote areas unserved by existing grid infrastructure, and appeal to parties looking for self-sufficiency and independence.

The formula is simple: If you can find out what is valuable to someone, then you have the key to motivating them. As a result, energy users who value reliability are the prime candidates for microgrids. Hit hard by a string of storms over the past 5 years, states on the east coast of the United States, including Connecticut and New York, have incentive programs to deploy microgrids to improve grid reliability. California recently put in place an energy storage program that could create financial incentives for microgrids. Island nations are also projected to be early customers of large-scale microgrids because they typically generate electricity with diesel generators, which is expensive. Navigant Research forecasts that the United States will fuel most growth in microgrids this decade, while Europe, which has a more reliable grid, will lag. It predicts growth from 685 MW last year to 5 GW by 2020.

1.4 Microgrid projects across the globe

In recent years, microgrids, being the core element of the smart grid evolution, have been a hot topic. According to a new report from Navigant Research, as of the fourth quarter of 2017, the research group's Microgrid Deployment Tracker (MDT) had identified 1869 projects—representing a total capacity of 20.7 GW—operating, under development, or proposed across 123 countries worldwide. That compares to 18 GW of microgrid capacity identified in the second quarter of 2016.

Across the globe, several microgrid projects have been implemented successfully in the advanced countries, while a few are in progress in developing countries too. The process is expected to move swiftly so that the smart grid can become a reality. There are a number of active microgrid projects around the world, namely in North America, the European Union, and South and East Asia, which are resulting in the improved operation of microgrids, and testing and evaluating microgrid demonstration systems. Globally, North America remains the foremost region for microgrid deployments, representing 69% of total installed capacity. Asia Pacific and Europe are also promising regions, sharing 19% and 12% respectively of total capacity [37]. Insistent smart grid deployment, emission reduction, and renewable source targets are fueling the demand for these technologies in Europe and North America.

Navigant Research noted that the increase in microgrid capacity was boosted by 60 new projects. These were mostly small installations, with the exception of a 2.2 GW microgrid installed at the Saudi Aramco gas-oil separation plant in Shaybah, Saudi Arabia, by Schweitzer Engineering Laboratories. That project—which is technically eight interconnected microgrids—is the largest single entry in Navigant's Tracker. Asian Pacific nations led all regions in microgrid deployments, with 8.4 GW of total capacity, though North America kept pace, with a total 6.97 GW. Navigant Research mentioned that while Asia Pacific and North America still account for more than three-quarters of all microgrid capacity in the Tracker, a major shift took place recently with the Middle East and Africa jumping to third place among all regions with more than 3 GW of total capacity. Europe also showed an unusual drop in capacity, with a total of 1.8 GW, due to a number of project updates where totals were adjusted downward to correctly represent their current status.

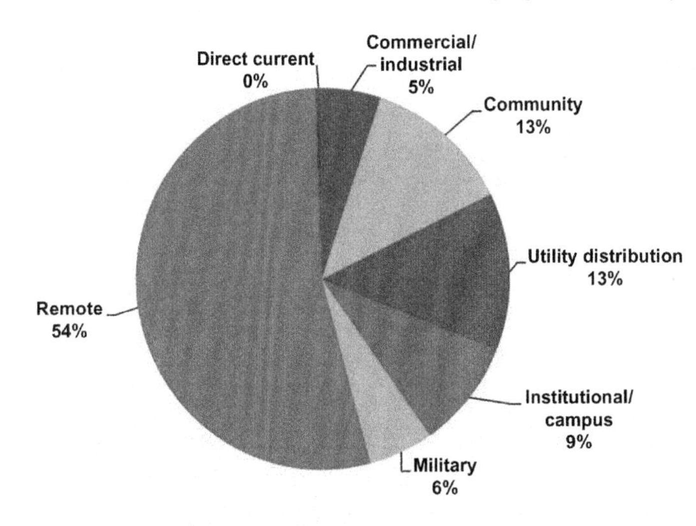

FIGURE 1.2 Total microgrid power capacity market share by segment, World Markets: 1Q 2016 [38].

Consequently, the Navigant Research group mentioned that while it has tracked the microgrid market since 2009, the microgrid deployment process is still rapidly taking hold worldwide. Navigant also noted that the microgrid market breaks into the following segments (with % of total power capacity as of the first quarter of 2016): remote (54%), utility distribution (13%), community (13%), institutional/campus (9%), military (6%) and commercial/industrial (5%) as shown in Fig. 1.2 [38].

1.4.1 Microgrid implementation in Ohio, United States

The pioneer and most well-known US microgrid research, development, and demonstration (RD&D) effort established in 1999 has been pursued under the Consortium for Electric Reliability Technology Solutions (CERTS). The CERTS microgrid was constructed at a site owned and operated by American Electric Power (AEP) near Columbus, Ohio. This test-bed as shown in Fig. 1.3 is a 480 Volt system, connected to the 13.8 kV distribution-voltage system through a transformer at the PCC. It has two main autonomous components, DG sources, and thyristor-based static switches [39].

The test-bed has three feeders, two of which have converter-based DG units driven by natural gas. One feeder has two 60 kW sources and another feeder has one 60 kW source which can be connected or islanded. The third feeder stays connected to the utility but can receive power from the microsources when the static switch is closed without injecting power into the utility. Two feeder branch circuits consist of sensitive loads and the third branch circuit has traditional nonsensitive loads. The objective of the test-bed is to demonstrate the system dynamics of each component of the CERTS microgrid.

1.4.2 Microgrid implementation in Canada Boston Bar—BC Hydro

The microgrid RD&D activities in Canada are focused on the development of control and protection strategies for autonomous microgrid operation at MVs. Two of the major

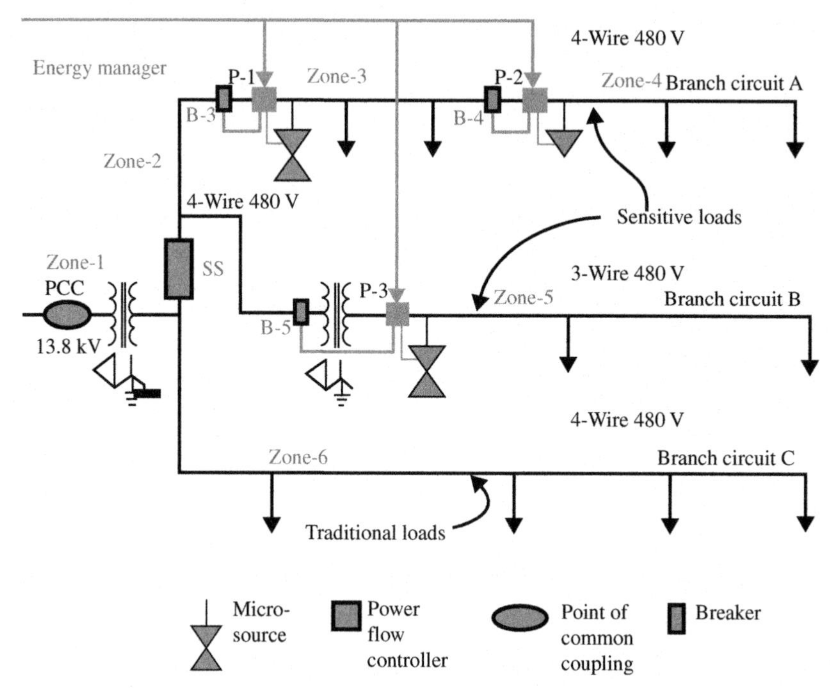

FIGURE 1.3 CERTS microgrid test-bed [39].

utility companies in Canada, BC Hydro and Hydro Quebec have implemented microgrid intentional islanding applications. The main objective of planned islanding projects is to reduce sustained power-outage durations and to enhance customer-based power supply reliability on rural feeders by utilizing an appropriately located independent power producer (IPP). Boston Bar town, a part of the BC Hydro rural area, is supplied by three 25 kV medium-voltage distribution feeders connected to the BC Hydro high voltage system through 60 km of 69 kV line. The Boston Bar IPP microgrid comprises of two 4.32 MVA run-of-river hydro power generators and is connected to one of the three feeders with a peak load of 3.0 MW, as shown in Fig. 1.4 [40].

Remote auto-synchronization capability was also added at the substation level to synchronize and connect the island area to the 69 kV feeder without causing load interruption. When a sustained power outage event, such as a permanent fault or line breakdown, occurs on the utility side of the substation, the main circuit breaker and feeder reclosers are opened. Then the substation breaker open position is telemetered via a leased telephone line used for communication between the generators' remote control site and the utility area control centre (UAAC) to the IPP operator [41].

1.4.3 Microgrid implementation in Greece

Within the frame of the European microgrids projects, several set-ups have been installed at different laboratories. The first European Union project funded by the

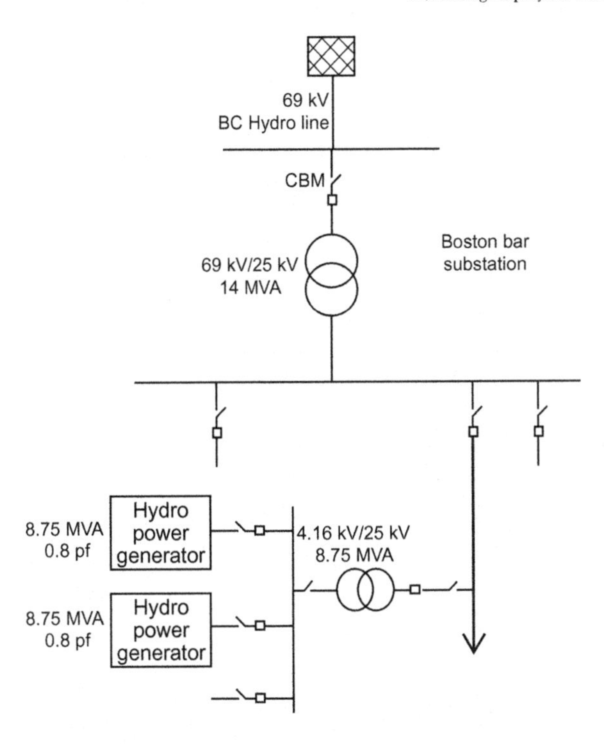

FIGURE 1.4 System configuration for the Boston Bar IPP and BC Hydro planned islanding site [40].

European Union was the Microgrids Project and it was undertaken by a consortium led by the National Technical University of Athens (NTUA). The microgrids project investigated a microgrid central controller (MCC) that promotes technical and economical operation, interfaces with loads and microsources and demand-side management, and provides set points or supervises local control to interruptible loads and microsources [42]. A pilot installation was installed in Kythnos Island, Greece, that evaluated a variety of DER to create a microgrid. Another project at NTUA is a laboratory-scale microgrid implementation as detailed further.

1.4.4 Microgrid implementation in Kythnos Island, Greece

There is one field demonstration of standalone DG in Greece, which possesses some features of a microgrid and it would be connected to the main grid in the near future. The system provides electricity to 12 houses in a small valley in Kythnos, an island in the cluster of Cyclades situated in the Aegean Sea. Fig. 1.5 shows the mini-grid layout for the settlement that is situated about 4 km from the MV grid line of the island [43]. The Kythnos mini-grid consists of a 10 kW solar PV capacity distributed in five smaller sub systems, a battery bank of 53 kWh capacity, a diesel generator of 5 kVA nominal output power, and three 4.5 kVA battery inverters to form the grid. Each house has an energy meter with 6 A fuse, as per the norms of the local electric utility public power corporation. Specially developed load controllers are installed at every house.

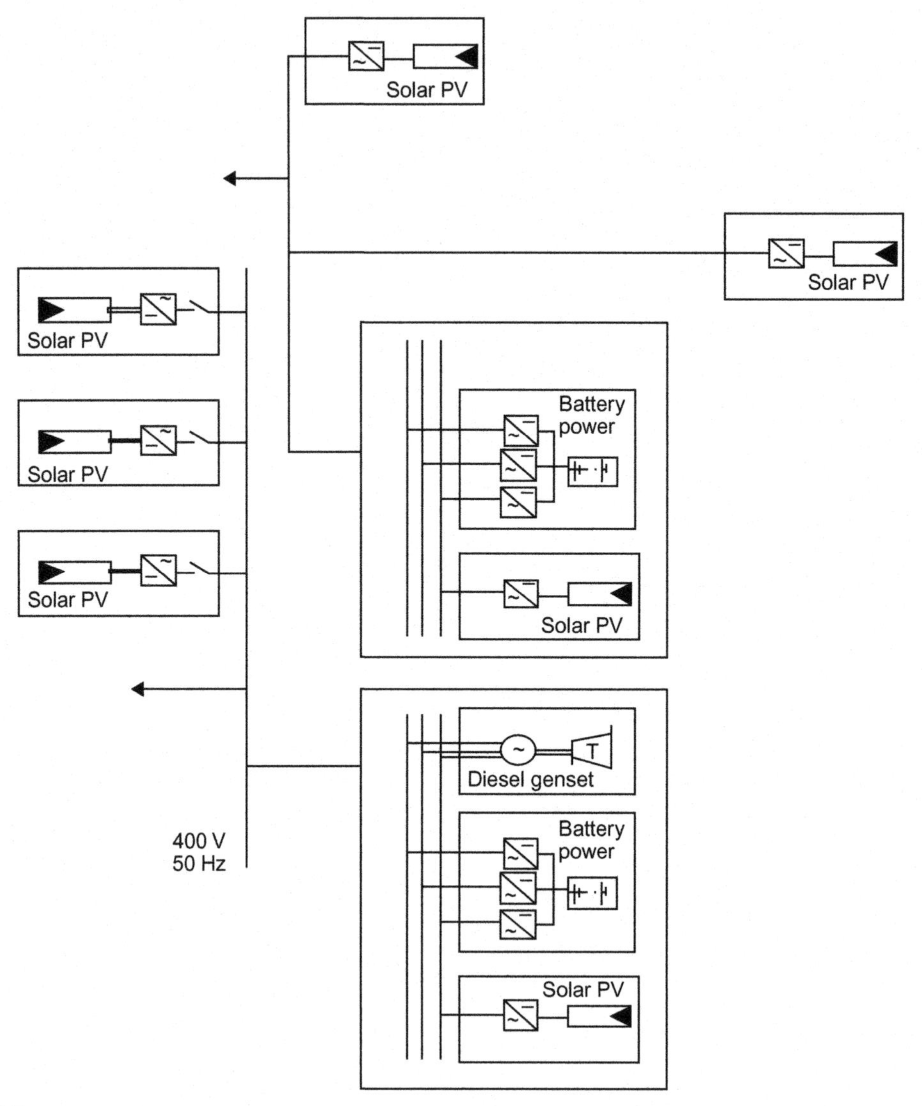

FIGURE 1.5 The Kythnos Island Microgrid—Greece [43].

1.4.5 Laboratory-scale microgrid implementation at NTUA—Greece

At the international level, the European Union has supported second major research efforts devoted exclusively to microgrids named "More Microgrid Projects." This project was executed to study alternative methods and strategies along with universalization and plug-and-play concepts. The demonstration site is an ecological estate in Mannheim-Wallstadt, Germany [42]. Continuing microgrid projects in Greece include a laboratory facility that has been set up at the National Technical University of Athens (NTUA), a

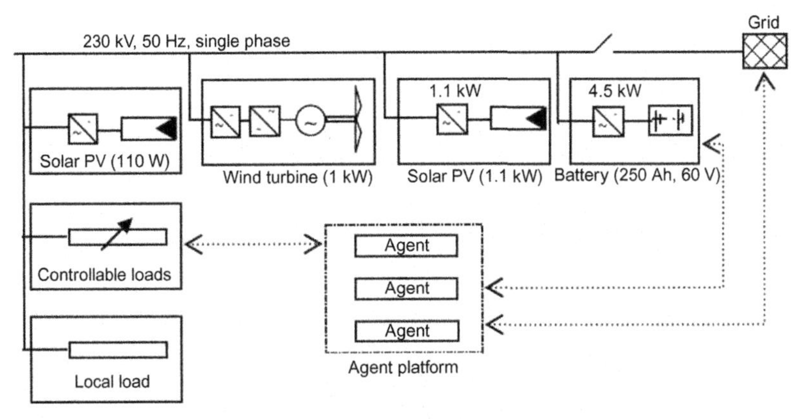

FIGURE 1.6 The Laboratory Microgrid Facility at NTUA [43].

specially designed single-phase system of the NTUA with agent control software with the objective to test small-scale equipment and control strategies for microgrid operation. NTUA microgrid test system shown in Fig. 1.6 consist of one wind turbine, two PV generators, battery storage of energy, and controllable loads. The battery is connected via a bidirectional Pulse Width Modulation (PWM) voltage source converter, which regulates the voltage and frequency when the system operates in the island mode. The battery inverter operates in voltage control mode (regulating the magnitude and phase/frequency of its output voltage) [43]. When the microgrid operates in parallel to the grid, the inverter follows the grid. Multiagent technology built on the Java Agent Development Framework (JADE) 3.0 platform has been implemented for the operation and control planning of the sources and the loads. The project has been completed successfully, providing several innovative technical solutions.

1.4.6 Microgrid implementation in Bronsbergen Holiday Park, Netherlands

In the Netherlands, one of the More Microgrids Projects is located at Bronsbergen Holiday Park, located near Zutphen. The park is electrified by a traditional three-phase 400 V network, which is connected to a 10 kV medium-voltage network via a distribution transformer located on the premises as shown in Fig. 1.7. The distribution transformer does not feed any low-voltage loads outside of the holiday park. Internally in the park, the 400 V supply from the distribution transformer is distributed over four cables, each protected by 200 A fuses on the three phases. It comprises 210 cottages, 108 of which are equipped with grid-connected PV systems of 315 kW, catering for a peak load of about 90 kW. The objective of this project is experimental validation of islanded microgrids by means of smart storage coupled by a flexible AC distribution system including evaluation of islanded operation, automatic isolation and reconnection, fault level of the microgrid, harmonic voltage distortion, energy management and lifetime optimization of the storage system, and parallel operation of converters [44].

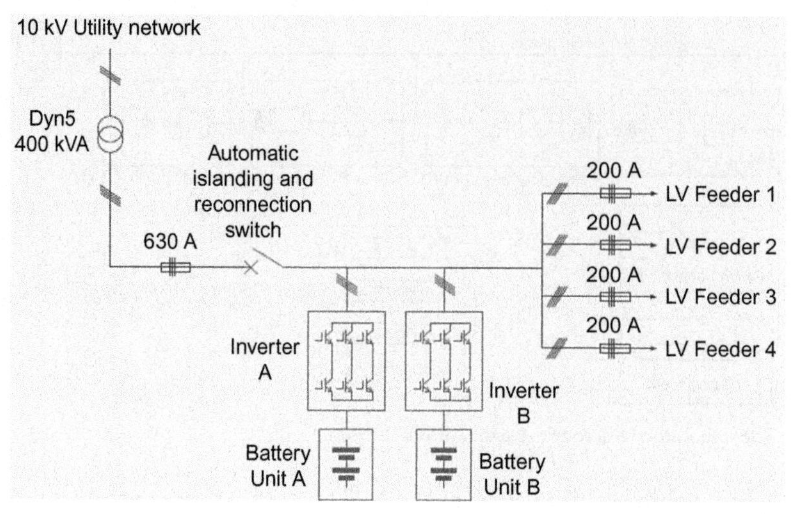

FIGURE 1.7 Schematic for the Bronsbergen Holiday Park Microgrid—Netherlands [44].

1.4.7 Microgrid implementation in DeMoTec—Germany

A comprehensive study on microgrid control methods has been performed at the Institut für Solare Energieversorgungstechnik (ISET), Germany. The design centre for modular systems technology (DeMoTec) microgrid at ISET, which is a general test site for DER has a total available generation capacity of 200 kW comprising a PV generator, a wind generator, two battery units, and two diesel generators. A number of loads with different priority levels and several automatic switches are there to divide the microgrid into three low-voltage islanded grids. A central crossbar switch cabinet connects all generators and loads to a local grid. Fig. 1.8 presents the diagram of the DeMoTec test microgrid.

To enable monitoring and control of the generators and of the operating states of the system a Supervisory Control and Data Acquisition System (SCADA) is deployed. The communication is done via a separate ethernet communication line and an XML-RPC communication protocol is used [45]. The DeMoTec promotes design, development and presentation of systems for the utilization of renewable energies and the rational use of energy. A DeMoTec master display is used to monitor the operations of a widely dispersed wind power plant system comprising about 80 representatively selected systems throughout Germany. Through this master display, isolated systems in Greece and Spain as well as the control of active low-voltage grids can be remotely monitored [46].

1.4.8 Residential microgrid implementation in Am Steinweg, Stutensee—Germany

One of the first pilot projects on microgrids with renewable energy in a residential neighborhood was carried out in the neighborhood Am Steinweg, Stutensee, a German village located about 15 minutes north of the Karlsruhe.

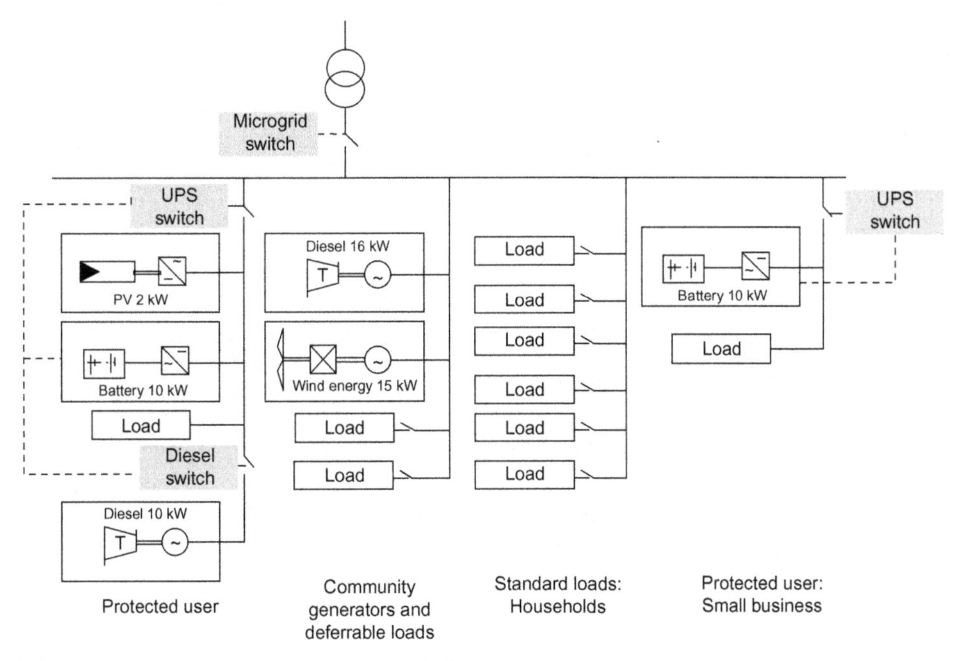

FIGURE 1.8 DeMoTec Microgrid Test System [46].

This is a three-phase low voltage grid with a neutral conductor, which is linked in one place by a 400 kVA transformer to the medium-voltage grid and has a circular structure. The energy sources in this microgrid are a CHP (with an optional electrical power of 28 kW), different PV installations (with a nominal power of 35 kWp) and a lead-free battery bank (880 Ah) with a bidirectional inverter. This inverter is designed for a power exchange of 100 kW. The batteries can deliver this power for half an hour. In total, 101 apartments are linked to this grid. The maximum active power through the transformer is determined as 150 kW. Fig. 1.9 shows the structure of the microgrid, with the division of loads and sources [43].

1.4.9 Microgrid implementation in Cesi Ricerca—Italy

The Cesi Ricerca DER test microgrid (Italy) is a low voltage (LV) microgrid, connected to the MV grid by means of a 800 kVA transformer. It comprises several generators with different technologies (renewable and conventional), controllable loads, and storage systems. This microgrid can provide electricity to the main grid with a maximum power of 350 kW. Fig. 1.10 presents the network configuration of Cesi Ricerca DER test microgrid [43].

It comprises the following DERs and storage systems:

- A hybrid energy system consisting of a PV plant (10 kW), a lead-acid storage, a diesel engine coupled with an asynchronous generator (7 kVA), and a simulated (8 kVA) asynchronous wind generator
- Five PV fields of different technologies for a total nominal power of 14 kW

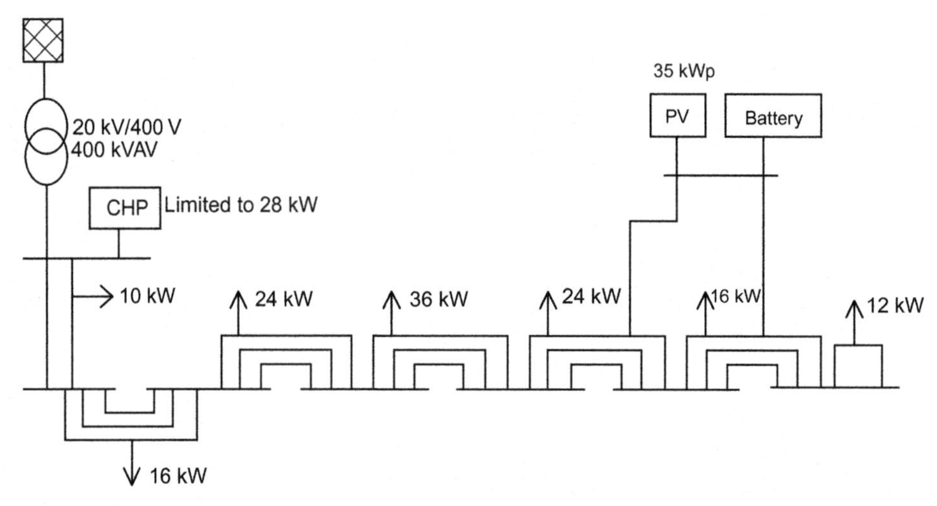

FIGURE 1.9 The residential microgrid of Am Steinweg in Stutensee—Germany [43].

- A solar thermal plant with a parabolic dish and a Stirling engine (10 kW)
- An ORC CHP system fueled by biomass (10, 90 kW)
- A CCHP plant with a gas microturbine (105, 170, 100 kW)
- A vanadium redox battery (42 kW, 2 hours)
- A lead acid battery system (100 kW, 1 hour)
- Two high-temperature Zebra batteries (64 kW, 30 minutes)
- A high-speed flywheel for power quality (100 kW, 30 seconds)
- A controllable three-phase resistive-inductive load (100 kW + 70 kVAR)
- A capacitive load and several R/L loads with local control (150 kVAR)

1.4.10 Microgrid implementation at the University of Manchester, United Kingdom

The hardware topology used in the University of Manchester microgrid/flywheel energy storage laboratory prototype is shown in Fig. 1.11. The overall microgrid test system is nominally rated at a 20 kVA, although the flywheel and power electronics are rated much higher at 100 kW connected with 0.4 kV mains supply of the laboratory, which is considered as the main grid. A synchronous generator coupled with an induction motor acts as the micro-source. A three-phase balanced load of 12 kW is connected at the end of the feeder. Control systems for real-time control of the microgrid hardware, using the Simulink/dSPACE control environment, was developed in 2005. The test-rig was designed to allow the investigation of power electronic interfaces for generation, loads, or energy storage. The AC/DC inverter, labeled 'flywheel inverter', can be configured in software to allow the interfacing of the flywheel storage system to the remaining microgrid unit. The microgrid may be operated in an islanded operation.

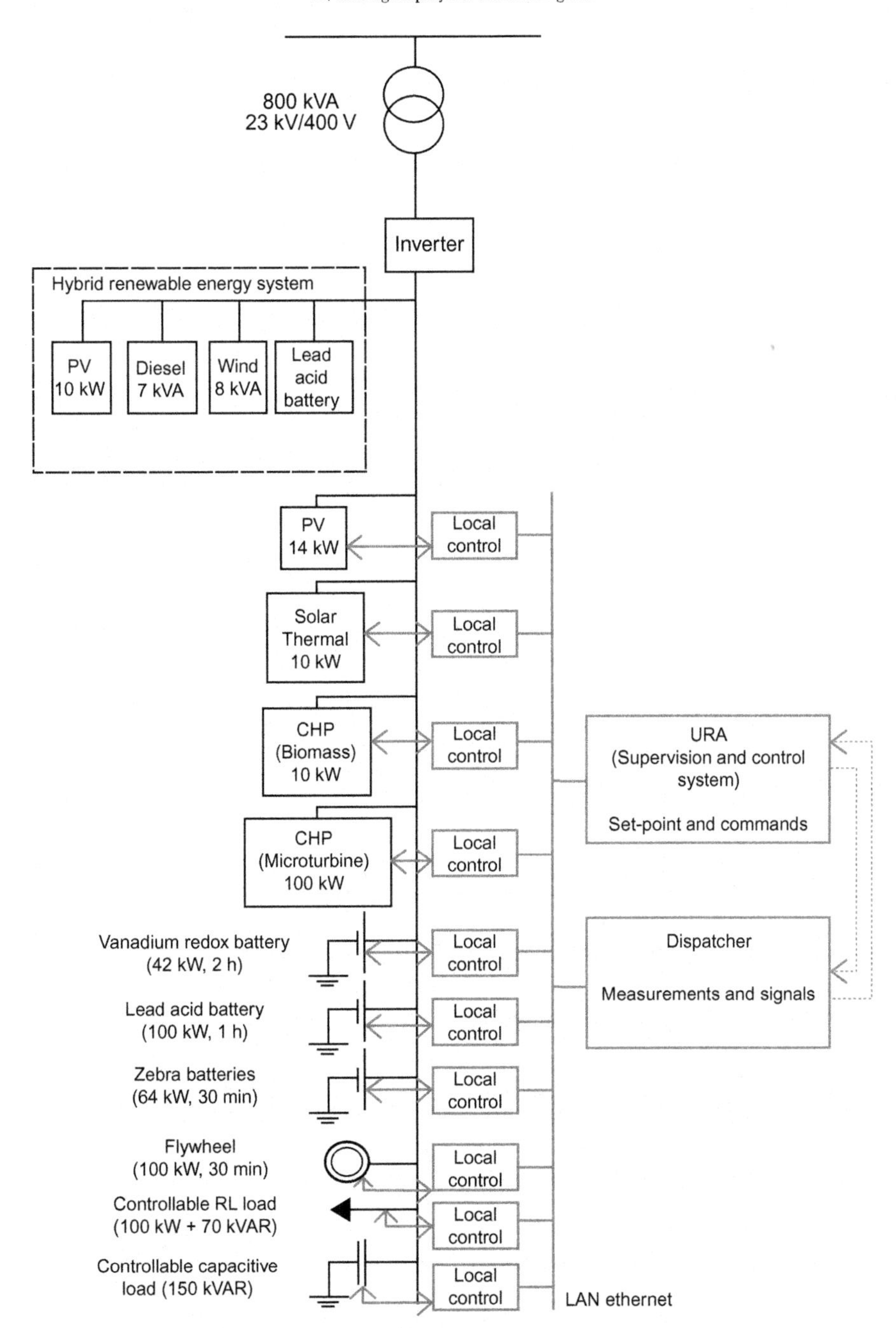

FIGURE 1.10 Cesi Ricerca, Italy Test Microgrid Network Configuration [43].

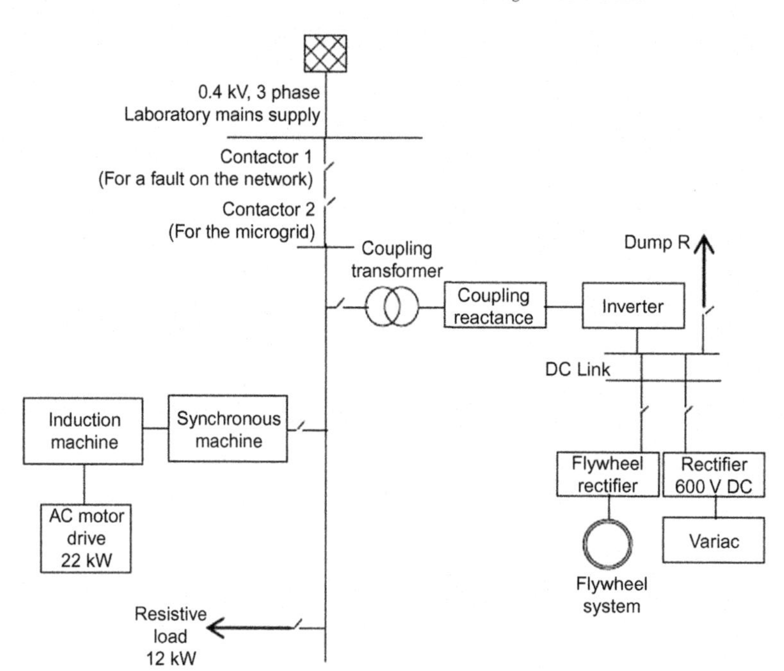

FIGURE 1.11 University of Manchester microgrid/flywheel energy storage laboratory prototype [43].

Breaker 1 is opened to emulate a loss of mains condition as shown in the Fig. 1.11. Breaker 2 is under the control of the microgrid controller system. Thus the onset of islanding, and resynchronization to mains, as well as mains-connected operation may be tested [43].

1.5 Description of a microgrid

Around the world, governments and industries are moving towards using cleaner energy sources and reducing environmental pollution. This has led to an increase in attention of DG using nonconventional and RESs, which are connected locally at the distribution system level. However, adverse impact on the grid structure and its operation, with increased penetration of DGs, is unavoidable. To reduce the impact of DGs and make conventional grids more suitable for their large-scale deployment, the concept of microgrids is proposed [44].

1.5.1 Definition of a microgrid

A microgrid is an interconnection of DGs, either a set of dispatchable generating units such as gas turbines and fuel cells or nondispatchable generators such as, wind turbines and solar PV units, integrated with electrical and thermal energy storage devices to meet customers' local energy needs, operating as a single system and small-scale, on low-voltage distribution systems, providing both power and heat. To ensure that the microgrid is operated as a single, aggregated system and meets power quality, reliability, and

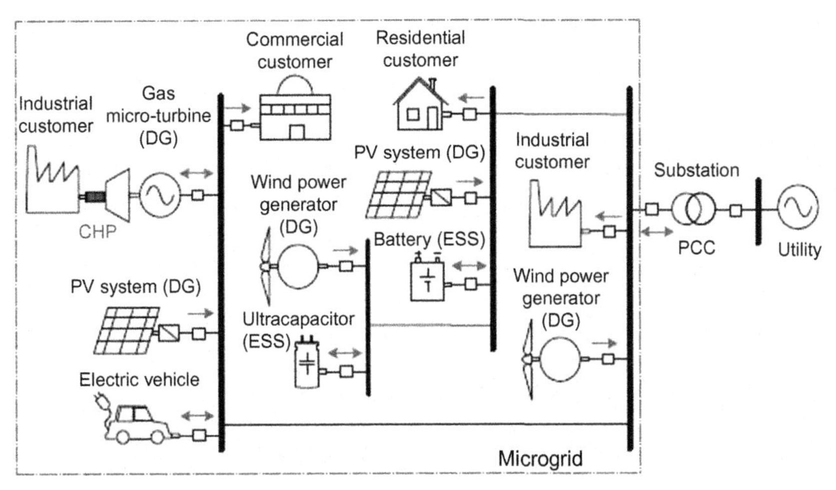

FIGURE 1.12 A typical microgrid structure integrating DERs and loads [4].

security standards, power electronic interfaces and controls need to be applied [9,47]. This control flexibility allows the microgrid to present itself to the main utility power system as a single controlled unit.

The MG serves a variety of customers, for example, residential buildings, commercial entities, and industrial parks, with the particularity that some loads (or all of them) may be controllable. A typical microgrid structure, including loads and DER units, is depicted in Fig. 1.12.

An MG is generally located downstream of the distribution substation. The electrical connection point of the MG to the utility grid constitutes the PCC. The internal coordination of groups of final consumers and DERs confined within the MG could be done independently from the distribution grid, so that the DSO need only consider the power exchange with the MG at the PCC. This allows the number of nodes under control by the DSO to be greatly reduced, while decision-making and top-level communication infrastructure is also simplified. The operation of MGs responds to multiple economic, technical, and environmental aims, by implementing smart energy management systems (EMS) and novel operation techniques based on ICTs.

The microgrid normally operates in a grid-connected mode through the substation transformer. However, it is also expected to provide sufficient generation and storage capacity, controls, and operational strategies to supply at least a portion of the load after being disconnected from the distribution system at the PCC and remain operational (after a smooth transition) as an autonomous (islanded) entity. This islanding capability reduces outages and allows service and power quality to be improved, therefore providing high reliability [48].

From the utility point of view, the application of microsources can potentially reduce the demand for distribution and transmission facilities. Clearly, DG located close to loads within the MGs can reduce power flows in transmission and distribution circuits with the consequent reduction of loss, also providing support to the utility network in times of stress by relieving congestion and aiding restoration after fault clearance [4].

To summarize, the basic concept of an MG is to aggregate and integrate DERs ideally near the end-users (loads) in order to provide them with an EPS characterized by the following functional and operational conditions:

- Efficient power production to meet the consumer's electricity demand.
- Energy management from supply and demand-side to achieve basic operation requirements such as power balance, voltage control, power quality, flexibility, and electrical safety, through the implementation of proper control techniques.
- Plug and play functionality, constituting a flexible system where new loads, DERs, and other devices can be implemented simply.
- Islanding capability, ensuring a safe, reliable, and prime quality energy supply.

1.5.2 Distributed generators

According to the Institute of Electrical and Electronics Engineers (IEEE), a DG is the generation of electricity by facilities sufficiently smaller than central generating plants as to allow interconnection at nearly any point in a power system. Generally, the power of the generator must be smaller than a few megawatts (10–50 MW) to be considered as a DG [10].

When it comes to MGs, usually smaller-scale DG systems are integrated, with installed capacities of tens to hundreds of kilowatts, such as microturbines, PV arrays, fuel cells, and wind turbines, among others. These units have emerged as a promising option to meet growing customer needs for electrical power, providing different economic, environmental, and technical benefits. The presence of generation close to demand potentially reduces losses and increases reliability. The electronic interfaces used by most DG systems allow power quality to be maintained and the overall efficiency to be improved by the application of control techniques based on EMSs. However, intermittent DERs, such as wind power generators and PV systems, may also introduce power quality issues which require the implementation of ESS. The economic advantages of DG respond to the lower capital investment for the construction of power systems, since distribution of generation units eliminates the extensive transmission systems. Finally, DG power systems provide environmental benefits as a result of offering a more efficient way of generating and distributing electricity, hence emphasizing RES deployment [4].

1.5.3 Energy storage devices

During islanded/isolated operation of microgrids, power generated from DG units cannot be instantly matched to load demands. A fundamental aspect of MGs' island/isolated mode of operation is the inclusion of ESS, which allows for the compensation of imbalances between generation and load to ensure quality of supply. ESS must be able to provide the amount of power required to balance the system disturbances and/or significant load changes. ESS implementation also mitigates the intermittency of RES, such as solar and wind energy, and it allows these distributed generators (together with the ESS) to operate as dispatchable units in order to provide additional power on request. In addition, ESS provides the energy requirements for a seamless transition between grid-connected and islanded autonomous operation modes. During grid-connected mode, ESS conveniently allows one to establish the optimal periods to interchange power with the distribution grid. Therefore the power and energy requirements of MGs are supported by ESS, thus raising the overall performance [49].

The storage options include batteries, flywheels, super (or ultra) capacitors, superconducting magnetic energy storage (SMES), compressed air, pumped hydroelectric and thermal energy storage. Electric vehicles (EV) are also seen as an alternative mobile option of energy storage as energy can be stored in their batteries when the demand and cost of electricity is low, and accessed later during high-demand periods [50].

1.5.4 Microgrid loads

A microgrid can serve electrical and/or thermal loads, although only electrical loads will be treated in this section. Electrical loads can be classified as critical and noncritical (or nonsensitive). Critical loads are those in which the electrical power must be maintained with a high quality and reliability, and thus cannot be interrupted. On the other hand, noncritical refers to those loads that can be disconnected from the energy supply for determined periods of time in order to maintain the microgrid operating conditions. The load management behavior of the microgrid mainly depends on the mode of operation and system requirements/market incentives, applying different strategies whether the MG operates in grid-connected or islanded mode.

In a grid-connected mode, the utility distribution system can compensate for any power discrepancy between demand/generated power within the microgrid to maintain the net power balance. Any difference could be also mitigated by load or generation shedding, when operational strategies or contractual obligations impose hard limits to the net import/export of power. Loadshedding is often required to maintain the power balance and consequently stabilize the microgrid voltage amplitude/angle during an autonomous mode of operation. In this sense, critical loads receive service priority through the implementation of a convenient operation strategy. Load control can also be executed to reduce the peak load and smooth out load profiles in order to optimize the performance of ESS units and dispatchable DG units [48].

1.5.5 Microgrid control

Advanced control strategies are vital for the realization of microgrids. In this sense, two different control approaches can be identified: centralized and decentralized. The implementation of a fully centralized control architecture becomes infeasible due to extensive communication and computation needs for the large number of controllable resources and rigorous performance and reliability requirements of MGs. On the other hand, a fully decentralized approach is not viable due to the strong coupling between units in the MG, which demands a minimum coordination and is impracticable when using only local variables. A compromise solution can be developed by applying a hierarchical control structure, consisting of three levels: primary, secondary, and tertiary [51].

The primary control deals with the inner control of the DG units, in order to keep the system stable using safe and fast control algorithms. Based on local measurements, primary control executes power sharing (and balance), islanding detection, and output control without using any critical communication link, which accomplishes the plug-and-play capability. In synchronous generators, the voltage regulator, governor, and the machine inertia itself perform power sharing and output control. Voltage source inverters (VSIs) used in electronic interfaces of DG require controllers to emulate the behavior of synchronous generators, which are composed of two stages: a DG power-sharing controller and inverter output controller. The latter is generally accomplished by an outer loop for voltage control and an inner loop for current settlement. Power sharing is performed by active power-frequency and reactive power-voltage droop controls simulating droop characteristics of synchronous machines [51].

The existence of line impedances drive load fluctuations to deteriorate the regulation of the microgrid voltage and frequency under droop control. Secondary control ensures that frequency and voltage deviations are restored towards zero after every change of generation or load inside the microgrid. Also referred to as the microgrid EMS, the secondary control is responsible for the secure, reliable, and economical operation of MGs in both grid-connected and islanded mode. This control level can be implemented in two different ways: centralized and decentralized. In general, centralized approaches are more suitable for isolated MGs with critical demand-supply balances and a fixed infrastructure, while decentralized approaches are associated with grid-connected MGs with multiple owners and a variable number of DER units. Secondary control operates on a slower time frame compared to primary control, determining the optimal (or near optimal) dispatch and unit commitment of DER units by three main methods: real-time optimization, expert systems, and decentralized hierarchical control [51].

Finally, the tertiary control regulates the power flow between the bulk power system and the MG at the PCC, which defines the long-term optimal set points depending on the requirements of the main grid, and also coordinates the clusters of MGs. This control level typically operates in the order of several minutes. Tertiary control can also be considered part of the main power grid.

1.6 Microgrids architectures

Thomas Edison built the first DC electricity supply system in September 1882 in New York City. Edison supported the DC distribution system, while George Westinghouse and Nikola Tesla promoted the AC system. Since the late of 19th century, AC has been the standard choice for commercial energy systems, based mainly on the ease with which AC voltage can be transformed into different levels, the capability of transmitting power over long distances, and the inherent characteristics of fossil energy-driven rotating machines with their synchronous generators.

However, DC distribution grids have shown a resurgence in recent years due to the development and deployment of RES based on DC power sources, and the rapid growth of DC loads which today constitute (together with AC loads with power electronic converters) the vast majority of loads in most power systems. DC distribution presents several advantages, such as a reduction in power losses and voltage drops, and an increase in the capacity of power lines, mainly due to the lack of reactive power flows, absence of voltage drops in lines reactance, and nonexistence of skin and proximity effects which reduce ohmic resistance of lines. As a result, its planning, implementation, and operation are simpler and less expensive [11].

In this context, microgrids involve both AC and DC components, and can be operated based on the principles of the AC power systems (i.e., AC microgrids) or DC power systems (i.e., DC microgrids), or a combination of both, through different architectures. MG architectures are mainly determined by the nature of the loads, the existing and planned distributed generators, the difficulties of building new electrical lines, the existing communications, the space to place energy storage devices and their specific power and energy requirements, among others. According to the literature, MG architectures can be divided into three main categories, namely: AC microgrid, DC microgrid and hybrid AC/DC microgrid.

1.6.1 Alternating current microgrid

Inspired by traditional EPSs, AC distribution is the most popular and commonly used structure for microgrid studies and implementations. By utilizing the existing AC network infrastructure (distribution, transformers, protections, etc.), AC microgrids are easier to design and implement and are built on proven and thus reliable technology. The first MG developed by the CERTS was formulated in 1998 as a cluster of microgenerators and storage with the ability to isolate itself from the utility seamlessly without interruption to the loads [52]. Based on the CERTS microgrid concept, an example of the AC MG architecture is shown in Fig. 1.13.

In the example, the MG has three AC feeders; two of them containing critical loads, DG and ESS, and the other one grouping noncritical loads. The MG is able to adapt generation and demand to any operating conditions by changing its topology through the circuit breakers. The connection of the MG to the distribution grid is managed by the static switch. This device can be operated to disconnect the MG when the quality of the electrical distribution grid is poor, leaving it in islanded operation mode. This maintains a high quality and reliable supply to the critical loads, which are fed both from the distributed generators and the energy stored in ESS devices. During a grid fault, the static switch is opened, as well as the circuit breaker of the third bus, in order to disconnect the noncritical loads from the grid to avoid their damage or malfunction [10].

In the AC microgrid architecture operated in grid-connected mode, the power flows directly from/to the grid, avoiding any series-connected converter and providing high reliability. The feeders have the same voltage and frequency conditions as the grid, so that the loads, generators, and energy storage devices must be grid-compliant. In fact, one of the main advantages of AC microgrid architecture is their compatibility with the existing electrical grid, which can be reconfigured to an AC microgrid scheme. One of the main drawbacks is the large amount of complex power electronics interfaces required (inverters and back-to-back converters) to synchronize DERs with the AC utility grid and provide high-quality AC currents without harmonics. The efficiency and reliability of the overall microgrid can be reduced, since complex electronic power converters present lower reliability than those with fewer components. Generally, an AC distribution system has more conversion steps than a DC system [11].

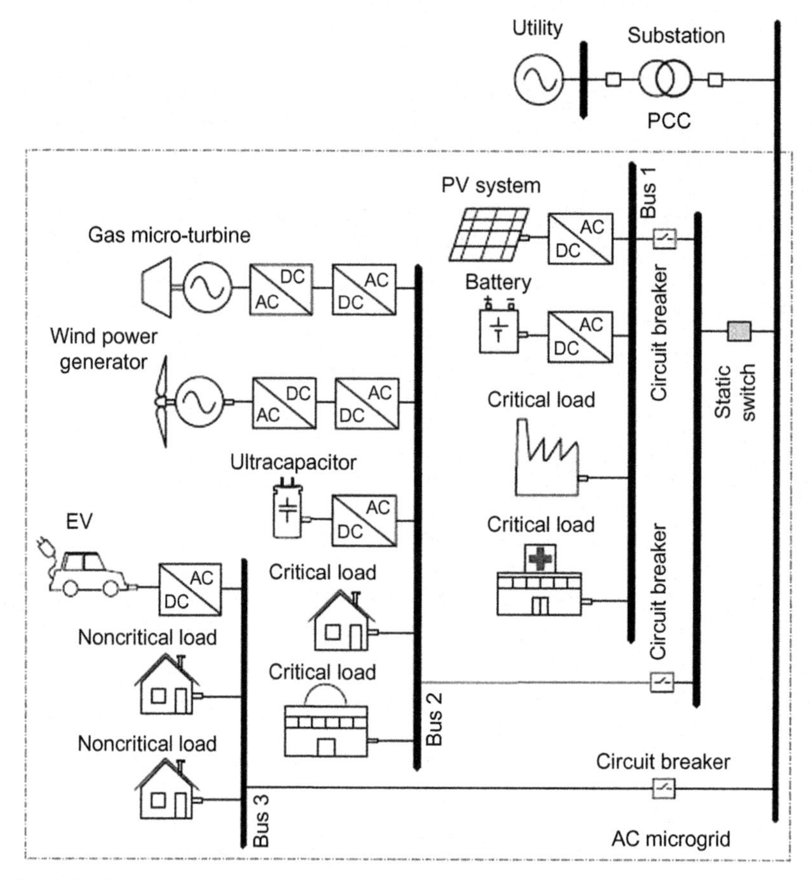

FIGURE 1.13 AC microgrid architecture with critical and noncritical loads [56].

1.6.2 Direct current microgrid

Most DER operate either natively at DC or have an intermediate DC link on their power electronic interface, whereas the end-point connection of ESS, such as supercapacitors, batteries, SMES, and fuel cells, is exclusively DC. On the other hand, many of today's consumer loads are DC supplied. According to some studies, nearly 30% of the generated AC power passes through a power electronic converter before it is utilized, with an amount of lost energy in wasteful conversions within the range of 10%–25% [53]. Therefore integrating these devices into DC microgrids through DC/DC converters becomes a smart choice not only in terms of increasing efficiency due to the reduction of conversion stages but also for achieving power quality with independence from the distribution grid.

An example of DC MG architecture is illustrated in Fig. 1.14 [10]. The main power electronic converter of this architecture is an AC/DC interface usually named as an interlinking converter (IC), which connects the DC MG to the AC grid at the PCC, after the voltage shift and galvanic isolation provided by the transformer. This converter must be

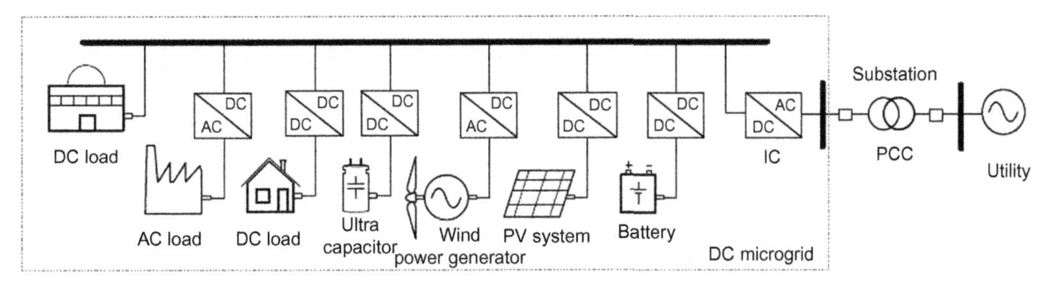

FIGURE 1.14 DC microgrid architecture [10].

bidirectional to allow for power exchange in both ways. Every DG and ESS device of the MG is linked to a DC bus with a determined voltage, most of them do it through a power electronic interface (DC/DC or AC/DC). The voltage regulation of the DC bus is done by the IC, providing a very high quality to the MG, regardless of the quality of the main grid.

Compared to AC MGs, DC MGs have a simpler structure, lower system costs and an overall improved efficiency. Grid synchronization of DERs, harmonics and reactive power flows are not the concern of this architecture, which simplifies its structure and control requirements. There are fewer losses in DC distribution lines compared to the AC distribution system, due to the absence of a reactive current component. Moreover, DC MGs have fault-ride-through capability and are not affected by blackouts or voltage sags which occur in the utility grid, due to the stored energy of the DC capacitors and the voltage control of the IC. On the other hand, DC MGs have some drawbacks such as the need to build DC distribution lines and the incompatibility with actual power systems. Also, the protection scheme for DC systems faces different challenges, mainly the immaturity of standards and guidelines and a limited practical experience. Special protection devices are needed for DC short-circuit currents interruption since there is no zero point crossing of the current wave. Another disadvantage of the DC architecture is that AC loads can't be directly connected to the microgrid and the voltage of DC loads is not standardized, so additional converters would be required. Nevertheless, the main weakness of this architecture is the series-connected IC handling the whole power flow from and to the distribution grid, since it reduces the reliability [54].

1.6.3 Hybrid alternating current/direct current microgrid

Combining the advantages of AC and DC architectures, hybrid AC/DC microgrids are gaining interest over the rest of the architectures, mainly because of the integration of two networks together in the same distribution grid, aiding the direct and efficient integration of both AC- and DC-based DERs and loads.

A typical hybrid microgrid structure is shown in Fig. 1.15, where the AC and DC grids can be distinguished, linked through the main AC/DC bidirectional converter [10]. AC loads are connected to the AC bus, whereas DC loads are connected to the DC bus, using a power converter to adapt the voltage level when necessary. DG and ESS units can be connected either to the AC or DC buses, with a minimum of conversion steps. The AC bus allows for the use of existing equipment; while the DC bus enables the use of a reduced

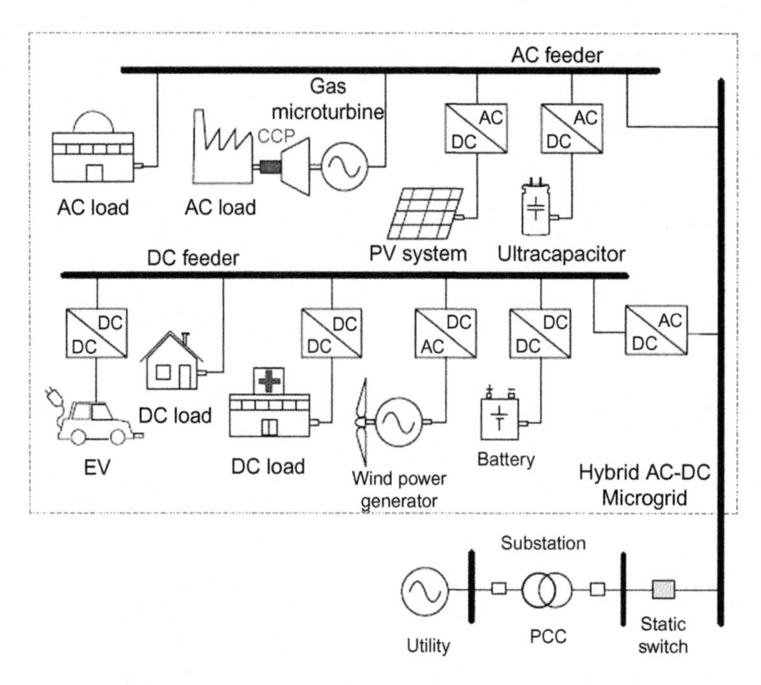

FIGURE 1.15 Hybrid AC/DC microgrid architecture [56].

number of simpler converters. Moreover, this architecture allows for the installation of sensible loads in the DC feeder, in combination with more robust loads installed in the AC feeder [10,55].

When dealing with integration issues, the hybrid AC/DC MG architecture simplifies the electronic interfaces. It also reduces the conversion stages and therefore the energy losses. The inclusion of every DC device on the DC bus, makes it easier to control harmonic injections into the AC side through the main converter, thus guaranteeing high-quality AC in the utility grid. With regard to the transformation of voltage levels, it can be performed in a simple manner on the AC-side with the use of conventional transformers, while on the DC-side this conversion is performed by electronic DC/DC converters.

A hybrid MG can be developed over an existing distribution grid, introducing a main AC/DC converter and a communication network for the connected devices. Although this makes the overall cost higher compared to AC microgrids due to this main power converter, the investment will be returned faster if the number of attached devices increases, as the number of total interface converters is reduced.

On the other hand, this architecture presents various weaknesses that need to be further investigated. When dealing with protection issues, AC equipment has reliable and widely studied technologies, while the DC-side of the MG still needs further research and developments in this regard. Furthermore, due to the main interface power converter, the reliability of the hybrid architecture is lower than that of the AC MG. Management is also more complex in a hybrid MG than in AC or DC microgrids, since stable and reliable power supply has to be ensured for both AC and DC buses through the application of appropriate control strategies for the connected devices. The major challenge of this architecture control is related to power balance when power exchange between the two networks (AC and DC) is required, what is still under study.

According to the main interface device and the interconnection to the utility grid, hybrid MGs can be classified in two main groups, namely coupled and decoupled AC configurations. At coupled AC topologies the AC bus of the microgrid is directly connected to the grid via a transformer and an AC/DC converter is used for the DC link. Instead, decoupled AC configurations are composed at least by an AC/DC and DC/AC stage; establishing no direct connection between the utility grid and the AC bus of the microgrid.

By comparing different microgrid architectures, pros and cons are given in Table 1.1.

TABLE 1.1 Advantages and disadvantages of alternating currecnt (AC) and direct current (DC) microgrids.

Performance	AC microgrid	DC microgrid
Conversion efficiency	Less efficient due to more power conversion stages.	Efficient due to fewer power conversion stages.
Topology	Relatively complex, large number of converters used.	Topology is relatively simple.
Controller complicity	Difficult, transient of AC systems are more complex than DC systems.	Easier than AC microgrid since topology less complicated.
Power quality (PQ) issues	Relatively complex voltage, frequency, phase angle and power factor are necessary to taken care.	Only voltage issue is concerned as PQ issue for DC microgrid, easier to provide high quality power to customers.
Standardization	Standards for AC microgrid are mature since existing standards for power systems are applicable for AC microgrid.	Standards for DC microgrid are still under construction.
Integration with alternative energy resources	Difficult, more conversion levels needed to integrate with DC DGs, AC current needs synchronized when integrating with high-frequency AC DGs.	Easier to integrate with PV, fuels cells, wind turbine and etc.
Interface with main grid	Can connect to main grid via PCC.	Bidirectional converters are needed to interface with main grid.
Protection	Can utilize existing protection technique from existing AC power system.	Still under research.
Stability	Less stable, frequency need to match when synchronizing, complex controllers, more conversion levels.	More stable, only voltage need to be controlled, less conversion levels increase system robustness.
Reliability	Lower than DC counterpart since more devices are used in AC microgrid which leads to higher failure rate of the whole System.	High
Cost	Relatively low cost with the same scale DC counterpart when utilizing existing devices like wires, generators and etc.	Relatively low cost comparing with the same scale AC counterpart when constructing a brand new microgrid.

1.7 Conclusion

MGs are envisioned as an attractive solution for integrating DG units in the utility grid, associated with a reduction on fossil fuel dependency, an increase in the efficiency of the overall electric grid and a substantial improvement in service quality and reliability seen by end customers. In this context, different MG architectures were presented in three main categories: AC, DC, and hybrid AC/DC. The AC microgrid is the most robust architecture due to the direct connection of loads to the utility in a grid-connected mode, and the fact that the whole MG is built on proven and reliable AC elements. In addition, they facilitate the implementation of the MG concept on the existing distribution networks, since they can be adapted to an MG scheme with a few modifications on their structure and components. This is the main reason why AC MGs are the most used architecture so far.

Reliability also relies on the energy conversion processes, and complex power electronic converters are more prone to failures than simpler ones. In this sense, hybrid AC/DC MGs present the lowest number of converters (and simpler ones), since DERs and loads can be connected to the AC or DC feeders depending on their characteristics. However, although the conversion processes of DC/AC and AC/DC are reduced compared with an individual AC or DC MG, many practical problems still exist for their implementation.

To obtain high-quality power for the MG loads, a DC MG is the most suitable option since the voltage of the DC bus is generated electronically, thus being independent from the main grid voltage and frequency, and their possible fluctuations. The main power converter needed to link the MG to the utility grid is nevertheless a weak point in this infrastructure, sacrificing reliability in order to enhance power quality. The integration of DC loads is simpler in this architecture, which makes it a great option to supply energy to modern electronic loads such as computers and servers. Another relevant aspect of the DC architecture is that it provides an easy management of the stored energy in ESS devices.

This new vision of power systems brought forward by the microgrid concept creates many new challenges that are multidisciplinary in nature. A robust and secure operation of such a complex and distributed system requires novel theoretical approaches regarding sensing, control, computational intelligence, software, and communication. The multiple software layers that the management and control of the distribution grid will require need also to be trustworthy, robust, flexible, user-friendly, and seamlessly integrated into the enormous databases that will be created. In summary, these issues are at the forefront of the research agenda in many disciplines and will need to be integrated to conceptualize the foundations of what is now called the smart grid.

Abbreviations

ADS	active distribution system
AEP	American electric power
CCHP	combined cooling, heat, and power
CERTS	consortium for electric reliability technology solutions
CESS	composite energy storage system
CHP	combined Production of heat and power
CSGs	conventional synchronous generators

DERs	distributed energy resources
DFIG	doubly fed induction generator
DG	distributed generation
DSO	distribution system operator
DVR	dynamic voltage restorer
EMS	energy management system
EMTDC	electro magnetic transient design and control
EPS	electrical power system
ESSs	energy storage systems
EV	electrical vehicle
FCLs	fault current limiters
GHG	green house gas
IC	interlinking converter
ICT	information and communication technologies
IEEE	institute of electrical and electronics engineers
IPP	independent power producer
ISET	Institut für Solare Energieversorgungstechnik
JADE	Java Agent DEvelopment Framework
MCC	microgrid central controller
MDT	microgrid deployment tracker
MG	microgrid
NTUA	National Technical University of Athens
PCBB	power converter building block
PCC	point of common coupling
PMSs	power management strategies
PQ	power quality
PSCAD	power system computer aided design
PV	photovoltaic
PWM	pulse width modulation
RES	renewable energy sources
SCADA	supervisory control and data acquisition system
SEPRI	Shandong Electric Power Research Institute
SMES	superconducting magnetic energy storage
STATCOM	Static Synchronous Compensator
UACC	utility area control centre
UC	unit commitment
VSIs	voltage source inverters

References

[1] M. Green, Community power, Nat. Energy 1 (3) (2016) 16014.

[2] A. Ipakchi, F. Albuyeh, Grid of the future, IEEE Power Energy Mag. 7 (2) (2009) 52–62.

[3] R.H. Lasseter, MicroGrids, in: IEEE Power Engineering Society Winter Meeting. Conference Proceedings (Cat. No.02CH37309), 2002, vol. 1, pp. 305–308.

[4] N. Hatziargyriou, Microgrid : Architectures and Control, John Wiley and Sons Ltd., IEEE Press, 2014.

[5] A. Timbus, A. Oudalov, C.N.M. Ho, Islanding detection in smart grids, in 2010 IEEE Energy Conversion Congress and Exposition, 2010, pp. 3631–3637.

[6] G. Platt, A. Berry, D. Cornforth, What role for microgrid? Smart Grid: Integrating Renewable, Distributed & Efficient Energy, Elsevier/Academic Press, 2012, pp. 185–207.

[7] M.Q. Wang, H.B. Gooi, Spinning reserve estimation in microgrids, IEEE Trans. Power Syst. 26 (3) (2011) 1164–1174.

[8] S.X. Chen, H.B. Gooi, M.Q. Wang, Sizing of energy storage for microgrids, IEEE Trans. Smart Grid 3 (1) (2012) 142–151.

[9] R. Lasseter, A. Akhil, C. Marnay, J. Stephens, J. Dagle, R. Guttromson, et al., Integration of distributed energy resources: the CERTS MicroGrid concept, Lawrence Berkeley Natl. Lab. (2002) 1–24.
[10] I. Patrao, E. Figueres, G. Garcerá, R. González-Medina, Microgrid architectures for low voltage distributed generation, Renew. Sustain. Energy Rev. 43 (2015) 415–424.
[11] J.J. Justo, F. Mwasilu, J. Lee, J.-W. Jung, AC-microgrids versus DC-microgrids with distributed energy resources: a review, Renew. Sustain. Energy Rev. 24 (2013) 387–405.
[12] M.Z. Kamh, R. Iravani, Unbalanced model and power-flow analysis of microgrids and active distribution systems, IEEE Trans. Power Deliv. 25 (4) (2010) 2851–2858.
[13] C.A. Hernandez-Aramburo, T.C. Green, N. Mugniot, Fuel consumption minimization of a microgrid, IEEE Trans. Ind. Appl. 41 (3) (May 2005) 673–681.
[14] Y. Uno, G. Fujita, R. Yokoyama, M. Matubara, T. Toyoshima, T. Tsukui, Evaluation of Micro-grid Supply and Demand Stability for Different Interconnections, in 2006 IEEE International Power and Energy Conference, 2006, pp. 612–617.
[15] J. Zhou, Z. Qi, Y. Zhao, W. Ye, Study of the active power exchanging between dynamic voltage restorer and the distribution power system, in: 2009 International Conference on Applied Superconductivity and Electromagnetic Devices, 2009, pp. 171–175.
[16] O. Vodyakho, C.S. Edrington, M. Steurer, S. Azongha, F. Fleming, Synchronization of three-phase converters and virtual microgrid implementation utilizing the Power-Hardware-in-the-Loop concept, in: 2010 Twenty-Fifth Annual IEEE Applied Power Electronics Conference and Exposition (APEC), 2010, pp. 216–222.
[17] N.J. Gil, J.A.P. Lopez, Exploiting automated demand response, generation and storage capabilities for hierarchical frequency control in islanded multi-microgrids, in: Proc. PSCC2008, 16th Power Syst. Comput. Conf., 2008, pp. 1–7.
[18] Y. Zoka, H. Sasaki, N. Yorino, K. Kawahara, C.C. Liu, An interaction problem of distributed generators installed in a MicroGrid, in: 2004 IEEE International Conference on Electric Utility Deregulation, Restructuring and Power Technologies, 2004, pp. 795–799.
[19] M. Wolter, H. Guercke, T. Isermann, L. Hofmann, Multi-agent based distributed power flow calculation, in: IEEE PES General Meeting, 2010, pp. 1–6.
[20] W.K.A. Najy, H.H. Zeineldin, W.L. Woon, Optimal protection coordination for microgrids with grid-connected and islanded capability, IEEE Trans. Ind. Electron. 60 (4) (2013) 1668–1677.
[21] X. Yu, A.M. Khambadkone, Multi-functional power converter building block to facilitate the connection of micro-grid, in: 2008 11th Workshop on Control and Modeling for Power Electronics, 2008, pp. 1–6.
[22] M.B. Delghavi, A. Yazdani, A unified control strategy for electronically interfaced distributed energy resources, IEEE Trans. Power Deliv. 27 (2) (2012) 803–812.
[23] T.L. Vandoorn, B. Meersman, L. Degroote, B. Renders, L. Vandevelde, A control strategy for islanded microgrids with DC-link voltage control, IEEE Trans. Power Deliv 26 (2) (2011) 703–713.
[24] H. Farhangi, Intelligent micro grid research at BCIT, in: 2008 IEEE Electr. Power Energy Conf. - Energy Innov., 2008.
[25] F. Katiraei, M.R. Iravani, Power management strategies for a microgrid with multiple distributed generation units, IEEE Trans. Power Syst. 21 (4) (2006) 1821–1831.
[26] M. Prodanovic, T.C. Green, High-quality power generation through distributed control of a power park microgrid, IEEE Trans. Ind. Electron. 53 (5) (2006) 1471–1482.
[27] Y. Zhao, L. Guo, Dynamical simulation of laboratory microgrid, in: 2009 Asia-Pacific Power and Energy Engineering Conference, 2009, pp. 1–5.
[28] H. Zhou, T. Bhattacharya, D. Tran, T.S.T. Siew, A.M. Khambadkone, Composite energy storage system involving battery and ultracapacitor with dynamic energy management in microgrid applications, IEEE Trans. Power Electron. 26 (3) (2011) 923–930.
[29] B. Brown, P.E., and S. D. E. Services, AC motors, motor control and motor protection, pp. 1–25.
[30] D.M. Vilathgamuwa, P.C. Loh, Y. Li, Protection of microgrids during utility voltage sags, IEEE Trans. Ind. Electron. 53 (5) (2006) 1427–1436.
[31] J. McDonald, Adaptive intelligent power systems: active distribution networks, Energy Policy 36 (12) (2008) 4346–4351.
[32] M. Fazeli, G.M. Asher, C. Klumpner, L. Yao, Novel integration of DFIG-based wind generators within microgrids, IEEE Trans. Energy Convers. 26 (3) (2011) 840–850.

[33] D. Lee, J. Park, H. Shin, Y. Choi, H. Lee, J. Choi, Microgrid village design with renewable energy resources and its economic feasibility evaluation, in: 2009 Transmission & Distribution Conference & Exposition: Asia and Pacific, 2009, pp. 1–4.
[34] R. Pawelek, I. Wasiak, P. Gburczyk, R. Mienski, Study on operation of energy storage in electrical power microgrid - modeling and simulation, in: Proceedings of 14th International Conference on Harmonics and Quality of Power - ICHQP 2010, 2010, pp. 1–5.
[35] P. Li, P. Degobert, B. Robyns, B. Francois, Implementation of interactivity across a resilient microgrid for power supply and exchange with an active distribution network, in: CIRED Seminar 2008: SmartGrids for Distribution, 2008, pp. 81–81.
[36] R. Aghatehrani, R. Kavasseri, Reactive power management of a DFIG wind system in microgrids based on voltage sensitivity analysis, IEEE Trans. Sustain. Energy 2 (4) (2011) 451–458.
[37] N. Research, Microgrid Deployment Tracker 2Q11, 2014. [Online]. Available: https://www.navigantresearch.com/. [Accessed: 23-Oct-2018].
[38] P. Asmus, M. Lawrence, Emerging Microgrid Business Models, Navigant (2016) 14.
[39] D.K. Nichols, J. Stevens, R.H. Lasseter, J.H. Eto, H.T. Vollkommer, Validation of the CERTS microgrid concept the CEC/CERTS microgrid testbed, in: 2006 IEEE Power Eng. Soc. Gen. Meet., 2006, pp. 1–3.
[40] B. Kroposki, R. Lasseter, T. Ise, S. Morozumi, S. Papathanassiou, N. Hatziargyriou, Making microgrids work, IEEE Power Energy Mag. 6 (3) (May 2008) 40–53.
[41] I. Mitra, T. Degner, M. Braun, Distributed generation and microgrids for small island electrification in developing countries : a review, Sol. Energy Soc. India 18 (1) (2008) 6–20.
[42] M. Barnes, A. Dimeas, A. Engler, C. Fitzer, N. Hatziargyriou, C. Jones, et al., Microgrid laboratory facilities, in: 2005 International Conference on Future Power Systems, 2005, pp. 1–6.
[43] N.W.A. Lidula, A.D. Rajapakse, Microgrids research: a review of experimental microgrids and test systems, Renew. Sustain. Energy Rev. 15 (1) (2011) 186–202.
[44] T.S. Ustun, C. Ozansoy, A. Zayegh, Recent developments in microgrids and example cases around the world—a review, Renew. Sustain. Energy Rev. 15 (8) (2011) 4030–4041.
[45] N. Hatziargyrioua, H. Asanob, R. Iravanic, C. Marnayd, An overview of ongoing research, development, and demonstration projects microgrids, IEEE Power Energy Magazine (2007) 78–94.
[46] D. Georgakis, S. Papathanassiou, N. Hatziargyriou, A. Engler, C. Hardt, Operation of a prototype microgrid system based on micro-sources quipped with fast-acting power electronics interfaces, in: 2004 IEEE 35th Annual Power Electronics Specialists Conference (IEEE Cat. No.04CH37551), pp. 2521–2526.
[47] N.D. Hatziargyriou, N. Jenkins, G. Strbac, J.A.P. Lopes, J. Ruela, A. Engler, et al., Microgrids - large scale integration of microgeneration to low voltage grids, CIGRE (2006).
[48] F. Katiraei, R. Iravani, N. Hatziargyriou, A. Dimeas, Microgrids management, IEEE Power Energy Mag. 6 (3) (2008) 54–65.
[49] M. Soshinskaya, W.H.J. Crijns-Graus, J.M. Guerrero, J.C. Vasquez, Microgrids: experiences, barriers and success factors, Renew. Sustain. Energy Rev. 40 (2014) 659–672.
[50] M.G. Molina, Emerging Advanced Energy Storage Systems: Dynamic Modeling, Control and Simulation, first ed, Nova Science Pub., Hauppauge, New York, 2013.
[51] D.E. Olivares, A. Mehrizi-Sani, A.H. Etemadi, C.A. Canizares, R. Iravani, M. Kazerani, et al., Trends in microgrid control, IEEE Trans. Smart Grid 5 (4) (2014) 1905–1919.
[52] Consultant report: integration of distributed energy resources. The CERTS MicroGrid concept, in: California Energy Commission, 2013.
[53] H. Gabbar, Smart Energy Grid Engineering., Elsevier Inc, 2016.
[54] E. Planas, J. Andreu, J.I. Gárate, et al., AC and DC technology in microgrids: a review, Renew. Sustain. Energy Rev. 43 (2015) 726–749.
[55] E. Unamuno, J.A. Barrena, Hybrid ac/dc microgrids—Part I: review and classification of topologies, Renew. Sustain. Energy Rev. 52 (2015) 1251–1259.
[56] C. Wang, Theoretical analysis and simulation Microgrid, 2013.

CHAPTER

2

Distributed energy resources and control

E.S.N. Raju P and Trapti Jain

Discipline of Electrical Engineering, Indian Institute of Technology Indore, Indore, India

2.1 Introduction

The conventional centralized power system, comprising a few large-scale generating power plants located far away from the end-use customer, has served our electricity needs for decades. However, with the rapid growth in electricity demand, problems have emerged with this conventional system, like the inefficient transmission of bulk power over long distances. New concerns facing the power industry include the constraints on the construction of new transmission lines, the liberalization of electricity markets, greenhouse gas emissions, and the production of nuclear waste by nuclear power plants. These issues forced the utilities and researchers to explore other possible solutions, paving the way for the evolution of DERs around the early 1990s [1–3]. The application of DERs is expected to provide benefits in terms of providing reliability and security of power, more economical power generation, a reduction in carbon dioxide emissions, and improved quality of power. However, the direct integration of DER units into the main grid through a power electronic interface (PEI) introduces the following issues [1–3]. The PEI for DERs can be direct current (DC)/DC, DC/alternating current (AC), AC/AC, or AC/DC/AC converters.

- DER units produce either a DC or a variable AC frequency. Therefore these DER units are interfaced to the distribution network or the local loads through a PEI. The direct integration of DER units at distribution voltage levels requires the development of new control strategies.
- DER units based on renewable energy sources (RESs) produce fluctuating active power due to their intermittent nature, which is problematic when DER units are connected directly to the main grid.

Distributed Energy Resources in Microgrids
DOI: https://doi.org/10.1016/B978-0-12-817774-7.00002-8

- The existing centralized power grid follows a multilevel power flow from transmission to the distribution network. The direct integration of DER units can cause power flow problems due to their behavior of injecting the power into the distribution network.
- In the existing, centralized power grid, the initial power balance for a new load is taken care of by the power stored in the large inertia of conventional synchronous generator. However, the direct integration of DER units with low inertia leads to power imbalances between the generation and the load due to lack of inertia.

The above issues differ from those found in conventional power generation sources. In addition to these, there are also a number of barriers in the form of technical, regulatory, and business issues when it comes to the direct integration of DER units into the main grid [1–3]. In order to overcome these limitations, Microgrids (MGs) are proposed by the Consortium for Electrical Reliability Technology Solutions in the United States [4].

A MG comprises low voltage (LV) or medium voltage (MV) distributed systems with distributed generation (DG) units, storage devices, distributed critical and noncritical loads, communication, control and automation systems, protective devices, and smart interconnecting switches [5]. MGs can be operated either in an islanded or a grid-connected mode of operation. The switching between the island and the grid-connected mode of operations is controlled by a static switch at the point of common coupling (PCC). The operation of MGs offers various environmental, economic, and technical benefits for the utility grid and customers as a whole. These benefits include:

- Reducing greenhouse gas emissions by utilizing less costly DGs based on RESs.
- More economical operation by reducing transmission and distribution (T&D) costs.
- Improved reliability by introducing a self-healing capability to the local distribution network.
- Better power quality by managing local loads.
- Higher flexibility, control, and efficiency of operation.
- Bidirectional power flow between the utility grid and the MG in the grid-connected mode of operation.
- Reliable and uninterrupted power supply to critical loads during main grid failures.
- The ability to disconnect from the utility grid in case of voltage fluctuations and upstream disturbances.

Moreover, control of DERs depends on the mode of operation of MGs. In view of the above aspects, this chapter presents the various types of DER technologies and their control at different hierarchical levels.

2.2 Distributed energy resources

DERs could be installed at electricity consumers' premises and/or electric utility facilities and consist of both DG units as well as distributed energy storage, as shown in Fig. 2.1. DERs can reduce physical and electrical distances between the generation and the load, thus, reduce the losses encountered in the T&D of power, minimize carbon emissions, and reschedule the establishment of new large generators and transmission lines.

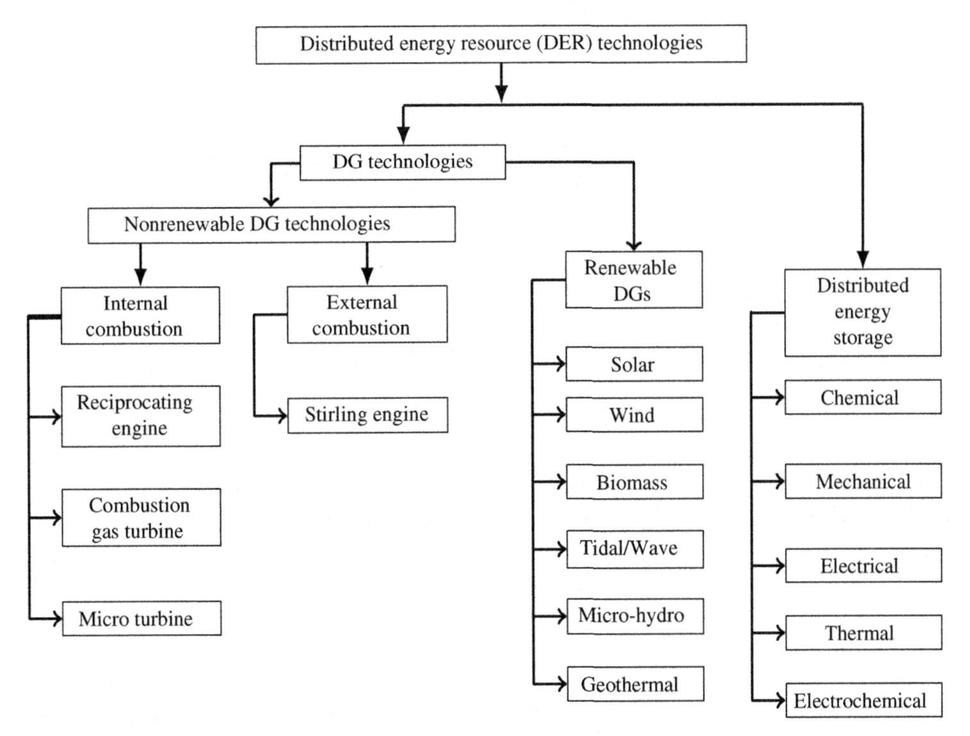

FIGURE 2.1 Distributed energy resource (DER) technologies.

In addition, DERs can also remove congestion from transmission and distribution lines, enhance grid voltage profile and power quality, and provide better usage of waste heat. Further, DERs require less time for their construction and deployment compared to huge power generation plants and T&D facilities. Moreover, DERs can provide power to remote locations, where T&D facilities are too costly to build and/or not available [3].

2.2.1 Distributed generators

Various definitions of DGs are summarized in the existing literature [2]. DG is defined as "generation of electricity by facilities that are sufficiently smaller than central generating plants so as to allow interconnection at nearly any point in the power system." The power generation from a DG unit is small-scale, ranging from 4 to 10,000 kW [1]. DG technologies are mainly categorized into conventional/nonrenewable energy and nonconventional/renewable energy types, requiring power electronic converters for their interface. The conventional or nonrenewable energy DG units are the dispatchable type and include reciprocating engines, combustion gas turbines (CGT), micro-turbines (MTs), and Stirling engines, whereas, the renewable energy or nondispatchable type DG units comprise solar photovoltaic (PV) and solar heating, wind/micro-wind, hydro/mini-hydro, geothermal, tidal/wave, and biomass, as shown in Fig. 2.1.

2.2.1.1 Reciprocating engines

Reciprocating engines were the first fossil-fuel-driven DG technologies developed more than 100 years ago and use diesel, natural gas, or waste gas as their fuel source. Currently, they are available from many manufacturers in all DG sizes ranging from less than 5 to over 5000 kW [6]. Further, reciprocating engines are primarily used in applications such as cogeneration, backup power and peaking power. Moreover, they offer low cost and good efficiency for DG applications. However, these engines require high maintenance and have high emissions in case of diesel-fueled units.

Fig. 2.2 shows the schematic operation diagram of a reciprocating engine, which operates in four cycles for power generation [6]. These four cycles include intake, compression, combustion, and exhaust. Fuel and air are mixed in the first intake cycle. In order to increase engine output, some engines are turbocharged or supercharged in which the intake air is compressed by a small compressor in the intake system. A fuel/air mixture is introduced into the combustion cylinder, then compressed as the piston moves towards the top of the cylinder. In diesel units, the air and fuel are introduced separately with fuel being injected after the air is compressed by the piston in the engine. As the piston nears the top of its movement, a spark is produced that ignites the mixture (in most diesel engines, the mixture is ignited by the compression alone). Dual fuel engines use a small amount of diesel pilot fuel in lieu of a spark to initiate combustion of the primary natural gas fuel [6]. The pressure of the hot combusted gases drives the piston down the cylinder. Energy generated in the moving piston is translated to rotational energy by a crankshaft. As the piston reaches the bottom of its stroke, the exhaust valve opens and the exhaust is expelled from the cylinder by the rising piston. Cogeneration configurations are also available with heat recovery from the gaseous exhaust and water as well as oil jackets.

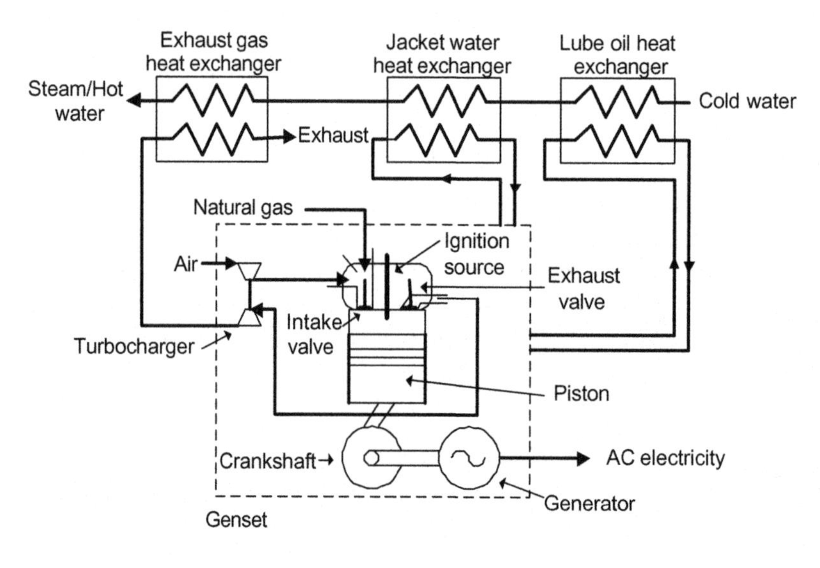

FIGURE 2.2 Schematic operation diagram of reciprocating engines [6].

2.2.1.2 *Micro-turbines*

An MT is a mechanism that uses the flow of a gas to convert thermal energy into mechanical energy. The technology used in the MT is derived from diesel engine turbochargers, aircraft auxiliary power systems, and automotive designs. MTs consist of a compressor, combustor, turbine, and generator, as shown in Fig. 2.3 [6]. The combustible (usually gas) is mixed in the combustion chamber with air, which is pumped by the compressor. This product makes the turbine rotate, which, at the same time, impulses the generator and the compressor. In the most commonly used design, the compressor and turbine are mounted above the same shaft of the electric generator. The compressors and turbines resemble automotive engine turbochargers, which are typically radial-flow designs. Most of the designs are single-shaft and use a high-speed permanent magnet generator for producing variable frequency and variable voltage AC power. A PEI (inverter/rectifier) is employed to produce 50/60 Hz AC or DC power. Most MT units are designed currently for continuous-duty operation and are recuperated to obtain higher electrical efficiencies. Moreover, MTs offer clean operation with low emissions and good efficiency. However, the costs to maintain them are high.

2.2.1.3 *Combustion gas turbines*

Historically, CGTs were developed as aero derivatives using jet propulsion engines as a design base. However, in oil and gas industries, they have been designed specifically for stationary power generation or for compression applications. CGTs range from 1 MW to over 100 MW [6]. They have low capital cost, low emission levels, heat recovery through steam, and infrequent maintenance requirements. However, they suffer from low electrical

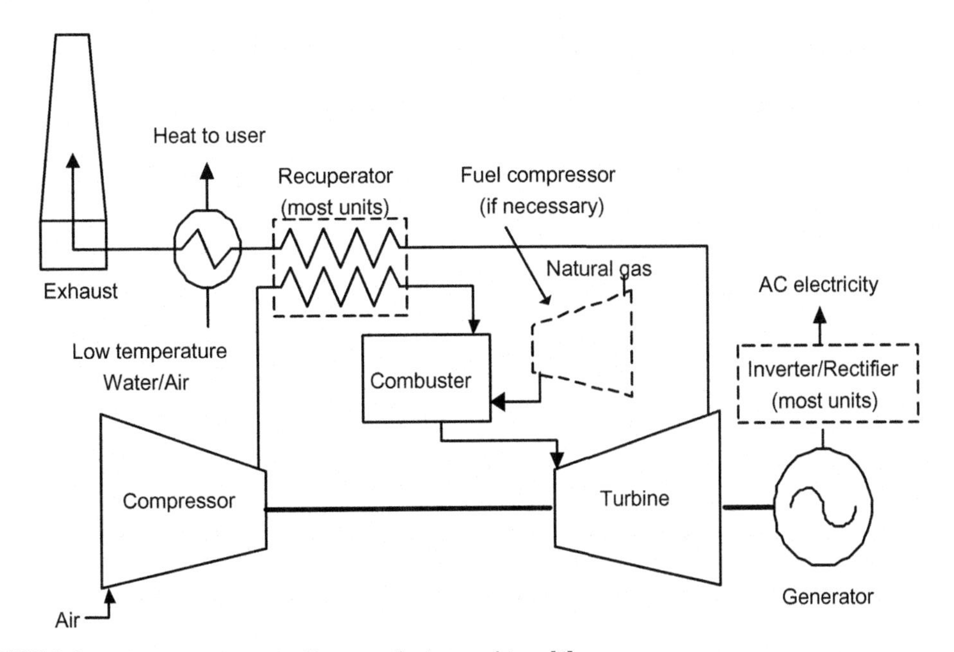

FIGURE 2.3 Schematic operation diagram of micro-turbines [6].

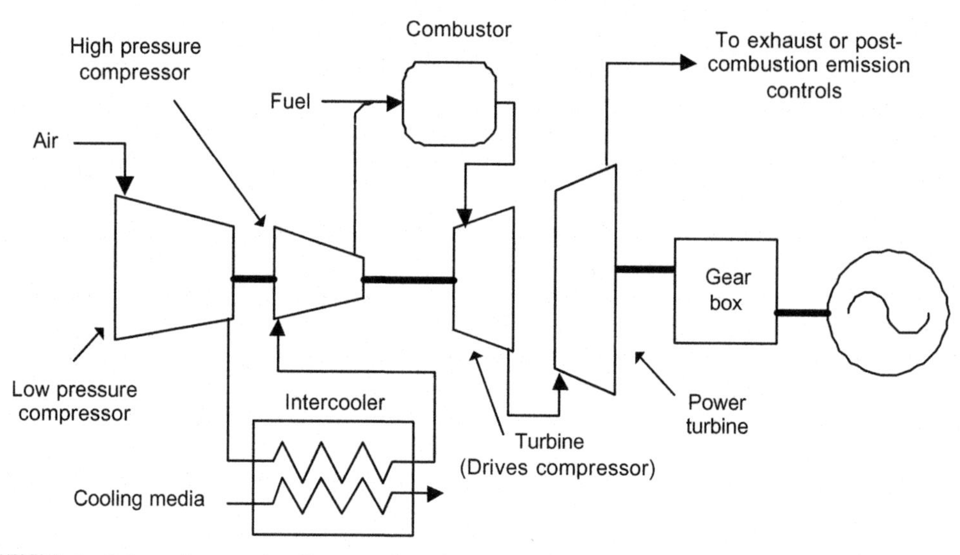

FIGURE 2.4 Schematic operation diagram of combustion gas turbines [6].

efficiency ratings. CGTs are used primarily in cogeneration and for peaking power applications. In a CGT, air is compressed and a gaseous or liquid fuel is ignited and the combustion products expand directly through the blades in a turbine to drive an electric generator, as shown in Fig. 2.4 [6]. Usually, CGTs have axial blades and are multiple-staged, whereas, smaller MTs are single-staged and have radial blades. Generally, the intercooler, shown in Fig. 2.4, is employed in larger units for economic benefits.

2.2.1.4 *Stirling engine*

The Stirling engine, invented by Robert Stirling in 1816, is a heat engine that is vastly different from the internal combustion engine. The Stirling engine uses the Stirling cycle, which is unlike the cycles used in internal combustion engines and has the potential to be much more efficient than a gasoline or diesel engine. The Stirling engine has six main components: containers, piston, displacer, crankshaft, flywheel, and external heat source, as shown in Fig. 2.5. The air at the bottom heats up, creating pressure on the small power piston, which moves up and rotates the wheel. The rotating wheel moves the big displacer down. The air cools down at the top, reducing the pressure and allowing the power piston to move down. This motion of the power piston moves the displacer upwards and the air at the bottom is heated again [7].

2.2.1.5 *Solar/photovoltaic*

PV cells convert the sun's energy directly into DC electricity, as shown in Fig. 2.6 [6]. In the PV system, semiconductive materials are used to construct the solar cells, which transform the self-contained energy of photons into electricity when they are exposed to sunlight. The cells are placed in an array that is either fixed or moving to keep tracking the sun in order to generate the maximum power. These PV systems are environmentally friendly without any kind of emission during operation. They are easy to use, simple in

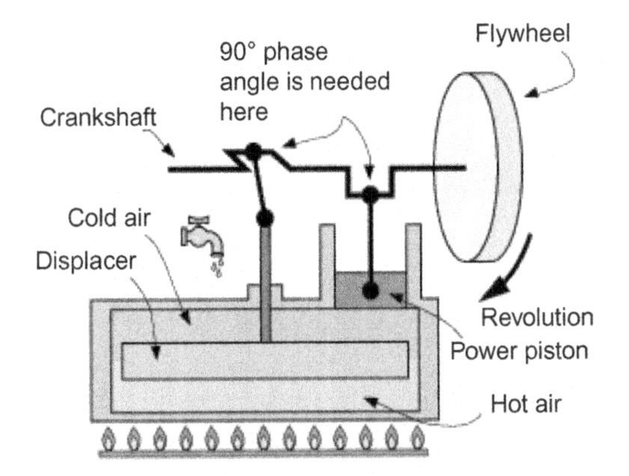

FIGURE 2.5 Schematic operation diagram of a Stirling engine [7].

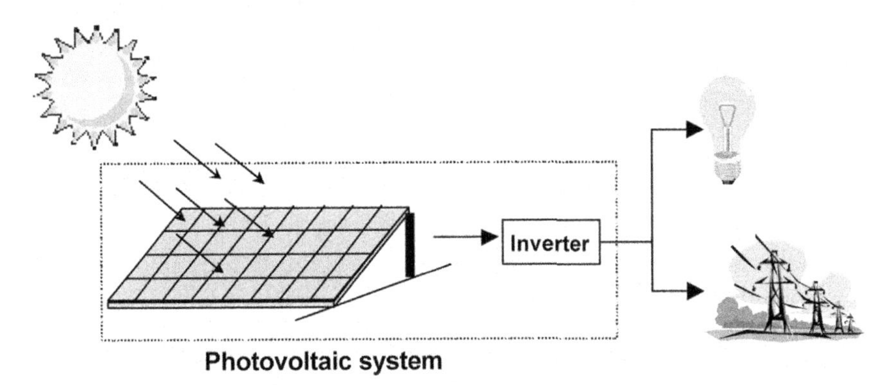

FIGURE 2.6 Schematic operation diagram of a photovoltaic (PV) system [6].

designs, and do not require any other fuel than sunlight. However, they need large spaces and have a high initial cost. A particular advantage of this technology is that it can be very effectively integrated into the structure of buildings. From a distribution system perspective, it is important to note that this technology requires its DC output to be converted to AC at the interface with the grid. This process introduces harmonic distortion to the grid and it is vital that the level of this distortion is controlled within limits set by industry standards.

2.2.1.6 *Wind turbines*

Wind turbines transform wind energy into electricity. Wind energy is a highly variable source, thus, it should be handled according to this characteristic. The principle of operation of a wind turbine is characterized by two conversion steps. Firstly, the rotor extracts the kinetic energy of the wind, changing it into mechanical torque in the shaft; and secondly, the generation system converts this torque into electricity, as shown in Fig. 2.7. In the most common system, the generator system gives an AC output voltage, which is dependent on the wind speed. Since the wind speed is variable, the generated voltage has

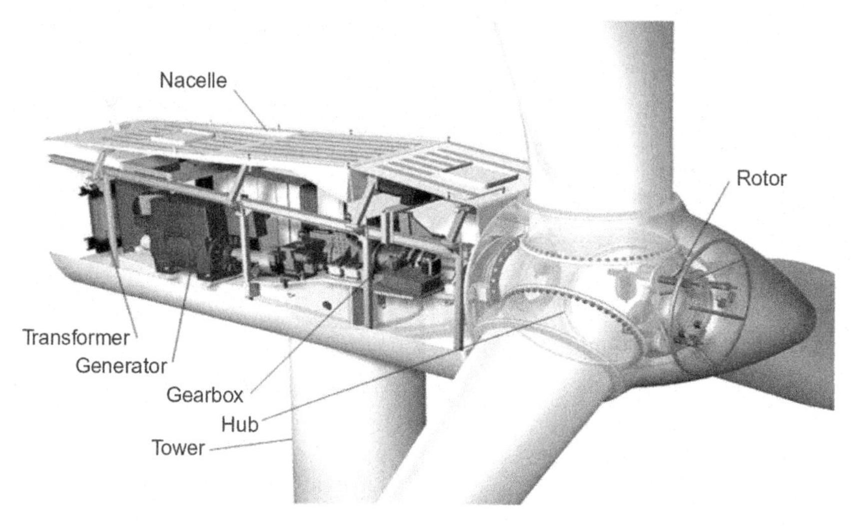

FIGURE 2.7 Schematic operation diagram of a wind turbine [6].

to be transferred to DC and back again to AC with the help of inverters. On the other hand, fixed-speed wind turbines can be connected directly to the grid. Wind generation challenges the MG system with a number of problems [6]. Firstly, the output of a wind farm is essentially uncontrollable, which causes an additional challenge to the control systems that ensure a balance between generation and demand and then constant system frequency. Secondly, most wind generators have different electrical characteristics compared to conventional synchronous generators and this has to be taken into account when connecting them to the T&D systems. However, it is possible to make wind generation a significant contributor to the electricity supply system since it is currently a leading renewable generating technology in many countries.

2.2.1.7 Micro-hydro distributed generation

Micro-hydro distributed generations (MHDGs) have been utilized as a renewable energy source for electricity generation. The major advantage of these MHDGs is that there are no requirements for the construction of huge dams and reservoirs. However, these DGs suffer from large variations in generation due to variable water flow caused by an uneven rainfall [8]. Fig. 2.8 shows the schematic operation diagram for MHDG. It can be seen that the fundamental components of MHDGs include the dam, penstock, turbine, tailrace, generator, and allied equipment. The head and flow rate of available water resources are two major components of MHDGs, which need careful attention before designing MHDGs [8]. The head is the vertical distance through which the water falls, while the flow rate is the volume of the water per unit of time. Electricity generation from MHDGs depends on the combination of suitable head and flow rates by using efficient and well-installed equipment. Generally, the efficiency of the MHDG is about 50% due to frictional losses and imperfection of turbines.

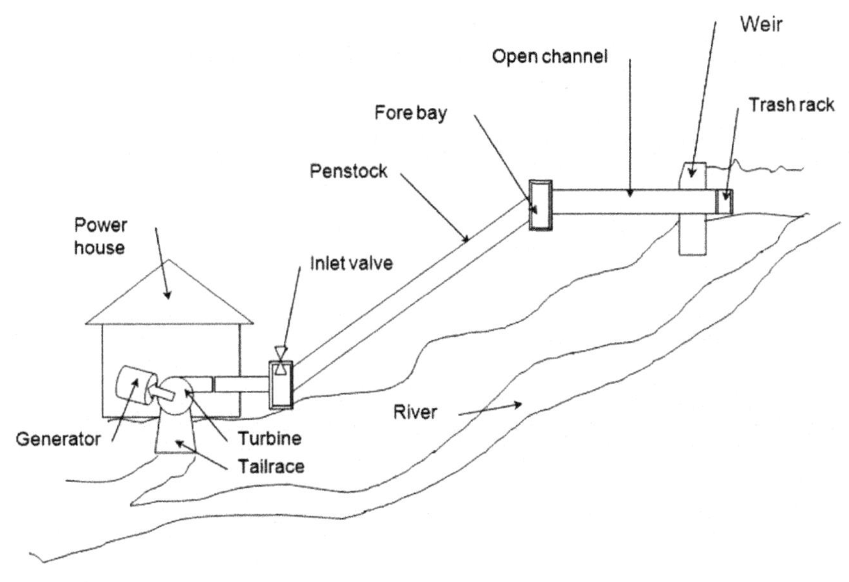

FIGURE 2.8 Schematic operation diagram of micro-hydro distributed generation [8].

2.2.1.8 *Other renewable distributed generations using biofuels and tidal/wave currents*

Biofuels include biomass and biogas. Biomass can be a wet or dry fuel derived from plant material and animal wastes. It can be combusted to generate electricity and/or heat in power plants with a wide range of outputs. Wet wastes can be fermented to produce a gas that can be used in gas engines or gas boilers. Biogas is essentially the same as the gas derived from wet biomass. The most common example of biogas is landfill gas. However, the same basic process that occurs in a landfill site can be reproduced in an anaerobic digester. This device allows organic material to decompose in the absence of oxygen in a controlled way, releasing methane. Both dry and gaseous biofuels can be used in gas engines (reciprocating) and gas turbines to produce electricity. The waste heat from this process can be captured to achieve cogeneration [3].

Tidal/wave currents are the ocean water mass response to tidal rise and fall. Tidal currents are generated by the horizontal movements of water, modified by seabed bathymetry, particularly near coasts or other constrictions, for example, islands. Tidal current flows result from the sinusoidal variation of various tidal components operating on different cycles. The energy production from tidal/wave currents is similar in many ways to the micro-hydro distributed generation. However, the difference is that the water flows in and out of the turbines in both directions instead of in just one forward direction.

2.2.2 Energy storage systems

Energy storage technologies allow electricity to be stored either directly or by employing a conversion process and are not strictly DG technologies [5]. However, they do have many characteristics in common with DGs and they are much talked about at present.

Energy storage systems (ESSs) are commonly implemented as the energy buffers in MGs due to the uncertain behavior of renewable energy-based DG units. Furthermore, ESSs are one of the most desirable solutions for maintaining the power balance in an MG, improve its stability, and tackling frequency and voltage variations. Depending on the form of energy used, ESSs technologies can be classified into chemical energy storage, mechanical energy storage, electrical energy storage, thermal energy storage, and electrochemical energy storage as described in the following subsection.

2.2.2.1 Classification of energy storage systems

ESSs technologies can be classified into the following categories [9].

- *Chemical energy storage*: electrical energy stored in the form of chemical energy.
  1. *Fuel cell*: This is an electrochemical cell that converts the chemical energy from a fuel into electricity through an electrochemical reaction of a hydrogen-containing fuel with oxygen or another oxidizing agent.
  2. *Electrolysis*: This involves the production, storage and utilization of hydrogen through the electrolysis of water.
- *Mechanical energy storage*: This is electrical energy stored in the form of mechanical energy. Major methods of mechanical storage are:
  1. *Flywheel energy storage*: This technique employs the mechanical energy of a spinning rotor to store energy.
  2. *Pumped hydro storage*: In this method, large amounts of water are pumped to an upper level, which is converted to electrical energy using a generator and turbine when there is a shortage of electricity.
  3. *Compressed air energy storage*: This system involves compressing air to a pressure of around 70 bar to store electrical energy.
- *Electrical energy storage*: This is usually implemented in applications that need a quick response.
  1. *Double layer capacitor*: It is also called ultracapacitor or supercapacitor storage. In this method, a dielectric gap between two conductors is employed. This technique has a high energy storage capability due to its high power ability.
  2. *Superconducting magnetic energy storage*: In this method, the voltage is stored in the superconducting coil after being switched to DC by an AC-DC converter.
- *Thermal energy storage*: The thermal storage method is based on converting energy either to ice or hot water.
  1. Geothermal energy storage systems,
  2. Solar-thermal energy conversion systems, and
  3. Phase-changing materials.
- *Electrochemical energy storage*: The chemical energy contained in the active material is converted directly into electrical energy.
  1. Batteries are an advanced technique for storing electrical energy in electrochemical form.
  2. Flow batteries use ions dissolved into liquid electrolytes.

ESSs can be utilized in MGs for various applications like energy arbitrage, peak shaving, load leveling, spinning reserve, voltage support, black start, frequency regulation, power quality, power reliability, time shifting, and capacity firming [9]. Electrochemical energy storage methods have been preferred for applications of load leveling, RESs integration, customer energy management, and ancillary services. Flywheel mechanical storage technologies have been used for emergency devices and low-energy applications. Moreover, electric vehicles (EVs) and plug-in hybrid electric vehicles are potential game-changers in future power systems due to their great flexibility in interacting with the power grid. Further details about energy storage technologies can be found in [9].

2.3 Control of distributed energy resources in microgrids

The main objectives of control of DERs in MGs are given as follows [10–18].

- To synchronize them with the utility grid.
- To maintain a power balance between the generation and the load, particularly, in the islanded mode of operation.
- To maintain stable voltage and frequency.
- To maintain quality of power injected into the utility grid.
- To maintain smooth power transfer between the MG and the utility grid for stable system operations under various operating conditions.
- To harness the maximum power from DER units based on various RESs.

The control of DERs depends on the mode of operation. In the grid-connected mode of operation, DC voltage is kept steady by the local controller of DER unit, whereas voltage and frequency of the AC bus are controlled by the utility grid's controller. In an islanded mode of operation, DC voltage may be controlled by the energy storage, while the voltage amplitude and frequency of the AC bus are controlled by the local controllers of the parallel-connected DER units. The control of DERs in MGs is performed on three levels: primary, secondary, and tertiary, as shown in Fig. 2.9. These control levels are based on their speed of response and the time frame in which they operate. They also differ in communication infrastructure requirements.

2.3.1 Primary level control of distributed energy resources in microgrids

The primary level control normally follows the set points given by upper-level controllers, which perform the control of local voltage, current, and power. It is responsible for load sharing among DER units and to maintain stability under islanded operation. It operates on timescales of the order of 10–100 ms.

In the case of AC microgrids (AC MGs), the most common primary control methods are decentralized droop control, active and reactive power control (PQ control), and voltage and frequency control (V/F control). The PQ control method, shown in Fig. 2.10, is applied in the grid-connected mode of operation, wherein, the reference values for the active and reactive power are given by the utility grid controller to the controllers of DERs [11]. It is used to generate the inductor reference currents, $\mathbf{i}^*_{\mathbf{Ldq}}$, with the help of the reference coupling inductance currents $\mathbf{i}^{\Sigma}_{\mathbf{dq}} = \mathbf{i}^*_{\mathbf{odq}} + \mathbf{i}_{\mathbf{Ldq}} - \mathbf{i}_{\mathbf{odq}}$ as given in Eq. (2.1).

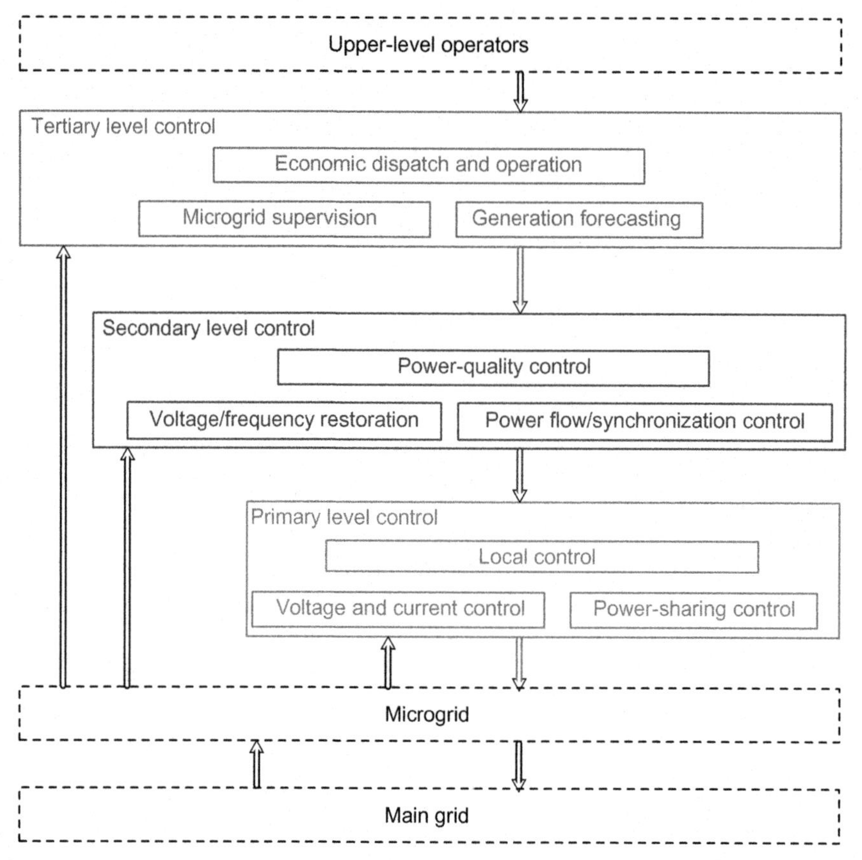

FIGURE 2.9 Control levels in microgrids.

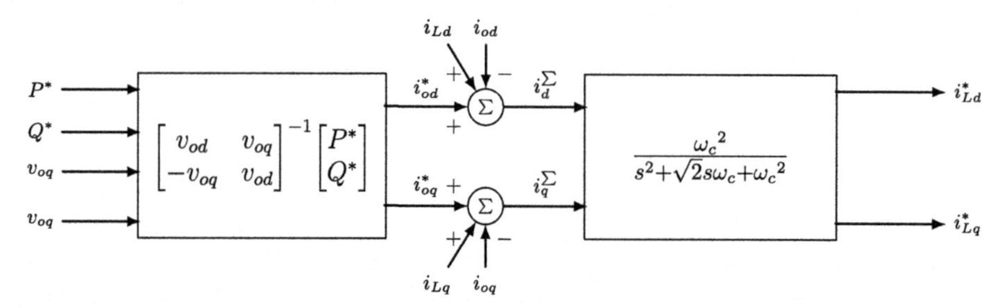

FIGURE 2.10 Power control (PQ) method.

$$\mathbf{i}^*_{\mathbf{Ldq}} = \frac{{\omega_c}^2}{s^2 + \sqrt{2}s\omega_c + {\omega_c}^2}(\mathbf{i}^*_{\mathbf{odq}} + \mathbf{i}_{\mathbf{Ldq}} - \mathbf{i}_{\mathbf{odq}}) \tag{2.1}$$

where ω_c is the cutoff frequency of the second-order Butterworth low-pass filter, the output reference currents, $\mathbf{i}^*_{\mathbf{odq}}$, are calculated using the reference powers, P^* and Q^*, and the output voltages, $\mathbf{v}_{\mathbf{odq}}$, as given in Eqs. (2.2) and (2.3).

$$i_{od}^{*} = \frac{v_{od}P* - v_{oq}Q*}{v_{od}^{2} + v_{oq}^{2}} \quad (2.2)$$

$$i_{oq}^{*} = \frac{v_{oq}P* + v_{od}Q*}{v_{od}^{2} + v_{oq}^{2}} \quad (2.3)$$

Droop control and V/F control methods are employed in the island mode of operation and provide load sharing among DER units without requiring time-critical communication links [12–18]. There are two types of droop control methods, namely P–f droop control and Q–V droop control, as shown in Fig. 2.11. The P-f control method controls the frequency by controlling the active power supplied by DER units, whereas, the Q–V droop control method controls the voltage magnitude by controlling the reactive power supplied by DER units. These methods can minimize fluctuations in the voltage and the frequency for small disturbances only. Droop control sets frequency and magnitude of output voltage according to the active and reactive power droop characteristics, respectively. This is achieved by emulating the operation of a conventional synchronous generator. The reference frequency (ω_{i-dr}^{*}) and voltage (v_{odi-dr}^{*} and v_{oqi-dr}^{*}) of ith DER unit are given in Eqs. (2.4) and (2.5), respectively.

$$\omega_{i-dr}^{*} = \omega_n - m_{Pi}P_i \quad \forall i \quad (2.4)$$

$$v_{odi-dr}^{*} = V_n - n_{Qi}Q_i \quad \text{and} \quad v_{oqi-dr}^{*} = 0 \quad \forall i \quad (2.5)$$

where ω_n and V_n stands for the nominal set point of frequency and d-axis output voltage, respectively; m_{Pi} and n_{Qi} represents static active and reactive power droop gains, respectively; P_i and Q_i are the average active and reactive powers of the ith DER unit, respectively. It should be noted that the q-axis component of voltage (v_{oqi}^{*}) is set to zero to facilitate voltage-oriented control. The static active and reactive power droop gains of the ith DER unit can be calculated for a given range of frequency and voltage magnitude as follows:

$$m_{Pi} = \frac{w_{max} - w_{min}}{P_{max\ i}} \quad \forall i \quad (2.6)$$

$$n_{Qi} = \frac{V_{odmax} - V_{odmnin}}{Q_{maxi}} \quad \forall i \quad (2.7)$$

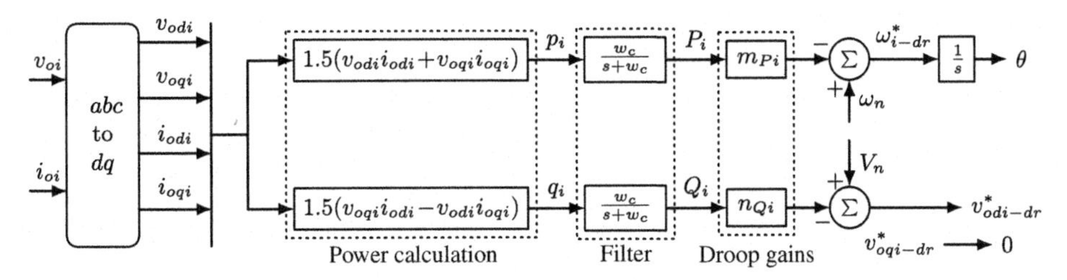

FIGURE 2.11 Power-sharing controller.

The average active and reactive powers utilized in Eqs. (2.4) and (2.5) are extracted from a low-pass filter, as given in Eqs. (2.8) and (2.9).

$$P_i = \frac{w_c}{s + w_c} \times \underbrace{1.5\,(v_{odi}i_{odi} + v_{oqi}i_{oqi})}_{\text{Instantaneous active power}} \quad \forall i \tag{2.8}$$

$$Q_i = \frac{w_c}{s + w_c} \times \underbrace{1.5\,(v_{oqi}i_{odi} + v_{odi}i_{oqi})}_{\text{Instantaneous active power}} \quad \forall i \tag{2.9}$$

where v_{odi}, v_{oqi}, i_{odi}, and i_{oqi} are the output voltages and currents of the ith DER unit in local dq reference frame, respectively, and w_c is the cutoff frequency of the low-pass filter.

In order to minimize large fluctuations in the voltage and frequency, the V/F control method is used. V/F control method uses proportional-integral (PI)/proportional–integral–derivative (PID) controllers to damp out the oscillations in the voltage and the frequency, as given in Eqs. (2.10) and (2.11) for the PI controller [12].

$$\Delta w = K_{pw}(\omega^*_{MG} - \omega_{MG}) + K_{iw}\int(\omega^*_{MG} - \omega_{MG})dt + \Delta ws \tag{2.10}$$

$$\Delta V = K_{pE}(V^*_{MG} - V_{MG}) + K_{iE}\int(V^*_{MG} - V_{MG})dt \tag{2.11}$$

where K_{pw}, K_{iw}, K_{pE}, and K_{iE} are the control parameters of the V/F control compensator, Δw and ΔV are frequency and voltage amplitude correction signals generated by the V/F control, respectively, and Δw_S is a synchronization term, which remains equal to zero when the grid is not present [12]. The frequency and voltage amplitude correction signals, Δw and ΔV, are responsible to compensate for deviations in frequency and voltage amplitude caused by the disturbances. The primary level controllers in AC MGs should be designed such that in an island mode of operation, it turns from PQ control to droop control for small disturbances and from droop control to V/F control for large disturbances.

In the case of DC MGs, the load current sharing is realized by an I–V droop controller, which is the most common primary control method [12]. The droop controller can be implemented by means of virtual resistance and can be expressed as:

$$v_{DCi} = v^*_{DC} - i_{DCi}R_{di} \quad \forall i \tag{2.12}$$

where v_{DCi} is the output voltage of each PEI of DER, v^*_{DC} is the reference value of the DC output voltage, i_{DCi} is the output current, and R_{di} is the virtual resistance. For large disturbances, the voltage control method, defined in Eq. (2.11), has been used to damp out the oscillations in the voltage.

2.3.2 Secondary level control of distributed energy resources in microgrids

The secondary level control appears on the top of primary level control and operates on a slower timescale of the order of 1–10 second. It ensures power quality by regulating the voltage and frequency deviations introduced by the primary level control and small-signal disturbances. The secondary level control can also be used to correct the load sharing

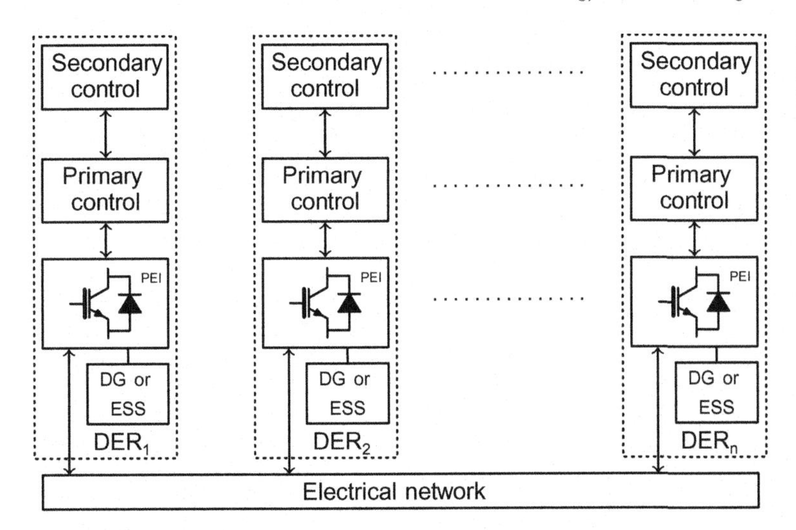

FIGURE 2.12 Schematic diagram of decentralized control of microgrids (MGs).

ratios done by the primary level control. It is also responsible for power flow/synchronization control. The secondary level control can be categorized further into decentralized, centralized, distributed multi-agent, and hierarchical control.

In a decentralized control of DERs in MGs, each primary control unit is controlled by its local secondary control unit, as shown in Fig. 2.12. The secondary control unit receives only local information and is not fully aware of either system-wide variables or actions of other secondary control units. The decentralized control of MGs provides total load sharing in a proper way as well as active damping of oscillations between the output filters. Furthermore, it ensures the stability of MGs on a global scale because of local measurements. However, decentralized control of DERs suffers from a lack of uniformity/consistency and coordination among DER units. To overcome these limitations, centralized controllers utilizing measurements based on either a communication link or without a communication link have been preferred. Centralized control of DERs in MGs, shown in Fig. 2.13, depends on the data gathered in a dedicated central secondary control unit that performs the required calculations and generates the control actions for all the primary control units at a single point. It requires extensive communications between the secondary and primary control units. The centralized controller determines the reference values for the primary control units, that is, output power and/or terminal voltage, for each DER unit.

Centralized and decentralized controllers have certain intrinsic disadvantages. For instance, they need a central computing and communication unit, which may suffer from single-point failures and poor control dynamics, respectively. In order to solve these problems, a distributed control approach has been applied in the secondary level control unit, as shown in Fig. 2.14. In this approach, every DER unit has its own local secondary control unit that can produce an appropriate control signal for the primary level control unit by using the measurements of the other DER units at each sample time. The main functions of the above centralized, decentralized, and distributed controllers for MGs are proper load sharing, coordination among DER units and voltage, as well as frequency regulation under small-signal disturbances. These requirements are of distinct timescales and

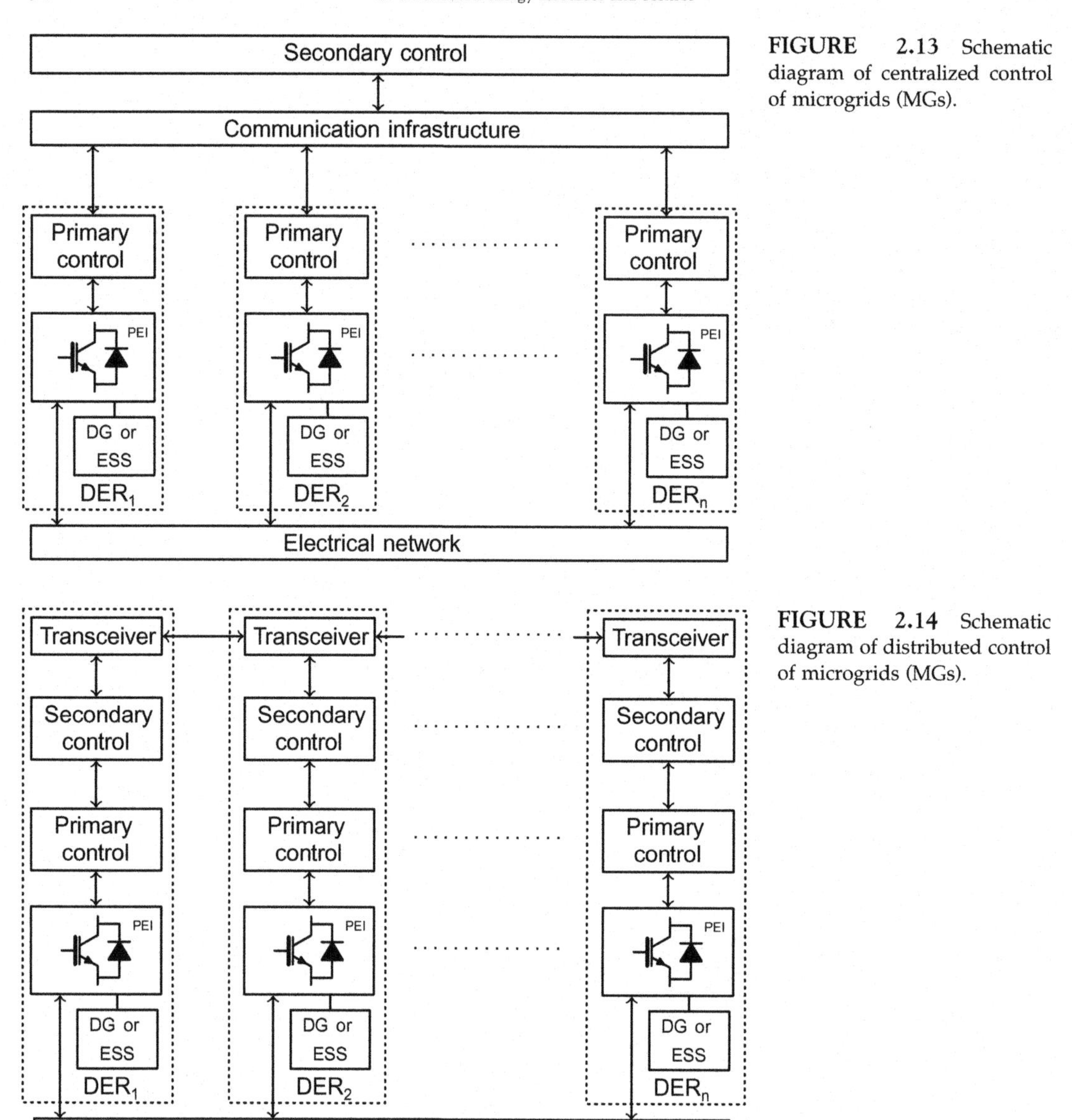

FIGURE 2.13 Schematic diagram of centralized control of microgrids (MGs).

FIGURE 2.14 Schematic diagram of distributed control of microgrids (MGs).

importance. Thus, to address the aforementioned requirements, a hierarchical control approach has been developed, as shown in Fig. 2.15.

2.3.3 Tertiary level control of distributed energy resources in microgrids

Finally, the tertiary level control is responsible for MG supervision, generation forecasting, economic dispatch and optimal power management. It also manages the power flow between the MG and the main power grid in the grid-connected mode of operation.

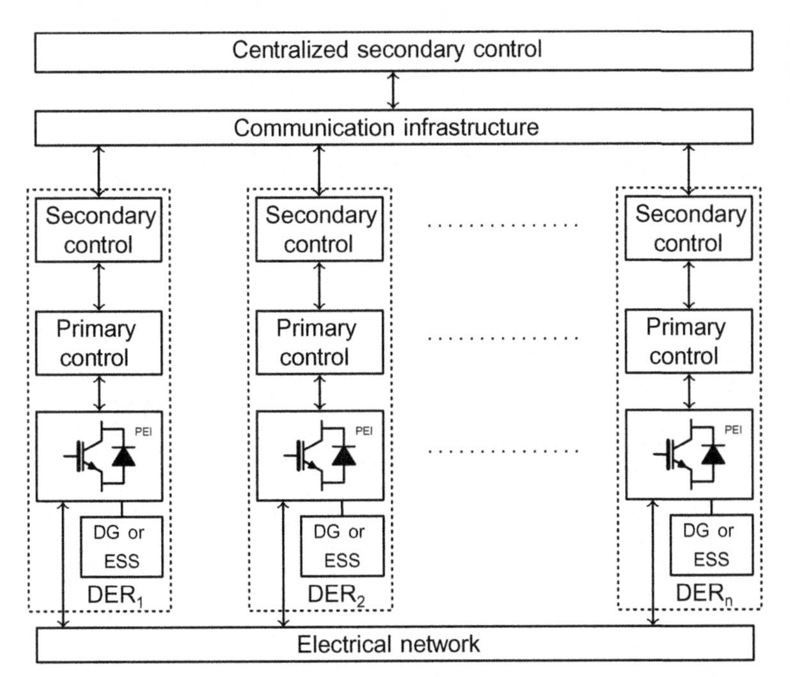

FIGURE 2.15 Schematic diagram of hierarchical control of microgrids (MGs).

The tertiary level control generally operates on a slow timescale based on a static power flow model in which references are updated every 15 minutes. Distribution system operators are the upper-level operators. They are responsible for operating and managing distribution networks operating at LV and MV levels.

2.4 Secondary level control of distributed energy resources in the typical alternating current microgrid test system

Fig. 2.16 shows a schematic diagram of a typical AC MG test system operating at a frequency of 50 Hz and voltage of 230 V per phase Root Mean Square. The AC MG system includes four inverter-interfaced distributed generation (IIDG) units, three lines, and locally connected loads, namely resistive (R)/inductive (RL), constant power load (CPL), rectifier interfaced active load (RIAL) and dynamic induction motor (IM) load. It also consists of secondary level decentralized and centralized controllers, and a combination of these two, that is, a two-level hierarchical controller. Each IIDG unit model consists of power processing and local control sections, as shown in Fig. 2.17. The power processing section includes a three-leg voltage source inverter, an output LC (inductor-capacitor) filter and a coupling inductor, while the control section can be divided into outer loop power-sharing controller and inner loop voltage and current controllers [13–15]. The outer loop power-sharing controller, shown in Fig. 2.11, sets the frequency and magnitude of the IIDG unit output voltage according to the active and reactive power droop characteristics, respectively. The inner loop voltage and current controllers, shown in Fig. 2.18, are designed to reject high-frequency disturbances and provide sufficient damping to the

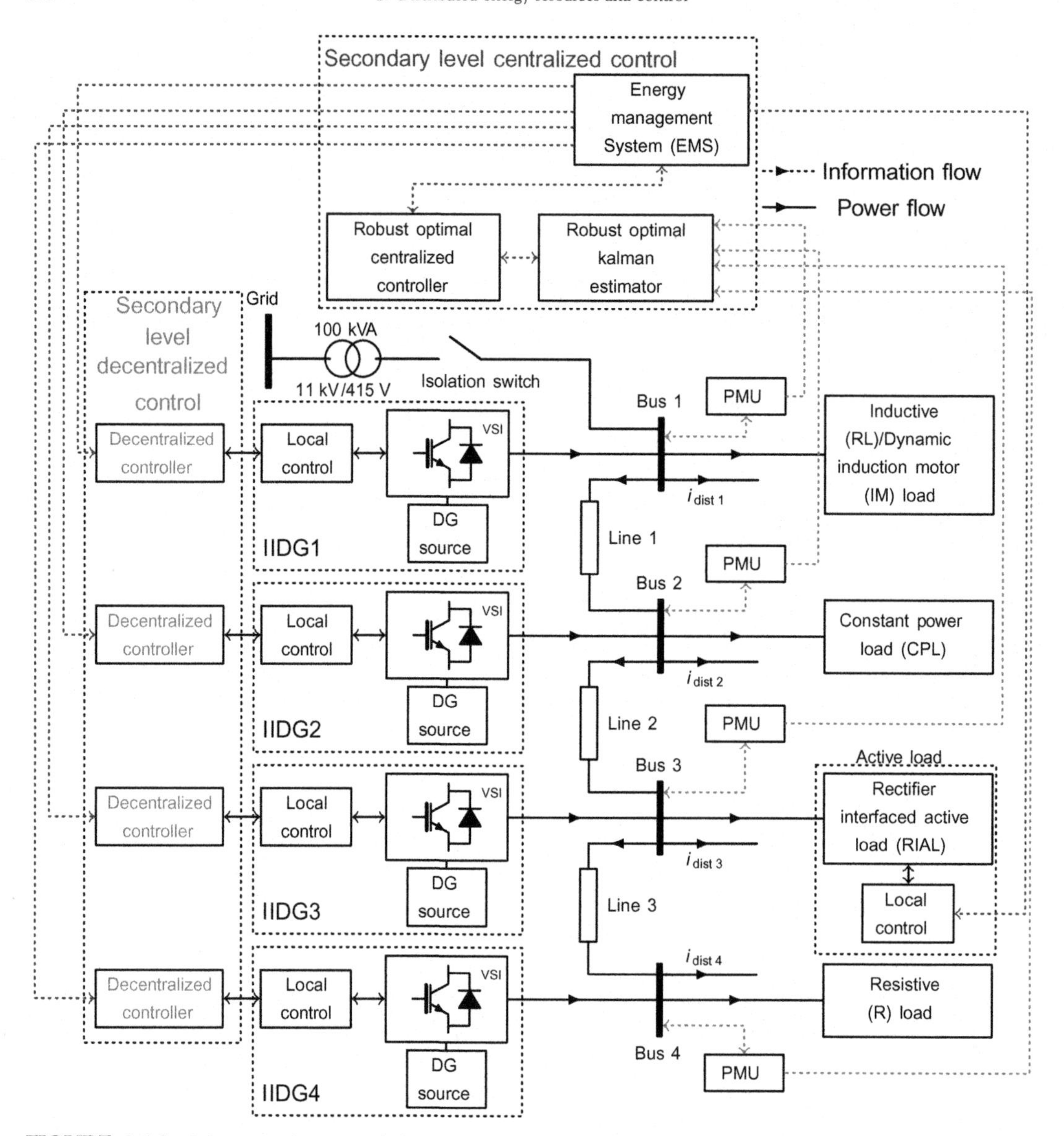

FIGURE 2.16 Schematic diagram of the studied islanded AC microgrid (ACMG) with static and dynamic loads.

output LC filter. The power processing and control sections of the RIAL are shown in Fig. 2.19 [14,15]. The control section can be divided into the outer DC voltage controller, which controls the DC voltage across the capacitor, C_{dc}, and the inner AC current controller, which controls the current through the inductor, L_f, as shown in Fig. 2.20. The MG is interfaced to the main utility grid bus through an isolation switch and a 415 V/11 kV step-up transformer at its PCC, which is bus 1. The AC MG can be operated either

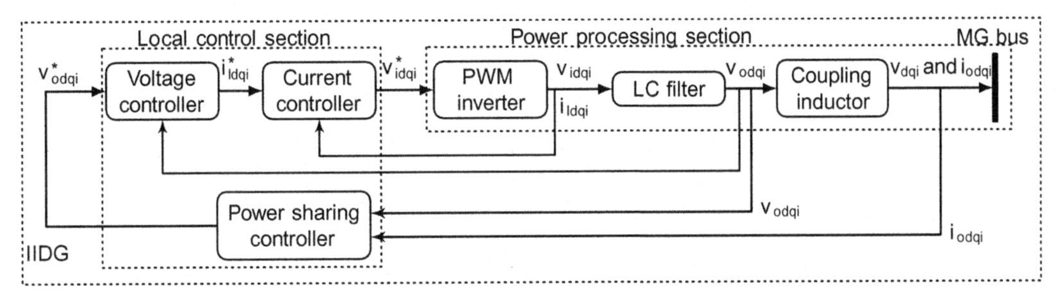

FIGURE 2.17 Power processing and control sections of an inverter-interfaced distributed generation (IIDG) unit.

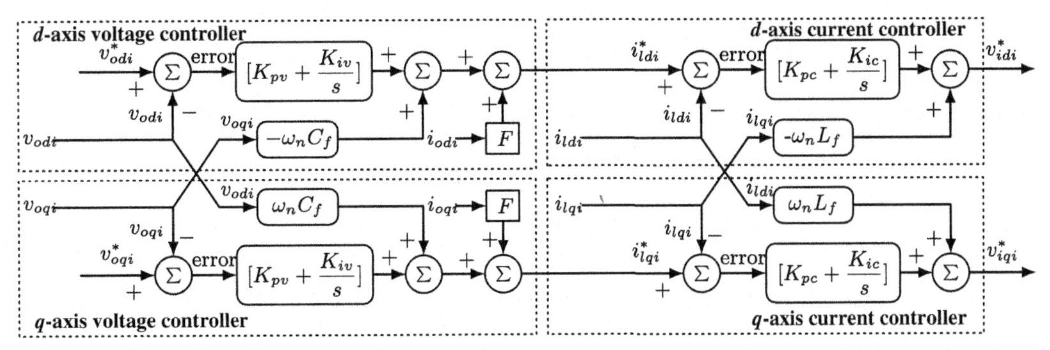

FIGURE 2.18 Voltage and current controllers of an inverter-interfaced distributed generation (IIDG) unit.

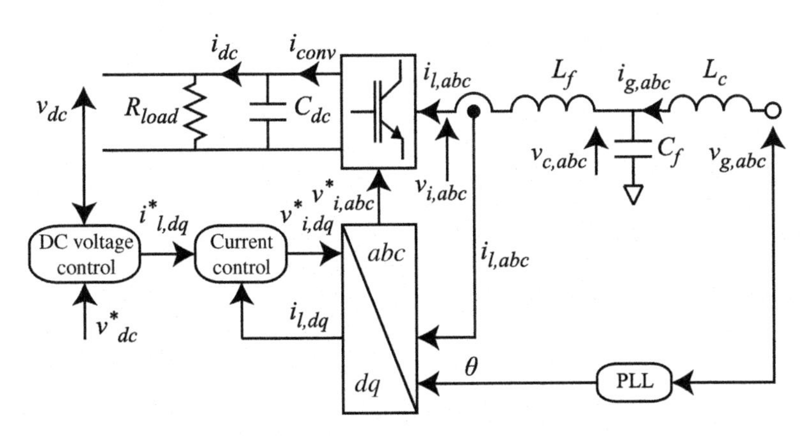

FIGURE 2.19 Rectifier interfaced active load circuit and control [15].

in islanded mode or grid-connected mode based on the status of isolation switch. The islanded operation is realized by opening the isolating switch, which disconnects the MG from the main grid, as shown in Fig. 2.16. In the island mode of operation, IIDG units are responsible for maintaining the system frequency and voltage along with

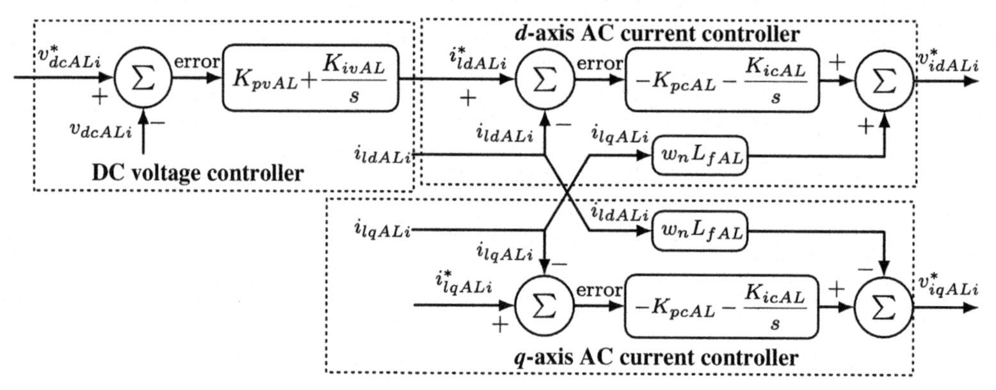

FIGURE 2.20 Direct current voltage and AC current controller of rectifier interfaced active load (RIAL).

meeting the total power demand. The parameters of the studied AC MG system are referred from Refs. [13–15].

The secondary level controllers, such as robust optimal decentralized supplementary control that loop around each IIDG unit [16], the robust optimal centralized controller [17], and the two-level hierarchical robust controller [18], have been designed for the stability enhancement of the studied AC MG system. It is worth mentioning that the decentralized and centralized controllers have been designed considering the equivalent static inductive (RL) load model of dynamic IM load at bus 1 (i.e., static loads), while the two-level hierarchical controller has been designed considering the dynamic IM load at bus 1 (i.e., static and dynamic loads). Time-domain simulations based on a linear model of the open-loop AC MG and closed-loop AC MG with these secondary controllers have been presented in Figs. 2.21, 2.22–2.24, respectively. The dynamic-step response of the deviation in frequency and voltage at all buses of the open-loop AC MG is shown in Fig. 2.21. It can be seen that the step response of the deviation in output variables increases continuously with respect to time, reflecting instability of the open-loop AC MG. It can be observed from Figs. 2.22–2.24 that the decentralized controller is faster than the centralized and the hierarchical controllers, whereas the hierarchical controller is faster than the centralized and slower than the decentralized controller. However, unlike the decentralized controller, the hierarchical controller has the ability to enhance the stability of MGs with static and dynamic loads. This can be verified from the time-domain simulation specification, that is, the settling time of the dynamic-step response of the deviation in frequency and voltage variables, given in Table 2.1.

Thus, it can be concluded that the hierarchical controllers are quite robust at dampening out the oscillations as well as settling quickly at a new operating point under the application of small load disturbances at different buses.

2.5 Conclusion

This chapter presents various types of DERs consisting of both DG units as well as distributed energy storages and their controls at different hierarchical levels. DG technologies, such as conventional/dispatchable types and renewable energy/nondispatchable

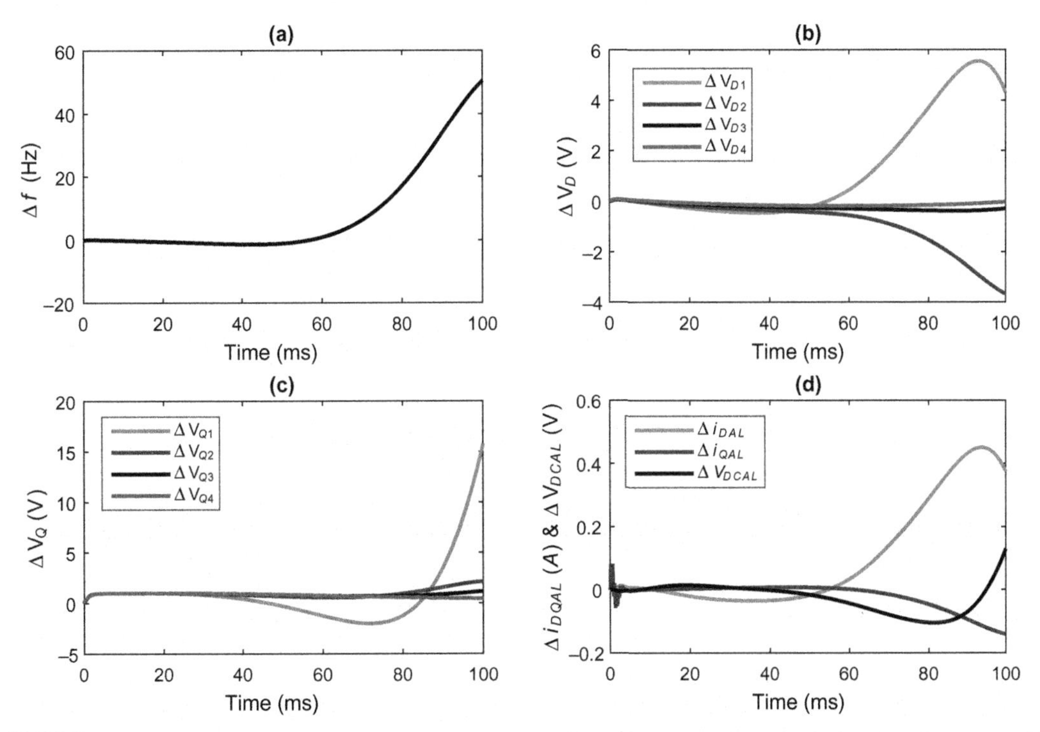

FIGURE 2.21 Response of the deviation in output variables of the open-loop-studied AC MG system. (A) Frequency. (B) D-components of bus voltages. (C) Q-components of bus voltages. (D) rectifier interfaced active load (RIAL) outputs.

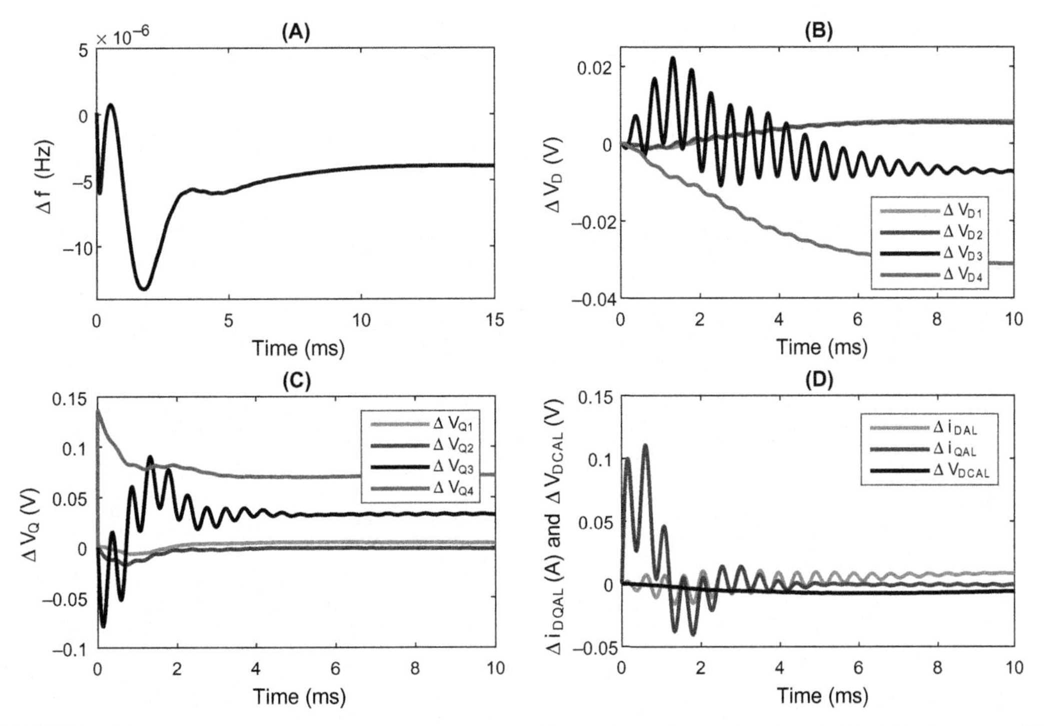

FIGURE 2.22 Response of the deviation in output variables with the decentralized controller. (A) Frequency. (B) D-components of bus voltages. (C) Q-components of bus voltages. (D) rectifier interfaced active load (RIAL) outputs.

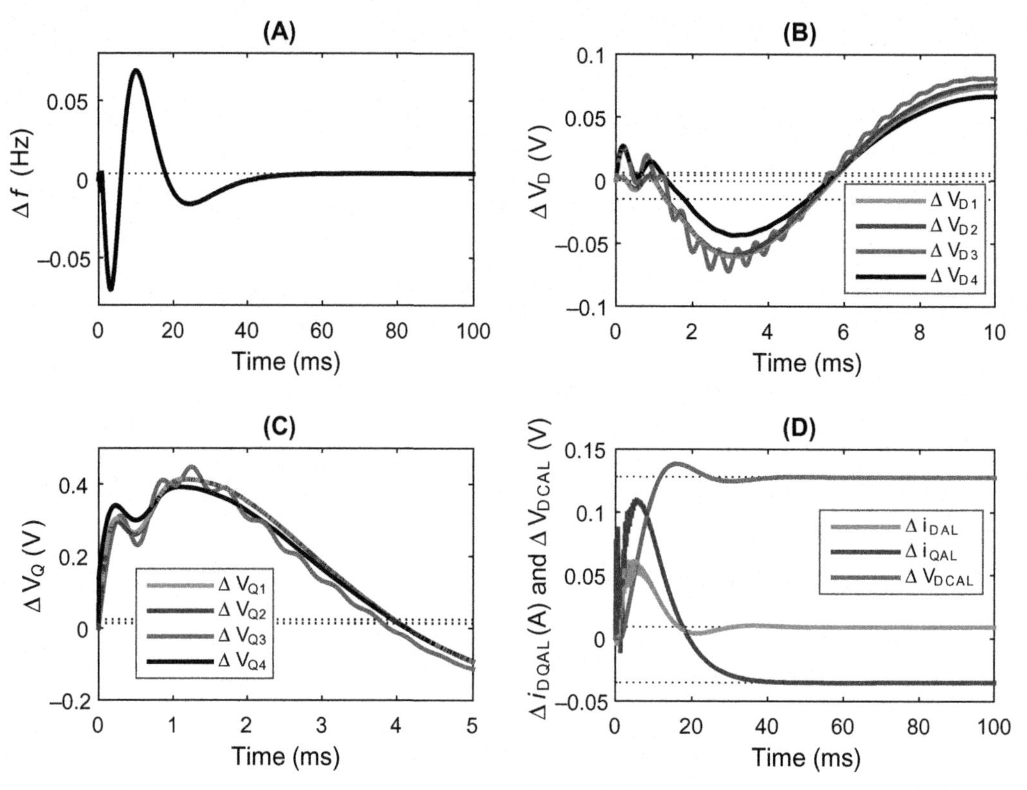

FIGURE 2.23 Response of the deviation in output variables with the centralized controller. (A) Frequency. (B) *d*-axis components of bus voltages. (C) *q*-axis components of bus voltages. (D) rectifier interfaced active load (RIAL) outputs.

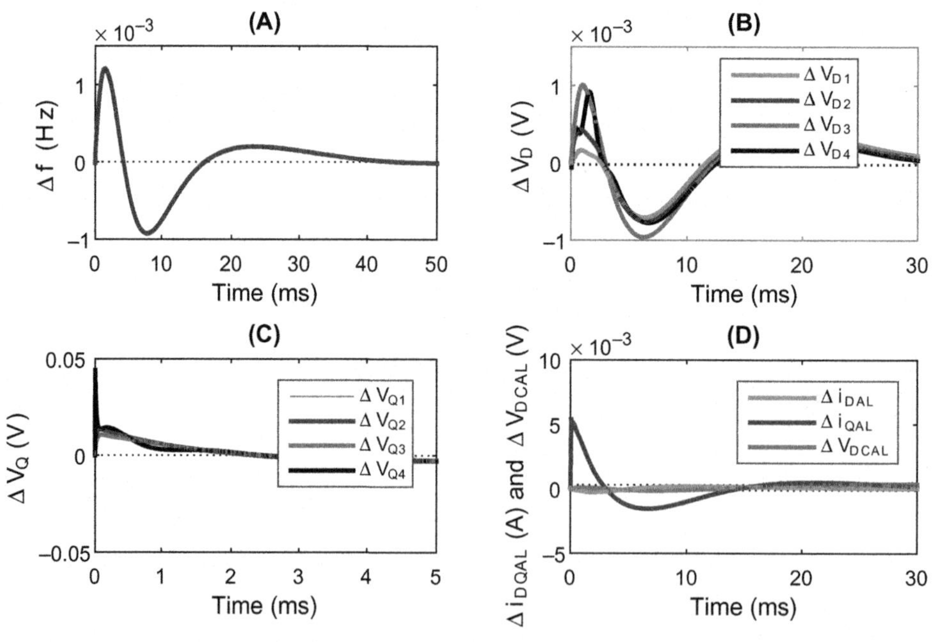

FIGURE 2.24 Response of the deviation in output variables with the two-level hierarchical controller. (A) Frequency. (B) *d*-axis components of bus voltages. (C) *q*-axis components of bus voltages. (D) Rectifier interfaced active load (RIAL) outputs.

TABLE 2.1 Performance comparison of the hierarchical controller with the decentralized and centralized controllers.

		With static loads		With static and dynamic loads
S. no.	**Deviation in output**	**With the decentralized controller: settling time (ms)**	**With the centralized controller: settling time (ms)**	**With the hierarchical controller: settling time (ms)**
1	Δf (Hz)	10.50	49.00	40.70
2	ΔV_1 (V)	8.35	40.60	38.30
3	ΔV_2 (V)	9.85	40.80	35.30
4	ΔV_3 (V)	11.35	39.80	33.80
5	ΔV_4 (V)	10.80	45.20	43.33

types, have been described. Furthermore, energy storage technologies, such as chemical energy storage, mechanical energy storage, electrical energy storage, thermal energy storage, and electrochemical energy storage, are discussed briefly. Finally, the results on the application of secondary level decentralized, centralized, and two-level hierarchical controllers to the typical AC MG system are analyzed. The results reveal that hierarchical controllers exhibit more robustness than decentralized and centralized controllers.

References

[1] R. Lasseter, et al., White paper on integration of distributed energy resources: the CERTS microgrid concept, LBNL-50829 Office of Power Technologies, the US Department of Energy, Berkeley, CA. Tech. Rep. DE-AC03-76SFW098, 2002.

[2] G. Pepermans, et al., Distributed generation: definition, benefits and issues, Energy Policy 33 (2005) 787–798.

[3] S. Chowdhury, S.P. Chowdhury, P. Crossley, Microgrids and Active Distribution Networks, Institution of Engineering and Technology, London, 2009.

[4] Jim Dyer, Consortium for Electric Reliability Technology Solutions (CERTS), March 2003.

[5] S. Parhizi, et al., State of the art in research on microgrids: a review, IEEE Access. 03 (2015) 890–925.

[6] Maine Public Utilities Commission and others, Assessment of Distributed Generation Technology Applications, Resource Dynamic Corporation, 2001.

[7] University Physics 10th Reference: edition, Addison Wisely, Young & Freedman. Available from: <http://www.physics.ubc.ca/outreach/web/phys420/index.php>.

[8] S. Sarip, et al., The potential of micro-hydropower plant for Orang Asli Community in Royal Belum State Park, Perak, Malaysia, in: Symposium on The 4th Royal Belum Scientific Expedition, 2016.

[9] X. Tan, Q. Li, H. Wang, Advances and trends of energy storage technology in microgrid, Int. J. Electr. Power Energy Syst. 44 (2013) 179–191.

[10] I.Y. Chung, et al., Control methods of inverter-interfaced distributed generators in a microgrid system, IEEE T. Ind. Appl. 46 (2010) 1078–1088.

[11] M.A. Hassan, M.A. Abido, Optimal design of microgrids in autonomous and grid-connected modes using particle swarm optimization, IEEE T. Power Electr. 26 (2011) 755–769.

[12] J.M. Guerrero, J.C. Vasquez, J. Matas, Hierarchical control of droop-controlled AC and DC microgrids—a general approach toward standardization, IEEE Trans. Ind. Electron. 58 (2011) 158–172.

[13] N. Pogaku, M. Prodanovic, T.C. Green, Modelling, analysis and testing of autonomous operation of an inverter-based MG, IEEE T. Power Electr. 22 (2007) 613–625.

[14] P.E.S.N. Raju, T. Jain, Impact of load dynamics and load sharing among distributed generations on stability and dynamic performance of islanded AC microgrids, Electr. Power Syst. Res. 157 (2018) 200–210.
[15] N. Bottrell, M. Prodanovic, T. Green, Dynamic stability of a microgrid with an active load, IEEE T. Power Electr. 28 (2013) 5107–5119.
[16] P.E.S.N. Raju, T. Jain, Optimal decentralized supplementary inverter control loop to mitigate instability in an islanded microgrid with active and passive loads, Int. J. Emerg. Electr. Power Syst. 18 (2017) 01–15.
[17] P.E.S.N. Raju, T. Jain, Robust optimal centralized controller to mitigate the small signal instability in an islanded inverter based microgrid with active and passive loads, Int. J. Elec. Power Energy Syst. 90 (2017) 225–236.
[18] P.E.S.N. Raju, T. Jain, A two-level hierarchical controller to enhance stability and dynamic performance of islanded inverter-based microgrids with static and dynamic loads, in: IEEE Trans. Ind. Inf. (early access), 2018. doi:10.1109/TII.2018.2869983

CHAPTER

3

Use of agents for isolated microgrids with frequency regulation

Juan M. Ramirez[1] *and Alejandra Pérez Pacheco*[2]

[1]Research Center and Advanced Studies, National Polytechnic Institute, Zapopan, Mexico

[2]Energy Control National Center, Peninsular Control Area, Mérida, Mexico

3.1 Introduction

The available research work on microgrids (MGs) is vast, ranging from power quality constraints, energy storage, islanding, and voltage control to frequency control.

The intermittent nature of the renewable energies employed in MGs may lead to large frequency fluctuations when the load frequency control capacity is not enough to compensate for the imbalance of generation and demand. To operate an MG, a balance between generation and load is required. Modifications of this balance generates a frequency deviation which, depending on the severity, could lead to cascading outages and/or undesirable load shedding [1–5]. The rate of change of frequency depends on the system inertia and, in an isolated MG, the frequency response tends to decay. A study on the Pennsylvania–New Jersey–Maryland Interconnection revealed that portfolios with a large percentage of renewable energy penetration and removal of baseload led to the lowest composite reliability indices [6]. In response, the Federal Energy Regulatory Commission (FERC) is considered to have an explicit requirement to provide frequency response for all new interconnected resources [4].

Several studies have been performed to mitigate the negative impact of renewable energy on the frequency response. This chapter focuses on the use of wind turbine (WT) since this technology has a significant percentage of penetration and is the fastest growing market [7,8]. In Ref. [9], the doubly fed induction generator (DFIG) control is modified to introduce the inertia response and it proves that with the new control the WT is able to provide kinetic energy. Then, the authors enhance the inertia contribution with the *deloading* capabilities by reducing the generator torque set point [10]. In Refs. [11,12], the authors develop a mathematical formulation for DFIG and fixed-speed WT that can be used in the power flow to calculate the steady-state frequency deviation.

Distributed Energy Resources in Microgrids
DOI: https://doi.org/10.1016/B978-0-12-817774-7.00003-X

Distributed generation and micro networks are widely related. It can be said that micro networks are a natural evolution of distributed generation. In fact, they have been defined as a solution which can take advantage of the potential of distributed generation by being integrated into the loads, where the set of loads and micro generators operates as a single system supplying power and heat. According to the review carried out, research regarding the optimization of hybrid generation systems and micro networks has focused mainly on the economic and environmental aspects, although restrictions that guarantee energy quality and stability are also important [13].

The design of isolated micro networks has been formulated as an optimization problem, which allows the appropriate sizes of the components to be obtained, costs to be minimized and the loss of power supply probability to be restricted [14]. Recently, different formulations have been made for the problem of dimensioning micro networks, taking into account hybrid generation and energy storage [15–17]. According to Ref. [18], sensitivity analysis has been revealed as a critical step in the planning of micro networks to develop a robust design in terms of economic feasibility.

Likewise, the operation of micro networks connected to the macro power system has been considered. In this sense, the power network has been modeled as a system of systems through a control strategy that incorporates distributed energy resources (DERs), storage devices, and loads; an objective function is used to minimize power exchanges between the micro networks and guarantee operation within technical limits [19].

Thus an energy management system (EMS) is required for the cooperative optimal operation of MGs [20]. A novel EMS based on a mixed integer linear programming framework is proposed in Ref. [21] to minimize operation costs. In Ref. [22], authors present an energy management strategy employing a stochastic programming model to minimize the expected operation cost. In Ref. [23], studies are conducted on EMS under the power-trading possibilities based on the noncooperative game theory.

Information technologies offer exciting opportunities to rethink the process of knowledge acquisition and achieve, among others, the following benefits: integration of media (text, audio, animation, and video), interactivity, and access to large amounts of information. To achieve these benefits, the use of computational agents is proposed. An agent is a hardware/software system that interacts with its environment (or other agents or humans), guided by one or several purposes, is proactive (reacts to events and sometimes anticipates making proposals), adaptable (can face situations novel), sociable (communicates, cooperates, or negotiates), and their behavior is predictable in a certain context.

In this chapter, some models presented in Ref. [12] are used on a decentralized optimization formulated by a multiagent system (MAS). The use of a decentralized approach ensures that the failure of an agent will not affect the entire system [24], where a backward–forward sweep method is applied and the MAS is able to find a solution even when one of them fails for several seconds. The communication between adjacent neighbors requires a minimum of information exchange, which proves to be useful, especially when not all the distributed DERs belong to the same utility and there is a need to maintain private information [25]. The optimization problem is separated into simpler subproblems and each agent works to achieve its specific objective while reaching a common goal. In Ref. [26], the authors integrate the MG market operations and implementation of DER, effectively demonstrating the autonomy and sociality that a MAS contributes to the power flow formulation in an MG. The IEEE Power Engineering Society's MAS Working Group did a compilation of the benefits of using a MAS in power engineering and explained them from two points: (1) construction of robust, flexible, and extensible systems and (2) the modeling approach [27]. This chapter uses the former.

The main contributions of this chapter deal with:

1. Abstraction of the optimal power flow problem for a belief–desire–intentions (BDI) architecture.
2. Formulation for a decentralized power flow using a MAS based on the Newton–Raphson method with a decoupled Jacobian.
3. Cooptimization for economic dispatching and frequency regulation.
4. The behavior analysis of DFIG WT under a droop control to contribute to the primary frequency response in interconnected and islanding mode.

3.2 Frequency control

One of the main indexes to evaluate the quality of the power supply is the frequency deviation. A healthy system maintains a stable frequency and when the system operates on a stressed mode it reflects itself in the frequency and voltage.

The key element for a steady operation is the balance between the generation and demand. When this balance is not met, it generates a frequency deviation.

Thus when supply exceeds the demand, the frequency changes to a value above the nominal (60 Hz) and, when the demand is higher than the power supply, the frequency decays. Fig. 3.1 mimics such balance.

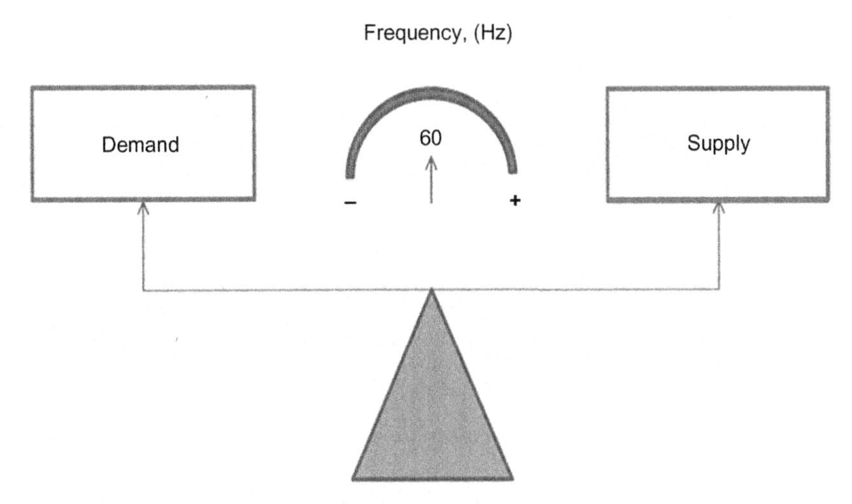

FIGURE 3.1 Generation/demand balance and frequency deviation.

In this context, there are three important concepts to address frequency response, inertial response, and primary frequency response. The first two are concerned with terms involved in frequency control, and the last is addressed in the following.

Frequency response is a measure of a system's ability to arrest and stabilize frequency deviations. This is affected by the collective response of generation and load in the system.

Inertial response is also named system inertia. The rotational masses of a generator have the capability to release or absorb kinetic energy and it influences how fast the frequency changes during an imbalance. The lower inertial response, the faster the rate of change of frequency during disturbances. The system inertia gives time for the primary frequency response to arrest the frequency deviation and stabilize the system [4].

3.2.1 Primary frequency response

Primary frequency response is the first stage of the frequency control, as shown in Fig. 3.2. It is a net of changes in the generation active power and power consumed by the loads in response to a frequency deviation. Primary frequency response is mainly provided by the autonomous actions of the turbines governors, while some of it is provided by frequency responsive loads [4].

Primary frequency control is the one in charge of restoring the system to a power equilibrium after a few seconds of the imbalance. To perform this action, the generating units are provided with a speed/droop control. When there is a load change in the system, the electrical torque of a generating unit changes, creating a mismatch between electrical and mechanical torque. This causes the unit to accelerate or slow down until the mismatch is canceled.

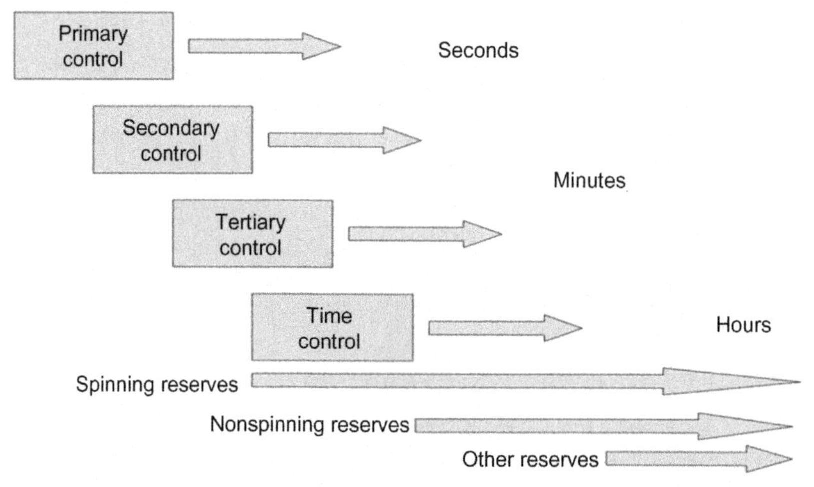

FIGURE 3.2 Frequency control continuum [4].

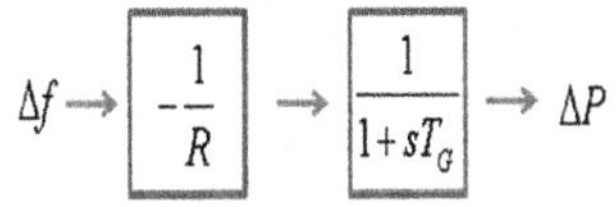

$\Delta f \rightarrow \boxed{-\frac{1}{R}} \rightarrow \boxed{\frac{1}{1+sT_G}} \rightarrow \Delta P$

FIGURE 3.3 Block diagram of a speed governor with droop [28].

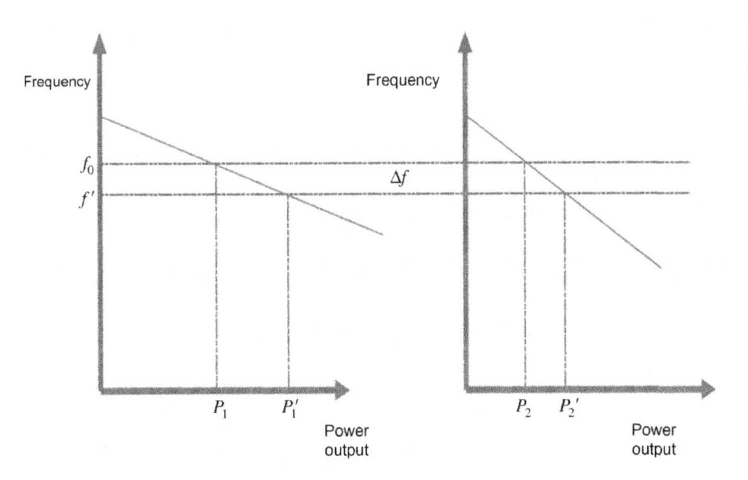

FIGURE 3.4 Ideal steady-state characteristics of governors with different droops [28].

3.2.2 Droop control

As explained in Ref. [28], the droop determines the steady-state speed versus load characteristics of the generation unit. The simplified block diagram of a governor is depicted in Fig. 3.3. It is a proportional controller with a gain of $1/R$ (droop). Δf is the frequency deviation and ΔP is the power output.

The droop characteristic allows sharing the loads between generators. Fig. 3.4 shows two generating units with different droop characteristics, initially working at nominal frequency f_0. When the units slow down due to an increase in load, the governors increase the output until reaching a new equilibrium in a new value of frequency f'. This means that using the droop control generates a steady-state frequency deviation which is then corrected by the secondary frequency control. The research in this chapter focuses on such deviation. In Refs. [28,29], the equation to calculate the variation in the power output of the ith synchronous generator, is the following [30]:

$$\frac{\Delta P_G}{P_{nG}} = -\frac{1}{\delta_G}\frac{\Delta f}{f_0} \tag{3.1}$$

where ΔP_G stands for the power output variation of the generator, P_{nG} is the rated power, δ_G its permanent droop, Δf becomes the frequency deviation, and f_0 the system's nominal frequency.

From Eq. (3.1) the concept declared at the beginning of this chapter can be proven: if the frequency deviation Δf is positive, the variation in the power output of the generator is negative. This implies that the supply exceeds the demand and, to attain an equilibrium, the generator reduces the output. The opposite with a negative frequency deviation is also true.

The final output of the synchronous generator is the initial set point plus the variation due to frequency change. Assuming that power output change is the difference between the setting point, P_{Gset}, and the new power after the change in frequency:

$$P_G = P_{Gset} - K\,\Delta f \tag{3.2}$$

$$K = \frac{P_{nG}}{\delta_G f_0} \tag{3.3}$$

where K is the generator's power frequency characteristic. Substituting Eq. (3.2) in the conventional power flow formulation results in the expression for a synchronous generator as used in the following:

$$\Delta P_k = \left(P_{Gset} - K\Delta f\right) - P_{LK} - P_k^{calc} \tag{3.4}$$

3.3 Wind power and frequency response

Because of the increased penetration of renewable energies, the system inertia is decaying. In the previous section, it is stated why this is problematic for the power system. Not only is there a lack of system inertia, but most renewable energies do not provide frequency response. One of the solutions is the *virtual inertia* in wind turbines [9–11]. The concept of *virtual inertia* in DFIG WT comes from the stored kinetic energy in the rotating mass of the blades [31].

For an air stream with density ρ flowing at variable speed ω through a cross-sectional rotor area **A**, the real mechanical power p^m output is given by:

$$p^m = \frac{1}{2}\rho A \omega^3 C_p \tag{3.5}$$

The fraction of power extracted by a real wind rotor may be determined by the performance coefficient C_p and is based on the modeling turbine characteristics [32]:

$$n_o = \mu_t \mu_g \tag{3.6}$$

The electrical power output (3.7) can be expressed by the product of the mechanical power (3.5) and the efficiency coefficient (3.6), which represents the efficiency of the turbine, μ_t, and the generator μ_g.

$$P_w = n_0 p^m \tag{3.7}$$

Fig. 3.5 shows a schematic of the DFIG. These turbines are a popular type of variable speed WT. When they are connected to a system they have a negligible effect on the system speed regulation because of the decoupling between the electrical and mechanical systems. However, they can have a negative impact if the baseload power plant is substituted by WT.

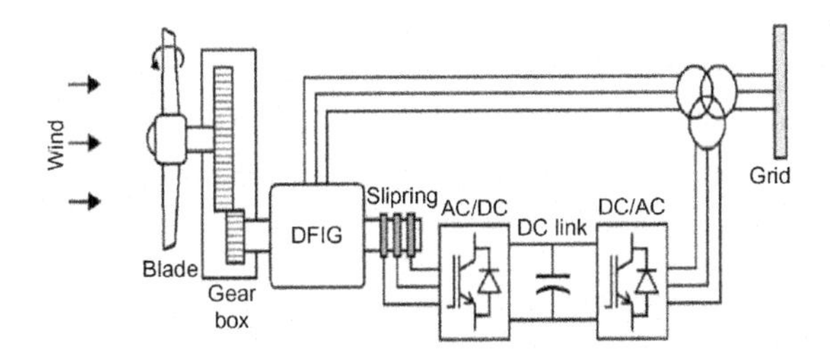

FIGURE 3.5 Schematic of a DFIG wind turbine [33].

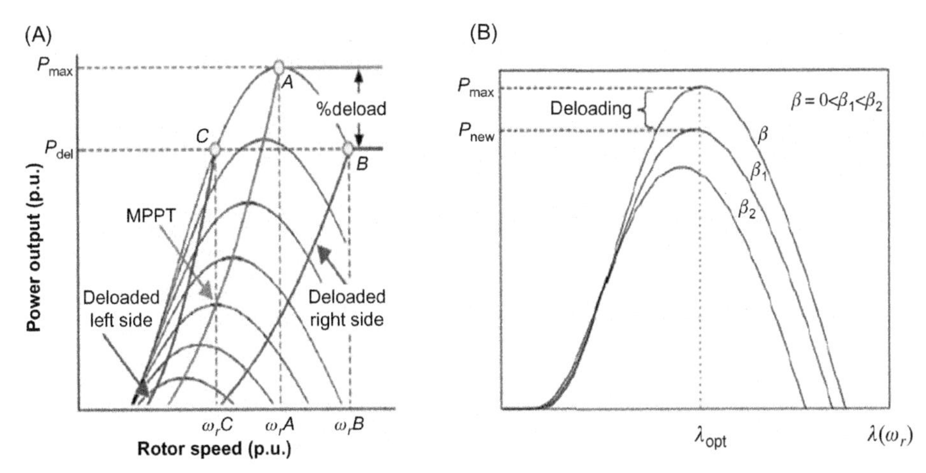

FIGURE 3.6 Deloading a wind turbine by: (A) shifting the operating point to the left or right [34] and (B) controlling the pitch angle [11].

Similarly to the synchronous generator, the DFIG's model assumes that it is possible to provide primary frequency regulation by means of a spinning reserve. The *deloading* factor k_d provides the wind turbine with this reserve (Fig. 3.6). The wind turbine output power, P_w, is calculated as explained in Ref. [33]:

$$P_{wset} = (1 - k_d)P_w \tag{3.8}$$

From Eq. (3.8), and assuming that the control works as a governor, the wind turbine power output with frequency control becomes:

$$P_{Gw} = (1 - k_d)P_w - K\Delta f \tag{3.9}$$

The mismatch expression used for the power flow formulation is:

$$\Delta P_k = (1 - k_d)P_w - K\Delta f - P_{Lk} - P_k^{calc} \tag{3.10}$$

It is noteworthy that the WT is only able to help in frequency deviation when this is negative. If the deviation is positive, then the supply needs to be curtailed further, but the wind turbine cannot do this anymore. Likewise, the turbine's droop characteristic is treated as static in this research. In Ref. [34], an adaptive droop is proposed, which helps with the stabilization of the turbines when the available wind power is below a certain threshold.

3.4 Example

To better understand how the frequency deviation affects the turbines, consider the following example: a wind turbine producing 0.3 p.u. of active power is supplying a load of $S = 0.5 + j0.2$ p.u. The small system is connected to the grid through a line with an impedance of $z = 0.005 + j0.05$ p.u. (Fig. 3.7).

Analyzing only the active power, since the frequency deviation depends on it, the generation has to be equal to the load (neglecting losses in the system):

$$P_G + P_W = P_L \tag{3.11}$$

where P_G is the power from the grid, P_W the one stemming from the wind turbine and P_L is the load. Substituting the corresponding equations for the synchronous generator (3.2) and the DFIG with frequency response (3.9) the following expression is obtained:

$$P_{G1set} - K_1\Delta f + (1 - k_d)P_w - K_w\Delta f = P_L \tag{3.12}$$

Using a permanent droop, δ, of 4% for both turbines, the corresponding power frequency characteristic, K, is 7.5 for the WT and 25 if we consider that the generator from the main grid is able to provide up to 1 p.u. of active power. For the *deloading* factor, k_d becomes 5%. Table 3.1 summarizes the characteristics of each turbine.

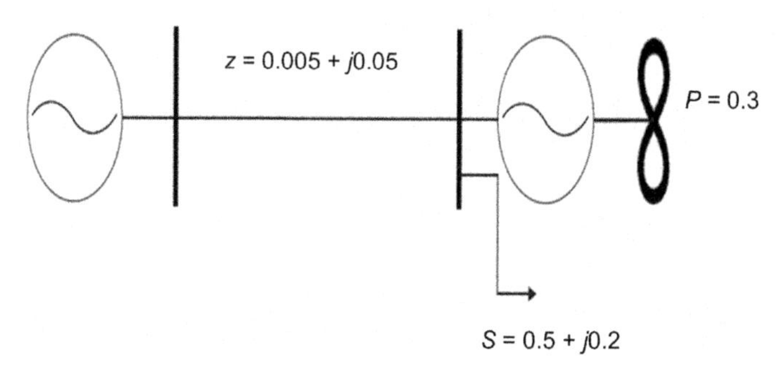

FIGURE 3.7 Test system.

TABLE 3.1 Turbine characteristics.

Element	SG	WT
P_n	1	0.3
Δ	4%	4%
K	25	7.5
k_d	NA	5%

Solving the frequency deviation in Eq. (3.12):

$$\Delta f = -\frac{P_L - P_{G1set} - (1 - k_d)P_w}{K_1 + K_w} \tag{3.13}$$

If the initial setting point of the synchronous generator is set to 0.2150, then the frequency deviation, Δf, is equal to 0. This means that the final output of the WT is the initial setting point 0.2850. Similarly, if the initial setting point, P_{G1set}, is less than needed, for example, 0.210 p.u., then the system has a frequency deviation of -0.0092 Hz. The final output of the WT is 0.2862 and 0.2138 for the main grid. As it can be observed from the results of this small test, the remaining missing supply is divided between both generators according to the capabilities and droop of each one.

3.5 Microgrids and the power flow formulation

The electric grid is no longer a one-way system and MGs are considered one of the most important research areas to make a successful transition toward the *smartgrid*. MGs are a combination of smart electric devices, power generation, and storage resources, connected to one or many loads that can operate in interconnected or islanding mode at will. Ironically, the future of the *smartgrid* lays upon an idea that appeared decades ago. Thomas Edison had the idea that the power system would involve small firms generating power for individual businesses. By 1886, 58 direct current (DC) MGs were installed but, with the alternating current enabling long-distance transmission of electricity, they were replaced by the larger utility monopolies. This monopoly lasted a century until the new wave of MGs appeared [35].

Depending on the size of the MG and the elements that compose it, there are also nanogrids and picogrids. According to Ref. [36], picogrids are responsible for peak shaving in households and they also carry out the load management side. Nanogrids serve single buildings and try to maximize the DER integration.

In 2015 about 0.1% of the total installed electric generating capacity in the United States was provided by MGs. They are mainly used in hospitals, military bases, and campuses because of their benefits: resiliency, efficiency, reliability, and cost saving [35].

3.5.1 Frequency in the microgrid

To operate an MG, a balance between generation and load is required. Modification of this balance generates a frequency deviation which, depending on the severity, could lead to cascading outages and/or undesirable load shedding [4]. Operating in interconnected mode, the MG is able to regulate the frequency mainly by the main grid; problems appear more commonly in the islanding mode, when the frequency regulation depends on the resources mixture of the MG.

The rate of change of frequency depends on the system inertia and, in an isolated MG, the frequency response tends to decay. One of the major components in an MG is the DER which, for the most part, does not participate in frequency response. On some days, the renewable energy penetration can exceed the 49%. When a fluctuation on frequency passes a threshold, the inverters usually disconnect from the system to avoid damages, which increases the frequency deviation.

Recently, new forms of contributing to the ancillary services, including frequency response, are under study. The California independent system operator (CAISO) conducted a study with the National Renewable Energy Laboratory (NREL) and First Solar to demonstrate the operating flexibility of a 300 MW solar photovoltaic (PV). The inverters followed the automatic generation control signals from CAISO, resulting in not only meeting the requirements for natural gas-fired plants but sometimes exceeding them [37].

The use of inverters to provide ancillary services in the MG is a new and promising trend. However, the development of frequency response in wind turbines is mature and well documented, which facilitates their integration into this research. Furthermore, wind availability is not limited to certain amount of hours during the day, like solar, providing a more reliable option.

3.6 Microgrid model

Consider an MG including solar and wind energy, energy storage system (ESS), dispatchable generators, and loads. First, the description of the general MG model is presented, and then the model representing each of the elements.

In the bus injection model, the complex net power S in a bus k can be defined as the difference between the generated power S_{Gk} and the demand S_{Lk},

$$S_k = S_{Gk} - S_{Lk} \tag{3.14}$$

Representing this net power in terms of voltages and admittances, it can be rewritten in active and reactive power:

$$\begin{aligned} P_k^{calc} &= V_k \sum_{j=1}^{n} V_j \left[G_{kj} \cos(\theta_k - \theta_j)\right] \\ Q_k^{calc} &= V_k \sum_{j=1}^{n} V_j \left[B_{kj} \sin(\theta_k - \theta_j)\right] \end{aligned} \tag{3.15}$$

Eq. (3.15) are the calculated powers in the bus, where G_{kj} and B_{kj} are the real and imaginary parts of the admittance matrix of a n bus system. V and θ represent the voltage magnitude and its phase angle at the bus, respectively. The mismatch between the calculated powers and the real ones can be expressed as:

$$\begin{aligned} \Delta P_k &= P_{Gk} - P_{Lk} - P_k^{calc} \\ \Delta Q_k &= Q_{Gk} - Q_{Lk} - Q_k^{calc} \end{aligned} \tag{3.16}$$

Minimizing the mismatch vector up to a predefined tolerance, yields the values of the voltage magnitude and phase angle for all the buses in the system.

3.6.1 Synchronous generator

A frequency-dependent model of synchronous generator corresponds to a generator that adjusts the active output power (P_G) by the prime mover's static response [24] (Eqs. 3.1–3.4).

3.6.2 Loads

Loads are modeled by the impedance-current-power (ZIP) model. The frequency dependency is introduced through factors K_q and K_p, depending if it is a reactive or an active load [38],

$$P_L = P_{Lset}\left(1 + K_p\Delta f\right)\left(a_1V^2 + a_2V + a_3\right) \quad (3.17)$$

$$Q_L = Q_{Lset}\left(1 + K_q\Delta f\right)\left(b_1V^2 + b_2V + b_3\right) \quad (3.18)$$

where V is the bus voltage, P_{Lset} and Q_{Lset} are the initial specified active and reactive power, respectively. Coefficients a_1, b_1, a_2, b_2, a_3, and b_3 represent the constant impedance, current, and power, respectively. Substituting Eqs. (3.17) and (3.18) in a power flow formulation, it is possible to take into account these types of loads:

$$\Delta P_k = P_{Gk} - P_{Lset}\left(1 + K_p\Delta f\right)\left(a_1V^2 + a_2V + a_3\right) - P_k^{calc} \quad (3.19)$$

$$\Delta Q_k = Q_{Gk} - Q_{Lset}\left(1 + K_q\Delta f\right)\left(b_1V^2 + b_2V + b_3\right) - Q_k^{calc} \quad (3.20)$$

3.6.3 Wind turbines and solar cells

Similarly to the synchronous generator, the DFIG's model assumes that it is possible to provide primary frequency regulation by means of a spinning reserve [9,10,12]. The *deloading* factor k_d provides the wind turbine with this reserve. The wind turbine output power, P_w, is calculated as explained in Ref. [39] (Eqs. 3.8–3.10).

3.6.4 Photovoltaic plants

Given temperature T and irradiance G, the output power p^p of a panel is related to the standard test conditions such as the maximum power correction factor for temperature γ, maximum power output p^p_{max}, temperature T_{STC} and irradiance G_{STC} [40]:

$$p^p = \frac{G}{G_{STC}} p^p_{max}[1 + \gamma(T - T_{STC})] \quad (3.21)$$

Thus regarding the PV panels, their available power becomes:

$$P_{PV} = n_{PV}p^p \quad (3.22)$$

where n_{PV} is the total number of PV panels and p^p is the output power of a single panel.

3.6.5 Storage devices

It is assumed that there are two types of ESS in the MG: electric vehicles (EVs) with only grid-to-vehicle capability, and a Li-ion battery. The state of charge (*SOC*) is modeled by Eq. (3.23), where P_{ch} and P_{dch} are charging and discharging power at time t, and B_{cap} is the battery capacity:

$$SOC_{avg}^{t+1} = SOC_{avg}^{t} + \frac{P_{ch}^{t} + P_{dch}^{t} n_b}{B_{cap}} \tag{3.23}$$

The efficiency, n_b, used for the EV is 92% [41] using a type two charger, while that of the stationary battery is 80%.

3.7 Decentralized power flow formulation, including frequency deviation

In Ref. [42], a power flow formulation including load and generator characteristics as well as system control devices is proposed. The power flow is solved by the Newton–Raphson method. For an n buses power system, where $l_1 \ldots l_l$ represent the PQ buses, the matrix structure of the system is as follows:

$$\begin{bmatrix} \Delta P_1 \\ \vdots \\ \Delta P_n \\ \Delta Q_{l1} \\ \vdots \\ \Delta Q_{ll} \end{bmatrix} = \begin{bmatrix} J_1 & J_2 & \frac{\partial P}{\partial \Delta f} \\ J_3 & J_4 & \frac{\partial Q}{\partial \Delta f} \end{bmatrix} \begin{bmatrix} \Delta\theta_2 \\ \vdots \\ \Delta\theta_n \\ \Delta V_{l1} \\ \vdots \\ \Delta V_{ll} \\ \Delta\Delta f \end{bmatrix} \tag{3.24}$$

3.7.1 Decentralized power flow

A system is split into two zones (Fig. 3.8), where bus k belongs to zone **A** and bus m to zone **B**, and it is assumed that this is the only connection between both zones. The Jacobian matrix of this system has the form:

$$\begin{bmatrix} \multicolumn{2}{c}{\mathbf{A}} & \begin{matrix} \frac{\partial P_k}{\partial \theta_m} & \frac{\partial P_k}{\partial V_m} \\ \frac{\partial Q_k}{\partial \theta_m} & \frac{\partial Q_k}{\partial V_m} \end{matrix} \\ \begin{matrix} \frac{\partial P_m}{\partial \theta_k} & \frac{\partial P_m}{\partial V_k} \\ \frac{\partial Q_m}{\partial \theta_k} & \frac{\partial Q_m}{\partial V_k} \end{matrix} & \mathbf{B} \end{bmatrix} \tag{3.25}$$

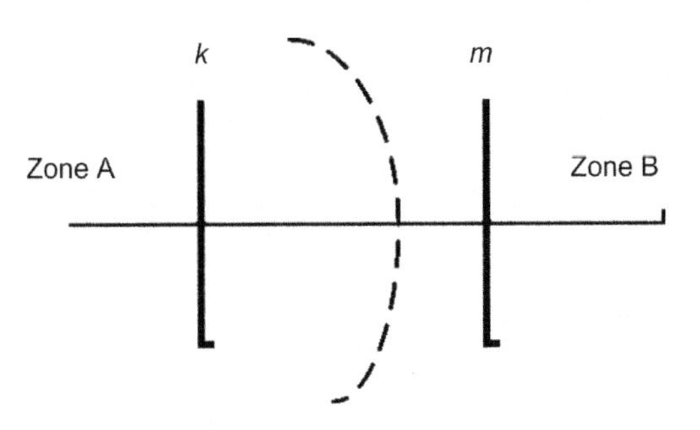

FIGURE 3.8 Breaking down the system into two zones.

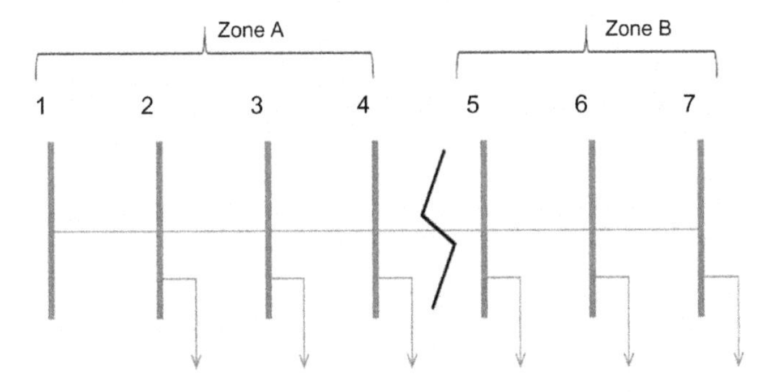

FIGURE 3.9 Seven buses test system.

Note that there are eight nonzero elements connecting both zones. Moving these elements outside the Jacobian matrix allows one to solve the two zones individually but in an iterative way, because each zone requires the last increment in the variables of the other one.

$$\begin{bmatrix} \Delta P_k - \dfrac{\partial P_k}{\partial \theta_m}\Delta\theta_m - \dfrac{\partial P_k}{\partial V_m}\Delta V_m \\ \Delta Q_k - \dfrac{\partial Q_k}{\partial \theta_m}\Delta\theta_m - \dfrac{\partial Q_k}{\partial V_m}\Delta V_m \\ \Delta P_m - \dfrac{\partial P_m}{\partial \theta_k}\Delta\theta_k - \dfrac{\partial P_m}{\partial V_k}\Delta V_k \\ \Delta Q_m - \dfrac{\partial Q_m}{\partial \theta_k}\Delta\theta_k - \dfrac{\partial Q_m}{\partial V_k}\Delta V_k \end{bmatrix} = \begin{bmatrix} \mathbf{A} & \mathbf{0} \\ \mathbf{0} & \mathbf{B} \end{bmatrix} \begin{bmatrix} \Delta\theta_k \\ \Delta V_k \\ \Delta\theta_m \\ \Delta V_m \end{bmatrix} \tag{3.26}$$

This approach has the following advantages: (1) none of the zones will be idle; (2) there is no need to duplicate the common variables; (3) the optimization can be done without using Lagrangian approaches; and (4) only one zone solves directly the frequency deviation but it gets updated in all the zones.

3.7.2 Example

Consider the test system depicted in Fig. 3.9, where buses 1–4 belong to zone A and buses 5–7 to zone B. Table 3.2 summarizes the data of the system per unit. Bus 1 is considered the slack bus.

TABLE 3.2 Line impedance and load value.

Element	Value
Line resistance	6:25e − 4
Line inductance	3.125e − 4
Load	0.95 + *j*0.31

Notice that there is only one line connecting both zones, this is between buses 4 and 5. Following the formulation that was presented before, it is possible to solve two separate systems:

$$\begin{bmatrix} \Delta P_2 \\ \Delta P_3 \\ \Delta P_4 - \frac{\partial P_4}{\partial \theta_5}\Delta\theta_5 - \frac{\partial P_4}{\partial V_5}\Delta V_5 \\ \Delta Q_2 \\ \Delta Q_3 \\ \Delta Q_4 - \frac{\partial Q_4}{\partial \theta_5}\Delta\theta_5 - \frac{\partial Q_4}{\partial V_5}\Delta V_5 \end{bmatrix} = \begin{bmatrix} \frac{\partial P_2}{\partial \theta_2} & \frac{\partial P_2}{\partial \theta_3} & \frac{\partial P_2}{\partial \theta_4} & \frac{\partial P_2}{\partial V_2} & \frac{\partial P_2}{\partial V_3} & \frac{\partial P_2}{\partial V_4} \\ \frac{\partial P_3}{\partial \theta_2} & \frac{\partial P_3}{\partial \theta_3} & \frac{\partial P_3}{\partial \theta_4} & \frac{\partial P_3}{\partial V_2} & \frac{\partial P_3}{\partial V_3} & \frac{\partial P_3}{\partial V_4} \\ \frac{\partial P_4}{\partial \theta_2} & \frac{\partial P_4}{\partial \theta_3} & \frac{\partial P_4}{\partial \theta_4} & \frac{\partial P_4}{\partial V_2} & \frac{\partial P_4}{\partial V_3} & \frac{\partial P_4}{\partial V_4} \\ \frac{\partial Q_2}{\partial \theta_2} & \frac{\partial Q_2}{\partial \theta_3} & \frac{\partial Q_2}{\partial \theta_4} & \frac{\partial Q_2}{\partial V_2} & \frac{\partial Q_2}{\partial V_3} & \frac{\partial Q_2}{\partial V_4} \\ \frac{\partial Q_3}{\partial \theta_2} & \frac{\partial Q_3}{\partial \theta_3} & \frac{\partial Q_3}{\partial \theta_4} & \frac{\partial Q_3}{\partial V_2} & \frac{\partial Q_3}{\partial V_3} & \frac{\partial Q_3}{\partial V_4} \\ \frac{\partial Q_4}{\partial \theta_2} & \frac{\partial Q_4}{\partial \theta_3} & \frac{\partial Q_4}{\partial \theta_4} & \frac{\partial Q_4}{\partial V_2} & \frac{\partial Q_4}{\partial V_3} & \frac{\partial Q_4}{\partial V_4} \end{bmatrix} \begin{bmatrix} \Delta\theta_2 \\ \Delta\theta_3 \\ \Delta\theta_4 \\ \Delta V_2 \\ \Delta V_3 \\ \Delta V_4 \end{bmatrix}$$

$$\begin{bmatrix} \Delta P_5 - \frac{\partial P_5}{\partial \theta_4}\Delta\theta_4 - \frac{\partial P_5}{\partial V_4}\Delta V_4 \\ \Delta P_6 \\ \Delta P_7 \\ \Delta Q_5 - \frac{\partial Q_5}{\partial \theta_4}\Delta\theta_4 - \frac{\partial Q_5}{\partial V_4}\Delta V_4 \\ \Delta Q_6 \\ \Delta Q_7 \end{bmatrix} = \begin{bmatrix} \frac{\partial P_5}{\partial \theta_5} & \frac{\partial P_5}{\partial \theta_6} & \frac{\partial P_5}{\partial \theta_7} & \frac{\partial P_5}{\partial V_5} & \frac{\partial P_5}{\partial V_6} & \frac{\partial P_5}{\partial V_7} \\ \frac{\partial P_6}{\partial \theta_5} & \frac{\partial P_6}{\partial \theta_6} & \frac{\partial P_6}{\partial \theta_7} & \frac{\partial P_6}{\partial V_5} & \frac{\partial P_6}{\partial V_6} & \frac{\partial P_6}{\partial V_7} \\ \frac{\partial P_7}{\partial \theta_5} & \frac{\partial P_7}{\partial \theta_6} & \frac{\partial P_7}{\partial \theta_7} & \frac{\partial P_7}{\partial V_5} & \frac{\partial P_7}{\partial V_6} & \frac{\partial P_7}{\partial V_7} \\ \frac{\partial Q_5}{\partial \theta_5} & \frac{\partial Q_5}{\partial \theta_6} & \frac{\partial Q_5}{\partial \theta_7} & \frac{\partial Q_5}{\partial V_5} & \frac{\partial Q_5}{\partial V_6} & \frac{\partial Q_5}{\partial V_7} \\ \frac{\partial Q_6}{\partial \theta_5} & \frac{\partial Q_6}{\partial \theta_6} & \frac{\partial Q_6}{\partial \theta_7} & \frac{\partial Q_6}{\partial V_5} & \frac{\partial Q_6}{\partial V_6} & \frac{\partial Q_6}{\partial V_7} \\ \frac{\partial Q_7}{\partial \theta_5} & \frac{\partial Q_7}{\partial \theta_6} & \frac{\partial Q_7}{\partial \theta_7} & \frac{\partial Q_7}{\partial V_5} & \frac{\partial Q_7}{\partial V_6} & \frac{\partial Q_7}{\partial V_7} \end{bmatrix} \begin{bmatrix} \Delta\theta_5 \\ \Delta\theta_6 \\ \Delta\theta_7 \\ \Delta V_5 \\ \Delta V_6 \\ \Delta V_7 \end{bmatrix}$$

Fig. 3.10 illustrates how the strategy works by a flowchart. The results and comparisons from a centralized approach and a decentralized one are presented in Table 3.3.

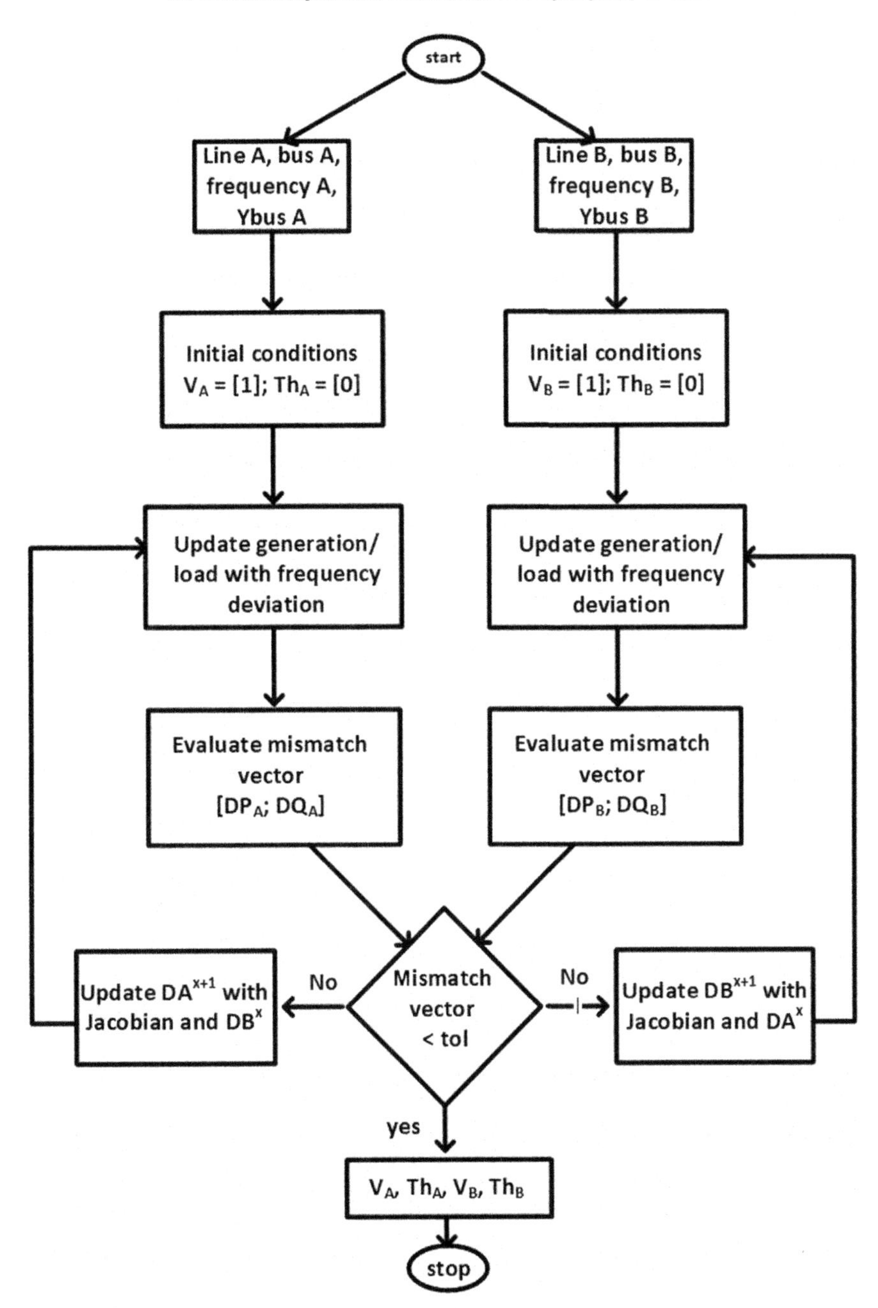

FIGURE 3.10 Flowchart of decentralized power flow.

TABLE 3.3 Results for the seven buses test system.

	Centralized	Decentralized
V_1	1.0000	1.0000
θ_1	0.0000	0.0000
V_2	0.9958	0.9958
θ_2	− 0.0356	− 0.0356
V_3	0.9923	0.9923
θ_3	− 0.0655	− 0.0655
V_4	0.9895	0.9895
θ_4	− 0.0896	− 0.0896
V_5	0.9874	0.9874
θ_5	− 0.1077	− 0.1077
V_6	0.9860	0.9860
θ_6	− 0.1198	− 0.1198
V_7	0.9853	0.9853
θ_7	− 0.1259	− 0.1259
Elapsed time	0.0110 s	0.0360 s

As can be seen, the results are equal but the computation time in the decentralized approach is higher. This method benefits from the radial distribution of the network. Having only one connection between the zones reduces considerably the amount of communication needed and the boundary variables. While the computation time is increased, it is also possible to use distributed computation that will reduce the work on one unit for large systems.

3.8 Optimal formulation and solution

In this chapter, the focus is on economical and frequency deviation optimization, first in connected mode with and without EV, and then in islanding mode using batteries.

3.8.1 Objective functions

There are two objectives taken into account: frequency deviation and economic dispatch. It is common to find them separately mainly because of the different timescales; thus a timescale suitable for both problems is used. Another solution is presented in Ref. [43] with a formulation in two different timescales; however, they do not address the conflict between objectives. Unlike other optimization problems, this multiobjective optimization searches the most preferred solution from a set of solutions found using the augmented ε constraint method [44]. Eq. (3.27) is the objective function considering the price of electricity c at time t and the power from/to the main grid P_G. A positive value for P_G means that the MG is buying energy from the main grid; on the other hand, a negative value represents revenues for the MG. The total number of steps in the optimization horizon is *nstep*:

$$min\, f_1(x) = \sum_{t=1}^{nstep} P_G^t \times c^t \tag{3.27}$$

The second objective is the frequency deviation:

$$min\, f_2(x) = \sum_{t=1}^{nstep} \Delta f^t \tag{3.28}$$

Thus it is assumed that for each analyzed time interval (some minutes long), frequency reaches a quiescent point in order to focus on economic aspects.

3.8.2 Constraints

The EMS must solve a constrained optimization problem taking into account the following: (1) limits in power generation (Eqs. 3.29 and 3.30), (2) maximum charging and discharging of batteries (Eqs. 3.31 and 3.32), and (3) frequency deviation limit (Eq. 3.33), that is,

$$P_{Gmin} \leq P_G \leq P_{Gmax} \tag{3.29}$$

$$Q_{Gmin} \leq Q_G \leq Q_{Gmax} \tag{3.30}$$

$$P_{chmin} \leq P_{ch} \leq P_{chmax} \tag{3.31}$$

$$P_{dchmin} \leq P_{dch} \leq P_{dchmax} \tag{3.32}$$

$$\Delta f \leq \Delta f_{max} \tag{3.33}$$

when the ESS is an EV there is an additional consideration: the EV needs to be fully charged before it is disconnected. Eq. (3.34) represents the energy required, E_{req}, to achieve this goal in the time horizon selected for the optimization:

$$\sum_{t=1}^{nstep} P_{ch}^t * \Delta t = E_{req} \tag{3.34}$$

If the MG includes batteries, the SOC's limits have to be monitored (Eq. 3.35):

$$SOC_{min} \leq SOC_{avg} \leq SOC_{max} \tag{3.35}$$

When the SOC reaches 5%, the battery will be inhibited for supplying energy. Similarly, if the SOC reaches 95% in order to avoid damages [5].

3.8.3 Jaya algorithm

Rao [45] developed an advanced optimization algorithm rooted on the teaching–learning-based optimization. This heuristic algorithm is named *Jaya*. It does not have the need to tune parameters and has a simple implementation. Eq. (3.36) is used to calculate the new values of the decision variables:

$$X_i = x_i + r_1\left(x_{i,best} - |x_i|\right) - r_2\left(x_{i,worst} - |x_i|\right) \tag{3.36}$$

where r_1 and r_2 are random numbers within the interval [0,1]. $x_{i,best}$ and $x_{i,worst}$ are decision variables' values that render the best and worst objective function. More information about the algorithm can be found in Ref. [46].

3.9 Multiagent system description

In general, cooperative learning is effective in domains where participants want to acquire skills such as planning, categorization, and memorization. The procedure consists of the participants learning the prerequisites of the topic to be taught and reinforcing the learning using a cooperative environment. There are several studies that suggest that the cooperative environment helps participants to understand complex tasks, where the domain of knowledge is complex, hierarchical, and requires deep knowledge of each level of the hierarchy. Currently, there are different trends to develop Computer Supported Collaborative Learning. A common factor in all of them is the use of Distributed Artificial Intelligence, which attempts to construct a model of the world populated by intelligent entities (agents) that interact through cooperation, coexistence, or competition in a distributed environment. MASs, where agents cooperate to carry out common objectives, is a technology that promises the solution to the problem of cooperation.

An intelligent agent displays flexible autonomy as identified by three properties. (1) Reactivity: the ability to respond to an event in a timely manner. (2) Proactiveness: the ability to take the initiative, as demonstrated through goal-directed behavior. (3) Social ability: the ability to interact with other agents in the environment.

The heart of an intelligent agent is its ability to reason about its goals and the state of the world and take actions to try to achieve its goals [47]. The choice of the reasoning technique employed is an open question, with many different solutions being proposed. These range from general-purpose reasoning such as that offered by artificial intelligence planners or the BDI framework to application-specific solutions such as neural networks or other pattern-recognition techniques trained to perform a specific operation.

The distributed power system architecture is perfectly suited for a MAS since they can construct a robust, flexible, and extensible system. Thus the concept of MAS is applied in those situations where different tasks have to be performed by different persons who or organizations which supposedly store private information not to be exposed to partner agents and have some individual goals or benefits to be fulfilled. In this sense, the management of an MG is an appropriate environment to use agents, since not all of the elements inside an MG belong to the same utility and their autonomy offers tangible benefits.

Authors followed a design strategy from a bottom-up perspective, considering what actions and messaging capabilities an agent should have, given its knowledge of the world. Likewise, it was verified that once implemented, agents will behave in the expected ways in order to continue robustly to meet its design objectives. Validation and verification of agent code is in many ways like assessing the correctness of any software system, with the possible interactions of various agents both simplifying and adding complexity to the task. Because agents operate as autonomous components, they can be thoroughly tested as stand-alone units, entirely analogously to unit testing of software classes [47].

In this application, the upper level agent coordinates with the lower level agents in a centralized manner to mitigate the redundancy, and the lower level agents cooperate with their neighboring agents to achieve the mutual support. The communication topology implies that the coordinated agent intercommunicates with every other agent, and each agent only exchanges the information with its neighboring agents. Here, it should be noted that a coordinated control mechanism is applied.

In this research, the Java Agent Development Framework [48] is used. This software complies with the standards set by the International Foundation of Intelligent Physical Agents, using Agent Communication Language. Previous work in MG optimization frequently assigns an agent to each DER [25], sometimes even one for each load in the system [49], but it is not always necessary to do this. If there is not going to be an auction among the DER to provide energy, a single agent can manage several sources and loads. In this chapter, no specific system partitioning was implemented; rather, it takes advantage of the radial scheme of the MG to split it into zones. Each zone has only one line connecting it with the adjacent zones. The proposed MAS only has two types of agents and uses the environment as a container and a means of communication for the agents. The architecture of each agent is presented in the following sections.

3.9.1 Zone and support agents

Zone agents have the BDI architecture [50] described in Fig. 3.11. The agent is initialized with a knowledge database describing the initial state of the environment. Every time the environment changes, the agent receives a message with new data and actualizes its beliefs according to the type of message that it receives. The desires of the agent are introduced to the objective functions or to the power flow problem. This depends on the initial goal of the agent and directly influences the intentions (actions) that the agent takes. The actions can be to apply the Jaya algorithm to find the optimal solution of the zone while cooperating with other agents, or to update the variables to find the power flow solution. Once the agent decides on an action, it sends the information with new changes in the environment to other agents until a common solution is found. The program for a power flow solution is written in Matlab and then wrapped in a method to use in Java. Documentation for the Matlab Java Builder can be found in Ref. [51]. One advantage of the proposed agent system architecture is the plug-and-play feature. Adding/removing elements or buses is quite simple as long as these do not affect the connections among zones. Even when new connections are created, it is still simple to introduce them, but keeping the connections intact facilitates the work. The database includes the structures presented in Fig. 3.12.

Support agents store the data from all zones in case a communication problem arises. The behavior of this agent is depicted in Fig. 3.13. Only one agent was used in this chapter; however, it is possible to include more if needed.

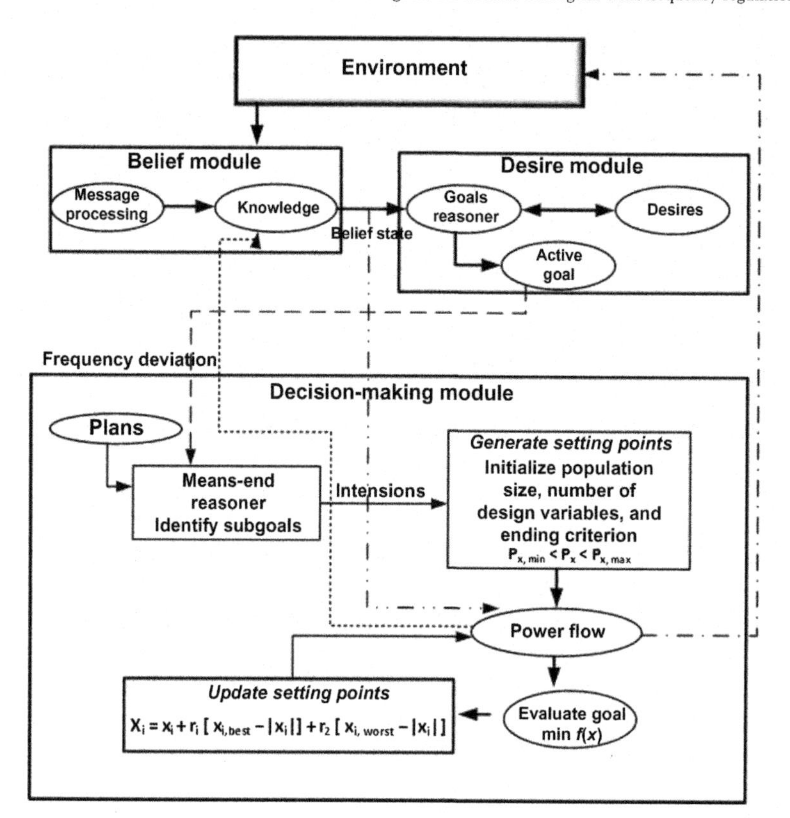

FIGURE 3.11 Architecture for the zone's agent.

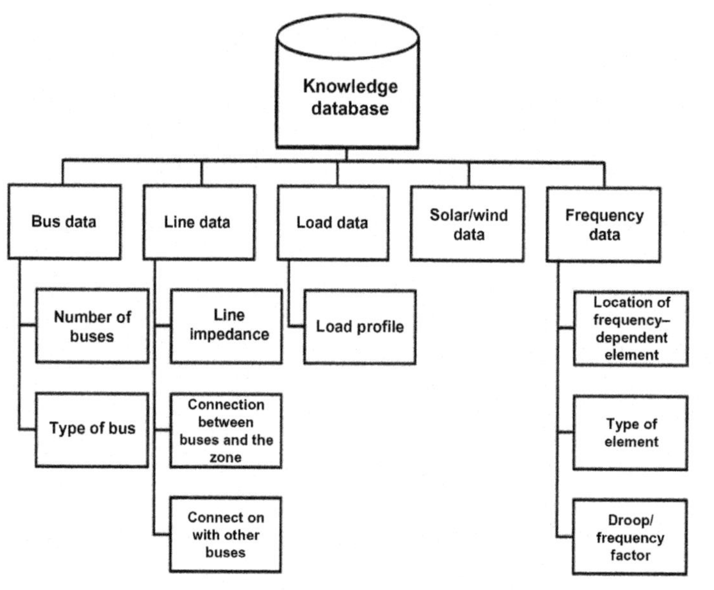

FIGURE 3.12 Initialization data for the zone's agents.

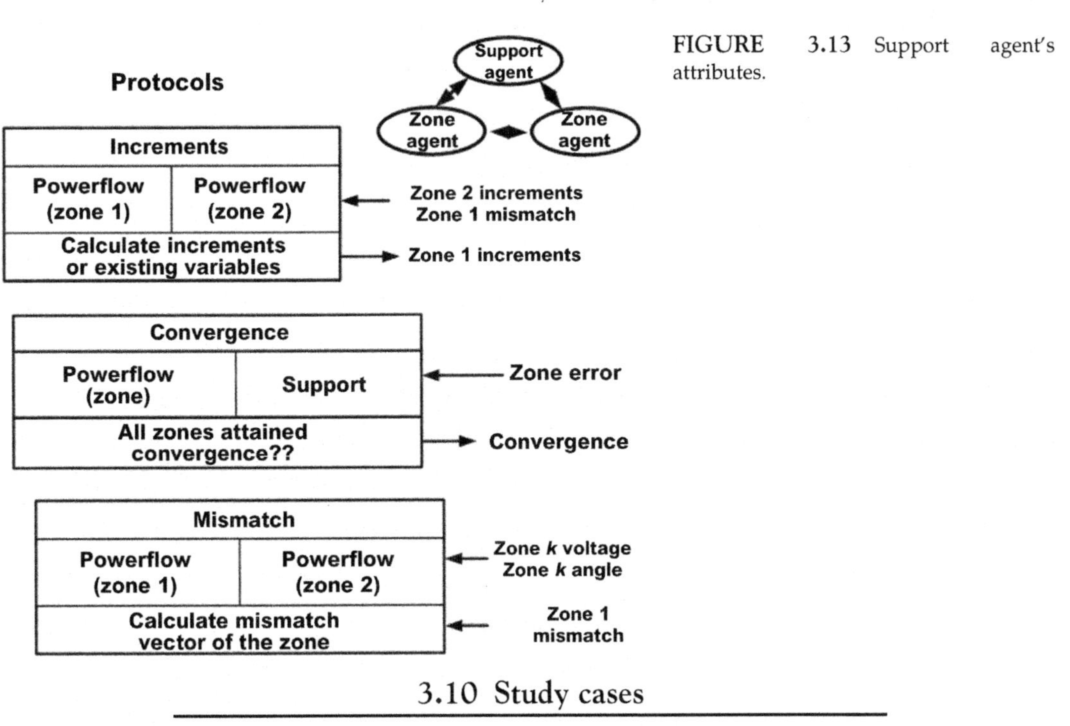

FIGURE 3.13 Support agent's attributes.

3.10 Study cases

Fig. 3.14 depicts the 10 buses MG, based on a benchmark proposed in Ref. [52]. Network, demand, and generation data are presented per unit. Data for wind speed and solar radiation in Fig. 3.15 is taken from the NREL [53]. The energy price in Fig. 3.16 is taken from the California ISO Open Access Same-time Information System [54]. The main grid is connected to the point of common coupling in bus 1 and there is no limit imposed to buy/sell energy from/to the grid. In order to assess the proposition, the optimization horizon is chosen to be equal to 18 time steps of 10 minutes (3 hours); chosen hours are from 8:00 to 11:00. The MG is divided in four zones, Table 3.4.

3.10.1 Conflicting objectives

Fig. 3.17 displays different solutions where the optimization is carried out using objective function f_1 (Eq. 3.17) as the main objective and the objective function f_2 (Eq. 3.18) as a constraint. The range for the frequency deviation lies within the interval $[-0.05, 0.05]$ Hz. As the frequency deviation varies from -0.05 to 0, the revenue of the MG decays, and once it reaches a positive deviation the revenue stops changing because the wind turbines do not help in frequency response anymore.

3.10.2 Operation without electric vehicle

When there are no EVs connected to the MG, the only zone that performs an optimization becomes the zone A and the decision variable is P_{Gset}. The permanent droop, δ, for

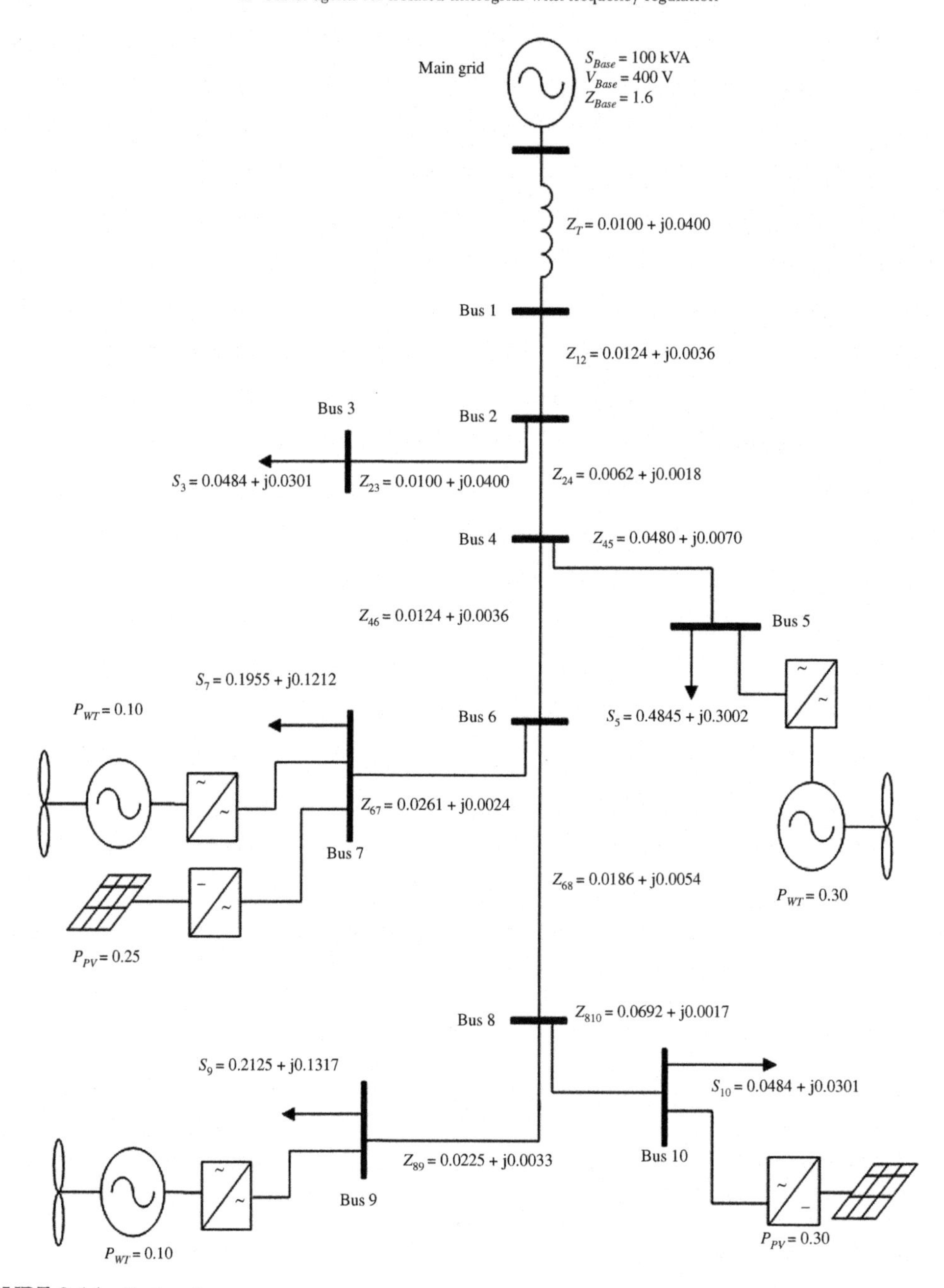

FIGURE 3.14 Test system.

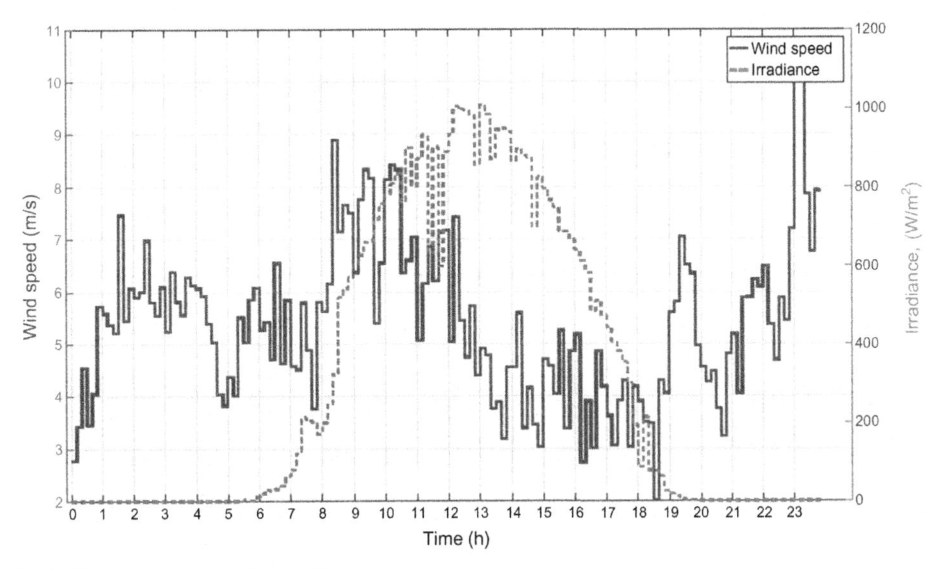

FIGURE 3.15 Solar and wind data [53].

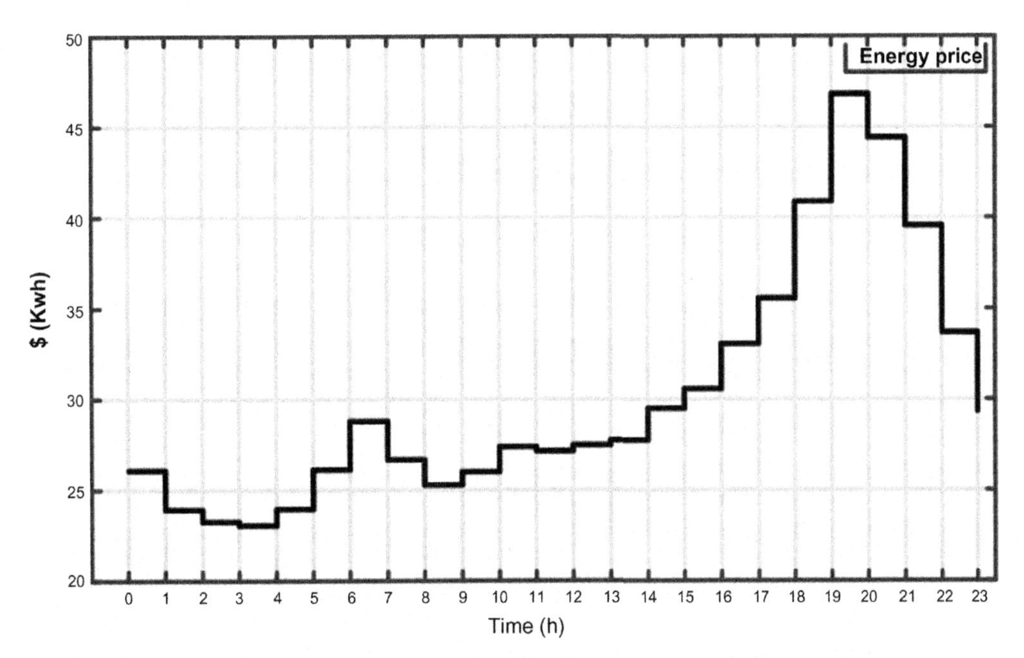

FIGURE 3.16 Energy price for one typical day [54].

TABLE 3.4 Buses and elements for each MG's zone.

Zone	Buses	Elements
A	1, 2, 3	Load, main generator
B	4, 5	Load, wind turbine
C	6, 7	Load, wind turbine, photovoltaic
D	8, 9, 10	Load, wind turbine, photovoltaic

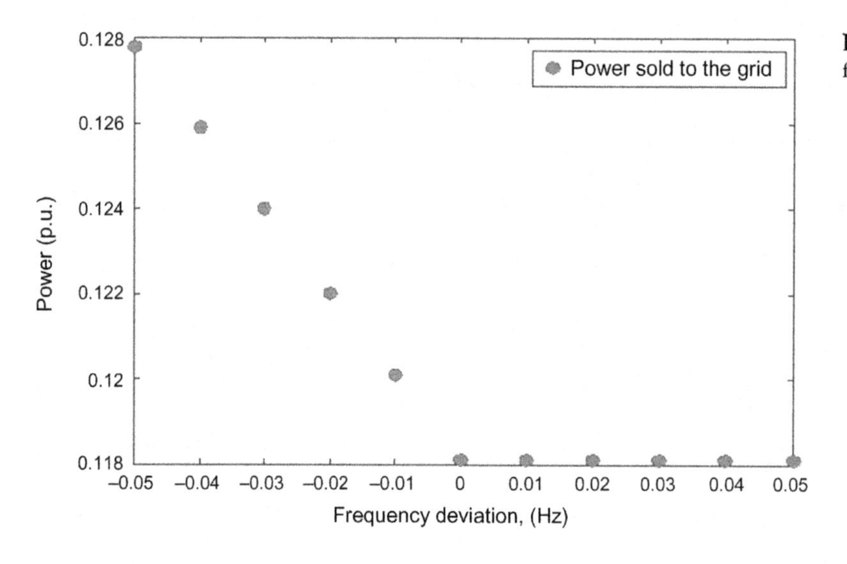

FIGURE 3.17 Pareto front.

generators is 4% to be within the advisory settings from the North American Electric Reliability Corporation [55]. Fig. 3.18 exhibits results obtained from the proposed algorithm. It is worth noting that when the DERs are not enough to supply the demand, the MG has to buy energy from the grid; otherwise, it sells the excess of energy. The frequency deviation (Fig. 3.18C) is inside the imposed constraint and, as expected, the agent tries to keep it in the negative end to maximize the revenues.

Table 3.5 summarizes the effects that the frequency deviation has on the generators. At 8:00 hours the final output of the WT, P_{Gw}, is equal to the maximum power it can deliver P_w; on the other hand, at 09:20 the frequency deviation is lower but the WT is not giving the maximum available power. This is especially important considering that, if there is a specific requirement to provide frequency response for all new interconnected resources, there needs to be a compensation for such service; otherwise it is not economically viable for the DER. Because the frequency deviation is always negative (meaning that there is more load than generation) when the MG is buying from the grid, the power P_G is bigger than the initial setting point P_{Gset}. Similarly, when the MG is selling, the power interchanged is lower than the initial setting point.

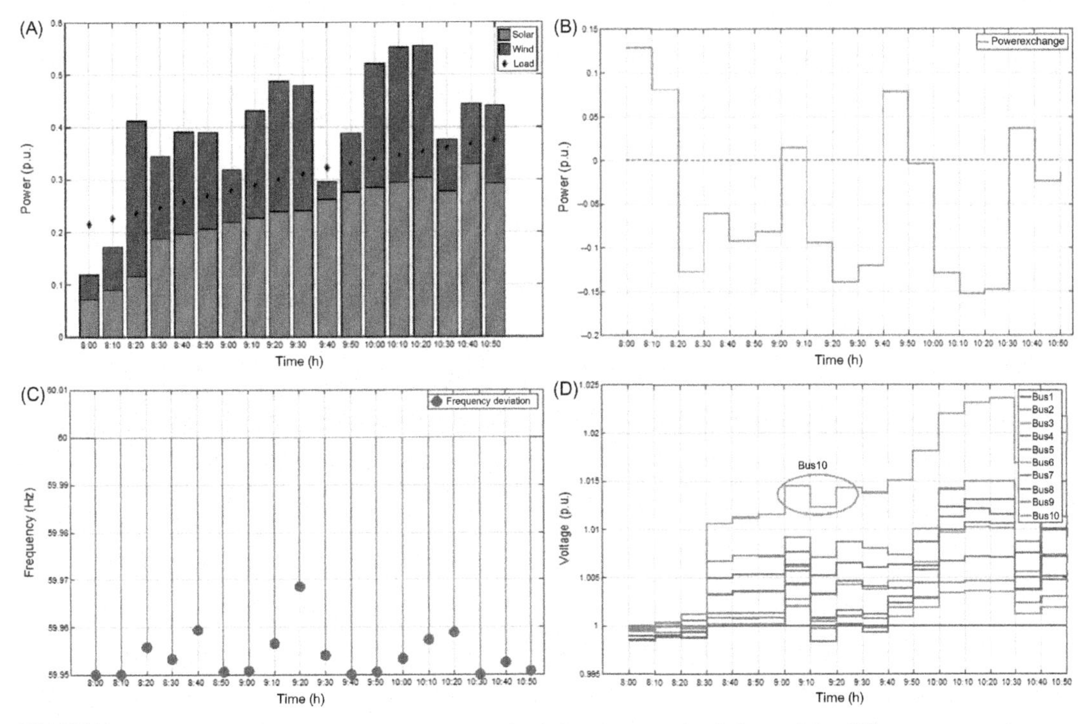

FIGURE 3.18 Results for the interconnected mode: (A) making up load demand by different sources (energy must be purchased from the network when the star is above the bars); (B) interchanged energy with the main grid (above the *dotted line*, purchased energy); (C) frequency deviation; and (D) voltage profile.

TABLE 3.5 Effect of frequency deviation on generators.

Parameter	08:00	09:20
Δf (frequency deviation, Hz)	− 0.050	− 0.0315
P_w (wind turbine output power, p.u.)	0.0271	0.1553
P_{wset} (wind turbine setting point, p.u.)	0.0244	0.1398
P_{Gw} (wind turbine output power with frequency control, p.u.)	0.0271	0.1437
P_{Gset} (synchronous generator setting point, p.u.)	0.1083	− 0.1518
P_G (synchronous generator active output power, p.u.)	0.1291	− 0.1387

3.10.3 Electric vehicle embedded

Introducing the EV only requires a change in the load data of the zone agent that has the charge station. In this case, the agent in zone D is the one chosen to have the charging station since the bus 10 is the one that presented a higher voltage in the previous results. Two EVs with different SOC at the starting time of the optimization were placed at bus 10 using a type two charger, with a maximum charging power rating of 6.66 kW.

Fig. 3.19 shows the attained results with the vehicles addition. In this application, optimization is carried out by two agents: the agent in zone A keeps the decision variable P_{Gset}, while the agent in zone D manages the charges of the EV using Eq. (3.24). The SOC of each vehicle changes each time accordingly to the charging at that time step (Fig. 3.20).

3.10.4 Microgrid without synchronous machine

When the MG is disconnected from the synchronous machine, the frequency control becomes more sensitive. Some of the solutions in islanding MGs include load shedding rules or demand-side management, which are not addressed herein. In order to maintain the frequency deviation between the desired limits (± 0.5% Hz) the WT located at buses 7 and 9 (Fig. 3.14) each has a maximum power output of 0.3 p.u. In this case a Li-ion battery with a capacity of 50 kW is connected to bus 4; results can be observed in Fig. 3.21. The main reason to use a stationary battery instead of an EV fleet is that at some steps the battery supplies energy to the grid and, since the EVs use Eq. (3.24) as a constraint, it is difficult to maintain the frequency deviation. The battery's charging/discharging is reflected in the SOC (Fig. 3.21B).

3.11 Discussion

Frequency variation in the MGs poses a significant challenge because MGs tend to demonstrate higher rates of mismatch between generation and demand; on the one hand, renewable energy is uncertain and intermittent, and on the other hand, the power consumption of the isolated community might vary frequently. As a result, taking proper care is vital to provide the scheduled frequency rate and compensate for the mismatch between power generation and consumption.

In view of the decentralized approach to the problem posed, this document proposes a coordinated, two-level control framework based on MAS to design the management of the MG. The top-level agent is used to handle the information in solving the optimization problem and the power flow problem. The lower level agents are designed to manage the data of the different zones that make up the MG. To demonstrate the validity of the proposed design method, the simulations are carried out in a system of 10 buses. The results show that the proposed method achieves satisfactory results in both the MG's energy and frequency management, under different operating conditions.

The process of developing a new agent depends on the description in a document of the agent's initial composition in terms of functionality, coordination protocols, and communication mechanisms with other agents. The proposed compositional architecture allows the agent to adapt to new requirements and modify its configuration at runtime without affecting the rest of the components and even the agent's ongoing activity, since each conversation is coordinated by independent connectors.

The MAS is programmed in a computer with Intel Core i7 4700HQ processor using Matlab 2016a, Java 7.1, and JADE 4.4. First, all the agents are located in one computer and then in two different computers to analyze the physically distributed benefits; in this case, the size of the system makes it hard to quantify the benefits. Nevertheless, it proves that it

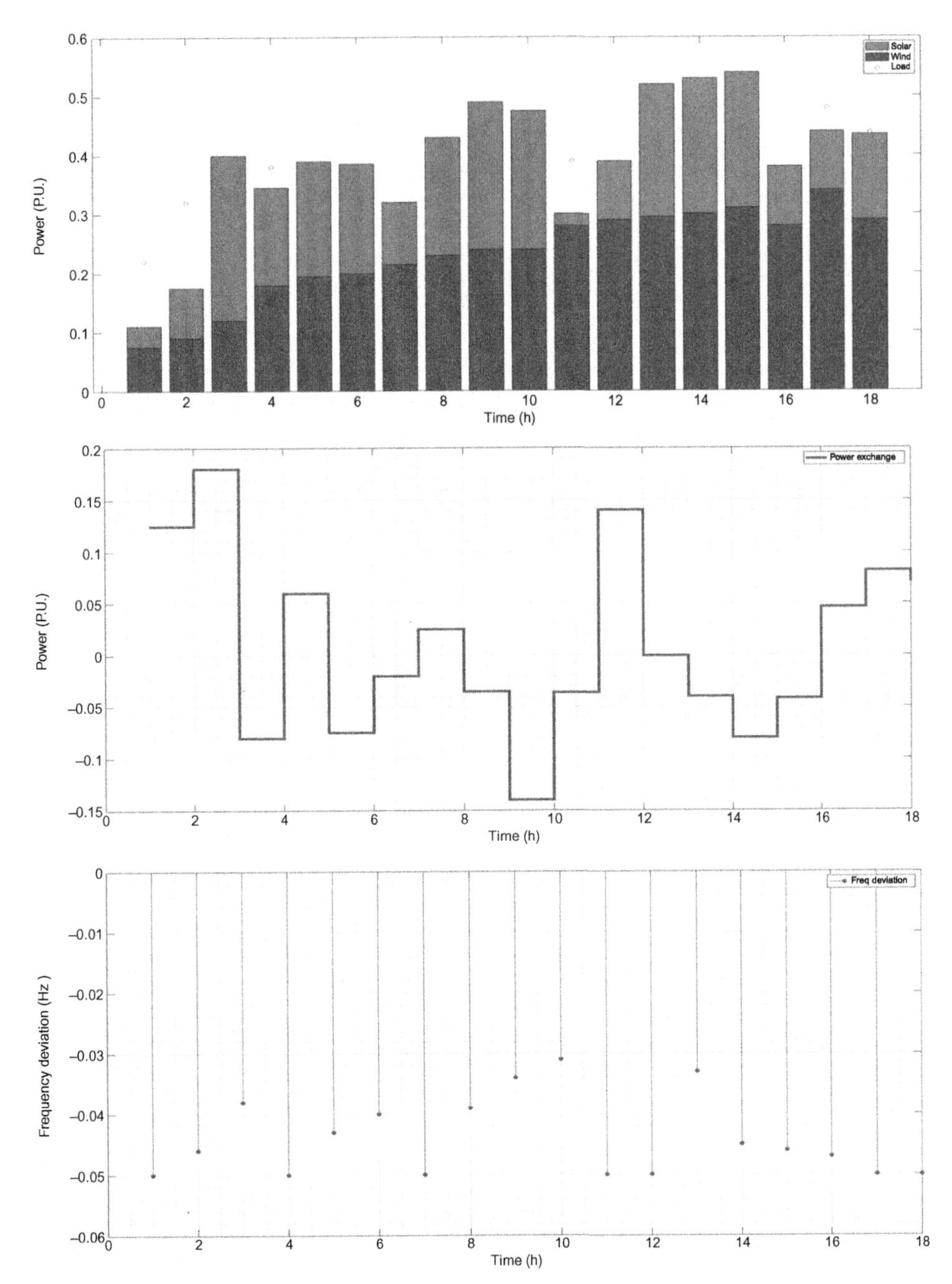

FIGURE 3.19 Results for the interconnected mode with EV: (A) making up load demand by different sources (energy must be purchased from the network when the star is above the bars); (B) interchanged energy (above the *dotted line*, purchased energy); and (C) frequency deviation.

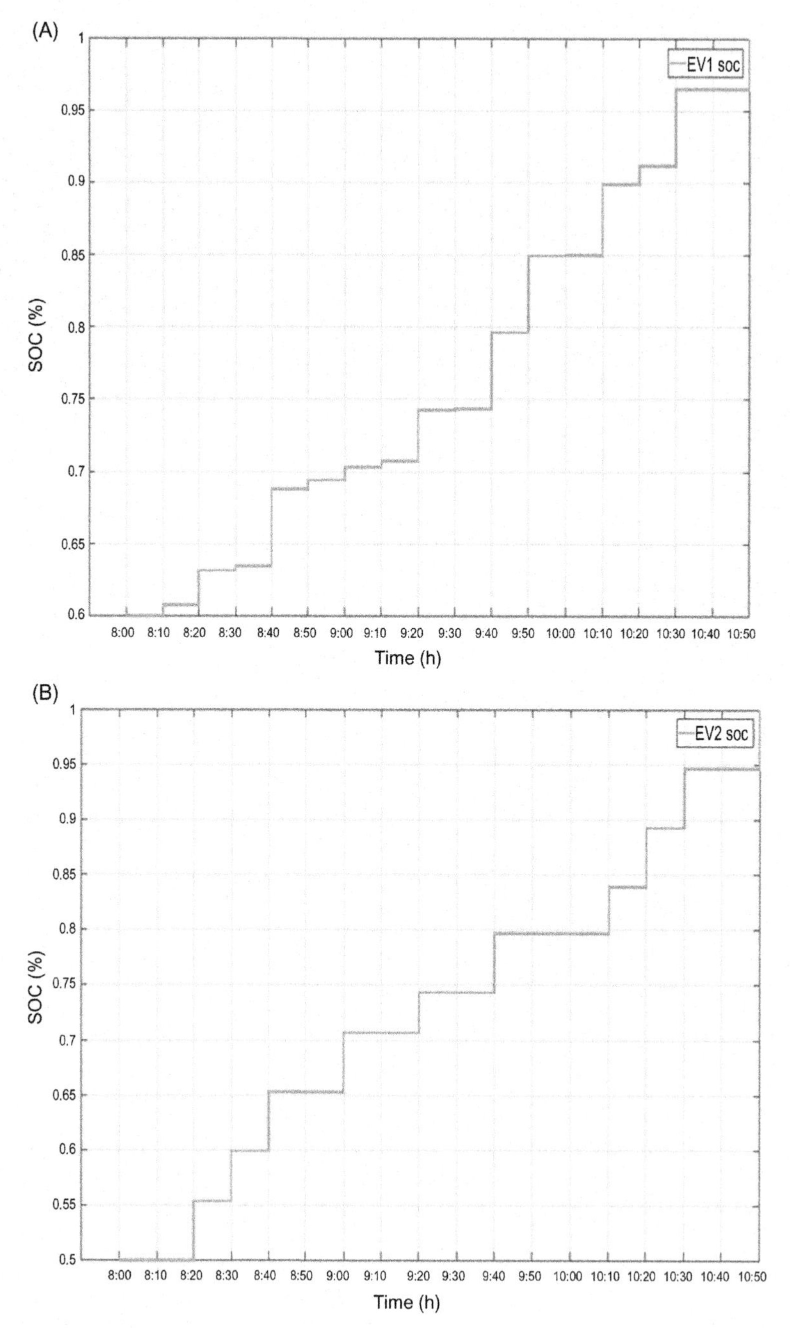

FIGURE 3.20 State of charge of each electric vehicle for every time step: (A) SOC of EV1 starting at 60% and (B) SOC of EV2 starting at 50%.

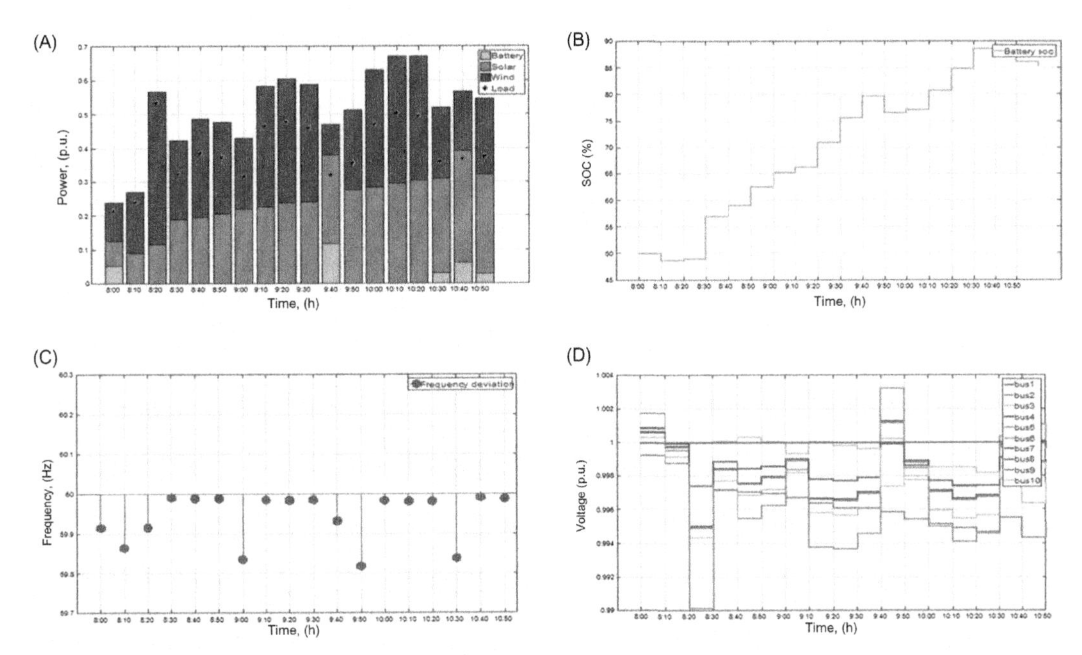

FIGURE 3.21 Results for the islanded mode with Li-ion battery: (A) making up load demand by different sources (energy may be sold to the network when the star is below the bars); (B) SOC of the Li-ion battery; (C) frequency deviation; and (D) voltage profile.

is possible to have agents in different computers with no difficulty of communication between them. This chapter uses the frequency deviation to provide an insight into how agents can be used in an MG with multiobjective optimization and emphasizes the need to use a set of Pareto solutions. Descriptions of the frequency-dependent models, decentralized power flow, and the abstraction of the problem are presented.

Optimal formulation rests on two objectives: frequency deviation and costs. Both functions are minimized using the heuristic algorithm Jaya, due to its simplicity, rapidity, and rendering of appropriate results. These indicate that frequency and voltages remain within the specified values both in isolated and connected operating mode while satisfying the demand. The use of the objective function related to costs allows that the load may be supplied from the cheaper source both with and without batteries. Thus the proposition is able to attain a satisfactory operation of the MG, both from a technical and an economical point of view.

Eqs. (3.4) and (3.10) indicate that the power programmed by the synchronous generator (representing the large system) and the wind turbine, respectively, are adjusted by the frequency deviation. Similarly, it is assumed that loads also vary with the value of the frequency (Eqs. 3.19 and 3.20). Thus the solution to the optimal problem can be visualized as a search for the best values of power injection, which satisfy the demand at the lowest cost, taking care of the limits or operational constraints (Eqs. 3.29–3.35). This search is achieved through multiple solutions of the power flow problem (3.26), until a solution that satisfies the requirements is achieved. Given the strategy proposed here, it is necessary that the support and zone agents share information during the search process, until reaching the objective.

From results, when the frequency deviation becomes negative (meaning that there is more load than generation) the MG is buying energy from the grid, and the power P_G is bigger than the initial setting point P_{Gset}. Similarly, when the MG is selling energy, the power interchanged is lower than the initial setting point. For the interconnected mode, when the load is above the clean generation, energy must be purchased from the large network, although by economy the frequency can be maintained at the lower limit 59.5 Hz. Otherwise, energy is sold to the large system, although by economy the frequency is not necessarily returned to 60 Hz, but this can be kept slightly deviated from the nominal one. When the MG operates in stand-alone mode, note that at some intervals the batteries must supply power to the MG in order to keep the frequency within acceptable limits. Voltage profile is maintained in standard conditions.

3.12 Conclusion

In this research, a MAS has been employed to optimally solve a management microgrid formulation. The agents presented here can decide on the course of action that will satisfy their objectives, which is the main difference with object oriented programming. In this application, the cooperation within the proposed MAS is by explicit design. The proposition has used the main advantages of a MAS: robustness and scalability.

Optimal formulation rests on two objectives: frequency deviation and costs. Both functions are minimized using the heuristic algorithm Jaya, due to its simplicity, rapidity, and rendering of appropriate results. These indicate that frequency and voltages remain within the specified values both in isolated and connected operating mode. The use of the objective function related to costs allows that load may be supplied from the cheaper source both with batteries and without them. Thus the proposition is able to attain a satisfactory operation of the MG both from a technical and an economical point of view.

References

[1] M.A. Hossain, H.R. Pota, M.J. Hossain, A.M.O. Haruni, Active power management in a low-voltage islanded microgrid, Int. J. Electr. Power Ener. Syst. 98 (2018) 36–47.
[2] T. Liu, X. Tan, B. Sun, Y. Wu, D.H.K. Tsang, Energy management of cooperative microgrids: a distributed optimization approach, Int. J. Electr. Power Energy Syst. 96 (2018) 335–346.
[3] M. Chen, X. Xiao, Hierarchical frequency control strategy of hybrid droop/VSG-based islanded microgrids, Electr. Power Syst. Res. 155 (2018) 131–143.
[4] US Federal Energy Regulatory Commission, Essential reliability services and the evolving bulk-power system-primary frequency response. Available from: <http://www.ferc.gov>, 2016.
[5] S. Chen, T. Zhang, H.B. Gooi, R.D. Masiello, W. Katzenstein, Penetration rate and effectiveness studies of aggregated BESS for frequency regulation, IEEE Trans. Smart Grid 7 (2016) 167–177.
[6] I. Prodan, E. Zio, A model predictive control framework for reliable microgrid energy management, Int. J. Electr. Power Energy Syst. 61 (2014) 399–409.
[7] A.C. Orrell, N.F. Foster, 2015 distributed wind market report, US Dept. Energy, Pacific Northwest Nat. Lab. Richland, WA, 2016.
[8] U. S. Department of Energy, Wind Vision: A New Era for Wind Power in the United States, US Dept. Energy, Oak Ridge, TN, 2015.

[9] J. Ekanayake, N. Jenkins, Comparison of the response of doubly fed and fixed-speed induction generator wind turbines to changes in network frequency, IEEE Trans. Ener. Convers. 19 (Dec. 2004).
[10] G. Ramtharan, N. Jenkins, J. Ekanayake, Frequency support from doubly fed induction generator wind turbines, IET Renew. Power Gen. 1 (1) (2007) 3.
[11] L. Castro, C. Fuerte-Esquivel, J. Tovar-Hernandez, Assessing the steady state post-disturbance condition of power systems including fixed speed and doubly-fed induction generators, in: 43rd North American Power Symposium, Boston, MA, 2011.
[12] L. Castro, C. Fuerte-Esquivel, J. Tovar-Hernandez, Solution of power flow with automatic load-frequency control devices including wind farms, IEEE Trans. Power Syst. 27 (4) (2012) 2186–2195.
[13] A.H. Fathima, K. Palanisamy, Optimization in microgrids with hybrid energy systems—a review, Renew. Sustain. Energy Rev. 45 (2015) 431–446.
[14] F.A. Bhuiyan, A. Yazdani, S.L. Primak, Optimal sizing approach for islanded microgrids, IET Renew. Power Gen. 9 (2) (2015) 166–175.
[15] Y. Ma, J. Ji, X. Tang, Triple-objective optimal sizing based on dynamic strategy for an islanded hybrid energy microgrid, Int. J. Green Energy 14 (2016) 310–316.
[16] J.P. Fossati, A. Galarza, A. Martín-Villate, L. Fontán, A method for optimal sizing energy storage systems for microgrids, Renew. Energy 77 (2015) 539–549.
[17] E. Ghiani, C. Vertuccio, F. Pilo, Optimal sizing and management of a smart microgrid for prevailing self-consumption, 2015 IEEE Eindhoven Power Tech. (2015) 1–6.
[18] C. Gamarra, J.M.J.M. Guerrero, Computational optimization techniques applied to microgrids planning: a review, renew, Sustain. Energy Rev. 48 (2015) 413–424.
[19] A. Ouammi, H. Dagdougui, R. Sacile, Optimal control of power flows and energy local storages in a network of microgrids modeled as a system of systems, IEEE Trans. Control Syst. Technol. 23 (1) (2015) 128–138.
[20] Q. Jiang, M. Xue, G. Geng, Energy management of microgrid in grid-connected and stand-alone modes, IEEE Trans. Power Syst. 28 (3) (2013) 3380–3389.
[21] A.C. Luna, N.L. Díaz, M. Graells, J.C. Vásquez, J.M. Guerrero, Cooperative energy management for a cluster of households prosumers, IEEE Trans. Consum. Electron. 62 (3) (2016) 235–242.
[22] W. Su, J. Wang, J. Roh, Stochastic energy scheduling in microgrids with intermittent renewable energy resources, IEEE Trans. Smart Grid 5 (4) (Nov. 2016) 1867–1883.
[23] Z. Min, S. Chen, L. Feng, H. Xiuqiong, A game-theoretic approach to analyzing power trading possibilities in multi-microgrids, Proc. CSEE 35 (4) (2015) 848–857.
[24] C.P. Nguyen, A.J. Flueck, A novel agent-based distributed power flow solver for smart grids, IEEE Trans. Smart Grid 6 (3) (2015) 1261–1270.
[25] G. Hug, S. Kar, C. Wu, Consensus + innovations approach for distributed multiagent coordination in a microgrid, IEEE Trans. Smart Grid 6 (4) (2015) 1893–1903.
[26] Y.S. Foo, H.B. Gooi, S.X. Chen, Multi-agent system for distributed management of microgrids, IEEE Trans. Power Syst 30 (1) (2015) 24–34.
[27] S. McArthur, E. Davidson, V. Catterson, A. Dimeas, N. Hatziargyriou, F. Ponci, et al., Multi-agent systems for power engineering applications—part I: concepts, approaches, and technical challenges, IEEE Trans. Power Syst. 22 (2007) 1743–1752.
[28] P. Kundur, Power System Stability and Control, McGraw-Hill, Palo Alto, CA, 1994.
[29] M. Okamura, Y. Oura, S. Hayashi, K. Uemura, F. Ishiguro, A new power flow model and solution method including load and generator characteristics and effects of system control devices, IEEE Trans. Power Appar. Syst. 94 (3) (1975) 1042–1050.
[30] G. Delille, B. Francois, G. Malarange, Dynamic frequency control support by energy storage to reduce the impact of wind and solar generation on isolated power system's inertia, IEEE Trans. Sustainable Energy 3 (4) (2012) 931–939.
[31] T. Littler, W. Du, Damping torque analysis of virtual inertia control for DFIG-based wind turbines, in: 5th International Conference on Electric Utility Deregulation and Restructuring and Power Technologies, Changsha, China, 2015.
[32] S. Heier, Grid Integration of Wind Energy Conversion System, second ed., John Wiley & Sons, Chichester, West Sussex, UK, 2006.
[33] A. Shrikant, Enabling Frequency and Voltage Regulation in Microgrids Using Wind Power Plants (M.Sc. thesis), Dept. Elect. Eng. Missouri Univ. Science and Technology, Rolla, MO, 2012.

[34] K. Vidyanandan, N. Senroy, Primary frequency regulation by deloaded wind turbines using variable droop, IEEE Trans. Power Syst. 28 (2) (2013) 837–846.
[35] M. Grimley, J. Farrell, Mighty Microgrids, ILSR Energy Democracy Initiative, Washington, DC, 2016.
[36] F. Martin-Martinez, A. Sanchez-Miralles, M. River, A literature review of microgrids: a functional layer, Renew. Sustain. Energy Rev. 62 (2016) 1133–1153.
[37] C. Loutan, V. Gevorgian, Using Renewables to Operate a Low-Carbon Grid: Demonstration of Advanced Reliability Services From a Utility-Scale Solar PV Plant, CAISO & NREL, Tempe, AZ, 2017.
[38] IEEE Task Force on Load Representation for Dynamic Performance, Load representation for dynamic performance analysis of power systems, IEEE Trans. Power Syst. 8 (2) (1993) 472–482.
[39] A. Morris, M. Ulieru, FRIEND: a human-aware BDI agent architecture, IEEE International Conference on Systems, Man and Cybernetics, Anchorage, AK, 2011, pp. 2413–2418.
[40] J. Momoh, Smart Grid Fundamentals of Design and Analysis, John Wiley & Sons, Hoboken, NJ, 2012.
[41] S. Papathanassiou, N. Hatziargyriou, K. Strunz, A benchmark low voltage microgrid network, in: Proc. CIGRE Symposium on Power Systems With Dispersed Generation, Athens, Greece, 2005.
[42] National Renewable Energy Laboratory, Renewable resource data center. Available from: <http://www.nrel.gov/rredc/>, 2016.
[43] D. Cai, E. Mallada, A. Wierman, Distributed optimization decomposition for joint economic dispatch and frequency regulation. IEEE Trans. Power Syst. 32 (6) (2017) 4370–4385.
[44] G. Mavrotas, Effective implementation of the ε-constrained method in multi-objective mathematical programming problems, Appl. Math. Comput. 213 (2009) 455–465.
[45] R.V. Rao, Jaya: a simple and new optimization algorithm for solving constrained and unconstrained optimization problems, Int. J. Indust. Eng. Comput. 7 (1) (2016) 19–34.
[46] R.V. Rao, Jaya-algorithm. Available from: <http://sites.google.com/site/jayaalgorithm>, 2016.
[47] V.M. Catterson, E.M. Davidson, S.D.J. McArthur, Practical applications of multi-agent systems in electric power systems, Euro. Trans. Electr. Power 22 (2012) 235–252.
[48] JADE, JAVA agent development framework. Available from: <http://jade.tilab.com>, 2017.
[49] M.H. Cintuglu, H. Martin, O.A. Mohammed, Real time implementation of multiagent-based game theory reverse auction model for microgrid market operation, IEEE Trans. Smart Grid 6 (2015) 1064–1072.
[50] A. Morris, M. Ulieru, FRIEND: a human-aware BDI agent architecture, in: IEEE International Conference on Systems, Man and Cybernetics, Anchorage, AK, 2011, pp. 2413–2418.
[51] The MathWorks Inc., MATLAB Builder for Java User's Guide. Available from: <www.mathworks.com>, 2006.
[52] S. Papathanassiou, N. Hatziargyriou, K. Strunz, A benchmark low voltage microgrid network, in: Proceedings of the CIGRE Symposium on Power Systems With Dispersed Generation, Athens, Greece, 2005.
[53] National Renewable Energy Laboratory, Renewable resource data center. Available from: <http://www.nrel.gov/rredc/>, 2016.
[54] California ISO, Open access same-time information system (OASIS). Available from: <http://oasis.caiso.com/mrioasis/logon.do>, 2017.
[55] North American Electric Reliability Corporation NERC, Industry advisory generator governor frequency response. Available from: <http://www.nerc.com/pa/rrm/bpsa/Pages/Alerts.aspx>, 2015.

CHAPTER

4

Su-Do-Ku and symmetric matrix puzzles—based optimal connections of photovoltaic modules in partially shaded total cross-tied array configuration for efficient performance

Rupendra Pachauri and Deepak Kumar

Electrical and Electronics Engineering Department, School of Engineering, University of Petroleum and Energy Studies, Dehradun, India

4.1 Introduction

The storage capacity of fossil fuels is scarce and limited restrained, researchers are exploring efficient alternative energy sources. In the context of the best available energy solution, photovoltaic (PV), wind turbine, and biofuel cell systems are leading renewable energy sources although each resource has its own environmental-based operating limitations [1,2].

PV technology faces abounding challenges due to various known and unacquainted causes, for example, technical and environmental pollutants. In the current scenario, the environmental limitations are rising exponentially due to dust and dirt accrual on the panel surface. The major causes of shadows on the PV plant are due to moving clouds, adjacent trees, and high-rise buildings in urban areas especially [3]. These shading causations account for the performance compromise of the PV system. Recently, the pursuance enrichment of PV technology has been a "hot" research area with the modification in electrical connections of

Distributed Energy Resources in Microgrids
DOI: https://doi.org/10.1016/B978-0-12-817774-7.00004-1

PV module gaining more popularity to raise the performance ability of PV array systems [4]. In an array, the PV modules are arranged in accustomed configurations such as series, parallel, series–parallel (SP), bridge-link (BL), total cross-tied (TCT), and honeycomb (HC) [5,6], but the adopted reconfiguration methodology of the electrically connected PV modules showed a performance enhancement when compared to the conventional approach.

4.2 Literature review

The profundity and depth of the literature review emphasizes the credibility of the current knowledge, including substantive findings in the area of partial shading on solar PV systems, studied to understand the research gap. The diversified review of PV array configurations in terms of their performance, reliability, accuracy, robustness, efficiency, and execution is procured. In the recent past, poly-researchers have focused on distinctive reconfiguration methodologies for performance enhancement of the PV system and comparing it with conventional configurations to show the effectiveness of proposed methods. In this chapter, the literature review section is subdivided into two categories to determine the concept of partial shading on new and advanced PV array configurations as (1) conventional PV array configurations and (2) shade dispersion–based PV array configurations.

4.2.1 Conventional photovoltaic array configurations

The performance investigation of asymmetrical 72×20 sized SP, TCT, BL configurations of PV array, considered three types of partial shading cases in the MATLAB/Simulink environment. Moreover, the TCT configuration had the best performance in terms of minimum power losses and highest fill factor (FF) for all the shading scenarios [5]. The authors of Ref. [6] investigated the behavior of MATLAB/Simulink-based electrical characteristics of PV interconnection schemes such as 9×4 and 6×6 sized SP, TCT and BL for performance comparison for fresh and dust accrual or shadow conditions of PV array in which the TCT configuration had the best performance among all discussed configurations.

In Ref. [7], passing clouds were considered as shadow conditions for performance investigations of series, SP, and TCT configured PV array. MATLAB/Simulink-based distinguished cloud positions were considered for extensive study, in which TCT had maximum power at global maximum power point (GMPP) and ameliorated FF. The authors of Ref. [8] studied the shading effect on 12 PV modules arranged in 2×6, 6×2, 4×3 and 3×4 sizes of each SP, BL, and TCT configuration, respectively. The sufficient shading test cases were carried out for the comprehensive investigations; and the TCT configuration had the best results compared to its counterparts. The authors of Ref. [9] designed a switching matrix based series, parallel, SP, TCT configurations, where a real-time experimental study of PV system reconfigurations was carried out and the performance was verified by the obtained results of considered configurations. In Ref. [10], a PV module comprising 20 solar cells was arranged in series and parallel configurations and performance investigations were carried out under the shaded area as 50%, 25%, 12.5%, and 5%, respectively. In Ref. [11], MATLAB simulation-based 3×2 sized series and SP connections were made in an execution investigation, and the hardware results were procured to determine the performance validation of the system. The performance parameters in terms of peak voltage and power were scrutinized for four different cases of partial shading test. The design of [12] was a damn clear paradigm of SP, BL, TCT, and HC configurations and investigated under four categories of shadow types: (1) short and narrow (978 W/m^2), (2) long and narrow (1017 W/m^2), (3) short and wide (1002 W/m^2), and (4) long and wide (1020 W/m^2) conditions. Moreover, TCT configuration had the best results among the 3×5 and 3×3 sized configurations. In Ref. [13], the authors studied a MATLAB/Simulink-based and experimental analysis of series-connected PV modules under the shading circumstances. The performance assessment was done and results were validated with the electrical behavior of $I-V$ and $P-V$ curves of the working system. The authors of Ref. [14] contemplated the mismatch losses of 6×3 sized SP, TCT, BL configured PV array under partially shading conditions (PSCs) resulting in superior performance of TCT over its counter-competitors. In Ref. [15], the performance of shaded 4×4 sized SP, TCT, and BL configurations was interrogated and the performances collated in terms of power, voltage at GMPP, and improvised FF. In Ref. [16], series, parallel, SP, TCT, and BL of configurations sized 2×6, 6×2, 2×4, 4×2, 3×4, 4×3, 4×6, 6×4, 3×3, and 4×4 were considered under 15 different types of random shading profiles. The electrical performance parameters, such as current, voltage, and power of all combinations, were achieved for each shading pattern for an extensive comparative study. In Refs. [17,18], a comprehensive study on 6×4 sized series, parallel, SP, TCT, BL, and HC configurations was carried for the probable shading scenarios. The performance assessment of the electrical characteristics of considered PV array configurations was done and compared for extensive study. The performance assessment of [2×12, 3×8, 4×6, 6×4, 8×3 and 12×2] sized SP, TCT configurations of PV array under PSCs has been carried out and it was observed that the TCT configuration demonstrated the best results in terms of the lowest relative power loss, maximum power, and voltage at GMPP. The performance of the SP and BL was highly reduced when compared to the TCT configurations [19] as investigated through the MATLAB/Simulink modeling of SP, TCT, and BL configurations under the shaded circumstances. In Ref. [20], the partial shading effect was analyzed on 4×4 sized SP, while TCT configurations were analyzed under horizontal and diagonal shading patterns. The obtained results of a laboratory-based experimental setup were compared and it was concluded that TCT configuration resulted in a satisfactorily efficient performance as minimum current loss and maximum FF. In Ref. [21], the 4×4 sized SP, BL, TCT, and HC configurations were considered for performance investigation and they were a 'cut above the rest' in terms of TCT configuration to other encountered configurations during different PSCs and they demonstrated a better response of (output) $I-V$ and $P-V$ curves.

The authors in Refs. [22,23] investigated the behavior of $I-V$ and $P-V$ curves of a PV array, arranged in various electrical plans, for example, series, parallel, and SP of three PV modules. The performance analysis has been affected under nonuniform and uniform solar irradiations, such as alteration in insolation due to passing clouds and dust accumulation using a MATLAB/Simulink environment.

4.2.2 Shade dispersion–based photovoltaic array configurations

Shade dispersion is a technique to scatter the effect of shade (previously at few positions of photovoltaic array) to the entire array to augment the magnitude of output power. The physical positions of the modules were rearranged as per the proposed reconfiguration schemes. In Refs. [24,25], 6 × 4 sized TCT, half-reconfigurable PV array, and fully reconfigurable PV array (FRPA) were investigated under four shading profiles. FRPA had maximum power of GMPP and the highest performance ratio (PR) as compared to other configurations. Refs. [26,27] showed 9 × 9 sized TCT, Su-Do-Ku, and optimal Su-Do-Ku PV array configurations which were considered for performance evaluation under a short and narrow shaded area. Moreover, a 36 × 36 sized PV array was built as a 4 × 4 array of 9 × 9 sized micro arrays and considered for performance evaluation under similar shading cases. The performance of optimal Su-Do-Ku was found best in terms of minimum mismatch loss. In Ref. [28], a 9 × 9 sized TCT, Su-Do-Ku, and genetic algorithm (GA)-based shade dispersion method for new configuration were compared to realize the power output (maximum) and yield an adequate voltage at GMPP for GA based proposed configuration. In Refs. [29,30,31], the performance comparison of 6 × 9, 4 × 4 sized TCT, SP-TCT, and Su-Do-Ku puzzle—based reconfigured TCT PV array configuration was done in terms of minimum power loss, improved FF, etc. In Ref. [32], the MATLAB/Simulink modeling of 4 × 5 and 4 × 9 sized configurations was reported for SP, TCT, BL, HC, hybrid SP-TCT, BL-TCT, and novel structure-1,2 puzzles pattern (NS-1 and NS-2)-based configurations. The achievements of all aforesaid configurations were inspected in terms of the smoothness behavior of $P-V$ curves, power loss, FF, and the effect of shade dispersion on the maximum power point (MPP) under different shading patterns. In Ref. [33], the performance of 5 × 7 sized TCT and physical reorientation of the modules, with fixed electrical connection (PRM-FEC) configurations, was investigated for the real environmental situation and the performance was validated by an experimental system. In Ref. [34], The SP, TCT, OTCT (optimal total cross-tied), NTCT (novel total cross-tied), BL, and HC array configurations were considered for investigation under the PSC, which recently proposed that NTCT had been shown to have better performance. In Refs. [35,36], the authors compared the performance of SP, BL, TCT, SP-TCT, BL-TCT, BL-HC, and novel magic square (MS) configurations of size 4 × 4. Based on the analysis, MS was found to be superior to the other considered configurations in terms of maximum power, voltage at GMPP, minimum power losses, and maximum FF.

4.2.3 Novelty of work

The above literature review motivates the authors to put forward the modeling of proposed Su-Do-Ku-total cross-tied (Su-Do-Ku-TCT) and symmetric matrix-total cross-tied (SM-TCT) configurations. Further, the performance of the proposed configurations was collated with the conventional TCT configuration. In addition to it, the execution of SM-TCT configuration was found to have a better performance matched to the existed TCT and new Su-Do-Ku-TCT configurations in terms of GMPP location, magnitude power losses, and degree of FF during all the PSCs. The key contributions of the work were highlighted as:

- The reconfiguration of PV modules of TCT array was carried out to design new configurations such as Su-Do-Ku-TCT, SM-TCT, and novel interconnection schemes used to disperse the shadow over the PV modules in an array.

- In accordance with shade dispersion, the SM-TCT configuration generated more potential power matched with the TCT and Su-Do-Ku-TCT array configurations.
- In order to verify the effectiveness of the advised configurations and extensive comparison with the TCT and Su-Do-Ku-TCT configurations for different shading portfolios, I–IV was carried out in terms of various performance parameters such as power, voltage at GMPP, power loss, FF, and performance augmentation.

4.3 Modeling of photovoltaic cell and array configurations

The depicted electrical-equivalent of Fig. 4.1 of a solar cell represents the ability to transform sunlight into electrical DC current through PV effect.

Cumulative equations for solar cell current (I_{cell}) can be expressed as:

$$I_{cell} = I_{ph} - I_D \tag{4.1}$$

$$I_{cell} = I_{ph} - I_o\left(\exp^{\left(\frac{qV_C}{AkT_C}\right)} - 1\right) \tag{4.2}$$

where $I_{ph}$is the photocurrent of solar cell (A), I_D is the diode current (A), I_o is the diode reverse saturation current (A), q is the electron charge (Coulomb), V_C is the cell voltage (V), A is the ideality factor, k is the Boltzmann's constant (J/K), and T_C is the cell temperature (°C).

The influential effect of primary environmental factors (irradiation and temperature) upon the nonlinear behavior of current–voltage was palpable. The simulation was accomplished for the available specifications of 50 W JIAWEI JW-50P model as given in Table 4.1.

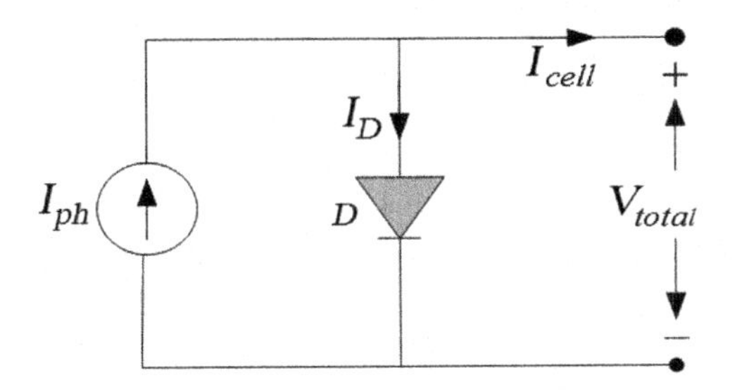

FIGURE 4.1 Solar cell equivalent circuit.

TABLE 4.1 Specifications of 50 W JIAWEI JW-50P PV module.

Electrical parameters	Specifications
Maximum power (P_m)	50 W
Voltage at maximum power (V_{mp})	17.8 V
Current at maximum power (I_{mp})	2.81 A
Open circuited voltage (V_{oc})	22.1 V
Short-circuit current (I_{sc})	3.15 A

11	12	13	14	15	16
21	22	23	24	25	26
31	32	33	34	35	36
41	42	43	44	45	46
51	52	53	54	55	56
61	62	63	64	65	66

(A)

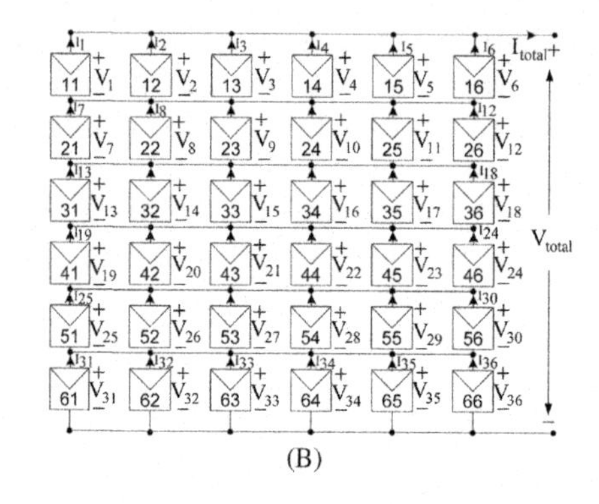

(B)

FIGURE 4.2 (A) TCT configuration and (B) electrical connections of PV modules in TCT configuration.

4.3.1 Total cross-tied configuration

The electrical connections of the PV modules in TCT configuration are shown in Fig. 4.2.

Considering the internal electrical interconnections of the TCT configuration, one can conclude that the summation of all the currents at different nodes and the voltages of parallel PV modules are expressed as:

$$V_{out} = \begin{cases} V_1 = V_2 = V_3 = V_4 = V_5 = V_6 \\ V_7 = V_8 = V_9 = V_{10} = V_{11} = V_{12} \\ V_{13} = V_{14} = V_{15} = V_{16} = V_{17} = V_{18} \\ V_{19} = V_{20} = V_{21} = V_{22} = V_{23} = V_{24} \\ V_{25} = V_{26} = V_{27} = V_{28} = V_{29} = V_{30} \\ V_{31} = V_{32} = V_{33} = V_{34} = V_{35} = V_{36} \end{cases} \tag{4.3}$$

$$I_{total} = \begin{cases} = I_1 + I_2 + I_3 + I_4 + I_5 + I_6 \\ = I_7 + I_8 + I_9 + I_{10} + I_{11} + I_{12} \\ = I_{13} + I_{14} + I_{15} + I_{16} + I_{17} + I_{18} \\ = I_{19} + I_{20} + I_{21} + I_{22} + I_{23} + I_{24} \\ = I_{25} + I_{26} + I_{27} + I_{28} + I_{29} + I_{30} \\ = I_{31} + I_{32} + I_{33} + I_{34} + I_{35} + I_{36} \end{cases} \tag{4.4}$$

$$V_{total} = \begin{cases} = V_1 + V_7 + V_{13} + V_{19} + V_{25} + V_{31} \\ = V_2 + V_8 + V_{14} + V_{20} + V_{26} + V_{32} \\ = V_3 + V_9 + V_{15} + V_{21} + V_{27} + V_{33} \\ = V_4 + V_{10} + V_{16} + V_{22} + V_{28} + V_{34} \\ = V_5 + V_{11} + V_{17} + V_{23} + V_{29} + V_{35} \\ = V_6 + V_{12} + V_{18} + V_{24} + V_{30} + V_{36} \end{cases} \tag{4.5}$$

Four shading cases, *I–IV*, considered in the investigation and the PV modules shaded by shading factor (α_s) can be expressed as:

$$\alpha_s = 1 - \frac{S_b}{S_x} \tag{4.6}$$

where S_x is the actual amount of irradiation incidence on PV module and S_b is the irradiance behind the shading.

For unshaded cells, the shaded current (I_{sh}) is defined as follows:

$$I_{sh} = \frac{S_x}{S_{STC}} \times I_{ph} \tag{4.7}$$

And for shaded cell, the current acquire I_{sh} as follows:

$$I_{sh} = (1 - \alpha_s)\frac{S_x I_{ph}}{S_{STC}} \tag{4.8}$$

In shading case-I, as shown in Fig. 4.7A, 12 cells were shaded in a row and the measured current can be derived as $I_1, I_2, I_3, \ldots, I_{36}$ from Eqs. (4.9) and (4.10) [21,37] as:

$$I_1, I_7, I_{13}, I_{19} = I_2, I_8, I_{14}, I_{20} = I_3, I_9, I_{15}, I_{21} = I_4, I_{10}, I_{16}, I_{22} = I_5, I_{11}, I_{17}, I_{23} = I_6, I_{12}, I_{18}, I_{24}$$
$$= I_{ph} - I_o\left[\exp\left(\frac{q(V_{1+7+13+19+\cdots\cdots+24})}{AkT_c}\right)\right] \tag{4.9}$$

$$I_{25}, I_{31} = I_{26}, I_{32} = I_{27}, I_{33} = I_{28}, I_{34} = I_{29}, I_{35} = I_{30}, I_{36}$$
$$= I_{ph} - \frac{12\alpha_S S}{1000} I_{ph} - I_o\left[\exp\left(\frac{q(V_{25+31,26+32,27+33,28+34,29+35,30+36})}{AkT_c}\right) - 1\right] \tag{4.10}$$

By substituting Eqs. (4.9) and (4.10) in Eqs. (4.3) and (4.4), a current–voltage relationship can be expressed as:

$$I_{total} = 2I_{ph} - \frac{12\alpha_S S}{1000} I_{ph} - 2I_o\left[\exp\left(\frac{q(V_{25+31})}{AkT_c}\right) - 1\right] \tag{4.11}$$

Considering Eq. (4.5), we can substitute V_{25+31} voltage in the Eq. (4.11) as follows:

$$I_{total} = 2I_{ph} - \frac{12\alpha_S S}{1000} I_{ph} - 2I_o\left[\exp\left(\frac{q(V_{total} - (V_{1+7+13+19}))}{AkT_c}\right) - 1\right] \tag{4.12}$$

Considering Eqs. (4.12) and (4.3), ($V_{1+7+13+19}$) can be expressed as follows:

$$V_{1+7+13+19} = \frac{AkT_c}{q} \ln\left(\frac{2I_{ph} - I_{total} - 2I_o}{2I_o}\right) \tag{4.13}$$

Substituting Eq. (4.13) in Eq. (4.12), the output current–voltage relationship is shown as:

$$I_{total} = 2I_{ph} - \frac{12\alpha_S S}{1000} I_{ph} - 2I_o \left[\exp^{\frac{q\left(V_{total} - \left(\frac{AkT_c}{q} \ln\left(\frac{2I_{ph} - I_{total} - 2I_o}{2I_o}\right)\right)\right)}{AkT}} - 1 \right] \tag{4.14}$$

$$I_{total} = 2I_{ph} - \frac{12\alpha_S S}{1000} I_{ph} - 2I_o \left[\exp^{\frac{\left(qV_{total} - \left(AkT_c \ln\left(\frac{2I_L - I_{total} - 2I_o}{2I_o}\right)\right)\right)}{AkT}} - 1 \right] \tag{4.15}$$

The mathematical modeling of TCT configuration for the rest of the three shading cases: III–IV can be done similarly as in case-I.

4.3.2 Su-Do-Ku-total cross-tied configuration

A Su-Do-Ku puzzle was considered with the objective to fill a 6×6 size grid with digits 1 to 6 such that each column–row had six 3×2 sub grids. In this chapter, the PV array comprised 36 PV modules with six rows and six columns. However, the physical placement of Su-Do-Ku-TCT is amended accordingly as the Su-Do-Ku puzzle pattern as shown in Fig. 4.3A and B.

4.3.3 Symmetric matrix-total cross-tied configuration

SM is a logic-based number placement matrix with a cyclical arrangement. These numbers are special because every row and column adds up to the last number and the obtained sum is the same (in all row and column cases). Moreover, any one diagonal digit repeats itself with the same number. The basic properties of a 6×6 sized SM configuration are shown in Fig. 4.4A–D.

From Fig. 4.4A–D, it is shown that the summation of each row and column is 21. Furthermore, the repetition of opposite 3×3 square matrix elements is existed, shown in Fig. 4.4D. Thus the properties of 6×6 SM are satisfied.

11	42	53	24	35	66
21	52	63	34	45	16
31	62	43	14	55	26
41	12	23	54	65	36
51	32	13	64	25	46
61	22	33	44	15	56

(A)

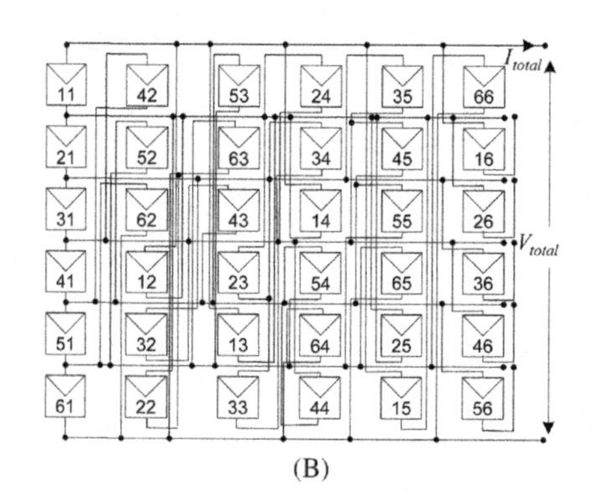

(B)

FIGURE 4.3 (A) Su-Do-Ku puzzle pattern and (B) electrical connections of PV modules in Su-Do-Ku-TCT configuration.

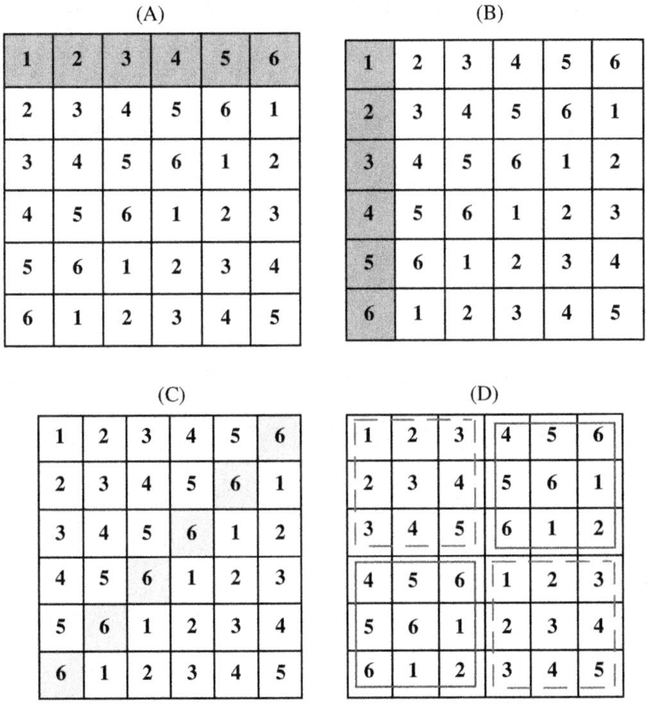

(A)

1	2	3	4	5	6
2	3	4	5	6	1
3	4	5	6	1	2
4	5	6	1	2	3
5	6	1	2	3	4
6	1	2	3	4	5

(B)

1	2	3	4	5	6
2	3	4	5	6	1
3	4	5	6	1	2
4	5	6	1	2	3
5	6	1	2	3	4
6	1	2	3	4	5

(C)

1	2	3	4	5	6
2	3	4	5	6	1
3	4	5	6	1	2
4	5	6	1	2	3
5	6	1	2	3	4
6	1	2	3	4	5

(D)

1	2	3	4	5	6
2	3	4	5	6	1
3	4	5	6	1	2
4	5	6	1	2	3
5	6	1	2	3	4
6	1	2	3	4	5

FIGURE 4.4 Properties of symmetric matrix: (A) row-wise sum; (B) column-wise sum; (C) repeated one diagonal digits; and (D) repeated SM submatrix.

4.3.3.1 Rules for a 6 × 6 sized symmetric matrix-total cross-tied configuration

For nth element corresponding to jth row and kth column of a pattern, n_{jk} is written as:

$$n_{jk}, \text{where} \begin{cases} j = \text{no. of row} \quad (j = 1, 2, \ldots, 6) \\ k = \text{no. of column} \; (k = 1, 2, \ldots, 6) \end{cases}$$

1. *Row-wise summation*

   The rule of row-wise summation in four different cases is shown in Fig. 4.4A and represented in Eq. (4.16) as:

$$\sum_{k=1}^{k=4} n_{jk} = \text{Summation for } j\text{th row } (j = 1, 2, \ldots\ldots 6) \tag{4.16}$$

2. *Column-wise summation*

   The rule of column-wise summation in four diversified cases is shown in Fig. 4.4B and represented in Eq. (4.14) as:

$$\sum_{j=1}^{j=4} n_{jk} = \text{Summation for } k\text{th column } (k = 1, 2, \ldots\ldots 6) \tag{4.17}$$

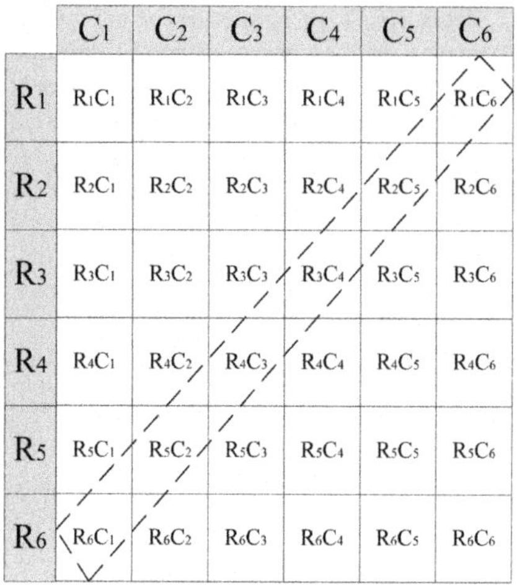

	C_1	C_2	C_3	C_4	C_5	C_6
R_1	R_1C_1	R_1C_2	R_1C_3	R_1C_4	R_1C_5	R_1C_6
R_2	R_2C_1	R_2C_2	R_2C_3	R_2C_4	R_2C_5	R_2C_6
R_3	R_3C_1	R_3C_2	R_3C_3	R_3C_4	R_3C_5	R_3C_6
R_4	R_4C_1	R_4C_2	R_4C_3	R_4C_4	R_4C_5	R_4C_6
R_5	R_5C_1	R_5C_2	R_5C_3	R_5C_4	R_5C_5	R_5C_6
R_6	R_6C_1	R_6C_2	R_6C_3	R_6C_4	R_6C_5	R_6C_6

FIGURE 4.5 Diagonally repeated number.

3. *Diagonally repeated number*

The rule of the diagonally repeated number in SM is shown in Figs. 4.4C and 4.5, and represented as shown in Eq. (4.18).

$$R_6C_1 = R_5C_2 = \cdots = R_1C_6 \tag{4.18}$$

By using Eqs. (4.16)–(4.18), the SM-TCT configuration of PV array is obtained.

The first digit is each module denotes the number of rows, while the second digit denotes the column of 6×6 PV array. TCT configuration of the PV array was rearranged on the basis of SM, which is shown in Fig. 4.6A–C.

Authors felt the need of a reconfigured methodology prerequisite for electrical interconnections among the PV modules for TCT configuration, such that a design of a PV array arranged in SM-TCT configuration could be assessed and the same was exhibited in Fig. 4.6A–C. This arrangement was framed using a proposed SM-TCT configuration without altering the electrical interconnections of the PV modules in an array under PSCs, where only the physical position of the PV modules varied. It is clear from Fig. 4.6C that the PV module number 22 (II row, II column) was located on the first row and second column physically and module number 32 (III row, II column) was located on the second row and second column in SM-TCT configuration.

4.4 Analysis of shading patterns

In this work, four more relevant and probable partial shading cases were considered in a 6×6 sized PV array [34]. The shading scenarios for TCT, Su-Do-Ku-TCT, and SM-TCT configurations were realized through the solar irradiation level of each PV module as scheduled in Fig. 4.7.

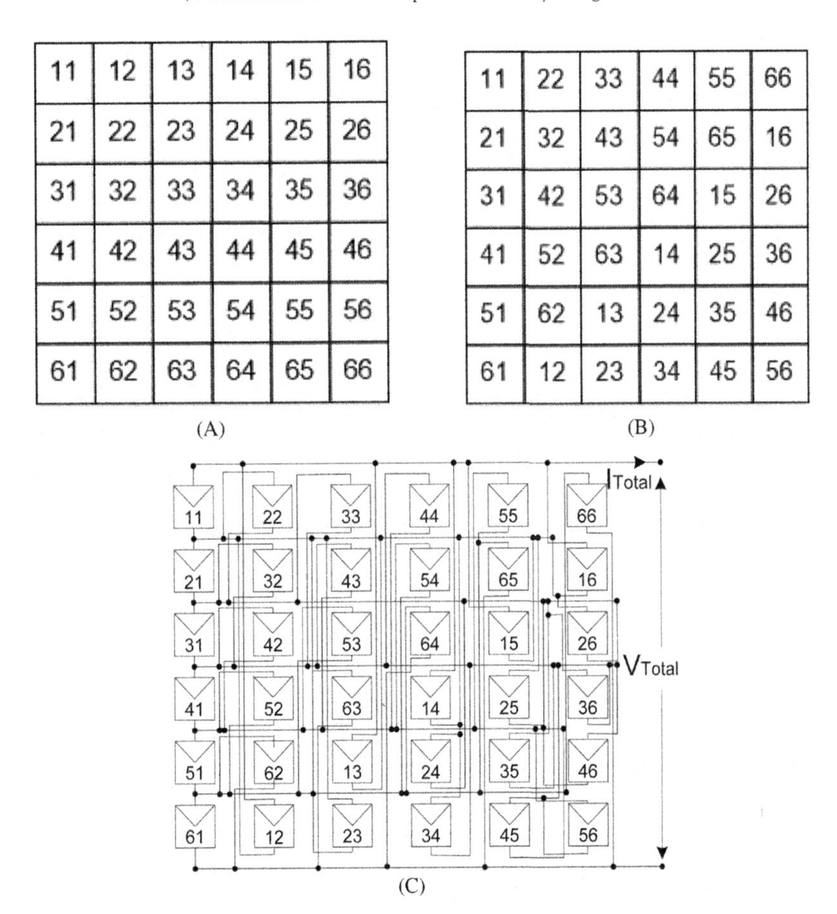

FIGURE 4.6 Methodology to reconfigure TCT into SM-TCT configuration: (A) TCT arrangement; (B) SM puzzle pattern; and (C) Electrical connections of PV modules in SM-TCT configuration.

4.5 Performance assessment of photovoltaic array configurations

The MPP of the array does not match that of the individual PV module under the shading profile, which is the actual cause of power losses. The misleading conditions of MPP tracking systems may be possible between local and global maxima points. Various performance parameters are investigated due to shading causes, demonstrated as $P-V$ and $I-V$ curves shown in Fig. 4.8.

The extreme probable power under PSCs is the summation of maximum magnitude of powers pertaining to singular PV modules while running the show independently under equal irradiance. Without shading, the maximum possible power of the array has always been higher than that of the peak value generated power under the shading conditions. The deviation between maximum power exhibited by the aforementioned array (with and without shading) is given by Eq. (4.19) as:

$$\text{Power loss} = P_{maximum} \text{ without shading} - P_{GMPP} \text{ under shading} \tag{4.19}$$

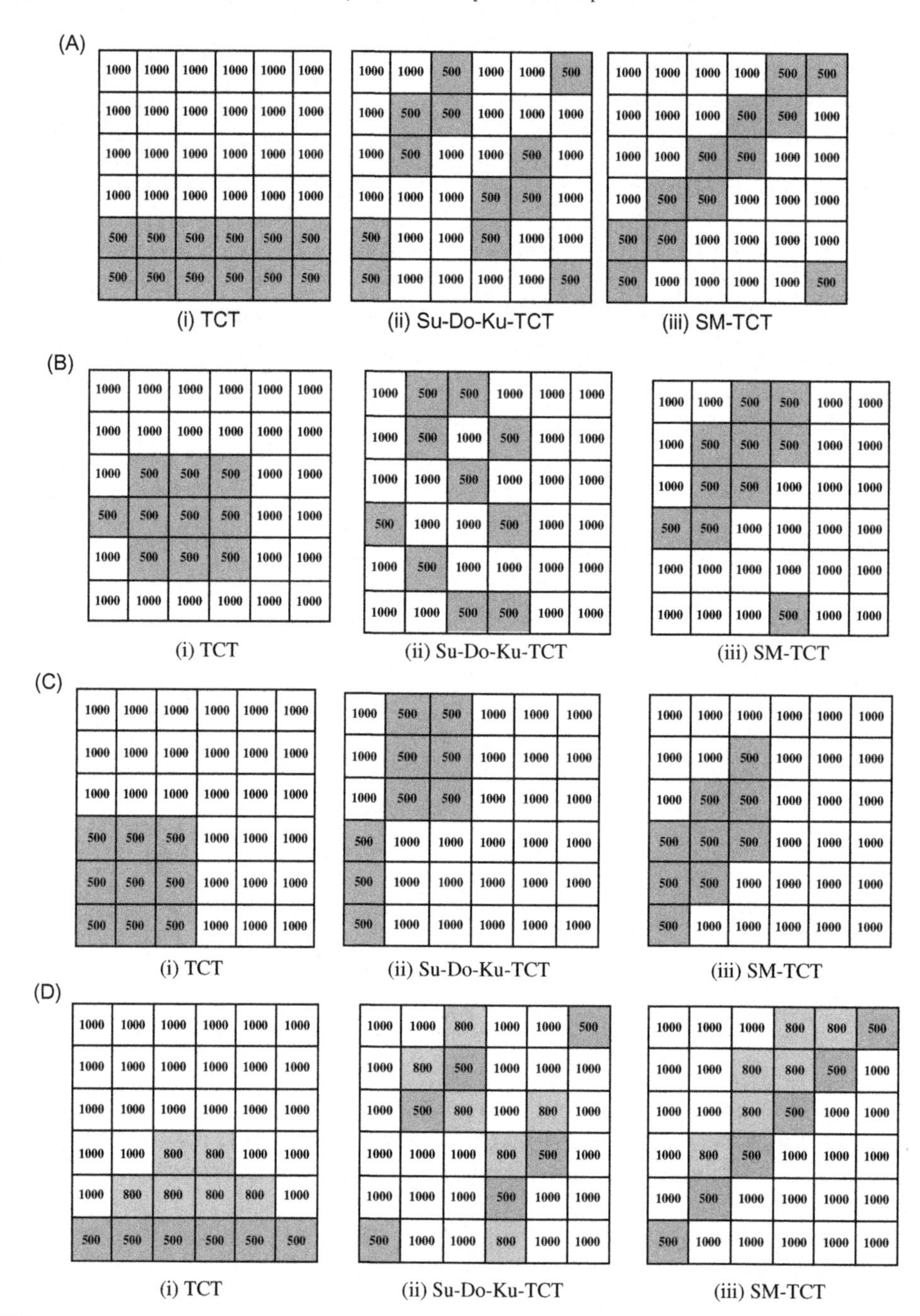

FIGURE 4.7 Probable shading cases: I–IV: (A) case-I: double row shading; (B) case-II: street light shape shading; (C) case-III: corner shading; and (D) case-IV: miscellaneous type shading.

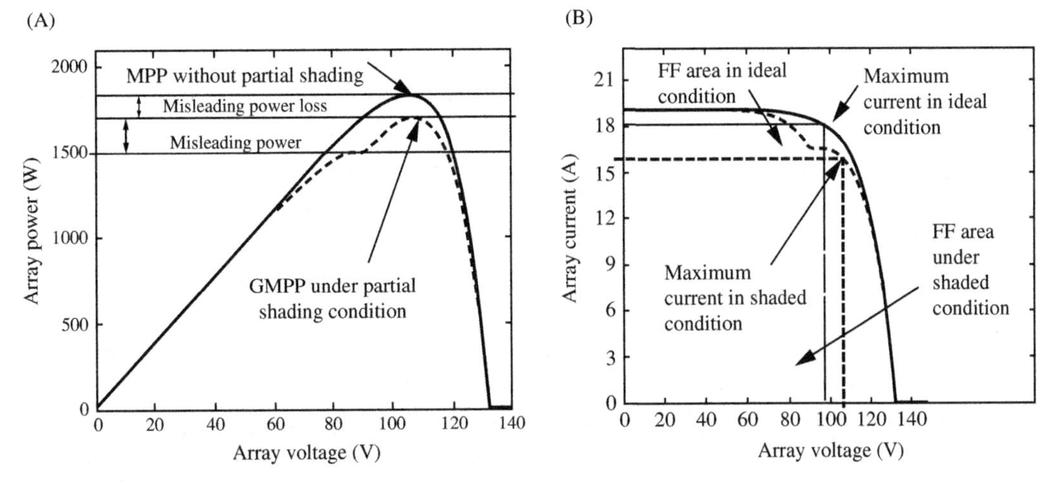

FIGURE 4.8 (A) $P-V$ and (B) $I-V$ curves under PSCs.

The performance assessment of the suggested configuration is appraised by four probable shading cases, shown in the previous section. The theoretical calculation of location regarding GMPP in the TCT and SM-TCT configurations are done. The DC current produced via entire PV modules under Standard Test Conditions (STC) is postulated to be I_m. So, the row currents in TCT connection for shading case-I are assessed as:

$$I_{r1} = I_{r2} = I_{r3} = I_{r4} = I_m + I_m + I_m + I_m + I_m + I_m = 6I_m \tag{4.20}$$

$$I_{r5} = I_{r6} = 0.5I_m + 0.5I_m + 0.5I_m + 0.5I_m + 0.5I_m + 0.5I_m = 3I_m \tag{4.21}$$

The generated current in all the rows is not similar and multiple peaks exist on the $P-V$ curves. The negligible amounts of variations are found in the voltages across each row. The voltage across the array is $V_{array} = 6 \times V_m$ and produced array power is expressed as:

$$P_{array} = V_{array} \times 6I_m \tag{4.22}$$

In the Su-Do-Ku-TCT configuration, the measurement of current in each row for shading cases-I is expressed in Eqs. (4.23)–(4.28) as:

$$I_{r1} = I_m + I_m + 0.5I_m + I_m + I_m + 0.5I_m = 5I_m \tag{4.23}$$

$$I_{r2} = I_m + 0.5I_m + 0.5I_m + I_m + I_m + I_m = 5I_m \tag{4.24}$$

$$I_{r3} = I_m + 0.5I_m + I_m + I_m + 0.5I_m + I_m = 5I_m \tag{4.25}$$

$$I_{r4} = I_m + I_m + I_m + 0.5I_m + 0.5I_m + I_m = 5I_m \tag{4.26}$$

$$I_{r5} = 0.5I_m + I_m + I_m + 0.5I_m + I_m + I_m = 5I_m \tag{4.27}$$

$$I_{r6} = 0.5I_m + I_m + I_m + I_m + I_m + 0.5I_m = 5I_m \tag{4.28}$$

Similarly, for SM-TCT configuration, the current in each row for shading case-I is given as:

$$I_{r1} = I_m + I_m + I_m + I_m + 0.5I_m + 0.5I_m = 5I_m \tag{4.29}$$

$$I_{r2} = I_m + I_m + I_m + 0.5I_m + 0.5I_m + I_m = 5I_m \tag{4.30}$$

$$I_{r3} = I_m + I_m + 0.5I_m + 0.5I_m + I_m + I_m = 5I_m \tag{4.31}$$

$$I_{r4} = I_m + 0.5I_m + 0.5I_m + I_m + I_m + 0.5I_m = 5I_m \tag{4.32}$$

$$I_{r5} = 0.5I_m + 0.5I_m + I_m + I_m + I_m + I_m = 5I_m \tag{4.33}$$

$$I_{r6} = 0.5I_m + I_m + I_m + I_m + I_m + 0.5I_m = 5I_m \tag{4.34}$$

It is to be observed that the row current in Su-Do-Ku-TCT and SM-TCT connection schemes is distinctive and obtained results are procured under all the shading cases: I–IV as shown in Table 4.2. The PV module's current and voltage under all four shading cases: I–IV for TCT, Su-Do-Ku-TCT and SM-TCT interconnection scheme is also depicted in Table 4.2. The improved power that is obtained is validated in all the four distinguished shading scenarios of PSCs for Su-Do-Ku-TCT and SM-TCT configurations, such that theoretical observations are also verified by the MATLAB simulation results [34].

4.6 Results and discussion

4.6.1 *P–V* and *I–V* curves of TCT, Su-Do-Ku-TCT, and SM-TCT configurations

In this section, the obtained electrical performance (*P–V* and *I–V* curves) for considering TCT, Su-Do-Ku-TCT, and SM-TCT configurations are discussed. In shading case-I, the array's GMPP for TCT configuration is obtained as 1080 W. The MPP of single unshaded (1000 W/m^2) and shaded (500 W/m^2) PV module is 50 and 24.5 W, respectively. In case-IV, additionally the solar irradiation level as 800 W/m^2 is considered for distinguished analysis. In Su-Do-Ku-TCT and SM-TCT interconnections, the array's GMPP are also 1129 W and 1250 W respectively, such that Fig. 4.9A exhibits the *P–V* curves of TCT, Su-Do-Ku-TCT, and SM-TCT configurations.

After an extensive investigation, it was seen that the SM-TCT interconnection scheme had the best performance among the TCT and Su-Do-Ku-TCT configurations in terms of sleekness of electrical curves. The smooth nature of the obtained curves abated the anticipation of erroneous tracking of MPP and the same was reflected by the *P–V* curves of three schemes of PV module arrangement for the acknowledged pattern of four shading profiles as shown in Fig. 4.9.

The *I–V* curves of TCT, Su-Do-Ku-TCT, and SM-TCT configurations for four instants shading cases-I–IV are shown in Fig. 4.10. The obtained efficient behavior of the *I–V* curve of SM-TCT had fewer undulations than TCT and Su-Do-Ku-TCT configurations. For all the considered shading cases-I–IV, the short-circuit (SC) current of TCT configuration is higher than the Su-Do-Ku-TCT and SM-TCT interconnections.

TABLE 4.2 Location of GMPP in TCT, Su-Do-Ku-TCT, and SM-TCT configurations under the considered four shading cases: I–IV.

	TCT				Su-Do-Ku-TCT				SM-TCT			
Partial shading cases	Orderly row current w.r.t. the bypassed panels		Array voltage	Array power	Orderly row current w.r.t. the bypassed panels		Array voltage	Array power	Orderly row current w.r.t. the bypassed panels		Array voltage	Array power
Case-I	I_{r6}	$3I_m$	$6V_m$	$18V_mI_m$	I_{r6}	$5I_m$	$6V_m$	$30V_mI_m$	I_{r6}	$5I_m$	$6V_m$	$30V_mI_m$
	I_{r5}	$3I_m$	$5V_m$	$15V_mI_m$	I_{r5}	$5I_m$	$5V_m$	$25V_mI_m$	I_{r5}	$5I_m$	$5V_m$	$25V_mI_m$
	I_{r4}	$6I_m$	$4V_m$	$24V_mI_m$	I_{r4}	$5I_m$	$4V_m$	$20V_mI_m$	I_{r4}	$5I_m$	$4V_m$	$20V_mI_m$
	I_{r3}	$6I_m$	$3V_m$	$18V_mI_m$	I_{r3}	$5I_m$	$3V_m$	$15V_mI_m$	I_{r3}	$5I_m$	$3V_m$	$15V_mI_m$
	I_{r2}	$6I_m$	$2V_m$	$12V_mI_m$	I_{r2}	$5I_m$	$2V_m$	$10V_mI_m$	I_{r2}	$5I_m$	$2V_m$	$10V_mI_m$
	I_{r1}	$6I_m$	V_m	$6V_mI_m$	I_{r1}	$5I_m$	V_m	$5V_mI_m$	I_{r1}	$5I_m$	V_m	$5V_mI_m$
Case-II	I_{r6}	$6I_m$	$6V_m$	$36V_mI_m$	I_{r6}	$5I_m$	$6V_m$	$30V_mI_m$	I_{r6}	$5.5I_m$	$6V_m$	$33V_mI_m$
	I_{r5}	$4.5I_m$	$5V_m$	$22.5V_mI_m$	I_{r5}	$5.5I_m$	$5V_m$	$27.5V_mI_m$	I_{r5}	$6I_m$	$5V_m$	$30V_mI_m$
	I_{r4}	$4I_m$	$4V_m$	$16V_mI_m$	I_{r4}	$5I_m$	$4V_m$	$20V_mI_m$	I_{r4}	$5I_m$	$4V_m$	$20V_mI_m$
	I_{r3}	$4.5I_m$	$3V_m$	$13.5V_mI_m$	I_{r3}	$5.5I_m$	$3V_m$	$16.5V_mI_m$	I_{r3}	$5I_m$	$3V_m$	$15V_mI_m$
	I_{r2}	$6I_m$	$2V_m$	$12V_mI_m$	I_{r2}	$5I_m$	$2V_m$	$10V_mI_m$	I_{r2}	$4.5I_m$	$2V_m$	$9V_mI_m$
	I_{r1}	$6I_m$	V_m	$6V_mI_m$	I_{r1}	$5I_m$	V_m	$5V_mI_m$	I_{r1}	$5I_m$	V_m	$5V_mI_m$
Case-III	I_{r6}	$4.5I_m$	$6V_m$	$27V_mI_m$	I_{r6}	$5.5I_m$	$6V_m$	$33V_mI_m$	I_{r6}	$5.5I_m$	$6V_m$	$33V_mI_m$
	I_{r5}	$4.5I_m$	$5V_m$	$22.5V_mI_m$	I_{r5}	$5.5I_m$	$5V_m$	$27.5V_mI_m$	I_{r5}	$5I_m$	$5V_m$	$25V_mI_m$
	I_{r4}	$4.5I_m$	$4V_m$	$18V_mI_m$	I_{r4}	$5.5I_m$	$4V_m$	$22V_mI_m$	I_{r4}	$4.5I_m$	$4V_m$	$18V_mI_m$
	I_{r3}	$6I_m$	$3V_m$	$18V_mI_m$	I_{r3}	$5I_m$	$3V_m$	$15V_mI_m$	I_{r3}	$5I_m$	$3V_m$	$15V_mI_m$
	I_{r2}	$6I_m$	$2V_m$	$12V_mI_m$	I_{r2}	$5I_m$	$2V_m$	$10V_mI_m$	I_{r2}	$5.5I_m$	$2V_m$	$11V_mI_m$
	I_{r1}	$6I_m$	V_m	$6V_mI_m$	I_{r1}	$5I_m$	V_m	$5V_mI_m$	I_{r1}	$6I_m$	V_m	$6V_mI_m$
Case-IV	I_{r6}	$3I_m$	$6V_m$	$18V_mI_m$	I_{r6}	$5.3I_m$	$6V_m$	$31.8V_mI_m$	I_{r6}	$5.5I_m$	$6V_m$	$33V_mI_m$
	I_{r5}	$5.2I_m$	$5V_m$	$26V_mI_m$	I_{r5}	$5.5I_m$	$5V_m$	$27.5V_mI_m$	I_{r5}	$5.5I_m$	$5V_m$	$27.5V_mI_m$
	I_{r4}	$5.6I_m$	$4V_m$	$22.4V_mI_m$	I_{r4}	$5.3I_m$	$4V_m$	$21.2V_mI_m$	I_{r4}	$5.3I_m$	$4V_m$	$21.2V_mI_m$
	I_{r3}	$6I_m$	$3V_m$	$18V_mI_m$	I_{r3}	$5.1I_m$	$3V_m$	$15.3V_mI_m$	I_{r3}	$5.3I_m$	$3V_m$	$15.9V_mI_m$
	I_{r2}	$6I_m$	$2V_m$	$12V_mI_m$	I_{r2}	$5.3I_m$	$2V_m$	$10.6V_mI_m$	I_{r2}	$5.1I_m$	$2V_m$	$10.2V_mI_m$
	I_{r1}	$6I_m$	V_m	$6V_mI_m$	I_{r1}	$5.3I_m$	V_m	$5.3V_mI_m$	I_{r1}	$5.1I_m$	V_m	$5.1V_mI_m$

SM-TCT, Symmetric matrix-total cross-tied; *TCT*, total cross-tied.

4.6.2 Power, voltage at global maximum power point, and power losses

Power and voltage at GMPP under all the shading cases are shown in Figs. 4.11 and 4.12. The TCT, Su-Do-Ku-TCT, and SM-TCT array GMPP without shading effect is observed as 1807 W. The relative loss (expressed in percentage) in power due to shading

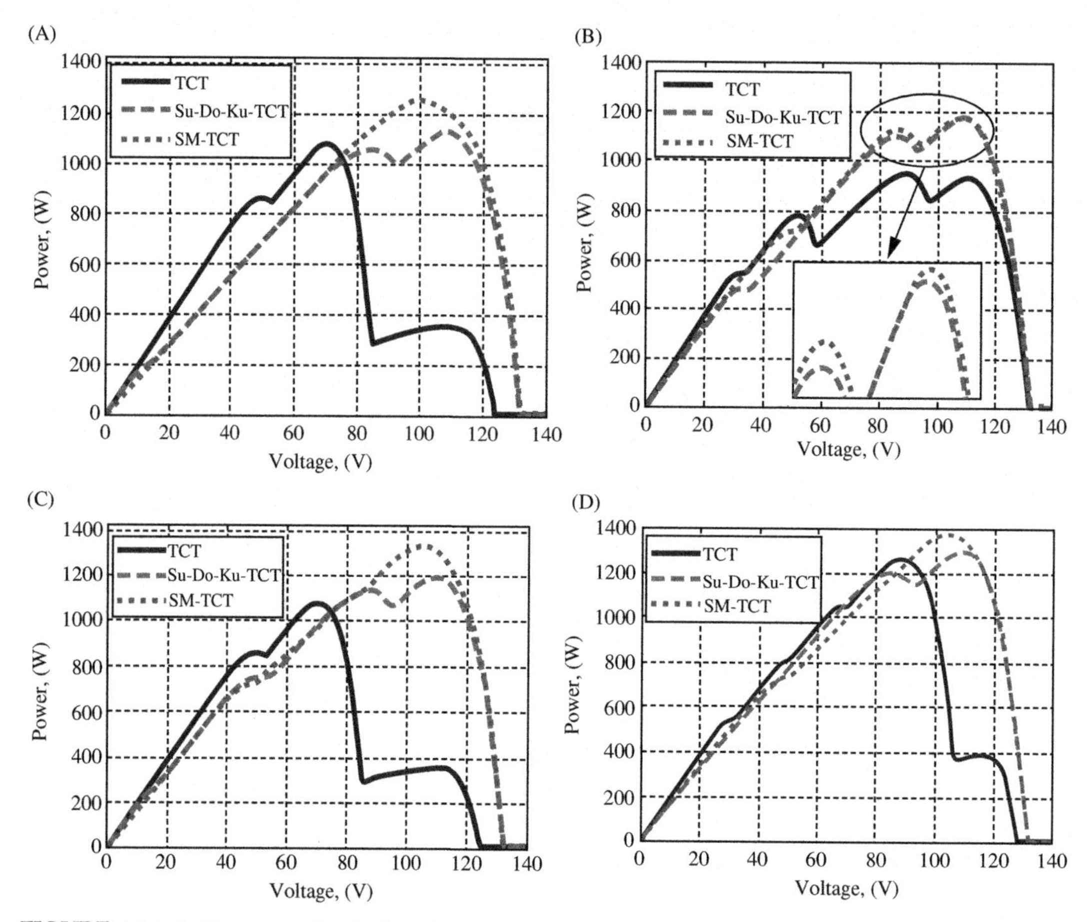

FIGURE 4.9 $P-V$ curves under the four shading cases: I–IV: (A) case-I; (B) case-II; (C) case-III; and (D) case-IV.

effect on TCT, Su-Do-Ku-TCT, and SM-TCT is shown in Table 4.2 and is equated using a bar chart representation in Fig. 4.13.

4.6.3 Fill factor

The FF is well defined as "the ratio of the power at GMPP to the product of open circuit voltage (V_{oc}) and SC current (I_{sc})". Explicitly, the FF is also signified as "the largest rectangle area which will fit inside the $I-V$ curve". These losses were introduced due to shading effect and they also reduce the FF [34]. FF can be evaluated using Eq. (4.35) as:

$$FF = \frac{\text{Power at } GMPP}{V_{oc} I_{sc}} \tag{4.35}$$

FF during various shading cases-I–IV, considered for TCT, Su-Do-Ku-TCT, and SM-TCT arrangement schemes is shown in Table 4.3 and compared using a bar chart as shown in Fig. 4.14.

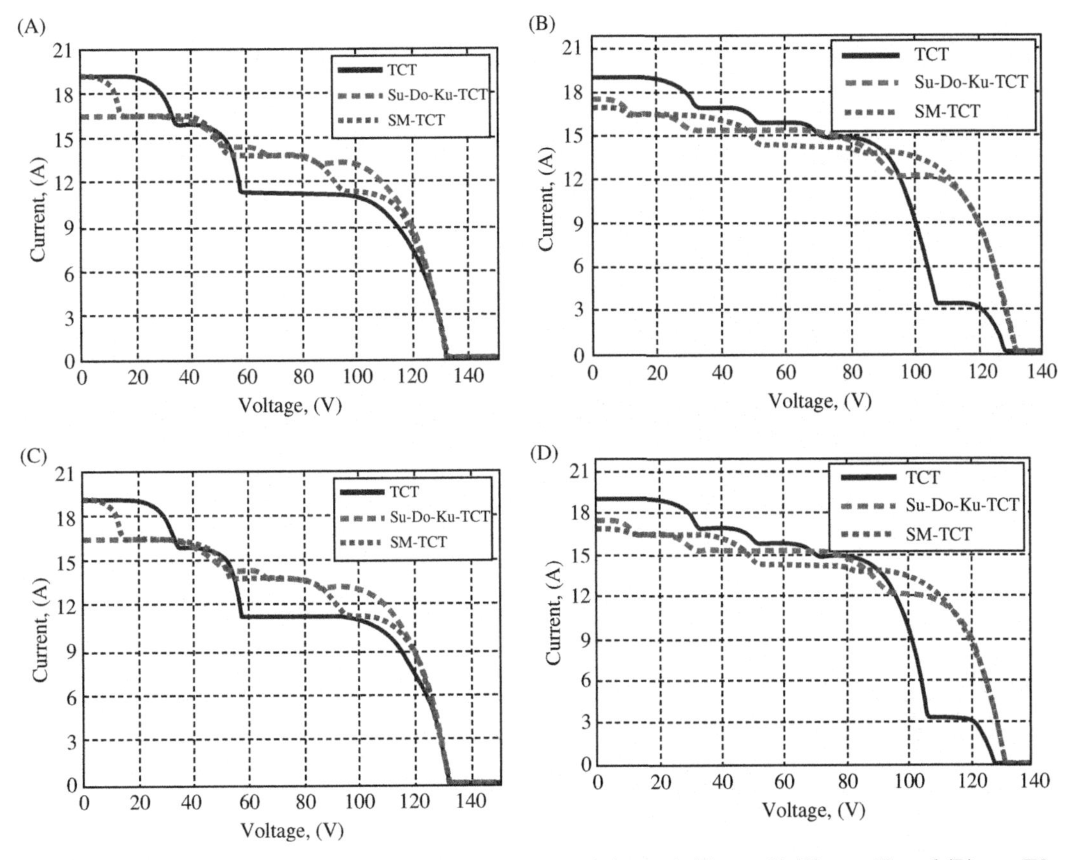

FIGURE 4.10 *I–V* curves under four shading cases: I–IV: (A) case-I; (B) case-II; (C) case-III; and (D) case-IV.

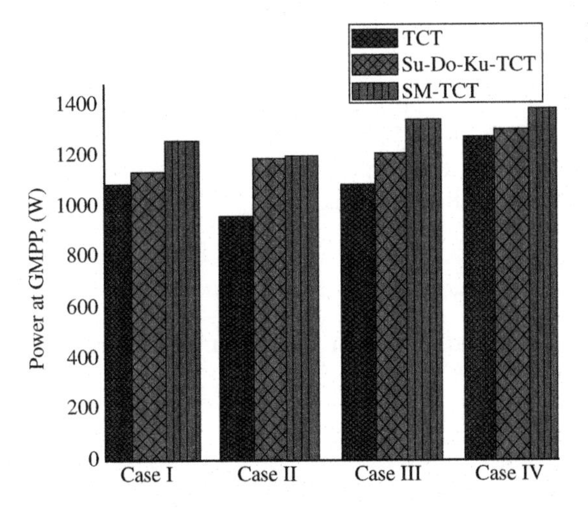

FIGURE 4.11 Power at GMPP.

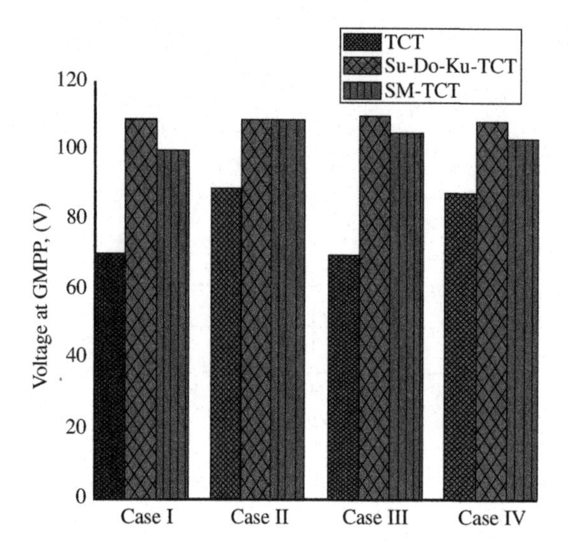

FIGURE 4.12 Voltage at GMPP.

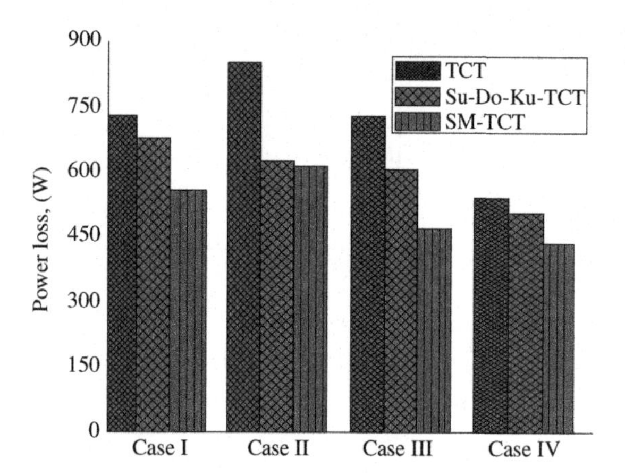

FIGURE 4.13 Power losses in TCT, Su-Do-Ku-TCT, and SM-TCT configurations.

4.6.4 Performance ratio

The increment in power at GMPP for all the configurations under shading cases: I–IV is called PR. In case-I, the PR increased from 59.76% in TCT, 62.47% in NTCT, and 69.17% in SM-TCT configurations. The acquired value of PR for considered shading scenarios is depicted in Table 4.3 and graphical representation is shown in Fig. 4.15.

4.6.5 Power enhancement

Due to shade dispersion, the increment in generated power of reconfigured PV array is called power enhancement (PE) and shown in Fig. 4.16 as:

$$\% PE = \frac{GMPP_{SM\text{-}TCT} - GMPP_{TCT}}{GMPP_{SM\text{-}TCT}} \times 100$$

TABLE 4.3 Power and voltage at GMPP, V_{oc}, I_{sc}, power loss, mismatch losses, FF, PR, and PE for TCT, Su-Do-Ku-TCT, and SM-TCT configurations under shading cases: I—IV.

Performance parameters under shading conditions	Case-I			Case-II			Case-III			Case-IV		
	TCT	Su-Do-Ku-TCT	SM-TCT	TCT	Su-Do-Ku-TCT	SM-TCT	TCT	Su-Do-Ku-TCT	SM-TCT	TCT	Su-Do-Ku-TCT	SM-TCT
Power at GMPP (W)	1080	1129	1250	956	1184	1195	1080	1201	1337	1267	1298	1372
Voltage at GMPP (V)	70.36	109.5	100.10	88.9	108.7	109	70.06	110	105	87.9	108.5	103.4
V_{oc} (V)	123.7	131.4	131.4	131	131.7	131.5	131.5	131.6	131.6	127.2	131.5	131.5
I_{sc} (A)	19.06	17.0	13.84	19.05	16.99	19.05	19.06	16.45	19.06	19.06	17.53	16.96
Power losses (W)	727	678	557	851	623	612	727	606	470	540	509	435
P_{loss}(%)	59.76	62.47	69.17	52.90	65.52	66.13	59.76	66.46	73.99	70.11	71.83	75.92
FF (%)	45.80	50.54	68.73	38.30	52.91	47.70	43.08	55.47	53.30	52.25	56.30	61.51
PR (%)	59.76	62.47	69.17	52.90	65.52	66.13	59.76	66.46	73.99	70.11	71.83	75.92
PE (%)	–	4.34	13.6	–	19.25	0.92	–	10.07	23.79	–	2.38	5.39
Best topology	*SM-TCT*			*SM-TCT/Su-Do-Ku*			*SM-TCT*			*SM-TCT*		

SM-TCT, Symmetric matrix-total cross-tied; *TCT*, total cross-tied.

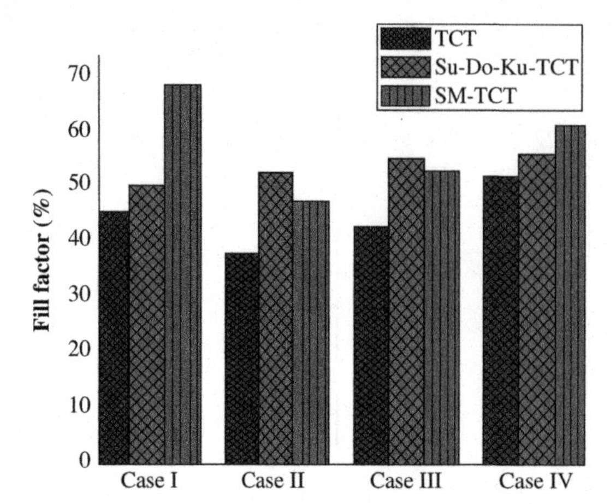

FIGURE 4.14 FF (%) for TCT, Su-Do-Ku-TCT, and SM-TCT configurations.

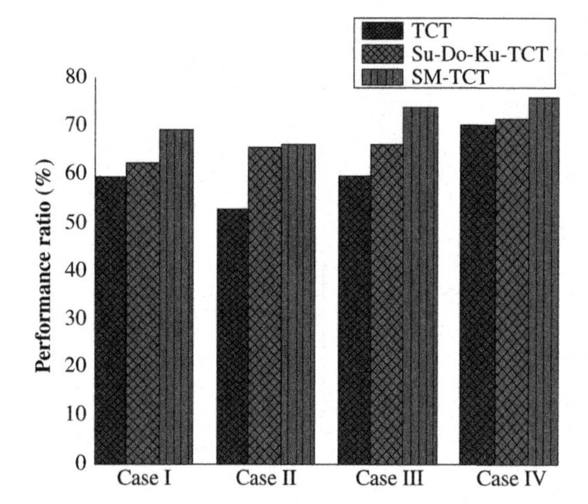

FIGURE 4.15 PR of TCT, Su-Do-Ku-TCT, and SM-TCT configurations.

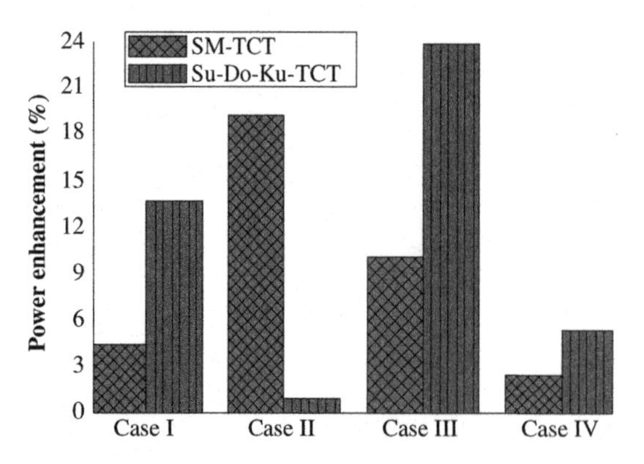

FIGURE 4.16 Power enhancement.

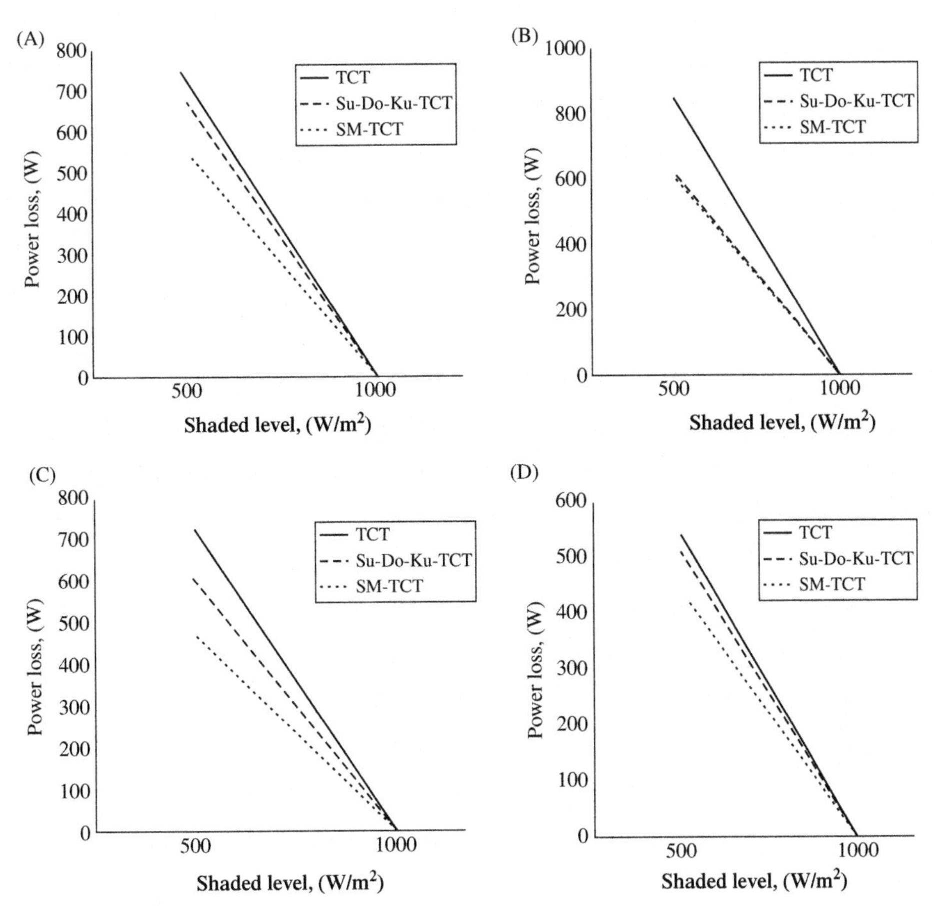

FIGURE 4.17 Shaded level effect on power loss under shading cases: I–IV: (A) case-I; (B) case-II; (C) case-III; and (D) case-IV.

4.6.6 Effect of shaded level on power loss

The proposed configuration SM-TCT is compared with the conventional TCT and Su-Do-Ku-TCT configurations in terms of power losses based on the shaded levels. The classical TCT and Su-Do-Ku-TCT configurations have maximum power losses as compared to the proposed one. The corresponding effect of shaded level on power losses is shown in Fig. 4.17A–D.

4.7 Conclusion

The authors have endeavored to propose and analyze Su-Do-Ku-TCT and SM-TCT configurations with that of existed TCT configuration of PV array for performance improvisation under considered partial shading cases-I–IV. The performance parameters such as power loss, PE, FF, and PR have been assessed and collated with classical TCT, Su-Do-Ku-TCT, and SM-TCT configurations. The results obtained from the SM-TCT configurations had fewer numbers of local maxima, and all the performance parameters were far better than TCT and Su-Do-Ku-TCT configurations. A simpler reconfiguration scheme can be employed to design arrays of any dimensions. Thus the proposed SM-TCT arrangement is helpful to design an efficient PV array for huge PV plants for maximum power generation.

References

[1] A. Woytea, J. Nijsa, R. Belmansa, Partial shadowing of photovoltaic arrays with different system configurations: literature review and field test results, Solar Energy 74 (2003) 217–233.

[2] D.M. Bote, E.C. Martin, Methodology for estimating building integrated photovoltaics electricity production under shadowing conditions and case study, Renew. Sustain. Energy Rev. 31 (2014) 492–500.

[3] D.L. Manna, V.L. Vigni, E.R. Sanseverino, V.D. Dio, P. Romano, Reconfigurable electrical interconnection strategies for photovoltaic arrays: a review, Renew. Sustain. Energy Rev. 33 (2014) 412–426.

[4] E.I. Batzelis, P.S. Georgilakis, S.A. Papathanassiou, Energy models for photovoltaic systems under partial shading conditions: a comprehensive review, IET Renewable Power Gener. 9 (4) (2015) 340–349.

[5] N.K. Gautam, N.D. Kaushika, An efficient algorithm to simulate the electrical performance of solar photovoltaic arrays, Energy 27 (2002) 347–361.

[6] N.D. Kaushika, N.K. Gautam, Energy yield simulations of interconnected solar PV arrays, IEEE Trans. Energy Convers. 18 (1) (2003) 127–134.

[7] D.D. Nguyen, B. Lehman, Modeling and simulation of solar PV arrays under changing illumination conditions, IEEE Conf. Comput. Power Electron. (2006) 295–299.

[8] E. Karatepe, M. Boztepe, M. Colak, Development of a suitable model for characterizing photovoltaic arrays with shaded solar cells, Solar Energy 81 (2007) 977–992.

[9] D. Nguyen, B. Lehman, An adaptive solar photovoltaic array using model based reconfiguration algorithm, IEEE Trans. Indust. Electron. 55 (7) (2008) 2644–2654.

[10] L. Gao, R.A. Dougal, Parallel connected solar PV system to address partial and rapidly fluctuating shadow conditions, IEEE Trans. Indust. Electron. 56 (5) (2009) 1548–1556.

[11] K. Ding, X.G. Bian, H.H. Liu, T. Peng, A MATLAB/Simulink based PV module model and its application under conditions of non-uniform irradiance, IEEE Trans. Energy Convers. 27 (4) (2012) 864–872.

[12] L.F.L. Villa, D. Picault, B. Raison, S. Bacha, A. Labonne, Maximizing the power output of partially shaded photovoltaic plants through optimization of the interconnections among its modules, IEEE J. Photovoltaics 2 (2) (2012) 154–163.

[13] L. Fialho, R. Melicio, V.M.F. Mendes, J. Figueiredo, M.C. Pereira, Effect of shading on series solar modules: simulation and experimental results, Procedia Technol. 17 (2014) 295–302.

[14] S. Moballegh, J. Jiang, Modeling, prediction, and experimental validations of power peaks of PV arrays under partial shading conditions, IEEE Trans. Sustain. Energy 5 (1) (2014) 293–300.

[15] S. Pareek, R. Dahiya, Output power maximization of partially shaded 4x4 PV field by altering its topology, Energy Procedia 54 (2014) 116–126.

[16] R. Ramaprabha, Selection of an optimum configuration of solar PV array under partial shaded condition using particle swarm optimization, Int. J. Electr. Comput. Energ. Electr. Commun. Eng. 8 (1) (2014) 89–96.

[17] F. Belhachat, C. Larbes, Modeling, analysis and comparison of solar photovoltaic array configurations under partial shading conditions, Solar Energy 120 (2015) 399–418.

[18] B. Celik, E. Karatepe, S. Silvestre, N. Gokmen, A. Chouder, Analysis of spatial fixed PV arrays configurations to maximize energy harvesting in BIPV applications, Renew. Energy 75 (2015) 534–540.

[19] H. Braun, S.T. Buddha, V. Krishnan, C. Tepedelenlioglu, A. Spanias, M. Banavar, et al., Topology reconfiguration for optimization of photovoltaic array output, Sustain. Energy Grids Networks 6 (2016) 58–69.

[20] A. Kumar, R.K. Pachauri, Y.K. Chauhan, Experimental analysis of SP/TCT PV array configurations under partial shading conditions, IEEE Conf. Power Electr. Intell. Control Energy Syst. (2016) 1–6.
[21] S. Mohammadnejad, A. Khalafi, S.M. Ahmadi, Mathematical analysis of total cross-tied photovoltaic array under partial shading condition and its comparison with other configurations, Solar Energy 133 (2016) 501–511.
[22] S. Pareek, R. Dahiya, Enhanced power generation of partial shaded photovoltaic fields by forecasting the interconnection of modules, Energy 95 (2016) 561–572.
[23] R.P. Vengatesh, S.E. Rajan, Analysis of PV module connected in different configurations under uniform and non-uniform solar radiations, Int. J. Green Energy 14 (13) (2017) 1507–1516.
[24] S. Vijayalekshmy, S.R. Iyer, B. Beevi, Comparative analysis on the performance of a short string of series-connected and parallel connected photovoltaic array under partial shading, J. Inst. Eng. (India): Ser. B 96 (3) (2014) 217–226.
[25] S.R. Potnuru, D. Pattabiraman, S.I. Ganesan, N. Chilakapati, Positioning of PV panels for reduction in line losses and mismatch losses in PV array, Renew. Energy 78 (2015) 264–275.
[26] M.Z.S. El-Dein, M. Kazerani, M.M.A. Salama, Optimal photovoltaic array reconfiguration to reduce partial shading losses, IEEE Trans. Sustain. Energy 4 (1) (2013) 145–153.
[27] B.I. Rani, G.S. Ilango, C. Nagamani, Enhanced power generation from PV array under partial shading conditions by shade dispersion using Su-Do-Ku configuration, IEEE Trans. Sustain. Energy 4 (3) (2013) 594–601.
[28] S.N. Deshkar, S.B. Dhale, J.S. Mukherjee, T.S. Babu, N. Rajasekar, Solar PV array reconfiguration under partial shading conditions for maximum power extraction using genetic algorithm, Renew. Sustain. Energy Rev. 43 (2015) 102–110.
[29] S. Vijayalekshmy, G.R. Bindu, S.R. Iyer, Performance improvement of partially shaded photovoltaic arrays under moving shadow conditions through shade dispersion, J. Inst. Eng. (India): Ser. B 97 (4) (2016) 569–575.
[30] A.S. Yadav, R.K. Pachauri, Y.K. Chauhan, Comprehensive investigation of PV arrays under different shading patterns by shade dispersion using puzzled pattern based Su-Do-Ku puzzle configuration, IEEE Conf. Next Gener. Comput. Technol. (2015) 824–830.
[31] A.S. Yadav, R.K. Pachauri, Y.K. Chauhan, Comprehensive investigation of PV arrays with puzzle shade dispersion for improved performance, Solar Energy 129 (2016) 256–285.
[32] P.S. Rao, P. Dinesh, G.S. Ilango, C. nagamani, Optimal Su-Do-Ku based interconnection scheme for increased power output from PV array under shading conditions, Front. Energy 9 (2015) 199–210.
[33] H.S. Sahu, S.K. Nayak, Extraction of maximum power from a PV array under non-uniform irradiation conditions, IEEE Trans. Electr. Devices 63 (12) (2016) 4825–4831.
[34] S. Vijayalekshmy, G.R. Bindu, S.R. Iyer, A novel Zig-Zag scheme for power enhancement of partially shaded solar arrays, Solar Energy 135 (2016) 92–102.
[35] N. Rakesh, T.V. Madhavaram, Performance enhancement of partially shaded solar PV array using novel shade dispersion technique, Front. Energy 10 (2015) 227–239.
[36] A.S. Yadav, R.K. Pachauri, Y.K. Chauhan, S. Choudhury, R. Singh, Performance enhancement of partially shaded PV array using novel shade dispersion effect on magic-square puzzle configuration, Solar Energy 144 (2017) 780–797.
[37] L.T.M. Tam, N.V. Duong, N.T. Tien, N.V. Ngu, A study on the output characteristic of photovoltaic array under partially shaded conditions, Appl. Mech. Mater. 472 (2014) 198–205.

CHAPTER

5

Dynamics of power flow in a stand-alone microgrid using four-leg inverters for nonlinear and unbalanced loads

Dhanashree Vyawahare

Electrical Design Group, Nuclear Power Corporation of India Limited, Mumbai, India

5.1 Introduction

In remote places, where grid supply is not available, an MG or distributed generation (DG) system using renewable energy sources transfers power through an inverter interface in stand-alone mode. In a stand-alone system, an unbalance in loads leads to lead to an unbalance in terminal voltages and requires unbalanced load compensators to maintain the balanced terminal voltages. Unbalanced voltages in general lead to more losses in motors due to negative sequence currents and may affect critical operations. Unbalanced load compensators are designed using three-leg inverters as well as four-leg inverters for three-phase, four-wire systems [1–3]. There are many nonlinear loads like single-phase large computer loads, variable frequency drives, etc. These are the main causes of harmonics in source currents. Different active filter designs have been proposed [4–6] to address the need of harmonics currents and to keep the current that is supplied by the grid or renewable energy source sinusoidal. The compensator designed in Ref. [7] uses shunt and series four-leg inverters which are required to be connected along with each individual DG system in an MG. For decades, such compensators have been used to tackle nonlinear and unbalanced load situations in MGs. But a compensator is a kind of 'local' circuit: operating near to the known unbalanced or nonlinear load fed by the renewable energy source. However, in distributed

Distributed Energy Resources in Microgrids
DOI: https://doi.org/10.1016/B978-0-12-817774-7.00005-3

systems, unbalance and nonlinearity in loads are not clearly defined and are not easily identifiable. Hence, local compensators have limitations in handling sporadic nonlinear and unbalanced loads in a typical distributed system.

Another approach to handle these sporadic unbalances and nonlinearities in loads is to make the system inherently robust for such random and small load changes using four-leg inverters. This second approach is addressed in detail in this chapter. Currently, four-leg inverters are widely used as power quality compensators for grid voltage unbalances for MG and DG systems in grid-connected mode [1,5–7], but the study of their use as distributed power interfaces has still not been investigated in detail. Three-leg inverters are being widely used as an interface in MGs, DG systems, and in the distributed operation of uninterrupted power supply (UPS) systems. In all the systems, mostly the loads are assumed to be linear and balanced. Three-leg inverters are not capable of controlling zero-sequence currents generated due to nonlinear and unbalanced loads. Hence, they are usually followed by a Δ-Y transformer to form a three-phase, four-wire low voltage (LV) distribution system (see Fig. 5.1). With nonlinear or unbalanced load, zero-sequence currents do not flow in the primary of Δ-Y transformer. However, they find a path through the neutral on secondary side leading to distorted and unbalanced voltages. Nevertheless this is the most popularly used topology for three-phase, four-wire DG systems. This is also the most common topology used in distributed UPS systems [8,9]. The inverter transformer (Δ-Y) provides a neutral connection for distribution. This transformer is usually designed such that winding provides filter inductance. Another common topology with three-leg inverters is to use a split DC link and connect the load neutral to the center of the DC link capacitors (see Fig. 5.2). In this case, for nonlinear and unbalanced loads, zero-sequence currents flow through the DC link capacitors. This produces ripples in the DC link voltage and hence capacitors of large sizes are required.

A four-leg inverter (see Fig. 5.3) is inherently capable of handling such loads [5,10]. The load neutral is connected to the center of the fourth leg. It controls the zero-sequence currents and thus avoids ripples in the DC link voltages. This is explained in detail in the closed-loop design of a four-leg inverter given in Section 5.2. An additional fourth leg also eliminates the need for an output transformer in LV and low power systems.

The issue of unbalanced loads in the stand-alone operations of renewable energy systems with a single four-leg inverter interface is discussed in detail in Refs. [12–14]. In Ref. [15], an MG consisting of two four-leg inverters employing a synchronous reference frame control strategy for positive, negative, and zero-sequence components of load

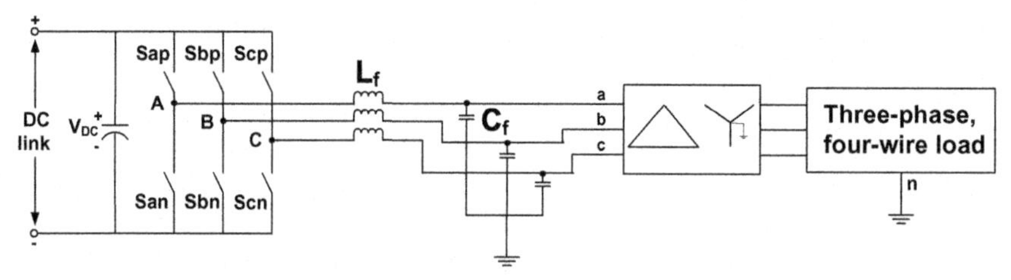

FIGURE 5.1 Three-leg inverter interface with a Δ-Y transformer.

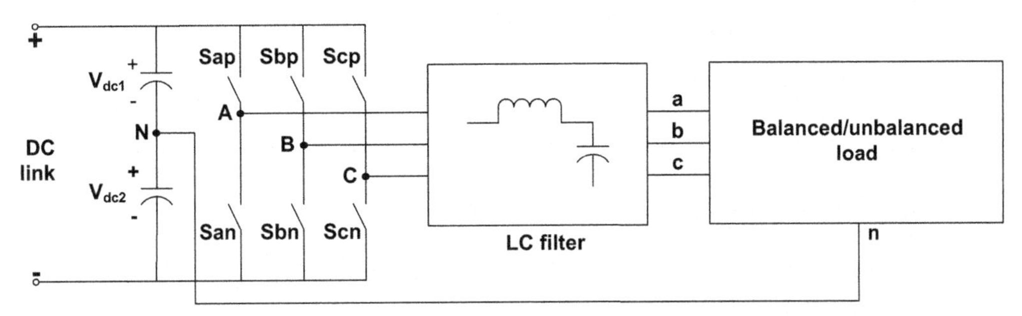

FIGURE 5.2 Three-leg inverter with a split direct current (DC) link capacitor.

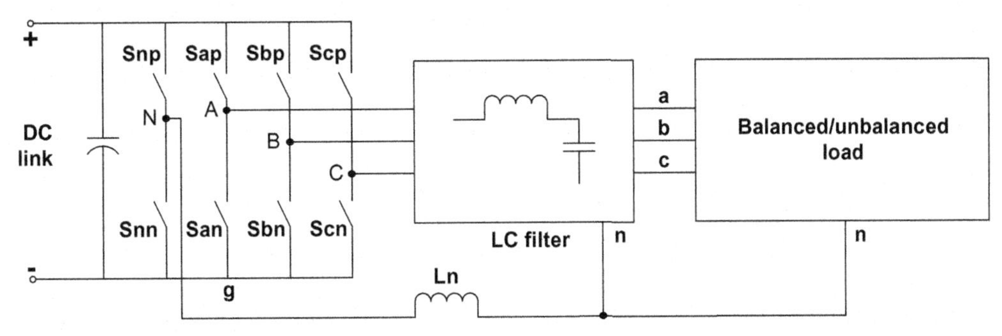

FIGURE 5.3 Four-leg inverter with a filter [11].

voltages is proposed. The concept of a decentralized operation of renewable energy sources using four-leg inverters as power interface to handle only unbalanced load situations and forming a four-wire MG is studied in Ref. [16]. In this chapter, a stand-alone MG is formed using four-leg inverters which caters to all the unidentified unbalanced as well as all types of nonlinear loads. It is shown that terminal voltages are maintained balanced and sinusoidal by each individual four-leg inverter.

The stand-alone operation is very similar to the parallel operation of inverters or UPS systems. In order to share the nonlinear load currents in proportion to the ratings of inverters, prior information about the harmonics present in the loads is necessary [17]. In Ref. [17], the knowledge of the presence of third or fifth harmonics in loads has been used to apply the droop controls corresponding to the harmonic frequencies. This results in an accurate sharing of nonlinear currents. However, as stated earlier, in MGs or DG systems, the variation of nonlinear and unbalanced loads is not known a priori and thus the load pattern is very random in nature. Hence, droop laws corresponding to only fundamental frequencies are applied. The droop control method leads to a desired decentralized operation. Since the droop corresponds to the fundamental frequency, it is observed that the average power corresponding to the fundamental frequency components of voltages and currents is shared according to the droop laws, but the instantaneous power due to harmonics in load currents does not follow the droop laws. Nevertheless, terminal voltages are maintained almost balanced and sinusoidal, not affecting the other three-phase balanced loads.

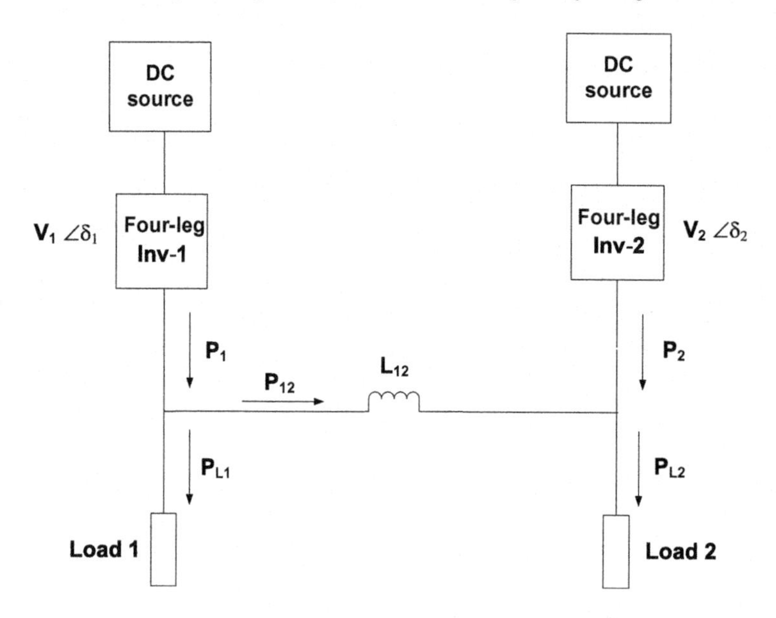

FIGURE 5.4 Stand-alone operation of two distributed generation (DG) systems.

As this chapter proposes a new idea of the MG using four-leg inverters to handle nonlinear and unbalanced loads, it becomes essential to study the power-flow dynamics and stability of the system as well. Power-flow and stability studies have been reported in literature for the decentralized operation of distributed sources or distributed UPS systems [18–22] assuming that the three-phase loads are 'linear and balanced'. In this work, for a proposed four-wire MG using four-leg inverters, a detailed study of the power-system dynamics and stability is carried out in the presence of 'nonlinear and unbalanced loads'. Power-flow analysis only for unbalanced load is already discussed in Ref. [16]. It is further extended for various types of standard and very common nonlinear loads. It can be noted that the power-flow analyses for unbalanced and balanced load conditions are the special cases of analysis done for a typical nonlinear load. The analysis is simplified due to balanced and sinusoidal voltages at the terminals maintained by a closed-loop system of each four-leg inverter. Also, droop laws govern the power sharing in a distributed system without any communication between the individual units. Decoupling between $P-f$ and $Q-V$ droop laws is ensured by considering the interconnecting line inductance to be sufficiently high for distantly placed DG units.

An MG consisting of two DG systems formed using four-leg inverters as shown in Fig. 5.4 is simulated and its laboratory implementation is carried out. The results for average power sharing and undistorted terminal voltages for various nonlinear and unbalanced loads are presented. Analysis is carried out for single-phase rectifier load on one of the phases. This emulates both the nonlinear and unbalanced load conditions. A nonlinear model representing the dynamics of power flow in interconnecting lines is developed. The developed model is validated using simulation data. An analysis of reactive power flow is also carried out.

The contributions of this chapter are:

1. Analysis of power flow and dynamics of the system for nonlinear and unbalanced load conditions.
2. Development and validation of nonlinear as well as linearized models for stability analysis using simulation data.
3. Development of laboratory prototype for a stand-alone MG using four-leg inverters as the interface.

This chapter is organized as follows. Section 5.2 explains the decentralized controller for each four-leg inverter. Section 5.3 presents the analysis of power flow in interconnecting lines for two DG systems for various nonlinear and unbalanced loads. A nonlinear model showing load-angle dynamics is derived for a typical load of single-phase diode bridge rectifier on one of the DG systems. This nonlinear dynamic model is validated using simulation data. In Section 5.4, linearization is carried out for the derived nonlinear differential equation for this typical nonlinear load and the small-signal stability of a two-DG system is studied. To take care of R-L loads, Q-V droops are also applied. This is discussed in brief in Section 5.5. Experimental results showing power sharing and balanced terminal voltages for the prototype microgrid are presented in Section 5.6. The conclusion is given in Section 5.7.

5.2 Closed-loop control design of four-leg inverter

To begin with, the controller design for a four-leg inverter MG interface is presented. The load neutral when connected to the center of the fourth leg of the inverter (Fig. 5.3) essentially works like three full bridge inverters with three-phase legs, sharing a common neutral conductor through the fourth leg. Consequently, the phase voltage at any switch position is $+V_{dc}$ or $-V_{dc}$.

The decentralized control system for each four-leg inverter, with *P-f* and *Q-V* droop laws and decoupled controllers for q–d–0-axes, is shown in Fig. 5.5. The switching of each four-leg inverter is controlled by a three-dimensional space vector modulation technique (3-D SVM) [11]. It has 15% more DC link utilization and low total harmonic distortion (THD) compared to other modulation techniques. The closed-loop system of a four-leg inverter is modeled as three independent decoupled circuits [10]. The control loop for a d-frame is shown in Fig. 5.6. Similar control loops are implemented for q-frame and 0-frame. Each controller has two loops, an inner capacitor current loop and an outer output voltage loop. Load currents act as a disturbance to the closed-loop system. V_{ref-q}, V_{ref-d}, and V_{ref-0} are derived from the droop laws as shown in Fig. 5.5. Capacitor current feedback is used. Since capacitor currents are proportional to the derivatives of capacitor voltages, they provide derivative control and the required damping aiding the PI regulator. The control parameters K_p, K_i, and S_a are derived using the classical root locus method. The values are chosen such that the controller is fast and robust in operation.

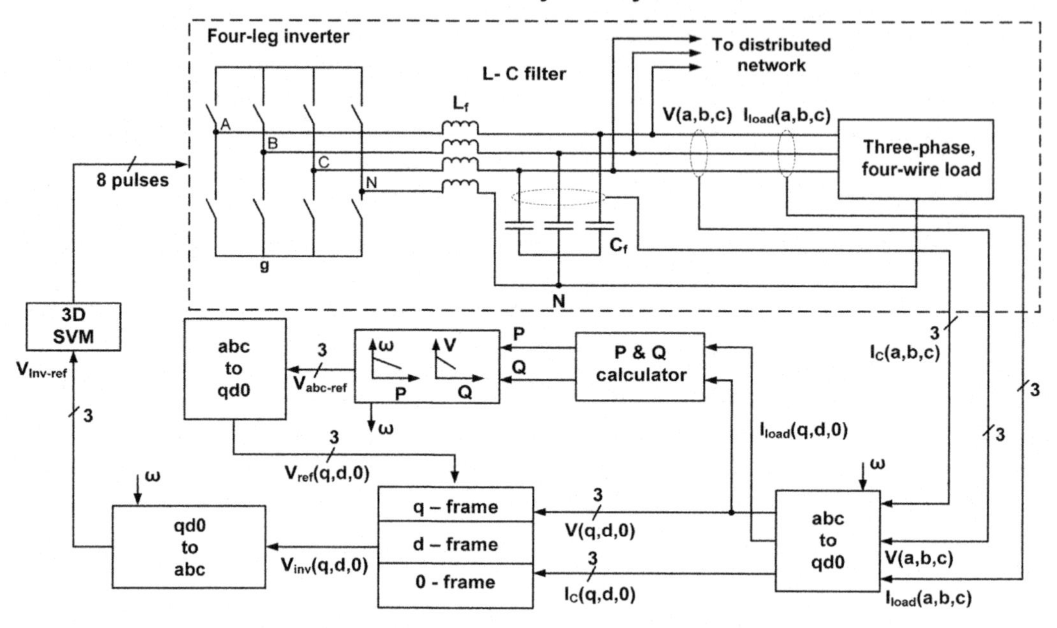

FIGURE 5.5 Decentralized closed-loop system of a single distributed generation (DG) unit.

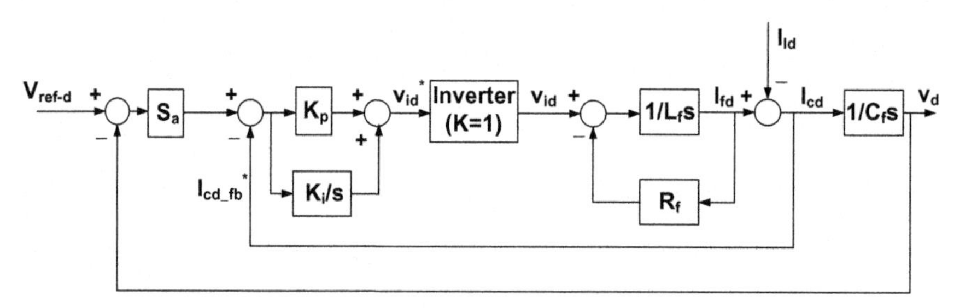

FIGURE 5.6 *d*-Frame controller [10].

5.2.1 Simulation of closed-loop operation

By carrying out the root locus exercise, controller parameters are selected as $K_p = 100$, $K_i = 50$, and $S_a = 4$. A DC link voltage is taken as 100 V. Since modulation index is kept at 1, corresponding output voltage after filtering is 40 V rms or 60 V peak value. Simulation results for the closed-loop operation of one four-leg inverter along with controller performance are given in Figs. 5.7 and 5.8.

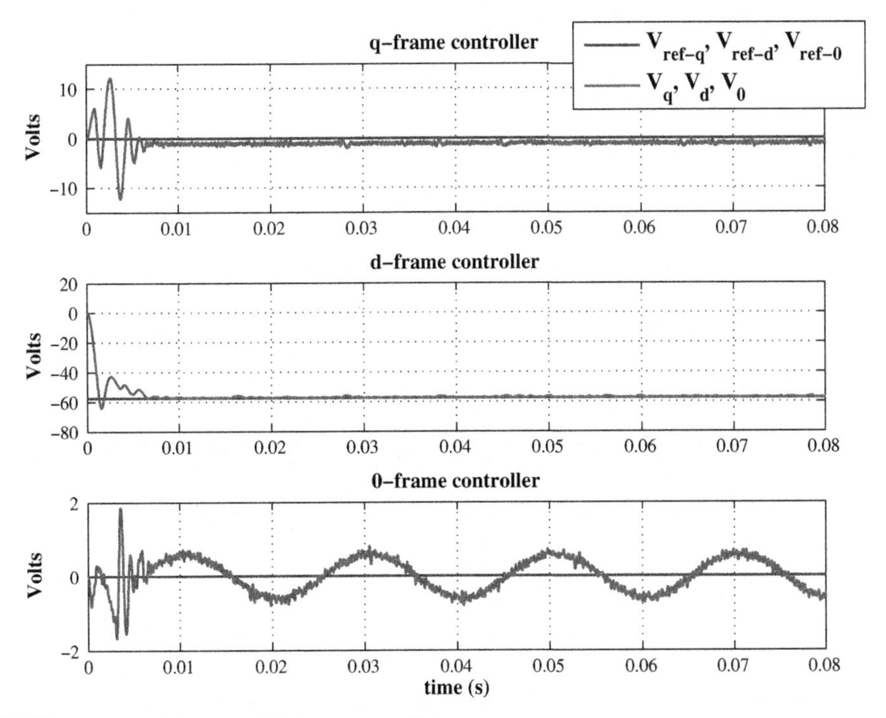

FIGURE 5.7 *q*-Frame, *d*-frame and 0-frame controller response.

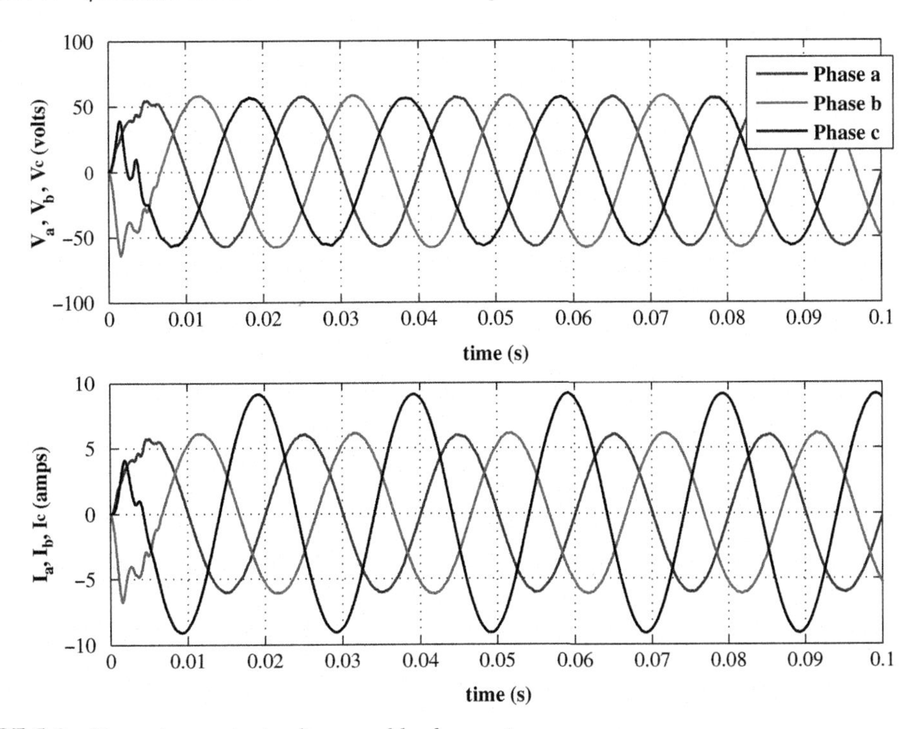

FIGURE 5.8 Three-phase output voltages and load currents.

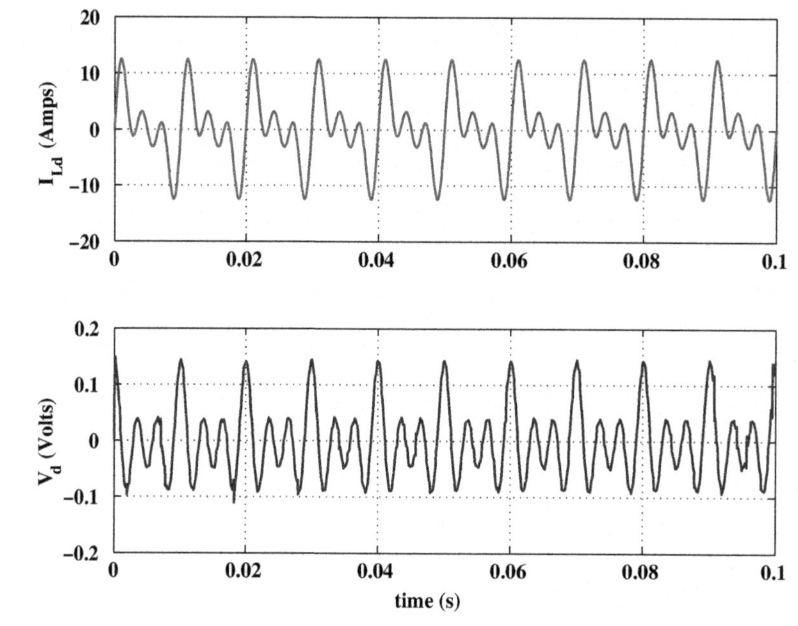

FIGURE 5.9 Perturbations in V_d due to I_{Ld} with harmonics.

For nonlinear and unbalanced loads, the load currents will have harmonics and negative sequence components. Hence, the effect of the disturbance signal I_{Ld} on the output V_d is analyzed for the designed controller parameter values. The corresponding transfer function is

$$\frac{V_d(s)}{I_{Ld}(s)} = \frac{(L_f s + R_f)s}{C_f s^2(L_f s + R_f) + (K_p s + K_i)(C_f s + S_a)}. \tag{5.1}$$

For a single-phase diode bridge rectifier load, the current phasor in stationary reference frame can be obtained in terms of phasors of all negative and positive sequence components of fundamental and harmonic currents as:

$$I = (I^p_{qd1}e^{j\omega t} + I^n_{qd1}e^{-j\omega t} + I^p_{qd3}e^{j3\omega t} + I^n_{qd3}e^{-j3\omega t} + I^p_{qd5}e^{j5\omega t} + I^n_{qd5}e^{-j5\omega t}). \tag{5.2}$$

This is discussed in detail in Section 5.3.1. In synchronous reference frame, I_{Ld} will thus contain 2ω, 4ω, and 6ω terms. Hence, an arbitrary current I_{Ld} (5.3) was applied to (5.1):

$$I_{Ld}(t) = 5(\sin 2\omega t + \sin 4\omega t + \sin 6\omega t) \text{ A}. \tag{5.3}$$

The corresponding perturbation in V_d is depicted in Fig. 5.9. Attenuation of the undesired effect of I_{Ld} on V_d because of the low gain offered by (5.1) is clearly visible. Thus, the designed controller tries to minimize the effect of nonlinear and unbalanced load current on the load terminal voltage.

5.3 Analysis of power flow with nonlinear and unbalanced loads

The power flow in interconnecting lines for a two-DG system for a variety of nonlinear and unbalanced loads is analyzed in a fashion similar to the classical load-flow study and uses terminal voltages and angles to determine the power flow in interconnecting lines. It helps in developing a linearized model for system dynamics and a stability study. For instantaneous power calculations, the three-phase quantities are transformed into a stationary reference frame using the transformation

$$\begin{bmatrix} f_\alpha \\ f_\beta \\ f_\gamma \end{bmatrix} = [T] \begin{bmatrix} f_a \\ f_b \\ f_c \end{bmatrix}, \tag{5.4}$$

where

$$T = \frac{2}{3} \begin{bmatrix} 1 & \cos\left(-\frac{2\pi}{3}\right) & \cos\left(\frac{2\pi}{3}\right) \\ 0 & \sin\left(-\frac{2\pi}{3}\right) & \sin\left(\frac{2\pi}{3}\right) \\ \frac{1}{2} & \frac{1}{2} & \frac{1}{2} \end{bmatrix}.$$

A four-leg inverter is capable of maintaining almost balanced undistorted output voltages at the nonlinear load terminals. These are first transformed into the stationary reference frame as

$$\begin{aligned} v_\alpha(t) &= v_\alpha, \\ v_\beta(t) &= \frac{(v_c - v_b)}{\sqrt{3}}, \\ v_\gamma(t) &= 0. \end{aligned} \tag{5.5}$$

A complex number representation of voltages is given as $\mathbf{V} = v_\alpha - jv_\beta$. Since voltages have only positive sequence components, the voltage phasor in stationary reference frame can also be represented as

$$\mathbf{V} = V_{qd}^p e^{j\omega t}, \tag{5.6}$$

where $V_{qd}^p = v_q + jv_d$. For analysis, different types of nonlinear and unbalanced loads are connected to one of the DGs. Most common nonlinear loads are single-phase bridge rectifier, three units of single-phase bridge rectifiers, and three-phase full-bridge rectifier. Expressions of power flow from DG-1 to DG-2 are derived for these loads in the next sections.

5.3.1 Single-phase diode bridge rectifier on one phase of DG-1

A diode bridge rectifier load is connected to one of the phases of a DG-1 system. This will create the situation of nonlinearity as well as unbalance in an MG. The currents in the interconnecting lines are

$$\begin{aligned} i_a(t) &= \mu I_1 \cos(\omega t - \phi_1) + I_3 \cos(3\omega t - \phi_3) + I_5 \cos(5\omega t - \phi_5) + \cdots, \\ i_b(t) &= I_1 \sin(\omega t - 2\pi/3 - \phi_1), \\ I_c(t) &= I_1 \sin(\omega t + 2\pi/3 - \phi_1). \end{aligned} \tag{5.7}$$

Where I_1 is the peak value of fundamental current when only a linear balanced load is connected and I_3 and I_5 are the peak values of harmonic currents. Due to the single-phase rectifier load on phase-a, the fundamental component will change and be taken care of by factor 'μ'. These line currents are transformed into a stationary reference frame using (5.4). In complex notation form, three-phase currents can be represented as a complex phasor:

$$\mathbf{I} = i_\alpha - j\ i_\beta. \tag{5.8}$$

Note that the component i_γ is not considered since, in instantaneous power-flow analysis, there is no corresponding v_γ component for terminal voltages [23]. Currents in a stationary reference frame can also be represented in terms of phasors of all negative and positive sequence components of fundamental and harmonic currents as

$$\mathbf{I} = (I^p_{qd1} e^{j\omega t} + I^n_{qd1} e^{-j\omega t} + I^p_{qd3} e^{j3\omega t} + I^n_{qd3} e^{-j3\omega t} + I^p_{qd5} e^{j5\omega t} + I^n_{qd5} e^{-j5\omega t}). \tag{5.9}$$

The coefficients for all sequence components, like I^p_{qd1}, I^n_{qd1} etc., are vectors. In general, they are represented as

$$f_{qdk} = f_{qk} + j\ f_{dk},$$

where k denotes the kth harmonic component. Superscript p and n are for positive and negative sequence components, respectively. Using (5.6) and (5.9), the apparent power flow in the line is calculated as:

$$S = (V^p_{qd} e^{j\omega_1 t})(I^p_{qd1} e^{j\omega_1 t} + I^n_{qd1} e^{-j\omega_1 t} + I^p_{qd3} e^{j3\omega_1 t} + I^n_{qd3} e^{-j3\omega_1 t} + I^p_{qd5} e^{j5\omega_1 t} + I^n_{qd5} e^{-j5\omega_1 t})^*. \tag{5.10}$$

The real part of (5.10) gives active power and the imaginary gives reactive power. Active power can be compactly expressed as

$$\begin{aligned} P_{12}(t) = P_0 &+ \sum_{k=1,3,5} (P^{nk}_c \cos((k+1)\omega_1 t) + P^{nk}_s \sin((k+1)\omega_1 t)) \\ &+ \sum_{k=3,5} (P^{pk}_c \cos((k-1)\omega_1 t) + P^{pk}_s \sin((k-1)\omega_1 t)). \end{aligned} \tag{5.11}$$

The power coefficients $P^{n1}_c, P^{n1}_s, Q^{n1}_c, Q^{n1}_s$, etc., are defined in Appendix 5.1, Power coefficients. In (5.11), P_0 is due to the fundamental positive sequence components of voltages and currents. It can be replaced by expression (5.12) which is usually the power flow in a balanced load condition.

$$P_0(t) = B_{12} \sin \delta_{12}, \quad \text{with} \quad B_{12} = \frac{3V_1V_2}{\omega_0 L_{12}}, \tag{5.12}$$

where V_1 and V_2 are inverter terminal voltages, ω_0 is the nominal frequency, and L_{12} is the inductance of line between DG-1 and DG-2. Substituting (5.12) in (5.11), P_{12} can be written as:

$$\begin{aligned} P_{12}(t) = B_{12} \sin \delta_{12} \quad &+ \sum_{k=1,3,5} (P_c^{nk} \cos((k+1)\omega_1 t) + P_s^{nk} \sin((k+1)\omega_1 t)) \\ &+ \sum_{k=3,5} (P_c^{pk} \cos((k-1)\omega_1 t) + P_s^{pk} \sin((k-1)\omega_1 t)). \end{aligned} \tag{5.13}$$

Power flow in interconnecting line P_{12} is thus the combination of power due to fundamental frequency positive sequence voltages and fundamental frequency as well as harmonic and unbalanced currents.

A similar expression is obtained for reactive power flow in interconnecting lines which is given as

$$\begin{aligned} Q_{12}(t) = Q_0 &+ \sum_{k=1,3,5} (Q_c^{nk} \cos((k+1)\omega_1 t) + Q_s^{nk} \sin((k+1)\omega_1 t)) \\ &+ \sum_{k=3,5} (Q_c^{pk} \cos((k-1)\omega_1 t) + Q_s^{pk} \sin((k-1)\omega_1 t)). \end{aligned} \tag{5.14}$$

5.3.2 Three single-phase diode bridge rectifiers as load

Three single-phase diode bridge rectifiers are connected to three phases of DG-1, and DG-2 continues to handle a linear balanced load. This will cause the following currents to flow in interconnecting lines.

$$\begin{aligned} i_a(t) &= I_1 \cos(\omega t) + I_3 \cos(3\omega t) + I_5 \cos(5\omega t) + \ldots, \\ i_a(t) &= I_1 \cos(\omega t - 2\pi/3) + I_3 \cos(3\omega t - 2\pi/3) + I_5 \cos(5\omega t - 2\pi/3) + \ldots, \\ i_c(t) &= I_1 \cos(\omega t + 2\pi/3) + I_3 \cos(3\omega t + 2\pi/3) + I_5 \cos(5\omega t + 2\pi/3) + \ldots \end{aligned} \tag{5.15}$$

Transformation of the interconnecting line currents with only dominant third and fifth harmonics in stationary reference frame is done using (5.4).

$$\begin{aligned} i_\alpha(t) &= I_1 \cos(\omega t) + I_3 \cos(3\omega t) + I_5 \cos(5\omega t), \\ i_\beta(t) &= -I_1 \sin(\omega t) - I_5 \sin(5\omega t). \end{aligned} \tag{5.16}$$

Note that the current is represented as (5.8). Currents transformed in stationary reference frame are represented in terms of phasors of all negative and positive sequence components of fundamental and harmonic currents as

$$\mathbf{I} = (I_{qd1}^p e^{j\omega t} + I_{qd3}^p e^{j3\omega t} + I_{qd5}^p e^{j5\omega t}). \tag{5.17}$$

Following steps similar to those in Section 5.3.1, the active and reactive powers are obtained as

$$P_{12}(t) = B_{12} \sin \delta_{12} + P_c^{p3} \cos 2\omega_1 t + P_s^{p3} \sin 2\omega_1 t + P_c^{p5} \cos 4\omega_1 t + P_s^{p5} \sin 4\omega_1 t, \tag{5.18}$$

TABLE 5.1 Power components for different nonlinear and unbalanced loads.

Load type	Coefficients present in P_{12}
Single-phase diode bridge-rectifier load	$P_0, P_c^{n1}, P_s^{n1}, P_c^{p3}, P_s^{p3}, P_c^{n3}, P_s^{n3}, P_c^{p5}, P_s^{p5}, P_c^{n5}, P_s^{n3}$
Three single-phase diode bridge rectifiers	$P_0, P_c^{p3}, P_s^{p3}, P_c^{p5}, P_s^{p5}$
Three-phase diode rectifier load and linear unbalanced load	$P_0, P_c^{n1}, P_s^{n1}, P_c^{p5}, P_s^{p5}, P_c^{n5}, P_s^{n5}$
Three-phase diode rectifier load	P_0, P_c^{n5}, P_s^{n5}
Linear unbalanced load	P_0, P_c^{n1}, P_s^{n1}
Linear balanced load	P_0

and

$$Q_{12}(t) = Q_0 + Q_c^{p3} \cos 2\omega_1 t + Q_s^{p3} \sin 2\omega_1 t + + Q_c^{p5} \cos 4\omega_1 t + Q_s^{p5} \sin 4\omega_1 t, \tag{5.19}$$

where the coefficients are the same as those given in Appendix 5.1, Power coefficients. It is noted here that since a three-phase load is balanced in nature, there are no fundamental negative sequence components in load currents. Oscillations in power are mainly due to harmonics in load currents.

5.3.3 Summary of power-flow analysis

Active power components for various combinations and types of nonlinear and unbalanced loads are derived using the above-described method and are summarized in Table 5.1.

From the above power-flow analysis, it is clear that the single-phase diode bridge rectifier load discussed in Section 5.3.1 is the load case that encompasses all other possible nonlinear and unbalanced load cases. Hence, further analysis is presented here for the single-phase diode bridge rectifier load.

5.3.4 Dynamics of power flow in interconnecting lines

In conventional power system, synchronous generators have a governing action to control frequency. Also the frequency goes down with an increase in load, but at the same time, the impedance offered by inductive loads also decreases, resulting in a slight improvement in frequency drooping. Thus, both the load-governing action and governing the action of each generator control the frequency. However, in MGs with inverter interfaces, a change in frequency can be achieved only by changing the reference in the controller. Hence, for a decentralized operation, P–f droop laws are applied to each four-leg inverter, which allow parallel operation. Analysis is started following the basic steps that are given in the conventional power system [24]. The method used is an extension of the power-flow analysis for conventional power systems.

The basic equation representing the power system's dynamics in terms of load angle δ_{12} is given as

$$\frac{d}{dt}\delta_{12}(t) = \omega_1(t) - \omega_2(t). \tag{5.20}$$

Note that δ_1 and δ_2 are respective voltage angles of DG-1 and DG-2 (see Fig. 5.4) and the load angle between the two is $\delta_{12}(t) = \delta_1(t) - \delta_2(t)$. The inverter frequencies ω_1 and ω_2 are calculated through P–f droops applied and expressed as

$$\omega_1(t) = \omega_0 - b_1(P_{R1} - P_1(t)), \quad \omega_2(t) = \omega_0 - b_2(P_{R2} - P_2(t)), \tag{5.21}$$

where ω_0 is the nominal frequency of the microgrid, P_{R1} and P_{R2} are rated power outputs for inverters, b_1 and b_2 are the droop coefficients and P_1 and P_2 are power outputs of DG-1 and DG-2. Droop coefficients are chosen to satisfy the condition

$$b_1 P_{R1} = b_2 P_{R2}.$$

From Fig. 5.4, P_1(t) and $P_2(t)$ can be written as

$$P_1 = P_{L1} + P_{12}, \quad P_2 = P_{L2} - P_{12}, \tag{5.22}$$

where P_{L1} and P_{L2} are loads connected to DG-1 and DG-2 respectively, and P_{12} is the power flow in the interconnecting line. From (5.22), it is clear that when the load is nonlinear or unbalanced, the active power supplied by the DG systems P_1 and P_2 will also contain similar harmonic components as in P_{12}. Using (5.20), (5.21), and (5.22), the nonlinear differential equation representing power flow dynamics is (without showing the time dependence of the parameters)

$$\frac{d}{dt}\delta_{12} = (b_1 + b_2)P_{12} + b_1 P_{L1} - b_2 P_{L2}. \tag{5.23}$$

Substituting P_{12} from (5.13), a general nonlinear model representing the dynamics of the power flow in a stand-alone MG is obtained as

$$\begin{aligned}\frac{d}{dt}\delta_{12} &= (b_1 + b_2)B_{12}\sin\delta_{12} + b_1 P_{L1} - b_2 P_{L2} \\ &\quad + (b_1 + b_2)\sum_{k=1,3,5}(P_c^{nk}\cos((k+1)\omega_1 t) + P_s^{nk}\sin((k+1)\omega_1 t)) \\ &\quad + (b_1 + b_2)\sum_{k=1,3,5}(P_c^{pk}\cos((k-1)\omega_1 t) + P_s^{pk}\sin((k-1)\omega_1 t)).\end{aligned} \tag{5.24}$$

This expression (5.24) represents the dynamics of an interconnected system. It shows that in an MG with DG systems, even in the case of nonlinear and unbalanced loads, the real power flow in interconnecting lines is still proportional to the difference of angles of the terminal voltage vectors (δ_{12}). The other oscillating components in power, which are due to harmonics and negative sequence currents, act as a disturbance. This model is useful for investigating the stability of the system.

5.3.5 Validation of nonlinear model

The nonlinear model (5.24) is validated against the simulation data. The DG system consisting of four-leg inverters and the single-phase diode bridge rectifier load on DG-1 is simulated. δ_{12} is obtained from the difference of load angles of DG-1 and DG-2 and is also obtained by solving the ordinary differential equation (ODE) (5.23). P_{L1} and P_{L2} are measured from the loads connected to DG-1 and DG-2. P_{12} is obtained by measuring the interconnecting line currents and voltages at DG-1 terminals as

$$P_{12} = Re(V_1 I_{12}^*). \tag{5.25}$$

At $t = 0.1$ seconds, two units are interconnected. For equal droop coefficients, the two units share an equal active load of 700 W. At $t = 0.2$ seconds the single-phase diode bridge rectifier load is turned on. The active power sharing between the two inverters is shown in Fig. 5.10. As discussed in Section 5.3, the power flow in the interconnecting line has second, fourth, and sixth harmonic components when the load is a single-phase rectifier. The Fourier analysis of P_{12} after $t = 0.2$ seconds, depicted in Fig. 5.11, confirms the presence of these harmonics derived in analysis.

Further the values of P_{L1}, P_{L2}, and P_{12} obtained from the simulation were used to solve (5.23) using a numerical ODE solver.

The time variation of δ_{12} calculated from load angles of terminal voltages and by solving a nonlinear model (5.23) numerically is shown in Fig. 5.12. As seen, plots for δ_{12} almost coincide and hence it can be said that the nonlinear model (5.24) is a fairly accurate representation of the δ_{12} dynamics for a two DG system feeding nonlinear and unbalanced loads.

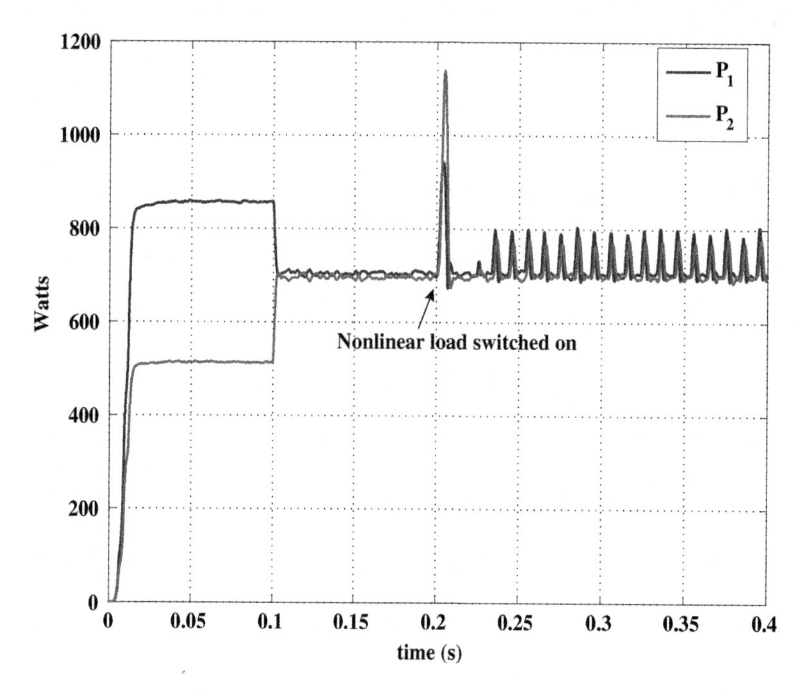

FIGURE 5.10 Power sharing between inverters.

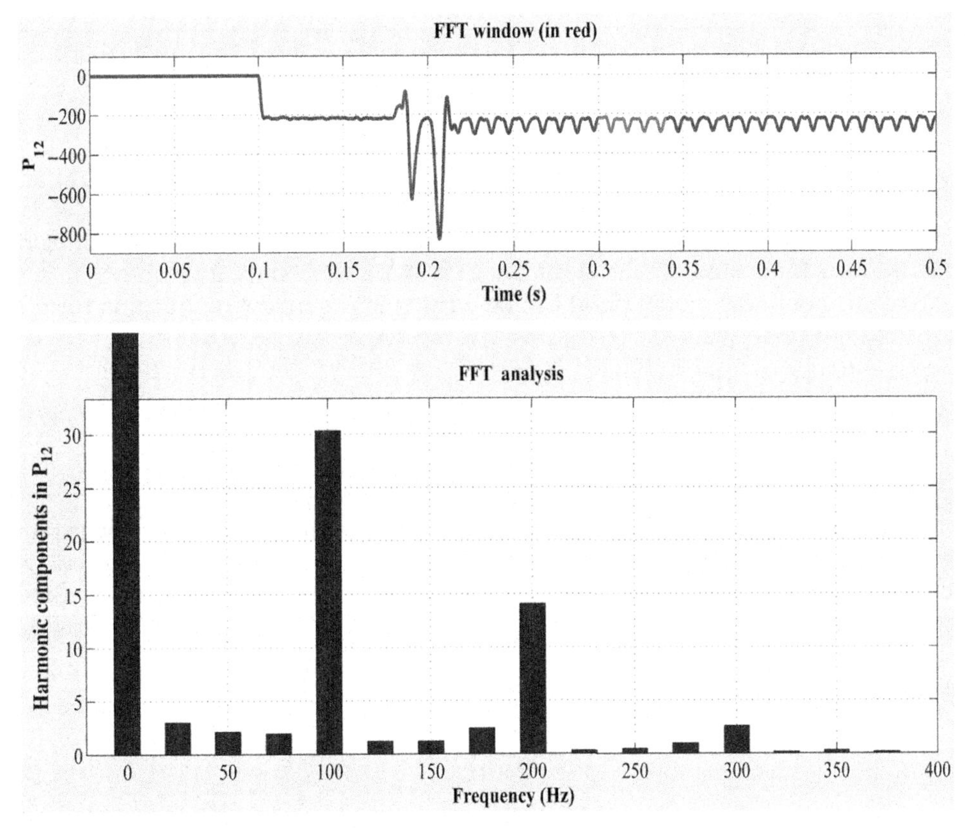

FIGURE 5.11 Fourier analysis of power in the interconnecting line (after $t = 0.2$ s).

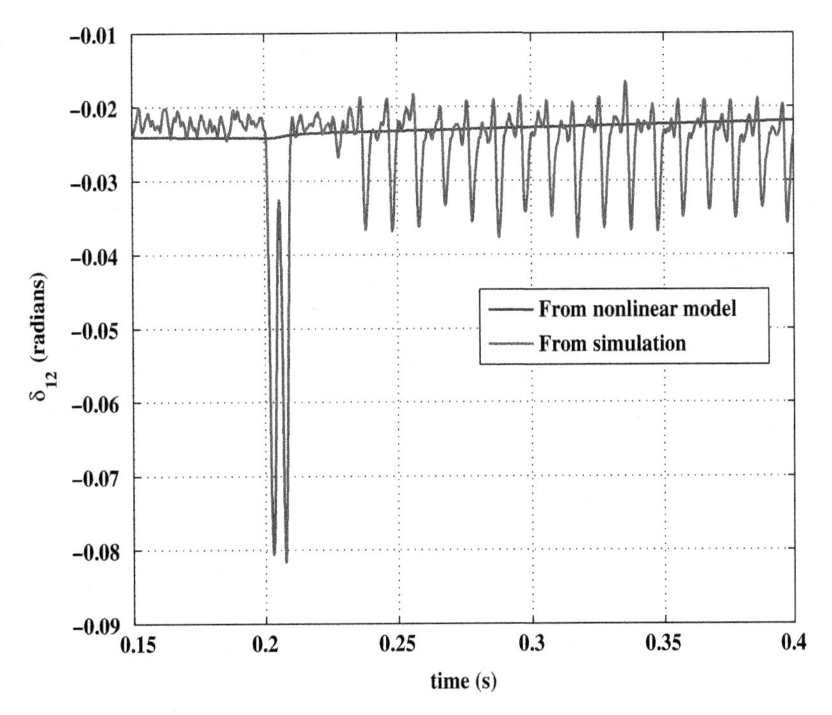

FIGURE 5.12 δ_{12} plots for nonlinear model for nonlinear loads.

5.4 Derivation of linearized model

From the nonlinear ODE (5.24) representing the system dynamics, it is very difficult to draw inferences regarding system stability. Also it is necessary to quantify the factors affecting the dynamics and stability in the presence of small nonlinear and unbalanced loads. These facts motivated the development of a linear version of (5.24).

The dynamic nonlinear model (5.24) can be written in the following function form:

$$\frac{d}{dt}\delta_{12} = f(\delta_{12}, P_{L1}, P_{L2}, P_c^{n1}, P_s^{n1}, P_c^{n3}, P_s^{n3} P_c^{p3}, P_s^{p3}, P_c^{n5}, P_s^{n5}, P_c^{p5}, P_s^{p5}, \omega_1, t) = f(\mathbf{M}). \tag{5.26}$$

For linearization, the operating point $\mathbf{M}^*$, given as

$$\mathbf{M}^*(\delta_{12}^*, P_{L1}^*, P_{L2}^*, P_c^{n1*}, P_s^{n1*}, P_c^{n3*}, P_s^{n3*}, P_c^{p3*}, P_s^{p3*}, P_c^{n5*}, P_s^{n5*}, P_c^{p5*}, P_s^{p5*}, \omega_1^*), \tag{5.27}$$

should be explicitly independent of time [25,26]. Hence, it is chosen to correspond to the steady-state balanced operation. Thus the system is assumed to be settled with a balanced load and a small perturbation in the form of nonlinearity or an unbalance in loads is created. The frequency at this operating point is close to nominal frequency, that is, $\omega_1^* = \omega_0$.

The function f ($\mathbf{M}$) is expressed in Taylor series at $\mathbf{M}^*$ as

$$\begin{aligned}\frac{d}{dt}\delta_{12} = {} & f(\mathbf{M}^*) + \left.\frac{\partial f}{\partial \delta_{12}}\right|_{\mathbf{M}^*}(\delta_{12} - \delta_{12}^*) + \left.\frac{\partial f}{\partial P_{L1}}\right|_{\mathbf{M}^*}(P_{L1} - P_{L1}^*) + \left.\frac{\partial f}{\partial P_{L2}}\right|_{\mathbf{M}^*}(P_{L2} - P_{L2}^*) \\ & + \sum_{k=1,3,5}\left(\left.\frac{\partial f}{\partial P_c^{nk}}\right|_{\mathbf{M}^*}(P_c^{nk} - P_c^{nk*}) + \left.\frac{\partial f}{\partial P_s^{nk}}\right|_{\mathbf{M}^*}(P_s^{nk} - P_s^{pk*})\right) \\ & + \sum_{k=3,5}\left(\left.\frac{\partial f}{\partial P_c^{pk}}\right|_{\mathbf{M}^*}(P_c^{pk} - P_c^{pk*}) + \left.\frac{\partial f}{\partial P_s^{pk}}\right|_{\mathbf{M}^*}(P_s^{pk} - P_s^{pk*})\right) \\ & + \left.\frac{\partial f}{\partial \omega_1}\right|_{\mathbf{M}^*}(\omega_1 - \omega_1^*) + \text{higher order terms.}\end{aligned} \tag{5.28}$$

The deviation variable for δ_{12} is defined as

$$\Delta\delta_{12} = \delta_{12} - \delta_{12}^*.$$

Other deviation variables are defined in Appendix 5.2, Coefficients of the linearized model.

The partial derivatives are evaluated at the operating point $\mathbf{M}^*$ and are given in Appendix 5.2, Coefficients of the linearized model. The equilibrium point $\mathbf{M}^*$ resembles to the balanced load condition and implies that

$$P_c^{n1*} = P_s^{n1*} = P_c^{p3*} = P_s^{p3*} = P_c^{n3*} = P_s^{n3*} = P_c^{p5*} = P_s^{p5*} = P_c^{n5*} = P_s^{n5*} = 0. \tag{5.29}$$

Thus the partial derivative $\left.\frac{\partial f}{\partial \omega_1}\right|_{\mathbf{M}^*}$ of $\Delta\omega_1$, is zero (see Appendix 5.2).

Noting $f(\mathbf{M}^*) = 0$, neglecting higher-order terms, and substituting partial derivative coefficients and deviation variables, the linearized ODE model for nonlinear and unbalanced load case is

$$\begin{aligned}\frac{d}{dt}\Delta\delta_{12} =& (b_1 + b_2)B_{12}\cos\delta_{12}^*\Delta\delta_{12} + b_1\Delta P_{L1} - b_2\Delta P_{L2}\\ &+ (b_1 + b_2)\sum_{k=1,3,5}\left(\cos((k+1)\omega_1 t)\Delta P_c^{nk} + \sin((k+1)\omega_1 t)\Delta P_s^{nk}\right)\\ &+ \sum\left(\cos((k+1)\omega_1 t)\Delta P_c^{pk} + \sin((k+1)\omega_1 t)\Delta P_s^{pk}\right).\end{aligned} \tag{5.30}$$

It is clear from the above equation that some coefficients are time-varying. It is worth mentioning here that the above linear model should be used for modeling situations when a 'small' nonlinear or unbalanced load is switched on in the presence of almost balanced linear loads in the MG. This scenario is very common in an MG.

The linearized model for unbalanced and balanced load cases can be derived along similar lines. For an unbalanced load case [16], there are only fundamental negative sequence currents and the linearized model can be written directly from (5.30) as

$$\frac{d}{dt}\Delta\delta_{12} = (b_1 + b_2)B_{12}\cos\delta_{12}^*\Delta\delta_{12} + b_1\Delta P_{L1} - b_2\Delta P_{L2} + \cos(2\omega_1 t)\Delta P_c^{n1} + \sin(2\omega_1 t)\Delta P_s^{n1}. \tag{5.31}$$

For balanced loads, the situation is much simpler and the linearized model is given as

$$\frac{d}{dt}\Delta\delta_{12} = (b_1 + b_2)B_{12}\cos\delta_{12}^*\Delta\delta_{12} + b_1\Delta P_{L1} - b_2\Delta P_{L2}. \tag{5.32}$$

The following observations are made from the models (5.24), (5.30), (5.31), and (5.32):

1. A balanced load situation is a special case of nonlinear and unbalanced load case.
2. From the nonlinear model (5.24), it is seen that δ_{12} depends on DG system frequency ω_1 (or ω_2). Thus, for a large unbalance, changes in DG system frequency affect the δ_{12} dynamics and consequently the power flow. However, in the linearized model (5.30) oscillations arising in ω_1 (or ω_2) due to load perturbations are not reflected in the dynamics of δ_{12} since the coefficient of $\Delta\omega_1$ is zero.

5.4.1 Small-signal stability analysis of the linearized model

In this section, a stability analysis of the linearized model for a nonlinear and unbalanced load situation is presented. Before proceeding to the unbalanced and nonlinear load case, the stability for a balanced load case is discussed. The dynamic model (5.32) for a balanced load case in state space form is

$$\Delta\dot{\delta}_{12} = (b_1 + b_2)B_{12}\cos\delta_{12}^*\Delta\delta_{12} + [b_1 \; -b_2]\begin{bmatrix}\Delta P_{L1}\\ \Delta P_{L2}\end{bmatrix} \tag{5.33}$$

As given in Refs. [18,20], this model is bounded-input-bounded-output (BIBO) stable as the coefficient $D_{f\delta12} = (b_1 + b_2)B_{12}\cos\delta_{12}^* < 0$ for negative droop coefficients b_1 and b_2.

Similarly the linear model for the nonlinear and unbalanced load (5.30) can be written in state-space model form $\dot{x} = Ax + Bu$ as shown in (5.34). Here C_k and S_k represent terms $\cos k\omega_1^* t$ and $\sin k\omega_1^* t$, respectively, with $k = 2, 4, 6$.

$$\Delta\dot{\delta}_{12} = (b_1 + b_2)B_{12}\cos\delta_{12}^*\Delta\delta_{12} + \begin{bmatrix} b_1 & -b_2 \end{bmatrix}\begin{bmatrix} \Delta P_{L1} \\ \Delta P_{L2} \end{bmatrix} + (b_1 + b_2)\begin{bmatrix} C_2 \\ S_2 \\ C_4 \\ S_4 \\ C_2 \\ S_2 \\ C_6 \\ S_6 \\ C_4 \\ S_4 \end{bmatrix}^T \begin{bmatrix} \Delta P_c^{n1} \\ \Delta P_s^{n1} \\ \Delta P_c^{n3} \\ \Delta P_s^{n3} \\ \Delta P_c^{p3} \\ \Delta P_s^{p3} \\ \Delta P_c^{n3} \\ \Delta P_s^{n3} \\ \Delta P_c^{n5} \\ \Delta P_s^{n5} \\ \Delta P_c^{p5} \\ \Delta P_s^{p5} \end{bmatrix}. \quad (5.34)$$

Thus, load power changes ΔP_{L1} & ΔP_{L2} and factors due to unbalance and nonlinearity in load, $\Delta P_c^{n1}, \Delta P_s^{n1}$, etc., can be treated as inputs (or disturbances) to the system. Stability is decided by the droop coefficients b_1 and b_2. Both these droop coefficients are negative and hence the linear models (5.30) or (5.34) are stable. Generalized state space representation of (5.34) is given as:

$$\dot{\mathbf{x}} = A\mathbf{x} + B_1\mathbf{u}_1 + B_2\mathbf{u}_2. \quad (5.35)$$

where, $\mathbf{u}_1$ corresponds to load changes and $\mathbf{u}_2$ corresponds to harmonics in P_{12} due to unbalance and nonlinearity in load. Terms B_2 and $\mathbf{u}_2$ are absent if loads are linear and balanced.

5.5 Reactive power sharing

In order to study load-angle dynamics, emphasis is given on active power-flow analysis. However, as the inductive load changes, terminal voltages should also change. Hence, the Q-V droop laws given in (5.36) are also applied to individual DG units.

$$V_1(t) = V_0 - c_1(Q_{R1} - Q_1(t)), \quad V_2(t) = V_0 - c_2(Q_{R2} - Q_2(t)), \quad (5.36)$$

where Q_{R1} and Q_{R2} are rated reactive power outputs of DG-1 and DG-2 respectively, c_1 and c_2 are droop coefficients, V_0 is the nominal system voltage and Q_1 and Q_2 are reactive power outputs of respective units. Due to load unbalance or nonlinearity, there are oscillations in reactive power flow Q_{12} in the interconnecting line as derived earlier in the power-flow analysis (5.14). Due to inductive output impedance of inverters and interconnecting line inductance P-f and Q-V decoupling is maintained.

5.6 Experimental results

An experimental set-up comprising two four-leg inverters feeding unbalanced loads was assembled in the laboratory to verify the preceding analysis. Fig. 5.13 shows the laboratory set-up and Table 5.2 gives the system details.

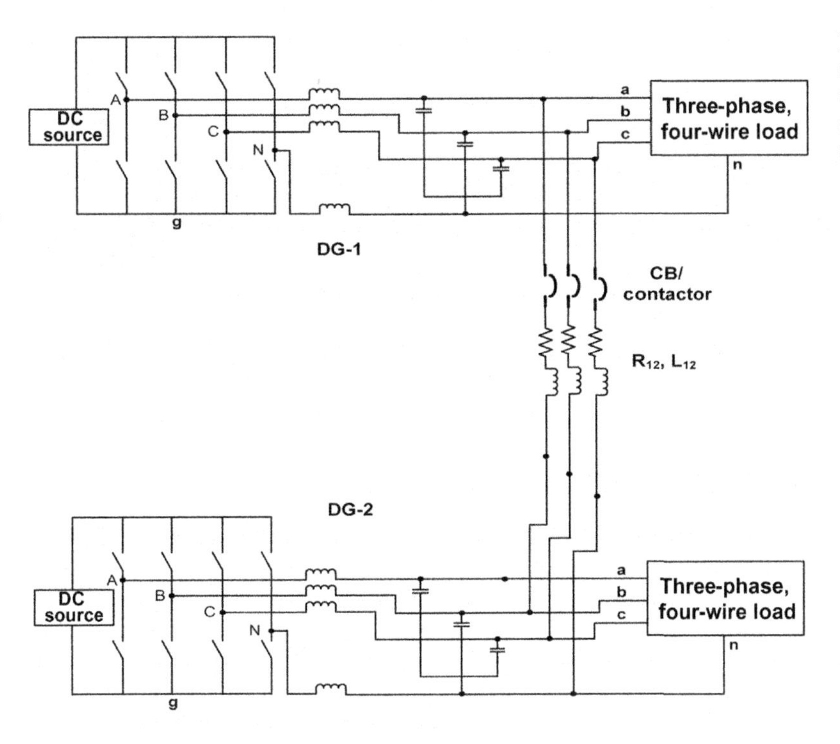

FIGURE 5.13 Detailed schematic diagram of distributed generation (DG) system using four-leg inverters.

TABLE 5.2 Data of two four-leg inverters.

Data	Inverter-1	Inverter-2
P_R	1000 W	1000 W
Q_R	800 VAr	800 VAr
b_1, b_2	−0.0002	−0.0002
f_0	50 Hz	50 Hz
c_1, c_2	−0.0002	−0.0002
V_0	92 V	92 V
Output voltage (ph) (rms)	65 V	65 V
L_f, L_{12}	3 mH	3 mH
C_f	200 μF	200 μF

In the laboratory, a renewable energy DC source is emulated using rectified output. Initial DC link voltage of each inverter is maintained at 210–220 V through a three-phase diode bridge rectifier. With an increase in load, the unregulated DC link voltage decreases and vice-versa. Each four-leg inverter is built with IGBTs rated for 35 A rms. The decentralized controller of each four-leg inverter is based on digital signal processor (DSP). Terminal voltages and load currents are attenuated to signals in the range of 0–5 V and are given to analog to digital converter (ADC). Calculations of active and reactive power, *P-f* and *Q-V* droop laws, decoupled closed-loop PI controllers, and SVM are performed by DSP. Finally pulses are generated by comparing SVM-generated signals and 5 kHz triangular carrier signal in FPGA. The sampling time used is 50 μs.

The nonlinear load used here is a set of three single-phase diode bridge rectifiers with a 680 μF capacitor at its output. Equal droops are selected so that the two units share the total load equally. P_1 and P_2 can also be expressed in the form of power supplied given in (5.11). After substitution in droop Eq. (5.21), it is seen that ω_1 and ω_2 are now time-varying with second or higher order even harmonic components superimposed on them. But due to small values of droop coefficients, it does not affect the sharing of 'average power' (P_0). It should be noted that P_1 and P_2, after sensing, were not passed through the low-pass filter (which is the usual practice in control system), so that their fast Fourier transform (FFT) analysis should reveal the harmonics present above the average value.

5.6.1 Single-phase diode bridge rectifier load on one phase of DG-2

Initially both the DG units are feeding resistive loads with $P_1 = 165$ W and $P_2 = 85$ W. The two units are then synchronized. For equal droops both share the total active load of 250 W equally (see Fig. 5.14). A single-phase diode bridge rectifier load connected to phase-a of DG-2 is switched on. It is observed that two inverters equally share the average increase in the load. The dominant harmonic components in FFT analysis (see Table 5.3) of

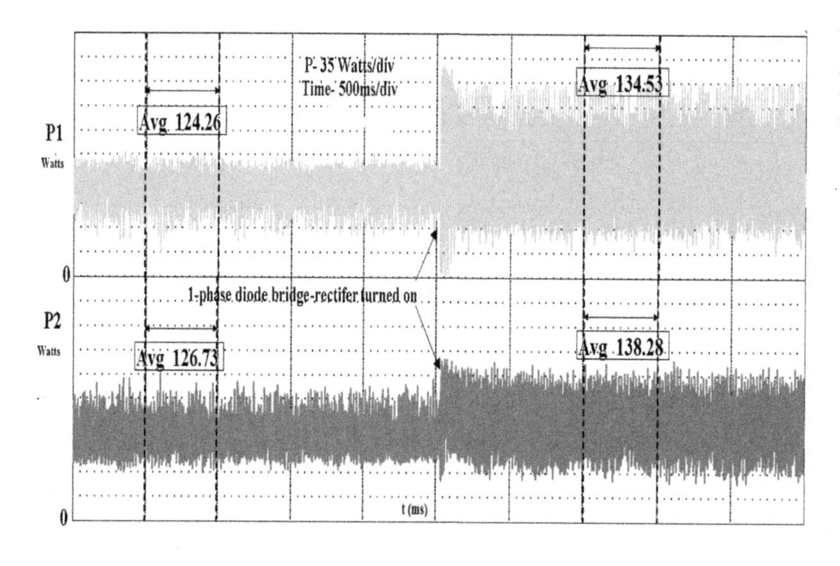

FIGURE 5.14 P_1 and P_2 during single-phase diode bridge-rectifier load switching.

TABLE 5.3 Harmonics in the P_1 and P_2 (single-phase diode bridge rectifier as load).

Harmonics in power	DG-1 (W)	DG-2 (W)
$P_0 = P\omega$	135.62	137.24
$P_{2\omega}$	43.28	16.52
$P_{4\omega}$	19.49	16.18
$P_{6\omega}$	27.91	12.51

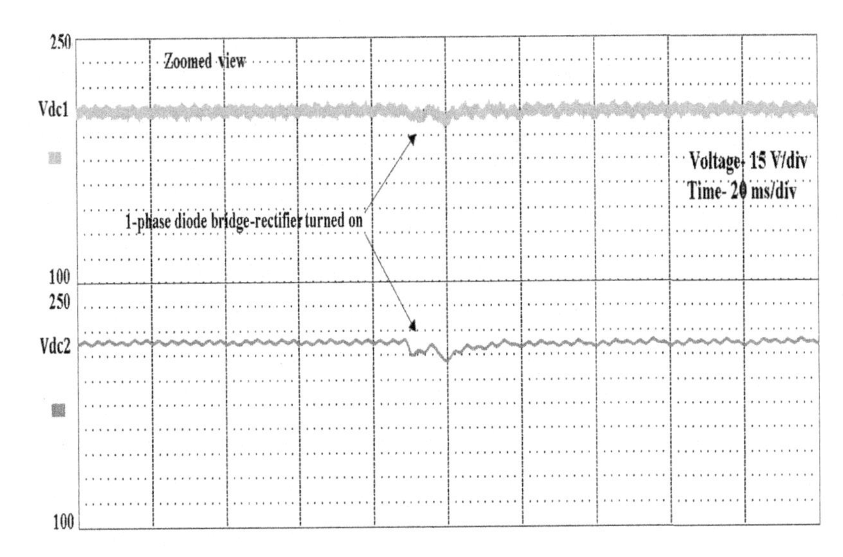

FIGURE 5.15 DC link voltages during single-phase diode bridge-rectifier load switching (zoomed in view).

data match with the theoretical analysis given in Section 5.3.1. The FFT analysis also shows that, although the average power is equally shared, harmonics in the power do not follow droop laws. The initial load connected to DG-1 is more and oscillating power components are also shared more by DG-1. This implies that harmonics in load currents after interconnection follow a low-resistance path.

The DC link voltages of inverters during the switching of diode bridge rectifiers are shown in Fig. 5.15. The FFT analysis is also carried out for the DC link voltages after the nonlinear and unbalanced load is switched on. It is clear from Table 5.4 that the DC link voltages are not affected by nonlinearity and unbalance in the load as in the case of three-leg split DC link capacitor topology. Terminal voltages are almost balanced and sinusoidal as shown in Fig. 5.16.

As per the IEEE Standard-519 [27] for systems with voltage levels equal to 69 kV and below, the maximum individual harmonic voltage should be limited to 3% and THD is limited to 5%. For a stand-alone system with an inverter at its output, standards which are applied generally for power UPS systems can also be referred. IEEE Standard-944 [28] recommends that for a stand-alone UPS system, the output voltage shall be a sinusoidal wave with no single harmonic component more than 3% rms, and a THD of no more than 5% of the magnitude of fundamental frequency component. Considering these

TABLE 5.4 Harmonics in DC link voltages (single-phase diode bridge rectifier as load).

Harmonics in direct current (DC) link voltages	V_{dc1}	V_{dc2}
DC	205	214
50	0.04	0.32
100	0.49	0.4
150	0.07	0.14
200	0.08	0.21
250	0.06	0.08
300	1.06	0.86

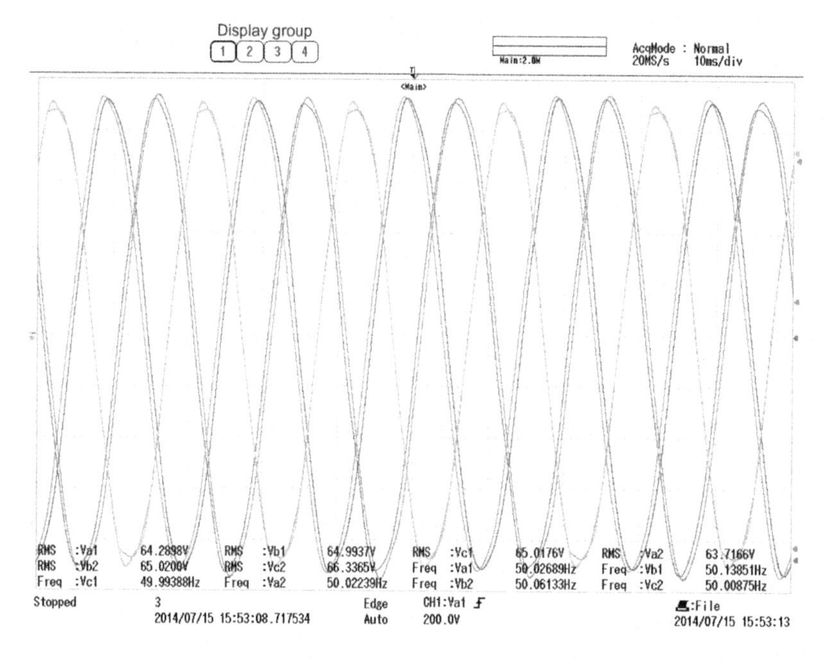

FIGURE 5.16 Terminal voltages during single-phase diode bridge-rectifier load switching.

standards, FFT analysis is carried out for the three-phase voltage data recorded from the power scope. Values of per phase %THD in voltages of DG-1 and DG-2 are tabulated in Tables 5.5 and 5.6. These are within the THD limits specified by the above standards.

5.6.2 Three single-phase diode bridge rectifier load on DG-2

This is a nonlinear but balanced type of load as analyzed in Section 5.3.2. Three diode bridges are turned on gradually and in each step it is observed that average power is

TABLE 5.5 THD calculations for DG-2 (for single-phase diode bridge rectifier on phase-b of DG-2).

Phases	%THD
a	2.43
b	1.97
c	2.02

TABLE 5.6 THD calculations for DG-2 (for single-phase diode bridge rectifier on phase-b of DG-2).

Phases	%THD
a	2.23
b	2.62
c	1.86

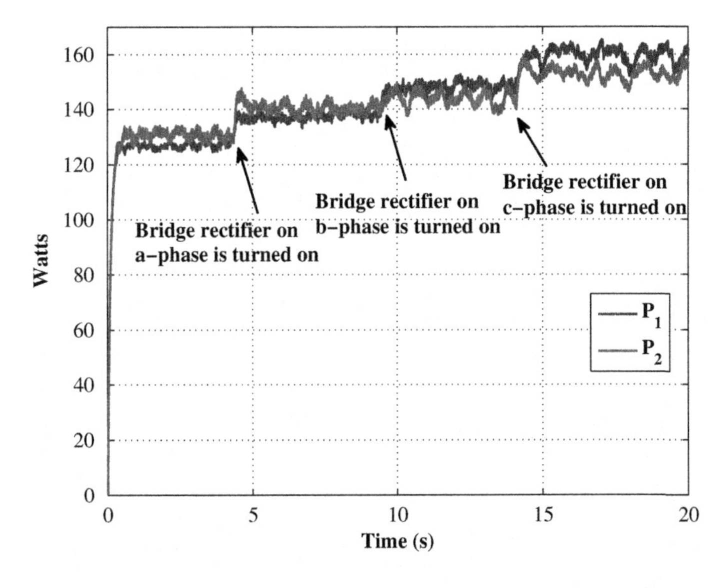

FIGURE 5.17 P_1 and P_2 during switching of three single-phase diode bridge rectifiers.

almost equally shared by the two units (see Fig. 5.17). Fig. 5.18 shows the average power sharing by the DG systems with all three single-phase rectifier bridges turned on. FFT analysis is carried out for P_1 and P_2 (see Table 5.7). The dominant harmonic oscillations in power are the same as derived in the theoretical analysis as given in Section 5.3.2. However the dominant harmonic components in P_1 and P_2 do not follow the droop laws (Table 5.8).

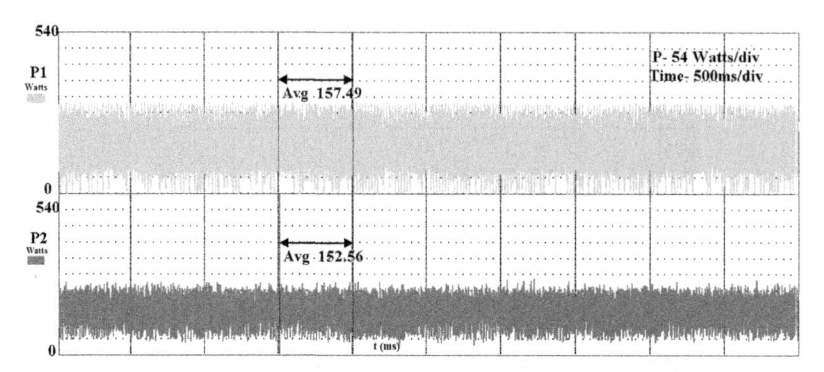

FIGURE 5.18 Power shared with three single-phase diode bridge rectifiers as loads.

TABLE 5.7 Harmonics in the P_1 and P_2 (three diode bridge rectifiers as load).

Harmonics in power	DG-1 (W)	DG-2 (W)
$P_0 = P_\omega$	155.44	154.19
$P_{2\omega}$	19.89	16.71
$P_{4\omega}$	6.25	10.14

TABLE 5.8 Harmonics in direct current (DC) link voltages (three diode bridge rectifiers as load).

Harmonics in DC link voltages	V_{dc1}	V_{dc2}
DC	200.46	212.11
50	0.24	0.21
100	0.59	0.33
150	0.13	0.29
200	0.1	0.19
250	0.016	0.06
300	0.87	0.79

Furthermore, FFT analysis is carried out for the DC link voltages. It is found that the DC link does not contain any dominant ripples. Since the %THD in output voltages is important for this type of load, FFT analysis is also carried out for the three-phase voltage data recorded from the power scope. Values of per phase %THD of DG-1 and DG-2 terminal voltages are given in Tables 5.9 and 5.10. These are less than the THD limits specified by the above standards.

TABLE 5.9 THD calculations for DG-1 (for three single-phase diode bridge rectifier on DG-2).

Phases	%THD
a	3.17
b	2.89
c	3.27

TABLE 5.10 THD calculations for DG-2 (for three single-phase diode bridge rectifier on DG-2).

Phases	%THD
a	4
b	3.86
c	3.13

It is worth emphasizing here that highly nonlinear and unbalanced loads are being applied to the DG system and there are no dedicated compensators near each DG unit. However, the small-signal analysis given in Section 5.4.1 is still applicable.

5.7 Conclusion

Nonlinearity and unbalance in electrical loads are very random in nature. These result in distorted terminal voltages which have an adverse effect on motor loads as well as on the switching devices of nonlinear loads. This important problem is addressed in this chapter in detail for stand-alone three-phase, four-wire MGs with DG systems. The idea proposed here is to make the system robust enough to handle small sporadic nonlinear and unbalanced loads. This is achieved by employing a four-leg inverter as an interface instead of the popularly used split DC link three-leg inverter or three-leg inverter followed by a Δ-Y transformer. A detailed analysis, simulation, and practical implementation of a stand-alone MG formed using two four-leg inverters is given. Experimental results obtained on a laboratory prototype of an MG for various types of nonlinear and unbalanced loads justify the correctness of the analysis.

Results for power sharing between the inverters in this type of load scenario show oscillations in instantaneous power due to harmonics and negative sequence components of load currents. But the average power supplied by them is always shared according to the droop coefficients. This is expected since droops are applied only for power corresponding to fundamental components of voltages and currents. Though the oscillating power is not shared as per the droop laws, the terminal voltages are almost balanced and THD is within the limits specified by IEEE standards without any additional load compensator. Thus, other balanced loads in the system are unaffected. In addition to this, experimental results clearly show that the DC link voltages are ripple-free and there is no need of large-size DC link capacitors as is the case of split DC link topology. Also, there is no need of a Δ-Y transformer, as the load neutral can be connected to the fourth neutral leg of a four-leg inverter. These merits justify the use of the four-leg inverter interface for three-phase, four-wire distributed system. However, for known large unbalanced or nonlinear loads, dedicated load compensators are always advisable.

Power-flow analysis and dynamics are studied in detail in the presence of nonlinear and unbalanced loads. This involves the derivation of a nonlinear as well as a linear model representing the system dynamics. The derived nonlinear model is complex and is validated using the simulation data. The nonlinear model for n-DG system is too difficult to enable the study of the system's dynamics. Hence, a linearized model is developed that helps to identify the factors which are responsible for the system's stability. It is clear from the linearized model that for a small unbalance or nonlinearity in loads, stability is still governed by droop coefficients similar to the classical power system and MGs using three-leg inverters. Oscillations in the instantaneous power flow only add disturbances to the system. This study emphasizes the use of four-leg inverters as an inverter interface in MGs instead of conventional three-leg inverters.

References

[1] Song, X., Wang, Y., Hu, W., Wang, Z., 2009. Three reference frame control scheme of 4 wire grid-connected inverter for microgrid under unbalanced grid voltage conditions. In: Twenty-Fourth Annual IEEE Applied Power Electronics Conference and Exposition, Washington, USA, February 2009, pp. 1301–1305.

[2] M. Savaghebi, A. Jalilian, J.C. Vasquez, J.M. Guerrero, Autonomous voltage unbalance compensation in an islanded droop-controlled microgrid, IEEE Trans. Power Electron. 60 (4) (2013) 1390–1402.

[3] M. Aredes, J. Halfner, K. Heumann, Three-phase four-wire shunt active filter control strategies, IEEE Trans. Power Electron. 12 (2) (1997) 311–318.

[4] S. Bhattacharya, T.M. Frank, D.M. Divan, B. Banerjee, Active filter system implementation, IEEE Trans. Ind. Appl. Mag. 4 (1998) 47–64.

[5] R.R. Sawant, M.C. Chandorkar, A multifunctional four-leg grid-connected compensator, IEEE Trans. Ind. Appl. 45 (1) (2009) 249–259.

[6] Kouzou, A., Rub, H.A., Mahmoudi, M.O., Boucherit, M.S., Knennel, R., 2011. Four wire shunt active filter based on four-leg inverter. In: International Aegean Conference on Electric Machines and Power Electronics and Electromotion, Istanbul, Turkey, September 2011, pp. 508–513.

[7] Y.W. Li, D.M. Vilathgamuwa, P.C. Loh, Microgrid power quality enhancement using a three-phase four-wire grid interfacing compensator, IEEE Trans. Ind. Appl. 41 (6) (2005) 1707–1719.

[8] White Paper on Comparing transformer-free to transformer-based UPS designs, 2012. Tech. Rep. Emerson Network Power (Liebert Corporation World Headquarters), Ohio, USA. Available from: <http://www.emersonnetworkpower.com/documentation/en-us/brands/liebert/documents/whitepapers/sl-24639-r01-12-final-web.pdf>.

[9] Rasmussen, N., 2011. White paper on the role of isolation transformers in data center UPS systems. Tech. Rep. 98, Schneider ElectricData Center Science Center, USA. Available from: <https://www.insight.com/content/dam/insight/en_US/pdfs/apc/apc-role-of-isolation-transformaers-in-data-center-ups-systems.pdf>.
[10] M.J. Ryan, R.W. De-Doncker, R.D. Lorenz, Decoupled control of four-leg inverter via a new 4 × 4 transformation matrix, IEEE Trans. Power Electron. 16 (5) (2001) 694–701.
[11] R. Zhang, V.H. Prasad, D. Boroyevich, F.C. Lee, Three-dimensional space vector modulation for four-leg voltage-source converters, IEEE Trans. Power Electron. 17 (3) (2002) 314–326.
[12] El-Barbari, S., Hofmann, W., 2000. Digital control of a four leg inverter for standalone photovoltaic systems with unbalanced load. In: Twenty-Sixth Annual Conference of IEEE, IECON, Nagoya, Japan, October 2000, pp. 729–734.
[13] I. Vechiu, O. Curea, H. Camblong, Transient operation of a four-leg inverter for autonomous applications with unbalanced load, IEEE Trans. Power Electron. 25 (2010) 399–407.
[14] Gannet, R.A., Sozio, J.C., Boroyevich, D., 2002. Application of synchronous frame controllers for unbalanced and non-linear load compensation in 4-leg inverters. In: Proceedings of the 17th Annual IEEE Applied Power Electronics Conference and Exposition, Dallas, TX, USA, March 2002, pp. 1038–1043.
[15] Hongbing, C., Xing, Z., Shengyong, L., Shuying, Y., 2010. Research on control strategies for distributed inverters in low voltage micro-grids. In: IEEE International Symposium on Power Electronics for Distributed Generation Systems, Hefei, China, June 2010, pp. 748–752.
[16] Vyawahare, D., Chandorkar, M.C., 2014. Power flow analysis in stand-alone 4-wire, 4-leg inverter micro-grid with unbalanced loads. In: Sixteenth European Conference on Power Electronics and Applications EPE-ECCE, Lappeenranta, Finland, August 2014.
[17] U. Borup, F. Blaabjerg, P.N. Enjeti, Sharing of nonlinear load in parallel-connected three-phase converters, IEEE Trans. Ind. Appl. 37 (6) (2001) 1817–1823.
[18] J.M. Guerrero, M.C. Chandorkar, T. Lee, P.C. Loh, Advanced control architectures for intelligent microgrids—part I: decentralized and hierarchical control, IEEE Trans. Ind. Electron. 60 (4) (2012) 1254–1262.
[19] J.M. Guerrero, P.C. Loh, T. Lee, M.C. Chandorkar, Advanced control architectures for intelligent microgrids—part II: power quality, energy storage, and AC/DC microgrids, IEEE Trans. Ind. Electron. 60 (4) (2012) 1263–1270.
[20] S.V. Iyer, M.N. Belur, M.C. Chandorkar, A generalized computational method to determine stability of a multi-inverter microgrid, IEEE Trans. Power Electron. 25 (9) (2010) 2420–2432.
[21] Chandorkar, M.C., Divan, D.M., 1996. Decentralized operation of distributed UPS systems. In: IEEE Conference on Power Electronics, Drives and Energy Systems for Industrial Growth, New Delhi, India.
[22] E.A.A. Coehelo, P.C. Cortizo, P.F.D. Garcia, Small-signal stability for parallel-connected inverters in stand-alone AC supply system, IEEE Trans. Power Electron. 38 (2) (2002) 533–542.
[23] Zhang, R., 1998. High Performance Power Converter Systems for Nonlinear and Unbalanced Load/Source (Ph.D. thesis), Virginia Polytechnic Institute and State University, Blacksburg, Virginia.
[24] H. Saadat, Power System Analysis., PSA Publishing, 2010.
[25] H.K. Khalil, Nonlinear Systems., Prentice Hall, 2002.
[26] M. Vidyasagar, Nonlinear Systems Analysis, SIAM, 2002.
[27] IEEE Std. 519–1992, 1993. IEEE recommended practices and requirements for harmonic control in electrical power systems, pp. 1–112.
[28] IEEE Std. 944-1986, 1996. IEEE recommended practice for the application and testing of uninterruptible power supplies for power generating stations.

Appendix 5.1 : Power coefficients

As discussed in Section 5.3, the power flow in line is oscillating in nature due to nonlinear and unbalanced loads. The coefficients of the active and reactive power are defined as follows:

$$P_0 = \frac{3}{2}(V_{qp}I_{qp} + V_{dp}I_{dp}), \qquad P_c^{n1} = \frac{3}{2}(V_{dp}I_{dn1} + V_{qp}I_{qn1}),$$

$$P_s^{n1} = -\frac{3}{2}(V_{dp}I_{qn1} - V_{qp}I_{dn1}), \quad P_c^{n3} = \frac{3}{2}(V_{dp}I_{dn3} + V_{qp}I_{qn3}),$$

$$P_s^{n3} = -\frac{3}{2}(V_{dp}I_{qn3} - V_{qp}I_{dn3}), \quad P_c^{p3} = \frac{3}{2}(V_{dp}I_{dp3} + V_{qp}I_{qp3}),$$

$$P_s^{p3} = \frac{3}{2}(V_{dp}I_{qp3} - V_{qp}I_{dp3}), \qquad P_c^{n5} = \frac{3}{2}(V_{dp}I_{dn5} + V_{qp}I_{qn5}),$$

$$P_s^{n5} = -\frac{3}{2}(V_{dp}I_{qn5} - V_{qp}I_{dn5}), \quad P_c^{p5} = \frac{3}{2}(V_{dp}I_{dp5} + V_{qp}I_{qp5}),$$

$$P_s^{p5} = \frac{3}{2}(V_{dp}I_{qp5} - V_{qp}I_{dp5}), \qquad Q_0 = \frac{3}{2}(V_{dp}I_{qp} - V_{qp}I_{dp}),$$

$$Q_s^{n1} = \frac{3}{2}(V_{qp}I_{qn1} + V_{dp}I_{dn1}), \qquad Q_c^{n1} = \frac{3}{2}(V_{dp}I_{qn1} - V_{qp}I_{dn1}),$$

$$Q_s^{n3} = \frac{3}{2}(V_{qp}I_{qn3} + V_{dp}I_{dn3}), \qquad Q_c^{n3} = \frac{3}{2}(V_{dp}I_{qn3} - V_{qp}I_{dn3}),$$

$$Q_s^{p3} = -\frac{3}{2}(V_{qp}I_{qp3} + V_{dp}I_{dp3}), \quad Q_c^{p3} = \frac{3}{2}(V_{dp}I_{qp3} - V_{qp}I_{d3}),$$

$$Q_s^{n5} = \frac{3}{2}(V_{qp}I_{qn5} + V_{dp}I_{dn5}), \qquad Q_c^{n5} = \frac{3}{2}(V_{dp}I_{qn5} - V_{qp}I_{dn5}),$$

$$Q_s^{p5} = -\frac{3}{2}(V_{qp}I_{qp5} + V_{dp}I_{dp5}), \quad Q_c^{p5} = \frac{3}{2}(V_{dp}I_{qp5} - V_{qp}I_{q5}).$$

Appendix 5.2 : Coefficients of the linearized model

In order to derive the linearized model for the nonlinear and unbalanced load cases, partial derivatives of (5.28) in Section 5.4 are evaluated at the balanced load operating point as follows:

$$\left.\frac{\partial f}{\partial \delta_{12}}\right|_{\mathbf{M}*} = (b_1 + b_2)B_{12} \cos \delta_{12}^*, \quad \left.\frac{\partial f}{\partial P_{L1}}\right|_{\mathbf{M}*} = b_1,$$
$$\left.\frac{\partial f}{\partial P_{L2}}\right|_{\mathbf{M}*} = -b_2, \quad \left.\frac{\partial f}{\partial P_c^{n1}}\right|_{\mathbf{M}*} = (b_1 + b_2)cos2\omega_1^* t,$$
$$\left.\frac{\partial f}{\partial P_s^{n1}}\right|_{\mathbf{M}*} = (b_1 + b_2)sin2\omega_1^* t, \quad \left.\frac{\partial f}{\partial P_c^{p3}}\right|_{\mathbf{M}*} = (b_1 + b_2)cos2\omega_1^* t$$
$$\left.\frac{\partial f}{\partial P_s^{p3}}\right|_{\mathbf{M}*} = (b_1 + b_2)sin2\omega_1^* t, \quad \left.\frac{\partial f}{\partial P_c^{n3}}\right|_{\mathbf{M}*} = (b_1 + b_2)cos4\omega_1^* t,$$
$$\left.\frac{\partial f}{\partial P_s^{n3}}\right|_{\mathbf{M}*} = (b_1 + b_2)sin4\omega_1^* t, \quad \left.\frac{\partial f}{\partial P_c^{p5}}\right|_{\mathbf{M}*} = (b_1 + b_2)cos4\omega_1^* t,$$
$$\left.\frac{\partial f}{\partial P_s^{n5}}\right|_{\mathbf{M}*} = (b_1 + b_2)sin4\omega_1^* t, \quad \left.\frac{\partial f}{\partial P_c^{n5}}\right|_{\mathbf{M}*} = (b_1 + b_2)cos6\omega_1^* t,$$
$$\left.\frac{\partial f}{\partial P_s^{n5}}\right|_{\mathbf{M}*} = (b_1 + b_2)sin6\omega_1^* t,$$

and

$$\begin{aligned}\left.\frac{\partial f}{\partial \omega_1}\right|_{\mathbf{M}^*} = &(b_1 + b_2)t(-2P_c^{n1*} \sin 2\omega_1^* t + 2P_s^{n1*} cos2\omega_1^* t - 2P_c^{p3*} \sin 2\omega_1^* t + 2P_s^{p3*} cos2\omega_1^* t \\ &+ (b_1 + b_2)t(-4P_c^{n3*} \sin 4\omega_1^* t + 4P_s^{n3*} cos4\omega_1^* t - 4P_c^{p5*} \sin 4\omega_1^* t + 4P_s^{p5*} cos4\omega_1^* t \\ &+ (b_1 + b_2)t(-6P_c^{n5*} \sin 6\omega_1^* t + 6P_s^{n5*} cos6\omega_1^* t).\end{aligned}$$

As mentioned earlier, the equilibrium point $\mathbf{M}^*$ resembles the balanced load condition, implying that the coefficients P_c^{n1*}, P_s^{n1*}, etc., are zero. As a consequence, $\left.\frac{\partial f}{\partial \omega_1}\right|_{\mathbf{M}^*}$ is zero.

Further, the deviation variables used in the linearized model (5.30) are defined as

$$\begin{aligned}
&\Delta\delta_{12} = \delta_{12} - \delta_{12}^*, && \Delta P_{L1} = P_{L1} - P_{L1}^*,\\
&\Delta P_{L2} = P_{L2} - P_{L2}^*, && \Delta P_c^{n1} = P_c^{n1} - P_c^{n1*},\\
&\Delta P_s^{n1} = P_s^{n1} - P_s^{n1*}, && \Delta P_c^{n3} = P_c^{n3} - P_c^{n3*},\\
&\Delta P_s^{n3} = P_s^{n3} - P_s^{n3*}, && \Delta P_c^{p3} = P_c^{p3} - P_c^{p3*},\\
&\Delta P_s^{p3} = P_s^{p3} - P_s^{p3*}, && \Delta P_c^{n5} = P_c^{n5} - P_c^{n5*},\\
&\Delta P_s^{n5} = P_s^{n5} - P_s^{n5*}, && \Delta P_c^{p5} = P_c^{p5} - P_c^{p5*},\\
&\Delta P_s^{p5} = P_s^{p5} - P_s^{p5*}, && \Delta\omega_1 = \omega_1 - \omega_1^*.
\end{aligned}$$

These deviation variables are used to derive the linearized model.

CHAPTER

6

Lithium-ion batteries as distributed energy storage systems for microgrids

Alberto Berrueta, Idoia San Martín, Pablo Sanchis and Alfredo Ursúa

Department of Electrical, Electronic and Communication Engineering, Institute of Smart Cities, Public University of Navarre, Pamplona, Spain

6.1 Introduction

A microgrid (MG) consists of a group of electricity generators and loads that typically operates connected to the main electricity grid, but can disconnect and operate autonomously. Among the virtues of MGs is their wide range of sizes. Small MGs can be built in single-family homes, with a domestic photovoltaic (PV) installation providing energy for home consumption. This is the case of an experimental installation located at the Public University of Navarra, Spain [1]. The energy is generated in this MG by means of a 4 kWp PV generator and a 6 kW wind turbine, and feeds the energy requirements of a four-member family home [2–4]. Larger MGs can be installed, for instance, in whole districts, such as the one located in the German town of Weinsberg, as shown in Fig. 6.1 [5]. This neighborhood comprises 23 households, the rooftops of which have an installed PV capacity of 145 kWp. When there is an excess of PV power production, dispatch management operates an air-to-water heat pump to heat water, thereby increasing the self-consumption. On the other hand, when there is a deficit of solar resources, a cogeneration unit is used as a back-up power source. Both centralized and distributed energy storage systems (ESSs) are key elements for the management, system integration, and increased self-sufficiency of this district. Given the distributed nature of renewable energies, these types of energy sources are commonly used to feed MGs. The most commonly used renewable sources (wind and solar PV) are nonmanageable, since the generation of electricity is subject to the availability of the natural resource (wind or solar radiation). Similarly, a great portion of the energy consumption in MGs is nonmanageable, given that it depends of the energy requirements of the MG's users.

Distributed Energy Resources in Microgrids
DOI: https://doi.org/10.1016/B978-0-12-817774-7.00006-5

FIGURE 6.1 Residential neighborhood in Weinsberg, Germany. The 23 households are linked by a microgrid with rooftop photovoltaic (PV) generation. *Source: Photo from Kaco New Energy, Integrated storage solution for large consumers. Available from: <http://kaco-newenergy.com/de/lounge/modellprojekt-weinsberg/> [5].*

An MG can operate both connected to the main electricity grid, as well as isolated (referred to as island mode), providing resiliency services to the customers connected to it. An ESS provides services to an MG, such as ensuring power quality, including frequency and voltage regulation, smoothing the output of renewable energy sources, and playing a crucial role in cost optimization. The role of an ESS is different in both operating modes. On the one hand, a grid-tied MG can exchange power with the main grid in order to meet its power demand. In such cases, an ESS allows a controlled power exchange in order to provide services to the main grid and support its proper performance. On the other hand, an ESS provides the required resiliency services to a stand-alone MG, that is, in a situation where there is a loss of utility service, the MG is able to provide the power required by the consumers [6]. Three main aspects are to be analyzed for the design of the ESS of an MG: (1) Is a soft transition from grid-tied to island mode required, or can the power supply be interrupted and restored after a few seconds? (2) Which of the loads are critical and need to be fed during the loss of utility service? and (3) What is the maximum time that the MG needs to hold the island mode? Finally, the role of the ESS during the transition between grid-tied and island mode is to allow a soft transition, with no power interruption. If the ESS has a fast response time and the control algorithm of the power converters is properly designed, a soft transition to island mode is possible, as shown in Fig. 6.2.

Among the vast variety of ESSs currently available in the market, Li-ion batteries offer advantageous characteristics to be used in MGs. Specially remarkable are their high-cycle efficiency, low maintenance requirements, low self-discharge, as well as their ability to supply both high power peaks during short times and low power during longer periods of time. Moreover, their lifetime and, therefore, the guarantee offered by most manufacturers, are increasing, while their price is dropping due to the economies of scale.

6.2 Evolution of lithium-ion batteries

Lithium was discovered in a mineral called petalite by Johann August Arfvedson in 1817, as shown in Fig. 6.3. This alkaline material was named lithion/lithina, from the

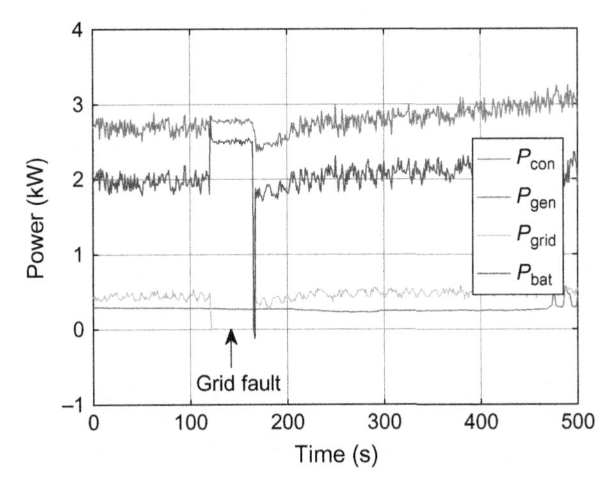

FIGURE 6.2 Grid fault measured in a domestic microgrid located at the Public University of Navarre (P_{grid} from $t = 122$ s to $t = 166$ s). Thanks to the fast response of the storage system (P_{bat}), the load power (P_{con}) was continuously met with no disconnection neither at the beginning nor at the end of the grid fault. P_{gen} represents the PV power generated in the microgrid.

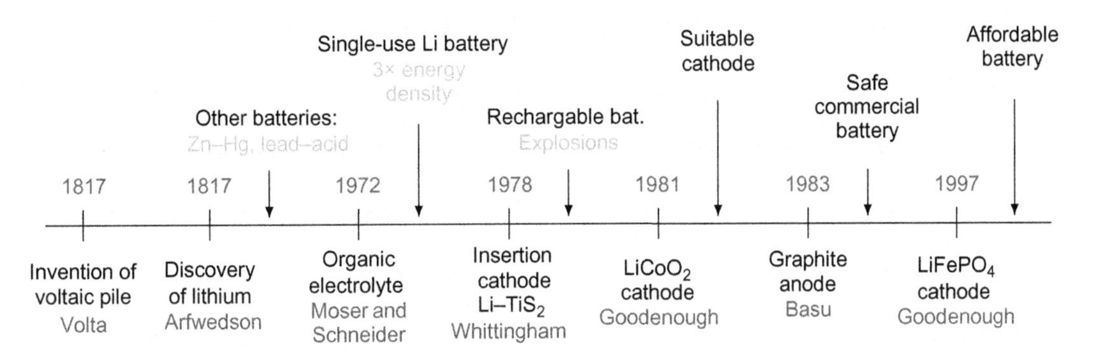

FIGURE 6.3 Main milestones during the early development of lithium batteries.

Greek word λιθοϛ (transliterated as lithos, meaning "stone"), to reflect its discovery in a solid mineral, as opposed to potassium, which had been discovered in plant ashes; and sodium, which was known partly for its high abundance in animal blood. Lithium was first isolated in 1821 by William Thomas Brande and Sir Humphrey Davy through the electrolysis of lithium oxide (Li_2O).

Some technical developments had to be tackled before lithium could be used as a battery material, given that traditional electrolytes were based on water, which has a decomposition voltage lower than the working voltage of a lithium battery. Therefore new organic electrolytes needed to be developed, the first of which was polyvinylpyridine. This first lithium battery has its genesis in research work done at the Jet Propulsion Laboratory and published in 1967 [7]. It was a primary battery (single-use, nonrechargeable battery). A few years later, in 1972, Moser and Schneider invented the lithium/iodine battery [8,9], which used a lithium metal anode and an iodine-based cathode. This movement from a zinc-based to a lithium-based battery achieved a five-times improvement of energy density (from 50 to 250 Wh/kg) [10]. The lithium-iodine battery was soon identified as a suitable power source for cardiac pacemakers [11], which had a tremendous impact on patient comfort, since its use in these devices resulted in reduced weight and an extension from two up to 10 years in the device's operational life.

Fig. 6.3 shows a chronology with the most important milestones for the development of lithium batteries. Under the timeline, the main discoveries and advances are shown, along with the main inventors. Above the line, the most remarkable dates are written in red color and the main characteristics of the technology development status during each period are summarized.

The fast and wide spread of consumer electronics during the 1970s stimulated an interest in moving to secondary, rechargeable systems. In theory, there was no difficulty on the anode side, since the ions formed from the lithium metal anode during discharge were expected to plate back to rebuild the metal anode during charge. Indeed, lithium deposition from carbonate-based solutions had been demonstrated already [12]. Therefore the research efforts were focused on cathode materials. The first insertion (or intercalation) cathode was proposed in 1978, which was titanium sulfide (TiS_2) [13]. Insertion electrodes are able, reversibly, to accept and release lithium ions from their open structure assuming various valence states, thereby allowing several charge-discharge battery cycles. The first commercial prototypes of a second lithium battery appeared in the late 1970s. However, some operational faults, including fire incidents, revealed problems concerning the battery anode. Due to the high reactivity of lithium metal, it reacts readily with the electrolyte, building the so-called solid–electrolyte interface (SEI). Irregularities at this layer can lead to uneven lithium deposition and dendrite formation, which may lead to a short circuit in the cells, causing thermal runaway, fire, or explosion.

The use of a reversible graphite electrode for the anode can be considered as the key advance that enabled the development of commercial lithium-ion batteries [14,15]. The choice of graphite is rather surprising, since it is unstable in the stability window of most common electrolytes. The electrolyte, in fact, decomposes at the graphite surface, forming an electronically blocking but ionically conductive layer that stops the decomposition while allowing the electrochemical process to continue. Therefore graphite is thermodynamically unstable but kinetically protected. This protective layer is known as the SEI. Graphite is nowadays the most commonly used anode material for lithium-ion batteries.

In contrast, a wide variety of cathode materials is used for Li-ion batteries. The cathode requirements (being capable of providing lithium ions and accepting them back in a reversible matter) were firstly fulfilled by $LiCoO_2$ (LCO), a material with a layered crystal structure discovered by Goodenough and co-workers in 1981 [16]. This is the cathode normally used nowadays for portable devices such as cell phones or laptops.

In the search for a cheaper, high-power Li-ion battery, Goodenough's group designed in 1997 a lithium iron phosphate ($LiFePO_4$) cathode [17], which is currently the flagship of the olivine components used as battery cathodes. This material is significantly cheaper than LCO, but suffers from a very high intrinsic resistance, which requires special electrode preparations involving sophisticated coating processes.

During the 21st century, an intensive research work was accomplished, discovering a wide number of cathode materials. Among them, the layered compound $Li(Ni_xMn_xCo_{1-2 \cdot x})O_2$

deserves special attention due to its widespread use. It exhibits outstanding electrochemical properties by combining the high capacity of $LiNiO_2$ and the thermal stability and low cost of manganese in $LiMnO_2$, and stable electrochemical characteristics of LCO [18].

The current need for long-lasting Li-ion batteries has spiked an interest in lithium titanium oxide (LTO) as the anode material to replace graphite. The negligible changes in the lattice structure upon accepting and releasing lithium ions and the lack of SEI interface due to the stability of this material inside the battery operation voltage, enlarge the lifetime and the reliability of the batteries. Companies like Leclanché are already selling LTO batteries.

The wide variety of materials available nowadays for lithium battery electrodes has led to the creation of an extensive catalogue of Li-ion batteries with different properties determined by the materials used and the manufacturing process.

6.3 Potential of lithium as energy storage material

Lithium is the lightest and smallest of all metals and has the lowest standard reduction potential. It is located in the top, left-hand corner of the periodic table, as shown in Fig. 6.4. Firstly, its small size eases its intercalation inside the crystal lattice of electrode materials, contributing to low mechanical stress and thereby minimizing cracks associated with charge–discharge cycles. Moreover, its low standard reduction potential gives rise to negative electrodes with a very low potential, which enlarges the battery working voltage (greater than 3 V) and its power and energy density. These properties make lithium a very attractive material for use as a battery active component.

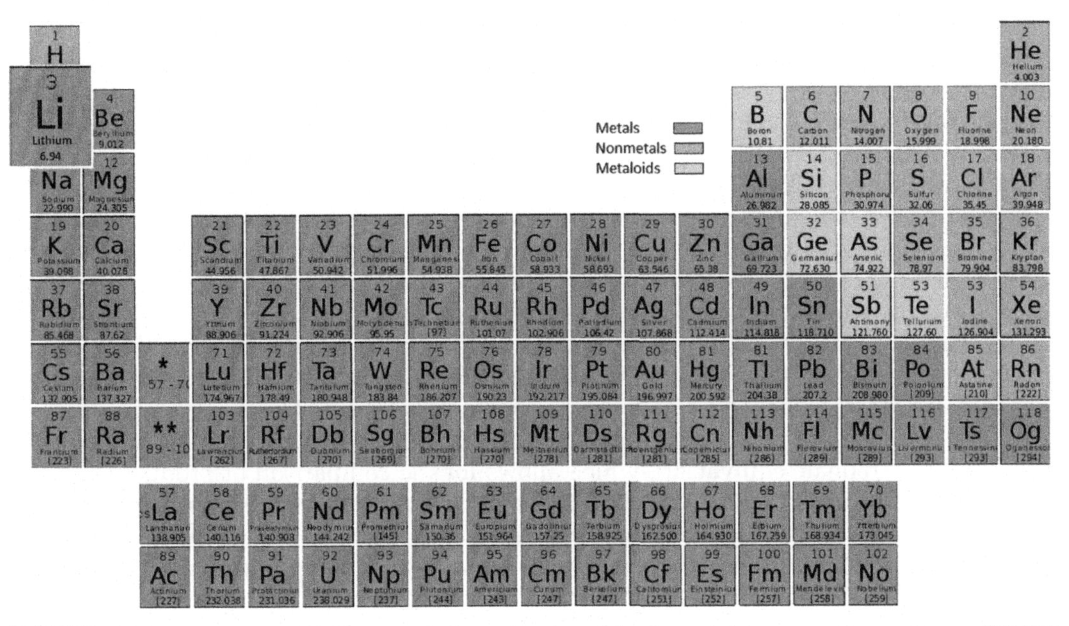

FIGURE 6.4 Periodic table of elements. Lithium is situated in the top, left corner. *Source: Dmarcus100 [CC BY-SA 4.0], from Wikimedia Commons.*

These characteristics are attractive for diverse industrial sectors, which has driven an increasing interest in Li-ion battery technology in recent years. During the decade of 2000s, the communication sector headed the Li-ion battery research and development aimed at replacing the previous Ni-Cd and Ni-MH batteries, which had poor functionality. During that decade, lithium batteries topped the portable device market, being currently the storage system used in virtually all electronic devices. During the decade of 2010s, the concept of electrochemical energy storage became compelling for the automotive sector, given the interest in electric vehicles (EVs). This is currently resulting in fast improvement of the technology, along with an unprecedented increasing demand. This improved functionality is catching the interest of a third main sector, of stationary applications, given than Li-ion batteries can be attractive for grid-connected applications such as domestic self-consumption, renewable energy generation improvement, electricity grid support, etc.

But, is there enough lithium available at a reasonable price to cover this increasing demand? Lithium is a sparse material, both in the universe and in the Earth's crust. Hydrogen and helium are estimated to make up roughly 74% and 24% respectively of all baryonic matter in the universe. The other elements comprise just 2% of ordinary matter. The material abundance in the universe follows a trend consisting of heavier nuclei being less common than lighter ones, given that shortly after the Big Bang (within a few hundred seconds) only light elements were produced in a process known as Big Bang nucleosynthesis. Specifically, only hydrogen, helium, and lithium were produced in this process. Heavier elements were mostly generated much later, inside stars. Lithium originated through three main processes: Big Bang nucleosynthesis (23% of the available lithium in the universe), the nuclear reactions that take place in dying low-mass stars (64%), and nuclear fission caused by cosmic rays (13%) [19].

Even though lithium nuclei are very light, this metal is an exception to the abovementioned abundance trend. The reason for this exception is that the nucleus of the lithium atom verges on instability. Actually, the two stable lithium isotopes found in nature have among the lowest binding energies per nucleon of all stable nuclides. Because of its relative nuclear instability, lithium is less common in the solar system than 25 of the first 32 chemical elements [20].

Interesting for industrial uses is the availability of lithium on the earth's surface. As a whole, the earth's crust contains approximately 20 parts per million of lithium, and the oceans contain 0.17 parts per million [21]. Lithium is, therefore, not a prevailing material, but it is more common than metals such as cobalt, tin, or lead, as shown in Fig. 6.5A (note the logarithmic scale). According to estimates of the US Geological Survey, there are 40 million tonnes of lithium mineable globally, 65% of that alone in the South American countries of Bolivia, Chile, and Argentina, as shown in Fig. 6.5B [22].

For standardization reasons, the unit used to refer to lithium production is tonnes of LCE (lithium carbonate equivalent), which allows for the comparison of production quantities of different lithium materials. In 2015, global lithium production was approximately 175,000 tonnes of LCE. According to projections, this amount will increase to 300,000 tonnes of LCE by 2020, as shown in Fig. 6.6, and continue growing to reach 700,000 tonnes of LCE in 2025 [22]. This increasing demand is rising the cost of lithium, as shown in Fig. 6.7A following typical demand-supply rules. However, this fact is not crucial for the price of Li-ion batteries, given the small amount of lithium required to build these

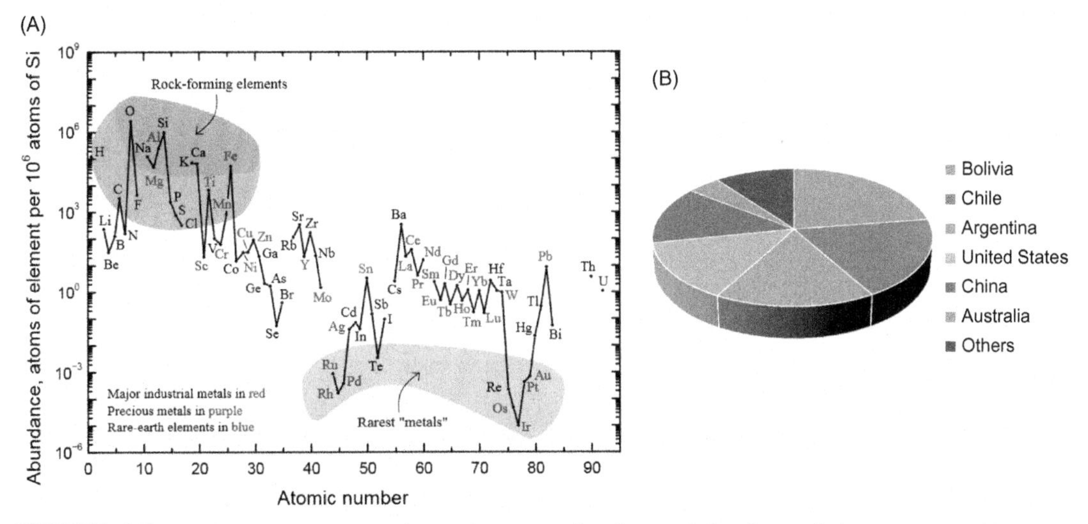

FIGURE 6.5 Lithium availability in the earth's crust: abundance of the chemical elements in earth's upper continental crust as a function of atomic number, reproduced from Wikipedia (A) and location of the main lithium reserves by country (B).

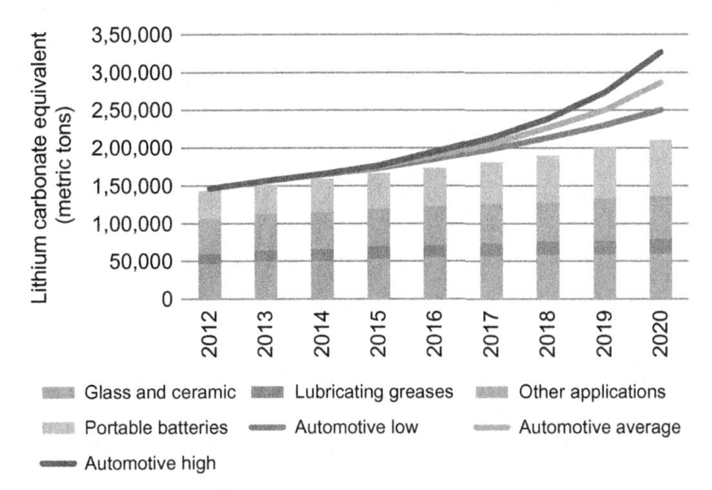

FIGURE 6.6 Global lithium production and consumer sector. Current data and prognosis for the future years. *Source: Data from Bloomberg (ALBEMARLE, Global lithium market outlook, in: Goldman Sachs HCID Conference, 2016, [23]).*

batteries. Actually, as mentioned in Table 6.1, 50 kg of LCE is required for an 85 kWh Tesla battery (this is equivalent to 110 g of Li per kWh). The lithium price of a battery is around €4.5 kW/h, which represents 2% of the overall battery price (roughly €212 kW/h).

Some concerns have been reported related to the availability of other materials required for the cathodes of Li-ion batteries. Most of them are related to cobalt, which is a key cathode component for portable electronics batteries. Cobalt is scarcer than lithium (see Fig. 6.5) and 65% of its global production comes from the Democratic Republic of the

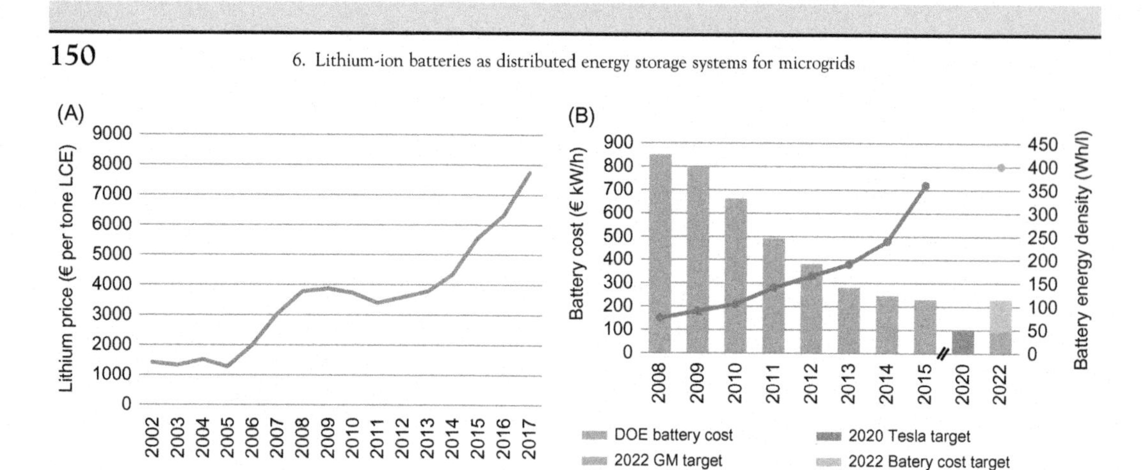

FIGURE 6.7 Price indicators related to Li-ion batteries: price of lithium from 2002 to 2017 in € per tonne LCE (A) and battery cost from 2008 to 2015, along with expected values for 2020 and 2022 in € per kWh (B) [24].

TABLE 6.1 Lithium carbonate equivalent required for the battery used in different devices [23].

Application	LCE
Cell phone	3 g
Notebook	30 g
Power tool	30–40 g
Hybrid electric vehicle 3 kWh	1.6 kg
Plug-in hybrid electric vehicle 15 kWh	11.8 kg
Electric vehicle 25 kWh	20 kg
Tesla 85 kWh	50.8 kg

Congo, a country that is extremely politically unstable and has deeply-rooted corruption structures. However, there is a wide variety of Li-ion batteries, as reported in the following sections, with different cobalt requirements (a high amount of cobalt is required for a LCO cathode used in a cell phone battery, but no cobalt is used in a $LiMn_2O_4$ cathode used for the first generation of Nissan Leaf batteries[1]). Therefore a cathode material is not as crucial as lithium for the Li-ion battery industry to expand, given the multiple cathodes that have already been designed and tested (cobalt, nickel, manganese, aluminum, etc.).

Having said this, although some materials required for Li-ion batteries are not broadly found in the earth's crust, the current increasing demand of these materials is still far from the global production capacity and from the global reserves. Actually, a demand of 300,000 tonnes LCE is predicted for the year 2020 while the reserves are estimated to be 40 million tonnes LCE, which means a 133-year availability at the production rate predicted

[1] In 2018, Nissan used its third-generation of Li-ion batteries for their electric vehicles, with an NMC cathode.

for 2020. If the production rate predicted for 2025 (700,000 tonnes LCE) is used for this calculation, 66 years of lithium supply are guaranteed. These numbers should be taken with care, given the fast-growing expectancies for this market during the upcoming years. However, based on these results, lithium can be considered, not as the definitive solution for energy storage, but as the best current and mid-term alternative while other technologies are being developed.

Actually, the effect of the increasing battery production is nowadays a significant reduction of Li-ion battery cost, given that scale economies, improvements of the manufacturing processes, and the discovery of cheaper battery materials have a bigger impact on battery cost than the increase in raw material prices. The battery price evolution from 2008 to 2015 is shown in Fig. 6.7B. A cost reduction from €850 kW/h in 2008 to €212 kW/h in 2015 is reported, which means a 75% price reduction in eight years. The US Department of Energy (DOE) has set a target for the year 2022 of €102 kW/h, which was considered by the International Energy Agency as a realistic value [24]. Two of the most remarkable companies involved in the sector of electrical mobility placed even more challenging targets, which consist of battery costs lower than €85 kW/h for year 2020 (Tesla) and 2022 (General Motors), as represented in Fig. 6.7.

Therefore rather than worrying about a lack of lithium for the increasing demand placed by the EV market in the short term, perhaps we should think about plausible shortages of rare earth materials. Such materials are used to build the permanent magnet for the electric motors. China controls about 95% of the global market for rare earth metals and expects to use most of these resources for its own production.

6.4 The cell, building block of a battery

The cell is the element in which the actual redox reaction takes place. It consists of two electrodes: an anode (negative) and a cathode (positive), separated by an ion-conductive membrane, as shown in Fig. 6.8. The main part of each electrode is the active material,

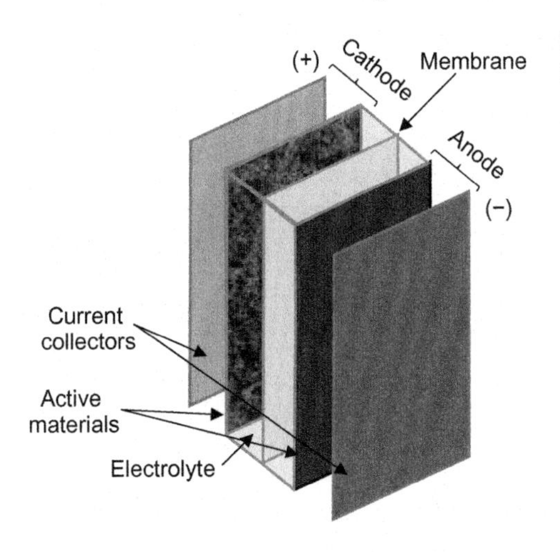

FIGURE 6.8 Schematic of the main components of a lithium-ion (Li-ion) cell.

which takes part in the electrochemical reaction and determines the battery performance. Besides the active material, each electrode has a current collector, which is a good electronic conductor material on which the active material is deposited. It provides the electrode with the required structural stability and conducts the electrons to the external electric circuit. Finally, both active materials and the membrane are immersed in an electrolyte, which allows the ion movement between both electrodes.

Li-ion cells are manufactured in three main shapes, which are cylindrical, prismatic, and pouch cells. Cylindrical cells, such as the one shown in Fig. 6.9A, have the cheapest manufacturing process. A continuous roll of electrode and membrane is wound up, and a metallic case is assembled in order to guarantee its tightness. These cells are named by a five-digit label, such as 18650. The two first digits indicate the diameter of the cell (18 mm), and the other three are its height in tenths of a millimeter (65 mm). The most common cylindrical cell is the 18650, but some companies are proposing 21,700 as the upcoming standard size for cylindrical batteries. The main drawback of cylindrical cells is their harder heat dissipation (explained with closer detail in Section 6.7), which limits their power density. Therefore the cell size is decided as a trade-off between an energy density as high as possible, while allowing a good heat transfer. Prismatic cells (Fig. 6.9B) are also manufactured with a continuous sheet of electrodes and a separator, which is wound up in a flattened cylinder and typically covered by a plastic or metallic case. This cell shape is more suitable for fitting into portable devices such as cell phones and tablets, and is the most common option for these applications. Finally, pouch cells, such as the one shown in Fig. 6.9C, are manufactured by the stacking of a number of electrodes and separators previously cut in the desired shape. The stack is then enclosed in a polymeric envelope. Even though these cutting and stacking processes lead to a higher manufacturing cost, pouch cells are commonly used due to the outstanding heat transmission properties that allow high power densities. Moreover, their polymeric enclosure prevents the risk of explosions which needs to be considered in the case of cylindrical and prismatic cells.

Besides the cell shape and size, there are diverse materials and manufacturing processes for each cell component. The leading choices, remarkable parameters, and main characteristics for each of them are covered in the following subsections.

FIGURE 6.9 Three cell shapes: 18650 cylindrical cell (A), prismatic cell (B) and pouch cell (C).

6.4.1 Negative electrode active material

The negative electrode of a lithium battery is often referred to as an anode, since the oxidation of the active material takes place during discharge. The anode material has an important influence on the battery performance, including safety, energy density, power density, and cycle life. These materials should fulfill the following conditions:

- Low potential versus Li/Li^+, which allows a larger cell potential when placed with a cathode.
- No significant change in the crystal structure should occur due to the intercalation and deintercalation of lithium ions, thereby leading to an extended battery lifetime.
- The reactions should be highly reversible, ideally with a coulombic efficiency $\eta_c = 1$.
- High diffusion of lithium ions through the crystal lattice.
- High electrode conductivity.
- High amount of charge stored per unit mass.

The characteristics, advantages, disadvantages, and challenges of the most used anode materials are covered in the following subsections and summarized in Table 6.2, where the variable C represents the material energy capacity, measured in mAh/g.

6.4.1.1 *Lithium metal*

As mentioned in Section 6.2, metallic lithium was used as anode for the first family of lithium batteries. Since it is purely lithium, the totality of the atoms can be ionized and used as charge carriers, providing an outstanding energy density (3860 Ah/kg [26]). Its electrochemical potential is also really low (−3.05 V vs standard hydrogen electrode [27]). These two characteristics yield an energy density as high as 600 Wh/kg on the cell level

TABLE 6.2 Main characteristics of anode materials.

Material	Theoretical capacity (mAh/g)	Real capacity (mAh/g)	Average potential (V)	Advantages	Issues
Li metal	3800	–	0	Energy density	High reactivity, dendrite growth, and fires, explosions
Graphite	372	~360	~0.1	Mature technology	Outside electrolyte stability
Cokes	–	~170	~0.15	High capacity and cycle life	High consumption of Li
Li_xSi_y	4200	~1000	~0.16	High capacity, low cost	Volume changes, nanostructures
Li_xSn_y	790	~700	~0.4	High capacity, low cost	Volume changes, nanostructures
$Li_4Ti_5O_{12}$ (LTO)	170	~145	~1.55	Low volume expansion, chemical, and thermal stability	Low specific capacity, and high voltage

Data from J.-K. Park, Principles and Applications of Lithium Secondary Batteries, John Wiley & Sons, Ltd., 2012 [18] and B. Scrosati, K. Abraham, W. van Schalkwijk, J. Hassoun, Lithium Batteries: Advanced Technologies and Applications, The ECS Series of Texts and Monographs, Wiley, 2013 [25].

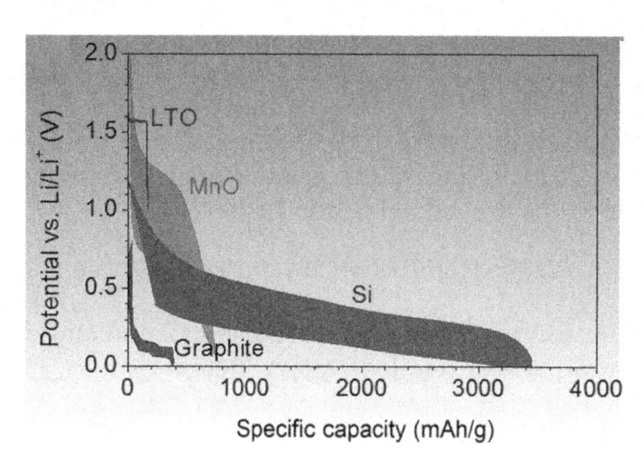

FIGURE 6.10 Potential and specific capacity of several anode materials at low charge/discharge rates to shown voltage hysteresis. *Source: Reprinted from N. Nitta, F. Wu, J.T. Lee, G. Yushin, Li-ion battery materials: present and future, Mater. Today 18 (5) (2015) 252–264, https://doi:10.1016/j.mattod.2014.10.040 [28], with permission from Elsevier.*

[26]. The safety problems associated with high lithium reactivity with air and water, as well as short-circuits, occurred as a consequence of dendrite formation on the metallic surface and made this anode unsuitable for practical batteries.

Some studies have been published attempting to stabilize lithium metal by coating the surface with polymeric or inorganic substances. Despite these efforts, lithium metal is not used nowadays as an anode material.

6.4.1.2 *Graphite*

Graphite is made of carbon atoms arranged in a planar, layered structure.[2] The electrochemical activity of carbon is due to the intercalation of Li between the graphene planes[3] (as much as one lithium atom per six carbon atoms can be stored). The main advantages of graphite are its low cost, abundance, low potential versus Li/Li^+ (see Fig. 6.10), high lithium diffusivity, high electrical conductivity, and low volume change during lithiation and delithiation [28].

For a closer analysis of the graphite anode, let's take a $Li_xC_6/Li_{1-x}CoO_2$ battery as an example, as the one shown in Fig. 6.11. Li_xC_6 denotes the graphite anode, in which one lithium reacts theoretically with six carbons, as shown in the equation below. The potential of pristine graphite is 3.0 V versus Li/Li^+, but rapidly declines when a small portion of lithium is intercalated and it ranges from 0.05 to 0.25 versus Li/Li^+ in the battery operation region.

$$Li_xC_6 \rightleftharpoons C_6 + x\ Li^+ + x\ e^-$$

[2] There are only three naturally occurring allotropes of carbon, which differ in terms of the structure and bonding of the atoms within the allotropes: diamond, which enjoys a diamond lattice crystalline structure; graphite, having a honeycomb lattice structure; and amorphous carbon (such as coal or soot), which does not have a crystalline structure.

[3] Note that graphite is made of hundreds of thousands of layers of graphene stacked together. The graphene used for some Li-ion batteries (sometime called graphene batteries) is separated single layers of graphite which are used as structural support for the active electrode materials. Therefore graphene is not the active material in graphene batteries.

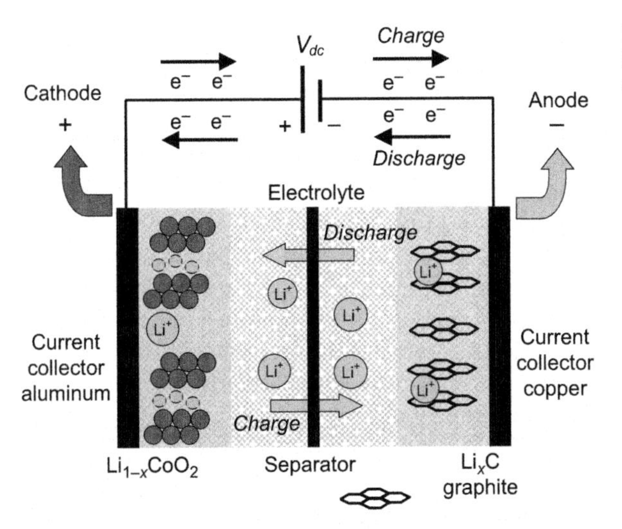

FIGURE 6.11 Movement of Li^+ in an electrolyte and insertion/extraction of Li^+ within electrodes in a lithium secondary battery.

Decomposition of the electrolyte occurs at the surface of the anode during charging, given that its reduction potential is higher than that of graphite. Such decomposition not only causes the formation of a SEI, but also suppresses the electron transfer reactions between the anode and the electrolyte, thus preventing further electrolyte decomposition. If the SEI is broken during normal battery operation and the anode gets in touch with the electrolyte, a new SEI is created to passivate the exposed area, thereby reducing the coulombic efficiency and battery lifetime, as explained in Sections 6.6.1 and 6.8.

Besides this SEI, the main drawbacks of graphite anodes are, firstly, their bad combination with the most common Li-ion battery electrolytes, which are based on propylene carbonate (PC). PC intercalates together with Li between the graphitic planes, leading to electrode exfoliation. Secondly, graphitic planes that build each particle are stacked in the same direction, leading to an uniaxial 10% strain along the edge planes due to lithium insertion and extraction, damaging the protective SEI. Some proposals to solve these problems suggest the use of amorphous carbon to protect the vulnerable edge planes from the electrolyte, thereby achieving a higher coulombic efficiency [29].

6.4.1.3 Amorphous carbon

Amorphous carbon behaves as small graphitic grains with disordered orientation, which makes them much less susceptible to electrolyte exfoliation because of their isotropic volume expansion, giving as a result a high capacity and high cycle life material. However, the high fraction of exposed edge planes increases the absolute quantity of SEI formed, reducing the coulombic efficiency during the first cycles and consuming an important part of the limited active lithium available in a fresh cell [28].

6.4.1.4 Lithium titanium oxide

LTO is a good anode material for low energy, but high power, high cycle life, and safe Li-ion batteries. Its combination of thermal stability, high rate, relatively high volumetric capacity and high cycle life compensates for the high cost of titanium, which increases the price of LTO batteries. However, this material has a larger potential versus Li/Li^+ (which reduces the cell voltage) and lower specific capacity than other materials, as shown in Fig. 6.10.

Its small volumetric change due to lithium insertion and extraction (0.2% change in volume [30,31]) reduces the strain of the electrode and the associated particle cracks. Moreover, its high potential prevents Li dendrite growth and electrolyte decomposition (since the electrolyte is stable at this potential). Unfortunately, surface reactions are not completely avoidable, suffering from gassing between the organic electrolyte and the LTO active material. Therefore carbon has been proposed as a particle coating [32,33], but this material catalyzes and accelerates electrolyte decomposition by the building of an SEI.

Even so, LTO anodes can last for tens of thousands of cycles, giving this electrode a distinct advantage over most other anode materials for high power and long life applications, especially when no high energy density is needed [28].

6.4.1.5 *Alloying materials*

Alloying materials refer to elements that electrochemically alloy and form compound phases with Li at a low potential. The most used alloying materials for Li-ion battery anodes are Si and Sn. These components can reach extremely high gravimetric capacity (see Li_xSi_y in Table 6.2), but usually suffer for extreme volume changes (several times its original volume) during lithiation and delithiation. This causes particles to fracture and lose electrical contact [34], shortening the cycle life and increasing the cell impedance. Some proposals related to particle encapsulation in a carbon shell or electrolyte additives to stabilize the SEI [35] have been published aimed at the improvement of these anode materials.

Silicon is the alloying material that has received the most attention due to its relatively low delithiation potential (~0.16 V), extremely high capacity (theoretically 4200 mAh/g, see Table 6.2 for comparison), abundance, low cost, chemical stability, and nontoxicity [28]. The alloy built with lithium is Li_xSi_y.

Tin (Li_xSn_y) has also been of some interest, given that it has similar properties to silicon, but with a higher conductivity that can counteract its lower gravimetric capacity and slightly lower cell voltage (see Table 6.2).

6.4.2 Positive electrode active material

The positive electrode of a lithium battery is often referred to as cathode, since a reduction of the active material takes place during discharge. As shown in Fig. 6.11, when the battery is being discharged, the cathode material is reduced by the electrons flowing from the cables, while the lithium ions are intercalated. Conversely, during the charging process, the emitted electrons and deintercalated ions are transported to the anode through the external circuit and the electrolyte by means of a nonspontaneous oxidation reaction.

A suitable cathode material is that which fulfils the following conditions:

- Reversible behavior and a flat potential with the intercalation–deintercalation of Li ions.
- It should be light and densely packed, and have high electrical and ionic conductivities for high power.
- Side reactions unrelated to Li-ion circulation should have a minimal weight in the total performance of the cathode.

TABLE 6.3 Main characteristics of cathode materials.

Material	Theoretical capacity (mAh/g)	Real capacity (mAh/g)	Avg. potential (V)	Advantages	Issues
LCO	274	~150	3.9	Field-tested operation	High cost, low thermal stability
LNO	275	215	3.7	Energy density, low cost	Thermal stability, Li blocking
NCA	200	~150	3.6	Long calendar life	Fast aging at high T
LMO	48	~130	4.0	Price, environmentally friendly	Reduced cycle life
NMO	280	180	3.8	Energy density, cost	Low Li diffusivity
NMC	278	170	3.7	High capacity	Relatively high price
LFP	170	~160	3.4	Low cost, safety, and environmental compatibility	Low electronic conductivity and potential

Data from J.-K. Park, Principles and Applications of Lithium Secondary Batteries, John Wiley & Sons, Ltd., 2012 [18]; B. Scrosati, K. Abraham, W. van Schalkwijk, J. Hassoun, Lithium Batteries: Advanced Technologies and Applications, The ECS Series of Texts and Monographs, Wiley, 2013 [25] and N. Nitta, F. Wu, J.T. Lee, G. Yushin, Li-ion battery materials: present and future, Mater. Today 18 (5) (2015) 252–264, https://doi:10.1016/j.mattod.2014.10.040 [28].

- Small changes in the crystal lattice volume in order to avoid active material desorption from current collectors.
- They should have electrochemical[4] and thermal[5] stability.

The development of Li-ion batteries has favored the rise of a wide variety of cathode materials, each of them having advantages and disadvantages, as detailed below and summarized in Table 6.3. Despite the continuous appearance of new families of cathode materials (conversion materials, fluorine and chlorine compounds, etc.), the most commonly-used options are intercalation compounds [28]. Intercalation materials have a crystal structure in which lithium ions can be reversibly intercalated and de-intercalated, provoking a small volume change. There are four crystal structures used for the cathode materials of Li-ion batteries, the main characteristics and representative materials of which are detailed below.

6.4.2.1 Layered structure compounds

The first-discovered and most-used structure is the layered lattice, which consists of several parallel layers between which the lithium ions are intercalated, as shown in LCO This is the first and most commercially successful Li-ion cathode material. It has a relatively high specific capacity (274 mAh/g), low self-discharge, high discharge voltage and good cycling performance. The major limitations are the high cost of cobalt, low thermal stability and fast capacity fade at high current rates or during deep cycling. Deep cycling (delithiation above 4.2 V, which means approximately 50% of Li extraction) induces lattice distortion from

[4] Electrochemical stability is achieved when reactions with the electrolyte are not spontaneous.

[5] Thermal stability refers to the exothermic release of oxygen when the cathode is heated above a certain point, resulting in a runaway reaction.

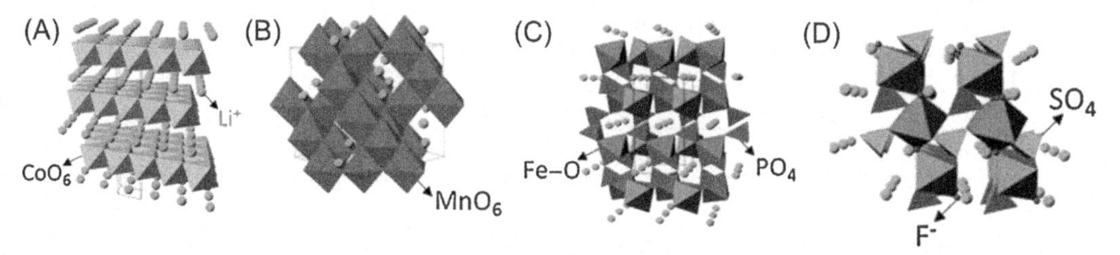

FIGURE 6.12 (A) These ions have a 2D mobility along the interlayer spaces, resulting in adequate ionic diffusivity in most of layered materials. Besides the traditional $LiCoO_2$ (LCO), there are other layered materials used as battery cathodes, most of them aimed at reducing cobalt costs: Representative crystal structures of cathode materials for lithium-ion batteries: layered $LiCoO_2$ (a), cubic $LiMn_2O_4$ spinel (b), olivine structured $LiFePO_4$ (c) and tavorite-type $LiFeSO_4F$ (d). Li ions are shown as light green spheres. *Reproduced from N. Nitta, F. Wu, J.T. Lee, G. Yushin, Li-ion battery materials: present and future, Mater. Today 18 (5) (2015) 252–264, https://doi:10.1016/j.mattod.2014.10.040 [28], with permission from Elsevier.*

hexagonal to monoclinic symmetry, deteriorating its cycling performance. Many different metals have been proposed as dopants, demonstrating promising but limited results [36].

$LiNiO_2$ (LNO) This material has the same crystal structure as LCO and a similar theoretical specific capacity of 275 mAh/g (see Table 6.3). However, pure LNO cathodes are not favorable, given than the Ni^{2+} ions have a tendency to substitute Li^+ sites during synthesis and delithiation, blocking the Li-diffusion pathways [37]. Moreover, it is even more thermally unstable than LCO. The partial substitution of Ni with Co is an effective way to reduce cationic disorder [38], and a small amount of Al can improve both thermal stability and electrochemical performance [39], building the NCA material explained hereafter.

$LiNi_{0.8}Co_{0.15}Al_{0.05}O_2$ (NCA) The high usable-discharge capacity and long storage calendar life of this material have made it a widespread cathode for commercial applications. For example, the cathode of the Panasonic cells manufactured for Tesla EVs is NCA. However, this material has been reported to have a severe capacity fade when subjected to high temperatures [40,41].

$LiMnO_2$ (LMO) This is a cheaper and less toxic option compared with previous layered materials. The lattice structure does not have a high stability, and therefore the layered structure tends to change irreversibly to a spinel structure. Moreover, the carbon anode impedance increases when Mn is dissolved in the electrolyte, a problem which is avoided with LTO anodes, given that there is no SEI layer in such anodes.

$Li(Ni_{0.5}Mn_{0.5})O_2$ (NMO) This cathode can maintain similar energy density to LCO, while reducing cost by using cheaper transition metals. However, it has a lower Li diffusivity, which may result in unappealing rate capability [42].

$LiNi_{0.33}Co_{0.33}Mn_{0.33}O_2$ (NMC) It has a similar achievable specific capacity to LCO and similar operating voltage, while having lower costs as a result of reduced Co requirements. This material has been reported to have a good cycle stability, and several rates of Ni, Co, and Mn have been proposed in order to achieve different characteristics. An even concentration gradient of these three transition metals in each material particle has been proposed in some research papers [43].

6.4.2.2 Spinel structure compounds

Spinel structures are cubic close-packed oxides with two tetrahedral and four octahedral sites per formula unit, where Li ions can be inserted, as represented in Fig. 6.12B. $LiMn_2O_4$ is a representative spinel active material.

$Li_2Mn_2O_4$ (LMO) Care should be taken with the acronym LMO, since it is sometimes used to refer to the layered $LiMnO_2$ cathode (p. 14). This paragraph concerns the spinel $Li_2Mn_2O_4$ cathode. It has several benefits related to the abundance, cost, and environmental friendliness of Mn. In this structure, Mn is located in the octahedral spinel sites and Li in the tetrahedral sites.

Therefore Li^+ can diffuse through vacant tetrahedral and octahedral interstitial sites in the 3D structure. However, this cathode suffers from insufficient long-term cyclability, which is believed to be originated by Mn dissolution and formation of tetragonal $Li_2Mn_2O_4$.

6.4.2.3 *Olivine structure compounds*

The olivine structure is a hexagonal, close-packed array of oxygen ions with half of the octahedral sites occupied with a metal and one-eighth of the tetrahedral sites occupied by a nonmetal, as represented in Fig. 6.12C. The representative olivine structure used as Li-ion cathode material is $LiFePO_4$.

$LiFePO_4$ (LFP) LFP has a good thermal stability and high-power capability. However, the major weakness is its low average potential. This issue is normally tackled by substituting oxygen by large polyanions (XO_4^{y-}) in the crystal lattice. The strong X—O bond reduces the Fe—O bond, increasing the ionization tendency of Fe^{3+}/Fe^{2+} and improving voltage [44,45]. Reduction in particle size, carbon coating, and cationic doping have proven to be effective in increasing the rate performance of this material [28].

6.4.2.4 *Tavorite structure compounds*

Tavorite materials have 1D diffusion channels, as represented in Fig. 6.12D. They have been proposed to exhibit low activation energies, allowing charge and discharge at very high rates, comparable to those observed in small olivine particles. Some of these materials contain vanadium, which raises concern about toxicity and environmental impact. The previously explained material families have been much more widely used than tavorite components [18,28].

6.4.3 Electrolyte

The electrolyte of a Li-ion battery is the medium for the movement of ions. Besides allowing the ion flow, an intercalation reaction takes place on the interface between the electrodes and the electrolyte. Therefore an electrolyte for Li-ion batteries should meet the following requirements [18]:

- *High ionic conductivity.* The movement of ions at the electrodes and diffusion within the electrolyte are especially important when the battery is rapidly charged or discharged.
- *High chemical and electrochemical stability toward electrodes.* Given the electrochemical reactions that take place in a battery, the electrolyte should be electrochemically stable within the potential range of redox reactions at both electrodes. Besides, it should be chemically stable toward the metals and polymers constituting the cathode, anode, and other parts of the cell.

- *Characteristics kept over a wide temperature range.* A common temperature for Li-ion batteries is considered to range from −20°C to 60°C, and a proper electrolyte should keep its characteristics through the whole range.
- *High safety.* Some materials used in electrolytes are flammable and may cause fires or explosions when heated to high temperatures during short circuits. Higher ignition points are favored, and nonflammable materials should be used when possible. The electrolyte should also have a low toxicity in case of leakage or disposal.
- *Low cost.* Given the fierce market competition for Li-ion batteries, expensive materials are unlikely to be adopted.

Four types of materials have been proposed to be used as electrolytes for Li-ion batteries. Their characteristics, advantages, and disadvantages are covered in the following subsections and summarized in Table 6.4. From these four types of electrolytes, liquid electrolytes are present in most of the Li-ion batteries that are nowadays on the market. On the other hand, solid polymers have not been commercially used. As explained below, gel polymers are used to manufacture the so-called lithium-polymer (LiPO) batteries and ionic liquids are a developing research area in which several promising results are being published.

6.4.3.1 *Liquid electrolytes*

Liquid electrolytes, consisting of lithium salts dissolved in organic solvents, are the most widely used electrolyte in commercially available batteries. Organic solvents are used instead of aqueous electrolytes because of the high potential of Li-ion batteries. The greater disadvantage of organic solvents is their low dielectric constant. A higher dielectric constant results in greater dissociation as it is inversely proportional to the coulombic force between cations and anions of the lithium salt. The working temperature is conditioned by the melting and boiling points of the solvent, which should remain liquid at any operating temperature and keep its ability to dissolve the lithium salt also at low temperatures. Some examples of solvents used in Li-ion batteries are ethylene carbonate (EC), PC, dimethyl carbonate, ethylmethyl carbonate, and other organic solvents.

TABLE 6.4 Comparative chart of electrolyte materials [18].

	Liquid electrolytes	**Ionic liquid electrolytes**	**Solid polymer electrolytes**	**Gel polymers electrolytes**
Composition	Organic solvents + Li salt	RT ionic liquids + Li salt	Polymer + Li salt	Org. solvents + polymer + Li salt
Ion conduct.	High	High	Low	Relatively high
Low-temp. performance	Relatively good	Poor	Poor	Relatively good
Thermal stability	Poor	Good	Excellent	Relatively good
Applicability	Most used	Promising future	Not used	Some pouch cells

Besides the solvent, attention should be paid to the dissolved lithium salt. Specifically, the salts with bigger ions are favored, given that Li salts having delocalized anions tend to dissociate more readily. However, an increase in ionic radius leads to less movement of anions, which can prevent the movement of lithium ions through the electrolyte.

6.4.3.2 Ionic liquid electrolytes

An ionic liquid refers to a salt in the liquid state, formed by organic cations and inorganic anions. In particular, those found in the liquid state at room temperature are called room-temperature ionic liquids. The main characteristics of ionic liquids that should be taken into account for their use as battery electrolytes are:

- They are liquid in a wide temperature range.
- They have high ionic conductivity due to the high concentration of ions.
- They are nonflammable and thermally resistant.
- They are chemically stable.
- They have a relatively high polarity and ionic conductivity.

Ionic liquids are a large family of materials that offer multiple options for the material design. Due to this issue, high expectations are put on ionic liquids to replace the problematic liquid electrolytes of Li-ion batteries, and intensive research on this topic has been made in recent years, as summarized by Efterkhari et al. [46] in their review paper. A promising possibility related to ionic liquids, which is still at an early stage of development, is supercooled ionic liquid crystals for fast ionic diffusion through the guided channels of a liquid-like medium. If this option is materialized, it will be a breakthrough for general electrochemistry, far beyond Li-ion batteries [46].

6.4.3.3 Solid polymer electrolytes

Solid electrolytes have several advantages over liquids, such as higher energy density and reliability, no risk of leakage, no release of combustible gases at high temperature, and no membrane requirement. Solid polymer electrolytes consist solely of a polymer and salt. The polymers should ease the lithium ion transport process and salt dissociation. Most research related to solid polymer electrolytes has focused on synthesizing new polymers to enhance ionic conductivity [47,48]. Ionic conductivity, mechanical strength, and electrode/electrolyte surface characteristics can be enhanced by adding inorganic particles to solid polymer electrolytes.

Despite such active research, Li-ion batteries with solid polymer electrolytes have yet to be fully commercialized. Compared to liquid electrolytes, solid polymer electrolytes have low ionic conductivity at room temperature, weak mechanical properties, and poor interfacial characteristics [18].

6.4.3.4 Gel polymer electrolytes

These electrolytes are formed by polymers, organic solvents, and lithium salts and are produced by mixing organic electrolytes with solid polymer matrices. Even though there is a polymeric phase, the ionic conductivity of gel electrolytes is between that of liquid and solid polymers, due to the electrolyte encapsulated in the polymer chains. Representative polymers used as matrices are polyacrylonitrile, polyvinylidene fluoride, and polymethyl methacrylate.

The lithium batteries with gel polymer electrolytes are called lithium-polymer batteries (LiPO batteries). The low ionic conductivity of gel polymers compared to liquid electrolytes is the main reason for the little advancement in the development of LiPO batteries. Poor interfacial contact with electrodes is another problem arising from gel polymer electrolytes, which has special significance when the battery is charged or discharged at high current rates.

6.4.4 Membrane

The membrane (or separator) is a nonactive material (it does not participate in the electrochemical reaction) which must be an electric insulator but ionic conductor. Together with the electrodes and the electrolyte, the separators play an important role in determining battery performance and safety [49]. A picture of a disassembled 40 Ah commercial Li-ion cell is shown in Fig. 6.13, where the membrane can be seen as a white, plastic film to separate both electrodes. The cell consists of 48 stacked electrodes connected in parallel to the connection taps (top part of the closed cell). An anode and a cathode are shown on the opened cell. The black part is the active material (graphite for the anode and NMC for the cathode), half of which was removed to show the copper and aluminum current collectors. The membrane, which is intercalated between the electrodes, is shown at the right of the opened cell.

The membranes are microporous polymer films with pores ranging from nanometers to micrometers. The most used material are polyethylene and polypropylene, which have various advantages, including outstanding mechanical strength, stability, and low cost [50,51]. These materials have a low thermal shut-down temperature (135°C–165°C), at which the membrane melts and the pores become blocked, restricting ion movement and thus improving battery safety.

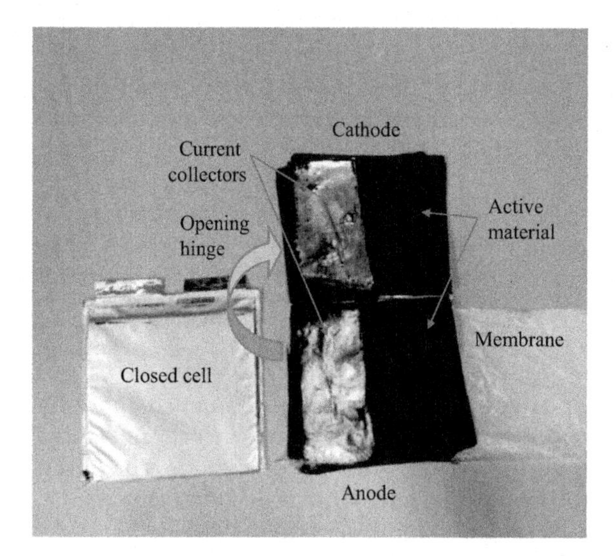

FIGURE 6.13 Disassembled 40 Ah pouch cell with graphite anode and NMC cathode.

The most important characteristics to take into account for membrane design are its thickness, electrical resistance, permeability, pore size, porosity, shut-down temperature, chemical stability, etc.

6.4.5 Current collectors

Current collectors are electrical conductors that supply or deliver electrons from the electrode active material. Given the required electronic conductivity, electrochemical stability, and mechanical strength, metals are commonly used as material for current collectors. Specifically, thin films (10–20 μm) of aluminum and copper are used at the cathode and anode, respectively. For the battery manufacturing, electrode-active materials are coated on these current collectors. In Fig. 6.13, the current collectors are made of aluminum (cathode current collector) and copper (anode current collector).

Copper is used as an anode current collector, given its low price and electrochemical stability at 0–3 V versus Li/Li^+, which is the working potential of carbon electrodes. However, copper cannot be used as a cathode current collector, given that its oxidation occurs at 3 V versus Li/Li^+ and cathodes have higher working potential. Taking this voltage limitation and cost into account, aluminum is the most appropriate material to be used as a cathode current collector.

6.5 Components of a battery system depending on its size

6.5.1 Single-cell batteries

The most elemental battery consists of a single cell that feeds an electronic device, as is usually the case for cell phones. Fig. 6.9B shows a Samsung cell phone battery, which is a prismatic cell with a rated voltage of 3.7 V and a rated capacity of 2.8 Ah. Besides the actual cell in which the energy is stored, these small batteries usually include a protection circuit that prevents damages in the battery due to overcurrent and external short-circuits. This protection circuit is usually included in the prismatic battery case.

6.5.2 Commercial, medium-size batteries

There are two possibilities to enlarge the energy capacity and the manageable power of a battery: increasing its capacity or its voltage. On the one hand, the capacity can be increased easily by manufacturing larger cells. The more electrode material is available for lithium insertion and deinsertion, the more energy can be stored in the cell. On the other hand, the cell voltage is defined by the equilibrium potential of the electrochemical reaction, which is around 3.7 V for Li-ion batteries. However, having a voltage higher than 3.7 V is a must for an efficient management of the battery power. For instance, the Tesla Powerwall battery, a suitable battery to be installed in an MG, offers a voltage range from 350 to 450 V, which allows it to be connected to several inverters available in the market. In order to offer this voltage, the American manufacturer builds this battery by series, connecting a number of cylindrical 18650 cells, such as the one shown in Fig. 6.9A. The Powerwall battery shown in Fig. 6.14 consists of 888 cells, assembled in eight paralleled series of 111 series-connected cells.

These batteries, built by several cells, require the installation of additional systems to guarantee the safe operation of the battery. Firstly, the voltage across each cell needs to be monitored continuously, given that an overvoltage or undervoltage in any of the cells can lead to a sudden breakdown and the subsequent destruction of the battery. Moreover, the battery current needs to be measured in order to avoid overcurrent performances that damage the battery. The temperature is also critical, given that Li-ion batteries have an operation temperature range that should be preserved. Batteries built by a large number of cells, such as the one shown in Fig. 6.14B, usually induce a temperature distribution along the battery that requires the use of a number of temperature sensors located in different parts of the battery.

The measurement of all the variables required to guarantee the safe operation of the battery is carried out by means of electronic cards, such as the one shown in Fig. 6.15A. These cards send the measured data to a microcontroller, also known as the battery management system (BMS), which guarantees the safe operation of the battery. The element that physically prevents the battery's unsafe operation is a safety relay. This is a normally open switch that is closed only when the BMS communicates that all the safety limits (voltage, current, temperature, etc.) are met. Besides the disconnection of any charge or load connected to the battery when safety limits are about to be surpassed, this safety relay should be able to protect the battery against overcurrent or even external short-circuits. In the case that this relay is not able to break a short-circuit, an additional fuse is usually installed in order to increase the safety of the battery. Fig. 6.15B shows one of these fuses, which is installed in a commercial, 133 V, 5.3 kWh battery pack.

FIGURE 6.14 Cell assembly of a Tesla Powerwall. This battery is built by 888 18650 cells.

(A)

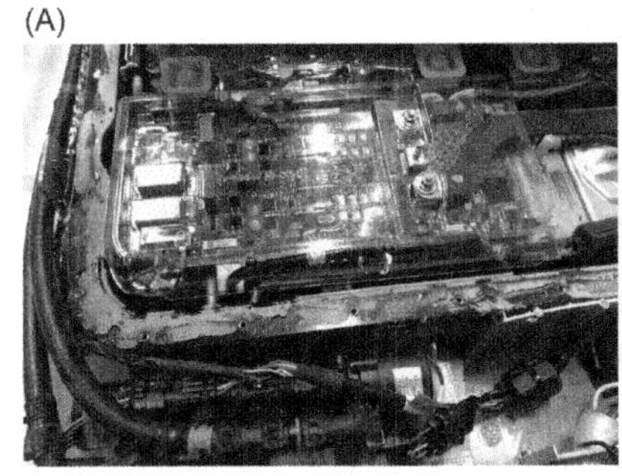

(B)

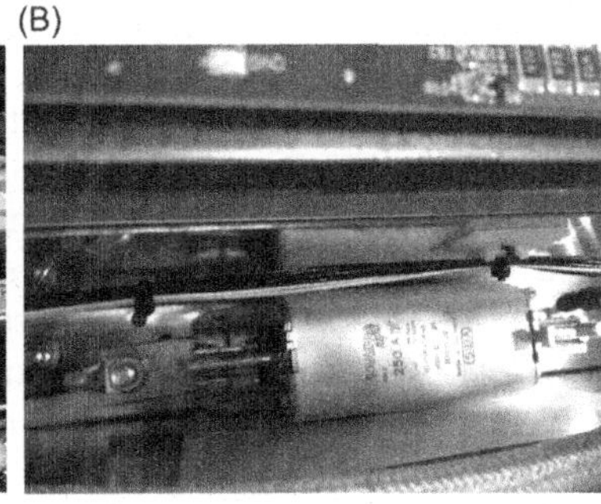

FIGURE 6.15 One of the electronic cards used in the Tesla Powerwall battery (A) and fuse of a 5.3 kWh lithium-ion (Li-ion) battery pack (B).

Given that a number of cells are connected in series in order to build these batteries, the current driven by each of the cells is forced to be the same. The small tolerance of the cell manufacturing process provokes a slight dispersion in the characteristics of these cells. This fact entails that the cells with poorer characteristics suffer deeper charge–discharge cycles, given that their capacity is lower. This leads to a faster aging of these cells, which, in turn, accelerates the cell capacity loss. This undesirable effect can be avoided by means of cell equalization. The BMSs of batteries with several series-connected cells should be able to balance the states of charge of the cells in order to achieve homogeneous aging phenomena. A typical equalization strategy, used by many BMSs, is so-called passive equalization, which consists of slow discharges of the cells with a higher state of charge (SOC) through small resistors located on the electronic cards.

Other main functions of the BMS are to figure out the battery SOC [52] and, based on it, provide the power limits of the battery at each time (which are function of SOC, operating temperature, state of health, etc.). Some BMSs also include a battery-load communication system that sends information about the battery state to the load connected to it.

6.5.3 Industrial, large battery systems

Larger battery systems, for instance distribution-grid-connected batteries that are installed in 20-feet containers, require further auxiliary systems. An important issue to take into account is the air conditioning inside the container. Containers are outdoor systems with a very poor thermal isolation that require air conditioning systems to guarantee a safe operation. Besides, they usually include fire-suppression systems, given that Li-ion batteries may release flammable gasses and a fire event can be dangerous. These auxiliary systems are controlled by the so-called energy management system, which takes care of the battery control in the level above BMS control.

The isolation capacity becomes relevant in large-size batteries, given that a series connection of cells generates greater voltage differences across the battery, which increase the risk of isolation failure. Besides, battery systems that manage voltage levels dangerous for people must have a ground-fault detection system to avoid accidents.

6.6 Electrical performance of Li-ion batteries

Most of the researchers divide the analysis of the electrical performance of Li-ion batteries into two parts [53,54]. Following this trend, Section 6.6.1 concerns thermodynamic processes, which analyze steady-state systems. Thermodynamic applied to batteries means open-circuit analysis (no current flow through the battery). The equilibrium potential in typical Li-ion batteries electrodes is addressed in this section, as well as the equilibrium between the anode and the electrolyte. This equilibrium has particular interest in Li-ion batteries, since the electrolyte is not stable at the working potentials of the negative electrode. The equilibrium is only possible through the formation of a so-called SEI which is a hot research topic for the manufacturing processes and which is of critical importance in the performance and aging of Li-ion batteries.

On the other hand, in Section 6.6.2, the most important electrochemical processes are gathered, that is, the effects that the electrochemical reaction has on the battery. Therefore electrochemical principles are applied when current is driven through the battery. The phenomena covered in this section are ohmic phenomena, related to the opposition of battery materials to the movement of electrons and ions; the electrochemical of the redox reactions driven by the Butler–Volmer equation, which is the causing of the so-called activation losses; and the effects of ion transport through the electrolyte and in the electrode, which is governed by diffusion laws and cause the diffusion (or concentration) losses. The result of these phenomena are overvoltages that lead to energy losses in the battery, as detailed in Section 6.6.2.

6.6.1 Thermodynamics

Thermodynamics is the branch of physics that addresses the fundamentals of the physical and chemical behavior of equilibrium matter [55]. Therefore thermodynamic laws are used to explain the state of a battery when it is kept at equilibrium, that is, when it is kept in open circuit and, therefore, neither electrons nor ions circulate through the system.

6.6.1.1 Equilibrium potential of insertion electrodes

The electrochemical reaction at the electrode–electrolyte interphase can be expressed as follows:

$$\mathrm{Li}^+ + \mathrm{e}^- + \theta_\mathrm{s} \rightleftharpoons \left[\mathrm{Li}^{+\delta} - \theta^{-\delta}\right]$$

where the expression $\left[\mathrm{Li}^{+\delta} - \theta^{-\delta}\right]$ represents the intercalated species. The equilibrium voltage of these kind of reactions can be expressed by means of the Nernst equation:

$$F \cdot v_{\mathrm{eq}} = F \cdot v_{\mathrm{eq}}^0 + R \cdot T \cdot \ln\left(\frac{x_\psi}{x_\varsigma}\right) \tag{6.1}$$

in which v_{eq}^0 is the equilibrium potential corresponding to the reference chemical potentials, and x_ς and x_ψ are the molar fractions of intercalated and deintercalated species respectively. Due to the pseudo binary interactions between Li ions and the host matrix during intercalation and deintercalation processes, Eq. 6.1 needs to be completed with a correction term:

$$F \cdot v_{\text{eq}} = F \cdot v_{\text{eq}}^0 + R \cdot T \cdot \ln\left(\frac{x_\psi}{x_\zeta}\right) + R \cdot T \cdot \ln\left(\frac{\gamma_\psi}{\gamma_\zeta}\right) \tag{6.2}$$

Since γ_j is related to Gibbs energy [56,57], the Redlich–Kister equation can be used to obtain the following expression for the equilibrium voltage:

$$v_{\text{eq}} = v_{\text{eq},c} - v_{\text{eq},a} = v_{\text{eq},c}^0 - v_{\text{eq},a}^0 + \frac{R \cdot T}{F} \cdot \ln\left(\frac{(1 - x_{\zeta,c}) \cdot x_{\zeta,a}}{x_{\zeta,c} \cdot (1 - x_{\zeta,a})}\right) + v_{\text{INT},c} - v_{\text{INT},a} \tag{6.3}$$

6.6.1.2 *Equilibrium between the anode and the electrolyte: solid–electrolyte interface*

The SEI is a passive layer created between the electrolyte and the graphite anode, the existence of which is crucial for the reversibility of Li-ion batteries. Without the SEI, the anode and electrolyte would react spontaneously and the battery would be destroyed. Actually, electrolyte solvents are unstable below 0.8 and above 4.5 V versus Li^+/Li in the presence of the electrode materials, which are strongly reducing/oxidizing [58], while the working potential of a carbon-based anode ranges from 0 to 1 V versus Li^+/Li [59].

Prior to its market release, each Li-ion battery undergoes a SEI-formation process, so that it has a stable SEI when acquired by the user. This formation process consists of a slow charge during which the electrolyte solvent degradation reaction takes place at the electrode–electrolyte interface, which often also involves the electrolyte salt and water impurity traces. The resulting insoluble products form a solid protective passivation layer adhering to the surface of the negative electrode, which is the SEI. Even though an ideal SEI would be built before the battery is put on the market and stay stable after that, there are irreversible side reactions that entail battery aging.

6.6.2 Electrochemistry

Electrochemistry is the study of electron transfer caused by redox reactions at the interface of an electron conductor, such as a metal, and an ionic conductor, such as an electrolyte. Therefore the electrochemical theory is the most suitable tool to explain the phenomena taking place in a battery when it is out of its thermodynamic equilibrium, that is, when it is being charged or discharged. The three main electrochemical phenomena that take place during the battery charging and discharging processes are described in the following subsections.

6.6.2.1 *Ohmic phenomena*

Ohmic losses represent the voltage drop due to the transfer of electrons in the electric circuit and the movement of ions through the electrolyte and membrane. These phenomena are determined, on the one hand, by the electronic conductivity of the electrodes and of the current collectors (usually copper and aluminum) and, on the other hand, by the ionic conductivity of the electrolyte and membrane [60]. Since the electrolyte does not store energy, its properties are unchanged for any value of SOC. However, the variation in the electrode lithium content leads to a conductivity change, making the ohmic phenomena SOC dependent. With regard to temperature, there are opposite effects over electronic and ionic conductivity. The electronic conductivity decreases for increasing temperature in most of conductive materials, whereas the ionic conductivity is increased by higher temperatures. Given that the effect over ionic conductivity is greater than that over electronic conductivity [18], ohmic losses decrease when the temperature is increased.

6.6.2.2 Polarizable electrodes

Any electrochemical reaction requires some energy to take place, which is called activation energy. This means that in a Li-ion battery the electrodes need to be polarized, that is, there has to be a voltage drop between each electrode and the electrolyte for the battery to charge or discharge. This voltage drop entails the need for increased energy for battery charging and reduced energy available when discharging. These power losses are known as activation losses or activation overpotential.

The Butler-Volmer Equation expresses the relationship between the faradaic current iF and the overpotential of an electrochemical reaction ($v_{elec} - v_{eq}$), based on the kinetics of the chemical reaction:

$$i_F = F \cdot S_{SEI} \cdot k^0 \cdot [c_O(0,t) \cdot exp(-\alpha \cdot f.(v_{elec} - v_{eq})) - c_R(0,t) \cdot \exp((1-\alpha) \cdot f.(v_{elec} - v_{eq}))] \quad (6.4)$$

where $f = F/(R \cdot T)$, S_{SEI} stands for the area of the electrode–electrolyte interphase, v_{eq} is the equilibrium electrode potential, v_{elec} the electrode potential when subject to a current, c_O and c_R the concentration of oxidant and reducing species respectively and α the apparent cathodic transfer coefficient.

The voltage difference required for the reaction to take place ($v_{elec} - v_{eq}$ in Eq. 6.4), combined with the faradaic current i_F, leads to power losses called activation losses.

6.6.2.3 Ion transport

Diffusion or concentration phenomena are described by Fick's laws of diffusion, which relate the diffusive flux, $\dot{m}$, to the concentration, x, through the Fick's coefficient, D, and the density, ρ. The first law, Eq. 6.5, states the relationship assuming a steady state; while the second law, Eq. 6.6, predicts how diffusion causes the concentration to change with time.

$$\dot{m} = -\rho \cdot D \cdot \nabla x \quad (6.5)$$

$$\frac{\delta x}{\delta t} = D \cdot \Delta x \quad (6.6)$$

Fick's diffusion coefficient, D, depends on the temperature, as expressed by the following equation:

$$D = D_0 \cdot \exp\left(-\frac{E_A}{R \cdot (T - T^0)}\right), \quad (6.7)$$

where D_0 and T^0 are the standard Fick's Diffusion Coefficient and temperature, respectively, and E_A is the activation energy of this process.

Diffusion phenomena have two main effects on the performance of a Li-ion battery. On the one hand, the ion diffusion through the crystal lattice in the electrodes creates a difference between the ion concentration at the electrode–electrolyte interphase, and the average electrolyte concentration, c_{Li}^{-}. Given the fact that the electrochemical reaction takes place on the interphase, the equilibrium voltage and the activation phenomena are defined by the concentration on this surface. This ion concentration gradient produces a concentration voltage drop during battery operation. On the other hand, ion diffusion through the membrane also has a major impact on the battery behavior, particularly during low temperature operation [57]. This process induces a voltage drop in the battery with characteristic dynamics, which is different from the electrode diffusion and needs to be analyzed separately. Both phenomena lead to energy losses in the battery and are called diffusion or concentration losses.

6.7 Thermal processes in Li-ion batteries

Heat generation is a fact in every energy system, and Li-ion batteries are not the exception. The inefficiencies of real processes prevent the total amount of energy being useful and a portion is dissipated as heat. This dissipative process takes place in the bulk of the battery cells, where the electrochemical reactions and the electron and ion movement go on. Therefore this energy must be conducted from the inner to the surface of the cell and finally dissipated to the ambient. As detailed in the following subsections, temperature has a remarkable influence on battery performance, and it can limit the battery functionality in some cases. Therefore a good understanding of the heat generation phenomena and suitable cell and battery pack designs that optimize thermal performance are important factors to achieve a good battery pack. Actually, Tesla Inc., a company that is taking great steps towards the Li-ion battery technical improvement and price decrease, owns over 40 patents directly related to thermal issues in such batteries, which include battery management strategies to limit thermal generation [61], heat dissipation [62,63], thermal event detection [64], avoidance of thermal propagation [65,66], etc.

The most remarkable issues about the influence that temperature has on Li-ion batteries, the thermal generation mechanisms, and the basics of thermal dissipation are covered in the following subsections.

6.7.1 Effect of temperature on battery performance

It is well known that the performance of Li-ion batteries reaches its maximum when operating at moderate, homogeneous ambient temperatures (between 20°C and 40°C). Both high and low temperatures have negative effects on their performance, and the optimal temperature is a trade-off between these issues that depends on the anode, cathode, and electrolyte components [67]. The pejorative effects that different thermal conditions have on battery performance are addressed in the following lines.

6.7.1.1 High temperature

Energy and power fade These are the two main battery aging phenomena explained in Section 6.8. Energy fade takes place when the active material inside the battery is turned into inactive material, reducing the battery capacity, which also cuts down the energy that can be stored in the battery. Power fade is related to impedance rise, which reduces the battery operating discharging voltage and increases its charging voltage.

Thermal runaway This is one of the most hazardous and undesirable processes that can take place in a Li-ion battery, and a significant amount of research and development efforts are aimed at the detection, prevention, and restraint of this effect. Actually, thermal runaway in two Li-ion batteries was the reason for the grounding of an entire fleet of Boeing 787 Dreamliners on January 2013, the first aircraft equipped with Li-ion batteries [68].

The thermal runaway event occurs in batteries when elevated temperatures trigger heat-generating exothermic reactions and the generated heat cannot be dissipated effectively, thereby activating further exothermic processes, as represented in Fig. 6.16. These exothermic reactions were reviewed by Spotnitz and Franklin [69], who identified the metastable components of the SEI to exothermically decompose at a temperature between 90°C and 120°C. After this breakdown, when the electrolyte is in contact with the anode, the electrolyte solvent starts to decompose at a temperature of 100°C, with the reaction peaking at 200°C, as shown in Fig. 6.16. The positive electrode can also react with the electrolyte or give off oxygen that reacts with the electrolyte in a highly exothermic reaction triggered only by high temperatures ($\approx$180°C, as shown in Fig. 6.16). In the presence of this released oxygen, organic electrolytes can combust once vaporized. By this point, the separator should have melted, which can cause battery short-circuits, which further increases the temperature. After these concatenated phenomena, the battery would probably have exploded and/or started burning.

The temperature at which the thermal runaway phenomena are triggered depends on a number of factors, such as the composition of the electrodes and electrolytes (and therefore the SEI) and the SOC (higher SOC has been proven to reduce the thermal runaway trigger temperature [70]).

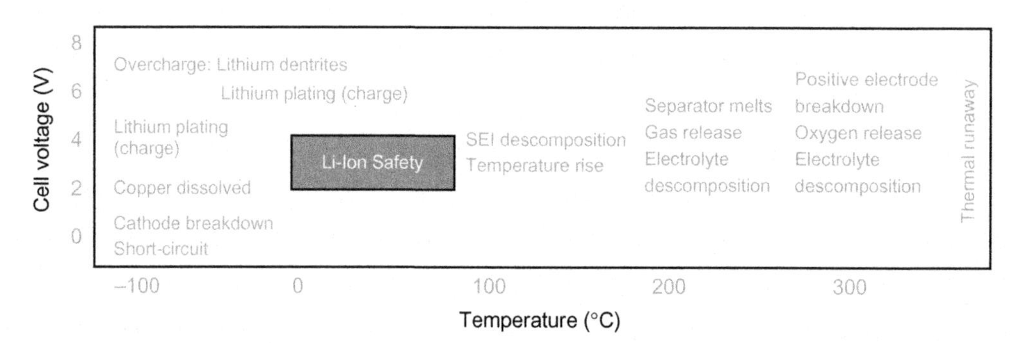

FIGURE 6.16 Lithium-ion cell operating window and destructive processes taking place out of this window.

6.7.1.2 *Low temperature*

Most of the batteries available in the market, including the Li-ion technologies, have reduced features when performing at low temperature, given the reduced ion conductivity of the involved materials and the lower reaction rate of the redox processes (which normally follow an Arrhenius trend) [71]. However, the requirements of the automotive sector, in which the batteries must be able to operate at subzero temperatures, encouraged the DOE to set a goal of 20% battery power performance with a temperature of −30°C. Therefore commercial hybrid EV manufacturers usually employ battery heating strategies to prevent battery degradation due to low temperatures.

Although low temperatures reduce the ionic conductivity of the SEI, and electrolyte and lithium ions have greater difficulties diffusing through the graphite anode, Zhang et al. [72] identified the poor charge transfer at the electrode–electrolyte interface as the performance limiting factor of Li-ion batteries at low temperatures.

Besides, when the lithium ions diffusion in the anode is deteriorated by low temperatures and their transport rate is kept high because of high current, metallic lithium is deposited on the surface of the SEI, which causes a new SEI formation process, thereby consuming active material and reducing the usable electrode–electrolyte contact area. As explained in Section 6.8, lithium deposition (or plating) is one of the most degrading effects concerning Li-ion batteries with carbonaceous anodes and can lead to the build-up of dendrites, which can eventually punch the separator and short circuit the battery.

6.7.2 Heat generation

The electrochemical phenomena described in Section 6.6.2 lead to internal overpotentials during charge and discharge processes which result in heat generation or absorption. For the thermal modeling of these effects, a common simplification is assuming an average heat capacity and isothermal cells, which are reasonable assumptions for typical battery applications. This heat generation rate is usually expressed as a simple function of the battery current (i) and voltage (v), operating temperature (T), and open-circuit voltage (v_{OC}) as follows:

$$\bar{\dot{Q}} = \frac{1}{v} \cdot \left(i \cdot (v - v_{OC}) + i \cdot T \cdot \frac{\partial v_{OC}}{\partial T} \right) \tag{6.8}$$

The first term of this equation is called irreversible or polarization heat. It is due to all the irreversible processes that take place in the battery. This term in always positive, since the working potential is higher than v_{OC} only when the battery is charging ($i > 0 \Leftrightarrow (v - v_{OC}) > 0$). The second term is the reversible or entropic heat, given that it is the consequence of reversible electrochemical processes. The entropic heat can have positive sign (heat dissipation) or negative (heat absorption) depending on the values of the current and entropic factor $\frac{\partial v_{OC}}{\partial T}$. The value of the entropic factor depends on the specific Li-ion battery technology and temperature, but has a typical shape with minimum, negative values at low SOC, maximum, positive values for medium SOC and average values for high SOC that have been published to be negative [73,74] or positive [75] depending on their referred study.

6.7.3 Heat dissipation

The heat generation in a battery as a consequence of the processes explained in Section 6.7.2 takes place in the whole bulk volume of the cell, leading to a temperature increase due to the absorption of this heat during the initial transient period. As the cell temperature increases, the heat transfer process from the inside to the ambient is also increased, until the steady-state operating conditions are reached, matching the amount of heat generated and evacuated [76]. The assumption of uniform heat generation along the volume is reasonable for Li-ion batteries, given the uniform manufacturing method consisting of the stacking of thin, uniform electrodes, current collectors, and separators. Under this assumption, a maximum temperature (T_{max}) is reached in the regions that are most separated from the outside of the cell, which are the central axis of a cylindrical cell (Fig. 6.17A), the symmetry plane of a pouch cell (Fig. 6.17B), and the central point of a prismatic cell (Fig. 6.17C). T_{max} has a main importance for the analysis of lithium-ion batteries, given that, as explained in Section 6.7.1, high temperatures have a negative influence on battery performance and lifetime. Moreover, if these maximum temperatures trigger a thermal runaway, a chain reaction will start, leading to the catastrophic failure of the cell.

The three fundamental mechanisms of heat transmission are conduction, convection, and radiation [76]. Given the small temperature difference between the batteries and the ambient, the radiated heat can be disregarded [77]. Therefore the generated heat is basically dissipated by means of conduction and convection.

The resolution of conduction and convection equations in complex environments is an arduous undertaking which concerns to Heat Transfer discipline. Some simplifications to this process, which are reasonable for common use of Li-ion batteries, are proposed hereinafter in order to provide a simplified solution of this engineering problem. With this purpose, a homogeneous and invariant heat conductivity λ and a uniform heat generation through the whole cell volume $\dot{q}$ are usually assumed. Let S be the external surface area of the cell, which is in contact with the ambient air, V the cell volume, and T_{sur} the surface temperature. The generated heat is transferred to the ambient through a convection process with a heat transfer coefficient h.

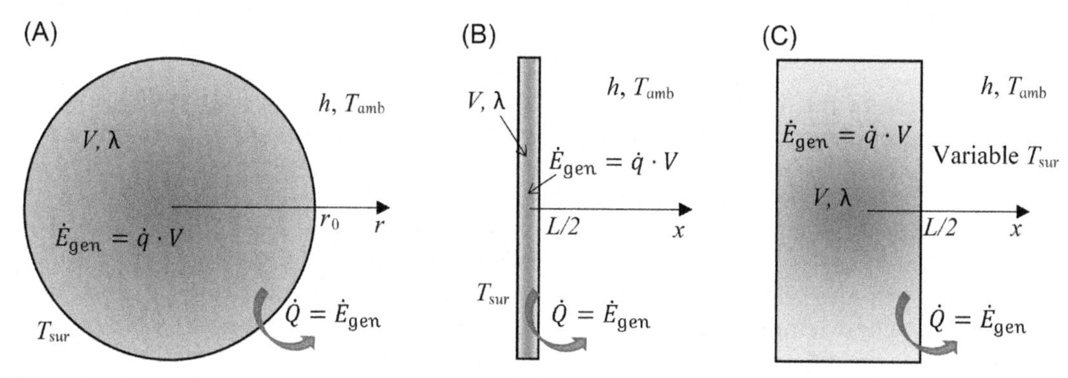

FIGURE 6.17 Schematic of thermal phenomena in different lithium-ion cells: cylindrical cell (A), pouch cell (B) and prismatic cell (C). The hottest areas are darker.

If the system has reached the steady state, the heat transfer rate from the cell to the ambient equals the total heat energy generated inside the cell, as represented in Fig. 6.17:

$$\dot{Q} = \dot{q} \cdot V \tag{6.9}$$

Using Newton's law of cooling:

$$T_{sur} = T_{amb} + \frac{\dot{q} \cdot V}{h \cdot S} \tag{6.10}$$

Eq. 6.10 has a straight solution if unidirectional heat flow can be assumed, which is realistic for cylindrical cells, the radius r_0 of which is much shorter than their length (radial flow, as represented in Fig. 6.17A) and pouch cells, the thickness L of which is much lower than their length and width (transversal flow, as represented in Fig. 6.17B). This 1D approximation provides worse results for prismatic cells, given than their width is long enough for the heat transferred through the top and bottom faces not to be negligible. The result of a pouch-type cell is:

$$T_{sur,pouch} = T_{amb} + \frac{\dot{q} \cdot L}{2 \cdot h} \tag{6.11}$$

and for the cylindrical case:

$$T_{sur,cyl} = T_{amb} + \frac{\dot{q} \cdot r_0}{2 \cdot h} \tag{6.12}$$

Given the spacial symmetry and the uniform generation, T_{sur} is constant for cylindrical and pouch cells. However, due to the 3D heat transmission, it is not the case for prismatic cells, as represented in Fig. 6.17. Note that these results are valid only under the abovementioned assumptions and once the steady state is reached.

6.8 Aging of lithium-ion batteries

The battery performance, based on the combination of two redox semireactions, seems straightforward at a first glance. However, there is an intrinsically complex system behind this simplicity. Besides the electrode-active materials and the electrolyte that allows the ion flow, several "inactive" components are required for the practical performance of a battery. Some of these components are separators to avoid short circuits between both electrodes, current collectors to gather or distribute the electrons along the electrodes, a polymeric film in which the electrolyte is embedded, conductive additives to enhance the conductivity of the electrodes, etc. All these contribute to battery function by maintaining its electrochemical characteristics and structure integrity [58].

The electrochemical reactions that are the basis of battery performance should ideally involve only the active materials, which are reversible and highly efficient. However, in a real system, all battery components can interact with one another, and many of these interactions are irreversible. Given this irreversibility, these interactions are considered aging mechanisms, since they entail an irretrievable reduction of battery capacity. Some of these processes are related to environmental variables, such as temperature, and conditions of use (discharge rate, charge protocol, depth of discharge, etc.).

As already covered in Section 6.4, there are different types of Li-ion batteries, with a number of different cathode and anode materials. Depending on these materials, particular parasitic reactions are induced and distinct aging mechanisms take place in the battery [78]. These reactions have macroscopic effects, with the most remarkable of them being battery capacity fade and impedance rise [79,80]. The macroscopic implications of battery aging on its performance are summarized below.

6.8.1 Aging effects on battery performance

All the phenomena described above have two measurable effects on the performance of Li-ion batteries, which are capacity fade and impedance rise (or power fade). The capacity fade is mainly due to the blocking of some electrode pores, or the lack of active material provoked by some of the factors described above. On the other hand, power fade is a consequence of a decrease in the electrical or ionic conductivity of any material present in the battery (especially the SEI layer), or a reduced surface area of the electrodes, which reduces the region at which the intercalation electrochemical reaction takes place.

The aging phenomena, which result in reduced battery capacity and power fade, are usually divided into two components for easier understanding and applicability. These two components are named cycle aging and calendar aging, and are assumed to be independent of each other, that is, the total capacity and power fade are a superposition of calendar and cycle aging phenomena [81]. Calendar aging refers to performance loss due only to the passing of time. This is the aging that a battery would suffer if it is stored at open circuit conditions, that is, with no use. Cycle aging denotes the changes suffered by the battery due exclusively to charge–discharge cycles [82]. Given that time goes by while the battery is cycled, the measurement of cycle aging is a challenging task, since it is always superimposed with calendar effects. Likewise, the measurement of the aging effects require the charging and discharging of the battery, which makes it difficult to measure the calendar effect avoiding the influence of measurement cycles. It has been proven already that the influence of cycle aging on battery capacity and power fade is much lower than that of cycle aging for normal battery operation [83]. There are different test protocols and measurement techniques proposed to separate both aging effects properly, like standard capacity and impedance measurement techniques [84], careful design of checkups for calendar aging characterization [83], standard charge-discharge cycles for cycle aging phenomena [85–87], aging acceleration techniques to shorten the required test time [88–90], etc.

6.9 Future perspectives of lithium-ion batteries

As explained at the beginning of the chapter, the first Li-ion battery boom took place in the 2000s and was related to portable devices. The second one is currently happening and is related to EVs and electrical mobility in general, and the third one is expected to happen during the following decades, with a rapid growth of stationary energy storage, especially small-scale batteries installed in households and PV systems. According to Bloomberg, Li-ion batteries for energy storage will become a €18 billion per year market by 2040 [91]. Due to this growing Li-ion battery demand, both in quantity and in technical requirements, important research efforts are being made at every level aimed at improving this technology and meeting market demands. Concerning engineering research in Europe aimed at optimizing the existing materials and getting the most out of each battery, a research project called *Batteries 2020* was recently concluded [92]. The Spanish IK4–IKERLAN research center led this project, which included first-level research partners from universities and companies such as the research group led by Prof. D.U. Sauer at the RWTH Aachen University, Aalborg University, Leclanché and Abengoa. This consortium was able to increase both the energy density and lifetime of large-format NMC pouch lithium-ion batteries towards the goals targeted for automotive batteries (250 Wh/kg at cell level and over 4000 cycles at 80% depth of discharge). Three strategies were tackled at the same time: (i) materials developments, specifically two new generations of NMC cathode materials were developed to improve the performance, stability, and cyclability of existing battery cells, (ii) better understanding of the aging phenomena, and (iii) reduction of battery cost by considering the second life of batteries already used in EV applications [93].

Great steps are being taken in the United States towards cheaper and more reliable Li-ion batteries. The engineering research done by Tesla and the construction of the factory known as Gigafactory 1 in Nevada are remarkable examples of the expected battery development during the upcoming years. The Tesla Gigafactory 1 is already operational while construction work is being completed at the Tahoe Reno Industrial Center, Nevada. Two construction phases are shown in Fig. 6.18, along with the final planned size. The factory started limited production of Powerwalls and Powerpacks in the first quarter of 2016 [94] using battery cells produced elsewhere, and began mass production of cells in January 2017 [95]. Given the expectations of building new gigafactories, Tesla began to refer to the

FIGURE 6.18 Gigafactory 1 construction phases. It doubled its size from July 2016 to January 2017, when 30% of the factory was completed. *Source: Planet Labs, Inc. [CC BY-SA 4.0], via Wikimedia Commons.*

SolarCity Gigafactory at Buffalo, New York, in February 2017, as Gigafactory 2. The projected Europe Gigafactory will be named either Gigafactory 4 or 5 and the location is expected to be announced in 2019.

At material level, there are several research lines aimed at improving battery materials. The major challenge is to move a step forward passing from the current intercalation chemistry to novel concepts that may lead to advanced batteries. Some of the most promising research lines are summarized below:

Lithium-metal Rechargeable lithium-metal batteries were manufactured before Li-ion batteries, but the dendrite growing prevented its widespread commercialization. In 2010, a trial Li-metal with a capacity of 300 Wh/kg was installed in an experimental electric vehicle. DBM Energy, the German manufacturer, claims 2500 cycles, short charge times and competitive pricing [96]. 300 Wh/kg would be the highest achieved energy density, given that NCA batteries in the Tesla S 85 has 250 Wh/kg and LMO batteries of the BMW i3 have 120 Wh/kg.

Lithium–air Similarly to fuel cells, the battery uses oxygen from the air to make the electrochemical reaction of the cathode, while the anode is made of lithium. The theoretical specific energy of Li–air batteries is 13 kWh/kg. If only one quarter of this theoretical value is reached, the combined battery energy density and high efficiency of an electric motor (~90%) could be on par with a gasoline tank (13 kWh/kg) and its low-efficient thermal motor (25%–30%).

However, the specific power of Li–air batteries is low, especially at cold temperatures. Air purity is also important for these batteries, and it may need to be filtered for its use in a battery. As a result, the battery may end up with compressors, pumps, and filters which increase its size and weight and consume part of the battery power. Another problem of Li–air batteries is the so-called sudden death syndrome due to the lithium peroxide films built by the reaction of lithium and oxygen, which prevents electron movement and results in an abrupt reduction in the battery's storage capacity. Research efforts are being made with additives that reduce the lithium peroxide formation and improve the cycle life.

Lithium–sulfur Li–S batteries are based on the dissolution of lithium from the anode and its incorporation into alkali metal polysulfide salts during discharge, and reverse plating to the anode while charging. These batteries do not need the intercalation material needed for Li-ion batteries. On the one hand, the intercalation graphite can be removed from the anode and metallic lithium can be used. On the other hand, the metal oxide cathode is replaced by cheaper and lighter sulfur. Each sulfur atom can host two lithium ions while the typical amount hosted by a Li-ion battery cathode material ranges from 0.5 to 0.7 lithium ions per host atom.

These batteries have a theoretical energy density on the order of 2600 Wh/kg, and a European H2020 Research Project ended in 2016 called ALISE reached a value of 500 Wh/kg^{-1} in a Li–S battery for cars developing new and optimized components regarding the anode, cathode, electrolyte, and separator [97]. Some issues that need to be solved to allow further development of Li–S batteries are the low conductivity and massive volume change of sulfur during charging and discharging. Carbon coating is being studied to improve the electrical conductivity of the sulfur cathode, but these issues have not yet been solved. Moreover, the achieved cycle life for Li–S batteries is still too short to allow its commercialization. While S and Li_2S_n are relatively insoluble in the electrolyte, many intermediate polysulfides are not, and their dissolution in the electrolyte causes irreversible loss of active sulfur. The use of a protective layer has been studied to improve this issue, but more research is still needed [98,99].

Solid-state lithium Solid-state technology replaces the liquid electrolyte of current Li-ion batteries with a solid polymer or a ceramic separator. A solid electrolyte allows alkali metals to plate and strip on both the cathode and the anode side without dendrites, which simplifies battery cell fabrication and eliminates the intercalation material, thereby reducing the mass of the battery.

Sodium-ion Na-ion batteries use the operation principles used by Li-ion, but using sodium ion as charge carriers. The advantages of Na-ion batteries are the low cost and abundance of Na compared to Li, and their ability to fully discharge with no damage to the cells, reducing the risk of short-circuits and ignition when batteries are stored (Li-ion batteries have a 30% of irreversible capacity that cannot be discharged).

A drawback of Na-ion batteries is their lower energy density, which may reduce its use in sectors where size and weight are not major issues, such as stationary energy storage and renewable energy plants. Moreover, the large expansion and contraction of electrode size during charge and discharge lead to a reduced lifetime of Na-ion batteries, and intensive research is aimed at improving the characteristics of this promising battery technology [100–103].

6.10 Results

After a careful reading of this chapter, the reader has a broad idea of the current state of the Li-ion batteries, an energy storage technology that is undergoing an unprecedented exponential growth. Special emphasis has been made in the chapter of the benefits that Li-ion batteries can bring to MGs. Specifically, the brief summary of the history from the beginning of this technology makes the reader conscious of the current state of the technology. The potential of lithium as an energy storage material is also analyzed in a section of the chapter in which the main advantages of lithium in the current technology scenario are presented. The amount of lithium required to manufacture a battery, the lithium reserves on earth, and the recent evolution and future perspective for Li-ion battery prices are covered in this section.

The next part of the chapter analyzes the building process of a battery, covering the various manufacturing materials that can be used. This leads to a classification of Li-ion batteries based on their materials. Advantages and disadvantages of each battery type are detailed in this section with the aim of contributing to the selection of the most suitable battery for each application. Besides the manufacturing materials, the remaining building components required for a commercial large-format battery are subsequently analyzed.

Thereafter, the electrical and thermal performance of Li-ion batteries are analyzed, leading to a discussion about the aging phenomena that determine the lifetime of the batteries. These sections provide useful tools for making the most of a battery installed in a particular system. Finally, the future perspectives of batteries (both Li-ion and after-lithium technologies) are analyzed in the chapter. This section includes ideas that can shed light on the future planning of companies and research institutions concerning electrochemical ESSs.

6.11 Conclusions

In the current scenario of both increasing electrification of the automotive sector and rising penetration of renewable energies, ESSs with high energy and power density, long lifetime, and high efficiency are required to allow these changes. Lithium-ion batteries and supercapacitors are put forward as suitable ESSs for these applications for being modular and affordable systems offering a wide range of design and management alternatives.

During the past decade, the consumer electronic sector brought about the development of Li-ion batteries, as this type of ESS has been the most common option for electronic devices during the current century. The recent rise of the EV market, requiring high energy density and an outstanding power–energy ratio, has led to fast technology development and a cost reduction of these batteries. Many technology outlooks predict Li-ion batteries to be a main option for electricity grid storage in future, given that the expected requirements coming from the increase of renewable energy generation share fit with the characteristics offered by Li-ion batteries.

The updated review of the current state of Li-ion batteries accomplished through this chapter, with special emphasis on its applicability to engineering systems, reveals the improving performance and cost reduction that this technology is undergoing. As a consequence, these ESSs are being installed in an increasing number of applications, using their potentials to improve the performance of the system.

Acknowledgments

We would like to acknowledge the support of the Spanish State Research Agency (AEI) and FEDER–UE under grants DPI2016-80641-R and DPI2016-80642-R; of Government of Navarra through research projects PI020 RENEWABLE STORAGE and 0011-1411-2018-000029 GERA; and the FPU Program of the Spanish Ministry of Education, Culture, and Sport (FPU13 /00542).

References

[1] J. Pascual, I. San Martín, A. Ursúa, P. Sanchis, L. Marroyo, Implementation and control of a residential microgrid based on renewable energy sources, hybrid storage systems and thermal controllable loads, in: 2013 IEEE Energy Conversion Congress and Exposition (ECCE), 2013, pp. 2304–2309, https://doi.org/10.1109/ECCE.2013.6646995.

[2] I. San Martín, A. Ursúa, P. Sanchis, Integration of fuel cells and supercapacitors in electrical microgrids: analysis, modelling and experimental validation, Int. J. Hydrog. Energy 38 (27) (2013) 11655–11671. Available from: https://doi.org/10.1016/j.ijhydene.2013.06.098.

[3] I. San Martín, A. Ursúa, P. Sanchis, Modelling of PEM fuel cell performance: steady-state and dynamic experimental validation, Energies 7 (2) (2014) 670–700. Available from: https://doi.org/10.3390/en7020670.

[4] A. Berrueta, I. San Martín, A. Hernández, A. Ursúa, P. Sanchis, Electro-thermal modelling of a super-capacitor and experimental validation, J. Power Sources 259 (2014) 154–165. Available from: https://doi.org/10.1016/j.jpowsour.2014.02.089.

[5] Kaco New Energy, Integrated storage solution for large consumers. Available from: <http://kaco-newenergy.com/de/lounge/modellprojekt-weinsberg/>.

[6] I.S. Martín, A. Berrueta, P. Sanchis, A. Ursúa, Methodology for sizing stand-alone hybrid systems: a case study of a traffic control system, Energy 153 (2018) 870–881. Available from: https://doi.org/10.1016/j.energy.2018.04.099.

[7] F. Gutmann, A.M. Hermann, A. Rembaum, Solid-state electrochemical cells based on charge transfer complexes, J. Electrochem. Soc. 114 (4) (1967) 323–329. Available from: https://doi.org/10.1149/1.2426586.
[8] J. Moser, Solid state lithium-iodine primary battery, US Patent 3,660,163 (May 2, 1972).
[9] J. Moser, A. Schneider, Primary cells and iodine containing cathodes therefore, US Patent 3,674,562 (July 4, 1972).
[10] W. Greatbatch, C. Holmes, The lithium/iodine battery: a historical perspective, PACE Pacing Clin. Electrophysiol 15 (11 II) (1992) 2034–2036. Available from: https://doi.org/10.1111/j.1540-8159.1992.tb03016.x.
[11] W. Greatbatch, J.H. Lee, W. Mathias, M. Eldridge, J.R. Moser, A.A. Schneider, The solid-state lithium battery: a new improved chemical power source for implantable cardiac pacemakers, IEEE Trans. Biomed. Eng. BME-18 (5) (1971) 317–324. Available from: https://doi.org/10.1109/TBME.1971.4502862.
[12] W.S. Harris, Ph.D. Thesis, University of California, Berkeley, UCRL-8381, 1958.
[13] M. Whittingham, Chemistry of intercalation compounds: metal guests in chalcogenide hosts, Prog. Solid State Chem. 12 (1) (1978) 41–99. Available from: https://doi.org/10.1016/0079-6786(78)90003-1.
[14] S. Basu, Ambient temperature rechargeable battery, US Patent 4,423,125, December 27, 1983.
[15] A. Yoshino, K. Sanechika, T. Nakajima, Secondary battery, US Patent 4,668,595, May 26, 1987.
[16] K. Mizushima, P. Jones, P. Wiseman, J. Goodenough, Li_xCoO_2 ($0 < x \leq 1$): A new cathode material for batteries of high energy density, Solid State Ionics 3 (Supplement C) (1981) 171–174. Available from: https://doi.org/10.1016/0167-2738(81)90077-1.
[17] A.K. Padhi, K.S. Nanjundaswamy, J.B. Goodenough, Phospho-olivines as positive-electrode materials for rechargeable lithium batteries, J. Electrochem. Soc. 144 (4) (1997) 1188–1194. Available from: https://doi.org/10.1149/1.1837571.
[18] J.-K. Park, Principles and Applications of Lithium Secondary Batteries, John Wiley & Sons, Ltd, 2012.
[19] K. Croswell, The Alchemy of the Heavens: Searching for Meaning in the Milky Way, Anchor Books, 1995.
[20] E. Vangioni Flam, M. Cassé, Cosmic lithium-beryllium-boron story, Astrophys. Space Sci. 265 (1) (1999) 77–86. Available from: https://doi.org/10.1023/A:1002197712862.
[21] J. Emsley, Nature's Building Blocks: An A–Z Guide to the Elements, Oxford Pakistan Paperbacks Series, Oxford University Press, 2001.
[22] AG, Lithium Report 2016, Report, Swiss Resource Capital AG, 2016.
[23] ALBEMARLE, Global lithium market outlook, in: Goldman Sachs HCID Conference, 2016.
[24] P. Cazzola, M. Gorner, Global EV Outlook 2016, Technical Report, International Energy Agency, 2016.
[25] B. Scrosati, K. Abraham, W. van Schalkwijk, J. Hassoun, Lithium Batteries: Advanced Technologies and Applications, The ECS Series of Texts and Monographs, Wiley, 2013.
[26] R. Korthauer, Handbuch Lithium-Ionen-Batterien, Springer, Berlin Heidelberg, 2013.
[27] W. Haynes, CRC Handbook of Chemistry and Physics, 93rd ed., Taylor & Francis, 2012.
[28] N. Nitta, F. Wu, J.T. Lee, G. Yushin, Li-ion battery materials: present and future, Mater. Today 18 (5) (2015) 252–264. Available from: https://doi.org/10.1016/j.mattod.2014.10.040.
[29] H. Nozaki, K. Nagaoka, K. Hoshi, N. Ohta, M. Inagaki, Carbon-coated graphite for anode of lithium ion rechargeable batteries: carbon coating conditions and precursors, J. Power Sources 194 (1) (2009) 486–493. Available from: https://doi.org/10.1016/j.jpowsour.2009.05.040.
[30] S. Scharner, W. Weppner, P. Schmid-Beurmann, Evidence of two-phase formation upon lithium insertion into the $Li_{1.33}Ti_{1.67}O_4$ spinel, J. Electrochem. Soc. 146 (3) (1999) 857–861. Available from: https://doi.org/10.1149/1.1391692.
[31] M. Wagemaker, D. Simon, E. Kelder, J. Schoonman, C. Ringpfeil, U. Haake, et al., A kinetic two-phase and equilibrium solid solution in spinel $Li_{4+x}Ti_5O_{12}$, Adv. Mater. 18 (23) (2006) 3169–3173. Available from: https://doi.org/10.1002/adma.200601636.
[32] Y.-B. He, M. Liu, Z.-D. Huang, B. Zhang, Y. Yu, B. Li, et al., Effect of solid electrolyte interface (SEI) film on cyclic performance of $Li_4Ti_5O_{12}$ anodes for Li ion batteries, J. Power Sources 239 (Supplement C) (2013) 269–276. Available from: https://doi.org/10.1016/j.jpowsour.2013.03.141.
[33] M.-S. Song, R.-H. Kim, S.-W. Baek, K.-S. Lee, K. Park, A. Benayad, Is $Li_4Ti_5O_{12}$ a solid-electrolyte-interphase-free electrode material in Li-ion batteries? Reactivity between the $Li_4Ti_5O_{12}$ electrode and electrolyte, J. Mater. Chem. A 2 (2014) 631–636. Available from: https://doi.org/10.1039/C3TA12728A.
[34] J. Wang, Y.-c K. Chen-Wiegart, J. Wang, In situ three-dimensional synchrotron X-ray nanotomagraphy of the (de)lithiation processes in tin anodes, Angew. Chem. Int. Ed. 53 (17) (2014) 4460–4464. Available from: https://doi.org/10.1002/anie.201310402.

[35] A. Bordes, K. Eom, T.F. Fuller, The effect of fluoroethylene carbonate additive content on the formation of the solid–electrolyte interphase and capacity fade of Li-ion full-cell employing nano Si-graphene composite anodes, J. Power Sources 257 (Suppl. C) (2014) 163–169. Available from: https://doi.org/10.1016/j.jpowsour.2013.12.144.

[36] J. Cho, Y.J. Kim, T.-J. Kim, B. Park, Zero-strain intercalation cathode for rechargeable Li-ion cell, Angew. Chem. Int. Ed. 40 (18) (2001) 3367–3369. Available from: https://doi.org/10.1002/1521-3773(20010917)40:18 < 3367::AID-ANIE3367 > 3.0.CO;2-A.

[37] A. Rougier, P. Gravereau, C. Delmas, Optimization of the composition of the $Li_{1-z}Ni_{1+z}O_2$ electrode materials: structural, magnetic, and electrochemical studies, J. Electrochem. Soc. 143 (4) (1996) 1168–1175. Available from: https://doi.org/10.1149/1.1836614.

[38] P. Kalyani, N. Kalaiselvi, Various aspects of LiNiO2 chemistry: a review, Sci. Technol. Adv. Mater. 6 (6) (2005) 689–703. Available from: https://doi.org/10.1016/j.stam.2005.06.001.

[39] C. Chen, J. Liu, M. Stoll, G. Henriksen, D. Vissers, K. Amine, Aluminum-doped lithium nickel cobalt oxide electrodes for high-power lithium-ion batteries, J. Power Sources 128 (2) (2004) 278–285. Available from: https://doi.org/10.1016/j.jpowsour.2003.10.009.

[40] I. Bloom, S.A. Jones, V.S. Battaglia, G.L. Henriksen, J.P. Christophersen, R.B. Wright, et al., Effect of cathode composition on capacity fade, impedance rise and power fade in high-power, lithium-ion cells, J. Power Sources 124 (2) (2003) 538–550. Available from: https://doi.org/10.1016/S0378-7753(03)00806-1.

[41] Y. Itou, Y. Ukyo, Performance of $LiNiCoO_2$ materials for advanced lithium-ion batteries, J. Power Sources 146 (1) (2005) 39–44. Available from: https://doi.org/10.1016/j.jpowsour.2005.03.091.

[42] E. Rossen, C. Jones, J. Dahn, Structure and electrochemistry of $Li_xMn_yNi_{1-y}O_2$, Solid State Ionics 57 (3) (1992) 311–318. Available from: https://doi.org/10.1016/0167-2738(92)90164-K.

[43] Y.-K. Sun, S.-T. Myung, B.-C. Park, J. Prakash, I. Belharouak, K. Amine, High-energy cathode material for long-life and safe lithium batteries, Nat. Mater. 8 (2009) 320–324. Available from: https://doi.org/10.1038/nmat2418N1.

[44] J. Tu, X. Zhao, G. Cao, D. Zhuang, T. Zhu, J. Tu, Enhanced cycling stability of $LiMn_2O_4$ by surface modification with melting impregnation method, Electrochim. Acta 51 (28) (2006) 6456–6462. Available from: https://doi.org/10.1016/j.electacta.2006.04.031.

[45] F. Lin, I. Markus, D. Nordlund, T.-C. Weng, M. Asta, H. Xin, et al., Surface reconstruction and chemical evolution of stoichiometric layered cathode materials for lithium-ion batteries, Nat. Commun. 5 (2014) 3529. Available from: https://doi.org/10.1038/ncomms4529.

[46] A. Eftekhari, Y. Liu, P. Chen, Different roles of ionic liquids in lithium batteries, J. Power Sources 334 (Suppl. C) (2016) 221–239. Available from: https://doi.org/10.1016/j.jpowsour.2016.10.025.

[47] D.-H. Kim, J. Kim, Synthesis of $LiFePO_4$ nanoparticles in polyol medium and their electrochemical properties, Electrochem. Solid-State Lett. 9 (9) (2006) A439–A442. Available from: https://doi.org/10.1149/1.2218308.

[48] H. Huang, S.-C. Yin, L.F. Nazar, Approaching theoretical capacity of $LiFePO_4$ at room temperature at high rates, Electrochem. Solid-State Lett. 4 (10) (2001) A170–A172. Available from: https://doi.org/10.1149/1.1396695.

[49] H. Lee, M. Yanilmaz, O. Toprakci, K. Fu, X. Zhang, A review of recent developments in membrane separators for rechargeable lithium-ion batteries, Energy Environ. Sci. 7 (2014) 3857–3886. Available from: https://doi.org/10.1039/C4EE01432D.

[50] P. Hohenberg, W. Kohn, Inhomogeneous electron gas, Phys. Rev. 136 (1964) B864–B871. Available from: https://doi.org/10.1103/PhysRev.136.B864.

[51] R.O. Jones, O. Gunnarsson, The density functional formalism, its applications and prospects, Rev. Mod. Phys. 61 (1989) 689–746. Available from: https://doi.org/10.1103/RevModPhys.61.689.

[52] A. Berrueta, I.S. Martín, P. Sanchis, et al., Comparison of state-of-charge estimation methods for stationary lithium-ion batteries, in: IECON 2016—42nd Annual Conference of the IEEE Industrial Electronics Society, 2016, pp. 2010–2015, https://doi.org/10.1109/IECON.2016.7794094.

[53] A. Berrueta Irigoyen, Energy Storage Systems Based on Lithium-Ion Batteries and Supercapacitors: Characterization, Modelling and Integration with Renewable Energies (Ph.D. thesis), Public University of Navarra, 2017.

[54] A. Berrueta, A. Urtasun, A. Ursúa, P. Sanchis, A comprehensive model for lithium-ion batteries: from the physical principles to an electrical model, Energy 144 (2018) 286–300. Available from: https://doi.org/10.1016/j.energy.2017.11.154.

[55] M.J. Moran, H.N. Shapiro, Fundamentals of Engineering Thermodynamics, John Wiley & Sons, Inc, 2004.

[56] E. Tatsukawa, K. Tamura, Activity correction on electrochemical reaction and diffusion in lithium intercalation electrodes for discharge/charge simulation by single particle model, Electrochim. Acta 115 (Suppl. C) (2014) 75–85. Available from: https://doi.org/10.1016/j.electacta.2013.10.099.

[57] J. Sabatier, J.M. Francisco, F. Guillemard, L. Lavigne, M. Moze, M. Merveillaut, Lithium-ion batteries modeling: a simple fractional differentiation based model and its associated parameters estimation method, Signal. Process. 107 (Suppl. C) (2015) 290–301. Available from: https://doi.org/10.1016/j.sigpro.2014.06.008.

[58] M.R. Palacín, A. de Guibert, Why do batteries fail? Science 351 (6273) 1253292. https://doi:10.1126/science.1253292.

[59] M.W. Verbrugge, B.J. Koch, Modeling lithium intercalation of single-fiber carbon microelectrodes, J. Electrochem. Soc. 143 (2) (1996) 600–608. Available from: https://doi.org/10.1149/1.1836486.

[60] A. Berrueta, V. Irigaray, P. Sanchis, A. Ursúa, Lithium-ion battery model and experimental validation, in: 17th European Conference on Power Electronics and Applications (EPE'15 ECCE-Europe), 2015, pp. 1–8, https://doi.org/10.1109/EPE.2015.7309337.

[61] K. Kelty, C. Kishiyama, S. Stewart, Low temperature fast charge of battery pack, US Patent 9,694,699, 2017.

[62] J. Mardall, P. Yeomans, Battery pack base plate heat exchanger, US Patent 13/736,217. 2014.

[63] J. Straubel, E. Berdichevsky, D. Lyons, T. Colson, M. Eberhard, I. Wright, et al., Battery mounting and cooling system, US Patent 9,065,103, 2015.

[64] F. LePort, Battery pack pressure monitoring system for thermal event detection, US Patent 9,083,064, 2015.

[65] P. Rawlinson, N. Herron, B. Edwards, G. Goetchius, Vehicle battery pack thermal barrier, US Patent 8,875,828, 2014.

[66] W. Hermann, Rigid cell separator for minimizing thermal runaway propagation within a battery pack, US Patent 8,481,191, 2013.

[67] T.M. Bandhauer, S. Garimella, T.F. Fuller, A critical review of thermal issues in lithium-ion batteries, J. Electrochem. Soc. 158 (3) (2011) R1–R25. Available from: https://doi.org/10.1149/1.3515880.

[68] N. Williard, W. He, C. Hendricks, M. Pecht, Lessons learned from the 787 dreamliner issue on lithium-ion battery reliability, Energies 6 (9) (2013) 4682–4695. Available from: https://doi.org/10.3390/en6094682.

[69] R. Spotnitz, J. Franklin, Abuse behavior of high-power, lithium-ion cells, J. Power Sources 113 (1) (2003) 81–100. Available from: https://doi.org/10.1016/S0378-7753(02)00488-3.

[70] S.A. Hallaj, H. Maleki, J. Hong, J. Selman, Thermal modeling and design considerations of lithium-ion batteries, J. Power Sources 83 (1) (1999) 1–8. Available from: https://doi.org/10.1016/S0378-7753(99)00178-0.

[71] G. Zhu, K. Wen, W. Lv, X. Zhou, Y. Liang, F. Yang, et al., Materials in-sights into low-temperature performances of lithium-ion batteries, J. Power Sources 300 (Suppl. C) (2015) 29–40. Available from: https://doi.org/10.1016/j.jpowsour.2015.09.056.

[72] S. Zhang, K. Xu, T. Jow, Electrochemical impedance study on the low temperature of Li-ion batteries, Electrochim. Acta 49 (7) (2004) 1057–1061. Available from: https://doi.org/10.1016/j.electacta.2003.10.016.

[73] S.J. Bazinski, X. Wang, The influence of cell temperature on the entropic coefficient of a lithium iron phosphate (LFP) pouch cell, J. Electro-chem. Soc. 161 (1) (2014) A168–A175. Available from: https://doi.org/10.1149/2.082401jes.

[74] J. Marcicki, X.G. Yang, Model-based estimation of reversible heat generation in lithium-ion cells, J. Electrochem. Soc. 161 (12) (2014) A1794–A1800. Available from: https://doi.org/10.1149/2.0281412jes.

[75] G. Vertiz, M. Oyarbide, H. Macicior, O. Miguel, I. Cantero, P.F. de Arroiabe, et al., Thermal characterization of large size lithium-ion pouch cell based on 1D electro-thermal model, J. Power Sources 272 (Suppl. C) (2014) 476–484. Available from: https://doi.org/10.1016/j.jpowsour.2014.08.092.

[76] Y. Cengel, Heat Transfer: A Practica Approach, McGraw-Hill, 2007.

[77] P. Gambhire, N. Ganesan, S. Basu, K.S. Hariharan, S.M. Kolake, T. Song, et al., A reduced order electrochemical thermal model for lithium ion cells, J. Power Sources 290 (Suppl. C) (2015) 87–101. Available from: https://doi.org/10.1016/j.jpowsour.2015.04.179.

[78] K. Jalkanen, J. Karppinen, L. Skogström, T. Laurila, M. Nisula, K. Vuorilehto, Cycle aging of commercial NMC/graphite pouch cells at different temperatures, Appl. Energy 154 (Suppl. C) (2015) 160–172. Available from: https://doi.org/10.1016/j.apenergy.2015.04.110.

[79] A. Berrueta, M. Heck, M. Jantsch, A. Ursúa, P. Sanchis, Combined dynamic programming and region-elimination technique algorithm for optimal sizing and management of lithium-ion batteries for photovoltaic plants, Appl. Energy 228 (2018) 1–11. Available from: https://doi.org/10.1016/j.apenergy.2018.06.060.

[80] A. Berrueta, J. Pascual, I. S. Martín, P. Sanchis, A. Ursúa, Influence of the aging model of lithium-ion batteries on the management of PV self-consumption systems, in: 18th IEEE International Conference on Environment and Electrical Engineering and 2nd IEEE Industrial and Commercial Power Systems Europe, EEEIC and I&CPS Europe, 2018.

[81] J. Schmalstieg, S. Käbitz, M. Ecker, D.U. Sauer, A holistic aging model for Li(NiMnCo)O_2 based 18650 lithium-ion batteries, J. Power Sources 257 (2014) 325–334. Available from: https://doi.org/10.1016/j.jpowsour.2014.02.012.

[82] S. Käbitz, J.B. Gerschler, M. Ecker, Y. Yurdagel, B. Emmermacher, D. André, et al., Cycle and calendar life study of a graphite |$LiNi_{1/3}Mn_{1/3}Co_{1/3}$ O_2 Li-ion high energy system. Part A: full cell characterization, J. Power Sources 239 (Suppl. C) (2013) 572–583. Available from: https://doi.org/10.1016/j.jpowsour.2013.03.045.

[83] J. Schmitt, A. Maheshwari, M. Heck, S. Lux, M. Vetter, Impedance change and capacity fade of lithium nickel manganese cobalt oxide-based batteries during calendar aging, J. Power Sources 353 (2017) 183–194. Available from: https://doi.org/10.1016/j.jpowsour.2017.03.090.

[84] J. Vetter, P. Novák, M. Wagner, C. Veit, K.-C. Möller, J. Besenhard, et al., Ageing mechanisms in lithium-ion batteries, J. Power Sources 147 (1) (2005) 269–281. Available from: https://doi.org/10.1016/j.jpowsour.2005.01.006.

[85] J. Wang, J. Purewal, P. Liu, J. Hicks-Garner, S. Soukazian, E. Sherman, et al., Degradation of lithium ion batteries employing graphite negatives and nickel–cobalt–manganese oxide + spinel manganese oxide positives: Part 1, aging mechanisms and life estimation, J. Power Sources 269 (2014) 937–948. Available from: https://doi.org/10.1016/j.jpowsour.2014.07.030.

[86] J. Purewal, J. Wang, J. Graetz, S. Soukiazian, H. Tataria, M.W. Verbrugge, Degradation of lithium ion batteries employing graphite negatives and nickel–cobalt–manganese oxide + spinel manganese oxide positives: Part 2, chemical–mechanical degradation model, J. Power Sources 272 (2014) 1154–1161. Available from: https://doi.org/10.1016/j.jpowsour.2014.07.028.

[87] S.F. Schuster, T. Bach, E. Fleder, J. Müller, M. Brand, G. Sextl, et al., Nonlinear aging characteristics of lithium-ion cells under different operational conditions, J. Energy Storage 1 (2015) 44–53. Available from: https://doi.org/10.1016/j.est.2015.05.003.

[88] W. Gu, Z. Sun, X. Wei, H. Dai, A new method of accelerated life testing based on the grey system theory for a model-based lithium-ion battery life evaluation system, J. Power Sources 267 (Suppl. C) (2014) 366–379. Available from: https://doi.org/10.1016/j.jpowsour.2014.05.103.

[89] D.I. Stroe, M. Świerczyński, A.I. Stan, R. Teodorescu, S.J. Andreasen, Accelerated lifetime testing methodology for lifetime estimation of lithium-ion batteries used in augmented wind power plants, IEEE Trans. Ind. Appl. 50 (6) (2014) 4006–4017. Available from: https://doi.org/10.1109/TIA.2014.2321028.

[90] T. Guan, P. Zuo, S. Sun, C. Du, L. Zhang, Y. Cui, et al., Degradation mechanism of $LiCoO_2$/mesocarbon microbeads battery based on accelerated aging tests, J. Power Sources 268 (Supplement C) (2014) 816–823. Available from: https://doi.org/10.1016/j.jpowsour.2014.06.113.

[91] New Energy Outlook 2017, Bloomberg new energy finance's annual long-term economic forecast of the world's power sector, Forecast, Bloomberg, 2017.

[92] EU, Batteries 2020, Online, 2016. Available from: <http://www.batteries2020.eu/index.html>.

[93] J.M. Timmermans, A. Nikolian, J.D. Hoog, R. Gopalakrishnan, S. Goutam, N. Omar, T. Coosemans, J.V. Mierlo, A. Warnecke, D.U. Sauer, M. Swierczynski, D.I. Stroe, E. Martinez-Laserna, E. Sarasketa-Zabala, J. Gastelurrutia, N. Nerea, Batteries 2020—lithium-ion battery first and second life ageing, validated battery models, lifetime modelling and ageing assessment of thermal parameters, in: 2016 18th European Conference on Power Electronics and Applications (EPE'16 ECCE Europe), 2016, pp. 1–23. https://doi.org/10.1109/EPE.2016.7695698.

[94] A. Johnston, Tesla starts off 2016 by producing and delivering powerwall, Online, 2016. Available from: <https://cleantechnica.com/2016/01/08/teslastarts-off-2016-producing-delivering-powerwall/>.

[95] T. Randall, Tesla flips the switch on the gigafactory, Online, 2017. Available from: <https://www.bloomberg.com/news/articles/2017-01-04/tesla-flips-the/switch-on-the-gigafactory>.

[96] I. Buchmann, Batteries in a Portable World, Cadex Electronics Inc., 2016.

[97] ALISE, Horizon 2020 Research Project, Online, 2017. Available from: <http://www.aliseproject.com/>.

[98] S. Jeong, Y. Lim, Y. Choi, G. Cho, K. Kim, H. Ahn, et al., Electrochemical properties of lithium sulfur cells using PEO polymer electrolytes prepared under three different mixing conditions, J. Power Sources 174 (2) (2007) 745–750. Available from: https://doi.org/10.1016/j.jpowsour.2007.06.108.

[99] M.M. Islam, V.S. Bryantsev, A.C.T. van Duin, ReaxFF reactive force field simulations on the influence of teflon on electrolyte decomposition during Li/SWCNT anode discharge in lithium–sulfur batteries, J. Electrochem. Soc. 161 (8) (2014) E3009–E3014. Available from: https://doi.org/10.1149/2.005408jes.

[100] H. Pan, Y.-S. Hu, L. Chen, Room-temperature stationary sodium-ion batteries for large-scale electric energy storage, Energy Environ. Sci. 6 (2013) 2338–2360. Available from: https://doi.org/10.1039/C3EE40847G.

[101] D.A. Stevens, J.R. Dahn, High capacity anode materials for rechargeable sodium-ion batteries, J. Electrochem. Soc. 147 (4) (2000) 1271–1273. Available from: https://doi.org/10.1149/1.1393348.

[102] H. Zhu, Z. Jia, Y. Chen, N. Weadock, J. Wan, O. Vaaland, et al., Tin anode for sodium-ion batteries using natural wood fiber as a mechanical buffer and electrolyte reservoir, Nano Lett. 13 (7) (2013) 3093–3100. Available from: https://doi.org/10.1021/nl400998t.

[103] I. Saadoune, S. Difi, S. Doubaji, K. Edström, P.E. Lippens, Electrode materials for sodium ion batteries: A cheaper solution for the energy storage, in: 2014 International Conference on Optimization of Electrical and Electronic Equipment (OPTIM), 2014, pp. 1078–1081. https://doi.org/10.1109/OPTIM.2014.6851038.

CHAPTER

7

Impact of dynamic performance of batteries in microgrids

M.C. Gustavo Pérez Hernández and Arturo Conde Enríquez

Department of Electrical Engineering, Autonomous University of Nuevo Leon, San Nicolás de los Garza, Nuevo Leon, México

7.1 Introduction

In recent years, the challenges of energy efficiency have evolved in many countries giving way to uncontrolled renewable energy sources in microgrids (MGs). The change in demand and the depletion of fossil fuel reserves have raised awareness in society and among energy organizations of the need to seek alternative solutions to this problem. Until a few years ago, the energy balance was controlled only by conventional generation sources (GENCO); thus, the safe and economical operation of the system was the exclusive responsibility of the utility provided by large conventional power plants. In this structure of vertical operation, the distribution networks were controlled by the electrical network satisfying their energy demand, and local control was made only for voltage regulation by connecting reactive elements in derivation. However, at present we have been working with noncontrollable energy sources with the support of storage systems for local control of demand.

Commonly the term distributed generation (DG) is used for the small-scale energy that is generated or produced near the points of consumption within a distribution network. Under the DG environment emerges the concept of an MG. An MG comprises several sources of generation such as: diesel plants, fuel cells, photovoltaic (PV), wind turbines, small hydraulic turbines, or storage systems that operate mostly interconnected to the main distribution network. These MGs can operate in isolation when a fault occurs or when they are interconnected in the main grid through a feeder with high impedance, and even interconnected through several feeders. These sources must interact with each other, behaving together as a consumer system or an energy producing system that includes a coordination of the control and protection devices, as well as the functionalities of a distribution management system (DMS).

DOI: https://doi.org/10.1016/B978-0-12-817774-7.00007-7

At present, the uncontrolled interconnection of DG sources in distribution grids has caused disturbances in the MGs manifesting in problems of frequency variation due to the natural energy imbalance in unregulated sources, in the low quality of energy due to variations in voltage due to the lack of availability of reactive sources, and severe distortion of the waveform due to the injection of nonfundamental frequency components. This results in a low quality of energy supply directly to consumers who are very close to unregulated sources. These MGs have undergone significant changes due to technological advances in their components and to the demands on the part of consumers, as well as to the reduction in cost of the devices that allow operating with the quality requirements for a better use of the electrical system.

Currently energy systems continue to rely on conventional generation units to provide energy storage. This means that, when a new load enters the grid, the initial energy balance must be satisfied with the support of the electrical grid, this change results in a small variation in the frequency of the system, which is quickly regulated by the reserve of the electrical grid. The integration of PV and wind systems represents an unfavorable condition for frequency control due to a lack of energy storage to mitigate this problem mainly in isolated MGs or with weak links to the utility. It is important to use storage devices that provide the balance of energy.

In this context, one of the applications that has proven quite useful is the use of storage systems in batteries to charge and discharge the energy when the grid requires it, thus improving the operation of the distribution grid. These storage systems have come a long way in intelligent distribution grids, and it has been shown in some works that, through proper programming, storage devices can store renewable energy when production is high or the price of energy is high, besides supporting the demand when there are high tariff conditions. In an MG that is isolated or with a weak link to the utility, the time response of the DG (microturbines, fuel cells, etc.) is slow, so it is convenient to have a storage system in order to provide a balance of energy in case of any disturbance or change in the load. These storage devices act as controllable AC voltage sources to cope with sudden changes in the system. However, all storage devices have their limitations regarding their storage capacity, times and limits of charge-discharge of energy, as well as their costs.

7.2 Microgrid operation

An MG, from a conceptual point of view is a medium-low voltage distribution system where the generation of energy in distributed form (DG), behaves as a system formed by many particular MGs attending to each demand in different points or nodes of a grid. This helps to reduce generation costs when considering renewable and nonrenewable energy sources interconnected near the charge. In addition, it means not having to rely on large conventional generators that imply high operating costs.

There are two modes of operation of an MG. The first is a mode connected to the grid, where it tries to satisfy the demand through the local generation of the system and the lack or excess of energy in the distribution grid can be absorbed or supplied by the main grid, in which the frequency is controlled by the utility. The second is in total or partial isolation, where the energy must be balanced between local generation and demand to maintain the frequency and stability of the system under optimal operating conditions. The representation of a distribution grid interconnected to the grid and in isolation is shown in Fig. 7.1.

Nowadays, research on the control of MGs has been exhausted due to the strong penetration of renewable sources that have caused disturbances in the distribution lines due to their own intermittence. Therefore the main variables to be controlled in an MG with renewable energy sources (RES) interconnected are the frequency and voltage.

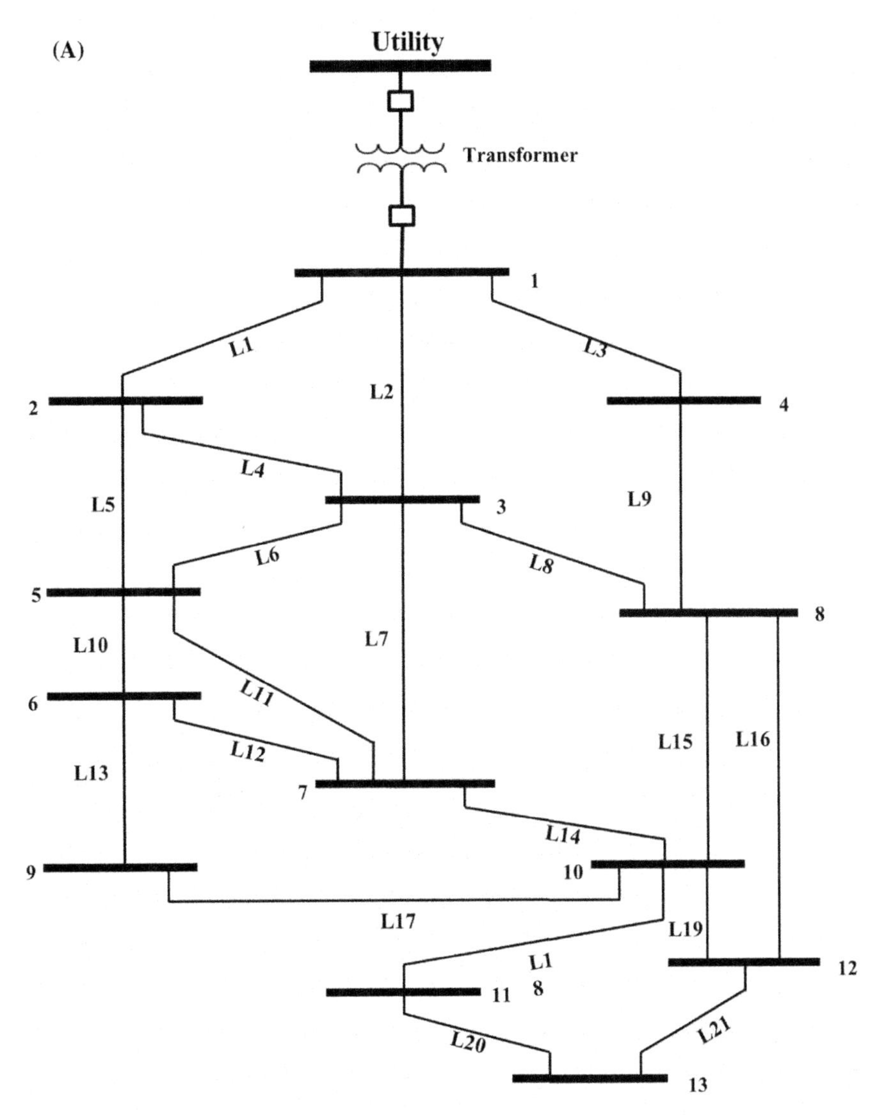

FIGURE 7.1 Mode of operation of a microgrid: (A) interconnected to the grid, (B) isolated from the grid.

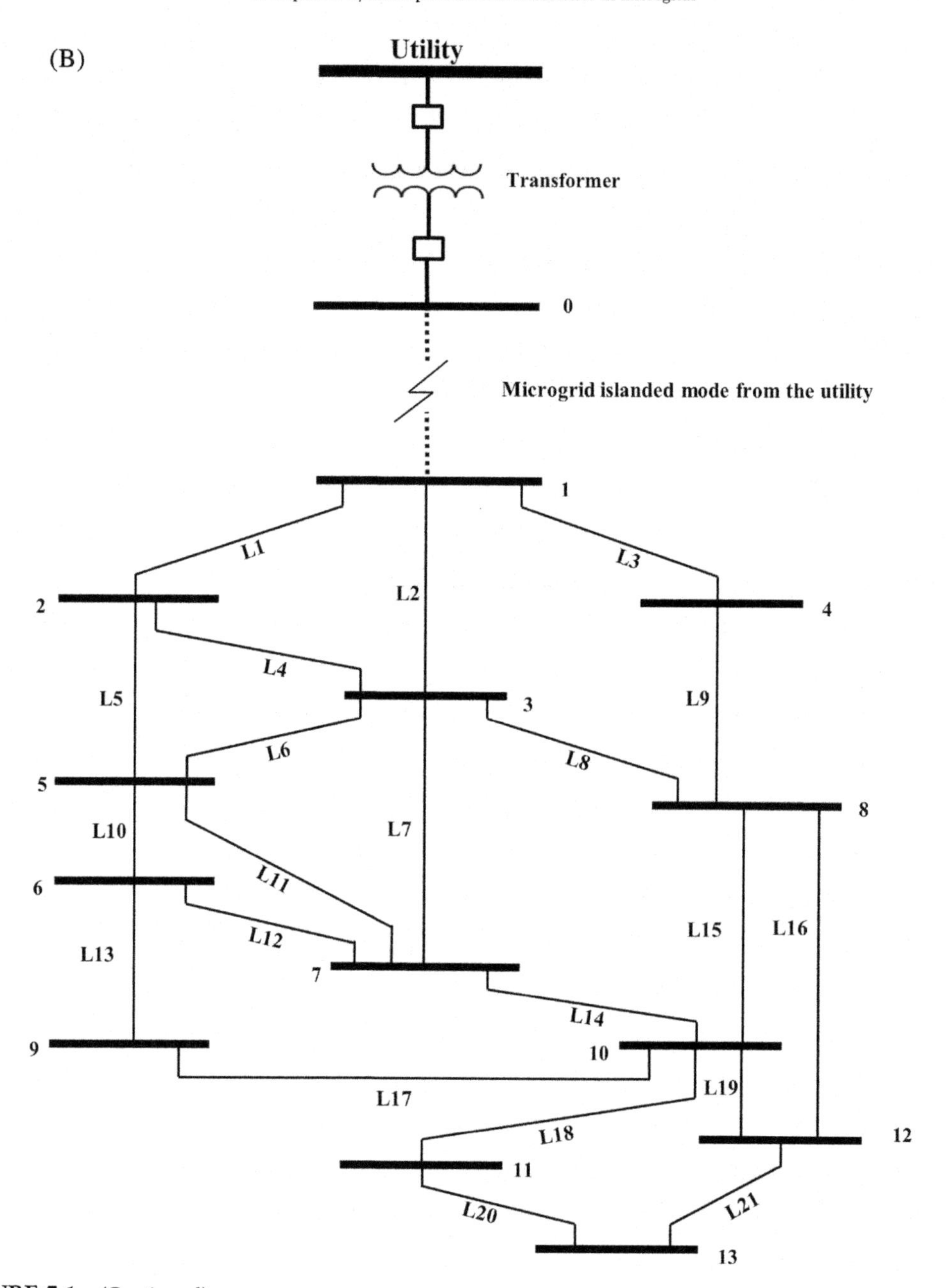

FIGURE 7.1 (Continued).

The reported works represent a trend towards self-sufficiency of the electric grids for the best use of intermittent energy, and an isolated operation architecture of the electric grid.

In the following Table 7.1, some works that have been carried out on MGs and their modes of operation are mentioned.

TABLE 7.1 Applications with microgrids (MGs).

	Operation mode/ system	Application/contribution
1	Grid-connected 5 bus [1]	They carry out the optimal programming of distributed generation and storage devices in intelligent networks, where each dispatchable device in the system finds its optimal schedule in equal-equal coordination with the neighboring devices.
2	Grid-isolated [2]	Minimization of the cost of the economic dispatch using renewable energies (solar, wind, and storage devices).
3	Community [3]	They propose an energy management scheme to maximize the use of solar energy in an intelligent microgrid (MG).
4	Grid-connected 7 bus [4]	They propose a hierarchical control to maximize the surplus of the distributed generation (DG), the surplus of the load and the surplus of the DG and load at the same time.
5	Grid-connected 33 bus [5]	They propose a methodology to optimize the dispatch of the response to demand in conjunction with distributed generation to provide energy and reserve by an operator minimizing operating costs.
6	Grid-connected 4 bus [6]	Management of real and reactive powers of the DG units interconnected electronically in a MG. Identify and investigate three strategies: voltage regulation, voltage drop characteristic, and load compensation of reactive power
7	Grid-connected 54 bus [7]	Control strategy to manage the local active/reactive power based on neural networks, able to regulate the voltage profiles at buses where the DG is connected, considering the restrictions of the capacity curve
8	Grid-connected 20 bus [8]	They describe an inverter model and a simplification of a spinning machine in MGs to maintain frequency stability and voltage stability through a control to size the storage system.

In this chapter, we present the dynamics of batteries located in an MG when it is interconnected through a feeder with large impedance making the distribution grid itself weak. The impact of these storage systems is analyzed through the battery bank (battery energy storage systems, BESS). Under these operating conditions, the operation dynamics are weighted, considering the charge and discharge times, as well as a finite number of operations. To determine the location and amount of energy that will be required from the BESS, a plan is made a day earlier where the availability of conventional energy is observed in conjunction with the wind and PV systems against the energy demand in the grid. Subsequently, the resulting frequency variation is evaluated and a DMS is proposed to reduce the frequency variations.

7.3 Battery energy storage systems dynamic model

Currently, energy systems continue with conventional generation units to provide inertia through their storage in their generation plants. This means that when a new load is connected to the grid, the initial energy balance must be satisfied with the inertia of the system and this change generates a small variation in the frequency of the system. Likewise, the integration of PV and wind systems has also been the cause of this problem due to the intermittency of said sources. In order to mitigate this problem, it is important to use devices that allow us to improve the operation of MGs, taking advantage of the available energy from either the utility grid or renewable sources. In this context, one of the applications that has proven quite useful is the use of storage systems in batteries to charge and discharge energy at times of interest to the system, thus increasing its efficiency.

It has been shown in some works that storage devices can store energy supplied by renewable sources when the production or price of energy be high.

Energy storage systems (ESS) are classified into mechanical, electrochemical, chemical, electrical, and thermal energy storage systems:

1. *Mechanical storage systems*. The most commonly used types of storage in this system are pumped hydroelectric power plants (pumped hydroelectric storage, PHS), compressed air energy storage (CAES), and flywheel energy storage (FES).
2. *Electrochemical storage systems*. This includes different existing battery technologies of practical use in the market such as lead acid, NiCd/NiMH, lithium ion, metal air, sodium sulfur, and sodium nickel chloride.
3. *Chemical energy storage systems*. The storage of chemical energy is more directed towards hydrogen and synthetic natural gas, classified as secondary energy, since it can be stored in large quantities.
4. *Electrical energy storage systems*. These include dual-layer capacitors or super capacitors, as well as magnetic energy storage superconducting systems that work in accordance with the electrodynamic principle.
5. *Thermal energy storage systems*. These store the available heat in various means in an isolated tank for different applications and especially for the generation of electricity. Thermal storage systems (TES) are implemented to overcome the mismatch between the demand and supply of thermal energy and, therefore, are important for the integration of RESs.

The type of storage systems most often used in an MG is that of batteries, since this system can operate bidirectionally with the grid, either as a load for the system or as an energy source, and this helps to compensate between the demand and conventional generation.

BESSs have many advantages in an MG with respect to renewable sources called unregulated, since they have a variety of distribution applications such as: frequency regulation, voltage regulation, storage of energy for contingencies, minimization of losses in the grid, backup of energy in case the grid is disconnected, rapid response to abrupt changes in demand, support for wind systems and PV by the intermittency in its generation, and in the market sector for the hourly sale of energy when the utility requires it.

7.3.1 State of the art

Some works are being devoted to the study of storage systems by pumping water, using renewable energy for their operation as is the case of [9], where they carry out a feasibility study and load sensitivity analysis for water systems pumping water with PV systems and with storage devices (batteries) or a diesel generator to obtain an optimal configuration that achieves a reliable system. In Ref. [10] batteries, are applied to an electrical system to ensure the safety, stability, and economic operation of the grid. Wind turbines are used to minimize the cost of the electricity grid with the BESS.

Nowadays, effort is being directed with regards to the development of wind power plants in remote rural communities, which are known as isolated electrical systems. These systems can be integrated with the cogeneration of diesel generators and wind turbines; however, these devices pose several technical challenges related to load balancing and frequency control. For this, it is necessary to include devices such as battery accumulators and flywheels of inertia to ensure a more stable operation and a higher level of penetration of wind energy is integrated into this system.

Energy storage systems have been used for different applications, both in MGs and in buildings as in Ref. [11], where they propose a multiobjective optimization model based on a demand response that integrates different sources of energy, such as PV panels, grid energy, combined cycle generation, storage devices such as batteries (BESS), and TES.

Currently, battery technology has been at the forefront of storage systems to improve performance and rapid response. In Europe, consumers have been using lithium-ion batteries to support PV systems with a rapid response in generation, and even for the price of energy for the purchase and sale of the same. As in Ref. [12], where they use a mixed integer problem to optimize the storage of batteries, considering actions of 15 and 60 minutes. In addition, they consider the limitation of a certain number of battery cycles to take into account their useful life, as well as a sensitivity analysis to identify the variability of the prices necessary to obtain profits in the operations of the expertise; so they implement a process of market prices with adjustable parameters. For each period they assume knowledge of future prices and do not consider additional operating costs or immediate energy costs.

Another application of BESS is the compensation of the errors in the forecasts of energy production that are in the market several hours before, as well as the optimization of the power and size of these devices operating in conjunction with the wind turbines as in Ref. [13] where they propose two strategies: in the first, the average wind power is considered as the power of dispatch (the one being generated) to minimize the capacity of the battery, and in the second apply two backup battery assemblies to avoid prolonged charging and discharging cycles where in each interval one of them is only charged and the other is only discharged, and thus make the battery life more efficient.

7.3.2 Model battery energy storage systems

The BESS model commonly used in Refs. [14–16] is determined by the states of charging and discharging in time intervals, considering some restrictions for its dynamics as maximum charge, minimum discharge, amount of charge-discharge power and times charging-discharging.

The BESS model is connected in the nodes according to the energy requirement for each of them and the model of the charge-discharge of the battery storage system is represented as follows:

$$P_{BAT}(t) = P_{BAT}(t-1) + \left[P_{Charge}(t) * \eta_C - P_{Discharge}(t) * \frac{1}{\eta_D}\right] * \Delta t \tag{7.1}$$

where η_C is the charge efficiency and η_D is the discharge efficiency. The efficiencies are related to the environmental temperature, the depth of charge, the depth of discharge, internal resistance and so on. The period of time is t; $P_{BAT}(t-1)$ is the power of the storage system on one preceding time point; $P_{Discharge}(t)$ is the released power in t period; $P_{Charge}(t)$ is the stored power in t period and Δt is the time interval.

The available energy of the BES system is limited by the storage power as represented in the following equation:

$$P_{BAT-min} \leq P_{BAT}(t) \leq P_{BAT-max} \tag{7.2}$$

where $P_{BAT-min}$ and $P_{BAT-max}$ are the minimum and maximum powers of BESS that you can have at each instant of time t.

During one time period t, the BESS can only be in one state:

$$\begin{aligned} P_{BAT-soc} &\geq P_{Charge}(t) \\ P_{BAT-soc} &\leq -P_{Discharge}(t) \end{aligned} \tag{7.3}$$

$P_{BAT-soc}$ is the states of the batteries of charge and discharge.

Also, in the BESS, it is important to consider both the times and the amount of power charged and released in each of the storage devices, according to the different existing technologies on the market.

Therefore in a state i the amount of power in each defined time interval is represented as follows:

$$\begin{aligned} P_{Charge-SOC}(t) &= \int_{t1}^{t2} P_{BAT} dt \\ P_{Discharge-SOC}(t) &= -\int_{t1}^{t2} P_{BAT} dt \end{aligned} \tag{7.4}$$

In some works, the aging of the batteries has been considered in order to optimize the duration or useful life of the different technologies of the BESS. Considering the above, the following equations represent the charging or discharging time of a BESS in a cycle of 24 hours.

$$\begin{aligned} T_c &= t_p * N_{charge} \\ T_d &= t_p * N_{discharge} \end{aligned} \tag{7.5}$$

Therefore the time constraint in which a BESS must be discharged and charged in a 24-hour cycle is determined by the following expression:

$$\begin{aligned} T_{ini} &\leq \sum_{t_p=1}^{Np} T_c \leq T_{fin} \\ T_{ini} &\leq \sum_{t_p=1}^{Np} T_d \leq T_{fin} \end{aligned} \tag{7.6}$$

Where t_p is the time at each moment during the 24 hours, N_{charge} and $N_{discharge}$ are the charging and discharging operations, T_c and T_d are the time of charging and discharging throughout the cycle of 24 hours, N_p the period of 24 hours, T_{ini} and T_{fin} start and end time respectively, which is the 24-hour period.

7.4 Selection of strategic points of the battery energy storage system to improve the regulation of energy in a system

With the high penetration of RESs in a distribution grid with conventional generation, the problem of frequency variation in the electrical system is presented, and the control of it becomes more complicated. Currently some research work has been geared towards regulating this frequency with the interconnection of BESS to solve this problem. However, in order to interpose the battery devices in an MG it is important to consider the sizing and proper location of the BESS, since otherwise this would impact on a storage capacity that is not necessary at the point of interconnection and consequently at very high economic costs.

Commonly the distribution grids are connected very close to the power substations (utility) or conventional sources ringed through normally open links with other substations, so, in case of any disturbance or failure in a circuit with a certain charge, it is possible to transfer that load to another circuit of the same substation (source) and even transfer that charge to another neighboring substation.

However, when an MG is very far from the substation and only has a main feeder with high impedance, this grid can be considered a weak MG. This grid becomes sensitive to any variation in voltage or frequency and could be at some point in island mode due to the disconnection of the same by some greater contingency. In the same way, we can say that within an MG there are also weak nodes, as well as instances in which the main feeder of that node is also powered by a driver with high impedance.

To operate a weak MG or in island mode requires the support of RESs interconnected according to the demands in each node and controlled by a DMS. Having intermittent renewable sources in an MG results in the variation of frequency and voltage, which is why it is necessary to intervene in battery storage systems to smooth the demand curve due to the intermittency of RES, to provide support in the event that the grid becomes island moded, to reduce the losses of the lines, and to correct the variation of frequency and voltage in the grid itself.

7.4.1 State of the art

In Ref. [17], they propose a two-stage stochastic optimization model to determine the optimal size of energy storage devices in a diesel-wind hybrid system, minimizing the total cost of the project, taking into account both the cost of operation and the cost in the diesel-wind power system.

Currently, several research projects are focusing on the location and optimal allocation of energy storage systems (ESS) in MGs as in Ref. [18], where they propose a methodology for the optimal allocation and economic functioning of ESS in MGs on the basis of a net present value. In this document, they propose a system with several optimal DG and energy storage devices available.

In Ref. [19], they focus on optimizing the size and location of the BESS to reduce the variation of the RES on the distribution side and control the frequency on the transmission side in the event of a failure. They are based on a heuristic method for the solution of optimal locations and capacities of BESS. On the storage side of transmission, they use two tools: Time Domain Power Flow (TDPF) to detect the optimal locations of the batteries and Complex-Valued Neural Networks for the sensitivity study. Combining TDPF and economic dispatch (DE) leads to the optimal size of BESS. In the storage part in distribution, the optimum amount of BESS is calculated to cut the peaks of the demand, and to smooth the charge curve.

Collapse in an MG or lack of stability means that the grid can operate in island mode, and for this it is important that it depends on the BESS to avoid a contingency; Thongchart [20] presents a method to evaluate the optimal size of BESS at a minimum cost through the use of frequency control based on particle swarm optimization of the microgrid, to assess the economic performance with different BESS technologies and perform comparisons with a typical grid including BESS.

7.5 Distribution management system

Changes in the distribution grids caused by the increase in demand, the concern for the burning of fossil fuels, and the high penetration of RESs have caused these MGs to become more complex in their operation, making them more complicated for energy operators to manage. Currently, some companies design tailored solutions to manage energy distribution; however, they have turned out to be very expensive due to the complexity of manipulating the different variables. Nowadays, intelligent grids are considered first-generation distribution grids, since they are implemented with an intelligent management system to monitor and regulate their operation.

A DMS is a set of software components that monitor and control the grid in order to optimize the behavior of an intelligent electricity grid and, therefore, allow companies to supply energy in a more reliable, efficient, and safe way [21]. In addition, it reduces interruptions, minimizes interruption time, maintains acceptable voltage levels, and reduces line losses. The operational changes that occur in the distribution grid are due to the continuous variations of energy dispatch and demand. Thus, connecting a large number of DG plants can cause a problem of regulation of critical voltage and variation of frequency in the distribution grids [22]. This means that the operational topology of a grid has to change from fixed to dynamic, since users will be able to store the unused generated energy in batteries, becoming a bidirectional grid exchanging energy in both directions and for this it is necessary to include a telecontrol and telecommunications system in the entire distribution grid to allow it operate dynamically in real time.

In order to achieve this optimal operation in the grid, continuous monitoring in real time, efficient control, and economic dispatch through an intelligent distribution management system is necessary, which must have a comprehensive solution based on efficient

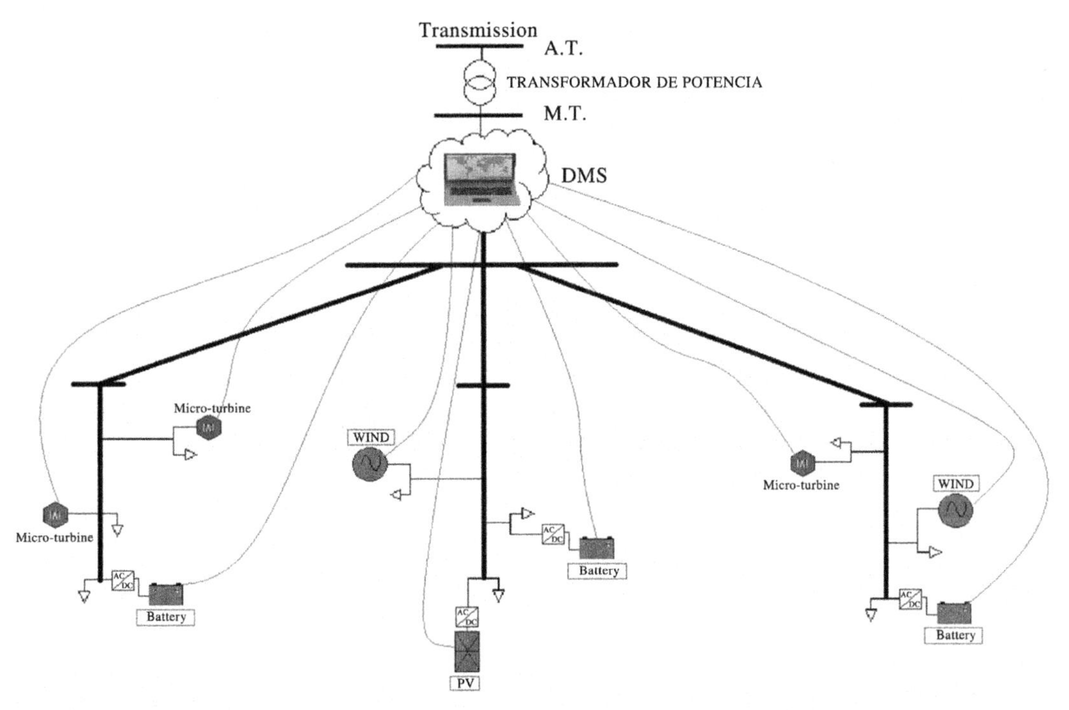

FIGURE 7.2 Diagram of a microgrid with a distribution management system (DMS).

algorithms and real-time information systems. These DMS algorithms for intelligent distributed generation systems must be calculated quickly, in order to analyze and make operational decisions in due time and in the appropriate manner [23]. A DMS system is shown in Fig. 7.2, where the red lines represent the communication between the DMS and each of the distributed generators.

7.5.1 Background

As mentioned before, electricity DGs have had a very important evolution in terms of the smart grid concept. These changes are mainly due to the substantial increase in distributed generation at a low scale (DGs) and to the operation in competitive environments in the market, which have led to the need to integrate entities such as Virtual Power Players (VPPs), which are generators and consumers. In Ref. [24], a methodology is proposed for a joint response to demand (DR) and distributed generation to provide power and reserve by a VPP that operates a distribution grid. The VPP [25] can add several types of distributed energy resources, such as DG, storage, DR, and electric vehicles. In Ref. [26], a totally distributed programming methodology is proposed based on an optimal control in discrete time called, primal-dual gradient descent. The goal is for the control center to be eliminated and for the optimal program for all devices to be found only through the iterative coordination of each device with its neighbor.

Some investigations propose a control of MG time intervals, as in Ref. [27] where they develop a new approach for the optimal energy management of electrical distribution in intelligent distribution grids. The novelty of the proposed approach lies in considering the optimal programming of the generation units of an automatic planning process in a dynamic, nondeterministic and not-totally-observable environment, which approaches the real conditions.

7.6 Control strategy with a distribution management system

The objective of a DMS in an MG is to control adequately all the variables that influence the proper operation of the distribution system, considering the behavior of the loads, the conventional generation, the operation of the renewable sources interconnected to the grid, and the availability of battery storage systems, to meet the criteria of quality, reliability, continuity, safety, and sustainability in the operation and control of the Electric System.

It is very common for distribution systems (medium voltage) not to have an energy management system within the grid itself, so it is herewith proposed that a control strategy should be included with a DMS to minimize the energy produced by conventional generators and maximize the storage energy in the batteries so it can be used in the moments when the grid requires it, or to store energy in the batteries when it is available through the grid itself.

The decisions made by the DMS in a MG are represented in the following proposed flow diagram (Figs. 7.3–7.5).

7.7 Results

To carry out the BESS control tests with the DMS, a distribution system consisting of a substation (conventional generation) was first considered, which provides active power with a capacity of 1100 kW and feeds an MG of 13 nodes. Node one is considered a slack node and the others as charge nodes, which have different charge profiles for each bus. The losses of the grid are taken into account for this system. The impedances of the distribution grid lines are marked in Fig. 7.6. This MG is made up of RESs such as PV and wind turbines with capacities and location in different nodes, as shown in Table 7.2.

Six different wind profiles and six different solar irradiation profiles were considered, to determine a single profile for each source located in the corresponding nodes. This takes into account that the profiles of the RES do not demonstrate the same behavior during the day.

The demands that were considered for this MG are shown in Table 7.3.

It starts with a day's energy planning to identify the weak nodes of the system, and thus determine the capacity and location of the batteries within the distribution system, and the time periods every 5 minutes for 24 hours.

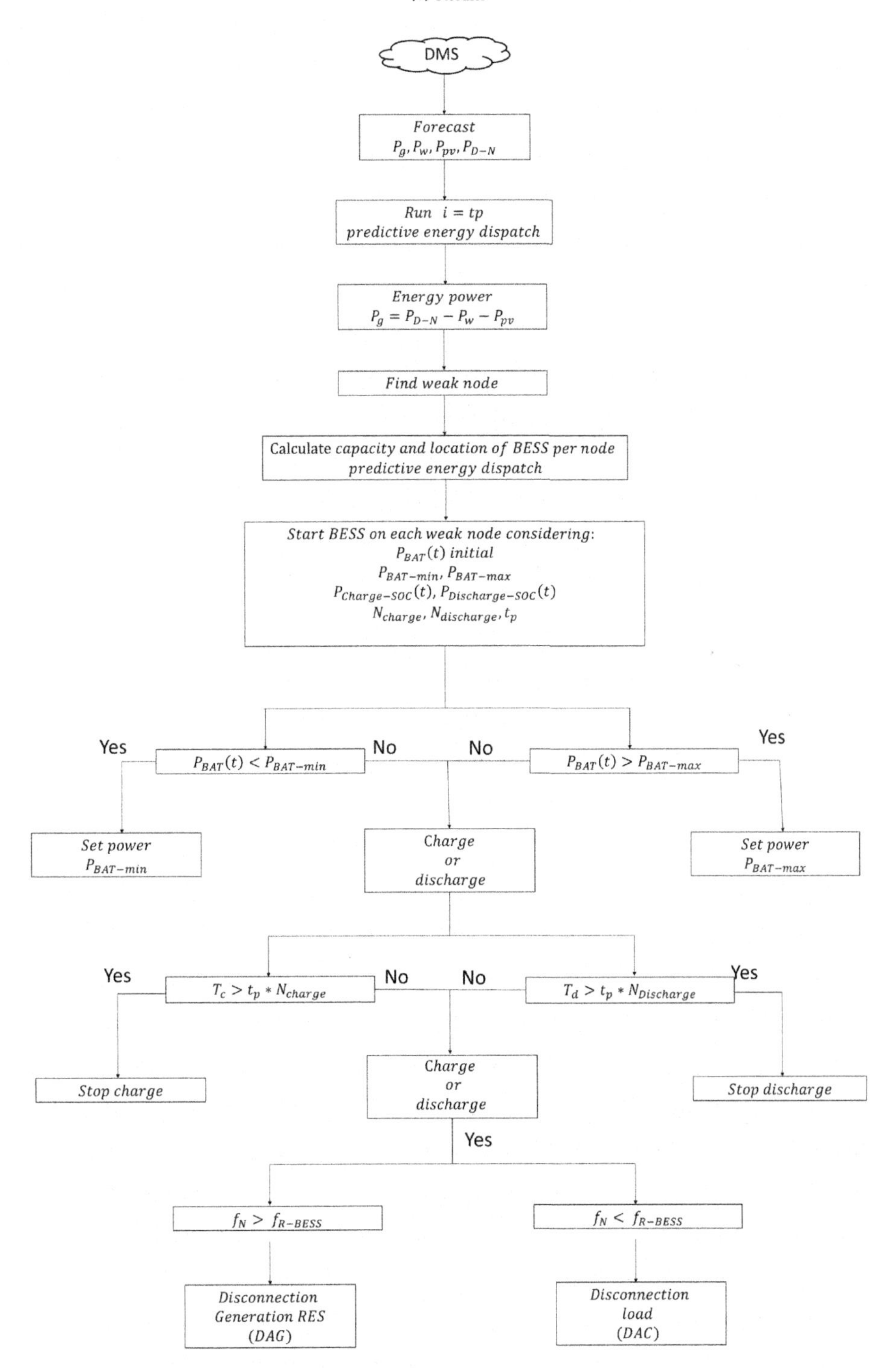

FIGURE 7.3 Distribution management system (DMS) flow chart.

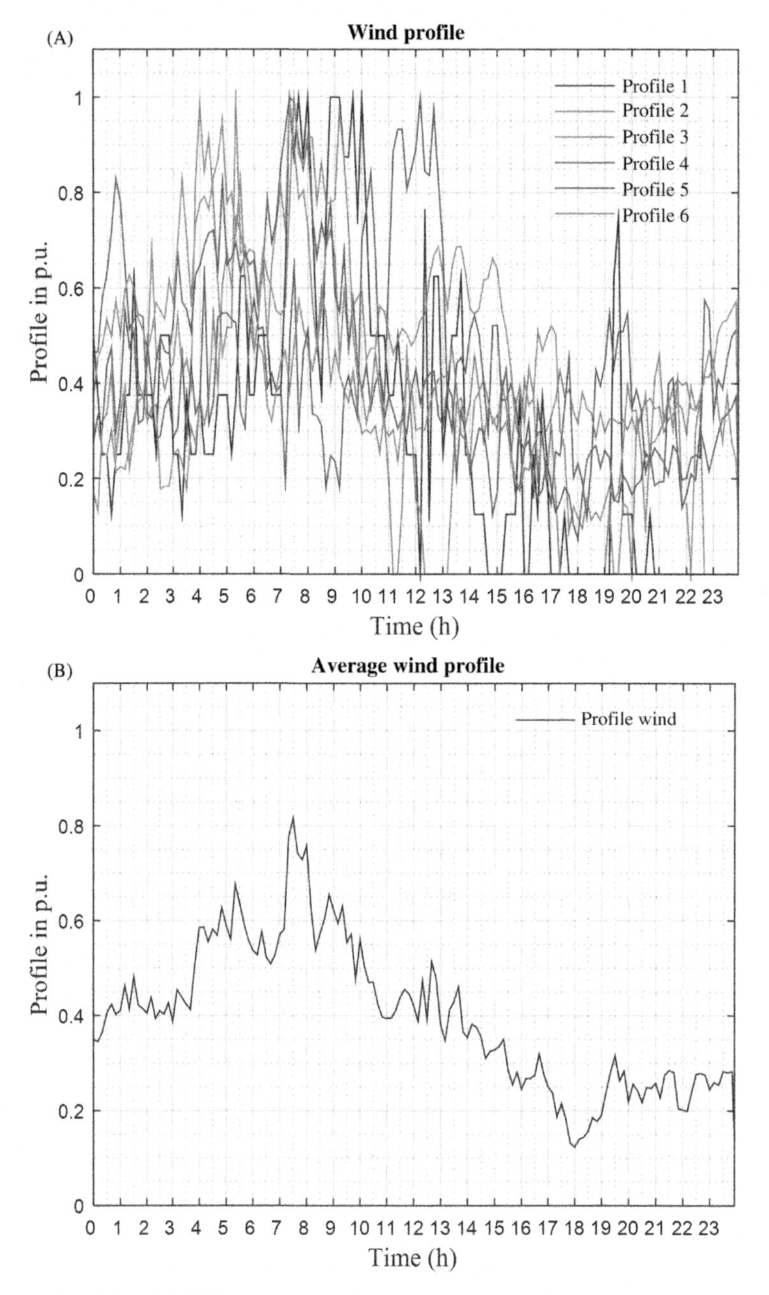

FIGURE 7.4 Wind profiles: (A) different profiles, (B) average wind profile.

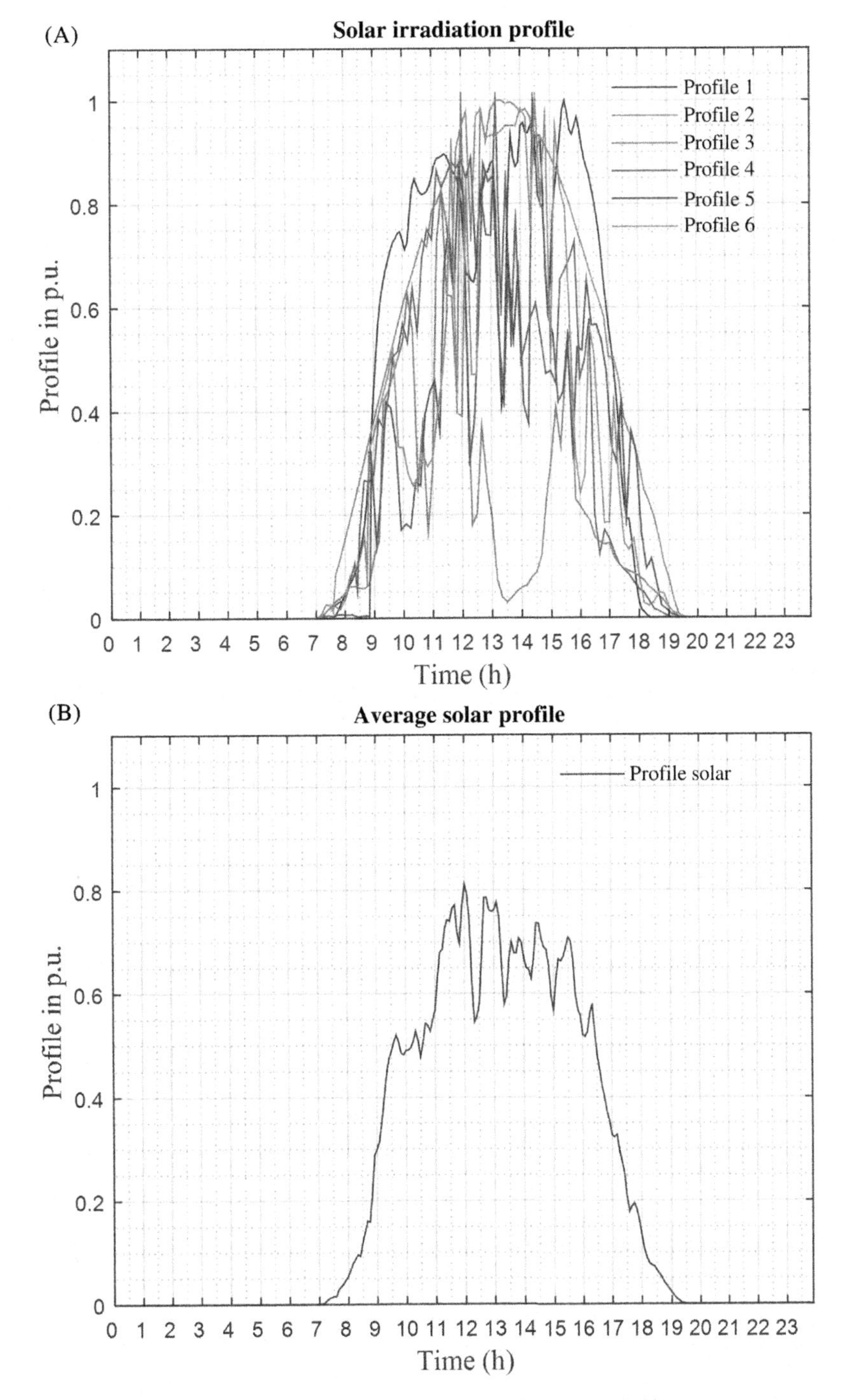

FIGURE 7.5 Solar irradiation profiles: (A) different profiles, (B) average profile.

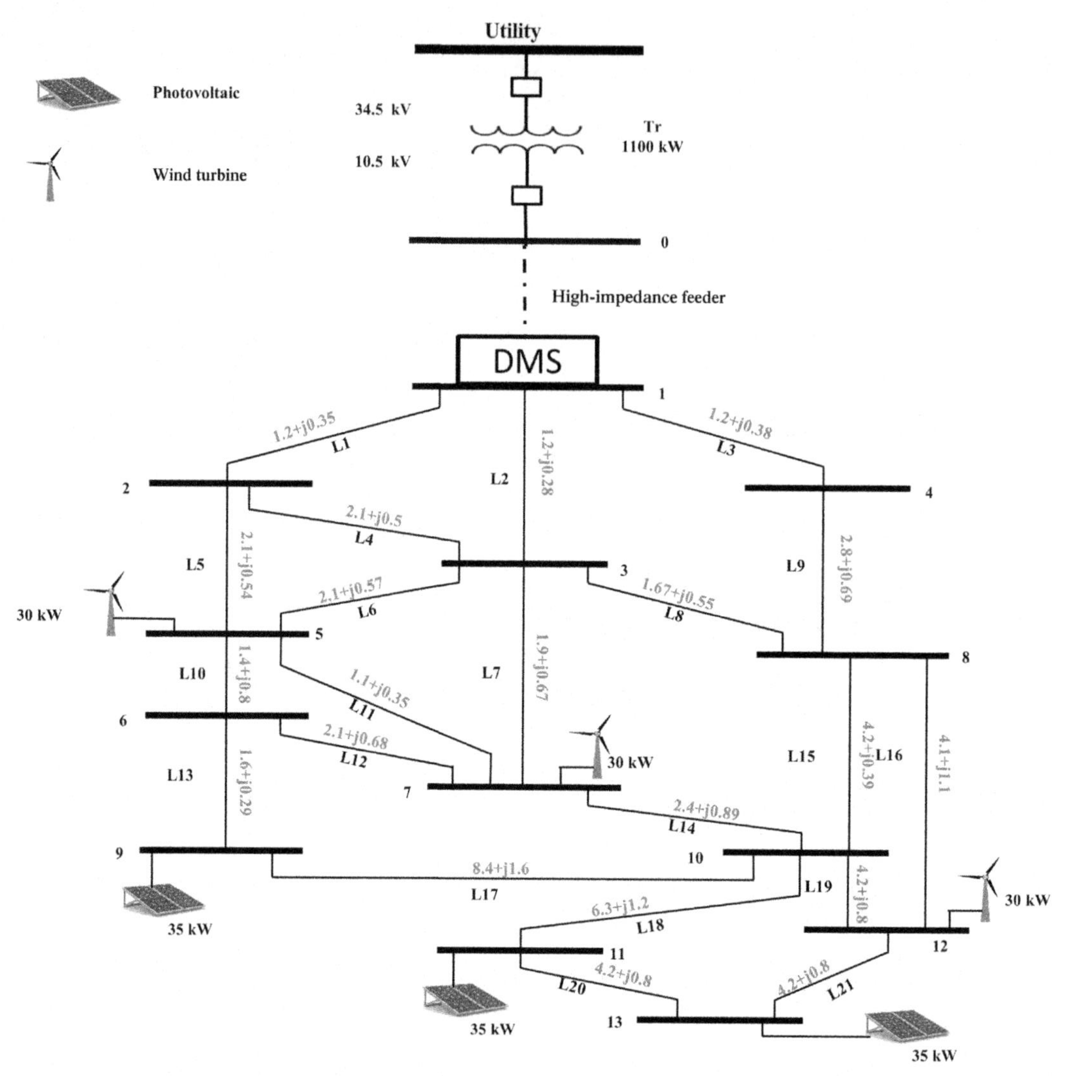

FIGURE 7.6 Distribution grid with renewable energy sources (RESs) and high impedance.

TABLE 7.2 Power capacities of the three MGs installed in the proposed grid.

Source (kW)	$N-5$	$N-7$	$N-9$	$N-11$	$N-12$	$N-13$
P_{wind}	30	30			30	
P_{pv}			35	35		35

TABLE 7.3 Loads in each of the nodes.

Node	$N-2$	$N-3$	$N-4$	$N-5$	$N-6$	$N-7$	$N-8$	$N-9$	$N-10$	$N-11$	$N-12$	$N-13$
Load	115	113	103	99	102	89	85	92	106	98	100	108

7.7.1 Microgrid connected to the main network with high impedance

Fig. 7.6 shows the distribution grid with high impedance and renewable sources interconnected in different nodes.

The following graphics show the demand and total generation of the system in weak MG mode, as well as the demand profiles in each of the nodes (Fig. 7.7).

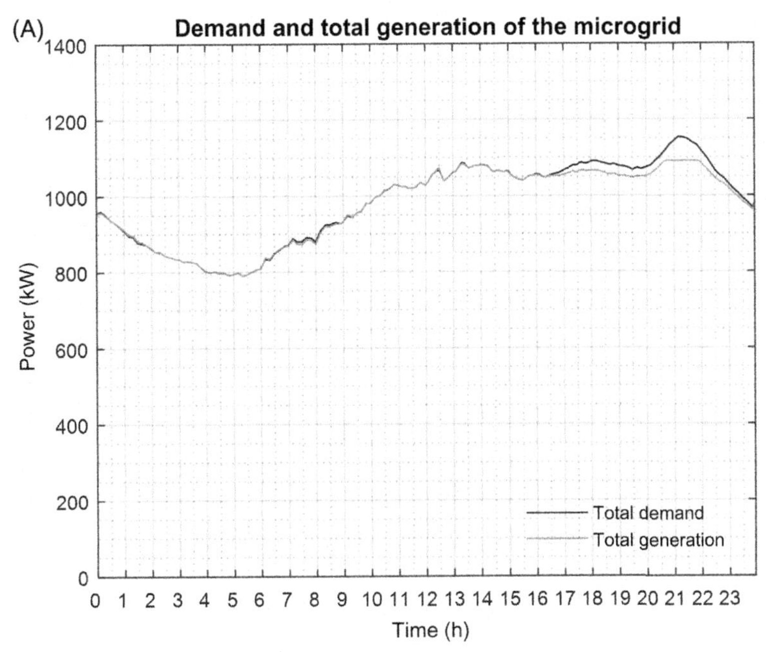

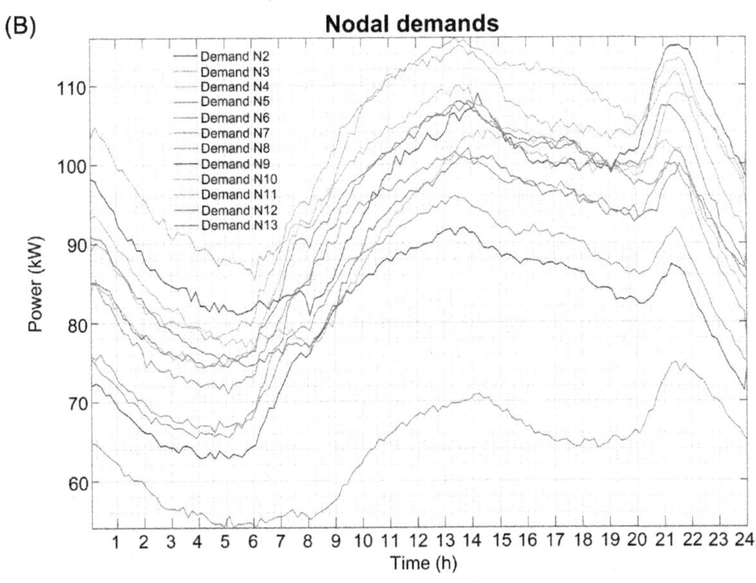

FIGURE 7.7 Behavior of the microgrid (MG): (A) generation and demand, (B) nodal demands.

7.7.1.1 Nodal simulations

Considering the aforementioned data and the grid with high impedance, the behavior of the grid in the nodes where the BESS are required is presented in Fig. 7.8.

As can be seen in Fig. 7.8, the energy required is in nodes 5, 7, 8, 9, 12, and 13 respectively, and the capacity of each of the BESS is determined by the maximum energy required in each one of the nodes as seen in the Table 7.4.

The location and capabilities of the BESSs are performed in the corresponding nodes in order to apply the dynamics in them. In this work, the following battery restrictions apply.

The capacity of the batteries is defined by the requirements of the planning the previous day, as shown in Fig. 7.8. In this case, the charging time (T_c) and discharge time (T_d) are defined as 480 minutes (8 hours) and 240 minutes (4 hours) respectively during a period of 24 hours, and depend on a number of charges and discharges during the day. This means that if the batteries operate the 480 minutes load distributed in a full day, they will stop charging even if the 24-hour period has not ended. And vice versa: if the discharge time defined by the technology type of the batteries is met, they will stop discharging even if the 24-hour period has not been completed. However, if one of the two conditions is met, the other will follow its dynamics until completing its operations or until the end of

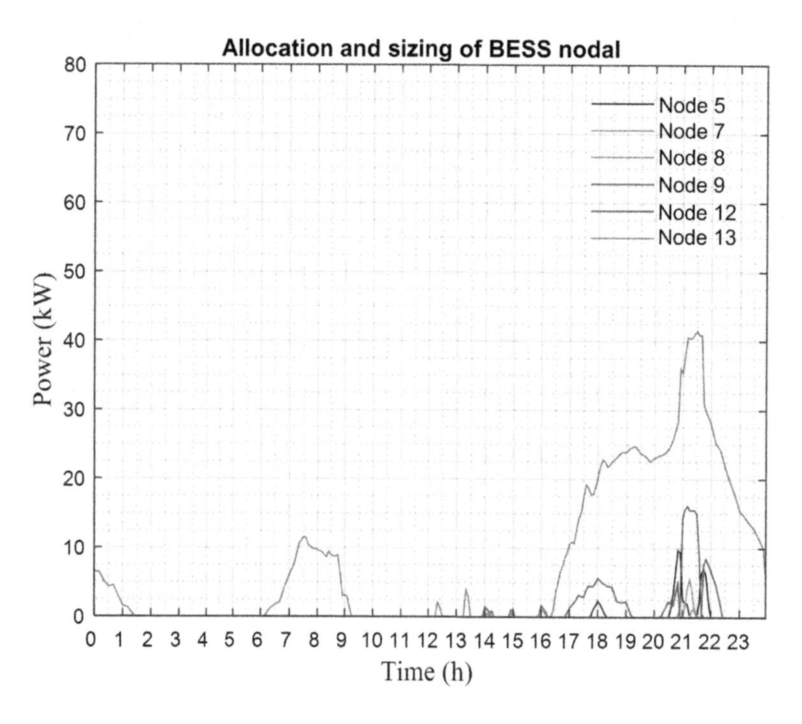

FIGURE 7.8 Location and sizing of the battery energy storage systems (BESS) nodales.

TABLE 7.4 Energy capacity required per node.

Node	5	7	8	9	12	13
Capacity (kW)	9.6572	5.54	1.18.21	16.1771	8.4442	41.4924

the full day with its dynamics. On the other hand, each time interval t_p, is defined by 5 minutes and the amount of charge and discharge was defined at 5 kW each time interval t_p. These parameters can change according to the type of battery used.

Some studies has been investigating the useful battery life by focusing on the depth of charge and discharge in them. In this model of batteries, a minimum amount of discharge and maximum charge is applied that can be varied depending on the type of batteries. In these simulations, the minimum and maximum storage capacities can be different in each node, according to what is required.

To observe the dynamics of the batteries in this scenario, we first choose node 5. Considering that the real demand has variations in time, we have the following:

In this node, it is observed that, due to the high impedance of the distribution grid, the energy balance is not met at certain times of the day (Fig. 7.9A). When including batteries with the aforementioned restrictions, it results in the following. It shows that the batteries operate by charging and discharging during the interval of 24 hours (Fig. 7.9B). It is observed that at approximately 12:30 hours, the BESS complies with the discharging operations and stops operating in that mode, and continues carrying out charge according to the same dynamics until the charging operations have been completed. For this nodal case, 96 charging operations were obtained ($T_c = t_p * N_{Charge}$), making a total of 480 minutes charge operation time and 48 discharging operations ($T_d = t_p * N_{Discharge}$), with a total discharge time of 240 minutes, throughout the 24-hour interval. It started with a capacity of 10 kW batteries, with a minimum discharge capacity of 10 kW and a maximum charge capacity of 150 kW (Fig. 7.10).

The following figure shows the contribution of BESS in node 5:

For this case the variation of the frequency is observed in the following graph:

In Fig. 7.11 the variation of the frequency in the indicated node is presented. Once the BESS is placed in the node to regulate the energy balance, since the battery's charging and discharging operations complied with the values set in the model, the BESS ceases to operate and the energy contributed by the BESS does not meet the energy balance at the end of the 24-hour interval. In this case, frequency variation is reflected in this node.

Considering the behavior in node 7 of the distribution grid, it is observed that the energy requirement in the node of Fig. 7.12A is approximately at 21 hours. Here, the dynamics of the battery remain in charged mode since the beginning of the 24-hour period since demand and generation meet the energy balance until 9:00 p.m., when the batteries are required to discharge. By no longer having an imbalance after discharging the BESS, it remains with that amount of energy until the start of the next 24-hour period.

In the case of node 7, the contribution of the BESS is minimal, as can be seen in the graph of Fig. 7.13. And the corresponding frequency curve is seen in Fig. 7.14.

As shown in Fig. 7.14, the variation of the frequency is minimal because the BESS covers the lack of energy in the node, as seen in Fig. 7.12B.

In Fig. 7.15, the behavior of the battery is according to the generation requirements to supply the demand in node 12. The BESS starts with the initial capacity set and stays for about 14 hrs, when it makes the power discharges to the grid. As the requirements are always of generation, the batteries reach their minimum discharge without covering all the amount of energy that is required. In this case, since there is no more energy available in this device, a charge disconnection must be performed, since the error difference

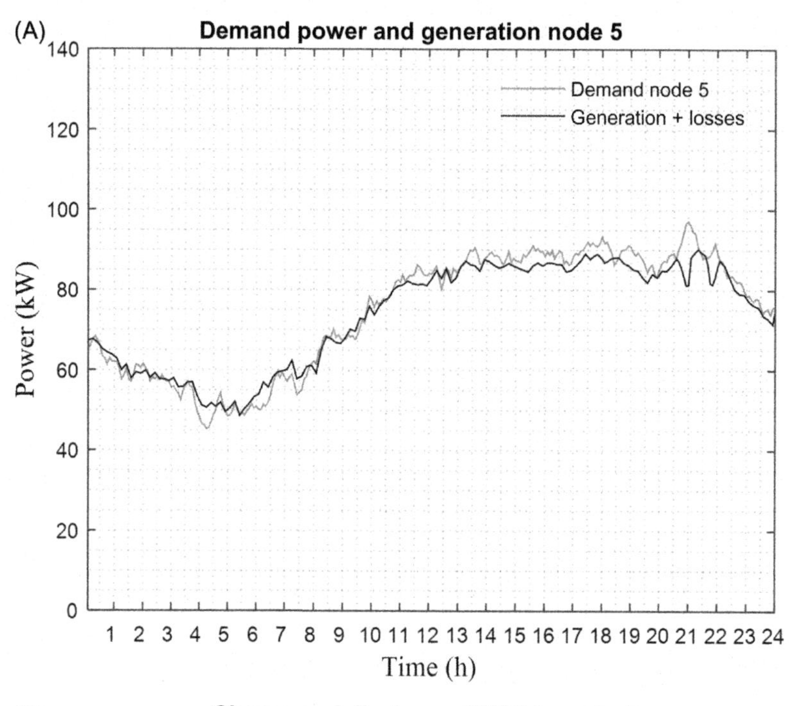

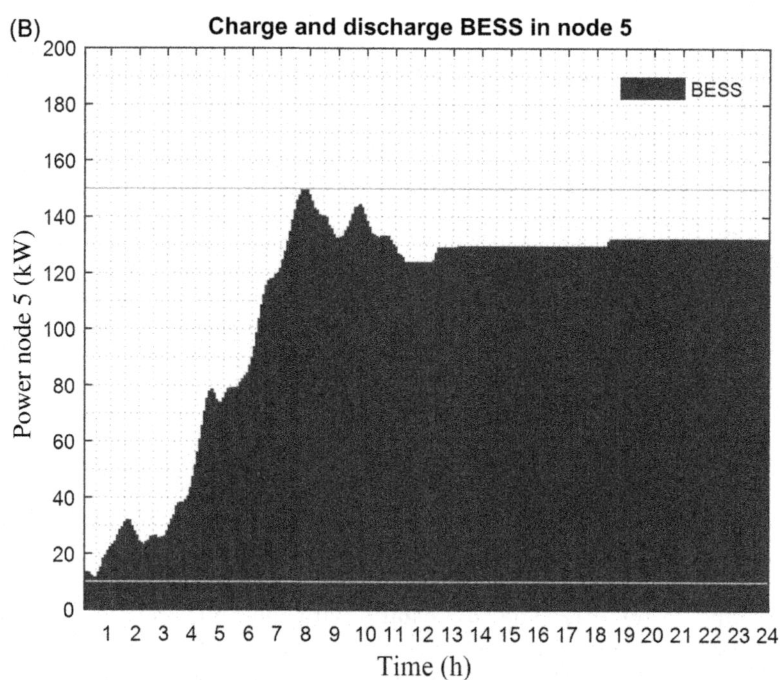

FIGURE 7.9 Dynamics node 5: (A) generation and demand, (B) battery behavior.

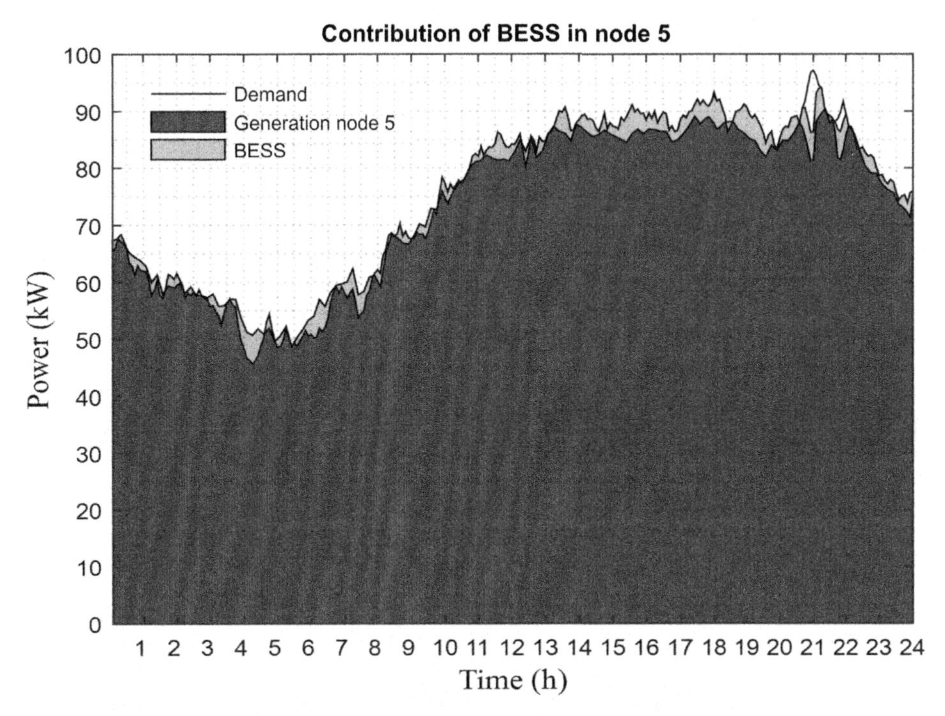

FIGURE 7.10 Contribution of power of generation and demand in node 5.

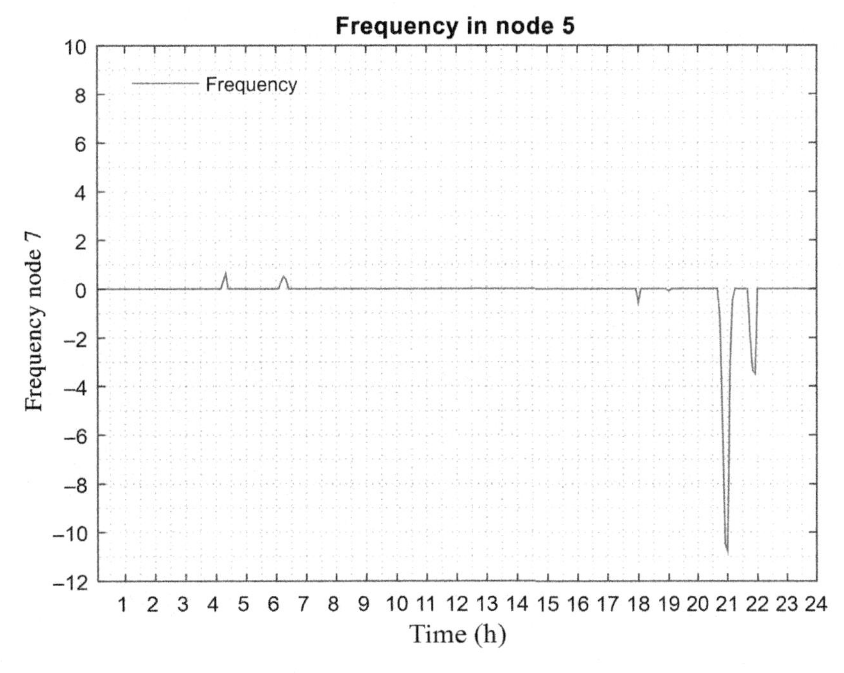

FIGURE 7.11 Frequency in node 5 with renewable energy source (RES).

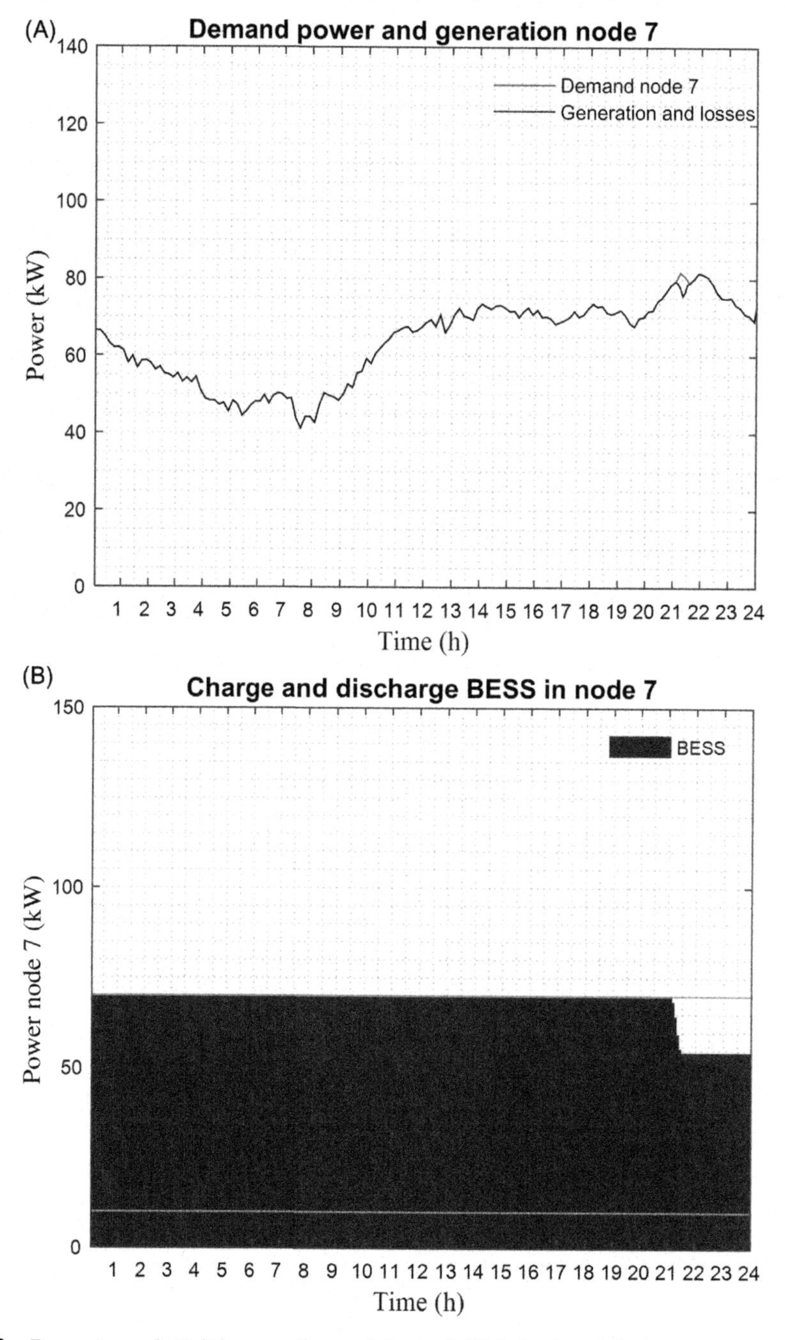

FIGURE 7.12 Dynamics node 7: (A) generation and demand, (B) behavior of the battery.

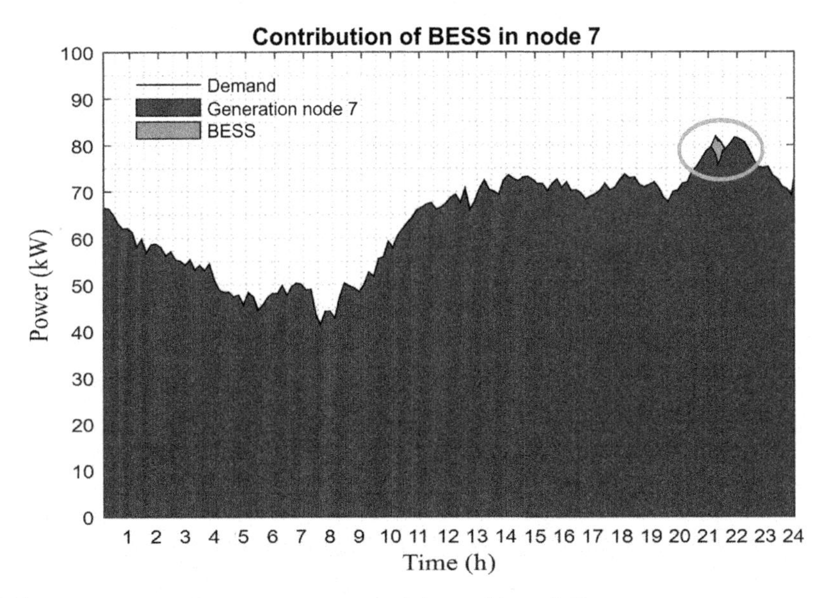

FIGURE 7.13 Contribution of power generation and demand in node 7.

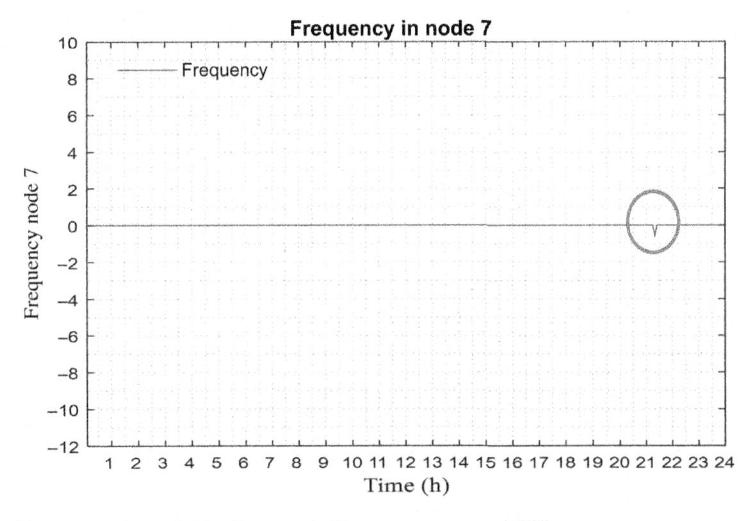

FIGURE 7.14 Frequency in node 7 with renewable energy source (RES).

between the total demand and the total generation is not zero and this implies variation of the nominal frequency of the system (f_N).

Likewise, the contribution of the conventional generation, the BESS and the demand curve in the aforementioned node are shown in Fig. 7.16.

In Fig. 7.17, the frequency is shown in node 12, where the variation of the same is shown due to the noncompensation of BESS in the node

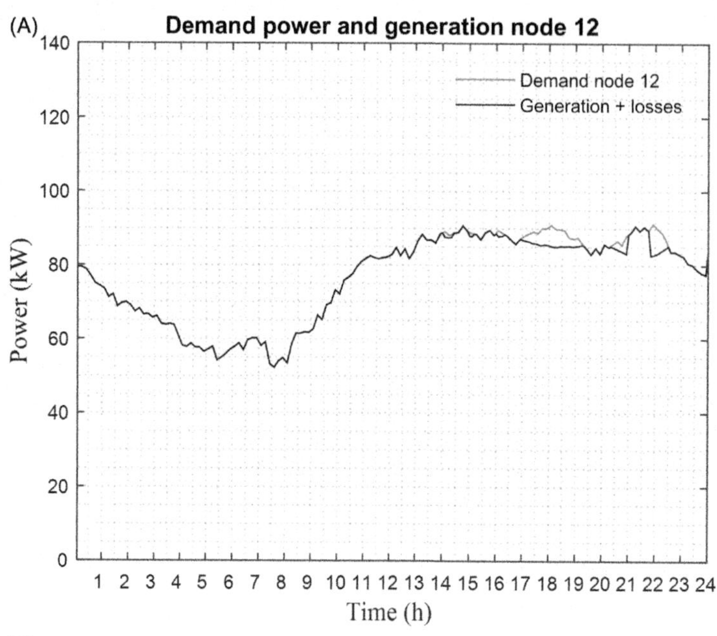

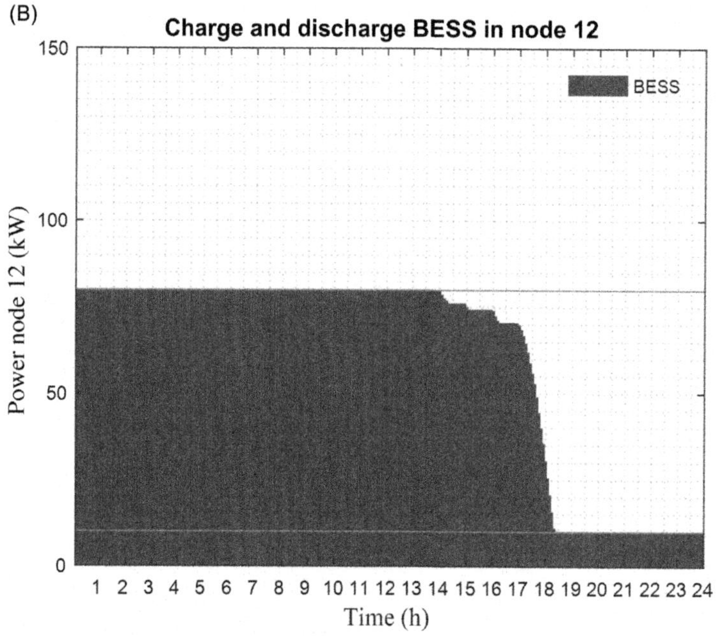

FIGURE 7.15 Dynamics node 12: (A) generation and demand, (B) behavior of the battery.

The highest energy requirement in the nodes of the proposed MG was node 13 in Fig. 7.18, in which demand and generation contributed positively and negatively throughout the 24-hour interval. This induces the BESS operating mode which charges and discharges that node. The charging time for this case was 205 minutes and the discharge

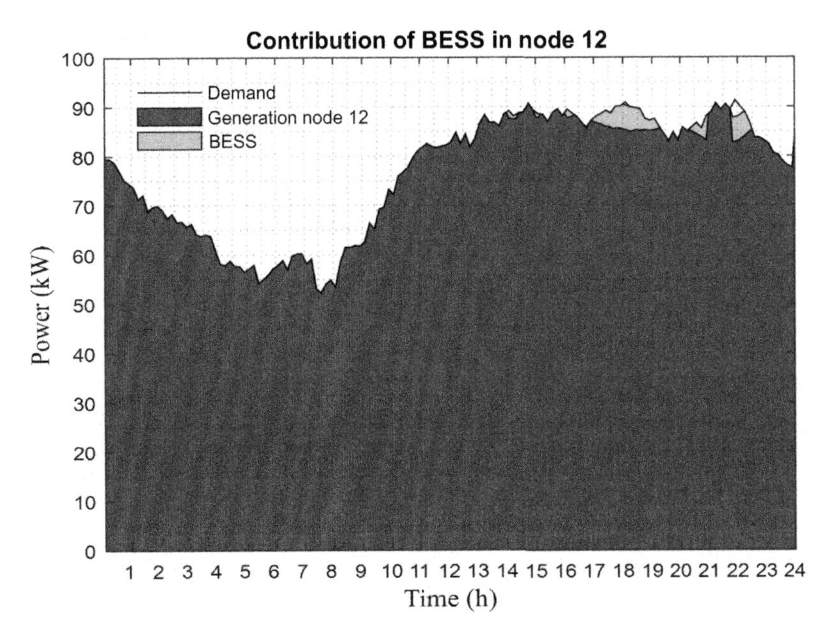

FIGURE 7.16 Contribution of power generation and demand in node 12.

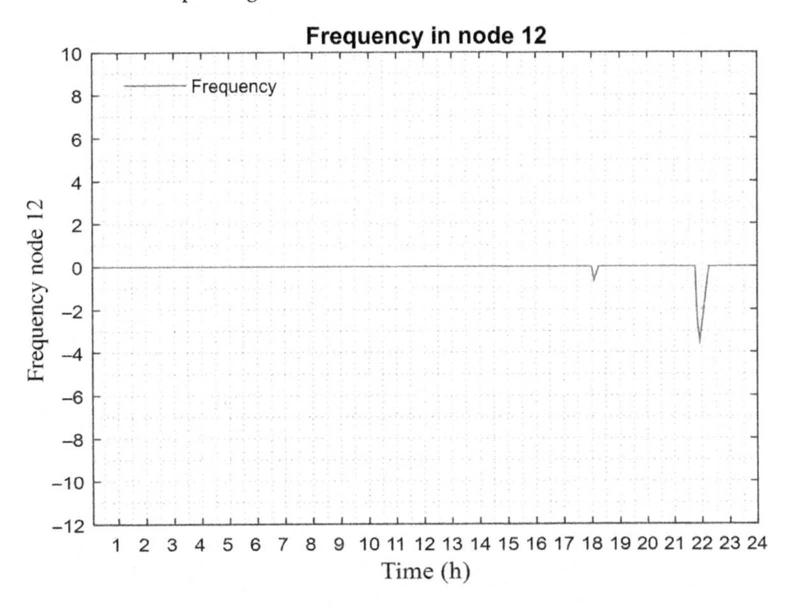

FIGURE 7.17 Frequency in node 12 with renewable energy source (RES).

time was 240 minutes, where this T_d, complied with the operations set in the battery model. So the discharge dynamics remained fixed after 17 hours.

As in the previous nodes, Fig. 7.19 shows the energy provided by the BESS with conventional generation and demand.

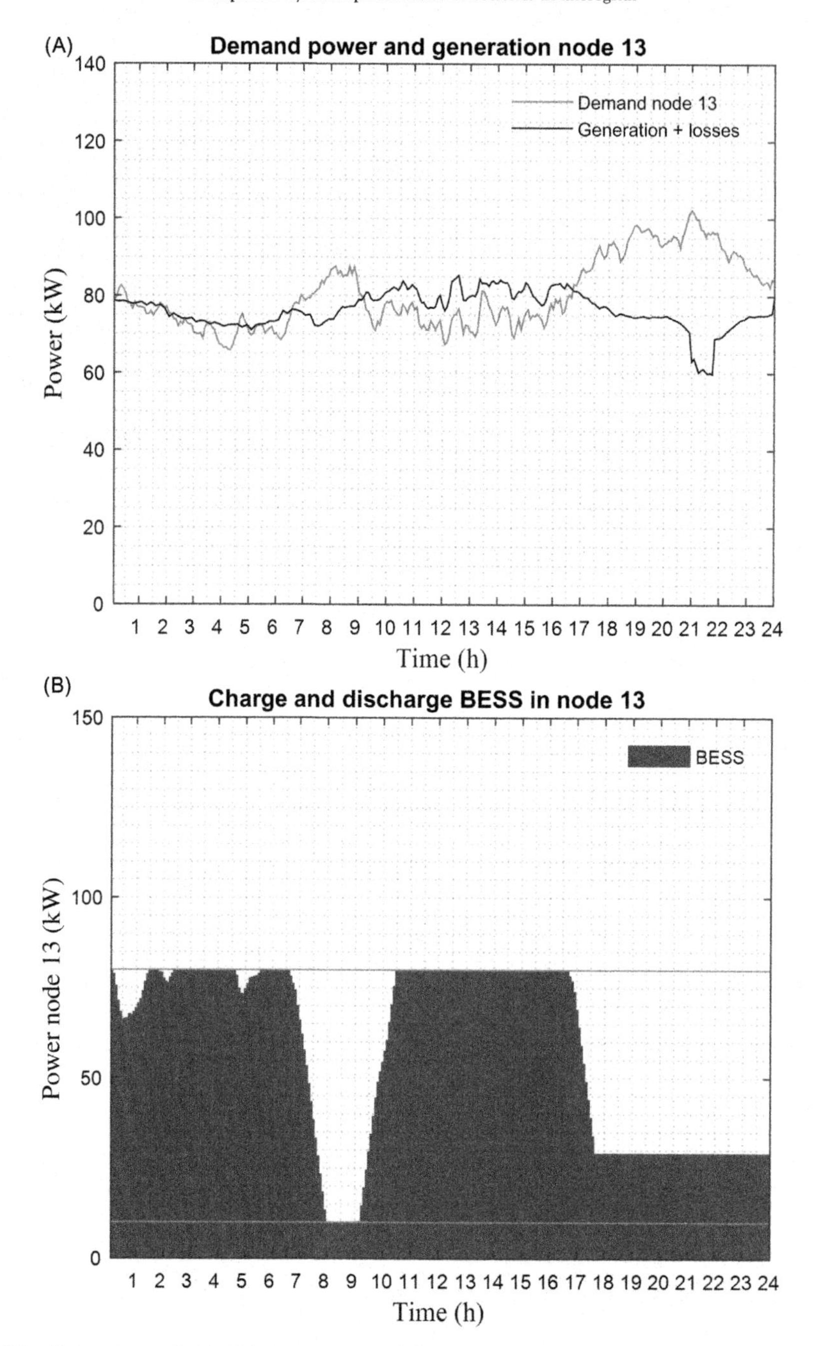

FIGURE 7.18 Dynamics node 13: (A) generation and demand, (B) behavior of the battery.

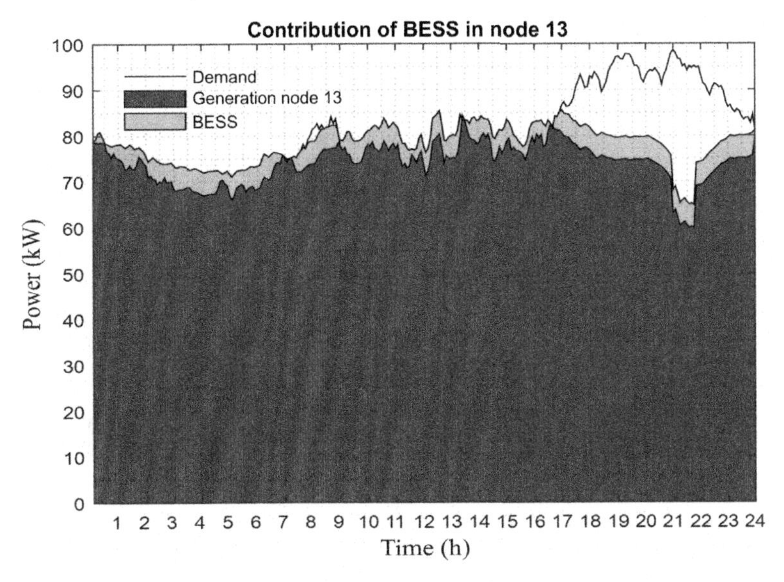

FIGURE 7.19 Contribution of power generation and demand in node 13.

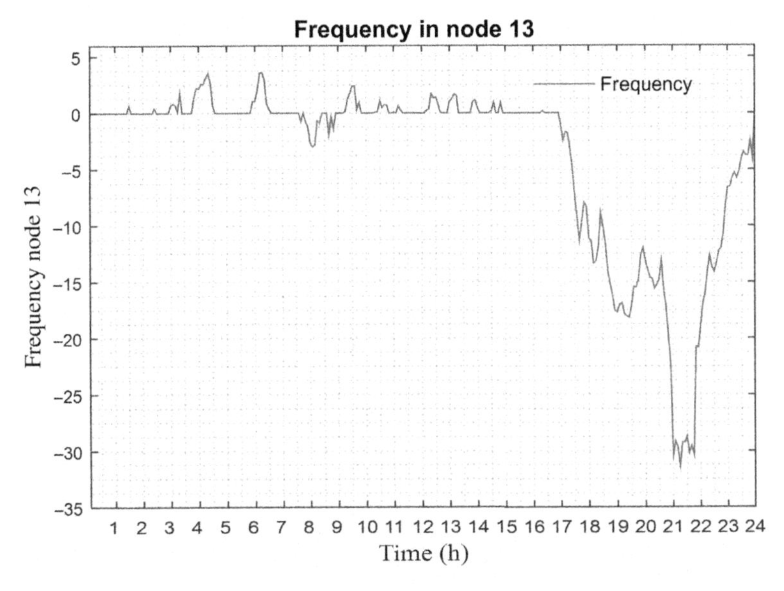

FIGURE 7.20 Frequency in node 13 with renewable energy source (RES).

Here the frequency has very variable dynamics with large negative magnitudes and small positivities (Fig. 7.20). When the demand increases, the frequency decreases in the same magnitude, and the DMS has to make the decision to cut the load in that node, so that the frequency takes nominal values.

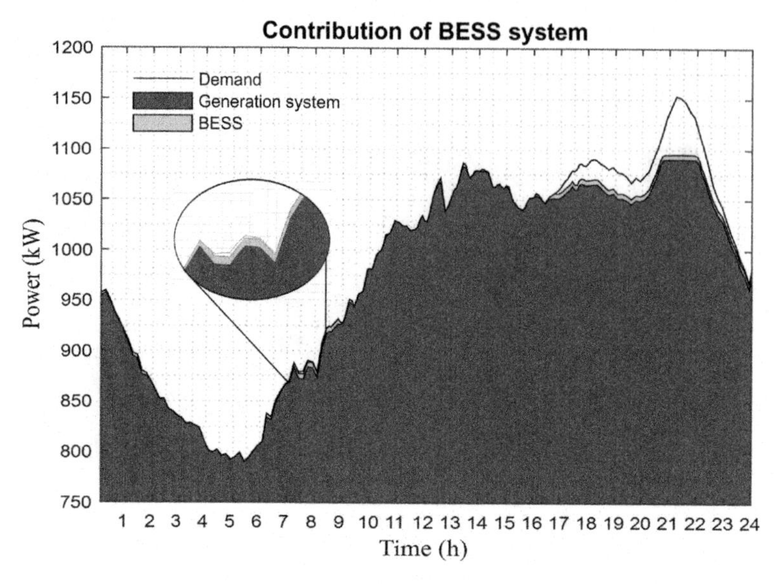

FIGURE 7.21 Contribution of power generation and demand in the system.

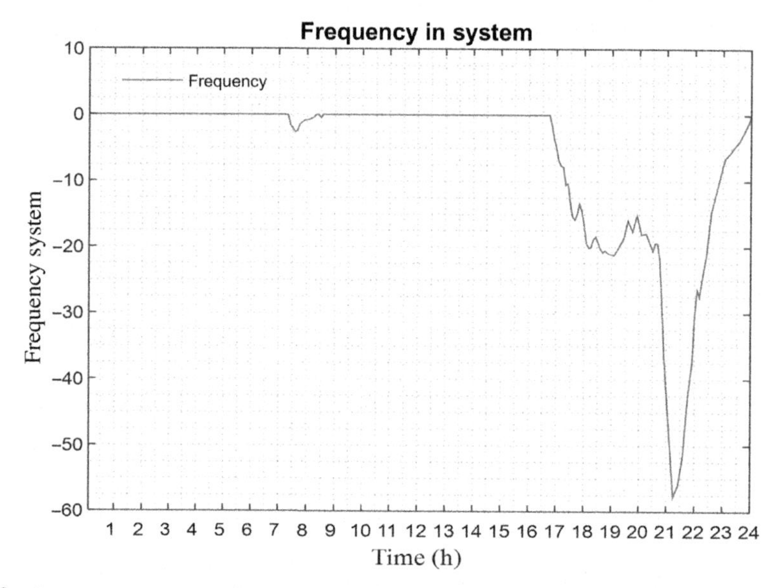

FIGURE 7.22 Frequency response of power generation and demand.

7.7.1.2 System simulations

The impact of BESS energy on the entire MG in general is presented in Fig. 7.21. Here is the contribution of the batteries in their entirety. It is observed that BESSs do not provide the necessary energy for the system to obtain the energy balance, so the DMS has to make appropriate decisions to level the frequency of the system. The frequency of the system as a whole is shown in Fig. 7.22.

7.8 Actions of the distribution management system

As observed in the previous simulations for the proposed MG, most of the nodes have energy requirements, so when applying the BESS to each of them, power stored in the battery is supplied and in some cases this is not enough energy and it is necessary to remove DMS decisions to the load-tripping scheme.

In this application, the performance of the DMS towards the BESS for the operation in each of the nodes is demonstrated. As seen in node 5, where the BESS charging and discharging operations are carried out before the end of the 24-hour period and therefore the load trip in that node is necessary. The same happens in node 12 and node 13, where the DMS decisions trigger the load because the batteries did not reach the energy balance, causing variation in the frequency. In the case of node 7, the energy requirement was very low and therefore the batteries managed to cover those power needs by balancing the frequency in that node.

7.9 Summary

Recent DMS research has focused on the management of distributed sources and, more specifically, on the control of frequency and voltage when interconnected with nonregulated sources. Today, this is a very important issue in electrical systems at the distribution level, since customers will also enter the game of buying and selling energy at different times. These customers play an important role in the electricity market because they have the potential to generate power from their own locations and send it to the grid when required, as well as being able to store surplus energy and sell it at convenient times for the system. This work showed the decisions made by the DMS on the interconnected batteries in each of the nodes where the frequency is very low due to the high demand in said nodes. We are assuming that the DMS is monitoring the behavior of the nodal frequency during all the time and, when this goes beyond the levels accepted by the interconnection code, the low-frequency-charge triggering scheme is enabled or, where appropriate, conventional generation or wind turbines provide input.

References

[1] N. Rahbari-Asr, Y. Zhang, M.-Y. Chow, Consensus based distributed scheduling for cooperative operation of distributed energy resources and storage devices in smart grids, Inst. Eng. Technol. 2016 J. 10 (5) (2015) 1268–1277.

[2] M. Bhoye, S.N. Purohit, I.N. Trivedi, M.H. Pandya, P. Jangir, N. Jangir, Energy Management of Renewable Energy Sources in a Microgrid using Cuckoo Search Algorithm, Conference on Electrical, Electronics and Computer Science (2016).

[3] W. Tushar, J.A. Zhang, C. Yuen, D.B. Smith, N. Ul Hassan, Management of renewable energy for a shared facility controller in smart grid, IEEE Access 4 (2016) 4269–4281.

[4] Y.S.F. Eddy, H.B. Gooi, S.X. Chen, Multi-agent system for distributed management of microgrids, IEEE Trans. Power Syst. 30 (1) (2015) 24–34.

[5] P. Faria, T. Soares, T. Pinto, T.M. Sousa, J. Soares, Z. Vale, et al., Dispatch of distributed energy resources to provide energy and reserve in smart grids using a particle swarm optimization approach, in: IEEE. IPP - Polytechnic Institute of Porto, Portugal, 2013.

[6] F. Katiraei, M.R. Iravani, Power management strategies for a microgrid with multiple distributed generation units, IEEE Trans. Power Syst. 21 (4) (2006) 1821–1831.

[7] V. Calderaro, G. Conio, V. Galdi, G. Massa, A. Piccolo, Active management of renewable energy sources for maximizing power production, Int. J. Electr. Power Energy Syst. 57 (2014) 64–72.

[8] H. Laaksonen, K. Kauhaniemi, 1 Voltage and Frequency Control of Low Voltage Microgrid With Converter-Based DG Units, 1, International Science Press, 2009.

[9] D.H. Muhsen, T. Khatib, H.T. Haider, A feasibility and load sensitivity analysis of photovoltaic water pumping system with battery and diesel generator, Energy Convers. Manag. 148 (2017) 287–304.

[10] N. Yan, Z.X. Xing, W. Li, B. Zhang, Economic dispatch application of power system with energy storage systems, IEEE Trans. Appl. Supercond. 26 (7) (2016).

[11] F. Wang, L. Zhou, H. Ren, X. Liu, S. Talari, M. Shafie-Khah, et al., Multi-objective optimization model of source–load–storage synergetic dispatch for a building energy management system based on TOU price demand response, IEEE Trans. Ind. Appl. 54 (2) (2018).

[12] D. Metz, J.T. Saraiva, Use of battery storage systems for price arbitrage operations in the15 and 60 min German intraday markets, Electr. Power Syst. Res. 160 (2018) 27–36.

[13] M. Gholami, S.H. Fathi, J. Milimonfared, Z. Chen, F. Deng, A new strategy based on hybrid battery–wind power system for wind power dispatching, IET Gen. Transm. Distrib. 12 (2017) 160–169.

[14] A. Castillo, X. Jiang, D.F. Gayme, Lossy DCOPF for optimizing congested grids with renewable energy and storage, in: American Control Conference (ACC), June 4–6, 2014. Portland, Oregon, USA, 2014.

[15] V.A. Boicea, Energy storage technologies: the past and the present, Proc. IEEE 102 (11) (2014) 1777–1794.

[16] B. Bahmani-Firouzi, R. Azizipanah-Abarghooee, Optimal sizing of battery energy storage for micro-grid operation management using a new improved bat algorithm, Int. J. Electr. Power Energy Syst. 56 (2014) 41–54.

[17] N. Nguyen-Hong, H. Nguyen-Duc, Y. Nakanishi, Optimal sizing of energy storage devices in isolated wind-diesel systems considering load growth uncertainty, IEEE Trans. Ind. Appl. 54 (3) (2018).

[18] C. Chen, S. Duan, T. Cai, B. Liu, G. Hu, Optimal allocation and economic analysis of energy storage system in microgrids, IEEE Trans. Power Electron. 26 (10) (2011) 2762–2773.

[19] M. Motalleb, E. Reihani, R. Ghorbani, Optimal placement and sizing of the storage supporting transmisión and distribution networks, Renew. Energy 94 (2016) 651–659.

[20] T. Kerdphol, K. Fuji, Y. Mitani, M. Watanabe, Y. Qudaih, Optimization of a battery energy storage system using particle swarm optimization for stand-alone microgrids, Electr. Power Energy Syst. 81 (2016) 32–39.

[21] V. Terzija, G. Valverde, D. Cai, P. Regulski, V. Madani, J. Fitch, et al., Wide-area monitoring, protection, and control of future electric power networks, Proc. IEEE 99 (1) (2011) 80–93.

[22] F. Bignucolo, R. Caldon, V. Prandonib, Radial MV networks voltage regulation with distribution management system coordinated controller, Electr. Power Syst. Res. 78 (4) (2008) 634–645.

[23] A.P. Sakis Meliopoulos, E. Polymeneas, Z. Tan, R. Huang, D. Zhao, Advanced distribution management system, IEEE Trans. Smart Grid 4 (4) (2013) 2109–2117.

[24] P. Faria, T. Soares, T. Pinto, T.M. Sousa, J. Soares, Z. Vale, et al., Dispatch of distributed energy resources to provide energy and reserve in smart grids using a particle swarm optimization approach, in: IEEE Computational Intelligence Applications in Smart Grid (CIASG), 2013, pp. 51–58.

[25] Z. Vale, T. Pinto, H. Morais, I. Praca, P. Faria, VPP's multi-level negotiation in smart grids and competitive electricity markets, in: IEEE Power and Energy Society General Meeting, July 24–29, 2011, pp. 1–8.

[26] N. Rahbari-Asr, Y. Zhang, M.-Y. Chow, Consensus-based distributed scheduling for cooperative operation of distributed energy resources and storage devices in smart grids, IET Gen. Transm. Distrib. 10 (5) (2016) 1268–1277.

[27] E.R. Sanseverino, M.L. Di Silvestre, M.G. Ippolito, A. De Paola, G.L. Re, An execution, monitoring and replanning approach for optimal energy management in microgrids, Energy 36 (2011) 3429–3436.

CHAPTER

8

Photovoltaic array reconfiguration to extract maximum power under partially shaded conditions

S. Saravanan[1], *R. Senthil Kumar*[1], *A. Prakash*[1], *T. Chinnadurai*[2], *Ramji Tiwari*[3], *N. Prabaharan*[4] *and B. Chitti Babu*[5]

[1]Department of EEE, Sri Krishna College of Technology, Coimbatore, India [2]Department of ICE, Sri Krishna College of Technology, Coimbatore, India [3]Department of EEE, Bharat Institute of Engineering and Technology, Hyderabad, India [4]School of Electrical and Electronics Engineering, SASTRA Deemed University, Thanjavur, India [5]Department of Electronics Engineering, Indian Institute of Information Technology, Design and Manufacturing, Chennai, India

8.1 Introduction

Nowadays, the world is facing high energy demands due to the increase of the human population. Most of the energy utilization to supply the growing population comes from fossil fuels. But this resource is becoming exhausted and is causing atmospheric pollution, among other problems. In order to address these problems, renewable energy sources (RES) like solar, wind energy, and fuel cells are being used. Within RES, energy is being produced predominantly from solar photovoltaic (PV) systems, because of the large amount of solar energy available to meet humanity's ever-increasing energy needs. Solar PV system-based power production has become famous and awareness of this is high among the general peoples. As such, governments provide grants aimed at pursuing technological developments to reduce the price per watt significantly of PV systems.

Despite some technological improvements, there are a few challenges remaining in PV power outcomes like the small conversion of efficiency, depending on varying irradiation

Distributed Energy Resources in Microgrids
DOI: https://doi.org/10.1016/B978-0-12-817774-7.00008-9

and temperature conditions [1]. The main issues that disturbs the energy harvest of PV system is loss due to divergence occurring in PV panels' electrical characteristics. The mismatch of the PV system is a result of the manufacturing systems employed, the aging of modules, poor soldering links, and Partial shaded (PS) or irregular irradiation conditions. From the above, PS conditions have created harmful consequences in terms of the yield efficiency of the PV system.

However the panels are sited as well as fixed after cautious planning, shading is inescapable in many circumstances due to space restrictions and also as a result of issues like clouds, buildings, trees, and snow. Moreover, inadequate maintenance may lead to dust settling on the panels, which creates an irregular response to radiation. The electrical performance characteristics of shaded panels differ from the unshaded panel, resulting in a mismatch of PV that rises with the intensity of shade. The PV energy harvest is reduced by this mismatch, and improving the PV outcome under such circumstances is a critical task as the decrease in the output is not only related to the shaded area but also depends on several other issues like array formation, the intensity of the shade, and the position of the shaded panel in the array [2]. When analyzing the effect of PS problems on the output of PV system, it is necessary to find suitable methods to counteract the shading effect.

When a PV panel is in a PS condition, it produces a minimum of photocurrent, while the drop is also related to the shading factor. If the PS problem occurs in a series PV string, it will cause the string to have current limitations. If the panel becomes shaded, it will be forced to conduct more current. As a result, the PV will operate in a reverse biased condition and thermal stress will increase across the shaded portion of panel. Moreover, confined hot spots will occur if the current is not equally distributed. To overcome this, development of the hotspot is necessary by using the bypass diode coupled with antiparallel to the PV panels to make the current flow along a different paths [3]. Even though the shaded portion of panels is secure from damage, the outcome of PV is not optimum. In addition to the bypass diodes, it leads to peaks in voltage-power characteristics and PV system of voltage–current (V-I) characteristics. Global Maximum Power Point Tracking (GMPPT) methods are involved in tracking the best point with the occurrence of multi-peaks in the PV system. Conventional tracking methods cannot be categorizing as Global Peak from the peak of local points. Various GMPPT methods contain many stages and different evolutionary and optimization algorithms have been offered successfully to track the global peak point [4].

The interconnection between the PV panels has an important effect on the output power under PS conditions. The above drawbacks are overcome by using PV array configurations schemes to reduce the power losses. In this survey, various PV array configuration strategy schemes such as Series (S), Parallel (P), Series-Parallel (SP), Bridged-Linked (BL), Honey-Comb (HC), Total-Cross Tied (TCT), Sudoku-configured PV array, Electrical Array Reconfiguration (EAR), and Adaptive Array Reconfiguration (AAR), Genetic Algorithm (GA), Particle Swarm Optimization (PSO), and Dominant Square (DS) are proposed [5–7].

In this chapter, several methods have been presented to decrease the effect of PS, including (1) establishment of bypass diodes, (2) employing global MPPT, (3) varying the architecture of power conditioning, (4) various reconfiguration schemes which extract maximum power, and (5) it is a challenging issue.

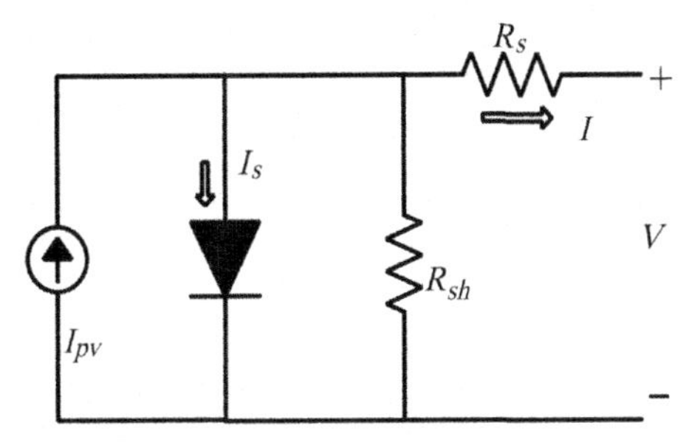

FIGURE 8.1 Solar cell equivalent circuit.

8.2 Photovoltaic panel modeling

A PV cell is made by a semiconductor pn junction diode, which converts solar energy into electricity. A large voltage is obtained when the cells are fixed in series, while a large current is obtained when the cells are fixed in parallel. Fig. 8.1 shows an equivalent circuit diagram of a single-diode PV panel. The modeling of a solar cell is defined by the voltage–current relationship within a PV system as follows [8].

$$I = I_{pv} - I_s\left(\exp\left(\frac{q(V + R_s I)}{N_s kTa}\right) - 1\right) - \frac{V + R_s I}{R_{sh}} \tag{8.1}$$

where I_{pv} is the PV current (A), q the electron charge (1.60217×10^{-19} C), I_s the saturation current (A), k the Boltzmann constant (1.38065×10^{-23} J/K), a the diode ideality constant, N_s the number of cells in series, R_s and R_{sh} the series and parallel resistances of cell (Ω), and T the temperature (K).

8.3 Configuration of bypass diodes

In general, as the climate changes from time to time, the solar panels experience changes due to shading. As a result, there is a consistent loss of power within the solar panels and these losses are eliminated by employing bypass diodes. Due to PS cells, the current is forced to flow through low-voltage cells. But in general, the current flows from high voltage to low voltage. Due to this abnormal condition, the solar panel becomes overheated and this leads to heavy power loss, making the solar panels perform as consumers instead of producers.

The arrangement of solar panels includes a junction box, in which the bypass diodes are fixed in parallel on the side of the cells. A minimum resistance path is provided for the available current to travel around the series of shaded solar cells. Due to this special arrangement, the heat gain is minimized and the current loss is reduced to a considerable level. The equation representing the bypass diode is given by the following:

$$I = I_0\left(e^{q(V - IR_s)/nk_BT} - 1\right) \tag{8.2}$$

TABLE 8.1 Parameters of a bypass diode.

Parameter	Value
I_0 (A)	4.32×10^{-10}
R_s (Ω)	0.32
N	1.52

The parameters of the bypass diodes are given in Table 8.1 [9].

Recently, many of the solar panels are designed with inbuilt bypass diodes. With the same arrangement, if more panels are wired together, connecting a bypass diode in parallel over the shaded panel will keep the current from flowing outside. This enables the entire system to reach maximum heat and at the same time inner bypass diodes act as active diodes. With a specific end goal to gauge the voltage over each bypass diode, a voltage sensor is utilized. For the most part in the cluster-based framework, bypass diodes are utilized to change the flow of the current created by the unshaded boards. In addition, the current will be restricted by the shaded panel's resistance [10]. Under various shading and normal conditions, the voltage varies from $-V_{oc}$ to 0.7 V and, at this stage, the bypass diode is in the forward biased mode (on state) carrying a voltage of 0.7 V with logic "0". The diode is in reverse biased mode (off state), when the voltage varies from 0 to $-V_{oc}$, with logic "1".

8.3.1 Blocking diodes

Blocking diodes are used to keep batteries from releasing in reverse through the solar panel boards during the evening. Current streams from high to low voltage, so on a bright day, the voltage of a panel board will be higher than the voltage of a profound cycle battery and this energy will normally spill out of the PV panel to the battery. However, at night, if the panel board is associated specifically to a battery, the voltage of the panel board will be lower than the voltage of the battery, so there is a probability of some retrogressive stream, which draws power out of the battery. It will not be as much as the stream in the day, however. Fig. 8.2 represents the bypass and blocking diodes in a PV panel, which are used to mitigate the hotspot issue.

8.4 Tracking of maximum power

In general, solar panels have nonlinear V-I characteristics, with a distinct power point at maximum level that depends on certain environmental factors, namely temperature and irradiation. Power electronics converters are employed to step up the voltage of the PV system using duty cycle control in addition to achieving MPPT. Some of these methods are discussed in the following section.

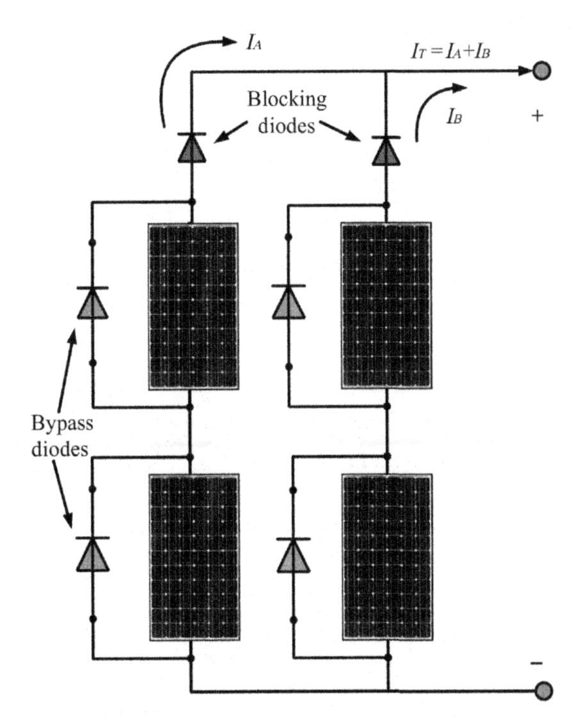

FIGURE 8.2 Placing of bypass and blocking diodes in a photovoltaic (PV) array.

8.4.1 Perturb and observe method

In general, the hill climbing method requires perturbation. But the Perturb and Observe (P&O) method, the operating voltage of the PV array needs to be perturbed continuously [11]. Bothering the duty cycle of the converter connected to the PV array will perturb the array voltage. Fig. 8.3 explains the power versus voltage characteristics of a PV array. From the curve, it is observed that the power increases as the voltage increases on the left part and the power decreases with an increase in voltage on the right part. The region where the curve's division point occurs is known as the Maximum Power Point (MPP).

Fig. 8.4 explains the algorithm for P&O. It is observed that the operating voltage V is perturbed with the MPPT cycle. V will continue to oscillate around the operating voltage once the MPP is reached. This causes a power loss which depends on the step width of a single perturbation [12]. With a large step width, there will be immediate response to sudden changes in the operating conditions. With a smaller step width, the losses under the stable conditions will be reduced. The drawback of the P&O algorithm is that if there is a sudden change in solar irradiance, P&O makes it appear as if the sudden change occurs due to the previous perturbation of the array operating voltage [12]. Fig. 8.5 represents the deviation of MPP with the P&O method under rapid solar irradiance changes.

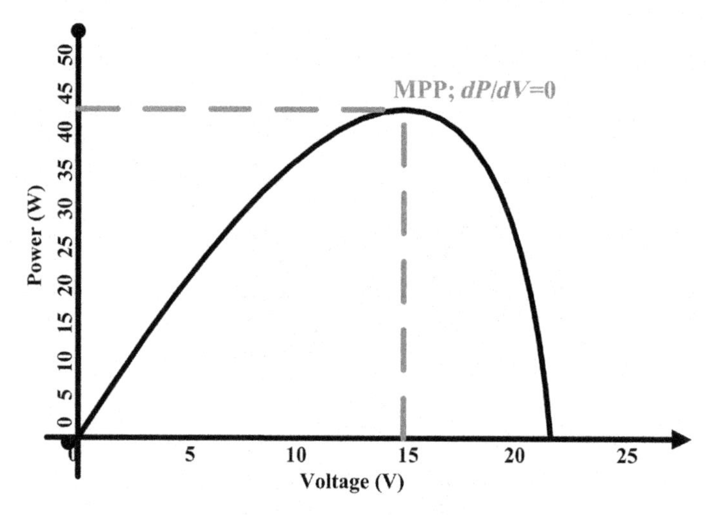

FIGURE 8.3 Power versus voltage curve at $(dP/dV) = 0$.

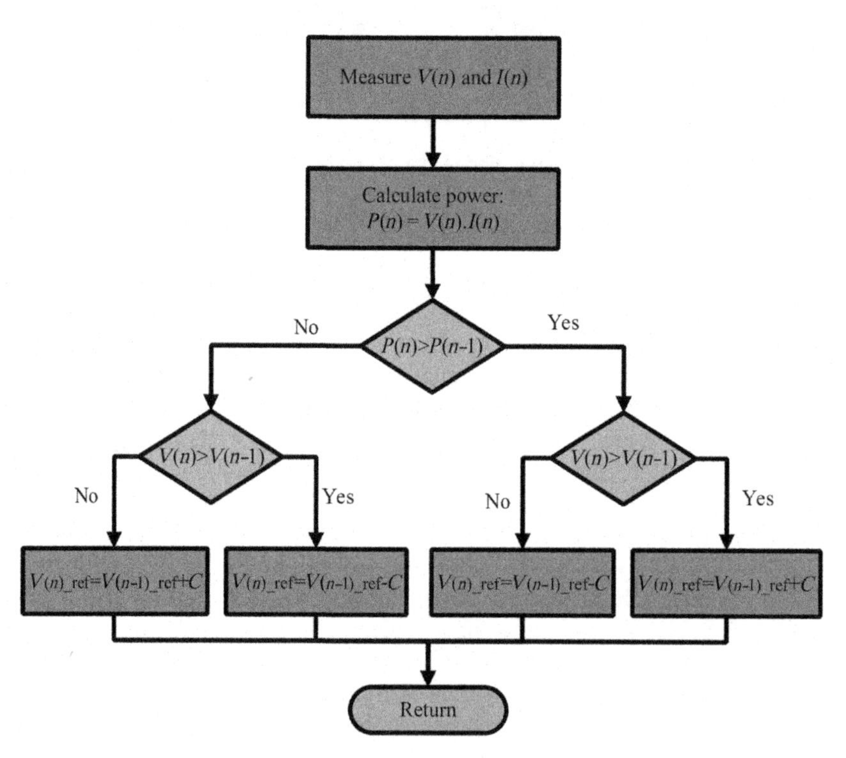

FIGURE 8.4 Flowchart for the Perturb and Observe algorithm.

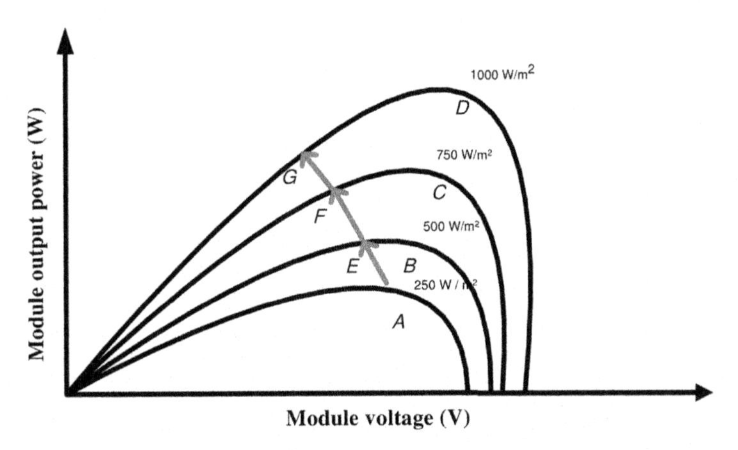

FIGURE 8.5 Deviation of Maximum Power Point (MPP) with Perturb and Observe (P&O) method.

8.4.2 Incremental Conductance Method

The Incremental Conductance Method (IncCond) method helps to overcome the disadvantages of the P&O method, where the slope of the PV array power curve is zero at MPP, positive on the left of the MPP and negative on the right of the MPP [13]. In this method, once the MPPT reaches the MPP, perturbation of the operating point is stopped. If this condition is not met, the direction in which the operating point to be perturbed can be calculated using the relationship between (dI/dV) and $(-I/V)$. This relationship is derived from the aforesaid conditions. This method can also track the rapidly increasing and decreasing irradiance conditions with higher accuracy than P&O. Mathematically, the conditions can be expressed as follows:

$$\left(\frac{dP}{dV}\right) = 0; \quad \text{at MPP} \tag{8.3}$$

$$\left(\frac{dP}{dV}\right) > 0; \quad \text{to the left of MPP} \tag{8.4}$$

$$\left(\frac{dP}{dV}\right) < 0; \quad \text{to the right of MPP} \tag{8.5}$$

Eqs. (8.3), (8.4), and (8.5) can be simplified using the following approximations:

$$\left(\frac{dP}{dV}\right) = d(IV) = I + \left(\frac{dI}{dV}\right) \tag{8.6}$$

$$\approx I + V\left(\frac{\Delta I}{\Delta V}\right) \tag{8.7}$$

From (8.7), (8.5) can be written as

$$\left(\frac{\Delta I}{\Delta V}\right) = \left(\frac{-I}{V}\right); \quad \text{at MPP} \tag{8.8}$$

$$\left(\frac{\Delta I}{\Delta V}\right) > \left(\frac{-I}{V}\right); \quad \text{to the left of MPP} \tag{8.9}$$

$$\left(\frac{\Delta I}{\Delta V}\right) < \left(\frac{-I}{V}\right); \quad \text{to the right of MPP} \tag{8.10}$$

From the above equations, it is very clear that the MPP is tracked by comparing the instantaneous conductance (I/V) to the incremental conductance ($\delta I/\delta V$). The flowchart for IncCond is shown in Fig. 8.6.

8.4.3 Load voltage and load current maximization

MPPT techniques maximize the power coming out of a PV array. When the PV array is connected to a PCU, maximizing the PV array power implies that the output power at the load of the PCU also increases. Conversely, maximizing the output power of the PCU also maximizes the PV array power, assuming a loss-less power conditioning unit (PCU) [14]. Loads are mostly of voltage-source type, current-source type, resistive type, or a combination of these. The V-I characteristics of these different loads are shown in Fig. 8.7. It can be inferred that for a voltage-source type load, the load current I_{out} should be maximized to reach the maximum power. For a current-source type load, the load voltage V_{out} should be maximized [15].

8.4.4 Fractional open circuit voltage (V_{oc}) and fractional short circuit current (I_{sc})

The relationship between V_{MPP} and V_{oc} of the PV array, under varying temperature levels, has given rise to the fractional V_{oc} method [14]. It is observed that

$$V_{MPP} \approx k_1 \times V_{oc} \tag{8.11}$$

where k_1 is the constant of proportionality and V_{MPP} is the panel voltage at MPP. Since k_1 is dependent on the characteristics of the PV array being used, V_{MPP} and V_{oc} for the specific PV array at different insolation and temperature levels has to be determined beforehand. The factor k_1 has been observed to be between 0.71 and 0.78.

Fractional I_{sc} results from the fact that, under varying insolation conditions, I_{MPP} is approximately linearly related to the I_{sc} of the PV array and is given by

$$I_{MPP} \approx k_2 \times I_{sc} \tag{8.12}$$

where k_2 is a proportionality constant and I_{MPP} is the panel current at MPP. Just like the fractional VOC technique, k_2 has to be determined according to the PV array that is being used. The constant k_2 is generally found to be between 0.78and 0.92. An additional switch usually is added to the PCU to periodically short the PV array for measuring I_{sc}.

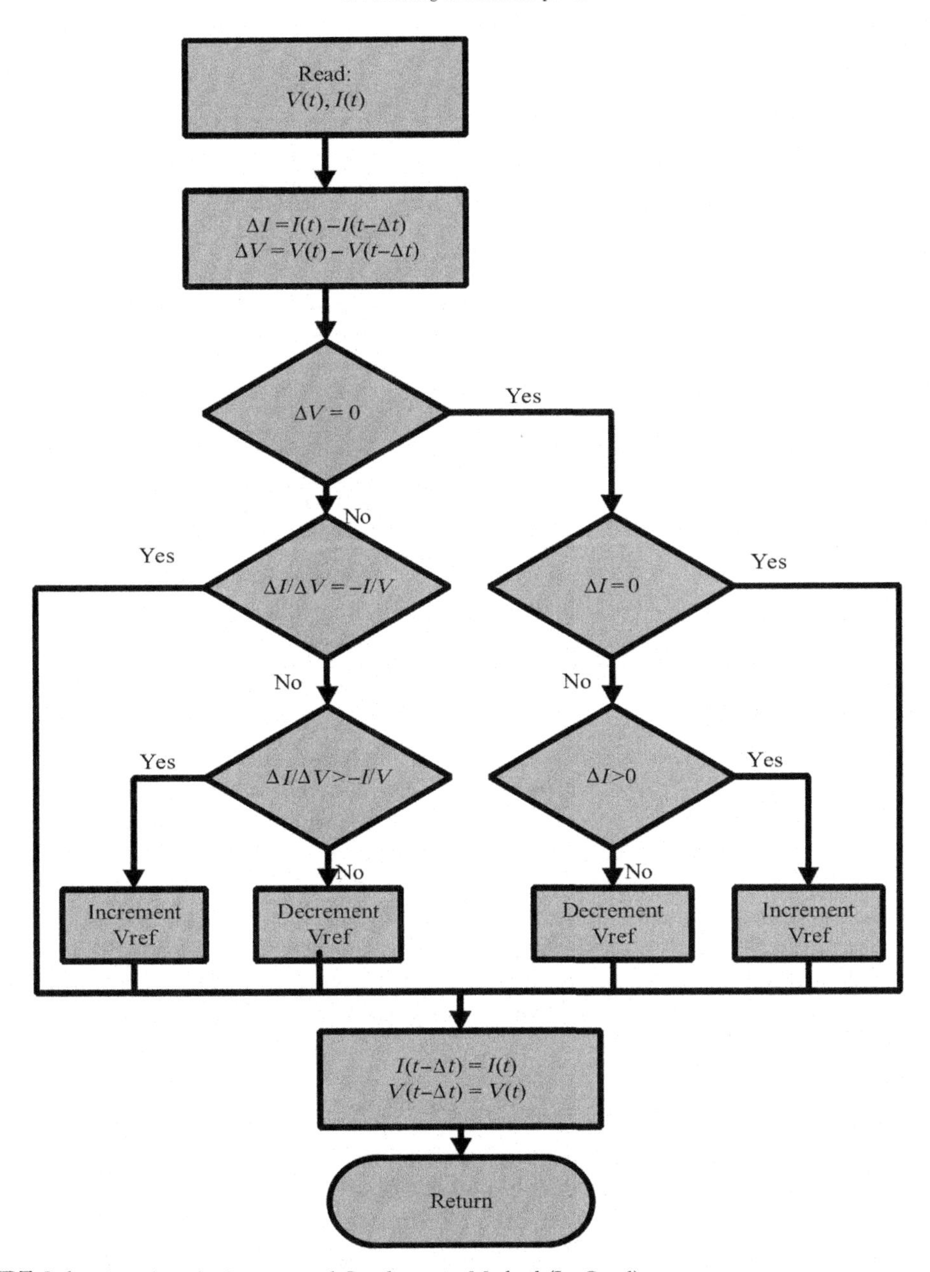

FIGURE 8.6 Flowchart for Incremental Conductance Method (IncCond).

8.4.5 *dP/dI* or *dP/dV*

In this method, the slope $(dP/dV \text{ or } dP/dI)$ of the power curve is computed and a feedback is provided to the PCU with some control to drive it to zero, since the slope at MPP is zero [16]. In this method, the sign of the slope is stored from the past few cycles.

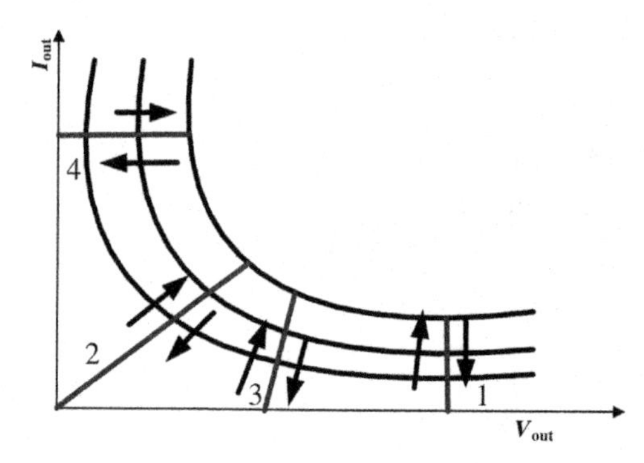

FIGURE 8.7 Different load types: 1, voltage source; 2, resistive source; 3, a combination of all three; 4, current source.

Based on these signs, suitable action is taken and the duty ratio of the PCU is either incremented or decremented to reach the MPP.

8.5 Photovoltaic system configurations

8.5.1 Central inverter

Normally, central inverters are placed away from harsh climatic conditions but near the electricity panel board. In the arrangement of the central inverter, energy from all the solar panels, which is in the form of DC, is collected together using a combiner box, after which it is converted to AC power. The output voltage obtained in this inverter is about as high as 600 V DC. Central inverters are used in industries of very large scale and the capacity is much larger than other forms of inverters (string and micro). A central inverter of rating 5 MW is required for generating power of 5 MW with the help of the PV system. The intended applications of the central inverter are large buildings, industries, and field establishments. The simple arrangement of a central inverter is depicted in Fig. 8.8.

8.5.1.1 Advantages of a central inverter

- Credibility: Central inverters have been around for quite a while now. With these devices, real-world field data is accessible with respect to its execution. This factor makes central inverters more popular than other inverters.
- Lower costs: Compared to micro and the string inverters, the cost of installing central inverters is considerably less per watt. This is mainly due to the fact that central inverters require fewer components.
- Reliability: As central inverters need to be maintained well, they are safeguarded within a protective environment. So, even though they may operate within very abnormal climatic conditions, there is little risk involved.

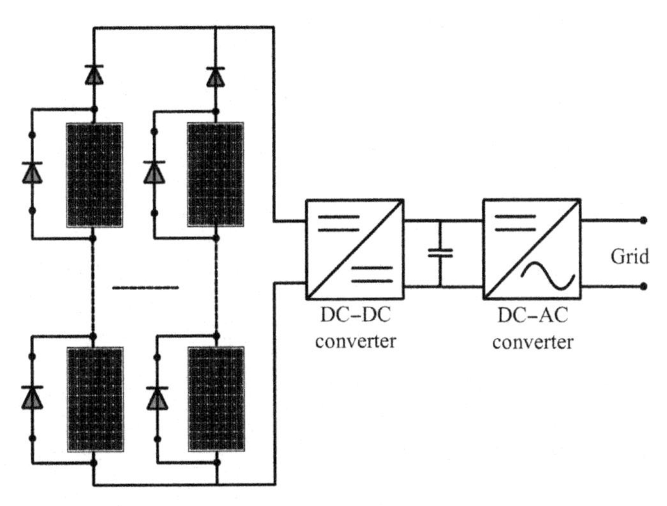

FIGURE 8.8 Diagram of a central inverter.

8.5.1.2 Disadvantages of a central inverter

- *Potential for a single point of failure:*
  If one panel becomes shaded or fails for some reason, the performance of the entire system is affected. If the panels are set up in series, they act as a single unit and the under-performance of one panel will affect the entire array. Reports from industrialists reveal the fact that 10% shading of a system leads to a decline of power output around 50%.
- *Higher risk factor:*
  The risk factor is higher in central inverters than the other inverter structures. High voltage DC is produced together when the panels are connected in series. High voltage is produced with the series connection of PV module and feeds DC power to the grid via grid-forming central inverters. Handling this situation can be a life-threatening task to both the owners and the end user.
- *Higher replacement cost:*
  When the centralized architecture of the central inverter fails, the cost incurred for the replacement of this inverter is very high when compared with micro and string inverters.

8.5.2 String inverter

String inverters convert DC to AC for *n* number of panels. This type of inverter is used in home and commercial solar power systems. Generally, string inverters are a large box structure, which is located at a distance slightly away from the solar array. With an efficiency rating ranging from 97.5 to 98.5, the number of string inverters might be more than one, depending on the installation size.

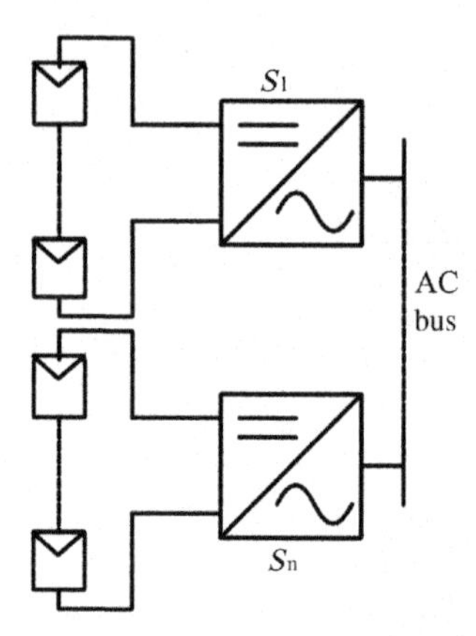

FIGURE 8.9 Diagram of a string inverter.

The power produced by all the panels is fed into a single inverter. From the inverter end, the electricity is distributed to the end users according to their requirements. Assume, for example, that there are 50 panels on the rooftop, arranged into 10 groups of 5 each. The 10 groups are the "strings" and they are connected in parallel to each other and, at the other end, a single inverter is connected to this arrangement. Fig. 8.9. depicts the structure of a string inverter.

Even though the string inverter is chosen based on the various types of application, residential apartments and huge commercial applications make use of this string inverter at a higher scale. String inverters may be chosen for structures such as roof installations with large surface areas and ground mounting, provided the panels are of same orientation. The string inverter acts as a mid-way between the micro and the central inverters. The arrangement procedures and the inverters mid-size makes the installation easier, while maintenance is also simpler compared to other types of inverters.

8.5.2.1 Advantages of a string inverter

- Flexible to design
- Higher efficiency
- Robustness
- Lower cost
- Capable of monitoring remote systems

8.5.2.2 Disadvantages of a string inverter

- Absence of maximum power point at panel level
- Absence of panel level monitoring

8.5.3 Microinverter

The structure of microinverter is very simple as it consists of very small box placed at the back or very close to the panel. As the design of the inverter is very small with regards to its size and rating, they are classified under small inverters.

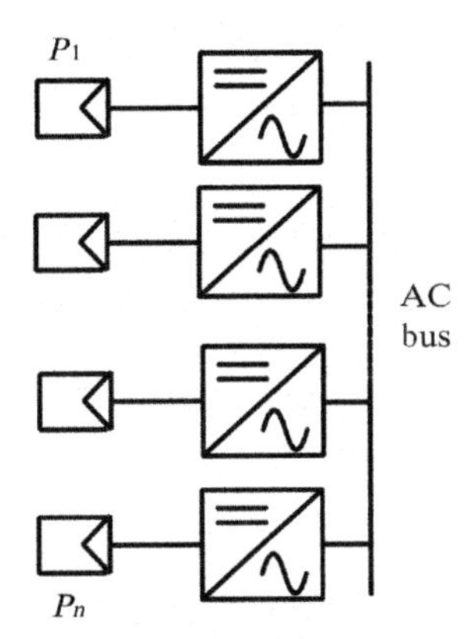

FIGURE 8.10 Diagram of a microinverter.

Microinverters are small inverters (both size-wise and rating-wise) that are designed to be attached to the back of each solar panel of the array. In some cases, they are attached to two solar panels instead of just one. With these, the direct current produced from the respective panels is inverted to alternating current and is then sent into the appliances. Fig. 8.10 shows the simple arrangement of a microinverter.

8.5.3.1 Advantages of microinverter

- MPPT at panel level is possible.
- Malfunction in single panel will not affect the performance of the entire system.
- Monitoring at panel level is possible.
- DC voltage of lower rating and increase in safety.
- Flexible in design and the modules can be moved to face all directions.
- Output obtained will be at the required level, as the shadowing on single module does not affects the entire string.
- Calculating the length of the string is not necessary.
- Whenever an update is required, it leads to the usage of different models.

8.5.3.2 Microinverter disadvantages

- The cost of the microinverter is twice the amount for that of string inverters, as the dollars per watt rate is high.
- System complexity in terms of installation.
- Heating issue at higher rate.
- As there are multiple units in a single array, the maintenance cost is higher for microinverters.

8.6 Reconfiguration strategies

A PV system is one of the renewable energy resources available in all areas. The main problem arising is partial shading where the irradiances of array modules are altered and lead to power losses and module degradation. The shadow module creates shadow patterns which can be classified as predictable and nonpredictable. Predictable patterns are based on the sun's position in the sky, as well as shading caused by buildings, chimneys, and trees. Predictable patterns also include shading from the front row of PV collectors and this is called mutual shading. Nonpredictable patterns are formed by nonpredictable parameters like dust, passing clouds, and the droppings of birds.

To overcome these issues, a PV array reconfiguration is the best solution to utilize the maximum solar power for shaded and unshaded modules effectively. Many reconfiguration methods have been practically exploited and are classified as:

- physical relocation,
- electrical rewiring, and
- electrical array reconfiguration.

With physical relocation, the position of the PV panels is interchanged physically to achieve equal irradiance and obtain the maximum power output in TCT interconnection PV array using the zig-zag method or Sudoku game theory. These methods can be applied with an even number of rows and columns (5×5), which has some limitation like considerable labor and excessive interconnection [17].

Electrical rewiring can be performed only once without physical relocation. For a large PV array, the additional rewiring causes complexity and difficulty in connection to utilize the maximum power of PV array.

The electrical array and DS scheme is one of the distinctive methods of PV array reconfiguration. The shade dispersal can be found by "$m \times n$" switching arrangements. Although the switching arrangements are effective, determining the combination of a suitable switching arrangement is difficult [18].

Hence the optimal switching techniques have been analyzed to identity the best combinations to track the maximum available power in a PV array.

1. Series (S) Array
2. Parallel (P) Array
3. Series-Parallel (SP) Array
4. Bridged-Linked (BL) Array
5. Honey-Comb (HC) Array
6. Total-Cross Tied (TCT) Array
7. Sudoku-Configured PV Array
8. Electrical Array Reconfiguration (EAR)
9. Adaptive Array Reconfiguration (AAR)
10. Genetic Algorithm
11. Particle Swarm Optimization
12. Dominant Square Method

8.6.1 Series array

The PV panels are fixed in series, as shown in Fig. 8.11, to obtain the required PV panel to mitigate the shading. If 24 panels are fixed in series, the array current, voltage, and power can be written as [19],

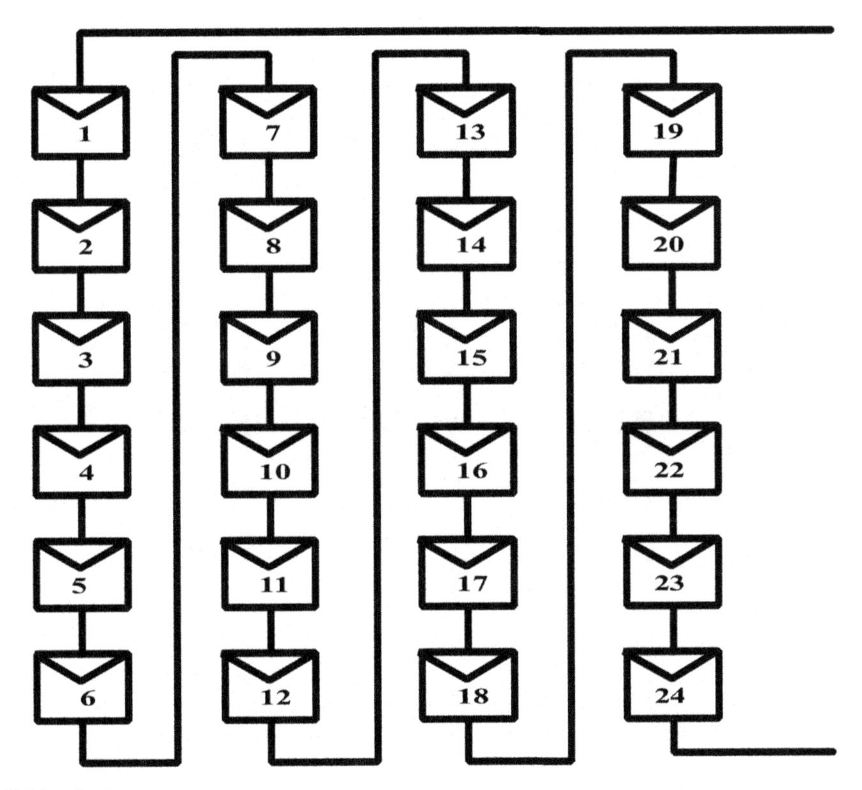

FIGURE 8.11 Series array.

$$I_{\text{array}} = I_{\text{string}} = I_{m1} = I_{m2} = \cdots = I_{24} = I_{24} \tag{8.13}$$

$$V_{\text{array}} = \sum_{i=1}^{24} V_{mi} = 24V_m \tag{8.14}$$

$$P_{\text{array}} = 24V_m I_m \tag{8.15}$$

The array current is the same as the panel current, while the array voltage is equal to the sum of panel voltages. If the impact of shading is extreme in a fixed series array, the shaded panel will enforce high current limitations. When panels are in a shaded condition in a series array, the particular panel current drops, as does the load current. In this instance, the panel is forced to perform under reverse bias and there is a need to install bypass diodes.

8.6.2 Parallel array

In order to avoid high current limitations and for portable PV systems, parallel-connected arrays are preferred as shown in Fig. 8.12. When the panels are connected in

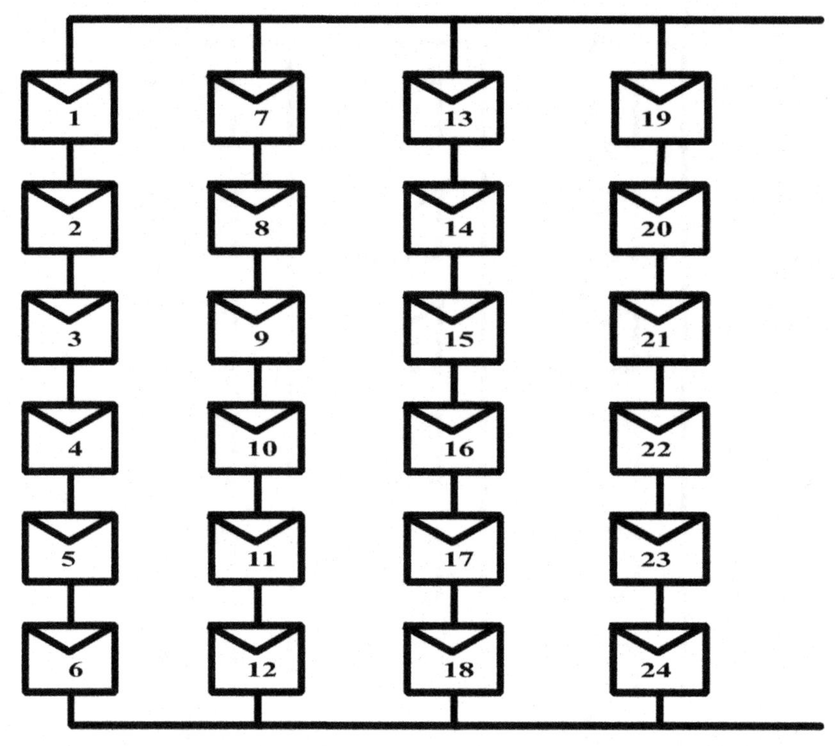

FIGURE 8.12 Parallel array.

parallel, the array current is the same as addition of all panel current and voltage across the panels are equal. The array current, voltage, and power can be written as,

$$I_{\text{array}} = \sum_{i=1}^{24} I_{mi} = 24I_m \tag{8.16}$$

$$V_{\text{array}} = V_{\text{string}} = V_{m1} = V_{m2} = \cdots = V_{24} = V_{24} \tag{8.17}$$

$$P_{\text{array}} = 24V_m I_m \tag{8.18}$$

8.6.3 Series parallel array

In order to form a PV module in this configuration, all the PV cells are electrically connected in serial and parallel combinations to form the PV array which is shown in Fig. 8.13. The panels are initially fixed in series and these strings are connected in parallel. According to requirements, the number of series strings and panels per string are connected.

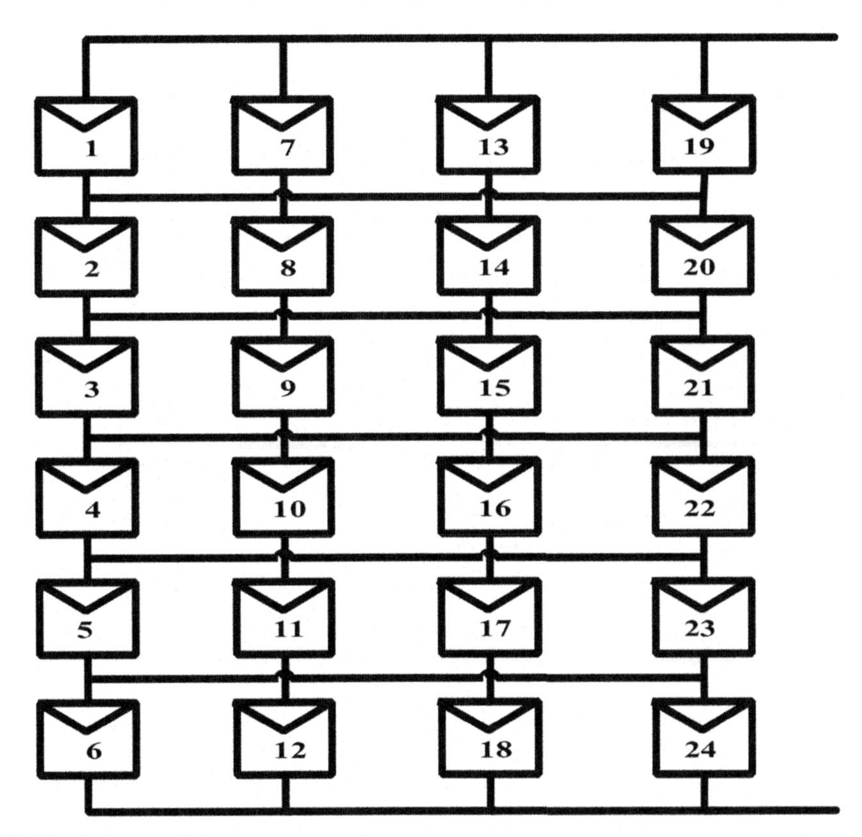

FIGURE 8.13 Series parallel array.

8.6.4 Bridged-Linked array

This approach involves a cross-tied topology, in which each module is connected in series as well as in parallel with many interconnections, as shown in Fig. 8.14. It consists of different panels in the form of bridge unit and cross ties are fixed between the bridges. When any module mismatches, all the modules performs identically but the generated array power is equal.

8.6.5 Honey-Comb array

The HC array is modified combination of BL with different sizes. This configuration consists of modules and submodules with cross ties, as shown in Fig. 8.15.

8.6.6 Total-Cross Tied configured scheme

By reconnecting the series-parallel configuration, the TCT configuration is attained. Each row of the junctions that are to be connected and the column modules are fixed in series and the row modules are fixed in parallel [20] as shown in Fig. 8.16. The current

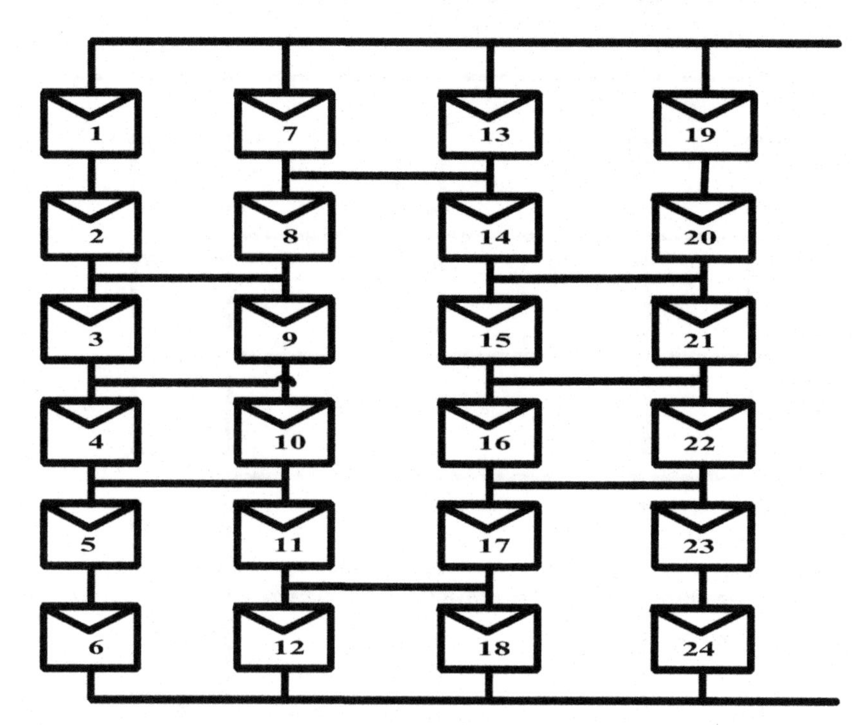

FIGURE 8.14 Bridged-Linked array.

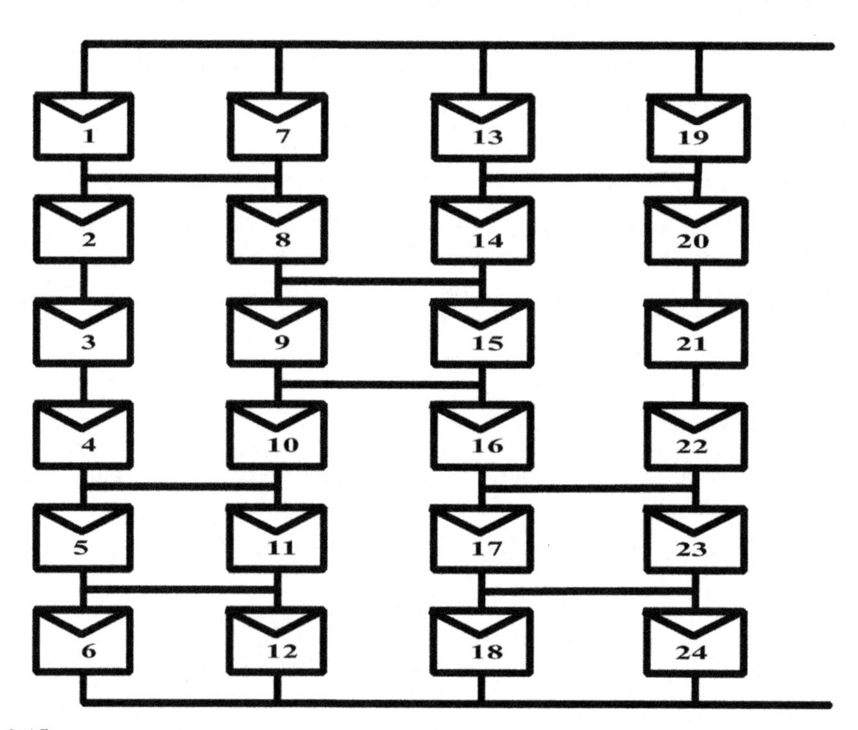

FIGURE 8.15 Honey-comb array.

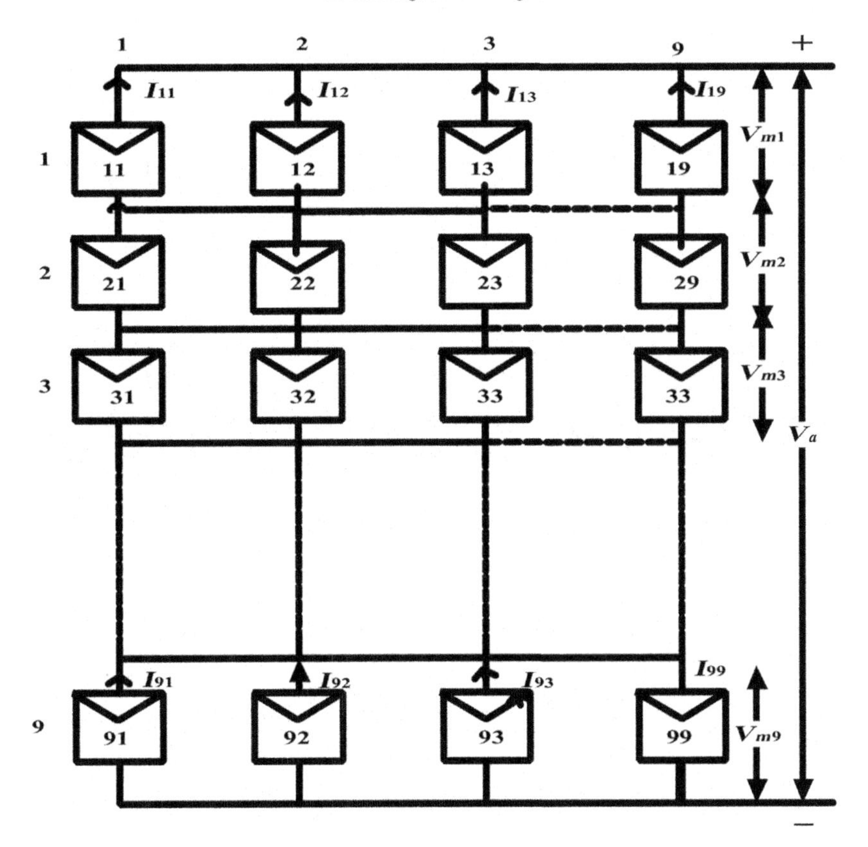

FIGURE 8.16 Total-Cross Tied (TCT) array.

produced in the module is proportional to the irradiance of the panel. The voltage is calculated by the summation of the voltages of all rows. Consider a 9×9 PV array for study, in which 81 panels are connected. The total current induced by the array at particular irradiance is given by [21],

$$I = KI_m \tag{8.19}$$

where $I_m$is the current generated by the module at standard irradiance G_0. The array of voltage is given by the total addition of the voltage in the entire nine rows. By using KVL,

$$V_a = \sum_{K=2}^{9} V_{mk} \tag{8.20}$$

where V_a is the PV array voltage, V_{mk} the panel voltage. Each node of current is calculated by KCL,

$$I_a = \sum_{c=1}^{9} (I_{rc} - I_{r+1}C) \quad \text{where } r = 1, 2, 3 \ldots \tag{8.21}$$

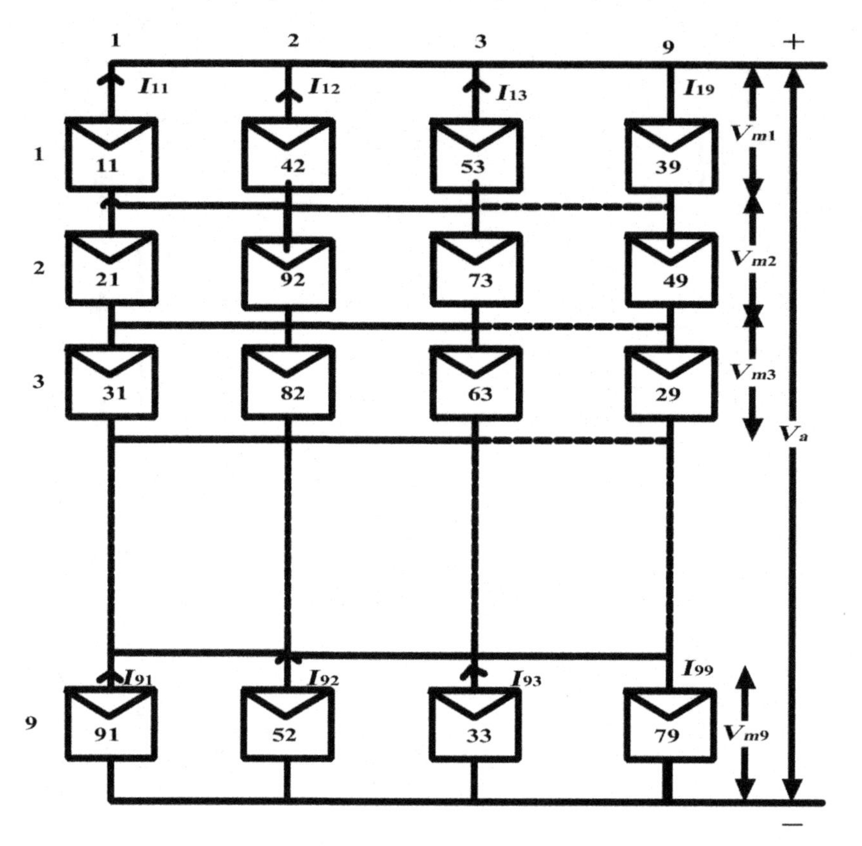

FIGURE 8.17 Sudoku-configured array.

8.6.7 Sudoku-configured photovoltaic array

Sudoku is a puzzle based on various combinations of placement of numbers. The 9×9 PV array with 81 solar cells has no repetition in number [22]. In this technique, the physical position of the modules is altered without varying the electrical wire connections among them. The uniform shading of dispersion is obtained in this Sudoku technique which is shown in Fig. 8.17. However, the physical relocation is a difficult and laborious task.

8.6.8 Electrical Array Reconfiguration

In this configuration method, all PV modules are assembled in a structure of a matrix of single string series-connected rows and parallel-connected modules per row [23]. The maximum available DC power depends on irradiance, temperature as well as their location in the structure of the matrix. A 9×9 PV array with 81 solar cells is shown in Fig. 8.18. To increase the current, more PV panels can be added in each row. To increase the voltage, more PV panels can be added in each column [24].

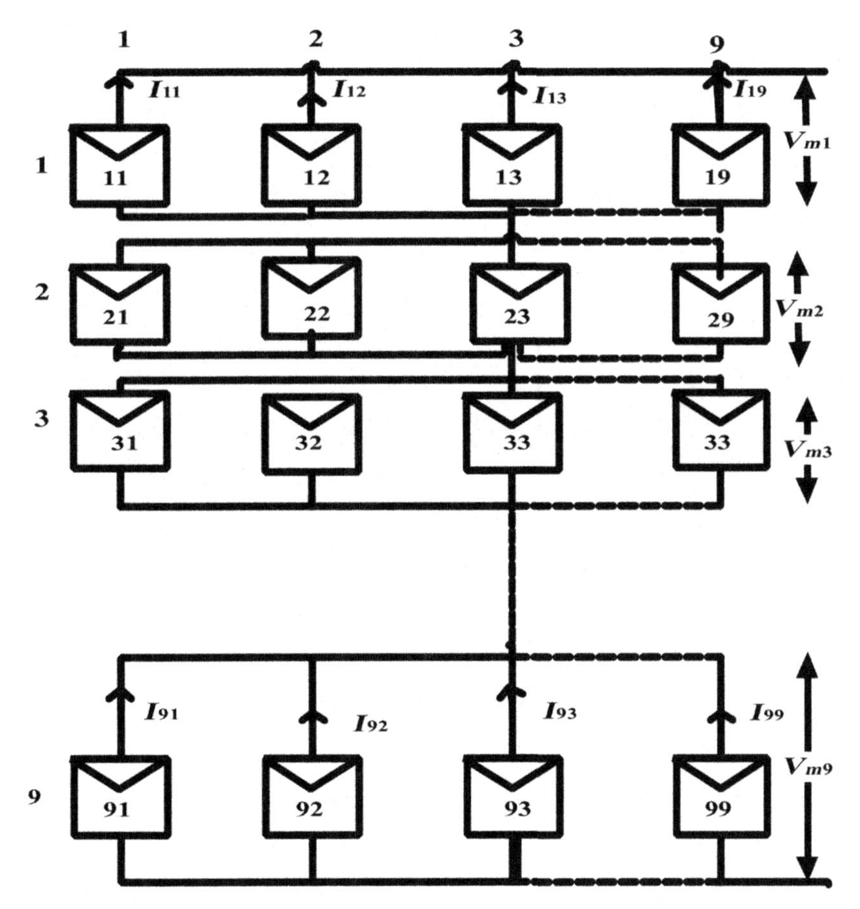

FIGURE 8.18 Electrical Array Reconfiguration (EAR).

$$\text{The number of configurations } N_{ci} = \frac{(m \cdot n)!}{m!(n!)^m} \tag{8.22}$$

where m is the number of series-connected rows, n the number of parallel-connected modules per row.

8.6.9 Adaptive Array Reconfiguration

An adaptive reconfiguration bank of solar PV array has a switching matrix to compute the maximum power under shaded conditions. It has two adaptive reconfiguration algorithms: bubble sort and the model reference approach. With the bubble-sort method, the PV array is adaptive and switches the array combination based on power calculations only once for every switching and then the next sort is implemented.

The flowchart for the control algorithm is given in Fig. 8.19. The second method, the model-based sorting algorithm, has fixed array and adaptive solar array capabilities. This

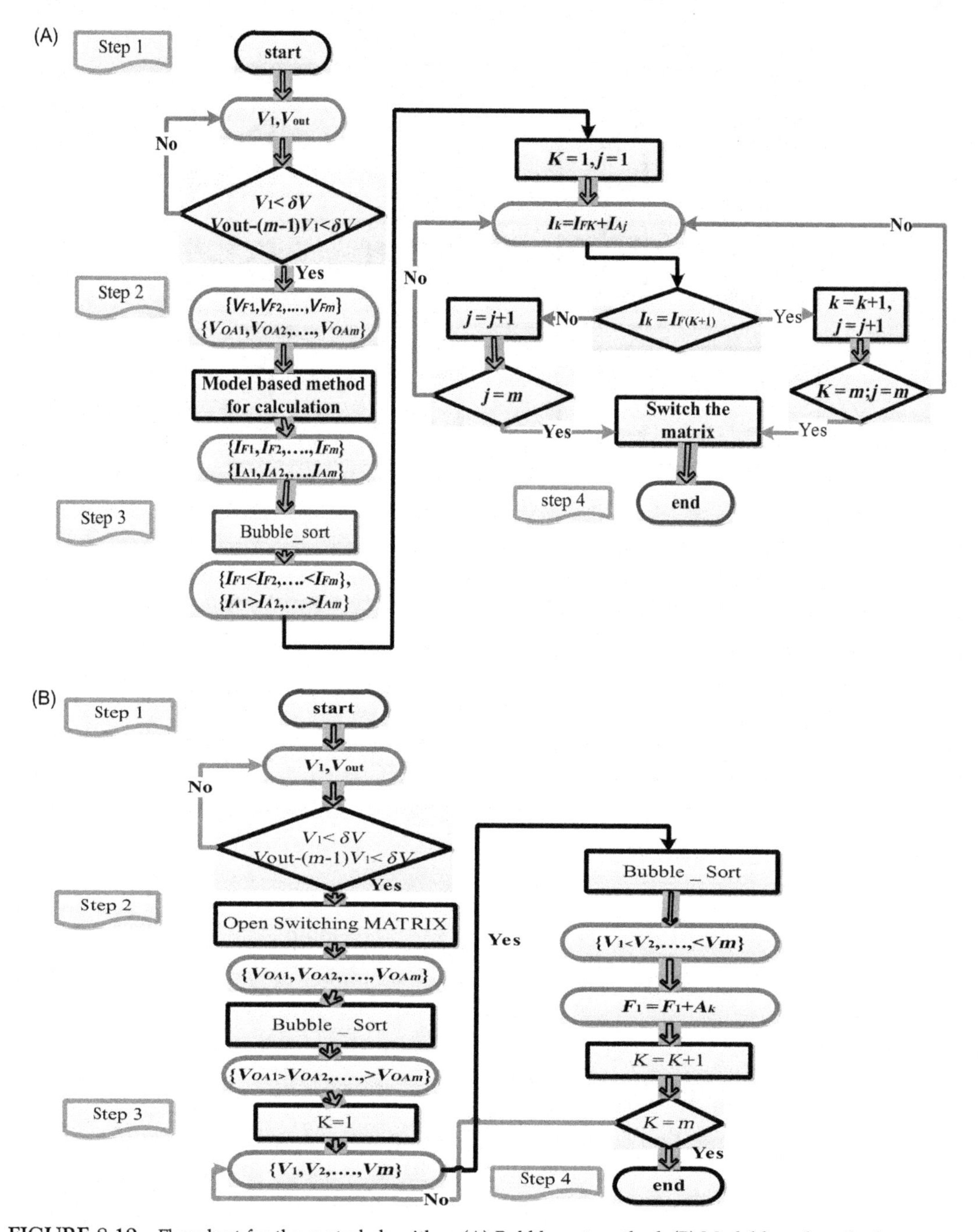

FIGURE 8.19 Flowchart for the control algorithm. (A) Bubble-sort method. (B) Model-based method.

method is used to predict the power of each row in a fixed solar array and, based on this power prediction, it can be switched simultaneously to an adaptive solar array connection. This algorithm depends on the connection between the illuminated PV cells or shaded rows of the adaptive banks to the fixed part of the shaded rows. Since the connection is in parallel, there is a very small range of increase in the current from the shaded adaptive cell. The major drawback of this method is the multiple number of switches required for switching the matrix.

8.6.10 Genetic Algorithm

The GA method, which is adaptable for various environmental conditions, helps to solve the issues encountered in PV array reconfigurations [25]. The standard deviation of row current is used to set the parameters to determine the optimal solution. The standard deviation is the measure of variation and dispersion amount for the set of values. When the standard deviation of the row current is low, then the data points converge to the average of the set. When the standard deviation of the row current is very high, then the data points are spread over the range of values.

In the PV-array-reconfiguration problem, the objective function is to minimize the standard deviation of row current which reduces the losses in mismatch and determines the maximum power under different environmental conditions. The objective function of array configuration is well defined and minimizes the standard deviation [26].

$$\text{Maximize (fitness }(i)) = \frac{1}{1+\sigma_i} \tag{8.23}$$

fitness(i) = fitness function of ith element in population

$$\sigma_i = \sqrt{\frac{1}{N}\left[\sum_{j=1}^{N}(I_j - I_m)^2\right]} \tag{8.24}$$

where N is total number of rows, σ_i is individual row current in standard deviation,

$$I_m = \tfrac{1}{N}\left[\sum_{j=1}^{N} I_j\right], \quad I_j \text{ is } j\text{th row current} \tag{8.25}$$

The GA approach has the drawback of large steps for the computational processes and a delay in convergence to find the shaded pattern. The following code represents the GA to reconfigure the system.

Step1: Initialize the parameters

Size of population, total number of iterations, irradiance pattern, crossover, probability of mutation

Step 2: Generate initial size of the population

$$\text{Evaluation fitness function} = \frac{1}{1+\sigma_i}$$

Step 3: GA Evaluation
For iteration = 1:N

- Selection of parents with the help of a roulette wheel selection.
- Crossover of parents with the help of probability of specified crossover.
- Mutation of offspring with the help of probability of specified mutation.
- Evaluation of new fitness function for the candidate.
- Selection of different individuals for next generation.

8.6.11 Particle Swarm Optimization (PSO)

PSO is used to perform on-time switching which handles the shading problem. The dispersing of shade can be done for a TCT-interconnected PV array using a PSO algorithm. The shading can be initialized randomly and row current is obtained for each row. Based on the row current, the velocity changes for each PV cells within the column arrangements. For each iteration, the row current is minimized. The difference in row current should be close to zero to locate global optimum solutions. When the solution is reached, the velocity change in each column is reduced based on velocity/number. Hence, the PSO method is efficient for performing on-time switching.

8.6.12 Dominant Square method

The DS method is based on the mathematical puzzle method and it adopts the position of number placements as number / alphabets to solve the given PV array matrix, that is, a11 alphabets are diagonally placed and certain rules of DS are followed for the initial position and diagonal position of the numbered alphabets [27]. All the alphabets are present in each row and columns. For example, the module number 12 represents its position as first row, second column in Fig. 8.20. After applying the DS method, PV panel number 32, originally positioned in the third row and second column, is relocated to the second row and second column in Fig. 8.21.

The strategy of various reconfiguration schemes is presented in Table 8.2.

A_{11}	A_{12}	A_{13}	A_{14}	A_{15}
B_{21}	B_{22}	B_{23}	B_{24}	B_{25}
C_{31}	C_{32}	C_{33}	C_{34}	C_{35}
D_{41}	D_{42}	D_{43}	D_{44}	D_{45}
E_{51}	E_{52}	E_{53}	E_{54}	E_{55}

FIGURE 8.20 Normal PV array matrix.

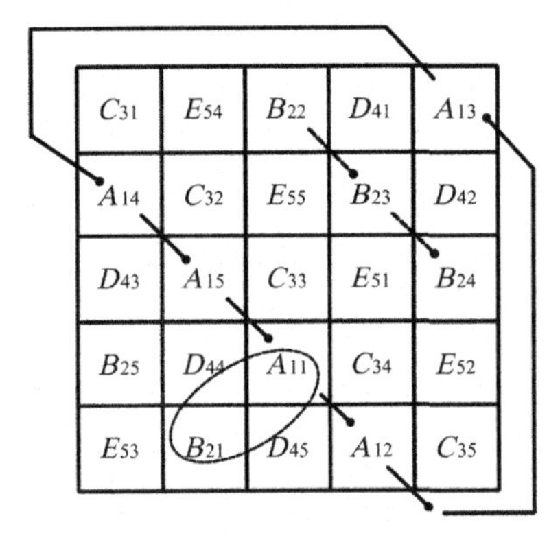

FIGURE 8.21 Dominant Square (DS) method.

TABLE 8.2 Strategy of reconfiguration schemes.

S. no.	Control algorithm	Strategy
1.	TCT-configured scheme	Series parallel connection
2.	Sudoku-configured scheme	Physical connection changed based on puzzle
3.	EAR	Irradiance equalization
4.	AAR	Adaptive bank
5.	GA	Row current minimization
6.	Particle Swarm Optimization (PSO)	Row current minimization
7.	DS	Row current minimization

8.7 Challenging issues

General challenges exist due to the position of nearby buildings, clouds, and trees, all of which can cause partial shading in PV arrays. In Germany, the energy loss occurs in the range of 5%–10% per annum due to the shade of nearby buildings. In Spain, the energy loss occurs in the range of 3%–6% per annum due to partial shading in PV farms [28] from the PV panels themselves.

Buildings, PV arrays, and trees are predictable shadow patterns. Passing clouds, snow, birds, bird litter, and dust are nonpredictable shadow patterns. The predictable patterns can be reduced by the proper design and location of the array arrangement with a proper distance between such obstacles to maximize the production of solar power.

The major challenges are:

- PV array connection in series experiences nonlinear internal resistance [29]

  In the series-connection of PV cells, the same current flows throughout the PV cells. Due to shading, less photon current is carried in the few PV cells and reverse bias happens in the shaded cells, which drains power from the fully illuminated cells. This issue leads to a hotspot problem if the system is not protected.

- Location and operating condition for MPP [30]
  When power generated by a PV cell varies continuously, the maximum power cannot generate in the PV cell. To extract the maximum power, the location and operating conditions of the array need to be addressed.
- Electrical monitoring [31]
  Data acquisition plays a major role in the collection of data to identify the fault, partial shade, connectivity and mismatch. DC parameters, AC parameters (current, voltage and power), temperature, wind speed, humidity, and irradiance are the parameters that need to be monitored for the generation of maximum power.
- Hardware complexity/switching matrix [32]
  For dynamic PV reconfigurations, the switching devices play an important role. The various switches like contactors, relays, SCR, MOSFET, and IGBT can be used based on their low cost, low maintenance, strong reliability, and its lifetime.

8.8 Conclusion

This chapter deals with the influence of PS conditions and its effect on PV array interconnections and bypass-diode-configuration-based PV systems. The different mitigation methods are described as a way of determining how to extract the maximum power from a PV system under various irradiation conditions and the restrictions of each method. Moreover, the effect of PS on energy yield and its requirement in the interconnection scheme and shade diffusion have also been presented. Reconfiguration array schemes and their switching combinations have given effective solution to track the maximum power under PS conditions.

References

[1] V.M. Gradella, G.J. Rafael, E.R. Filhol, Modelling and circuit–based simulation of photovoltaic arrays, Braz. J. Power Electron. 14 (1) (2009) 3–45.

[2] M.C. Alonso-García, J.M. Ruiz, F. Chenlo, Experimental study of mismatch and shading effects in the I-V characteristic of a photovoltaic panel, Sol. Energy Mater. Sol. Cells 90 (3) (2006) 329–340.

[3] S. Silvestre, A. Boronat, A. Chouder, Study of bypass diodes configuration on PV panels, Appl. Energy 86 (9) (2009) 1632–1640.

[4] J.Y. Hyok, J.D. Yong, K.J. Gu, K.J. Hyung, L.T. Won, W.C. Yuen, A real maximum power point tracking method for mismatching compensation in PV array under partially shaded conditions, IEEE Trans. Power Electron. 26 (4) (2011) 1001–1009.

[5] O. Bingol, B. Ozkaya, Analysis and comparison of different PV array configurations under partial shading conditions, Sol. Energy 160 (2018) 336–343.

[6] F. Belhachat, C. Larbes, Reducing partial shading power loss with an integrated smart bypass, Sol. Energy 103 (2015) 134–142.

[7] D.S. Pillai, J.P. Ram, M.S.S. Nihanth, N. Rajasekar, A simple, sensorless and fixed reconfiguration scheme for maximum power enhancement in PV systems, Energy Conver. Manage. 172 (2018) 402–417.

[8] V.M. Gradella, G.J. Rafael, E.R. Filhol, Comprehensive approach to modelling and simulation of photovoltaic arrays, IEEE Trans. Power Electron. 24 (5) (2009) 1198–1208.

[9] E.D. Dorado, J. Cidrás, C. Carrillo, Discretized model for partially shaded PV arrays composed of PV panels-with overlapping bypass diodes, Sol. Energy 157 (2013) 103–115.

[10] D.P. Winston, B.P. Kumar, S.C. Christabel, A.J. Chamkha, R. Sathyamurthy, Maximum power extraction in solar renewable power system - a bypass diode scanning approach, Comput. Electr. Eng. 70 (2018) 1–15.
[11] D. Peftitsis, G. Adamidis, A. Balouktsis, A new mppt method for photovoltaic generation systems based on hill climbing algorithm, in: Proceedings of the 18th International Conference on Electrical Machines, 2008, pp. 1–5.
[12] H. Knopf, Analysis, simulation and evaluation of maximum power point tracking (MPPT) methods for a solar powered vehicle (M.Sc.), Portland State University, 1999.
[13] Z. Xuesong, S. Daichun, M. Youjie, C. Deshu, The simulation and design for mppt of pv system based on incremental conductance method, WASE Int. Conf. Inform. Eng. 2 (2010) 314–317.
[14] T. Esram, P. Chapman, Comparison of photovoltaic array maximum power point tracking techniques, IEEE Trans. Energy Con. 22 (2) (2007) 439–449.
[15] M.A. Masoum, H. Dehbonei, E.F. Fuchs, Theoretical and experimental analyses of photovoltaic systems with voltage and current-based maximum power point tracking, IEEE Power Eng. Rev. 22 (8) (2002) 62.
[16] S. Chiang, K. Chang, C. Yen, Residential photovoltaic energy storage system, IEEE Trans. Ind. Electron. 45 (3) (1998) 385–394.
[17] T.S. Babu, J.P. Ram, T. Dragičević, Particle swarm optimization based solar PV array reconfiguration of the maximum power extraction under partial shading conditions, IEEE Trans. Sustain. Energy 9 (1) (2018) 74–85.
[18] D. Nguyen, B. Lehman, An adaptive solar photovoltaic array using model-based reconfiguration algorithm, IEEE Trans. Ind. Electron. 55 (7) (2008) 2644–2654.
[19] S. Malathy, R. Rama Prabha, Comprehensive analysis on the role of array size and configuration on energy yield of photo voltaic systems under shaded conditions, Renew. Sustain. Energy Rev 49 (2015) 672–679.
[20] H. Brauna, S.T. Buddha, V. Krishnana, Topology reconfiguration for optimization of photovoltaic array output, Sustain. Energy, Grids Net. 6 (2016) 58–69.
[21] D. Picault, B. Raison, S. Bacha, J. Aguilera, J. De La Casa, Changing photovoltaic array interconnections to reduce mismatch losses: a case study, in: Proceeding on International Conference on Environment and Electrical Engineering.
[22] B.I. Rani, G.S. Ilango, C. Nagamani, Enhanced power generation from PV array under partial shading conditions by shade dispersion using Su Do Ku configuration, IEEE Trans. Sustain. Energy 4 (3) (2013) 594–601.
[23] G.V. Quesada, F.G. Gispert, Electrical PV array reconfiguration strategy for energy extraction improvement in grid-connected PV systems, IEEE Trans. Ind. Electron. 56 (11) (2009) 4319–4331.
[24] D. Nguyen, B. Lehman, An adaptive solar photovoltaic array using model-based reconfiguration algorithm, IEEE Trans. Ind. Electron. 55 (7) (2008) 2644–2654.
[25] S.N. Deshkar, S.B. Dhale, J.S. Mukherjee, T.S. Babu, N. Raja Sekar, Solar PV array reconfiguration under partial shading conditions for maximum power extraction using genetic algorithm, Renew. Sustain. Energy Rev. 43 (2015) 102–110.
[26] A. Harrag, S. Messalti, Adaptive GA-based reconfiguration of photovoltaic array combating partial shading conditions, Neural Comput. App. 30 (4) (2018) 1145–1170.
[27] B. Dhanalakshmi, N. Rajasekar, Dominance square based array reconfiguration scheme for power loss reduction in solar PhotoVoltaic (PV) systems, Energy Conver. Manage. 156 (2018) 84–102.
[28] M.Z. Shams El-Dein, M.M. Kazerani, M.M.A. Salama, Optimal photovoltaic array reconfiguration to reduce partial shading losses, IEEE Trans. Sustain. Energy 4 (1) (2013) 145–153.
[29] R. Ramaprabha, B.L. Mathur, A. Comprehensive, Review and analysis of solar photovoltaic array configurations under partial shaded conditions, Int. J. PhotoEnergy (2012) 1–16.
[30] V. Vaidya, D. Wilson, Maximum power tracking in solar cell arrays using time-based reconfiguration, Renew. Energy 50 (2013) 74–81.
[31] D.L. Manna, V.L. Vigni, E.R. Sanseverino, V.D. Dio, P. Romano, Reconfigurable electrical interconnection strategies for photovoltaic arrays: a review, Renew. Sustain. Energy Rev. 33 (2014) 412–442.
[32] E.R. Sanseverino, T.N. Ngoc, M. Cardinale, V.L. Vigni, D. Musso, Dynamic programming and Munkres algorithm for optimal photovoltaic arrays reconfiguration, Sol. Energy 122 (2015) 347–358.

CHAPTER

9

Communications and internet of things for microgrids, smart buildings, and homes

Gianluca Fadda, Mauro Fadda, Emilio Ghiani and Virginia Pilloni

Department of Electrical and Electronics Engineering, University of Cagliari, Cagliari, Italy

9.1 Introduction

Nowadays, new business opportunities are becoming available in the global market of energy production, distribution, and utilization. Indeed, empowering technologies such as smart meters, smart sensors, and smart control systems, internet of things (IoT) communications, and Cloud platforms are allowing the growth of several kinds of energy management applications, in domestic, commercial and industrial sectors (smart homes/buildings).

The current industrial revolution is going to change, completely, the way in which energy is generated, delivered, and used but, at the same time, the electric industry faces many challenges, mainly related to the need to balance supply and demand as grid complexity grows and to keep the whole network under a suitable level of cyber security.

Under this scenario, this chapter refers to the general topic of applying the IoT paradigm to energy management in microgrids (MGs), smart buildings, and homes.

9.2 The transition toward the smart grid

Several power grid evolution strategies are still under study due to many crucial challenging aspects that are evolving at the same time but in different ways. Fig. 9.1 shows the evolutionary character of smart grids (SGs) [1].

Distributed Energy Resources in Microgrids
DOI: https://doi.org/10.1016/B978-0-12-817774-7.00009-0

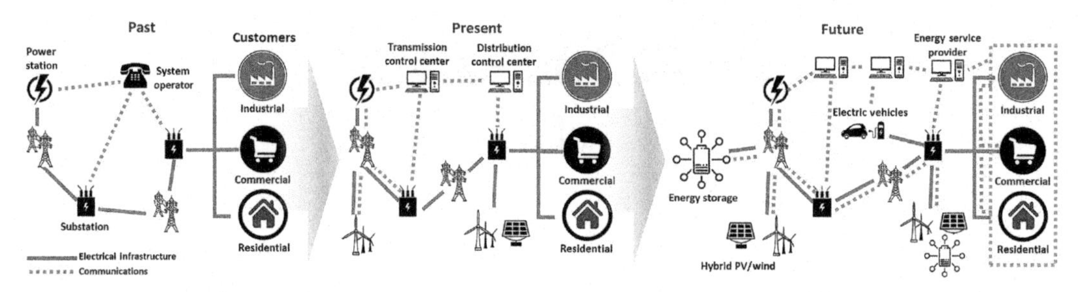

FIGURE 9.1 The evolutionary process of smart grids. *Source: Adapted from N. Andreadou, M.O. Guardiola, G. Fulli, Telecommunication technologies for smart grid projects with focus on smart metering applications, Energies (2016) [1].*

Indeed, the continuous growth of renewable energy sources (RESs) of different types must fit within the requirements imposed in terms of emission reduction and efficient energy usage [1]. The SG of the future will exploit modern and reliable telecommunication technologies to implement energy management systems able to cope with all the new scenarios.

Monitoring and control of the energy generation, as well as transmission and distribution, are the main issues when dealing with the "smartization" of the power grid [2], requiring the interaction between different disciplines with a coordinated design of all the layers in the SG architecture [3], with interactive information and technology (ICT) devices and smart meters (SM) for empower an advanced metering infrastructure.

From a general point of view, three different kind of networks should be considered, depending on the specific function domain of the power grid: the transmission of the electrical energy refers to the high voltage (HV) network, while its distribution is enabled through the medium voltage (MV) network, and the low voltage (LV) network provides such energy to end users [1].

This chapter focuses on some key aspects regarding some LV and MV network arrangements and management system architectures, like the ones related to MGs, smart homes, and buildings.

The US Department of Energy (DOE) defines an MG as follows: "A MG is a group of interconnected loads and DERs within clearly defined electrical boundaries that acts as a single controllable entity with respect to the grid. A MG can connect and disconnect from the grid to enable it to operate in both grid-connected or island mode." As such, the possibility of having more load/customers within the MG boundary is intrinsically intended. Then, among the various definitions and arrangements, an MG can be used, for instance, like an integrated energy system located downstream of a main distribution substation through a point of common coupling (PCC) containing production and consumption users [4] but the concept of an MG can be applied also to the small power system destined to feed smart homes and buildings.

A possible system architecture referring to the MG concept, for MV networks, is shown in Fig. 9.2. An MG can support two different operating modes: when is running as "grid-connected," the MG is linked to the main grid through the distribution substation transformer, while when running as "islanded" (or autonomous), it results in being isolated from the main grid, typically during a blackout or brownout [5,6].

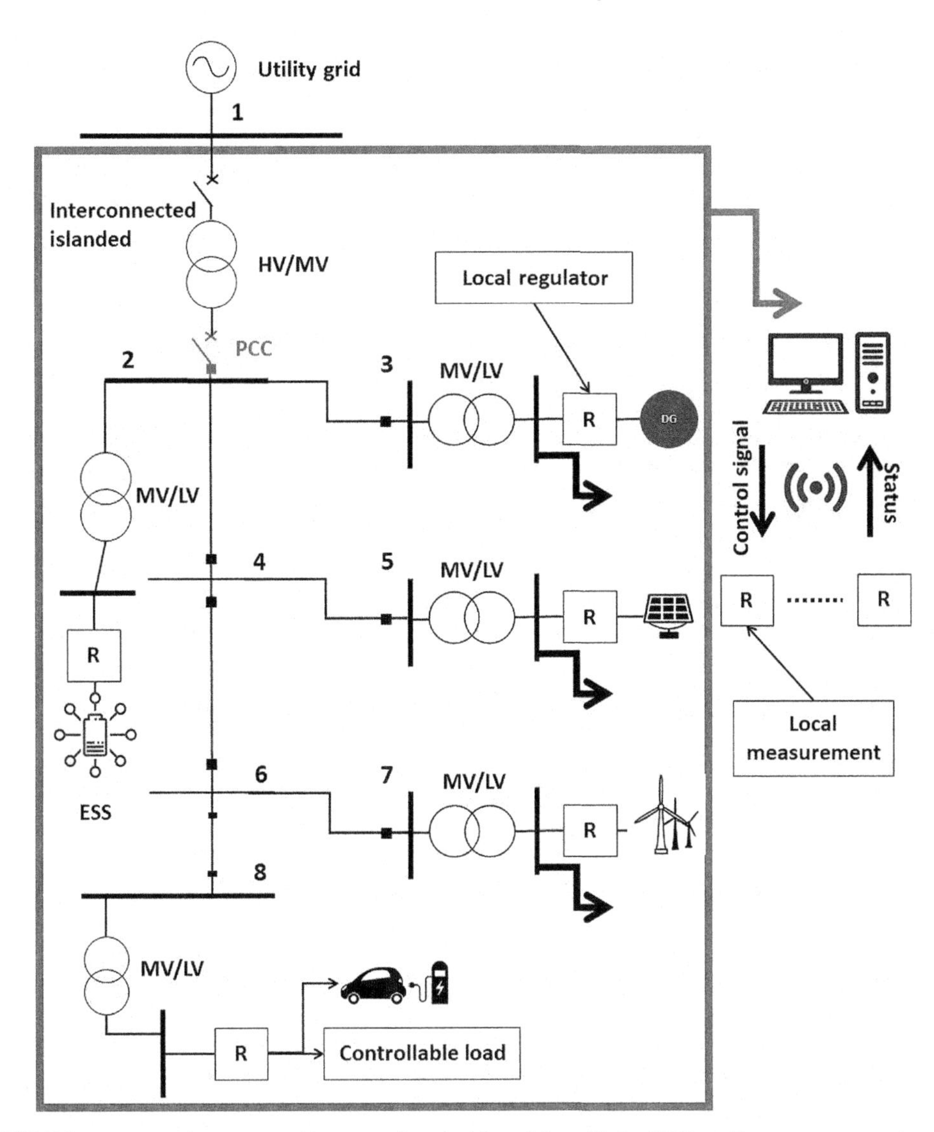

FIGURE 9.2 Microgrid system architecture. *Source: Adapted from W. Su, J. Wang, Energy management systems in microgrid operations, Electr. J. (2012) [7].*

All the most common MG arrangements consider five different typical components [4,8]:

1. PCC is the connection point for the power production, distribution network, and customer interface.
2. Distributed generation (DG) usually denotes a small-scale electric power supply directly connected to the distribution system at or near the load feeder, which supplies power in an intermittent way, through one or more RES, based on the network demand.

3. Energy storage systems (ESS), which may allow the implementation of energy buffering strategies when the energy price from the main grid is cheaper or an over-generation from the local DGs occurs. ESS can also be employed as an additional power generator during peak demand periods.
4. Controllable loads, such as plug-in electric vehicles (EVs) or thermostatically controlled loads, which are smart things able to modify their own electric energy usage based on real-time set points.
5. Critical loads, such as schools or hospitals, to which MG have to guarantee the highest levels of power supply availability and reliability.

The SG transition for MGs requires the distribution of the communication, computation, and storage resources and services to be on or close to devices and systems in the control of end-users, as well as a distribution management system (DMS) as fundamental part of the advanced grid modernization, for controlling all the distributed devices and components, including protection, intelligent electronic devices (IED) and microprocessor-based controllers of power system equipment. For instance, by exploiting the concept of "fog computing," introduced by Cisco in 2014 and claimed as "a standard that defines how edge computing should work, and which facilitates the operation of compute, storage, and networking services between end devices and cloud computing data centers" [9]. The edge computing, often just referred to as edge, "brings processing close to the data source, and it does not need to be sent to a remote cloud or other centralized systems for processing" [10].

In July 2018, the IEEE Standards Association approved the OpenFog Reference Architecture, developed by the OpenFog Consortium, an association of high-tech industry companies and academic institutions, including Cisco Systems, Intel, Microsoft, Princeton University, Dell, and ARM Holdings, as an official standard, under the name of IEEE 1934 [11]. The IEEE 1934 is a technical reference framework, "designed to enable processing to be distributed across things-to-cloud continuum" [12].

From a general point of view, all smart-grid scenarios considered in this chapter exhibit the same three-layer-architecture composed by the smart things tier (i.e., the lower layer), the fog tier (i.e., the middle layer), and finally the macrostation tier (i.e., the upper layer) as shown in Fig. 9.3.

The middle layer, namely the "Fog tier," could include all the network edge devices, such as smart meters, MGs, or any other smart appliance which could implement an energy load balancing application, by switching in a suitable way among alternative distributed energy resources (DERs) at the lowest layer, namely the "smart thing tier," based on data generated by their grid sensors and devices. In terms of communication signals, based on the received information, the so-called "fog collectors" at the fog tier, make actuations at the SG tier, process data so as to drop those references that have to be consumed locally, and send the filtered data to the higher layer, specifically the "macrostation tier," for management purposes, that is, visualization and reporting for real-time or transaction analytics [13].

The fog computing paradigm enables the implementation of applications of big data (BD) analytics in real-time, since by this way a dense distribution of data-collection points can be managed. The smart thing tier refers to machine-to-machine (M2M) interactions, which allow local devices, that is, protections or controllers, to communicate with a remote

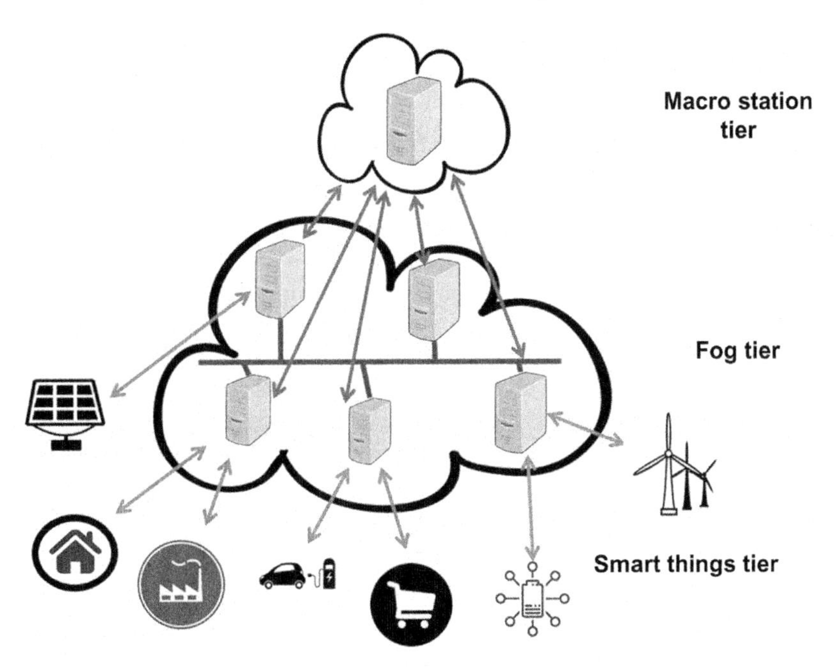

FIGURE 9.3 Fog computing architecture in a SG.

system and for making a suitable response to a particular event or situation within a processing time from milliseconds to seconds, that is in real-time [14]. The fog tier and the macrostation tier refer to human-to-machine-interactions, that is, visualization and reporting, but also supervision and control of systems and processes. Such communication exchange over the Fog platform can span from seconds to minutes for real-time analytics, and up to days for transactional analytics. Due to this fact, the Fog must support different kinds of storage, from transient at the lowest tier to semi-permanent at the highest tier. Then, when dealing with the internet of everything for streaming and real applications implemented by a fog computing approach, the following classification of interactions can be considered:

- M2M: Any machine can send/receive data to/from any other one, by exploiting the networking capabilities of IoT and sensors.
- People-to-Machine: Any person can send/receive data to/from any other person or machine, and the connection relies on capabilities in data and analytics.
- People-to-People: People can exchange data through a cooperation mechanism.

Smart local grid (SLG) is another emerging concept, referred to a network and communication infrastructure consisting of multiple MGs, aiming to improve their effectiveness and reliability at a local level, by allowing communication among all devices [15]. By applying fog computing to SLG, the bandwidth and latency issues can be highly reduced by keeping costs to the minimum. Indeed, by allowing M2M direct connections, SLG enables real-time decisions without the need for a data exchange with the Cloud [16]. SLG is a smart network where all the devices are connected to the Cloud through open communication standards, but they are also autonomous in making decisions when some changes force them to reply in real time. The opportunity for devices to communicate with the Fog allow them to solve problems with a higher complexity. A classification of the application technologies is summarized in Table 9.1.

BD, Big data; *QoS*, quality-of-service.

The fog computing paradigm is in charge of supplying new SG models to manage response demand. The main advantage is the communication overhead reduction between devices compared to fully distributed SG models, where there is no possibility of improving power consumption due to the lack of sharing computing and communication information between devices and users.

A Cloud Computing approach in which all users (i.e., supplier and customer) are continuously connected to the Cloud can improve the development of centralized demand response management algorithms [17]. Considering macrogrids and MGs as Fog devices may reduce excessive and unnecessary communications between end users, typically due

TABLE 9.1 Classification of the application technologies.

Technology	Applications	Fog computing applications	Smart grid features	Big data
Energy management	• MG management • Dynamic demand response operated within the MG • Real-time monitoring on applications for SG	• Data metric communication implementing private fog for small-size networks • Dynamic bandwidth increasing for fog applications to avoid congestions • Fog MG to MG interaction • Demand response model definition	• MG management • Dynamic pricing	• High QoS for real-time BD applications
Information management	• Smart meter data streams in Cloud • Dynamic data center operation	• Guaranteed work-flow latency and processing rates with the help of Fog data optimization • Dynamic pricing model in SG architectures according to load on Fog Data Services • Adequate data transfer framework from users to Fog and vice versa	• Cost optimization • Data Storage and processing	• For BD processing with Platform-as-a-Service and Infrastructure-as-a-Service
Security	• Security and protection system for electric power information • Privacy preserving over encrypted metering data for SG	• Fog as software as a service for data privacy issues in large scale deployment of SG • Security mechanisms definition while using fog computing applications • Effective and efficient security and privacy policies to support increasing data from smart meters	• Data security and privacy • Threat detection • Cyber security	• Important to ensure that all technology and application components include and maintain acceptable levels of security and privacy mechanisms

to distance in a fully distributed model. The idea is to allow customers to communicate frequently with close Fog devices, allowing them to interact periodically with the Cloud.

Another proposed solution in SG applications is demand response management [18], where a Fog device is in charge of coordinating a mutual power exchange between MGs as well as MGs and the main grid. A new power management algorithm has been proposed in Ref. [18] to minimize loss power, create exchange pairs among MGs and prioritize communications. In the demand response management solution, two main layers can be identified. The first one is related to the consumers connected to the same Fog device and the way to obtain local information from it. Likewise, in the second layer, several Fog devices are attached to the same Cloud server. Fog devices can also be interconnected and, thanks to the collected information, they are able to group themselves in order to minimize power losses and consequently limit communication cost.

Another interesting solution has been proposed in Ref. [19], based on two main steps: the first one is a classification step in which heterogeneous EVs are dynamically grouped; then each sub-group is scheduled in terms of charging demand considering a sliding-window iterative approach.

9.3 Internet of things concept for microgrids, smart homes, and buildings

Nowadays, modern systems generate energy in a distributed manner, making management very difficult in terms of security, stability, and reliability. For MGs, the main contribution of energy is provided by Power Electronic-based (PE-based) energy sources. The main complexity is how to manage the controllability of DGs that influence thresholds of variables defined for a distributed system. Energy sources need to be deeply analyzed in terms of their:

- Stochastic trend: Nonlinear energy generation function is typically due to natural occurrences, such as the weather or atmospheric agents. In these cases, mathematical algorithms can help to optimize the management of energy.
- Controllability: Its variability can reach very high levels depending on the source nature, configuration, and size of the system.
- Emission awareness: Considering RES, they implicitly show an emission-aware nature, so can help to improve the environment.

Not all energy sources are "equivalent," especially considering transmission and conversion. Typically, consumers use electrical energy, both for its easy way to be managed and transferred, and for the presence of a detailed infrastructure. Nevertheless, consumers can behave differently considering their needs. Energy management could significantly diverge in terms of utility type (e.g., residential or industrial users) and its interpretation and evaluation are very difficult both for users and operators. A punctual self-organization of loads could significantly improve energy utilization, but a similar solution is more expensive and looks undesirable from the user's point of view.

In this context, the new IoT paradigm can help one to find innovative solutions to improve the management of energy systems. All objects (i.e., devices, sensors, etc.) can cooperate in collecting information to analyze consumers' behavior over time. Then, after defining a data processing system, all this information can be interpreted and converted in practical actions to curtail or shed loads based on user preferences. Another application used by system operators for load control acts on time-shiftable and thermostatically controllable loads using incentive-based offers.

A communication system is able to collect, process, and save information from each object. In the system infrastructure, it is necessary to have a decision-maker that, after sensing and processing procedures, is able to transmit commands to low-level components of the energy system. To do this, a two-way bidirectional communication system is necessary to allow all components in the energy system to exchange information. The stochastic nature of loads needs to be taken into account in designing the communication system, selecting as best as possible the components with a higher level of authority, which can be a very difficult choice considering the presence of many independent devices in a modern system [20].

Security is another main topic to be managed in an IoT architecture and this becomes even more complex in energy systems. Integrating wireless and wired communication networks in a modern energy system increases vulnerability due to the accessibility of information. For these reasons, a trusted privacy management system needs to be carried out in order to allow both consumers and providers to control their privacy and use a permission-based procedure. Many service providers can access different types of information and share them with third parties, which in turn raises issues about the privacy of the end-users. This is possible only with the end user's permission.

In fact there are energy systems where some components can change position, as in the case of a SG with EVs, in which the system load profile and scheduling strategies [21,22]. EVs need to communicate with local and central controllers, making crucial M2M communications. In the near future, the increase in the number of "green" loads will drastically change the behaviors of consumers and, consequently, energy consumption. It is expected that smart EVs will substitute traditional cars in future [23,24].

Considering the IoT perspective and the security of the system, two kinds of information can be "defined":

- The first one is typically related to the privacy of users and they don't want to share the position of EVs in a particular moment.
- The second one is related to warnings about the security of users' information, for example, protect against economic losses.

Energy systems are divided into four categories as indicated in Ref. [25] and each one presents a different security model: appliance group, monitor group, central controller, and interface with users. The main objective is to find a solution in terms of computational complexity, memory, and data integrity to reach the highest level of energy efficiency.

The main challenge is to find the best compromise in terms of exchanged information. Here, it needs to be considered that, while a greater information load exchanged between components would bring a considerable improvement in terms of energy consumption, at the same time the same greater information load would make the system more vulnerable from external attacks.

IoT systems, where the number of heterogeneous objects is very high, cannot use traditional methods to control devices from cyber attacks but, on the other hand, different security approaches can be adopted considering distributed and decentralized architectures characterized by several levels. In this context, machine learning methods can be used to improve security thanks to data collected by different sources located in different places even if considering the same physical layer [26].

Energy management systems and the strategies behind their design are driven by the objectives that consumers and providers desire to reach in terms of:

- reducing energy losses;
- maximizing the profit (i.e., for providers);
- minimizing voltage deviation;
- earning money (i.e., for consumers); and
- minimizing costs.

Therefore, both sides may act to limit divergencies in searching for a common solution. A new paradigm of energy systems has been carried out to allow energy customers to produce and provide their own energy to a distribution system. In this case, customers are called prosumers [27]. Thanks to bidirectional connections, prosumers are able to exchange also data with the system.

In a context in which smart buildings and smart cities are improving their performance thanks to the internet, the need to design an IoT-based platform to ensure efficient, secure, economical, and reliable energy distribution has already become a must.

Currently, two energy management strategies are commonly recognized as the best solutions:

- passive scheme: able to operate autonomously; and
- interactive scheme: local and global system information is shared between nodes with the objective of finding the best operating point.

The second scheme, the interactive power/energy management system (IP/EMS), needs more intelligence to integrate different data formats and technologies. It is useful to consider this heterogeneity both for the computational characteristics of each device and the constraints of each node.

Three different communication schemes can be performed in an IP/EMS:

- Centralized: A centralized controller is in charge of finding the optimum point for the objective, so that most end users do not have the authority to manage energy.
- Decentralized: Each end user can decide individually to optimize both production and consumption of energy. The main disadvantage is the higher complexity of the communication architecture due to more data transactions between nodes. In fact, in a scheme like this, nodes share data with neighbors to take the best decision.
- Hybrid.

9.4 Energy management in microgrids

9.4.1 The role of the energy management system

An MG EMS is the control hierarchy adopted to ensure the optimal management of the energy supplied by the DG units to the loads, and to manage the resynchronization response of the system when the MG switches between interconnected and islanded modes, based on the real-time operating conditions of MG components and the status of the system [5].

A power system based on a smart MG enables end users to play an active role in the energy demand-supply market, through the integration of controllable loads and renewable sources. The dynamics of demand and supply sides follow different variation criteria. As a consequence, the power supply may be unable to satisfy the total demand during peak periods, and a suitable control structure able to manage this issue needs to be implemented. In order to satisfy the needs of users, the missing power could come from the MG's storage devices or from other MGs or the macrostation. All these ways of power transmission generate energetic losses, which could be higher or lower, depending on the distance between the two nodes of exchange. Hence, the maximum efficiency in the power exchanges is another aspect which must be considered by the energy management system. Several algorithms and methods have been proposed in order to solve various issues, related to the energy management of MGs, including those previously mentioned [7].

On the basis of the methodology applied for implementing the control strategy, MG architectures can refer to a different network topology. The three main categories found in the literature are centralized, decentralized, and distributed control approach.

9.4.2 Centralized, decentralized and distributed energy management system

In a centralized approach, as shown in Fig. 9.4, the overall control is performed by a main unit, typically implementing a Supervisory Control and Data Acquisition (SCADA) system, which processes all the information received from the various devices over the

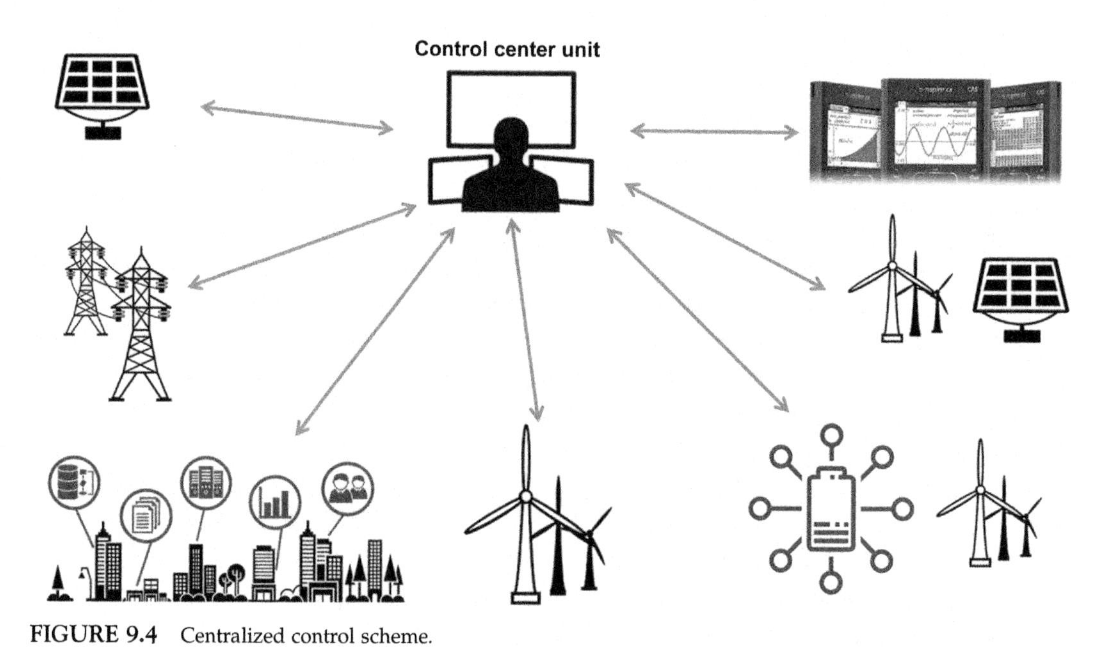

FIGURE 9.4 Centralized control scheme.

network and sends the control signals to all the agents (i.e., power generators, consumers, or prosumers of a MG) suitable to perform management of the generated power and its economic dispatch.

From a communication point of view, such an approach requires a bi-directional link for each of the agents controlled by the central SCADA system. Although centralized methods are currently in use in many power systems, such a control strategy is not the best approach for the implementation of future MGs, because of its limitations in terms of scalability, security, and effectiveness [4].

In a decentralized approach, presented in Fig. 9.5, each agent is able to perform its own control strategy, based on the local values of electrical parameters such as voltage, current, and frequency, in order to reach better results in terms of profit or stability.

Thus fewer communication channels are needed compared to those of the centralized methodology, since in a decentralized strategy an information exchange among agents, or between agents and the main SCADA, is required, except for some signals in particular that are sent and received between the leader agents and their local center. Any lack of communication links means that the global optimization and stability of the whole system cannot be guaranteed [5]. On the other hand, the privacy protection policies for the agents become more secure and reliable [28–30]. Due to its structure, a decentralized system results in better stability compared to a centralized system with the same connections. Indeed, in a decentralized system, when some connections among leaders and other agents are lost, at least some local sub-systems can still keep their stability.

A performance comparison of centralized and decentralized methods applied to a small-scale power system has been discussed in Ref. [31]. In both cases, an improvement in terms of transient stability has been provided.

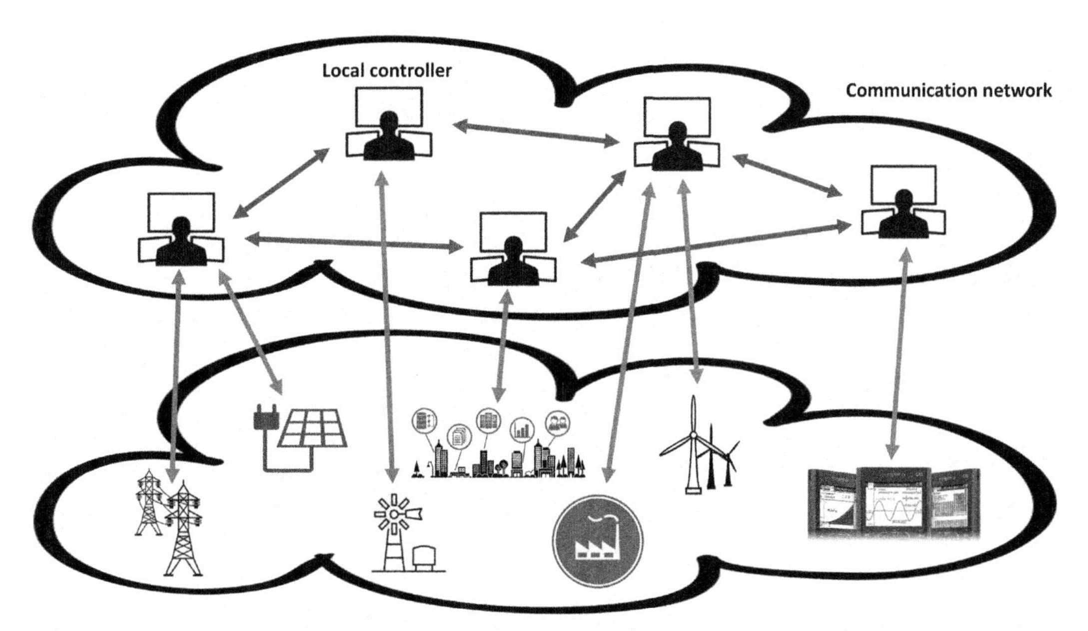

FIGURE 9.5 Decentralized control scheme.

In the distributed approach, each agent exploits the local information provided by its neighbors in order to achieve the global optimal point. This information sharing is performed by agents through local, bi-directional communication links.

The distributed control scheme overcomes one of the main limits of the decentralized method, since it can achieve global optimization as in the centralized approach, by allowing users to share their knowledge. Other important advantages of such methodology are plug-and-play and the fact that the power supply is unaffected by changes in the grid topology. In Fig. 9.6 an example of an MG implemented by using a distributed approach is shown.

Several distributed algorithms were studied and applied in the literature to MG and power system structures in order to solve management problems, mainly related to the economic dispatch of the energy [32–36]. A typical scenario refers to different roles in the electrical market, that is, producers, consumers, and prosumers, each one having its own private cost function which consequently sets the proper business strategy. A distributed approach appears to be the best solution to preserve the privacy policies needed to enable fair competition among different kind of subjects acting on the same market with different

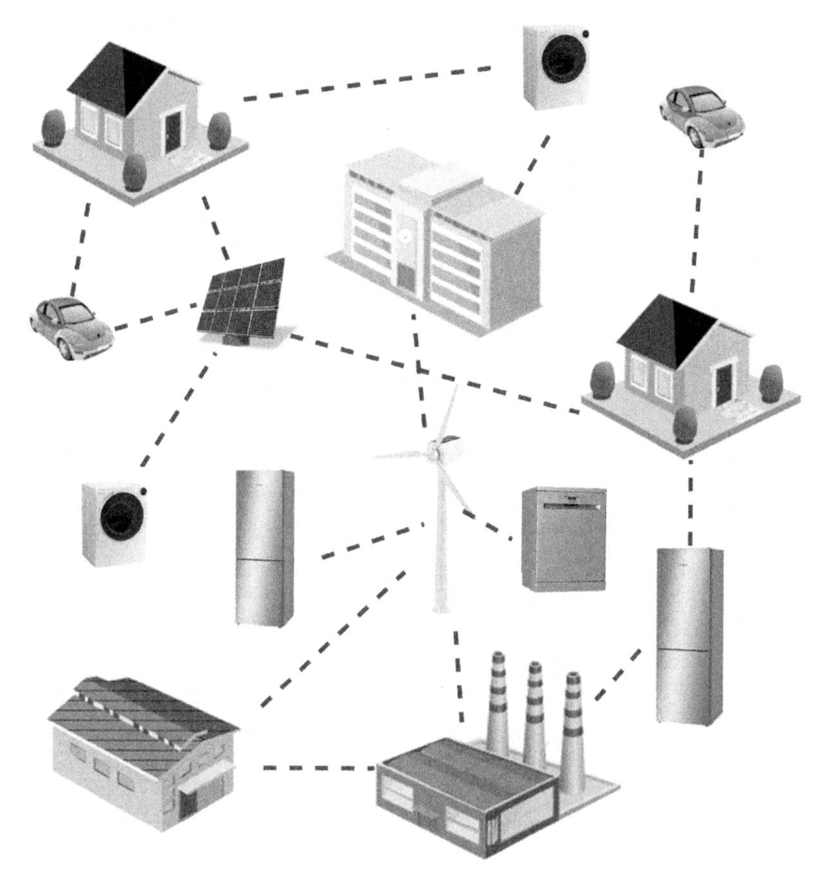

FIGURE 9.6 Microgrid (MG) controlled with distributed approach.

goals [5]. Table 9.2 shows the comparison of advantages and disadvantages of the three control schemes suggested in Ref. [5].

9.4.3 Energy management system in smart grid applications

Several energy management approaches which exploit a suitable interaction among these different approaches have been proposed in the literature [8,15,37]. This is typically a challenging task as the MG application scenarios refer to the energy dispatching problem, due to the intermittent nature of the power generation and the dynamic nature of the power absorption [15].

Demand Response (DR) and Demand Side Management (DSM) are two typical energy management strategies applied in SG applications, both designed to minimize the energy displacement or optimize the energy consumption through the alignment of demand-and-response dynamics to the current price of electricity, for example, by applying peak shaving, which is commonly used to reduce peak demand, by shifting power usage to off-peak hours and lowering total energy consumption.

Two market mechanisms, namely incentive-based and price-based, are typically considered when dealing with the implementation of a demand-response program. More complex and innovative market-based mechanism, such as time-of-use (TOU) pricing, critical peak pricing (CPP), extreme day CPP, or real-time pricing can be implemented through the DSM framework [5], the architecture of which is shown in Fig. 9.7.

The support technologies that enable the DSM strategy, such as wide area network (WAN) and home area network (HAN), are all bi-directional communication channels.

The term Virtual Power Plant (VPP) is frequently used when referring to an MG, representing a software system that enables the automatic and remote dispatch and

TABLE 9.2 Comparison of centralized, decentralized, and distributed control method.

Control type	Pros	Cons
Centralized	• Easy to implement • Easy to maintenance in case of single point failure	• Computational burden • Difficult to expand • Single point of failure (highly unstable) • High level of connectivity required
Decentralized	• Only local information • Equipped with control island-area also without leaders • Parallel computation	• Absence of communication links between agents • Moderate scalability
Distributed	• Easy to expand (high scalability) • Low computational cost (parallel computation)	• Needs synchronization • Time-consuming for local agents to reach consensus • Convergence rates affected by communication network topology • Two-way communication infrastructure need • Cost to upgrade on the existing control and communication

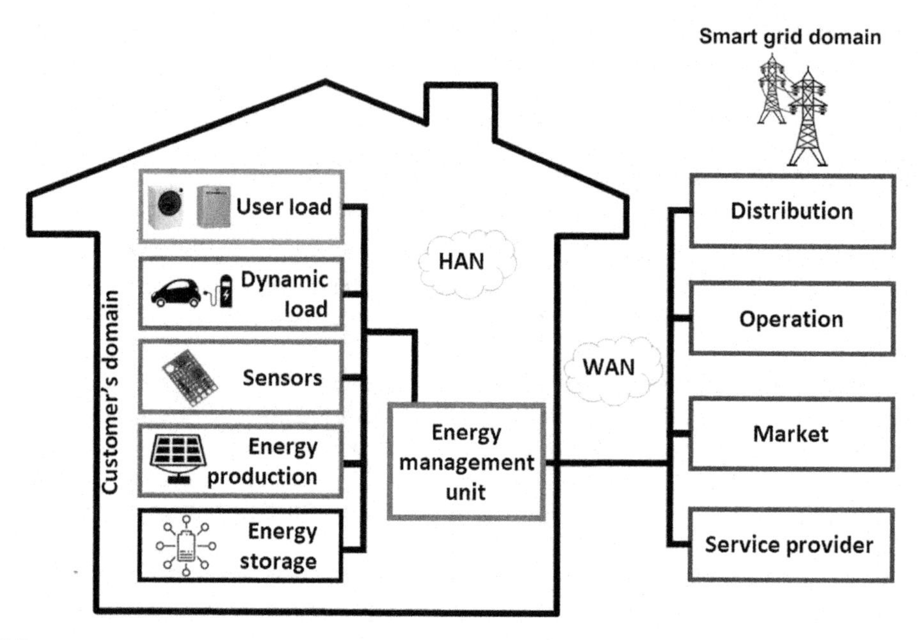

FIGURE 9.7 Architecture of Demand Side Management (DSM) framework.

optimization of power generation, demand, and storage in a single and secure web-connected way. VPP enables the concept of Internet of Energy, by allowing grid networks to customize supply-and-demand services for different customers, which maximizes revenues for both end users and distribution utilities [6]. A schematic diagram which highlights the versatility of a VPP is shown in Fig. 9.8. VPPs overcome the geographical boundaries and the static set of resource limits of a typical MG.

Sensors, smart meters, and data centers need to be connected to exchange collected information. For this reason, a two-way SG communication and information infrastructure is essential, and may be characterized by:

- interoperability,
- reliability,
- flexibility,
- scalability, and
- safety.

Moreover, taking into account the heterogeneity of devices in an SG, resulting in different communication requirements (i.e., frequency range, throughput, latency, etc.), a secure two-way backbone needs to be optimized to meet all of them [9,38]. Transmission and storage of information are highly critical and need to be kept as secure as possible to prevent cyber-attacks. Many cyber security solutions have been proposed and clearly described in the literature to make SG networks and devices safe [39,40].

SG communications can be categorized as HAN, neighborhood area network, and WAN in terms of the covered geographical region, as depicted in Fig. 9.9.

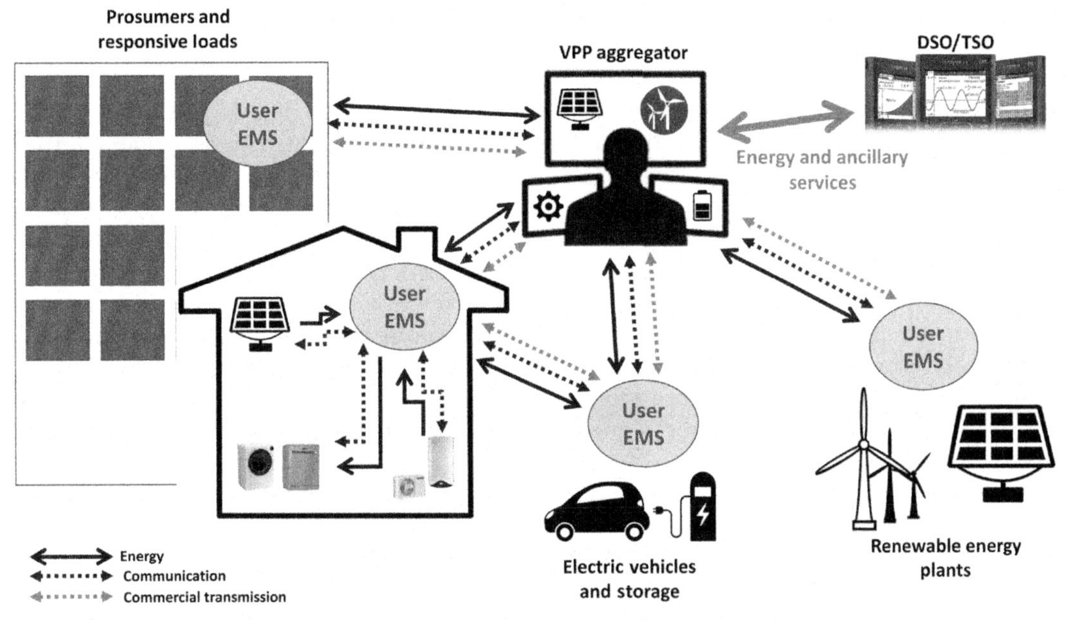

FIGURE 9.8 Diagram displaying Virtual Power Plant (VPP) versatility.

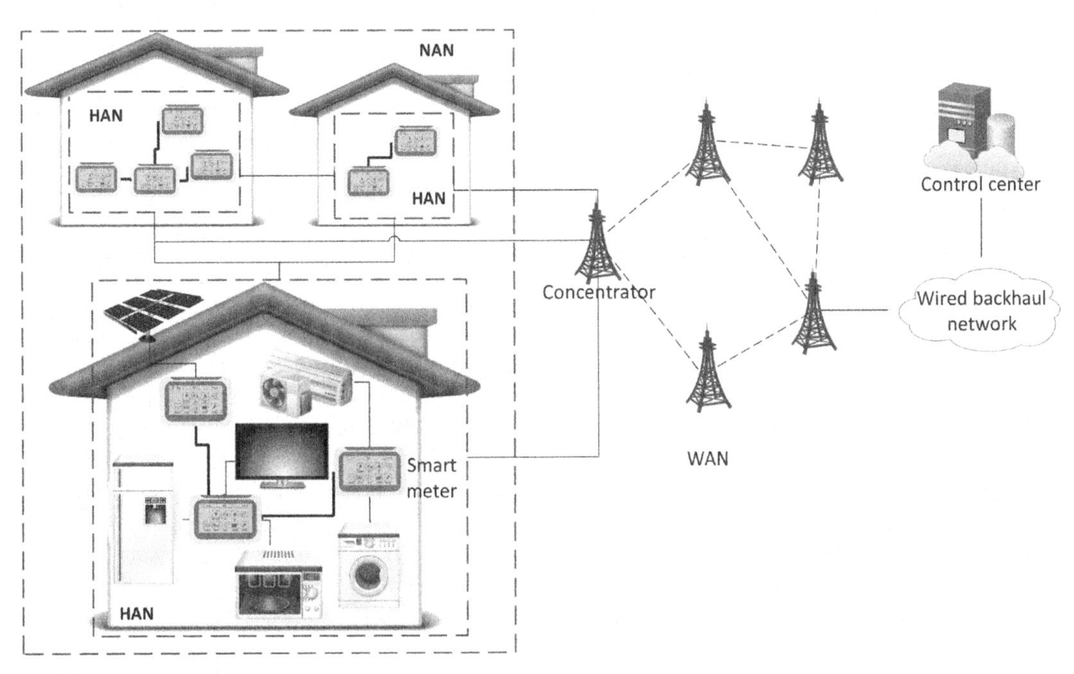

FIGURE 9.9 The architecture of the smart grid (SG) communication layer.

Typically, SG infrastructure comprises power line communication (PLC), wireless communication, cellular communication, and finally internet-based virtual private networks (VPN) [41].

PLC combines the electrical power distribution line with the data signal [42,43]: PLC acts by modulating carrier signals directly on the power cables. PLC represents an alternative broadband networking infrastructure without installing dedicated and expensive network wires such as twisted pair cables or fiber optics. The main disadvantage of PLC is that it suffers from multiple types of interference: since the power wiring is unshielded and untwisted, it acts as an antenna, thus causing an interference to the existing users of the same frequency band, as well as receiving interference from radio signals. Improper wiring and circuit breakers along the line can cause noise and connection interruptions, respectively. Moreover, when the interruption of electric service occurs, there may also be the interruption of the communication channel between the ICT devices. Furthermore, security issues can arise since the signal carried along the cables could be easily intercepted and, thus, cryptography is needed. Nevertheless, PLC's high availability, cost-effectiveness, and ease of installation and management have contributed to its widespread usage. There are two major PLC technologies that operate at different bandwidths: narrowband PLC (NB-PLC) and broadband PLC (BB-PLC) [44,45].

NB-PLC, which operates in the 3÷148.5 kHz frequency range, is characterized by low data rates (up to 500 kbps) and long distances (more than 150 km). It is mostly used for appliance control. BB-PLC is used for applications that require higher data rates (up to 200 Mbps on very short distances). It operates in the HF/VHF frequency bands (1.8–250 MHz) and it is not able to travel more than 1.5 km. It is typically used for in-home networking applications requiring a high data rate.

It is worth mentioning that fiber optic technology can be used in a software defined network (SDN) as wired communication media [46]. Since it provides high data rates and ensures reliable communications, it is mostly used to connect concentrators and provide communication between concentrators and control centers. It has many advantages such as data transmission over long distances and low losses, but it is much more expensive compared to PLC solutions and it is better suited to backbone or substation connections.

Along with PLC, the main communication media in the future are expected to be wireless media, because of their typically low installation costs and the network's easy design. In this regard, WiMax [47] is an option for wireless communication media within neighborhood area networks (NANs) and local area networks (LANs) [48], and its performance in an active DN management is assessed in Ref. [49]. It is defined by the IEEE 802.15.6 communication standard for broadband wireless communications. It works in the microwave frequency band at 2÷66 GHz, and it is characterized by low latency (lower than 100 ms round trip time), lower deployment, and reduced operating costs with respect to fiber optics, and the availability of traffic management tools to ensure a good quality-of-service (QoS) [50]. Furthermore, data exchange is secured by appropriately designed protocols, such as advanced encryption schemes.

Wireless sensor networks (WSNs) can be used widely for load monitoring and local component interactions (e.g., for distributed network management applications) because of their ease of installation and management, along with the low complexity and low deployments costs. The reference standard for low-rate, low-power WSNs is the IEEE

802.15.4 [48], whose most used protocol is the ZigBee. It operates at different frequency bands: 868 MHz in Europe, 915 MHz in the United States, and 2.4 GHz in the rest of the world. It is characterized by low data rates (up to 250 kbps) and low communication ranges (up to 100 m).

The latest mobile communication technologies are the most promising choice for communication between SMs/IEDs and the DMS/EMS since they support wide coverage, low latency, high throughput, and QoS differentiation [51]. In particular, 5G/long term evolution (LTE) networks are envisioned to be highly used in the near future, thanks to their high data rates, high availability, and low energy consumption [52–55]. In high-density device grids, such as SDNs, where a large number of devices are connected to a network, 5G facilitates the data exchange from the remote devices, with statistical observations and analysis, allowing their analysis in real-time to optimize the energy distribution [56]. Such requirements are guaranteed with 5G, which has very large bandwidth, a high-speed connection, and low-energy consumption [57]. Despite the frequency bands not having been assigned yet, it is supposed to operate in the mm-wave band.

Machine-type communication (MTC) is defined as the data exchange and processing among machines with minimum intervention from humans. Different perspectives of MTC challenges are defined and attempts are made to overcome them in the Third Generation Partnership Project (3GPP) by different solutions and technologies like LTE [58] and low power wide area network (LPWAN) [59] introduced as NarrowBand IoT [60].

An internet VPN converts an established public network into a high-speed private network to carry its traffic. There are lots of technologies proposed in the literature, such as internet protocol security (IPsec) and multiprotocol label switching (MPLS), both based on overlay and peer-to-peer VPN architectures [61,62]. As a drawback, overlay models require fully meshed circuits for optimal routing and it is difficult to size intersite circuit capacity. Peer-to peer models also need complex filters and all VPN routes are carried in the service provider. Unlike MPLS, IPsec suffers from scalability problems in the stage of managing large VPNs and it does not provide a robust connection due to the deficiency of predictable communication performance over the internet.

9.4.4 Issues emerging in sensing, measurement, control, and automation technologies

A smart meter is an advanced energy meter that can [63–65]: provide real-time energy consumption and power quality monitoring; provide information on prices; enable demand-response integration through dynamic tariff schemes; support appliance energy management; and detect safety-and-security hazards such as blackouts, brownouts and electricity thefts. Even though their use represents a number of opportunities, it also introduces some issues to be taken into consideration [66].

- Scalability: All the elements of an MG network need addresses and identity numbers. Furthermore, the amount of data logs to be stored is proportional to the number of smart meters installed in the network. Therefore, scalability problems arise when the number of customers, and thus smart meters, increases [67,68]. This is particularly true if their DERs are also installed and interconnected [69,70].

- Uncertainties: Benefits introduced by SGs may not be measurable and/or readily evident (e.g., savings resulting from new infrastructure or peak energy reduction). As a result, electricity distributors may find difficulties in passing on their savings to retailers, or from retailers to consumers [71]. Other causes of uncertainty are represented by core permeability, core losses, leakage reactance, winding resistance, and iron-cored current transformers, which present magnetizing current, magnetic saturation, flux leakage, and eddy current heating [63,72]. All these elements can introduce measurement errors which affect the DSM performance.
- Safety: The presence of users can be inferred from energy consumption and appliance usage data [73]. Furthermore, various malicious actions may be generated [74,75]: production of fake smart meter readings, manipulation of energy costs and/or computations for energy management, smart meter control, smart meter cloning, and the violation of users' smart meter to get access to their home network and secure applications.
- Costs: Although DSM enables a reduction in bills for customers and peak demand for providers, it also increases the operational complexity of the system without reducing the overall appliance energy consumption [76,77]. The amount of data that is exchanged among network elements and computed to operate a DSM may be a cause of network, computational, and memory overloads [78,79]. Furthermore, not only does DSM require a great effort from customers, who have to keep themselves informed about DR programs and frequent price changes, but it is also expensive to purchase and install smart meters [80,81]. As far as providers are concerned, it is not clear who should be in charge of promoting DR programs. In addition, the reduction of peak demand reduces the income due to peak energy cost. On the other hand, the rebound of energy usage after high price times can cause new, unexpected peaks [82]. Finally, the DR infrastructures require an initial investment that may be difficult to recover [83].
- Standardization: The different rates of regional development hinder the definition of a unified standard policy system for DSM [84]. Additionally, the hardware installed and its related software are typically vendor-dependent [85].

9.5 Smart energy management system architectures

The smart energy management system (SEMS) exploits the interoperability of various wired and wireless communication technologies, with a suitable software layer, to offer a complete solution for the automation of tasks, such as energy measurement or device control [7]. A generic SEMS architecture is shown in Fig. 9.10.

The key components of the system are:

1. the data collection engine, which enables in a scalable way the acquisition, aggregation, decoding, check, validation, and storage of heterogeneous data coming from different metering systems, through custom interfaces called "adapters";

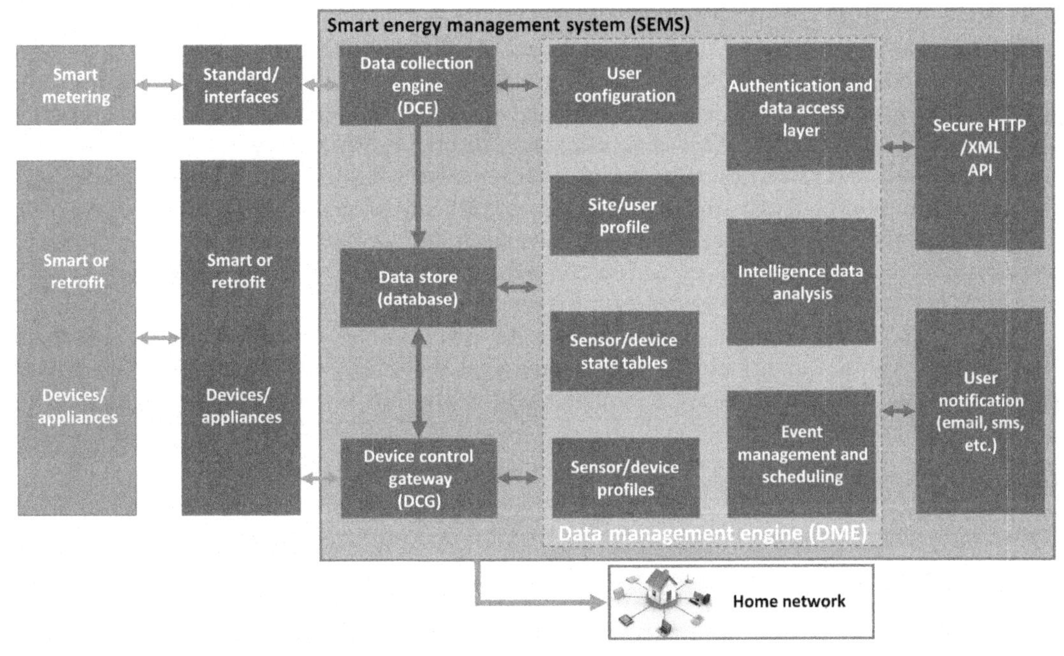

FIGURE 9.10 Smart energy management system architecture.

2. the device control gateway, which enables the control and automation capabilities of the EMS, by supporting several smart appliance control technologies and protocols, through individual plug-in adapters for connection to devices and sensors;
3. the data management engine which enables secure interaction between the user interface (UI) components and the energy monitoring and control functionalities; and
4. the local and remote UI, which is built by using the data access and device control API provided by the energy management system, by using cross-platform XHTML and AJAX web technologies to enable access through any kind of web-enabled devices, such as PCs, workstations, touch screens tablets, personal digital assistants and smartphones. The UI allows users to get information in an easy and detailed way, providing also reporting tools and rule-based notifications or alert messages.

In SDNs, energy consumption and RES production are measured by smart metering systems [86]. The data gathered from the SM are sent through wired or wireless networks to the EMS, which may be also responsible for billing-related functionalities. Based on the level of intelligence and on the provided functionalities, the metering system can be classified as one of the following [86–88]:

- automated meter reading (AMR);
- automatic meter management (AMM); and
- advanced metering infrastructure (AMI).

In AMR, the data acquired from SMs are sent to a central system every hour, day, week, or year. The central system is responsible for analyzing data and performing billing procedures based on real-time consumption data, rather than on an estimated energy consumption. AMM extends AMR functionalities by allowing bidirectional data exchange, therefore enabling limited forms of meter control. Thanks to AMM, demand response programs offering time-based rates, can be implemented to encourage users to save energy during peak-demand times. AMI is an evolution of both AMR and AMM toward IoT systems. AMR and AMM can be considered as subsystems of the AMI. The AMI is defined as the combination of hardware (e.g., SMs, communication media) and software (e.g., the data management system) used to ensure measurement, storage, and processing of user's consumption data. AMI provides various interfaces to interact with both users and service providers by using ICT technologies, summarized in Table 9.3.

Fig. 9.11 shows the typical architecture of an AMI [86,89] able to integrate SMs with other objects inside a smart home, such as energy-consuming appliances and RES power generation [90]. These objects are connected within the HAN. Smart homes belonging to

TABLE 9.3 Characteristics of main communication technologies used in advanced metering infrastructure (AMI).

Technology	Standards	Data rate	Frequency band	Communication range	Network
NB-PLC	• IEC 61334 • G3-PLC • PRIME • ITU-TG. HNEM • IEEE P1901.2	• Single carrier: tens of kbps • Multicarrier: <500 kbps	• 3÷148.5 kHz (EU: CENELEC band 3–148.5 kHz)	• >150 km	• NAN, LAN, WAN
BB-PLC	• IEEE 1901 • ITU-T G.9960/61 • HomePlug	• <200 Mbps	• 2÷86 MHz	• <1.5 km	• HAN
Fiber optics	• IEEE 802.3ah • ITU T G.983/984	• <10 Gbps	• ~186÷236 THz	• <60 km	• WAN
WSN	• IEEE 802.15.4 (ZigBee)	• <250 kbps	• 2.4 GHz • EU: 868 MHz • United States: 915 MHz	• <1600 m	• HAN, NAN
WiMAX	• IEEE 802.16	• <1 Gbps	• Typically 2.3, 2.5, and 3.5 GHz	• Good: 0÷30 km Bad: 30÷100 km	• NAN, LAN, WANN
Mobile communication	• 2G, 3G, 3.5G, 4G, 4.5G • 5G (expected)	• <1 Gbps	• Typically 700, 850, 1800, 1900, 2100, 2300, 2600 MHz	• Good: 0÷30 km Bad: 30÷100 km	• LAN, WAN

(Continued)

TABLE 9.3 (Continued)

Technology	Standards	Data rate	Frequency band	Communication range	Network
Thread	• IEEE 802.15.4	• 250 kbps	• 2.4 GHz	• 30 m	
Z-wave	• Z-Wave	• 100 kbps	• 900 MHz	• 30 m	
LoRa	• LoRa	• 100 kbps	• 863, 915 MHz	• >10 m	
Sigfox	• Sigfox	• 10, 100 kbps	• 863, 915 MHz	• >10 m	
Bluetooth	• IEEE 801.15.1	• 1 Mbps	• 2.4 GHz	• 10 m	
Satellite	• IEEE 521	• 1 Mbps	• 30÷300 GHz	• 6000 km	

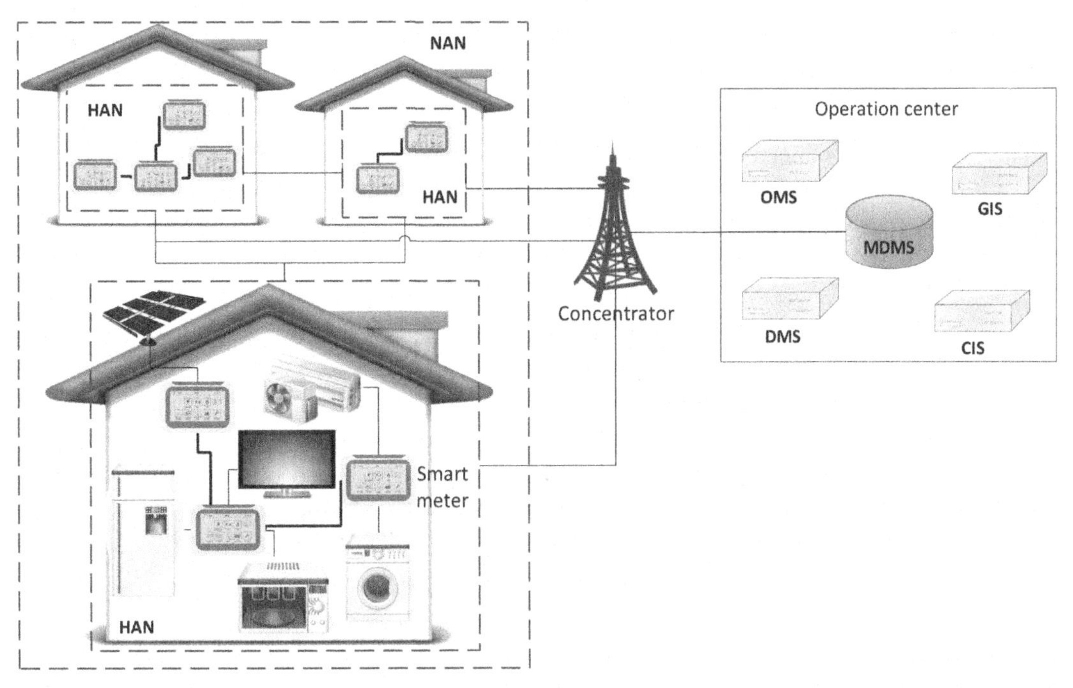

FIGURE 9.11 Basic elements of an advanced metering infrastructure (AMI).

the same neighborhood can cooperate, for example sharing the power generated by a photovoltaic plant. In this case, they form a NAN.

Data collected from HANs and NANs are sent to the closest concentrator, which forwards them to the closest operation center [89]: if only one central operation center is responsible for the whole system, the AMI is said to be centralized; if each LAN is served by one operation center, the AMI is distributed. Each distributed operation center is locally responsible for analyzing and managing data, as well as for making decisions. Only summaries and required integrated information are sent to the central operation center through a backbone network. This distributed approach offers multiple advantages with respect to the centralized one: not only does it reduce communication overheads and save on communication resources, but it also improves stability and resilience to noise and interferences, thanks to shorter communication distances [55,91].

The core of an operation center is the metering data management system (MDMS) [86,89,92]. It is responsible for analyzing and storing metering data, controlling SMs remotely, and performing billing operations. Furthermore, the MDMS provides interfaces with the following functional elements of the operation center:

- The outage management system, which detects, manages and registers power outages.
- The geographic information system, which provides geographic information about the location of the elements of the SDN.
- The consumer information system, which manages information about the consumer, such as consumption rates and billing-related data. It enables the development of new products and services, based on the consumer profile.
- The data management system, which provides control, management and forecasting functionalities.

The whole AMI is based on data collected from SMs and sent through wired and wireless communication media to operation centers, where they are stored and processed so that decisions can be made and sent back to the consumer's premises.

9.6 Case study: communication and internet of things from smart energy management

In order to demonstrate how communication and IoT can improve energy management in MGs, a smart home energy management (SHEM) system based on a profile characterization of the involved users' appliances is analyzed hereinafter [93,90]. The aim of such a system is to shift in a dynamic way the tasks of controlled appliances to lower the overall energy cost of a household, while also exploiting RES. The main benefits introduced by it are:

- the joint optimization of DER and grid network power usage in a neighborhood, taking into consideration uncertainties of the former and cost variations of the latter in a real-time manner;
- the classification of appliances pertaining to different classes and with the most diverse usage patterns; and
- the fact that users are differentiated and clustered based on their appliance usage preferences and needs.

The reference scenario is that of a group of houses such as a block or a condominium, namely the Cooperative Neighborhood. The rationale behind considering such a neighborhood is that in case the energy produced by RESs in a smart home at a given moment

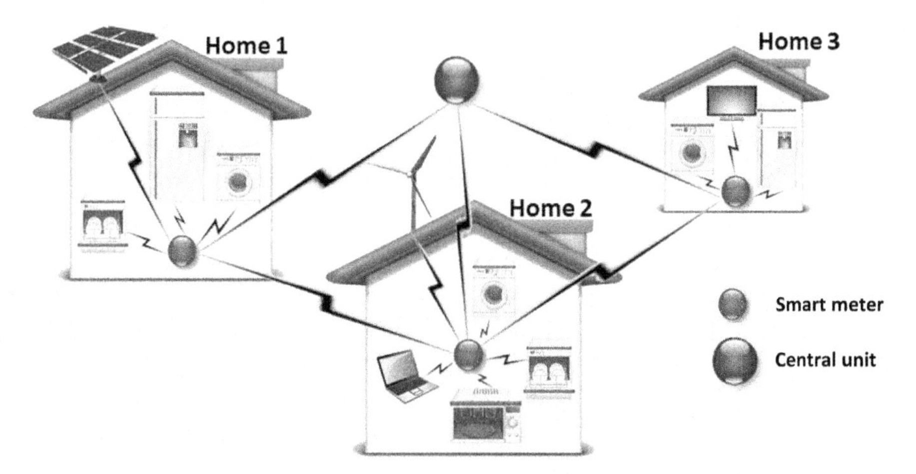

FIGURE 9.12 Reference case study scenario.

cannot be partially or entirely used by loads in the same home, this energy can be transferred to one of the neighbors [83,84].

Considering Fig. 9.12, inside each house there are appliances that consume energy. On the other hand, power supplies such as the electric grid, solar panels, and micro wind turbines provide energy that can be used to run appliances. Smart meters and actuators are associated with these appliances to monitor their energy consumption/production and control their activation/deactivation. The appliances are divided into four groups, based on their characteristics and requirements:

- G1: not controlled loads, that is, small loads such as lights[1] and smartphone chargers, and not controlled high loads such as freezer and fridge[2];
- G2: switching controlled high loads, for example, washing machines and clothes dryers;
- G3: thermostatically controlled high loads, that is, appliances that are controlled by a thermostat such as Heating Ventilation and Air Conditioning systems and water heaters; and
- G4: supplies such as solar panels and micro wind turbines.

At first, when a new appliance is plugged in a HAN, information related to the appliance's characteristics and the tasks it can perform will be detected by smart meters and sent to a central unit that connects all the households in the neighborhood. Users' habits, that is, how family members usually use appliances, are monitored and sent to the central unit as well. Based on this information, a profile of their energy consumption habits, namely a user profile, is associated with the users. If, for example, the house is empty during working hours, it is unlikely that appliances such as the TV or lights are turned on during this time span. This information is used as input to the energy management

[1] Although not controlled, presence sensors can be used for automatically switching lights on and off so as to save energy when nobody is in a room.

[2] Freezers and fridges account for approximately 10% of the total household consumption [28].

algorithms in the SHEM system, which will decide the best scheduling for each controlled appliance.

Consider the appliances (or energy sources) in the entire cooperative neighborhood indexed with $i \in \{1, 2, \ldots, I\}$ and the homes indexed with $h \in 1, 2, \ldots, H$. Each house's smart meter, namely ESM_h, is enhanced with storage and processing capabilities. Accordingly, it stores the key parameters about appliance i, depending on which group it belongs to, as illustrated in Table 9.4.

The SHEM system studied in this section is designed to perform three basic functions:

- It monitors and analyses users' habits with reference to appliance usage. Based on this information, a user profile is created.
- It detects power surplus due to RESs production and distributes this power to the houses of the same neighborhood, with the aim of maximizing its consumption.

TABLE 9.4 Key parameters of appliances for the smart home energy management (SHEM) system.

Type	Parameter	Description
G1	G_h^1	Set of appliances of G1 for home h
	$x_i(t)$	State (on/off) for appliance i at time t
	P_i^{cons}	Power consumed by appliance i
	$Pr_i(t)$	Probability that appliance i is on at time t
G2	G_h^2	Set of appliances of G2 for home h
	$x_i(t)$	State (on/off) for appliance i at time t
	P_i^{cons}	Power consumed by appliance i
	t_i^{exec}	Time needed by appliance i to perform its tasks
	$t_i^{\min ST}$	Minimum starting time for appliance i
	t_i^{DL}	Latest deadline for appliance i
G3	G_h^3	Set of appliances of G3 for home h
	$x_i(t)$	State (on/off) for appliance i at time t
	P_i^{cons}	Power consumed by appliance i
	$t_i^{exec}(T_i^{\exp})$	Time needed by appliance i for the temperature to reach the expected temperature $T_i^{\exp}$
	$[T_i^{\min}, T_i^{\max}]$	User's preferred temperature interval
G4	G_h^4	Set of RESs of G4 for home h
	$X_i(t)$	State (on/off) for RES i at time t
	P_i^{prod}	Power produced by RES i at time t
	$\Pr_i(t)$	Probability that RES i has power to deliver at time

- It sets the most convenient starting time of controllable appliances so that they are turned ON when it is more convenient, according to TOU tariffs and RES energy production. In order to accomplish this function, two algorithms are used:
  - The cost saving task scheduling (CSTS), which schedules tasks characterized by a high-power load in off-peak times, considering the user profile; and
  - The renewable source power allocation (RSPA), which dynamically shifts tasks in order to maximize the use of renewable energy that is made available by neighbors.

The sequence of steps to be performed is shown in Fig. 9.13.

As soon as appliance i placed in home h needs to start, it sends an activation request to ESM_h. If appliance i is not controllable or it is not a supplier (i.e., it belongs to G_h^1 or G_h^4) it just needs to notify the ESM_h that it is changing state ($x_i(t) = ON$) for the whole duration

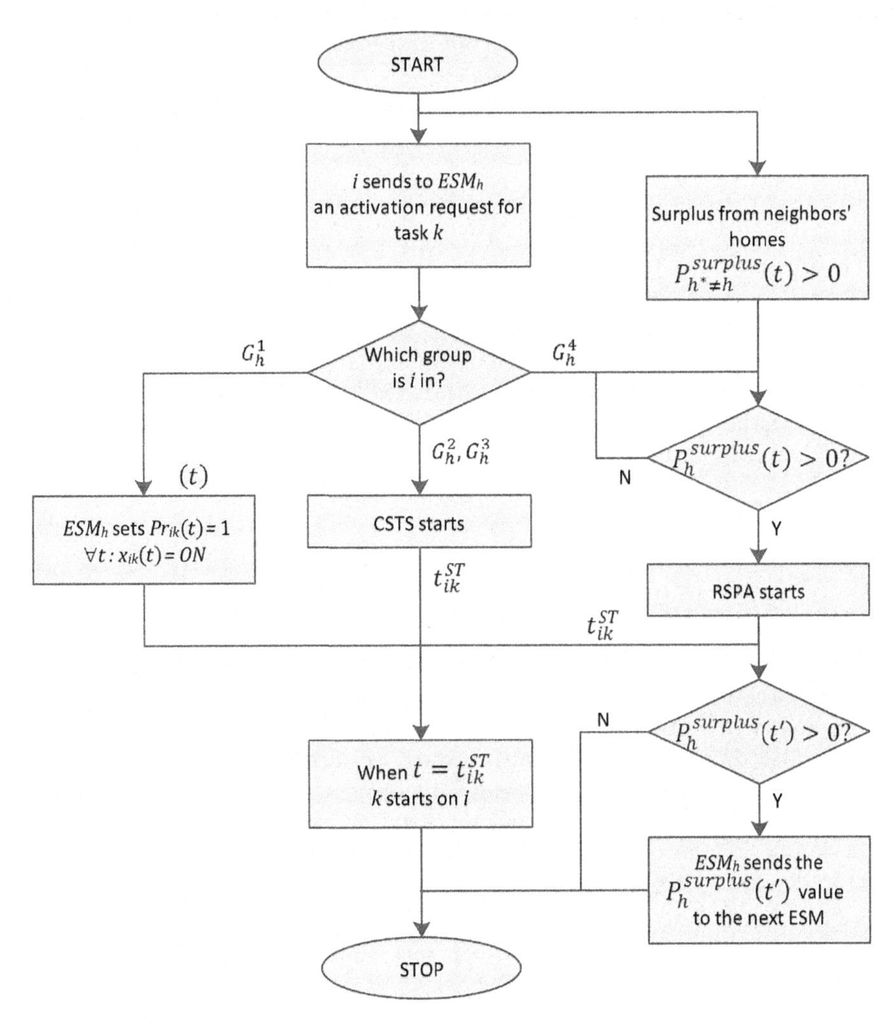

FIGURE 9.13 Task assignment steps.

of the task. ESM_h sets its probability to be on to 1 accordingly. When appliance i stops, it informs ESM_h, which sets $Pr_i(t)$ to its probability to turn on again, according to the user profile. Its power consumption and duration values are monitored and sent to the central unit, which analyses them and updates the user profile accordingly.

If appliance i is a controllable consumer, that is, it belongs to G_h^2 or G_h^3, the CSTS algorithm is started. CSTS is a centralized algorithm that is performed by the ESM to postpone the starting time t_i^{ST} of G_h^2 and G_h^3 appliances, so that their tasks are executed, if possible, during off-peak hours, when the electricity charge is lower.

For G_h^2 appliances, which are controlled according to users' time preferences, the user can set the minimum starting time t_i^{minST} and the deadline t_i^{DL} when the task needs to be carried out. For G_h^3 appliances, which are controlled according to users' temperature preferences, the user can set the preferred temperature interval, that is, T_i^{min} and T_i^{max}. Therefore, the starting time t_i^{ST} is computed by the CSTS according to the user preferences, provided that the available power P^{max} is not exceeded by the simultaneous usage of the appliances that made an activation request.

The *cost contribution* of appliance $i \in G_h^2$ starting at time t_i^{ST} and ending at time $t_i^{END} = t_i^{ST} + t_i^{exec} \leq t_i^{DL}$ is defined as in (9.1):

$$C_i^{G2}\left(t_i^{ST}\right) = P_i^{cons} \cdot \int_{t_i^{ST}}^{t_i^{END}} \phi(t)dt \tag{9.1}$$

where $\phi(t)$ is the electricity tariff at time t.

For an appliance $i \in G_h^3$, two cases need to be distinguished: if, at the current time t^{cur}, the temperature $T_i^{cur} = T(t^{cur})$ is outside the preferred temperature interval $T_i^{cur} \notin \left[T_i^{min}, T_i^{max}\right]$, the appliance needs to start immediately, and thus its starting time is set to $s_i^{ST} = t^{cur}$, and the system is only left to decide the optimal ending time t_i^{END}; otherwise, the system decides both the starting and ending times. Therefore, the cost contribution of appliance $i \in G_h^3$ starting at time t_i^{ST} and ending at time t_i^{END} is evaluated as in (9.2):

$$C_i^{G3}\left(t_i^{ST}, t_i^{END}\right) = P_i^{cons} \cdot \int_{t_i^{ST}}^{t_i^{END}} \phi(t)dt \tag{9.2}$$

where $t_i^{END} = t_i^{ST} + t_i^{exec}\left(T_i^{exp}\right)$, with $t_i^{exec}\left(T_i^{exp}\right)$ defined as the time needed for the temperature to reach $T_i^{exp} \in \left[T_i^{min}, T_i^{max}\right]$. The value $t_i^{exec}\left(T_i^{exp}\right)$ depends on the characteristic parameters of the appliance. The computation of its value falls outside the scope of this paper and will not be explained here. The reader can find the details in Refs. [90,93,94].

Let Λ_h be the array of appliances $i \in G_h^2, G_h^3$ that made an activation request, either because a new task of a G2 appliance has to start, or because the current temperature T_i^{cur} of a G3 appliance is not inside the preferred temperature interval. The problem to be solved by the CSTS algorithm is expressed by (9.3):

$$\min \sum_{i \in \Lambda_h} \sum_{t,t'} C_i^{G2}(t) y_i(t) + C_i^{G3}(t,t') y_i(t) y_i(t') \tag{9.3}$$

$$st.\ y_i\left(t = t_i^{ST}\right) = 1,\ y_i\left(t \neq t_i^{ST}\right) = 0$$

$$y_i\left(t' = t_i^{END}\right) = 1,\ y_i\left(t' \neq t_i^{END}\right) = 0$$

where the constraints guarantee that the cost is only considered if t is the starting time and t' is the ending time for node i. The starting and ending times of the appliances are therefore scheduled according to the CSTS results.

When a supplier (i.e., an appliance that belongs to G_h^4) starts producing some power, or a power surplus coming from neighboring houses is detected by ESM_h, the ESM computes the $P_h^{surplus}(t)$ value of the power surplus related to house h at time t. $P_h^{surplus}(t)$ takes into account all the power surplus contributions that are made available by the neighbor houses along with the power supplied by G_h^4 appliances, and it is decreased by the power consumed by the appliances inside home h if they are on, according to (9.4):

$$P_h^{surplus}(t) = \sum_{h* \neq h} P_{h*}^{surplus}(t) - \sum_{i \in G_h^1} P_i^{cons} \cdot Pr_i(t) - \sum_{i \in \{G_h^2, G_h^3\}} P_i^{cons} \cdot x_i(t) + \sum_{i \in G_h^4} P_i^{prod}(t) \tag{9.4}$$

Whenever $P_h^{surplus}(t) > 0$ is verified, ESM_h broadcasts this information to the appliances it controls. If there is any G_h^2 or G_h^3 appliance that is waiting to turn on and its power consumption is lower than the available surplus power, RSPA is started. Since the cost to consume the power produced by RESs is 0, the benefit of the appliance being turned on when surplus power is available is proportional to its power consumption and in inverse proportion to the time before its deadline. Indeed, when referring to energy-cost saving, appliances that consume more power P_i^{cons} have the priority to be scheduled when it is more convenient, that is, when surplus power is available. Furthermore, priority needs to be given to tasks characterized by closer deadlines. Calling t the current time, tasks with closer deadlines have higher values of the ratio $t/(t_i^{DL} - t_i^{exec})$. Summarizing, the benefit value is defined as in (9.5):

$$b_i(t) = P_i^{cons} \cdot \frac{t}{t_i^{DL} - t_i^{exec}} \tag{9.5}$$

The RSPA algorithm assigns the surplus power to the appliances characterized by the highest benefit values. The available surplus power is then updated by subtracting the power consumption of the appliances that have been selected, and the tasks are started on the selected appliances. If there is any surplus power still available, it is sent to the closest ESM.

9.7 Conclusion

The most interesting approaches referring to state-of-the-art technologies in the integration among power, control, and communication systems in MGs, smart buildings, and smart homes are presented and discussed in this chapter. It describes in detail the enabling models and technologies, which are able to manage heterogeneous DERs, for the purposes of enhancing the integration and coordination of small-scale power generation for self-consumption, as well as for trading energy or selling ancillary services on the open market and to distribution system operators.

References

[1] N. Andreadou, M.O. Guardiola, G. Fulli, Telecommunication technologies for smart grid projects with focus on smart metering applications, Energies 9 (2016) 375.

[2] A.R. Di Fazio, T. Erseghe, E. Ghiani, M. Murroni, P. Siano, F. Silvestro, Integration of renewable energy sources, energy storage systems, and electrical vehicles with smart power distribution networks, J. Ambient Intell. Human. Comput. 4 (6) (2013) 663–671.

[3] E. Ghiani, et al., A multidisciplinary approach for the development of smart distribution networks, Energies 11 (2018) 2530.

[4] H. Pourbabak, T. Chen, B. Zhang, W. Su, Control and energy management system in microgrids, in: Clean Energy Microgrids, The Institution of Engineering and Technology (IET) 2017.

[5] P. Asmus, Microgrids, virtual power plants and our distributed energy future, Electr. J. (2010).

[6] R.H. Lasseter, MicroGrids, in: 2002 IEEE Power Eng. Soc. Winter Meet. Conf. Proc. (Cat. No. 02CH37309), 2002.

[7] W. Su, J. Wang, Energy management systems in microgrid operations, Electr. J. (2012).

[8] H.T. Haider, O.H. See, W. Elmenreich, A review of residential demand response of smart grid, Renew. Sustain. Energy Rev. (2016).

[9] V.C. Güngör, et al., Smart grid technologies: communication technologies and standards, IEEE Trans. Ind. Inform. (2011).

[10] Edge computing vs. Fog computing: Definitions and enterprise uses. [Online]. Available from: <https://www.cisco.com/c/en/us/solutions/enterprise-networks/edge-computing.html>.

[11] M. Chiang, T. Zhang, Fog and IoT: an overview of research opportunities, IEEE Internet Things J. (2016).

[12] P.O. Ostberg et al., Reliable capacity provisioning for distributed cloud/edge/fog computing applications, in: EuCNC 2017 - European Conference on Networks and Communications, 2017.

[13] New IEEE 1934TM standard delivers framework for developing applications and business models enabled by fog computing. Available from: <https://standards.ieee.org/news/2018/ieee1934-standard-fog-computing.html>.

[14] I. Stojmenovic, Fog computing: a cloud to the ground support for smart things and machine-to-machine networks, in: 2014 Australasian Telecommunication Networks and Applications Conference, ATNAC 2014, 2015.

[15] H.S.V.S.K. Nunna, A.M. Saklani, A. Sesetti, S. Battula, S. Doolla, D. Srinivasan, Multi-agent based Demand Response management system for combined operation of smart microgrids, Sustain. Energy Grids Netw. (2016).

[16] J. Byun, I. Hong, S. Park, Intelligent cloud home energy management system using household appliance priority based scheduling based on prediction of renewable energy capability, IEEE Trans. Consum. Electron. (2012).

[17] Z.M. Fadlullah, D.M. Quan, N. Kato, I. Stojmenovic, GTES: an optimized game-theoretic demand-side management scheme for smart grid, IEEE Syst. J. (2014).

[18] C. Wei, Z.M. Fadlullah, N. Kato, I. Stojmenovic, On optimally reducing power loss in micro-grids with power storage devices, IEEE J. Sel. Areas Commun. (2014).

[19] R. Jin, B. Wang, P. Zhang, P.B. Luh, Decentralised online charging scheduling for large populations of electric vehicles: a cyber-physical system approach, Int. J. Parallel Emergent Distrib. Syst. (2013).

[20] M. Tahanan, W. van Ackooij, A. Frangioni, F. Lacalandra, Large-scale unit commitment under uncertainty, 4OR (2015).

[21] D. Van Der Meer, G. RamChandra Mouli, G. Morales-España, L. Ramirez Elizondo, P. Bauer, Erratum to: energy management system with pv power forecast to optimally charge evs at the workplace (IEEE Transactions on Industrial Informatics (2018) 14:1 (311-320), IEEE Trans. Industr. Inform. (2018).

[22] K. Mahmud, G.E. Town, S. Morsalin, M.J. Hossain, Integration of electric vehicles and management in the internet of energy, Renew. Sustain. Energy Rev. (2018).

[23] M. Fadda, M. Murroni, V. Popescu, Interference issues for VANET communications in the TVWS in urban environments, IEEE Trans. Veh. Technol. (2016).

[24] P. Angueira, M. Fadda, J. Morgade, M. Murroni, V. Popescu, Field measurements for practical unlicensed communication in the UHF band, Telecommun. Syst. (2016).

[25] T. Song, R. Li, B. Mei, J. Yu, X. Xing, X. Cheng, A privacy preserving communication protocol for IoT applications in smart homes, IEEE Internet Things J. (2017).
[26] A. Buczak, E. Guven, A survey of data mining and machine learning methods for cyber security intrusion detection, IEEE Commun. Surv. Tutorials (2015).
[27] R. Zafar, A. Mahmood, S. Razzaq, W. Ali, U. Naeem, K. Shehzad, Prosumer based energy management and sharing in smart grid, Renew. Sustain. Energy Rev. (2018).
[28] L. Gan, U. Topcu, S.H. Low, Optimal decentralized protocol for electric vehicle charging, IEEE Trans. Power Syst. (2013).
[29] Y. Guo, J. Xiong, S. Xu, W. Su, Two-stage economic operation of microgrid-like electric vehicle parking deck, IEEE Trans. Smart Grid (2016).
[30] Y. He, B. Venkatesh, L. Guan, Optimal scheduling for charging and discharging of electric vehicles, IEEE Trans. Smart Grid (2012).
[31] T. Senjyu, et al., Power system stabilization based on robust centralized and decentralized controllers, 2005 Int. Power Eng. Conf. (2005).
[32] R. Olfati-Saber, J.A. Fax, R.M. Murray, Consensus and cooperation in networked multi-agent systems, Proc. IEEE (2007).
[33] W. Ren, R.W. Beard, E.M. Atkins, A survey of consensus problems in multi-agent coordination, in: Proceedings of the 2005, American Control Conference, 2005, vol. 3, 2005, pp. 1859–1864.
[34] R. Mudumbai, S. Dasgupta, B.B. Cho, Distributed control for optimal economic dispatch of a network of heterogeneous power generators, IEEE Trans. Power Syst. (2012).
[35] Z. Zhang, M.Y. Chow, Incremental cost consensus algorithm in a smart grid environment, IEEE Power and Energy Society General Meeting 2011, San Diego, CA, 2011, pp. 1–6.
[36] N. Rahbari-Asr, U. Ojha, Z. Zhang, M.Y. Chow, Incremental welfare consensus algorithm for cooperative distributed generation/demand response in smart grid, IEEE Trans. Smart Grid (2014).
[37] E. Karfopoulos, et al., A multi-agent system providing demand response services from residential consumers, Electr. Power Syst. Res. (2015).
[38] E. Ancillotti, R. Bruno, M. Conti, The role of communication systems in smart grids: architectures, technical solutions and research challenges, Comput. Commun. (2013).
[39] X. Fan, G. Gong, Security challenges in smart-grid metering and control systems, Technol. Innov. Manag. Rev. (2013).
[40] W. Li, X. Zhang, Simulation of the smart grid communications: challenges, techniques, and future trends, Comput. Electr. Eng. (2014).
[41] V.C. Gungor, F.C. Lambert, A survey on communication networks for electric system automation, Comput. Netw. (2006).
[42] Y. Kabalci, A survey on smart metering and smart grid communication, Renew. Sustain. Energy Rev. (2016).
[43] K. Sharma, L.M. Saini, Power-line communications for smart grid: progress, challenges, opportunities and status, Renew. Sustain. Energy Rev. 67 (2017) 704–751.
[44] Y. Yoldaş, A. Önen, S.M. Muyeen, A.V. Vasilakos, İ. Alan, Enhancing smart grid with microgrids: challenges and opportunities, Renew. Sustain. Energy Rev. 72 (2017) 205–214.
[45] S. Galli, A. Scaglione, Z. Wang, Power line communications and the smart grid, 2010 First IEEE International Conference on Smart Grid Communications (2010) 303–308.
[46] M. Kuzlu, M. Pipattanasomporn, S. Rahman, Communication network requirements for major smart grid applications in HAN, NAN and WAN, Comput. Netw. 67 (2014) 74–88.
[47] M. Lixia, M. Murroni, V. Popescu, PAPR reduction in multicarrier modulations using genetic algorithms, in: Proceedings of the International Conference on Optimisation of Electrical and Electronic Equipment, OPTIM, 2010.
[48] A. Mahmood, N. Javaid, S. Razzaq, A review of wireless communications for smart grid, Renew. Sustain. Energy Rev. 41 (2015) 248–260.
[49] G. Celli, P.A. Pegoraro, F. Pilo, G. Pisano, S. Sulis, DMS cyber-physical simulation for assessing the impact of state estimation and communication media in smart grid operation, IEEE Trans. Power Syst. 29 (5) (2014) 2436–2446.
[50] M. Anedda, G.-M. Muntean, M. Murroni, Adaptive real-time multi-user access network selection algorithm for load-balancing over heterogeneous wireless networks, IEEE International Symposium on Broadband Multimedia Systems and Broadcasting, BMSB 2016, Nara, 2016, pp. 1–4.

[51] L. Militano, M. Nitti, L. Atzori, A. Iera, Enhancing the navigability in a social network of smart objects: a Shapley-value based approach, Comput. Netw. 103 (2016) 1–14.
[52] H.A. Foudeh, A.S. Mokhtar, Automated Meter Reading and Advanced Metering Infrastructure projects, in: 2015 JIEEEC 9th Jordanian International Electrical and Electronics Engineering Conference, JIEEEC 2015, 2016.
[53] G.C. Madueño, J.J. Nielsen, D.M. Kim, N.K. Pratas, Stefanović, P. Popovski, Assessment of LTE wireless access for monitoring of energy distribution in the smart grid, IEEE J. Sel. Areas Commun. 34 (3) (2016) 675–688.
[54] M. Garau, M. Anedda, C. Desogus, E. Ghiani, M. Murroni, G. Celli, A 5G cellular technology for distributed monitoring and control in smart grid, in: 2017 IEEE International Symposium on Broadband Multimedia Systems and Broadcasting (BMSB), 2017, pp. 1–6.
[55] N. Saputro, K. Akkaya, Investigation of smart meter data reporting strategies for optimized performance in smart grid AMI networks, IEEE Internet Things J. 4 (4) (2017) 894–904.
[56] N. Panwar, S. Sharma, A.K. Singh, A survey on 5G: the next generation of mobile communication, Phys. Commun. 18 (2016) 64–84.
[57] J. Montalban, et al., Multimedia multicast services in 5G networks: subgrouping and non-orthogonal multiple access techniques, IEEE Commun. Mag. (2018).
[58] C.O. Nnamani, C.L. Anioke, C.I. Ani, M. Anedda, M. Murroni, Load-shared redundant interface for LTE access network, in: IEEE International Symposium on Broadband Multimedia Systems and Broadcasting, BMSB, 2017.
[59] M. Anedda, C. Desogus, M. Murroni, D.D. Giusto, G.-M. Muntean, An energy-efficient solution for multi-hop communications in low power wide area networks, in: 2018 IEEE Int. Symp. Broadband Multimed. Syst. Broadcast., 2018, pp. 1–5.
[60] X. Lin, et al., Positioning for the internet of things: a 3GPP perspective, IEEE Commun. Mag. (2017).
[61] R. Cohen, On the establishment of an access VPN in broadband access networks, IEEE Commun. Mag. (2003).
[62] H.Y. Xu, The research of building VPN based on IPsec and MPLS technology, Commun. Comput. Inform. Sci. (2011).
[63] J. Ekanayake, K. Liyanage, J. Wu, A. Yokoyama, N. Jenkins, Smart Grid: Technol. Appl. (2012).
[64] F. Benzi, N. Anglani, E. Bassi, L. Frosini, Electricity smart meters interfacing the households, IEEE Trans. Ind. Electron. (2011).
[65] V.C. Gungor, et al., Smart grid and smart homes: key players and pilot projects, IEEE Ind. Electron. Mag. (2012).
[66] I. Colak, S. Sagiroglu, G. Fulli, M. Yesilbudak, C.F. Covrig, A survey on the critical issues in smart grid technologies, Renew. Sustain. Energy Rev. (2016).
[67] S.S.S.R. Depuru, L. Wang, V. Devabhaktuni, Smart meters for power grid: challenges, issues, advantages and status, Renew. Sustain. Energy Rev. (2011).
[68] M. Erol-Kantarci, H.T. Mouftah, Smart grid forensic science: applications, challenges, and open issues, IEEE Commun. Mag. (2013).
[69] I.-K. Song, W.-W. Jung, J.-Y. Kim, S.-Y. Yun, J.-H. Choi, S.-J. Ahn, Operation schemes of smart distribution networks with distributed energy resources for loss reduction and service restoration, IEEE Trans. Smart Grid (2013).
[70] I.K. Song, S.Y. Yun, S.C. Kwon, N.H. Kwak, Design of smart distribution management system for obtaining real-time security analysis and predictive operation in Korea, IEEE Trans. Smart Grid (2013).
[71] M.P. McHenry, Technical and governance considerations for advanced metering infrastructure/smart meters: technology, security, uncertainty, costs, benefits, and risks, Energy Policy (2013).
[72] C.R. Bayliss, B.J. Hardy, Current and voltage transformers, Transmission Distrib. Electr. Eng. (2012) 157–159.
[73] T. Krishnamurti, et al., Preparing for smart grid technologies: a behavioral decision research approach to understanding consumer expectations about smart meters, Energy Policy (2012).
[74] S.M. Amin, Smart grid: overview, issues and opportunities. Advances and challenges in sensing, modeling, simulation, optimization and control, Eur. J. Control (2011).
[75] K. Sharma, L. Mohan Saini, Performance analysis of smart metering for smart grid: an overview, Renew. Sustain. Energy Rev. (2015).

[76] G. Strbac, Demand side management: benefits and challenges, Energy Policy (2008).
[77] M.E. Khodayar, H. Wu, Demand forecasting in the smart grid paradigm: features and challenges, Electr. J. (2015).
[78] D.E. Allen, A. Apostolov, D.G. Kreiss, Automated analysis of power system events, IEEE Power Energy Mag. 3 (5) (2005) 48–55.
[79] K.M. Muttaqi, J. Aghaei, V. Ganapathy, A.E. Nezhad, Technical challenges for electric power industries with implementation of distribution system automation in smart grids, Renew. Sustain. Energy Rev. (2015).
[80] J.H. Kim, A. Shcherbakova, Common failures of demand response, Energy (2011).
[81] K. Spees, L.B. Lave, Demand response and electricity market efficiency, Electr. J. (2007).
[82] L. Gelazanskas, K.A.A. Gamage, Demand side management in smart grid: a review and proposals for future direction, Sustain. Cities Soc. (2014).
[83] J. Wang, C.N. Bloyd, Z. Hu, Z. Tan, Demand response in China, Energy (2010).
[84] Z. Ming, S. Li, H. Yanying, Status, challenges and countermeasures of demand-side management development in China, Renew. Sustain. Energy Rev. (2015).
[85] L. Zhu, D. Shi, X. Duan, Standard function blocks for flexible IED in IEC 61850-based substation automation, IEEE Trans. Power Deliv. (2011).
[86] Y. Kabalci, A survey on smart metering and smart grid communication, Renew. Sustain. Energy Rev. 57 (2016) 302–318.
[87] H.A. Foudeh, A.S. Mokhtar, Automated meter reading and advanced metering infrastructure projects, in: 2015 9th Jordanian International Electrical and Electronics Engineering Conference (JIEEEC), 2015, pp. 1–6.
[88] D. Rua, D. Issicaba, F.J. Soares, P.M.R. Almeida, R.J. Rei, J.A.P. Lopes, Advanced Metering Infrastructure functionalities for electric mobility, 2010 IEEE PES Innovative Smart Grid Technologies Conference Europe (ISGT Europe) (2010) 1–7.
[89] J. Jiang, Y. Qian, Distributed communication architecture for smart grid applications, IEEE Commun. Mag. 54 (12) (2016) 60–67.
[90] V. Pilloni, A. Floris, A. Meloni, L. Atzori, Smart home energy management including renewable sources: a QoE-driven approach, IEEE Trans. Smart Grid 9 (3) (2018) 2006–2018.
[91] M. Nitti, V. Popescu, M. Fadda, Using an IoT platform for trustworthy D2D communications in a real indoor environment, IEEE Trans. Netw. Serv. Manag. (2018).
[92] J. Zhou, R.Q. Hu, Y. Qian, Scalable distributed communication architectures to support advanced metering infrastructure in smart grid, IEEE Trans. Parallel Distrib. Syst. 23 (9) (2012) 1632–1642.
[93] A. Floris, A. Meloni, V. Pilloni, L. Atzori, A QoE-aware approach for smart home energy management, in: 2015 IEEE Glob. Commun. Conf., 2015.
[94] A. Pilloni, A. Pisano, E. Usai, Parameter tuning and chattering adjustment of super-twisting sliding mode control system for linear plants, in: VariableStructure Systems (VSS), 2012 12th International Workshop on, 2012, pp. 479–484.

CHAPTER

10

Communications, cybersecurity, and the internet of things for microgrids

Ajit Renjit

Electric Power Research Institute (EPRI), Palo Alto, CA, United States

10.1 Introduction

Microgrids (MGs) are small autonomous power systems that can operate when the broader grid is down, creating resiliency and providing energy in remote areas and following natural disasters. Most MG demonstration projects across the world focused on the deployment of distributed energy resource (DER) and the communication integration of these resources with the MG management system or an MG controller. Many of these projects involved the integration of inverter-based systems [1], such as solar photovoltaic (PV) and energy storage systems and noninverter-based system like diesel or natural gas gensets and combined heat and power units, including diverse sizes and manufacturers. To enable stable and resilient operation of an MG, communication protocols and the physical networks to carry the messages between DER and the MG controller are essential in MG implementations.

10.2 Background information on typical microgrid architectures

A typical MG architecture as shown in Fig. 10.1 illustrates two levels of communication.

1. At the device-level between the local MG controller and individual DER in the dashed line
2. At the MG controller-level between DER management system (DERMS) [2] and the local MG controller in the thick continuous line.

Distributed Energy Resources in Microgrids
DOI: https://doi.org/10.1016/B978-0-12-817774-7.00010-7

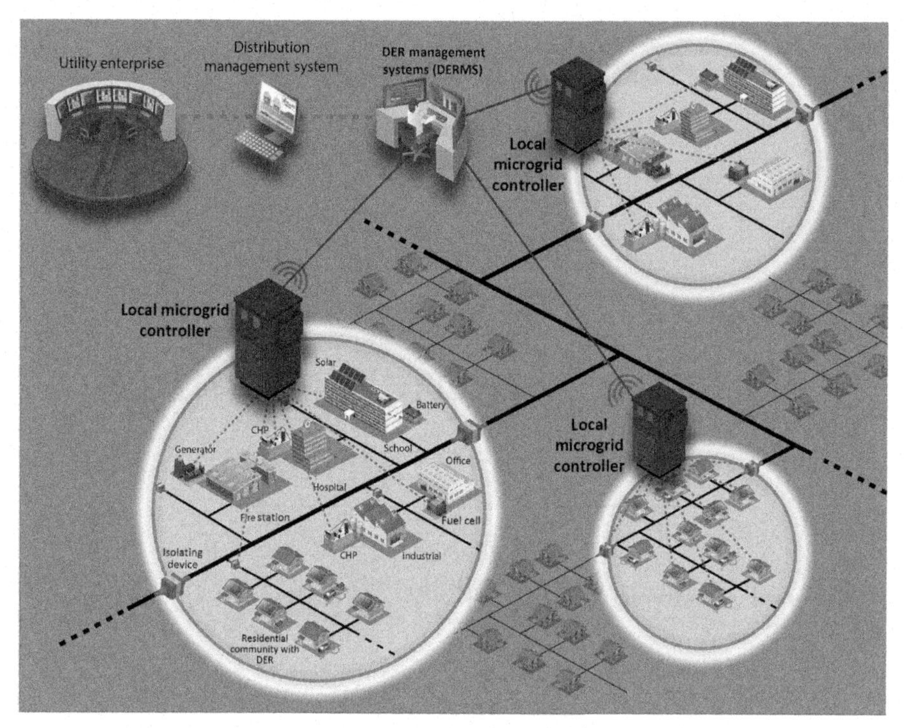

FIGURE 10.1 Typical microgrid architecture illustrating communication between different actors—utility DERMS, local MG controller, and DER.

10.2.1 Controller-level and device-level interfaces

It is important to note the key differences between the two interfaces—controller-level and device-level—to understand the architectural significance.

1. Number of interactions: A key difference between the controller-level and device-level functions is the number of interactions required to maintain a sustained grid service. At the device-level, these interactions happen more often with a large number of devices which are grouped in alignment with the power system needs. Since many of these devices are variable energy resource (e.g., solar) achieving a service in a stable and sustained fashion requires very frequent monitoring and adjustment of device settings.
2. Device-type: The services provided by DER are device-type specific since not all DER types support the same function. However, the services offered by an MG controller, the group managing entity, is agnostic of the type of downstream DER.
3. Simple versus complex functions: The grid supportive functions and the associated settings at the device-level are very granular and contain a range of information unlike the controller-level functions. A reactive power request at the group-level can be requested by a simple reactive power group setpoint whereas the device-level functions (e.g., volt-var) require much more granular information.
4. Net-result oriented versus direct functions: Finally, the instructions at the group-level are net-result oriented while the device specific functions are explicit instructions. When

an upstream entity (e.g., DERMS) requests a specific service from the downstream entity (e.g., MG controller), the DERMS does not care about the individual DER setpoints and functions enabled by the microgrid controller to provide the requested service. MG controllers are control systems with intelligence that can render requested services from DER-groups in creative ways that optimize the service to the utility. Contrary to controller-level functions, device-level functions are explicit with direct instructions.

Before selecting the relevant communication protocol upstream and downstream of the MG controller, it is important to define the messages or information model that should be carried by the protocol. To develop the information model, the different functions or modes of operation supported by the DER and the MG controller need to be defined first.

The first section of this chapter covers the information model at the device-level by identifying functions supported by the DER. These functions vary depending on the type of DER and their role in supporting different applications in an MG. Different protocols that can be used to carry the device-level messages, the implementation methods, and the level of cybersecurity supported are discussed. The final section of this chapter will focus on communication protocols that enable the utility integration of MGs. These are business-to-business (B2B) protocols that act as the interface between the utility operations and the MG controller, enabling several MG functions like islanding, re-synchronization, voltage support, load relief, etc.

10.3 Common distributed energy resource functions and communication protocols within a microgrid

In this section, the messages or information model (in dashed line) required to communicate between the local MG controller and the individual DER within the MG are explained. To define these messages, it is important to identify the functions supported by the DER for various MG applications (Table 10.1, Fig. 10.2).

TABLE 10.1 Applicable device-level functions for different types of distributed energy resource (DER).

Sl. no.	Device-level function	PV	Energy storage	Rotating machinery-type DER
1	Limit maximum real power	✓		✓
2	Active power smoothing control		✓	
3	Peak power limiting function		✓	✓
4	Active power setpoint		✓	✓
5	Reactive power setpoint	✓	✓	✓
6	Fixed power factor	✓	✓	✓
7	Voltage and frequency master—isochronous (grid-forming DER)		✓	✓
8	Voltage and frequency master—droop (grid-forming DER)		✓	✓

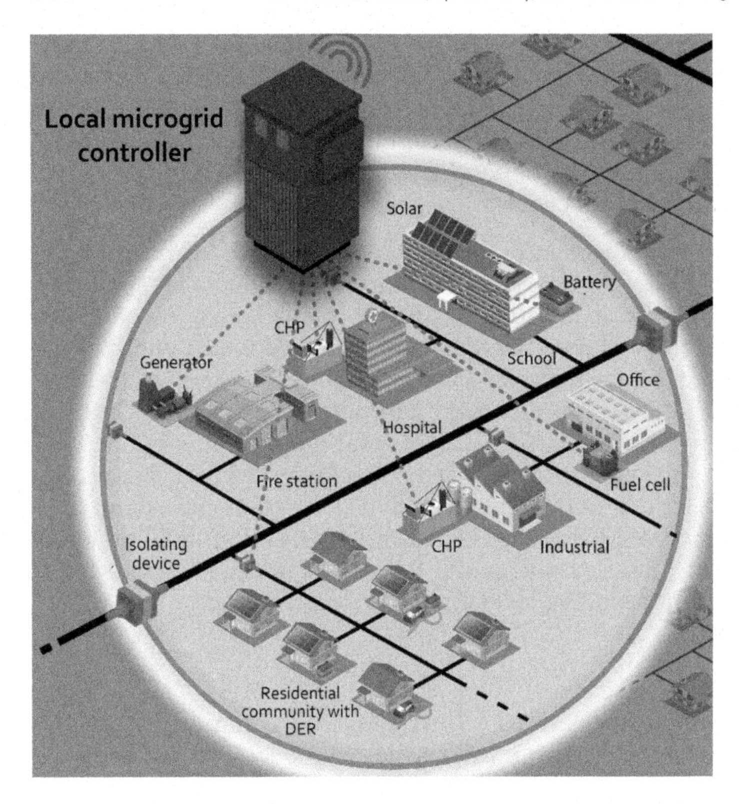

FIGURE 10.2 Microgrid architecture illustrating communication between the local MG controller and DER.

10.3.1 Common distributed energy resource functions

10.3.1.1 Limit maximum real power

10.3.1.1.1 Purpose

This function [3–5] is intended to provide a flexible mechanism through which the power either in or out of a DER might be self-limited, if so desired. This includes generations from a PV system or the charging and discharging of an energy storage system. This function could be used to limit the maximum output from PV to prevent the overloading of local assets, such as transformers. It can also be used to manage PV output when optimizing a diverse DER mix for increased usage of renewables and decreased usage of DER that emits CO_2. It is proposed that the Limit DER Power Output Function be percentage based, according to the WMax and WChaMax capability of the device. The effect of this setting is illustrated in Fig. 10.3.

10.3.1.1.2 Message definition

- *Read maximum output power setting*: A query to read the present setting as a percent of WMax_Output.
- *Set maximum output power*: A command to set the maximum generation level as a percent of WMax_Output. Percentage-based settings allow communication to large groups of devices of differing sizes and capacities.

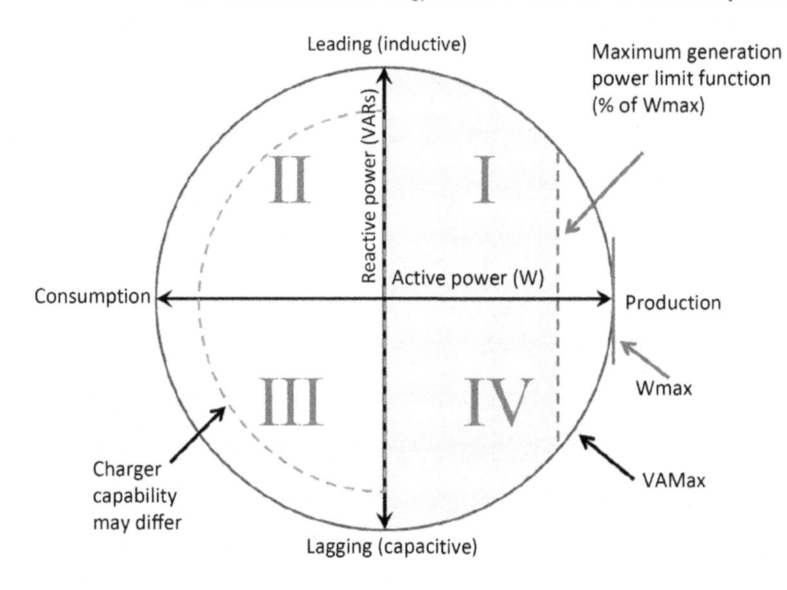

FIGURE 10.3 An example of maximum generation settings.

- *Read maximum input power setting*: A query to read the present setting as a percent of WMax_Input.
- *Set maximum input power*: A command to set the maximum generation level as a percent of WMax_Input. Percentage-based settings allow communication to large groups of devices of differing sizes and capacities.

10.3.1.2 Active power smoothing control

10.3.1.2.1 Purpose

This function is intended to provide a flexible mechanism through which highly intermittent renewable power generated by a renewable energy resource (e.g., PV or wind) is compensated with energy storage power to achieve stable combined power, if so desired. This is a common use case in MGs with energy storage and a significant amount of PV. The function enables the MG operator to specify an allowed ramp rate for the PV output. The operation can provide regulated ramp rate at the PV output or a PV plus load output.

10.3.1.2.2 Message definition

- *Smoothing target SOC*: The system tries to keep the energy storage SOC at this target level while keeping the overall power ramp rate within the limits.
- *Positive ramp rate limit (kW/min)*: The ramp-up limit that enables a smooth output at the point of common coupling (PCC) when the power output of the PV plant instantaneously increases.
- *Negative ramp rate limit (kW/min)*: The ramp-down limit that enables a smooth output at the PCC when the power output of the PV plant instantaneously decreases.
- *Reference power measurement point*: The measurement point that measures power output against which the energy storage conducts a ramp rate control operation (Fig. 10.4).

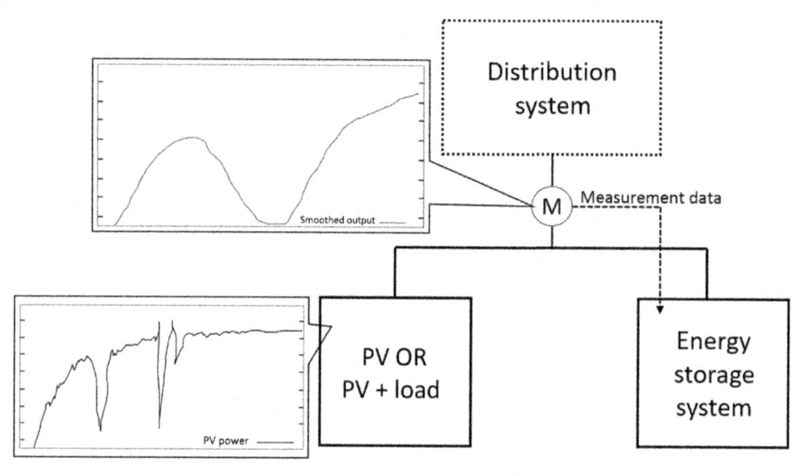

FIGURE 10.4 An example of the implementation of the active power-smoothing control system.

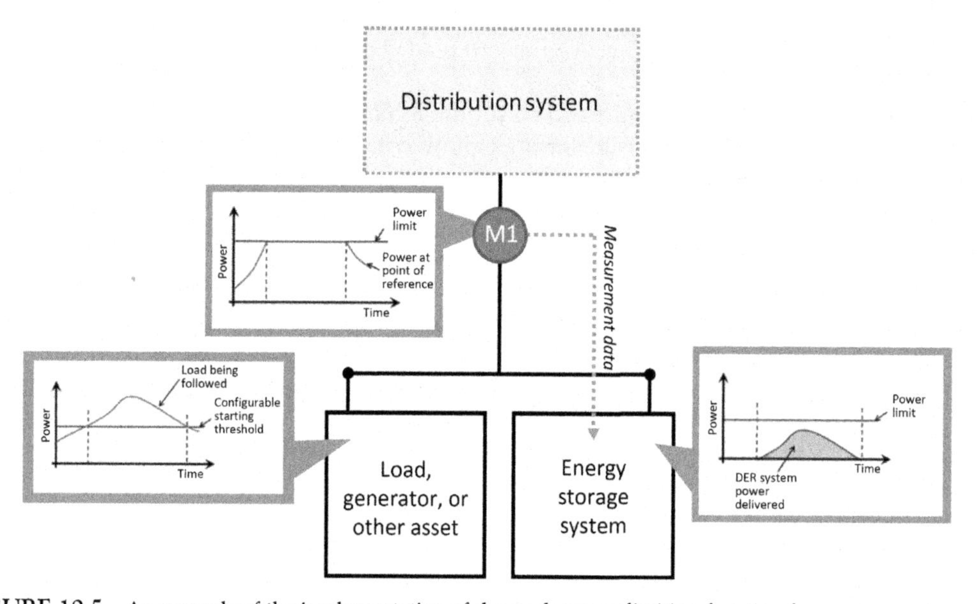

FIGURE 10.5 An example of the implementation of the peak-power limiting function for an energy storage.

10.3.1.3 Peak power limiting function

10.3.1.3.1 Purpose

This function is intended to provide a flexible mechanism through which inverters, such as those associated with energy storage systems, may be configured to provide a peak power limiting function. This function involves the variable dispatch of energy in order to prevent the power level at some point of reference from exceeding a given threshold (Fig. 10.5).

10.3.1.3.2 Message definition

Peak power limit: This is the target power level limit, expressed in watts.

Reference power measurement point: This is the power measurement in watts which the DER is using as the reference for peak power limiting. From the perspective of this function, this quantity is read-only. It is the responsibility of the DER manufacturer and user to configure and establish how the DER acquires this measurement.

10.3.1.4 Active power setpoint

10.3.1.4.1 Purpose

This function is intended to provide a simple mechanism through which the active power of DER may be directly managed. This can be achieved by updating the charging and discharging of energy storage systems or the real power capability of rotating machinery-type generators. This function assumes that the intelligence which determines the active power setpoint resides at the MG controller, and that the DER (to the extent possible) follows the requests it is given.

10.3.1.4.2 Message definition

Set active power setpoint: The MG controller controls the DER to operate at constant real power levels. For energy storage, this is determined by the charge/discharge rate. This setting is provided as a percentage between +100% (discharging) and −100% (charging). For rotating machinery-type generators, this value will be always positive.

Minimum reserve percentage (percentage of usable capacity): A reserve percentage that the user may define if desired. Rotating machinery-type generators will not generate below this amount (i.e., fuel level) until the reserve percentage is changed. The storage system will not discharge below this amount (i.e., state of charge) until the reserve percentage is changed.

Maximum reserve percentage (percentage of usable capacity): A reserve percentage that the user may define if desired. The storage system will not charge above this amount until the reserve percentage is changed.

Set maximum storage charge rate in watts: The maximum power rate at which the storage unit may be charged, in watts.

Set maximum storage discharge rate in watts: The maximum power rate at which the storage unit may be discharged, in watts.

10.3.1.5 Reactive power setpoint

10.3.1.5.1 Purpose

This function is intended to provide a simple mechanism through which reactive power of DER may be directly managed. This can be achieved by updating the charging and discharging of energy storage systems or the reactive power capability of rotating machinery-type generators. This function assumes that the intelligence which determines the reactive power setpoint resides at the MG controller-level, and that the DER (to the extent possible) follows the requests it is given.

10.3.1.5.2 Message definition

Set reactive power setpoint: The MG controller controls the DER to operate at constant reactive power levels. For energy storage, this is determined by the charge/discharge rate. This setting is provided as a percentage between +100% (discharging) and −100% (charging). For rotating machinery-type generators, this value will be always positive.

Minimum reserve percentage (percentage of usable capacity): A reserve percentage that the user may define if desired. Rotating machinery-type generators will not generate below this amount (i.e., fuel level) until the reserve percentage is changed. The storage system will not discharge below this amount (i.e., state of charge) until the reserve percentage is changed.

Maximum reserve percentage (percentage of usable capacity): A reserve percentage that the user may define if desired. The storage system will not charge above this amount until the reserve percentage is changed.

Set maximum storage charge rate in volt ampere reactives (VARs): The maximum power rate at which the storage unit may be charged, in VARs.

Set maximum storage discharge rate in VARs: The maximum power rate at which the storage unit may be discharged, in VARs.

10.3.1.6 Fixed power factor control

10.3.1.6.1 Purpose

This function [6] is intended to provide a simple mechanism through which the power factor of a DER may be set to a fixed value.

10.3.1.6.2 Message definition

Set power factor – positive current, power flow to grid: A command to set the power factor. Typically provided as a number between +1.00 and −1.00, each of which results in zero VARs. A setting of zero may not be used.

Set power factor – negative current, power flow to DER: A command to set the power factor. Typically provided as a number between +1.00 and −1.00, each of which results in zero VARs. A setting of zero may not be used.

10.3.1.7 Voltage and frequency master—isochronous (grid-forming distributed energy resource)

10.3.1.7.1 Purpose

This function is intended to provide a simple mechanism through which the DER can be switched from a grid-following function to a grid-forming function. One of the active functions in the grid-forming mode of operation is the isochronous voltage and frequency master. This function is used for MG applications when acting as the primary grid-forming generator where the DER under this mode of operation should be the largest generator on the MG.

10.3.1.7.2 Message definition

Set gridform-ISO mode: This enables the grid-forming, isochronous functionality in the DER regulating to voltage and frequency references.

Set nominal voltage and frequency: These parameters are used to set the nominal voltage and frequency of the MG.

10.3.1.8 Voltage and frequency master—droop (grid-forming distributed energy resource)

10.3.1.8.1 Purpose

This function is intended to provide a simple mechanism through which the DER can be switched from a grid-following function to a grid-forming function. One of the active functions in the grid-forming mode of operation is the droop-enabled voltage and frequency master. This function is used for MG applications when the DER under this mode of operation parallels with other grid-forming DERs or can be its own grid-forming unit with droop enabled.

In this mode, the DER regulates the MG voltage and frequency based on the active and reactive power requirement in the system. The regulated frequency and voltage are a function of the defined droop voltage and frequency. A basic illustration of the functionality can be seen in Fig. 10.6.

10.3.1.8.2 Message definition

Set gridform—VF mode: This enables the grid-forming, droop functionality in the DER regulating to voltage and frequency references set by the MG controller.

Droop slope: Defined rate of change of frequency and voltage as a function of real and reactive power, respectively.

Deadband: The frequency and voltage ranges where the DER output power is constant.

Set nominal voltage and frequency: These parameters determine the nominal voltage and frequency of the MG at which they are regulated.

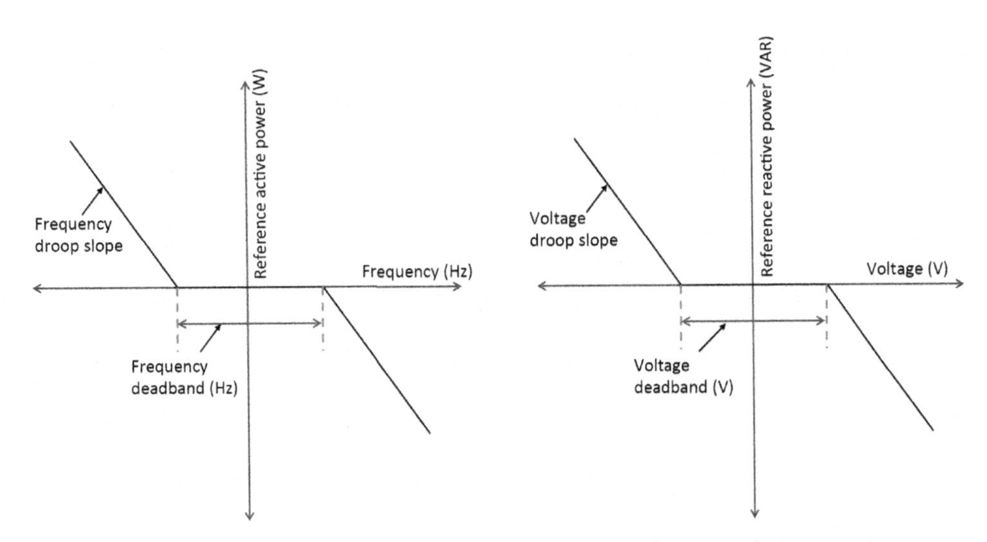

FIGURE 10.6 An example of the implementations of the voltage and frequency droop curves.

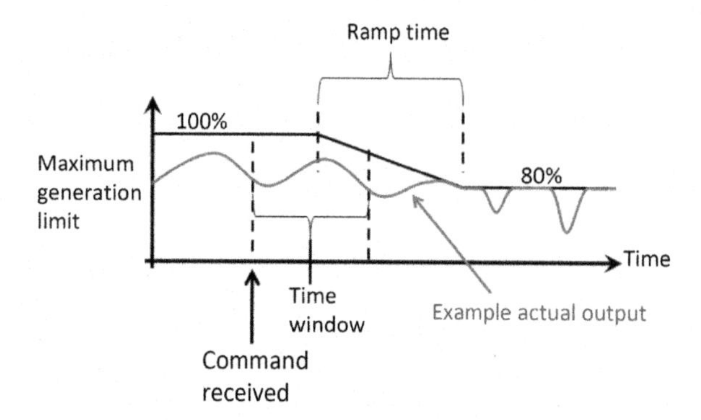

FIGURE 10.7 Example of function settings.

10.3.1.9 Common messages across all functions

The following information exchanges are associated with this function:

- *Time window*: A time in seconds, over which a new setting is to take effect. For example, if the time window is set to 60 seconds, then the DER would delay a random time between 0 and 60 seconds prior to beginning to make the new setting effect. This setting is provided to accommodate communication systems that might address large numbers of devices in groups.
- *Reversion timeout*: A time in seconds, after which a setting below 100% expires and the device returns to its natural "WMax, delivered" limits. Reversion Timeout = 0 means that there is no timeout.
- *Ramp time—output increasing*: A time in seconds, over which the DER linearly places the new limit into effect when increasing output. For example, if a device is operating with no limit on watts generated (i.e., 100% setting), then it receives a command to reduce to 80% with a "Ramp Time" of 60 seconds, then the upper limit on allowed watts generated is reduced linearly from 100% to 80% over a 60-second period after the command begins to take effect.
- *Ramp time—output decreasing*: A time in seconds, over which the DER linearly places the new limit into effect. For example, if a device is operating with no limit on watts generated (i.e., 100% setting), then it receives a command to reduce to 80% with a "ramp time" of 60 seconds, then the upper limit on allowed watts generated is reduced linearly from 100% to 80% over a 60-second period after the command begins to take effect.
- *Ramp time—input increasing*: A time in seconds, over which the DER linearly places the new limit into effect when decreasing input. Only applies to energy storage. (optional).
- *Ramp time—input decreasing*: A time in seconds, over which the DER linearly places the new limit into effect when decreasing output. Only applies to energy storage (optional) (Fig. 10.7).

10.3.2 Distributed energy resource communication protocols in microgrids

This section describes different communication protocols [7] that can be used to carry the messages between an MG controller and DER.

10.3.2.1 Modbus

Protocol overview: The Modbus protocol was developed in 1979 by Modicon, incorporated for industrial automation systems and Modicon programmable controllers. It has since become an industry standard method for the transfer of discrete/analog I/O information and register data between industrial control and monitoring devices. Recently, Modbus has also become a dominant choice for communications in DER because of its simplicity, stability, and support. However, for MG applications, without a standardized approach, DER manufacturers are currently using customized register maps, requiring engineers to modify products and/or control systems to match these custom mappings.

Modbus devices communicate using a master-slave (client-server) technique in which only one device (the master/client) can initiate transactions (called queries). The other devices (slaves/servers) respond by supplying the requested data to the master, or by taking the action requested in the query. A slave is any peripheral device (e.g., PV, energy storage, gensets, or other measuring devices) which processes information and sends its output to the master (e.g., an MG controller) using Modbus. Masters can address individual slaves, or can initiate a broadcast message to all slaves. Slaves return a response to all queries addressed to them individually, but do not respond to broadcast queries. Slaves do not initiate messages on their own, they only respond to queries from the master.

Cybersecurity: Security will vary by choice of transport layer. Modbus TCP supports transport layer security (TLS), however not all DER manufacturers support it. Wireless applications may require special attention, however Wi-Fi and ZigBee both support security. It is important to note that Modbus is often used within the DER plant and is therefore often protected by physical security.

10.3.2.2 DNP3

Protocol overview: IEEE 1815, also known as Distributed Network Protocol 3 (DNP3), was created to standardize communications between substation equipment, remote terminal units, International Electrotechnical Commission (IECs), and control systems. The protocol is largely used in utility and energy industries including water, waste water, transportation, and the oil and gas industry. It was created in 1993 and is largely based on work in IEC 61850. DNP3 was a big step forward from Modbus because it supported new, important features to support communication to devices over long distances. It preserves chains of events, event storage, delivery confirmation, and event timestamping; none of which are in Modbus. It also supports event reporting and storing information collected for later transmission. This contrasts to Modbus where only instantaneous data is transmitted, without a timestamp, and without means for event reporting. DNP3 is similar to Modbus in that it can be implemented using a standardized points configuration (application note) or a customized points configuration for the specific application.

In DNP3 networks there are masters and outstations. Masters are management systems and outstations are remote devices or terminals. The master can send a request to the outstations and they respond with the requested data. Outstation units also support event-based reporting. In this case, outstation units can be configured to generate reports to the master when specific conditions are met. This is often used for alarms or other critical information. Otherwise all data is stored and timestamped on the outstation unit until the master requests it.

Cybersecurity: DNP3 supports secure authentication to enforce role-base access. It can also make use of TLS when implemented on IP-based networks, though it does not include encryption as part of the specification. DNP3 can be implemented with or without security so if secure authentication is needed, Secure Authentication version five (or later) should be requested in the request for proposal (RFP) process.

10.3.2.3 IEC 61850

Protocol overview: IEC 61850 [8] is a standard that defines data models, exchanges, and events between power systems substations. This standard can be mapped to a number of legacy protocols such as manufacturer message specification, generic object oriented substation events, and sampled measured values. However, IEC 61850 was not designed for serial communication protocols as it is intended to run over ethernet networks. This should have a positive impact on the cost and operation of power system components at the application layer and above. Leveraging the services of this standard is proposed to achieve reliability and security. Services in this standard are: retrieving the device description, fast and reliable host-to-host status information, exchange of status, reporting data or sequence of events, data logging, retrieving sample values from sensors, time synchronization, and file transfer for online configuration of components. Table 10.2 shows the structure of the IEC 61850 standard.

Communication protocols define how data bits are transmitted on the transmission medium. However, they do not define the data organization in an IED or device with communication capability. For MG components, each DER has a unique functionality, and a subset of data attributes may not be present in all DERs. Each IED manufacturer may have different naming conventions for data. The IEC 61850 data-naming conventions are based on a power systems context, which helps ensure interoperability between devices from various vendors used in the same communication system.

Logical nodes consist of data elements and methods that are related to device functionality, and each data element has a unique name that is defined by the standard. One or multiple logical nodes form a logical device. The services of IEC 61850 form a functional MG communications system. Transmitting via ethernet and mapping legacy protocols into the standard simplifies the design of the communication system, and reduces the cost of integrating multivendor installations. Using IEDs to translate power system component protocols into object-oriented data representations permits blending devices controlled with a variety of communication protocols within the MG.

Cybersecurity: IEC 61850 does not discuss the technical aspects of cybersecurity. In reality, the security corresponding to IEC 61850 is entrusted to another standard, the IEC 62351.

TABLE 10.2 List of applicable device-level protocols.

Protocol	Transport	Physical
IEC 61850	TCP/IP	Ethernet
Modbus	TCP/IP	Ethernet
	N/A	RS 485
DNP3	TCP/IP	Ethernet

10.4 Common controller functions and communication protocols for utility integration of microgrids

From the utility perspective, it is not practical to have each MG controller independently define the grid-services or protocols that they can provide. In a high DER penetration scenario, there could be potentially thousands of MGs that may offer such services and it is not reasonable to expect distribution control strategies that deal uniquely with each type of offered service. Likewise, from the vendor's perspective, it is not practical to have each utility independently define the services or protocols that they can utilize. There are many utilities, and vendors need volume and consistency in the market in order to provide quality products at feasible costs (Fig. 10.8).

Standard service/function definitions and protocols for MG controllers exist and are being actively improved and maintained. Utilities engaged in MG projects are encouraged to consider and build upon these standards, offering improvements and extensions as learning occurs. Table 10.3 provides a concise summary of applicable functions and their functional definitions between the MG controllers and upstream utility control systems.

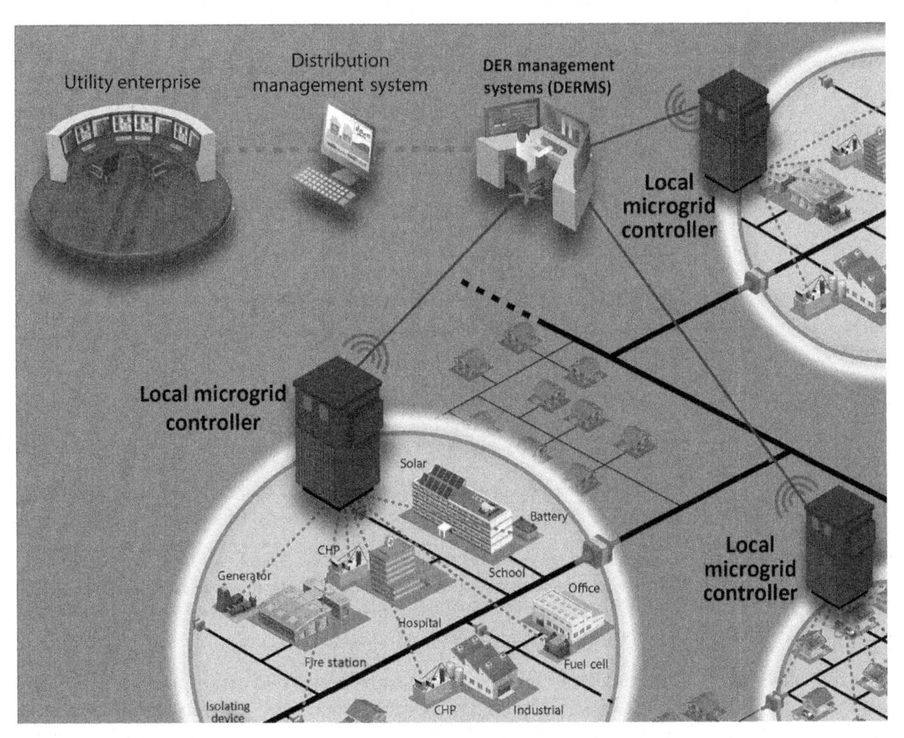

FIGURE 10.8 MG architecture illustrating controller-level communication between the utility back-end, DER management systems (DERMS), and the local MG controller.

10.4.1 Standard functional definitions at the microgrid controller interface

TABLE 10.3 List of applicable functions and their functional definitions at the controller-level.

Sl. no	Function	Definition
1	MG capability discovery	The purpose of this function is to read/report the capabilities of an MG. This function is specifically focused on as-built or installed capability (e.g., static, nameplate, nonvariable quantities), not real-time status data which is time-variable in nature.
2	MG status monitoring	The purpose of this function is to read/report the present status of an MG. In this context, "status" refers foremost to the present value and range of adjustability of real and reactive power levels. Depending on devices and the specific protocol mapping, a wide range of additional group-status-related parameters may also be exchanged using this function.
3	MG forecasting	The purpose of this function is to support the exchange of forecasts of MG availability. Specifically, this method addresses forecasts for the availability of real and reactive power from a DER group. In future additions, forecasts could be made available in a similar fashion for any service that the DER group can provide.
4	MG maximum real power limiting	The purpose of this function is to request that the maximum real power (generated or absorbed) be limited to the specified level.
5	MG ramp rate limit control	The purpose of this function is to manage the maximum ramp rates at the MG level.
6	MG real power dispatch	The purpose of this function is to request/dispatch real power from an MG. This method has two forms: 1. A request that the real power for the group be set to a specified level. 2. A request that the real power for the group be raised/lowered by a specified amount.
7	MG reactive power dispatch	The purpose of this function is to request/dispatch reactive power from an MG. This method is in the form of a request that the reactive power for the group be set to a specified level.
8	MG voltage regulation function	This is a control function by which DER support for various voltage needs may be requested. Requesting entities (e.g., utilities) could specify a target voltage or an increase/decrease adjustment. Requests could be made at the MG level. The entity providing this service or the MG controller could use a variety of settings of individual DER in order to provide this service.
9	Provide price to MG	This is a function by which energy price signals can be provided to a group of DER (e.g., via an MG controller). This function is not a bid/offer mechanism but implies that such markets (or other price-determining mechanisms), if they exist, have been conducted.
10	Request cost of service from MG	This function exchanges the cost of a specified service from an MG. This information could be, for example, exchanged prior to actionable services being requested or rendered in order to aid in decision making.

10.4.2 Standard protocol capability at microgrid controller interface (to utility)

Communication protocol standards have also been developed to support MG controller interfaces to utility control systems. The encodings continue to be improved and may or may not be supported in given products. For both scalability and sustainability, communication protocol standards should be required at MG controller interfaces. The standard information model for supporting communication between the controller and upstream utility control systems like DERMS has been defined by the standard IEC 61968-5. Using this information model, several B2B protocols that acts as the interface between utility operations and the MG controller. Major protocols in this space are the popular internet-based protocols like IEEE 2030.5 and OpenADR. An overview of IEEE 2030.5, its features and the associate's cybersecurity is explained in the section below. At this level, protocols are categorized based on several functions including:

- *Extensibility*—The ability of the communication protocol to expand indefinitely to include more MGs, different types of MGs (i.e., based on operation and ownership model) in a cohesive way.
- *Interoperability*—Use of standard-based and open communication technologies that support heterogeneous communications.
- *Efficiency*—Using low overheads in terms of network bandwidth, CPU, RAM and programming memory essential for IoT devices.
- *Security*—Ability to secure, cohesively, the networks end-to-end for diverse devices and technologies.

10.4.2.1 IEEE 2030.5

Protocol overview: IEEE 2030.5 [9] is an application layer specification formerly referred to as SEP 2.0 (Smart Energy Profile 2.0). It was developed as a communication protocol to securely integrate consumer's smart devices into the smart grid including smart loads, electric vehicles, and DERs. The protocol reduces communications architectural challenges by using the widely used IP at the internet layer and supporting a variety of protocols at the physical layer (including ethernet, wi-fi, powerline communications and different low-power radio technologies).

IEEE 2030.5 uses a client-server network architecture. The server hosts the necessary DER resources (e.g., volt/var curve) for the client which are accessed through methods called "polling" and "subscription/notification." The most common and simpler way to retrieve data from the server is by polling. IEEE 2030.5 clients use representational state transfer or RESTful webservices to access the resources from the server. This includes mechanisms like GET, HEAD, PUT, POST, and DELETE. IEEE 2030.5 uses extensible markup language (XML) for encapsulating the commands and data. Schemas specify how to format the XML files so they can be recognized by others. This is different from the other DER protocols (DNP3 and Modbus). DNP3 and Modbus standardize the association of specific points with specific data. Only the data, not the meaning of the data, is sent. In XML, the content is human-readable because measurement values and metadata are included. The protocol uses IP at the internet layer and supports a variety of protocols at the physical layer (including ethernet, wi-fi, powerline communications and different low-power radio technologies). This reduces communication architectural challenges for utilities to talk to consumer devices, including residential PV and energy storage.

Cybersecurity: A complete implementation of an IEEE 2030.5 communication stack also includes all the mandated cybersecurity features in the standard. This ensures all transactions between clients and servers to be secured using HTTP over TLS (also called HTTPS). All IEEE 2030.5 devices use digital certificates to authenticate their identity. Once authenticated by a server, the device can access different resources in the server based on its identity and permissions associated with that identity. All data transactions between the server and device are encrypted at the transport layer using a secure cipher suite.

10.5 Conclusions

In order to enable open communication standards for DERs operating in MGs, a common set of functions for DER should be identified and adopted. Standard communication protocols can only result in interoperability between DER and MG controllers. To realize the complete potential of renewables and energy storage in MGs, standards will be required to enable a wide range of device types, sizes, and brands to operate together, providing uniform services in response to uniform controls.

References

[1] A.A. Renjit, A. Mondal, M.S. Illindala, A.S. Khalsa, Analytical methods for characterizing frequency dynamics in islanded microgrids with gensets and energy storage, IEEE T. Ind. Appl. 53 (3) (2017) 1815–1823.
[2] B.K. Seal, A.A. Renjit, B. Deaver, Understanding DERMS, technical publication, Electric Power Research Institute (EPRI), Report no. 3002013049, June 2018.
[3] B.K. Seal, B. Ealey, Common functions for smart inverters: 4th edition, technical publication, Electric Power Research Institute (EPRI), Report no3002008217, December 2016.
[4] IEEE Std 1547-2018, IEEE Standard for Interconnection and Interoperability of Distributed Energy Resources with Associated Electric Power System Interfaces.
[5] California Rule 21. 2018. Available from: <http://www.cpuc.ca.gov/Rule21/>.
[6] A.A. Renjit, A.M. Huque, B.K. Seal, B. Ealey, Voltage regulation support from smart inverters, technical publication, Electric Power Research Institute (EPRI) Journal, 2017.
[7] B. Ealey, A. Renjit, W. Johnson, Protocol reference guide overview of application layer protocols for DER and DR, technical update, Electric Power Research Institute (EPRI), Report no3002009850, December 2017.
[8] A. Bani-Ahmed, L. Weber, A. Nasiri, H. Hosseini, Microgrid communications: state of the art and future trends, in: Renewable Energy Research and Application (ICRERA), International Conference on 2014, pp. 780–785.
[9] P2030.5, IEEE Standard for Smart Energy Profile Application Protocol, in: IEEE Std 2030.5–2018 (Revision of IEEE Std 2030.5–2013), 21 Dec. 2018, pp. 1–361.

Further reading

B.K. Seal, G. Grey, G. Horst, J. Simmins, Common Functions for DER Group Management, third ed., EPRI, Palo Alto, CA, 2016. 3002008215.

CHAPTER

11

Transmission system-friendly microgrids: an option to provide ancillary services

Thomas Krechel[1], Francisco Sanchez[1], Francisco Gonzalez-Longatt[1], Harold Chamorro[2] and Jose Luis Rueda[3]

[1]Centre for Renewable Energy Systems Technology – CREST. Loughborough University, Loughborough, United Kingdom [2]KTH Royal Institute of Technology, Stockholm, Sweden [3]Section Intelligent Electrical Power Grids, Department of Electrical Sustainable Energy, Delft University of Technology, Stockholm, Sweden

11.1 Introduction

The classical concept of a microgrid (MG) is defined as a set of interconnected distributed energy resources (DERs) capable of providing sufficient and continuous energy to a significant portion of internal load demand. The MG has been intensively researched during the last decade, primarily as an alternative mechanism to support the main supply grid and provide it with resilience [1].

The concept of an MG is not new and the first attempt to define it can found on Consortium for Electric Reliability Technology Solutions (CERTS) MicroGrid White Paper (2001), created by Lasseter et al. [2]. That white paper recognized the significant potential of smaller DERs (<100 kW/unit) to meet customers' and the power utility's needs. The report recognized the MG as an organized set of DERs.

The concept of an MG has evolved over time and has also been adapted to specific interests and locations. The European vision of an MG and its applicability is found in many EU research projects [3–5]. The European vision of an MG includes the penetration of DERs (microturbines, fuel cells, photovoltaics (PV), etc.) in low voltage (LV) distribution systems together with storage devices (flywheels, energy capacitors, and batteries) and

Distributed Energy Resources in Microgrids
DOI: https://doi.org/10.1016/B978-0-12-817774-7.00011-9

flexible loads. The MG can be operated in a non-autonomous way if interconnected to the grid, or in an autonomous way if disconnected from the main grid. The operation of micro-sources (μS) in the network can provide distinct benefits to the overall system performance if managed and coordinated efficiently. Modern MGs can be found as purely direct current (DC), alternating current (AC), or hybrid AC/DC schemes; however, as far as this chapter is concerned, the use of pure AC signals (fundamental frequency) are considered in the MG concept.

In this chapter, the MG is defined as a cluster of μS and loads operating as a single controllable system that provides power to its local area. With the light provided by the dawn of smart grids (SG), together with the high penetration of cybernetical systems in MGs, MGs might be characterized as the "building blocks of smart grids" [6,7], as they provide a very promising and novel network structure.

MGs typically include several DER technologies as in the case of solar PV (which outputs DC power) or microturbines (high-frequency AC power) [8]. Those technologies typically require power electronic converters (PECs) to provide an interface, such as a DC/AC or DC/AC/DC converter, to connect the primary source with the main power grid.

The power electronic interfaces provide an excellent mechanism to enable the controllability of features to the μS; especially considering the technological advances reached with the development of SGs.

The power inverter, or simply an inverter, is a frequently used PEC used in μS to provide interfacing between DC and AC.

Power inverters can play an essential role in the MG control when multilayered management systems are implemented. The inverter plays a critical role in frequency and voltage control in islanded MGs, and also, facilitates participation in black-start strategies.

MGs feature special control requirements and strategies to perform local balancing and to maximize their economic benefit.

The new IEEE Standard 1547-2018 [9], IEEE Standard for Interconnection and Interoperability of DERs with Associated Electric Power Systems Interfaces, overcomes the limitations of the low penetration of power converter-based DER in the electrical system, and adds advanced features on the DER, like smart features for solar PV inverters, including the following [10]:

- voltage ride-through,
- frequency ride-through,
- voltage support,
- frequency support, and
- ramp rates.

Standards are a key enabler to the deployment of MGs, and the associated DER within them. The IEEE Standard 1547-2018 [9] created a new tendency in the PEC industry, the so-called "smart inverter." The IEEE standard defines many features that should be included in a power inverter in order to be named smart, especially the digital architecture, bidirectional communications capability, and robust software infrastructure.

Modern MG controls must deliver special functional features [11]: (1) humility to represent the MG to the utility grid as a single self-controlled entity so that it can provide frequency control like a classical synchronous generator; (2) control the active power flow in the

MG in order to avoid power flow exceeding cable/line ratings; (3) control capability to regulate voltage and frequency within acceptable power-quality bounds during islanding operations and to dispatch resources to maintain an energy balance; and (4) provide mechanisms to safely reconnect and resynchronize the MG with the main grid, when required.

There are numerous benefits to implementing a standard for MG controllers. Such standards can facilitate the deployment and adoption of MG concepts by the utility industry; reduce system integration costs; help resolve in-the-field interoperability issues to reduce time to deployment; and lower technical barriers to advanced applications for MGs.

An MG is basically a group of interconnected loads and DERs within a clearly defined electrical boundary. The MGs have to act as a single controllable entity with respect to the utility grid and, depending on the operational mode, they can be connected and disconnected from the utility grid, enabling the MG to operate in both grid-connected or islanded modes. In this chapter, the grid-connected operational mode is only considered (see Fig. 11.1).

Fig. 11.1 shows a basic schematic diagram of a representative structure of an MG as considered in this chapter. The boundary between the utility grid and the MG is defined by the low side of the step-down transformer and indicated as the connection to the distribution grid at the point of interconnection (POI). The POI is a very important point in the MG because it is where all the interactions with the utility grid take place, encompassing the majority of the standards [9,12,13]. There are many topologies on the design of an MG, and it is possible to have multiple connections, including an alternate grid utility connection at the substation. However, an MG will have only one connection with the utility grid, which will be closed at a given time [12].

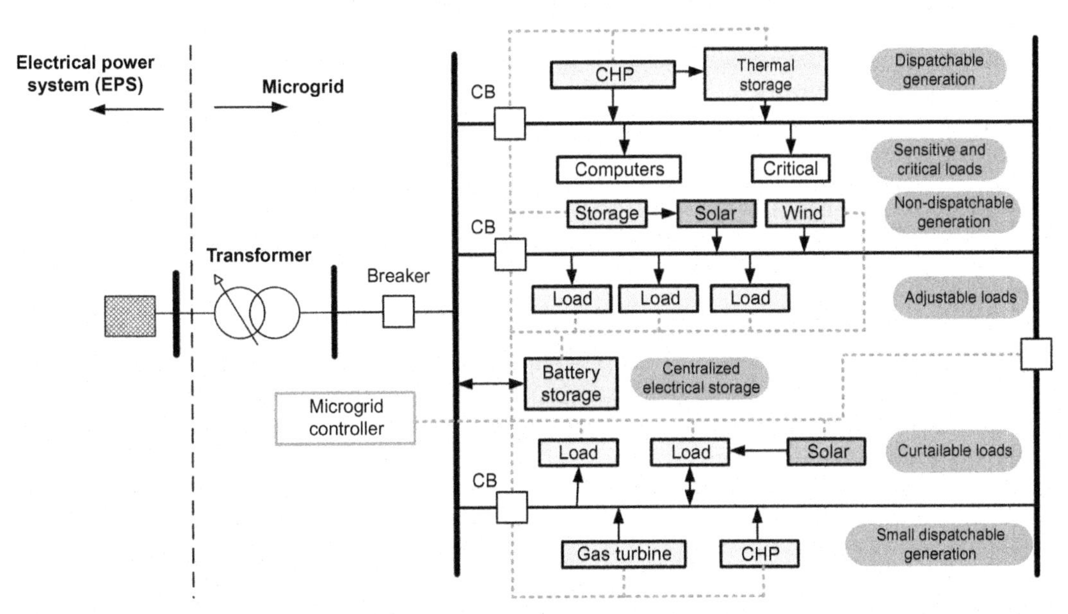

FIGURE 11.1 An accurate representation of a microgrid structure and its components. *Taken from IEEE Power and Energy Society, 2017. IEEE Standard for the Specification of Microgrid Controllers and the model figures has been modified for this specific chapter.*

A very detailed discussion of planning, designing, and operating the MG is given in IEEE Std. 1547.4 [10]. However, this standard does not address issues related to the power exchanges between the MG and the utility grid.

The modern IEEE Std. 2030.7, IEEE Standard for the Specification of Microgrid Controllers [11] is a crucial document that defines the operation and control of the MG. It introduces the fundamental concept of the MG energy management system (MEMS) and includes the specifications of the control functions that define the MG as a system that can manage itself, and operate autonomously or grid-connected, and seamlessly connect to the utility grid. The system also enables the MG to be disconnected from the utility grid to exchange power and supply ancillary services.

It is clear that the controllability of the MG is increasing and the potential value to provide support to the distribution network operators is a real option, especially enhancing the operation and increasing the resilience.

Today, the digitalization of the power systems creates a clear path to overcoming barriers and allowing a very active and dynamic interaction between transmission system operators (TSOs) and DSOs. One of these trends is the increasing volume of distributed generation (DG) being connected to the main grid and, more important, enhancing the services offered by the TSOs and DSOs. The active distribution network is a reality and has the potential of unfolding a massive synergy of DSO services to support TSOs.

Interactions between TSOs and DSOs could help minimize many power system problems, by Ref. [14]:

- Relieving the congestion at the transmission-distribution interface (with the potential to defer infrastructure investment);
- Relieving the congestion of transmission lines and distribution lines;
- Providing fully managed voltage support (TSO $\leftrightarrow$ DSO);
- Balancing power supply;
- (Anti)Islanding, resynchronization and black-start; and
- Offering coordinated protection.

There are many demonstratives projects on possible architectures for optimized interaction between TSOs and DSOs, as Ref. [15] shows with a summary of a demonstration project focused on the coordinated provision of power balancing, congestion management, and voltage regulation.

A collaborative approach to TSOs and DSOs involves the coordinated operation of reactive power controllers in order to keep a suitable control of the voltages of the system. MGs, with a high penetration of DER, can provide voltage support to the whole system by adjusting the reactive power flows at their interface and taking advantage of DG's reactive power capability.

The European grid code, the Demand and Connection Code (DCC) [16], establishes the new distribution systems and requests the provision of technical capacity to restrain the reactive power flowing upwards in the transmission system at low active power consumption, that is below 25% of their maximal power import capacity. Also, the DCC mentions other solutions may exist and can be considered, provided that TSOs and DSOs jointly assess both the technical and economic benefits through common analysis [17].

This chapter is dedicated to introducing the concept of grid-friendly or smart-converters and how they can be used to create a fully controllable MG able to provide auxiliary services, specifically voltage control, to the transmission system: the so-called "transmission system-friendly microgrid." This chapter presents the main challenges and the solution of a Volt-Var control on converter-dominated MGs, using the concept of grid-friendly or smart-converters with wide area control. This chapter dedicates a section to the current numerical results of the proposed control algorithm and its implementation using Python and simulations using DIgSILENT PowerFactory to demonstrate the suitability of the proposed transmission system-friendly MG.

11.2 Transmission system-friendly microgrid

This chapter uses the concept of grid-friendly or smart-converters and proposes an optimized mechanism to enable a fully controllable MG to provide Volt-Var auxiliary services to the transmission system; it is the development of a transmission system-friendly MG.

A traditional power system uses transformers equipped with an on load tap changer (OLTC) to maintain control of the high voltage (HV) level by the means of extra high voltage (EHV)/HV while the EVH level is managed through a classic Volt-Var control (shunt capacitors/reactors, transformer tap changers, synchronous generators, synchronous condensers, FACTS, for example, STATCOM, SVC, and HVDC) (see Fig. 11.2).

The effectiveness of the reactive power measures to control voltages in power systems is well-documented in the literature [18]. In fact, the influence of the reactive power from

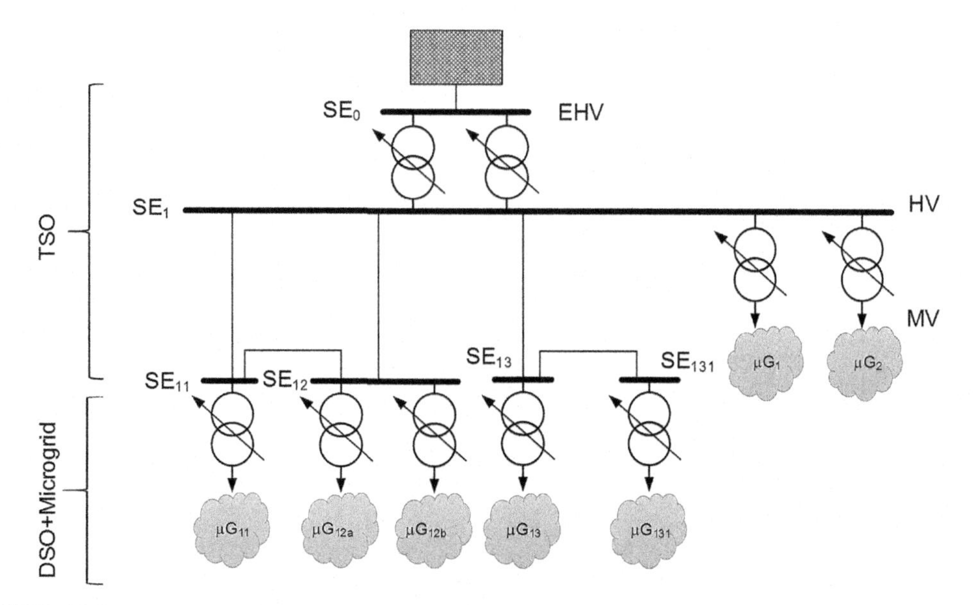

FIGURE 11.2 Illustrative scheme of UHV/high voltage (HV)/medium voltage (MV) + microgrid (MG) system. Transmission System Operators (TSO)–Distribution System Operator (TSO–DSO) boundaries are shown.

MV networks on the HV voltage is roughly proportional to the short-circuit power of the substation short-circuit current (SCC).

Traditionally, additional reactive power sources/controls are provided in MV/HV, including:

- Distribution voltage/var control;
- Voltage regulators;
- Transformer tap changers;
- Shunt capacitors/reactors;
- Series capacitors/reactors;
- D-STATCOM.

The technological advances on DER technologies and their interface to the grid are making the power converter-based DER a very attractive solution to provide Volt-Var control and positively impact the distribution systems. Modern inverter-interfaced DERs have the potential to improve customer load voltage, voltage regulation, voltage flicker, and other related issues.

DERs provides a double mechanism to manage voltages inside the DSO + MG system. Power converter-based DERs allow fine control of the power generation and control (if appropriate controls are applied), allowing voltage band management through changes in the system. Also, power electronic control allows a decoupling between active and reactive power is making it more attractive to use inverters to control voltage fluctuations.

Fig. 11.3 shows the illustrative case of a simple MG connected to an MV system, the relationship between the voltages (U_1 and U_2) and active (P_L, P_G) and reactive power (Q_L, Q_G, Q_C):

$$U_2 \approx U_1 + \frac{R_{line}\overbrace{[P_{DER} - P_L \pm \Delta P_{control}]}^{\text{Active power}} + X_{line}\overbrace{[Q_{DER} - Q_L \pm \Delta Q_{control}]}^{\text{Reactive power}}}{U_2} \tag{11.1}$$

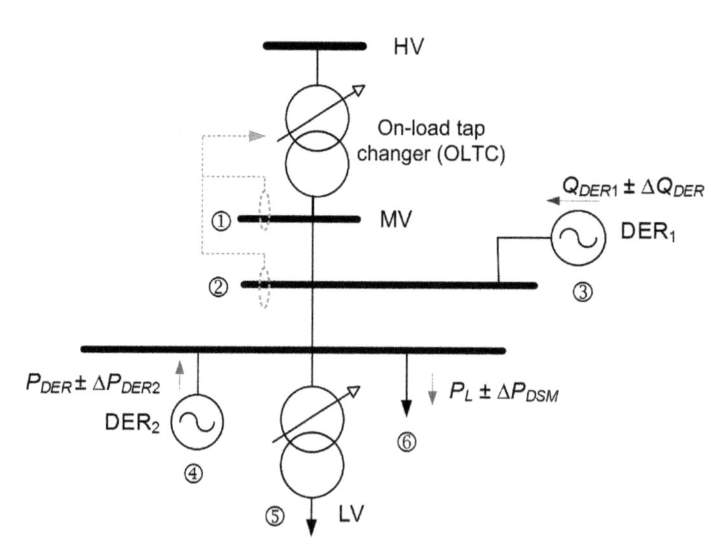

FIGURE 11.3 Mechanism of the influence of voltage control: ① and ② on load tap changer (OLTC), ③ and ④ distributed energy resource (DER) active and reactive power generation/consumption, ⑤ adjustable transformer (low voltage, LV), ⑥ active demand-side management.

TABLE 11.1 Effect of line resistance (R_{line}) and reactance (X_{line}) on the voltage (U_2).

Voltage level	R/X ratio	Voltage level influenced by
Transmission system (HV)	$R_{line} << X_{line}$	Reactive power
MV distribution	$R_{line} < X_{line}$	Active and reactive power
LV distribution	$R_{line} > X_{line}$	Active power

HV, high voltage; *MV*, medium voltage; *LV*: low voltage.

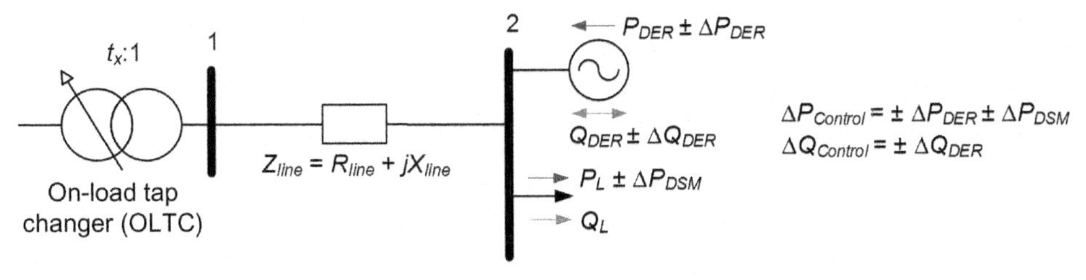

FIGURE 11.4 Illustrative representation of a simple microgrid and its component.

where $\Delta P_{control}$ and $\Delta Q_{control}$ are the changes on the smart-converter active and reactive power generation/consumption. Table 11.1 shows a summary of the variables influencing the voltage control at HV, MV, and LV.

It is clear that controlling the active and reactive generation/consumption of a smart-converter allows control of the voltage inside the MG. Fig. 11.4 shows an illustrative example where the active power generation of a smart-converter and a controllable load (active power consumption) are used for voltage management; for illustrative purposes the reactive power is not considered.

Fig. 11.5 shows the effect of acting on the active power in the MG as a precise mechanism of controlling the voltage at the LV side of the step-down transformer. However, it is clearly a limitation on this approach as local voltage control is easily enforced in a smart converter without the use of complicated and expensive communication. But, using local voltage control at the converters, may increase MG voltages, making the OLTC act by compensating for HVs during light load/heavy generation conditions. As a consequence, there is the potential risk of the OLTC over-acting and negatively affecting the power quality and life of the step-down transformer. The effect of the smart inverter and control strategies on the smart-converter and OLTC control is depicted in Fig. 11.6.

The transmission system-friendly MG is an MG with two main elements:

- *Smart-converters*: DERs are interfaced to the grid by smart-converters, each one of which is enabled with digital architecture, bidirectional communications capability, and robust software infrastructure.
- *Wide-Area Volt-Var control*: The wide-area controller monitors and controls beyond the traditional boundary between the utility grid and the microgrid (POI). In fact, the controller considers the OLTC transformers at the interface between the TSOs and the

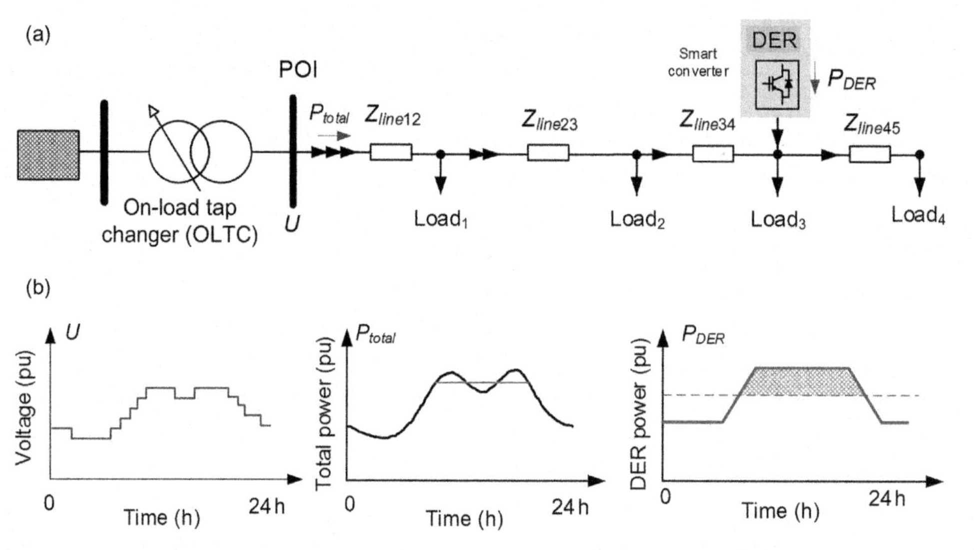

FIGURE 11.5 An illustrated example of voltage band management through changes in power generation/consumption. (A) Representative radial configuration of an MG + DSO, where a smart-converter is integrated. (B) (*from left to right*) Representative LV, active power generation of the smart-converter, load demand (P_L).

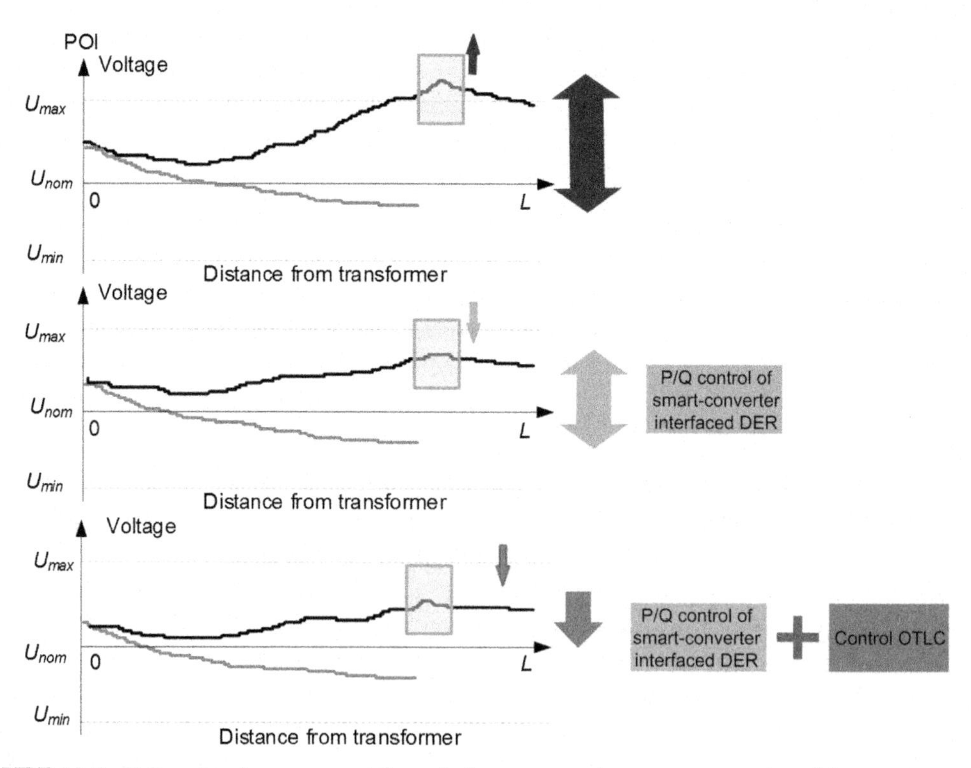

FIGURE 11.6 Voltage band management through changes on active power generation of the smart-converter, load demand (P_{load}), and its effect on the load tap changer (OLTC).

DSOs. Monitoring the voltages at the MG and the distribution system, the controller is able to control the voltage (and thereby the load power). Details of the Wide-Area Volt-Var control are discussed in the next subsection.

The smart-converter at the grid-friendly device is able to help regulate voltage by absorbing and injecting reactive power from the grid by using the Volt-Var control method for voltage regulation.

The primary advantage of using a smart-converter to regulate voltage is its ability to shift power quickly as it is a power electronic device. Typical applications of smart-converters are based on a local voltage controller. This chapter goes beyond this by proposing a novel, optimized Volt-Var controller.

11.2.1 Wide-Area Volt-Var controller

The IEEE Std. 2030.7, IEEE Standard for the Specification of Microgrid Controllers [11] establishes the main requirements for an MG controller. It must include the control functions that define the MG as a system that can manage itself, operate autonomously or in grid-connected fashion, and connect to and disconnect from the main distribution grid for the exchange of power and the supply of ancillary services. Fig. 11.7 shows a functional framework of the typical MG control system as specified on IEEE Std. 2030.7.

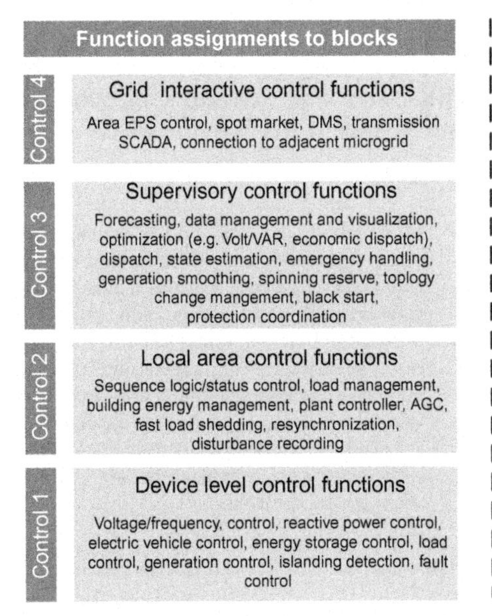

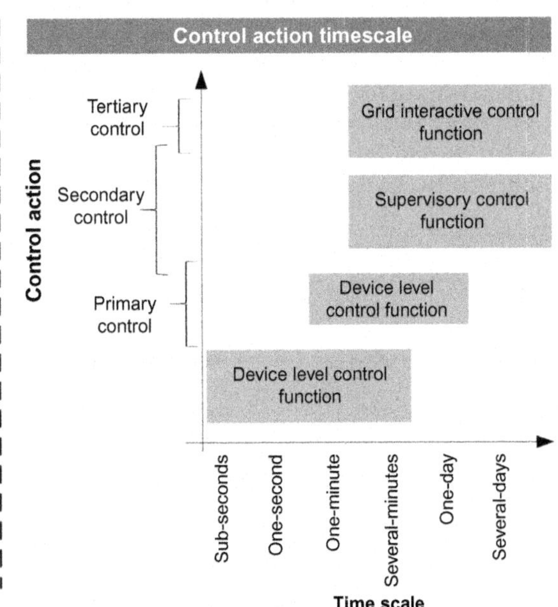

FIGURE 11.7 A graphical description of the functional framework for microgrid (MG) control system as specified on IEEE Std. 2030.7. *Modified from Bernabeu, E., Hoke, A., Walling, R., Zhou, G., 2018. Application of IEEE Std 1547 and Operating Experiences of Grid-Supportive Distributed Energy Resources PJM Engagement in the Application of IEEE 1547.*

The hierarchical levels of control for MGs imposed by IEEE Std. 2030.7 may be categorized as primary, secondary, and tertiary [11]:

- Primary control is the lowest hierarchical control level and it is entirely based on local measurements and includes control actions such as islanding detection, output control, and power sharing (and balance control).
- Secondary control, the *microgrid* energy management system (mEMS), is responsible for MG operation in either a grid-connected or islanded mode.
- Tertiary control is the highest level of control and sets long-term and "optimal" set points, depending on the host grid's requirements.

The MG control included in the IEEE Std. 2030.7 has a very important constraint. While, for purposes of interconnection, an MG shall present itself to the grid to which it is connected as a single controllable entity, the controller is designed to meet the MG requirements at the POI with the MG. As a consequence, controllers based on IEEE Std. 2030.7 limit the synergy MG + DSO/TSO.

In this chapter, the core of the transmission system-friendly MG is the Wide-Area Volt-Var controller. Fig. 11.8 illustrates the main aspects of this controller regarding measured signals and control actions. It must be noticed that controls affect the OLTC transformers at the interface between the TSOs and the DSOs.

One of the main characteristics of the Wide-Area Volt-Var controller is the time scale of the decision process. Although the controller can work in real time, the proposed approach has been limited to a time scale similar to that provided by the supervisory control and data acquisition (SCADA), 5–15 minutes resolution.

The monitoring component of the Wide-Area Volt-Var controller is responsible for taking voltage measurements at each load node in the MG, and also extending the voltage measurement beyond the POI by taking voltage measurements at the LV side of the step-down substation transformer.

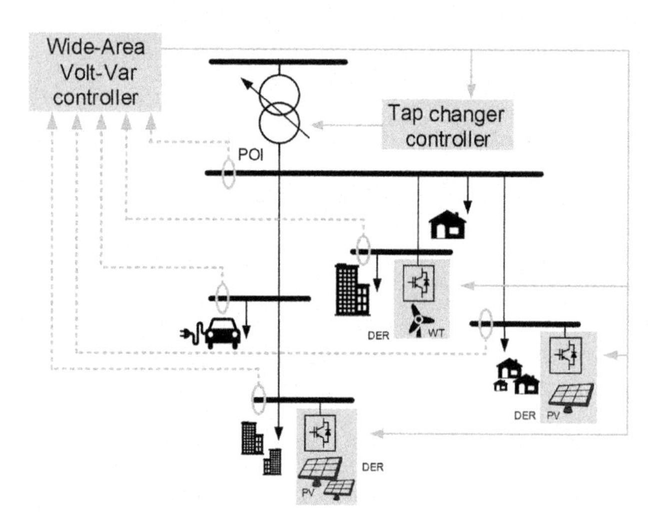

FIGURE 11.8 The traditional SCADA system provides an effective monitoring and acting platform to collect, effectively, measurements and status data from the process and remotely intervene in the process, by using a centralized and reliable platform. General scheme of the proposed Wide-Area Volt-Var controller. *Dotted lines* are used to represent measurement signals, while *solid lines* represent control actions taken by the controller.

11.2.2 Methodology

The power converter-based MGs provide enormous opportunities for them to be used as an essential resource enabling their services to help the operation of the transmission system. In this book, the concept of grid-friendly or smart-converters is proposed as an optimized mechanism to enable a fully controllable MG to provide Volt-Var auxiliary services to the transmission system; it is the development of a transmission system-friendly MG.

The DER technologies and their interface to the grid (smart-converter) are very useful mechanisms for providing Volt-Var control, which can impact positively on distribution and transmission systems.

This section presents a methodology to enable the smart converter installed inside an MG to provide voltage control with the Volt-Var control, thus reducing the active power losses in the MG. All these advantages are reached by optimally adjusting the reactive power injection of the smart-converter and considering settings of the interface transformers.

From a mathematical point of view, an optimization problem aims to find the best solution from all feasible solutions.

In this chapter, the problem of reducing the steady-state active power losses $P_{losses} = f(\mathbf{x})$ inside the MG is formulated as a minimization optimization problem.

$$\min[P_{losses}] = \min_{\mathbf{x}} \left[f(\mathbf{x})\right] \tag{11.2}$$

The optimization problem is solved by taking decision variables, $\mathbf{x} = [\ x_1\ x_2, \ldots, x_n]^T$ in the way to allow the calculations the reactive power output of a smart-converter at the DERs ($Q_{G,DER} = [Q_{G,DER1}, Q_{G,DER2}, \ldots, Q_{G,DER,Nder}]^T$) as well as the tap settings of the transformers ($\mathbf{TRX_{tap}} = [TAP_1, TAP_2, \ldots, TAP_{Ntrx}]^T$) connected to the MG so that the system's active power losses are minimized, while keeping the bus voltage inside the quality limits ($U_{min} \le U \le U_{max}$).

The objective function $f(\mathbf{x})$ in Eq. (11.3) has been modified explicitly to include the active power losses and voltage control. As a consequence, an *Exterior Penalty Function* method is applied to include the voltage constraints and transform the objective function to an augmented objective function set for the subsequent execution of an unconstrained optimization algorithm.

The augmented objective function, $F(\mathbf{x},k_p)$, which takes into account the main objective function $f(\mathbf{x})$ and the penalty function $g(\mathbf{x},k_p)$, is defined in Eq. (11.3).

$$F(\mathbf{x},k_p) = f(\mathbf{x}) + g(\mathbf{x},k_p) \tag{11.3}$$

where the decision variables (x_i) are limited to:

$$x_i^{low} \le x_i \le x_i^{high} \forall i \in [1, \ldots, n] \tag{11.4}$$

where $\mathbf{x}$ is a vector of the decision or control variables $\{x_i\}$. In this particular problem, the control variables are the reactive power output of the smart-converters installed at the DER and the tap position at the transformer's interface.

The objective function $f(\mathbf{x})$ represents the total system active power losses $f(\mathbf{x}) = P_{losses}$ and $g(\mathbf{x}, k_p)$ is a penalty function that depends on the system voltages.

The coefficient k_p represents a penalty multiplier. The penalty function $g(\mathbf{x}, k_p)$ is defined in 11.5.

$$g(\mathbf{x},k_p) = k_p \left[\sum_{i=1}^{N} h_i(\mathbf{x}) \right]^2 \tag{11.5}$$

$$h_i(\mathbf{x}) = \begin{cases} 0, & \mathbf{x}^{\mathbf{low}} \leq \mathbf{x} \leq \mathbf{x}^{\mathbf{high}} \\ 1, & \text{otherwise} \end{cases} \quad \forall i \in [1, \ldots, N] \tag{11.6}$$

where $h_i(x)$ is an auxiliary function that has a value of zero if the voltage at bus i is within the allowed range and has a value of one if the voltage is above or below these limits and N is the total number of buses in the system.

The number of buses with a voltage outside the allowable range is squared, multiplied by the penalty factor, and then added to the objective function. By squaring the number of buses with a voltage outside the acceptable range, more weight is assigned to the penalty function, thereby increasing the objective function in a higher proportion. Nonetheless, the solution obtained might entail a higher value of system losses.

A small penalty factor may produce a solution close to the constraint limits. The value assigned to the penalty factor can be set constant throughout the optimization process, or it can be set so that it varies with each iteration from an initial maximum to a final minimum, thereby unfeasible solutions are more heavily penalized at the beginning of the process.

The methodology proposed in this chapter is summarized and depicted in the flowchart shown in Fig. 11.9.

11.3 Simulation and results

This section is dedicated to illustrating the proposed methodology and assessing the performance on improving the active power losses of a typical European distribution system considering the integration of low-carbon technologies: electric vehicles (EVs), wind turbines, and PV systems.

11.3.1 Test system

A typical European distribution system is selected for illustrative and performance assessment of the proposed methodology. The CIGRE Task Force C6.04 created the original version of the test system and presented it in the report titled "Benchmark Systems for Network Integration of Renewable and Distributed Energy Resources" [19].

The MV distribution network benchmark was derived from a real, physical MV distribution network located in southern Germany. This network supplies a small town and the surrounding rural area. The CIGRE Task Force C6.04 made an effort to reproduce the

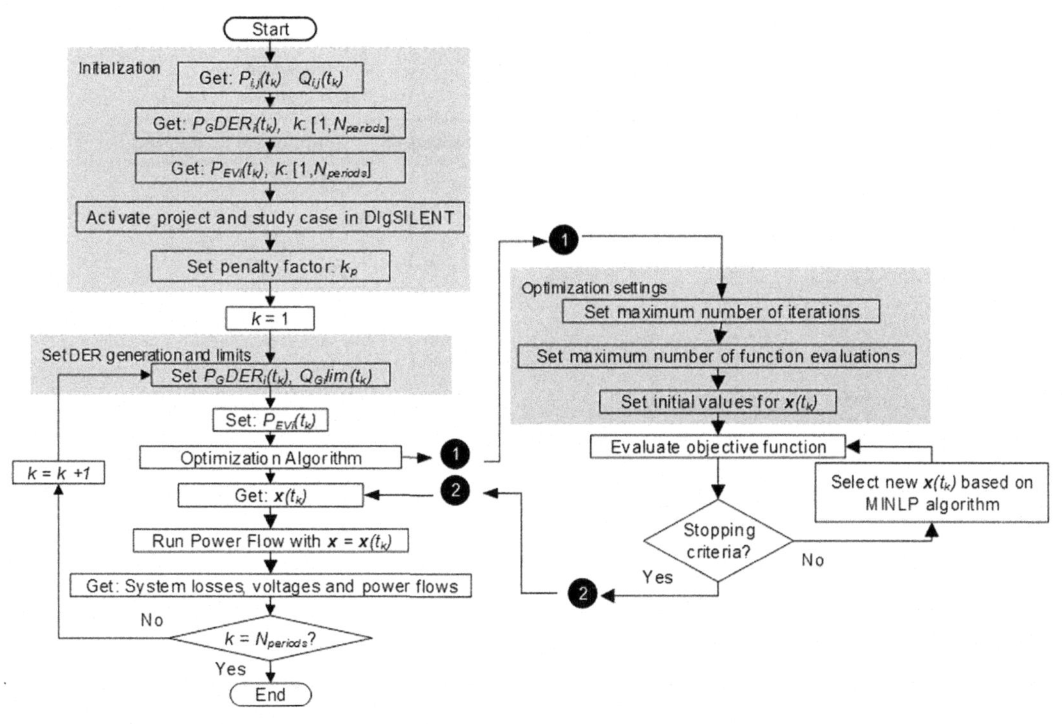

FIGURE 11.9 Illustrative flowchart summarising the proposed methodology of transmission system-friendly micro-grid (MG).

original data when compared with this original network. The number of nodes for the benchmark network was reduced to enhance user-friendliness and flexibility, while adequately maintaining the realistic character of the network [19].

Real distribution networks can be imbalanced, especially at LV lateral along the MV line. The European MV distribution network benchmark does not explicitly include imbalances; consequently, the network is assumed to be symmetrical and balanced.

The test system consists of two typical European MV distribution feeders (named Feeder 1 and 2 and highlighted on Fig. 11.11). The feeders are two 20 kV, 50 Hz, three-phase feeders, and the test system includes switching devices allowing the feeders to be operated in radial or meshed topology. In this chapter, the "S2" and "S3" switches are considered open and "S1" is closed, as a consequence, the feeders form a close ring when interconnected by line 14-8. The feeders are fed from a 110-kV sub-transmission network, where two step-down transformers (110/20 kV) are used to feed the feeders separately. The authors have selected the European MV distribution feeders because they can be used for DER integration studies.

The European MV distribution feeders are provided with three-phase lumped loads, the coincident peak loads for each bus of the test system is presented in Table 11.2. The loads are assumed to be symmetrical and therefore equal in all three phases. The loads used to represent the equivalent feeders are placed in buses 1 and 12.

TABLE 11.2 Main load parameters of *Test System* (20 kV, 50 Hz)

	Apparent power S_{max} (kVA)		Power factor $pf = \cos\phi$	
Bus	**Residential**	**Commercial /industrial**	**Residential**	**Commercial /industrial**
1	15,300	5,100	0.98	0.95
2	–	–	–	–
3	285	265	0.97	0.85
4	445	–	0.97	–
5	750	–	0.97	–
6	565	–	0.97	–
7	–	90	–	0.85
8	605	–	0.97	–
9	–	675	–	0.85
10	490	80	0.97	0.85
11	340	–	0.97	–
12	15,300	5280	0.98	0.95
13	–	40	–	0.85
14	215	390	0.97	0.85

The maximum apparent power (S_{max}) presented in Table 11.2 together with the power factor ($pf = \cos\phi$) of the commercial/industrial and residential customer can be used to calculate the maximum active and reactive power consumption at each bus ($P_{max,i}$ and $Q_{max,i}$ $\forall$ I = 1, ..., $N_{bus} = 14$).

The test system is designed to reproduce the performance of commercial/industrial and residential customers throughout 24 hours. As a consequence, the daily loadability profile is used to calculate the maximum active and reactive power at each bus and at any time of the day.

Fig. 11.10 show the daily loadability profiles $S[\%]$ of commercial/industrial and residential customers throughout 24 hours. The loadability pattern is used to calculate the appropriate load at each bus at each time period of the day.

The maximum active and reactive power consumption of the load installed at the ith bus at the time t_k are calculated as:

$$P_{load,i}(t_k) = S_{max,i} \cdot pf^{type}(t_k) \cdot S^{type}_{max}(t_k) \tag{11.7}$$

$$Q_{load,i}(t_k) = S_{max,i} \cdot \left\{ \sqrt{1 - \left[pf^{type}(t_k)\right]^2} \right\} \cdot S^{type}_{max}(t_k) \tag{11.8}$$

where $S_{max,i}$ is obtained from the Table 11.2 and $S^{type}{}_{max}(t_k)$ is obtained from Fig. 11.10.

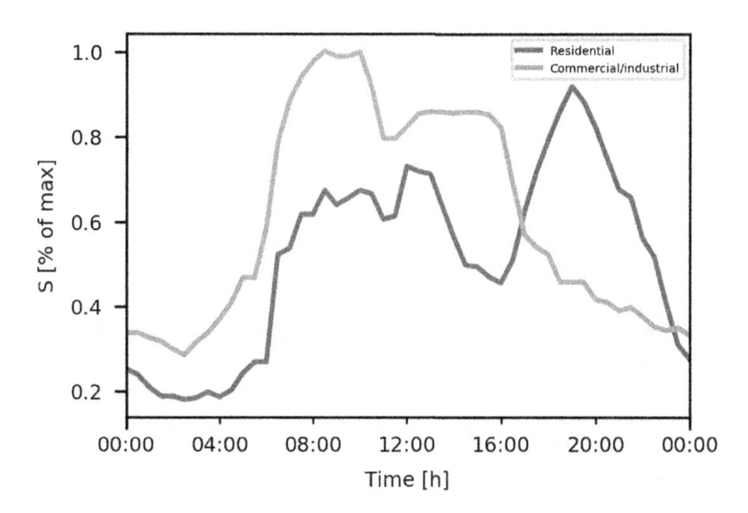

FIGURE 11.10 Daily demand profiles for residential and commercial/industrial loads.

TABLE 11.3 Details of the main electrical parameters of distributed energy resource (DER) units.

Bus	DER type	P_{max} (kW)
3	Photovoltaic	20
4	Photovoltaic	20
5	Photovoltaic	30
6	Photovoltaic	30
7	Wind turbine	1500
8	Photovoltaic	30
9	Photovoltaic	30
10	Photovoltaic	40
11	Photovoltaic	10

In this chapter, the model has been modified slightly to allow the integration of low-carbon technologies: PV, wind power, and plug-in EVs. The PV and wind turbine units are implemented as generation units considering the stochasticity of the primary energy resources. A single DER unit is integrated per bus, and the maximum active power (P_{max}) of each unit is detailed in Table 11.3.

The test system showed in Fig. 11.11 has been modified to accommodate the DER detailed in Table 11.2. The new network is shown in Fig. 11.12.

11.3.2 Electric vehicle clusters

The test system was improved by including clusters of EVs; the idea behind this is to include the effect of the low-carbon technology on the active power losses of the test system, which adds stress to the system.

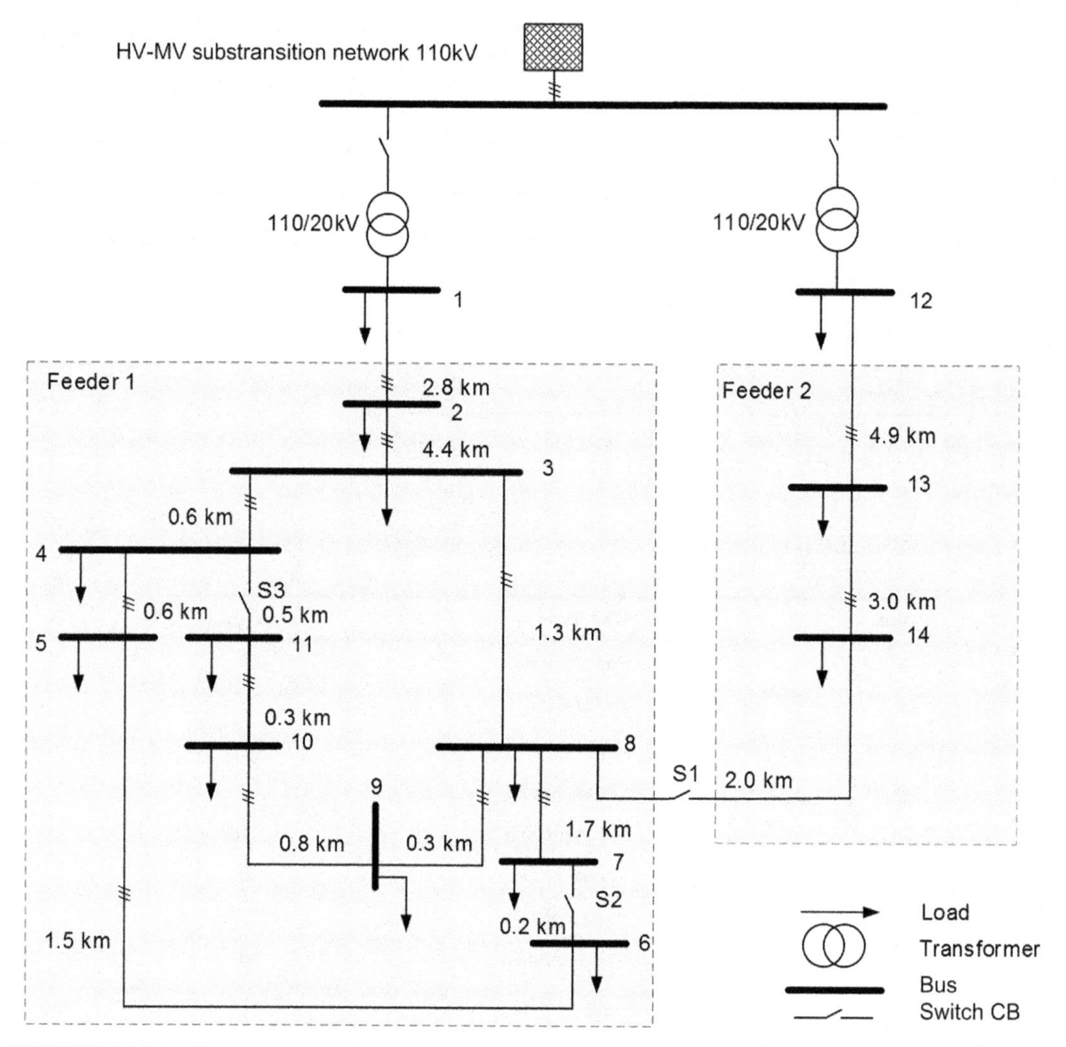

FIGURE 11.11 Illustrative single-line diagram depicting the topology of a European medium voltage (MV) distribution network benchmark. *Taken from CIGRE, 2014. Benchmark Systems for Network Integration of Renewable and Distributed Energy Resources, April 2014 and slightly modified for this specific chapter.*

In this chapter, the EV demand is calculated by using the model presented in Ref. [20]. This model can be configured to accommodate the nature of the type of electric charge pattern. Consequently, the EV demand has been designed to reproduce three charging patterns:

- Uncontrolled or also known as "dumb charging" in which the EV is connected as soon as it is back from its daily trip, typically at around 6 or 7 pm. This constitutes the worst case from the point of view of system operation because the peak load from EV charging tends to occur at the same time as the overall system demand peak.

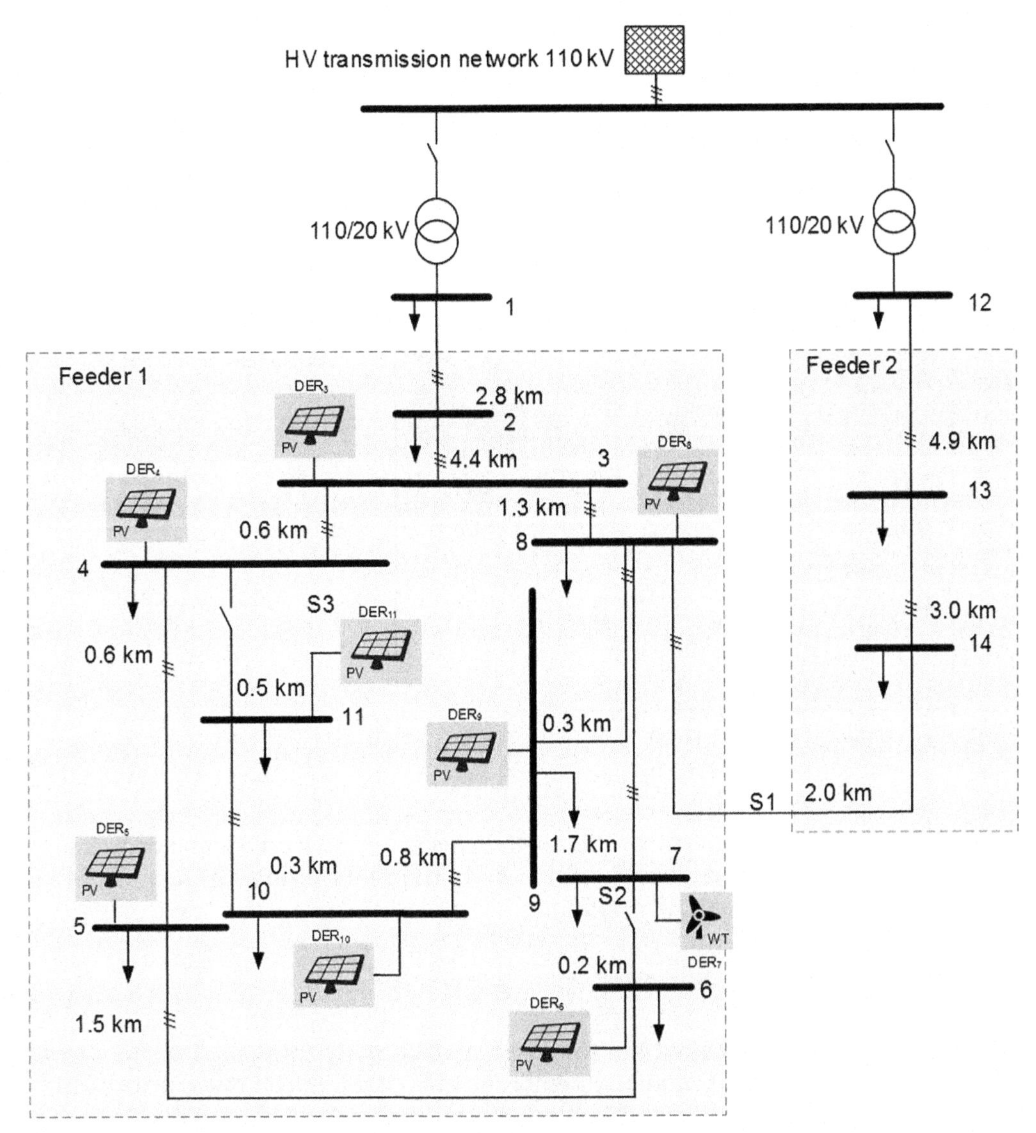

FIGURE 11.12 Illustrative single-line diagram depicting the test system and the integration of distributed energy resource (DER).

- Dual tariff or "off-peak" charging, in which the user is encouraged to charge during a specific time window (typically corresponding to system valley hours).
- Intelligent or "smart charging" in which a control system manages the EV charging according to system demand as well as energy prices.

Three EV clusters are connected to the test system specifically at buses 7, 11, and 13. For illustrative purposes, each EV cluster is connected to a different bus and is modeled with varying proportions of dumb, smart and off-peak charging as detailed in Table 11.4.

TABLE 11.4 Summary of the simulated electric vehicle (EV) pattern scenarios.

Charging scenario	Dumb charging (%)	Smart charging (%)	Off-peak charging (%)	EV cluster connection
A	20	70	10	Bus 7
B	70	20	10	Bus 11
C	50	50	0	Bus 13

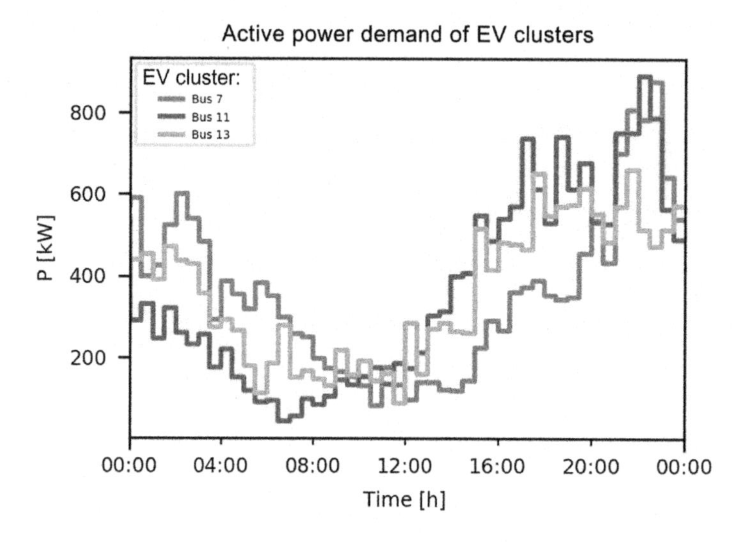

FIGURE 11.13 Patterns of lumped electrical demand of charging stations of electric vehicle (EV) clusters.

Fig. 11.13 shows the simulated profile of the cluster of electrical demand of charging stations for the EV clusters defined in Table 11.4. The daily profile (24 hours) has been simulated considering a 30-minute resolution.

11.3.3 Implementation

The authors have implemented the proposed methodology (presented in the previous section) in a very complex co-simulation environment. DIgSILENT PowerFactory 2018 SP 3 has been used as the primary modeling and power system simulation tool.

The test system has been modeled using DIgSILENT PowerFactory and the power system analysis command ComLfd has been used to obtain the steady-state conditions of the test system, including the active power losses (P_{losses}), voltages (U_i), and power flow (P_{ij}). The power system simulation software has been controlled by using Python API, and the Python environment has been used to develop the whole methodology, where the optimization problem is solved efficiently by the open-source C++ code, Bonmin.

The power demand, DER generation profile and EV charger pattern have been precalculated, and all the input data is stored using Microsoft Excel Files, then the data is read using a specific purpose-developed Python Script, all the data has been considered for a single day, 24 hours, considering a 30-minute resolution.

The vector **x** defines the control variables in the optimization process. In this specific chapter and the test system, the decision-variable vector (**x**) comprises the reactive power output of the wind turbine $\mathbf{Q_{G,WT}}$, the reactive power generation of the smart-converters connected to the PV units $\mathbf{Q_{G,PV}}$ and the tap changers of both transformers $\mathbf{TRX_{tap}} = [TAP_1, TAP_2]^T$), as a consequence the vector is defined as:

$$\mathbf{x} = \left[\mathbf{Q_{G,WT}} \quad \mathbf{Q_{G,PV}} \quad \mathbf{TRX_{tap}} \right]^T \tag{11.9}$$

where

$$\mathbf{Q_{G,WT}} = [Q_{GWT}]$$
$$\mathbf{Q_{G,PV}} = \left[Q_{G,PV3} \quad Q_{G,PV3} \quad Q_{G,PV5} \quad Q_{G,PV6} \quad Q_{G,PV8} \quad Q_{G,PV9} \quad Q_{G,PV10} \quad Q_{G,PV11} \right]^T$$
$$\mathbf{TRX_{tap}} = \left[TAP_1 \quad TAP_2 \right]^T$$

In the case of the transformer taps, it is considered that each transformer has ± 10 load-changing taps and that each tap corresponds to 0.625% of the transformer-rated voltage (tap position zero).

For the DER, the reactive power output is constrained inside the safety operative limits of each unit, which are given by the fixed maximum apparent power and the variable active power output at each period as defined in the following equation:

$$Q_{GDER,i}^{\lim} = \pm \sqrt{\left[S_{G,i}\right]^2 - \left[P_{GDER,i}\right]^2} \tag{11.10}$$

The objective function considers the total power losses as well as the voltage profile for all system buses. These constraints require a calculation of the nonlinear network, power-flow equations. Moreover, the transformer taps can only take integer values. Therefore the equations formulated in Eqs. (11.2)–(11.10) correspond to a mixed-integer nonlinear programming (MINLP) optimization problem. In this chapter, the penalty factor was set constant with a value of 100.

By formulating the problem as an MINLP, more exact results can be obtained as the full complexity of the network is considered and no rounding is necessary to obtain the optimum transformer tap settings. The MINLP optimization model formulated above is solved efficiently by the open-source C++ code Bonmin [21].

11.3.4 Summary of simulation cases

The proposed methodology is illustrated and its performance assessed by using four simulation scenarios. A summary of them is presented in Table 11.5 and simulation results are explained in the next subsections.

11.4 Base case

Considering Fig. 11.14, it is evident that due to the low demand both from residential and commercial/industrial consumers in the early hours of the day (from 00:00 to 06:00) the voltages are higher than 1.0 pu throughout the system.

TABLE 11.5 Transformer tap settings for the different simulation cases.

Case	Transformer 0–1 tap setting (TAP_1)	Transformer 0–12 tap setting (TAP_2)
Base case	10	5
Case 1	10	5
Case 2	10	5
Case 3a	Variable	Variable
Case 3b	Variable	Variable

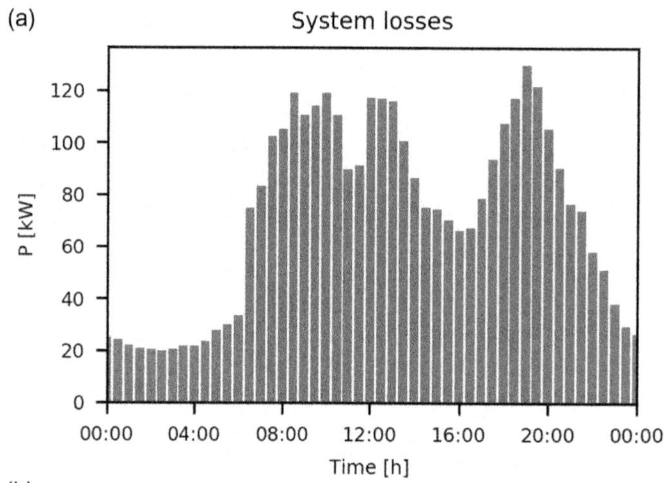

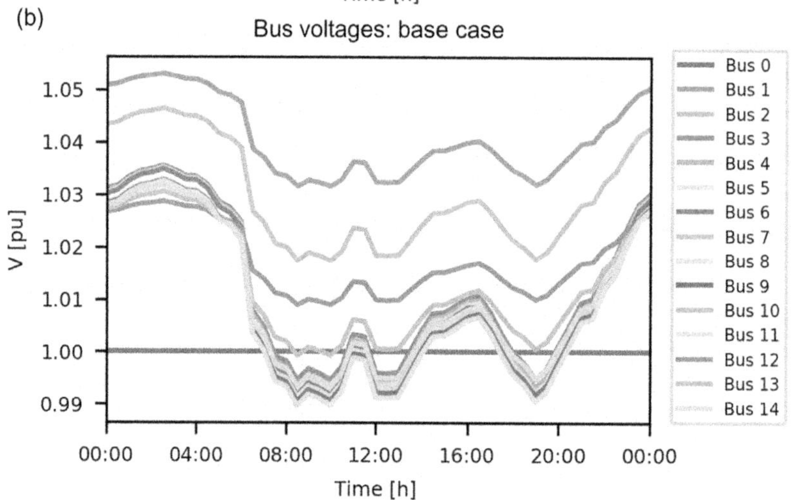

FIGURE 11.14 *Base case* (A) system losses, (B) system voltages.

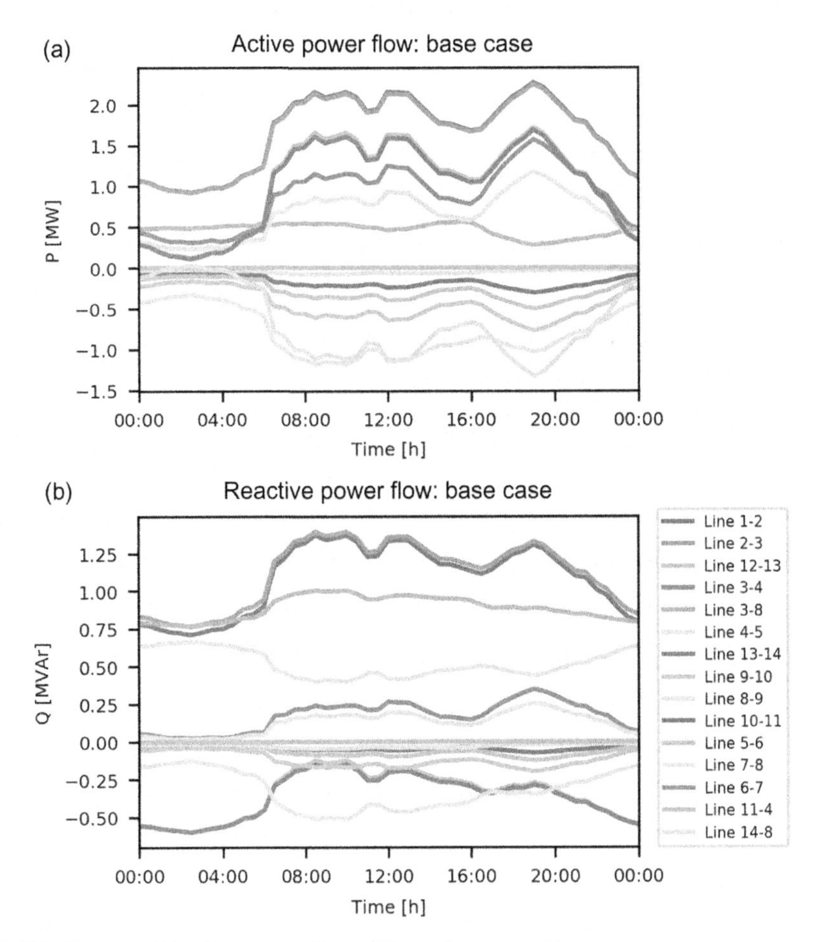

FIGURE 11.15 *Base case* (A) active power flows, (B) reactive power flows.

In this period, the voltage of bus 1 rises above 1.05 pu. However, after 07:30, overall voltages begin to descend and reach a minimum value of 0.989 pu in bus 11 at 08:30.

The voltage profile improves after 12:00 following a reduction in residential demand, however, after 16:00 they decrease sharply and reach a new minimum of 0.991 pu at 19:00. System losses increase sharply at around 07:30 due to increased demand and the maximum value is 130.05 kW at 19:00.

Fig. 11.15 shows the active and reactive power flows through the network for the base case.

Lines 1-2 and 2-3 on Feeder 1 carry most of the active and reactive power through the system. Active and reactive power flows through lines 11-4 and 6-7 is zero because switches S2 and S3 are open.

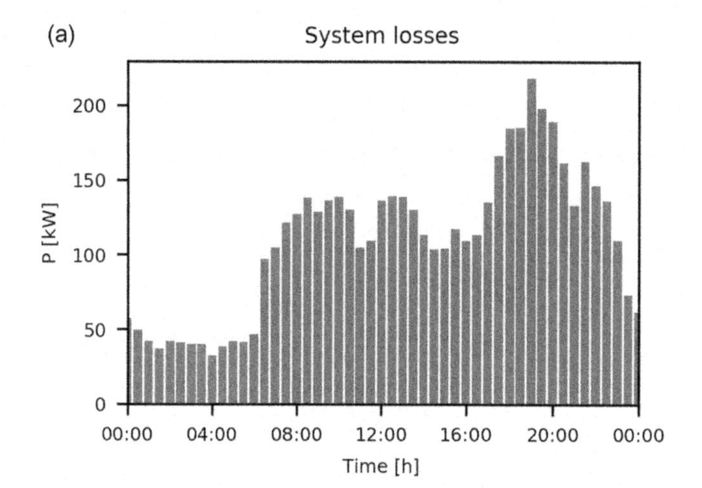

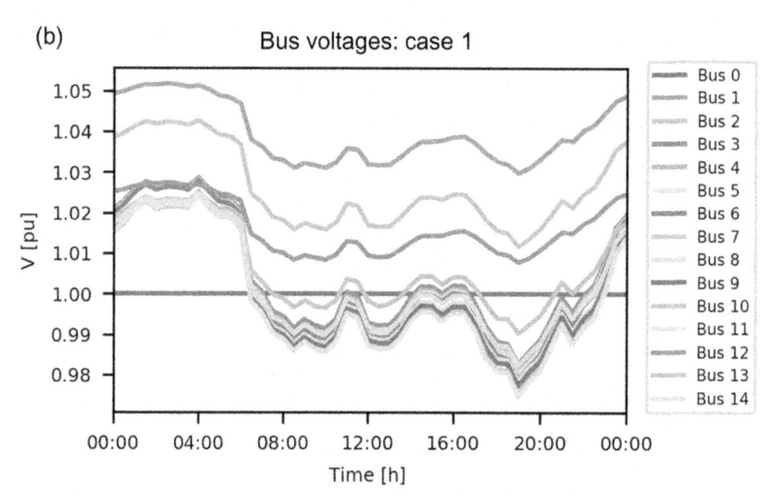

FIGURE 11.16 Case 1 (A) system losses, (B) system voltages.

11.5 Case 1: inclusion electric clusters

The extra EV demand causes voltages to drop further during the evening peak-demand period as compared with the base case. It remains, however, 5% above the nominal voltage at the beginning and at the end of the day for bus 1.

The voltage drops further down at the time of the most massive EV demand. The minimum voltage, in this case, reaches 0.975 pu in bus 11 at 19:00. As expected from the additional EV demand, the flows of both active and reactive power increase (Fig. 11.16, Fig. 11.17).

11.6 Case 2: integration of distributed energy resource

This simulation case includes the integration of DER, as indicated in the modeling section, and the maximum active power of each unit is detailed in Table 11.3. The wind

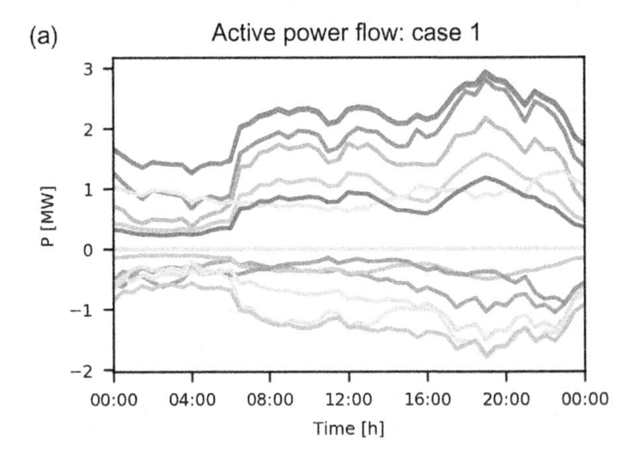

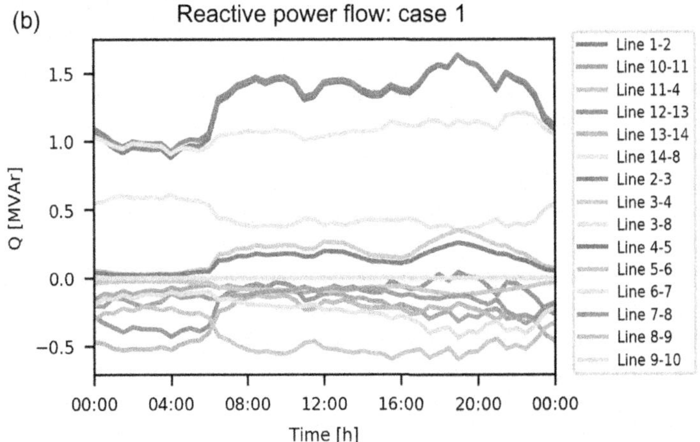

FIGURE 11.17 Case 1 (A) active power flows, (B) reactive power flows.

turbine active power production is presented in Fig. 11.18. For this case, the power factor of the DER units is set at 0.8 (Fig. 11.19).

By including the DERs, the system losses are reduced and the voltage profiles are improved in the middle part of the day. Notably, the voltage in all buses is within 5% of the nominal value during the day, compared to the previous cases without DERs. The reduction in system losses can be understood more readily by observing the flows of active power shown in Fig. 11.20A and B. For example, the active power flow in Line 2-3 is reduced by almost 20% concerning Case 1. The DERs introduced, namely, the wind turbine and PV units supply a percentage of the demand locally, thereby reducing power lost in transmission.

11.7 Case 3: optimization

Two scenarios are considered for this case. In Scenario A, the acceptable voltage deviation is limited to $\Delta U = \pm 5.0\%$ of the nominal value and in Scenario B, it is limited to $\Delta U = \pm 2.0\%$ of the nominal value of 1.0 pu (i.e., $0.98 \leq U_i \leq 1.02$ pu).

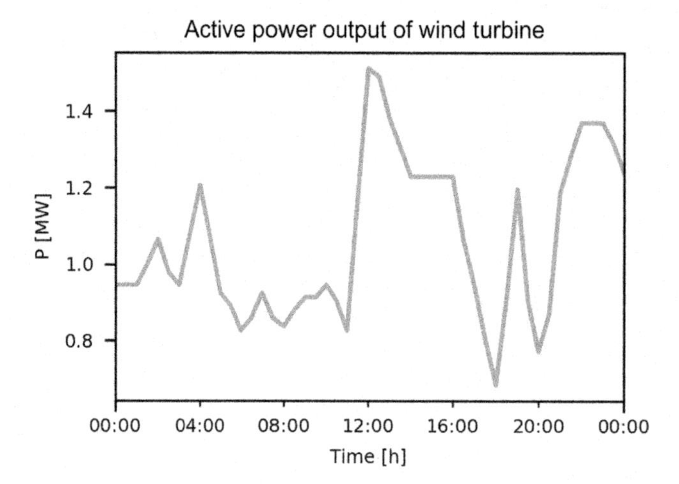

FIGURE 11.18 Active power output of wind turbine and photovoltaics (PV) units.

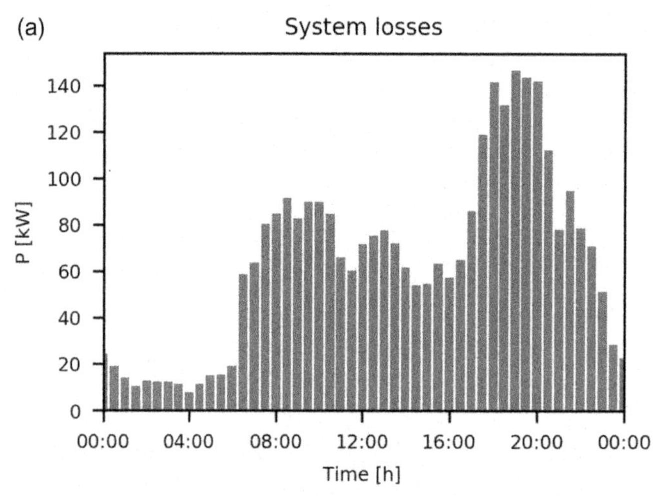

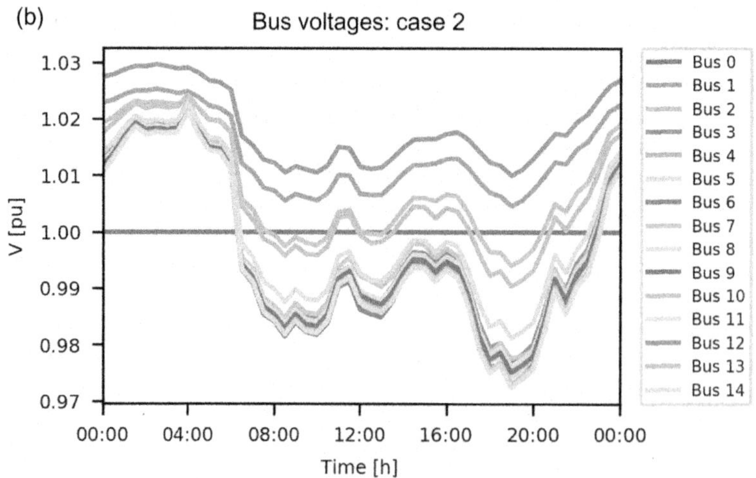

FIGURE 11.19 Case 2 (A) system losses, (B) system voltages.

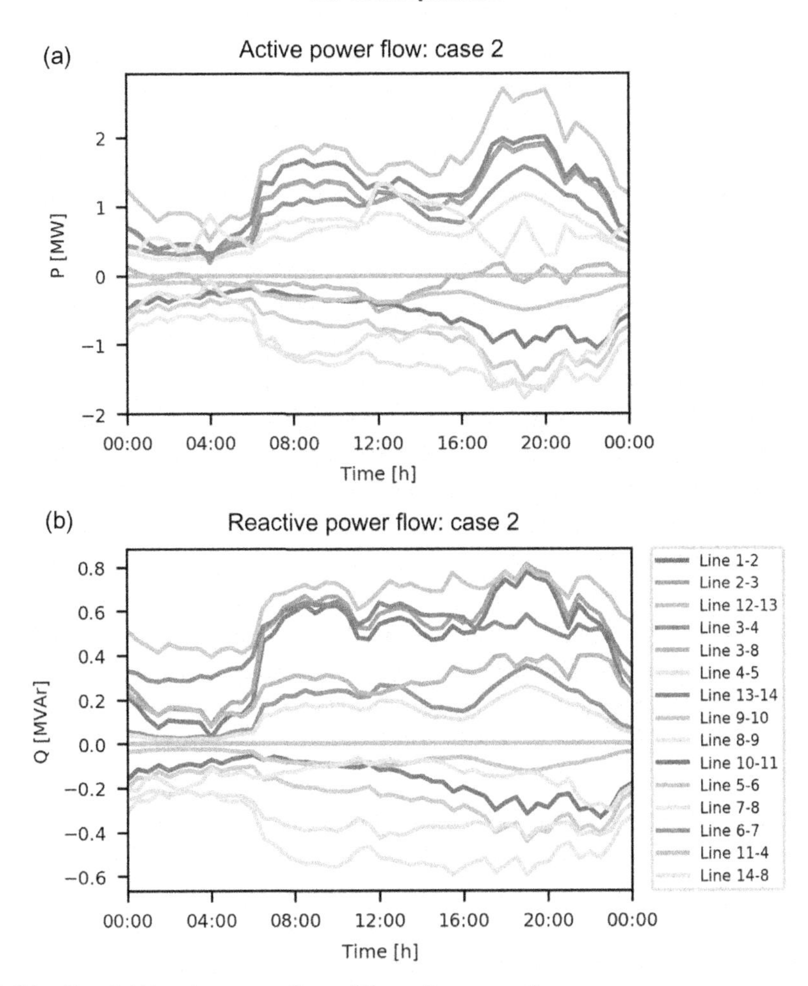

FIGURE 11.20 Case 2 (A) active power flows, (B) reactive power flows.

Fig. 11.21 shows the system voltage profile for each bus in Case 3a and the overall system losses for Cases 1, 2 and 3a.

In Case 1, with EV clusters and no DERs, the losses range from 32.39 kW at 04:00 to 218.26 kW at 19:00 (peak load). The inclusion of DER in Case 2 effectively reduces the power that needs to be transmitted from the main substation and therefore has the effect of significantly reducing system losses which vary from 7.76 kW at 04:00 to 146.5 kW at 19:00.

Finally, in Case 3a, the reactive power of the DER and the operation of the transformer tap changers are optimized to reduce system losses as well as to keep the voltage profile within 5% of the nominal values. In this case the system losses at 04:00 are 6.07 kW and the losses at 19:00 are 127.58 kW (Figs. 11.22, 11.23).

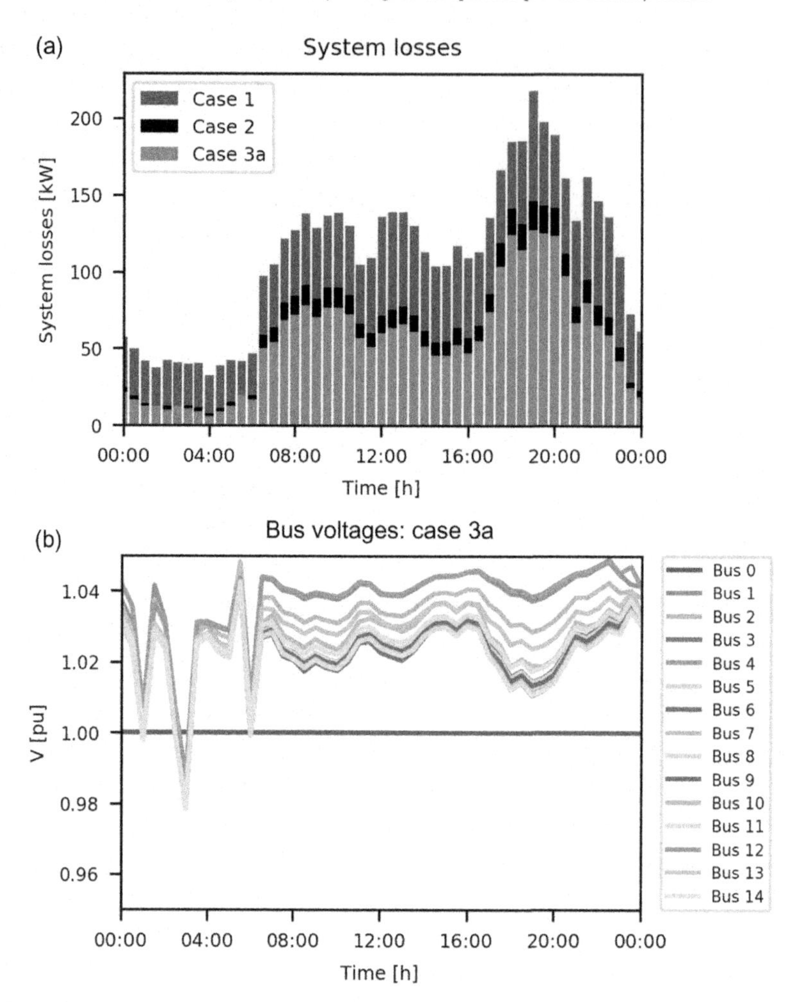

FIGURE 11.21 Case 3a (A) system losses, (B) system voltages.

In this scenario, the acceptable range of system voltages is narrowed to tolerate just a 2% deviation from the nominal value.

As before, solutions of the optimization routine that produce voltages outside this range are more heavily penalized.

As shown in Fig. 11.24, a more stringent permissible voltage range has the consequence of increasing system losses during the peak-demand periods as lower voltages lead to higher currents and consequently more losses.

Fig. 11.25 presents the reactive power output of the wind turbine and the total PV generation units for both scenarios of Case 3.

In Case 3a, the reactive power output from the wind turbine is always positive, except at 01:30 when it absorbs 0.32 MVAr.

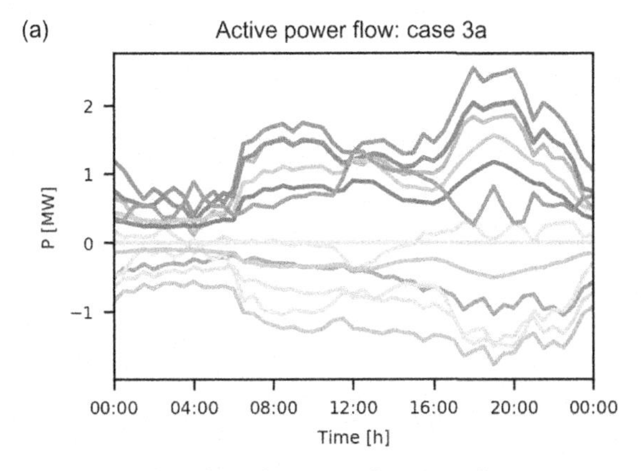

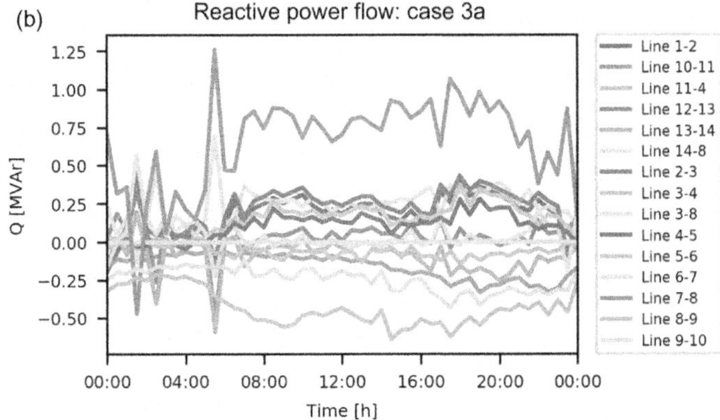

FIGURE 11.22 Case 3a (A) active power flows, (B) reactive power flows.

After this, the output stabilizes to a positive exported reactive power with an average of 0.84 MVAr.

In Case 3b, on the other hand, the reactive power output from the wind turbine experiences more substantial fluctuations throughout the day. For instance, at 19:30 it exports 1.33 MVAr and at 20:00 it imports 0.48 MVAr.

The individual reactive power output from the PV unit connected at bus 10 is shown in Fig. 11.26. It absorbs reactive power at nearly the maximum rate until 07:00 when it begins to inject it to the network. The maximum reactive power output that each individual PV unit can inject or absorb depends on the active power that it is injecting to the system at each moment as well as on the maximum apparent power of the unit.

Fig. 11.27 shows the tap position of both transformers. The tap positions are changed accordingly to keep the voltage within the acceptable range during the day.

Higher tap values are necessary from 07:00 until 16:00 to cover the increase in the residential load and from 16:00 to around 23:00 because of the added demand from EV charging.

In Scenario A, the taps of both transformers are set at their maximum value from 06:30 until 22:30, whereas in Scenario B, with more stringent voltage limits, the operations of the tap changers increase significantly thereby reducing the useful life of the units.

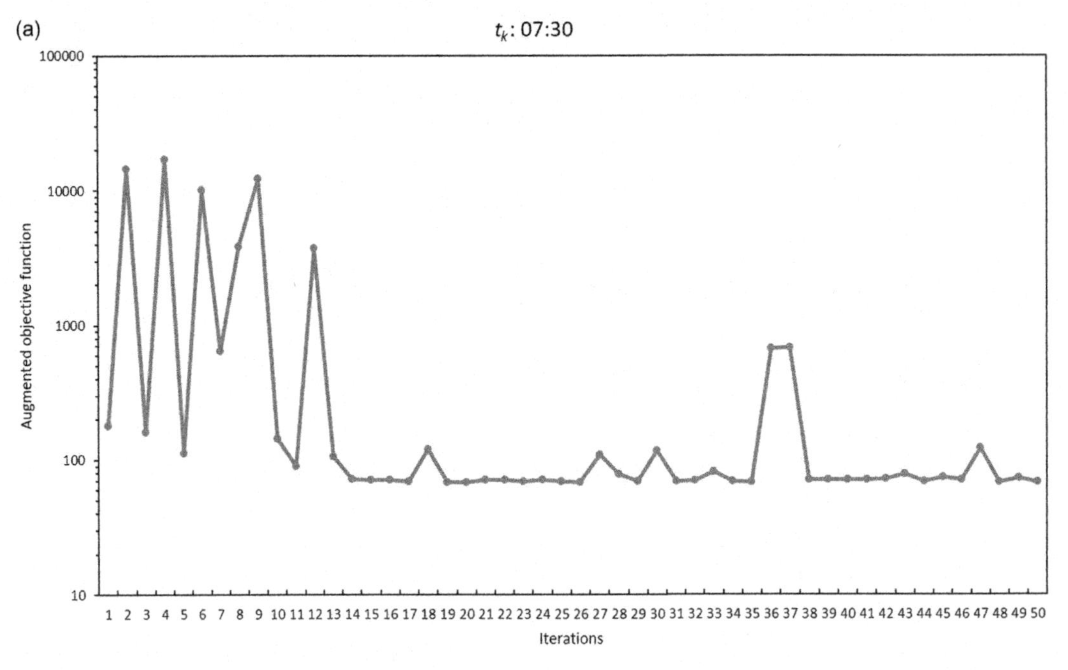

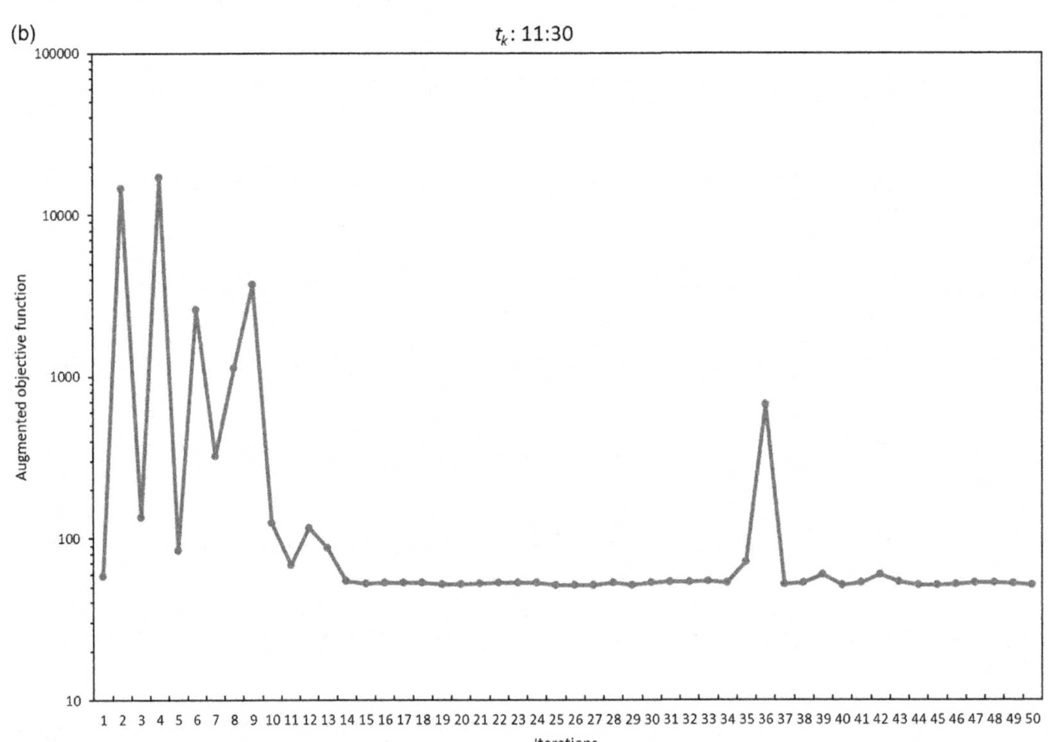

FIGURE 11.23 Performance of the optimization function (A) $t_k = 07{:}30$, (B) $t_k = 11{:}30$.

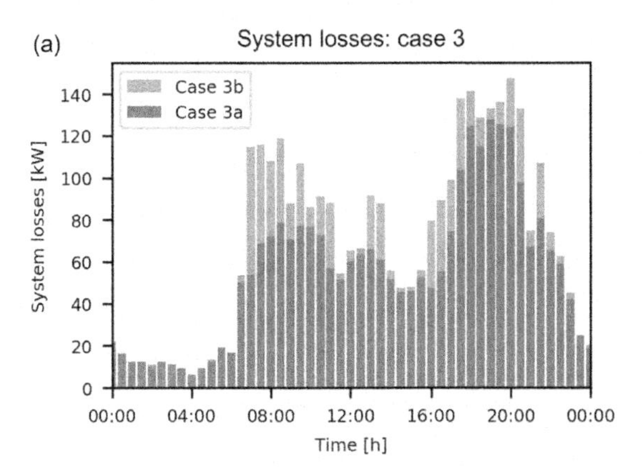

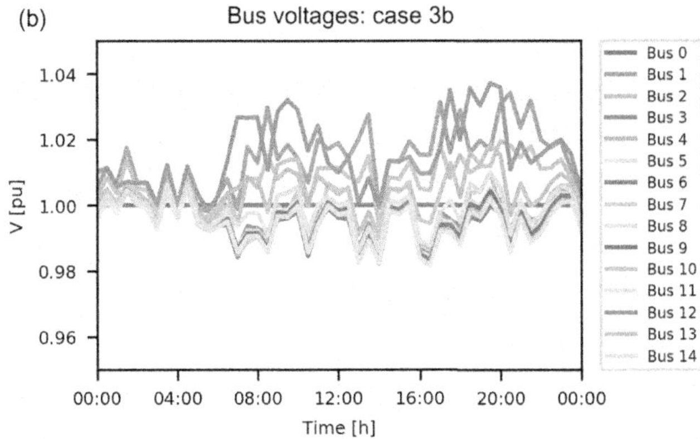

FIGURE 11.24 Case 3b (A) system losses, (B) system voltages.

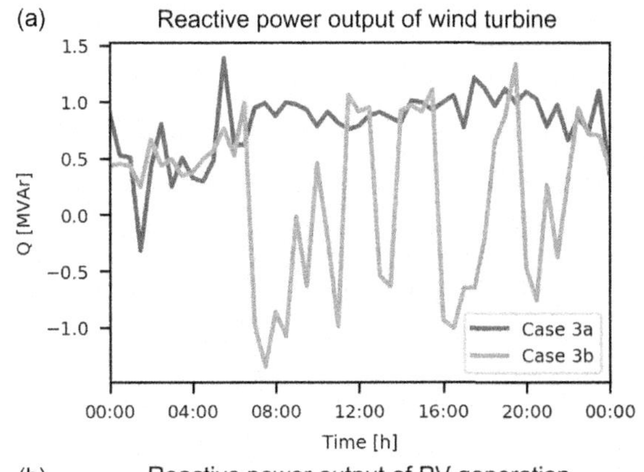

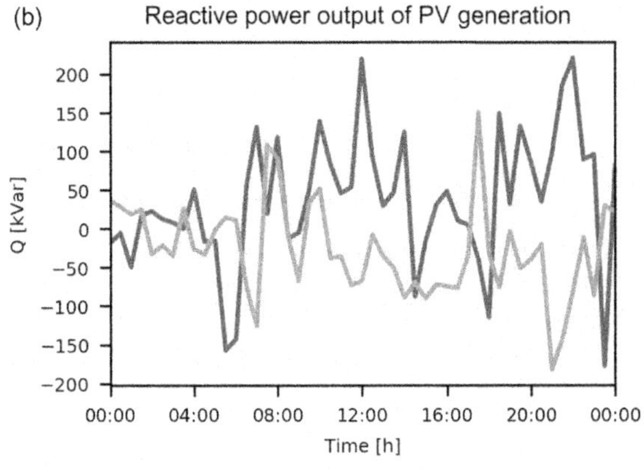

FIGURE 11.25 Case 3. (A) Reactive power output of wind turbine, (B) reactive power output of photovoltaics (PV) units.

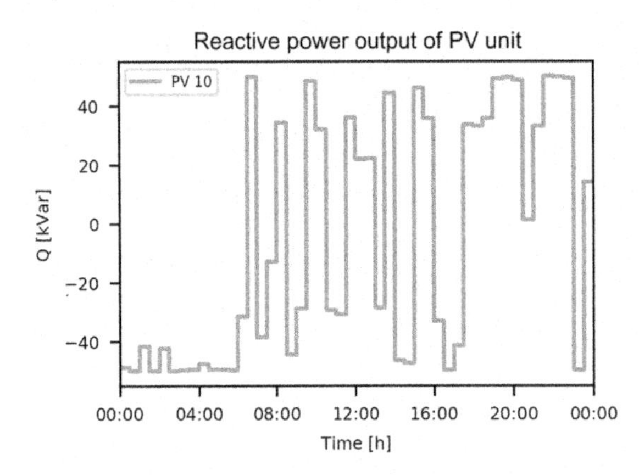

FIGURE 11.26 Reactive power output of photovoltaics (PV) 10.

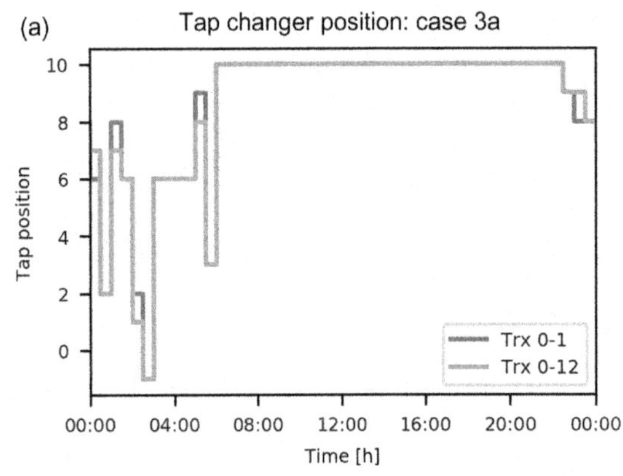

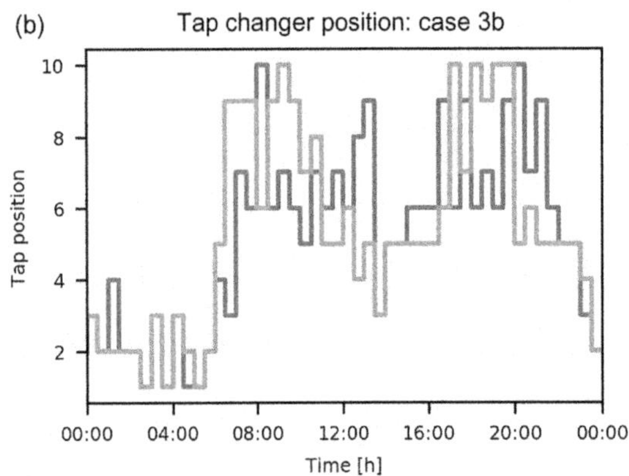

FIGURE 11.27 (A) Transformer tap positions for Case 3a, (B) transformer tap positions for Case 3b.

11.8 Conclusions

The most recent developments on grid-friendly or smart-converters, allow the PECs to perform additional tasks that could help the TSO with the operational problems: voltage control, low SCCs, etc. This chapter is dedicated to introducing the concept of grid-friendly or smart-converters and how they can be used to create a fully controllable MG able to provide auxiliary services to the transmission system: the so-called "transmission system-friendly MG." This chapter introduced the concept of grid-friendly or smart-converters and their main capabilities, the functionalities of which are enhanced by wide area control and the concept is a transmission system-friendly MG. Also, the chapter presented numerical results of the proposed control algorithm and its implementation using Python, while simulations using DIgSILENT PowerFactory demonstrated the suitability of the proposed transmission system-friendly MG.

References

[1] F., Gonzalez-Longatt, B.S., Rajpurohit, J.L.R., Torres, S.N., Singh, Simulation platform for autonomous smart multi-terminal DC micro-grid, in: 2016 IEEE Innovative Smart Grid Technologies - Asia (ISGT-Asia), 2016, pp. 630–635.

[2] Lasseter, R., et al., Consortium for Electric Reliability Technology Solutions White Paper on Integration of Distributed Energy Resources. 2001.

[3] Microgrids, 2018. (Online). Available from: <http://www.microgrids.eu/default.php> (accessed 13.10.18).

[4] Interflex – Home, 2018. (Online). Available from: <https://interflex-h2020.com/> (accessed 13.10.18).

[5] Home – Green Energy Storage, 2018. (Online). Available from: <http://www.greenenergystorage.eu/> (accessed 13.10. 18).

[6] A., Oudalov, T., Degner, F., van Overbeeke, J.M., Yarza, Microgrid: architectures and control, in: Microgrids Architectures and Control Wiley, 2003, pp. 1–24 (Chapter 2).

[7] R.H., Lasseter, MicroGrids. In: 2002 IEEE Power Engineering Society Winter Meeting Conference Proceedings (Cat. No.02CH37309), vol. 1, 2002, pp. 305–308.

[8] F.G.-L., Rajeev Kumar Chauhan, Bharat Singh Rajpurohit, S.N.S., DC microgrid in residential buildings, in: DC Distribution Systems and Microgrids, 2018, p. 437.

[9] IEEE PES Industry Technical Support Task Force, 2018. Impact of IEEE 1547 Standard on Smart Inverters.

[10] E., Bernabeu, A., Hoke, R., Walling, G., Zhou, 2018. Application of IEEE Std 1547 and Operating Experiences of Grid-Supportive Distributed Energy Resources PJM Engagement in the Application of IEEE 1547.

[11] A. Hirsch, Y. Parag, J. Guerrero, Microgrids: a review of technologies, key drivers, and outstanding issues, Renew. Sustain. Energy Rev. 90 (2018) 402–411.

[12] IEEE Power and Energy Society, 2017. IEEE Standard for the Specification of Microgrid Controllers.

[13] IEC 61727:2004 | IEC Webstore | invertor, smart city, LVDC, 2018. (Online). Available from < https://webstore.iec.ch/publication/5736> (accessed 13.10. 18).

[14] The International Smart Grid Action Network and Clean Energy Ministerial, 2014. TSO–DSO interaction: an Overview of Current Interaction Between Transmission and Distribution System Operators and an Assessment of Their Cooperation in Smart Grids.

[15] G. Migliavacca, et al., SmartNet: H2020 project analysing TSO–DSO interaction to enable ancillary services provision from distribution networks, CIRED Open Access Proc. J. (1) (2017) 1998–2002.

[16] Demand Connection Code,(Online). Available from: <https://electricity.network-codes.eu/network_codes/dcc/>, 2018 (accessed 13.10.18).

[17] J. Morin, F. Colas, X. Guillad, J.-Y. Dieulot, S. Grenard, Joint DSO–TSO reactive power management for an HV system considering MV systems support, CIRED Open Access Proc. J. (1) (2017) 1269–1273.

[18] T. Van Cutsem, C. Vournas, Voltage Stability of Electric Power Systems, Kluwer Academic Publishers, Boston, 1998.

[19] CIGRE, 2014. Benchmark Systems for Network Integration of Renewable and Distributed Energy Resources, April, 2014.

[20] F., Sanchez, F., Gonzalez-Longatt, D., Bogdanov, Impact assessment of frequency support by electric vehicles: Great Britain Scenario 2025, in: XXth International Symposium on Electrical Apparatus and Technologies (SIELA 2018), 2018, pp. 1–8.

[21] P. Bonami, et al., An algorithmic framework for convex mixed integer nonlinear programs, Discret. Optim. 5 (2) (2008) 186–204.

CHAPTER

12

Energy management of various microgrid test systems using swarm evolutionary algorithms

Bishwajit Dey, Kumar Shivam and Biplab Bhattacharyya

Department of Electrical Engineering, Indian Institute of Technology (Indian School of Mines), Dhanbad, India

12.1 Introduction

An MG comprises a low voltage (LV) system along with distributed energy resources (DERs), which include conventional generators, renewable energy sources (RES), storage devices, and flexible loads. DERs, such as micro-turbines (MTs), fuel cells (FC), wind turbines (WTs), and photovoltaic (PV) systems in addition to storage devices such as flywheels, batteries, energy capacitors etc., are all used in an MG. An MG has two modes of operation: namely, islanded mode and grid-connected mode and hence it is of benefit to both the grid and the customer. The primary MG control, also known as the coordinated control, is used to optimize the allocation of power among DERs, while reducing the cost of producing the energy as well as reducing the emissions.

Literatures reveals that authors used conventional ways in which to do economic scheduling and to determine the optimum allocation of generators, satisfying both demand and generation constraints [1,2]. A fuzzy interval optimization approach to solve the Environmental/Economic Dispatch (EED) problem with uncertain parameters within the constraints and the objective functions is studied by Ref. [3]. Authors in Refs. [4–6,7] used the hybrid and modified versions of various soft computing techniques such as particle swarm optimization (PSO), fireworks algorithm, and imperialist competitive algorithm etc. to minimize the MG cost and emission of an LV MG comprising a micro-turbine FC, PV, WT, and battery storage. There are several methods for the optimal sizing of MGs. Some authors use artificial intelligence (AI) tools such as genetic algorithm (GA) and PSO, while others use the rule-based method and optimal global methods available [8–13]. A fuzzy

Distributed Energy Resources in Microgrids
DOI: https://doi.org/10.1016/B978-0-12-817774-7.00012-0

interval optimization approach to solving the EED problem with uncertain parameters in the constraints and the objective functions is studied in Refs. [14,15] and uses iterative and AI-based methods for optimizing a hybrid PV system. Similarly in Refs. [16,17], the GA finds the optimal configuration of the hybrid system and optimizes the operation strategy using each of the optimal configurations. Reference [18] uses PSO to reduce the generation cost of an MG system comprising an MT, FC, PV, WT and an energy storage device backed up by the utility. Authors used the differential evolution technique to solve an economic and emission dispatch problem of an MG using combined heat and power in Ref. [19]. The day-ahead forecasted value of PV and WT along with the real-time market prices were considered while diminishing the MG generation cost in Ref. [20] and both emission and MG generation cost in Ref. [21]. The harmony search algorithm was used to minimize the MG system generation cost involving penetrable PV, MT, and FC in Refs. [22,23] investigates a hybrid dynamic economic emission dispatch problem involving thermal, WT, and PV generation systems on IEEE 30 bus, six-unit and IEEE 57 bus, seven-unit test systems. The author implemented the imperialist competitive algorithm (ICA) in Refs. [24,25] for optimal scheduling of the DERs of a virtual power plant, thus minimizing the generation cost of the system.

This chapter uses a dedicated and powerful whale optimization algorithm (WOA) technique to diminish the net operating cost of three grid-connected MG test systems mathematically formulated in the next section and comprising PV, WT, FC, MT, and energy storage devices. Section 12.2 of this chapter discusses the formation of objective functions of the MG including the constraints involved. Sections 12.3, 12.4, and 12.5 provides detailed knowledge of the grey wolf optimization (GWO) algorithm, symbiotic organism search (SOS), and WOA used in the chapter. Furthermore, Section 12.6 discusses the results obtained and analyzes a comparative study among the results of other literatures. The chapter concludes in Section 12.7.

12.2 Energy management problem formulation

The MG considered in this work is a LV grid-connected MG. It may consist of an MT, a proton-exchange membrane FC, an energy storage system, a PV system, and a wind farm all connected to the utility. The overall operating cost of an MG comprises the sum of distributed generation (DG) fuel costs and utility sharing prices. Hence, the cost objective function emphasizes on optimal sizing of power from DG sources. Such a fitness function can be defined as follows:

$$MinF(x) = \sum_{t=1}^{24} \sum_{k=1}^{n} a_k * P_k(t) + b * P_{ess}(t) + c * P_{Grid}(t) \tag{12.1}$$

where $k = 1,2,3,\ldots,n$ and $t = 1,2,3,\ldots$ 24 are the total number of DG sources and hours of operation respectively. $P(t)$ is the power output of the DG source at tth hour, $P_{ess}(t)$, and $P_{Grid}(t)$ are the power output of the energy storage device and the grid at tth hour. a_k, b, and c are the fuel cost coefficients of the kth DG source, energy storage system, and the real-time market price of the grid (utility) respectively.

The above-mentioned objective function in (12.1) is bound to some equality and inequality constraints as explained below:

12.2.1 Generation limit

The hourly power outputs of all the DG sources, including the storage device and utility, should fall between its maximum and minimum operating range.

$$P_{k/Grid,min}(t) \leq P_{k/Grid}(t) \leq P_{k/Grid,\max} \tag{12.2}$$

where $P_{k,min}(t)$ is the minimum output power and $P_{k,max}(t)$ is the maximum output power of the DG sources at the tth hour respectively.

12.2.2 Storage device limits

The rate of charging and discharging of storage device of test system 2 for each time interval are limited as follows

$$W_{ess,t} = W_{ess,t-1} + \eta_{charge} P_{charge} \Delta t - \frac{1}{\eta_{discharge}} P_{discharge} \Delta t \tag{12.3}$$

$$\begin{cases} W_{ess,\ \min} \leq W_{ess,t} \leq W_{ess,\max} \\ P_{charge,t} \leq P_{charge,\max} \\ P_{discharge,t} \leq P_{discharge,\max} \end{cases} \tag{12.4}$$

where $W_{ess,t}$ and $W_{ess,t-1}$ are the energy stored in the battery at tth and $(t-1)$th hour respectively. P_{charge} and $P_{discharge}$ are the allowable rate of charge and discharge during a definite period, say Δt. The suffixes 'min' and 'max' denote the minimum and maximum of the respective parameter.

12.2.3 Equality constraint between supply and demand of power

The total generation of power from the DGs in the MG during every hour including the RES must fulfill the total load demand of the grid. Transmission losses are neglected as the MG considered is a small LV one. Therefore

$$\sum_{k=1}^{n} P_{k/Grid}(t) + P_{ess}(t) + P_{PV}(t) + P_{WT}(t) = P_{load}(t) \tag{12.5}$$

Three different MG test systems are studied to exploit the availability and the dependency on the renewable energy sources and the grid that are acting as DG sources for the MG. For MG test system 1, there are two cases which are discussed below:

Case 1: This case takes into consideration that the grid (utility) maintains a buying/selling mechanism with the MG. In other words, in the case of excess power produced by the DG sources of the MG, excess power is sold to the grid, while power is also bought from the grid, when the DG sources cannot provide the load demand of the MG.

Case 2: In the second case, the grid is not allowed to import power from the MG. In simple words, the grid backs up the MG within hours of insufficient power but does not buy power, even if the MG has excess power.

Likewise, MG test system 2 and MG test system 3 are also RES-integrated MGs, which are highly dependent on the grid to satisfy their load demand when and if it is not available from DG sources. The objective is the energy management and optimal sizing of these DG sources and the grid, as well to minimize the MG cost.

The proposed WOA-based approach is implemented to find the optimal sizing of the DG sources in all the three different cases.

12.3 Grey wolf optimization algorithm

The GWO algorithm, first formulated by Mirjalili [26], is a new swarm intelligence technique based on the characteristic behavior of grey wolves (*Canis lupus*), which prefer to live in a pack. This algorithm focuses on their leadership and hunting abilities. An average pack of 5–12 grey wolves has a precise dominant hierarchy. The dominant wolf, Alpha, which becomes the leader of the pack, is responsible for making trivial decisions for the entire pack and has to be the most efficient one in managing the group. However, it is not necessarily the strongest member of the pack. The second best candidate to become Alpha if the dominant wolf dies or becomes too old is called Beta. The primary function of Beta is to maintain discipline in the pack, providing advice and feedback to Alpha. The least important wolf plays the role of scapegoat and is marked as Omega. Omega is the lowest-ranked wolf in the pack. The category of wolf other than Alpha, Beta, or Omega is designated as Delta. Scouts, hunters, caretakers, sentinels, and elders fall into this category.

In this particular algorithm, the most appropriate solution is known as Alpha (α), followed by Beta (β), and Delta (δ). Remaining solutions are assumed to be Omega (ω). The ω-wolves come after these three categories of wolves. While hunting, the wolf pack tends to encircle its prey. The encircling behavior of the pack of wolves is formulated as:

$$\vec{D} = \left|\vec{C}.\vec{X}_p(iter) - \vec{X}(iter)\right| \tag{12.6}$$

$$\vec{X}(iter+1) = \vec{X}_p(iter) - \vec{A}.\vec{D} \tag{12.7}$$

$\vec{X}$ and $\vec{X}_p$ are the vectors marking the position of grey wolf and prey respectively and *iter* is the current iteration count. Coefficient vectors $\vec{A}$ and $\vec{C}$ are numerically calculated as:

$$\begin{aligned} \vec{A} &= 2 \cdot \vec{a} \cdot \vec{r}_1 - \vec{a} \\ \vec{C} &= 2 \cdot \vec{r}_2 \end{aligned} \tag{12.8}$$

The vectors $\vec{r}_1$ and $\vec{r}_2$ are randomized between 0 and 1 and vector $\vec{a}$ is linearly decreased from 2 to 0 over the course of iterations. The members of the pack update their positions around the prey in any random location and these are expressed mathematically by Eqs. (12.6) and (12.7). In order to depict the hunting behavior of grey wolves mathematically, we suppose that the whereabouts of the prey is already known to Alpha (best candidate solution), Beta, and Delta. Three best solutions are marked and the remaining search agents are asked to update their positions accordingly.

$$\left.\begin{aligned}\vec{D}_\alpha &= \left|\vec{C}_1\vec{X}_\alpha - \vec{X}\right| \\ \vec{D}_\beta &= \left|\vec{C}_2\vec{X}_\beta - \vec{X}\right| \\ \vec{D}_\gamma &= \left|\vec{C}_3\vec{X}_\gamma - \vec{X}\right|\end{aligned}\right\} \tag{12.9}$$

The location of the grey wolf during hunting is denoted by (12.10).

$$\left.\begin{aligned}\vec{X}_1 &= \vec{X}_\alpha - \vec{A}_1.(\vec{D}_\alpha) \\ \vec{X}_2 &= \vec{X}_\beta - \vec{A}_2.(\vec{D}_\beta) \\ \vec{X}_3 &= \vec{X}_\gamma - \vec{A}_3.(\vec{D}_\gamma)\end{aligned}\right\} \tag{12.10}$$

Eq. (12.11) is used to update the positions of the grey wolf.

$$\vec{X}(iter + 1) = \frac{\vec{X}_1 + \vec{X}_2 + \vec{X}_3}{3} \tag{12.11}$$

If $\left|\vec{A}\right| > 1$, then wolves are diverged from the current prey to find a better prey and when $\left|\vec{A}\right| < 1$, the wolves are forcefully attracted towards the prey. At the end of every iteration, α, β, and δ wolves perform their position updating towards the probable location of the prey. The GWO algorithm concludes when the maximum number of iterations is attained.

The flowchart of the GWO algorithm is given below:

12.4 The symbiotic organisms search algorithm

SOS is a relatively new, powerful and meta-heuristic evolutionary algorithm applied to optimize many mathematical and engineering problems [27]. It works by simulating the symbiotic strategies acquired by the organisms among themselves to survive and sustain in the ecosystem. The fact that SOS doesn't require any algorithm specific parameters makes it superior to many other meta-heuristic algorithms. The symbiotic relationships that are found in nature are of three types, namely mutualism, commensalism, and parasitism. These relationships are further formulated below and the SOS algorithm is developed as under:

12.4.1 Mutualism phase

In the mutualism phase of SOS, both the species involved in the relationship benefit. One of the best illustrations of mutualism observed in nature is the relationship between honey bees and flowers. The honey bees accumulate nectar from flowers and turn it into honey and hence are benefitted from the flowers. In this process the bees also carry the pollen grains from one flower to another and thus assist in pollination. This phase can be mathematically developed by the following equations:

$$X_{inew} = X_i + rand(0,1) * (X_{best} - Mutual\_Vector * BF_1) \tag{12.12}$$

$$X_{jnew} = X_j + rand(0,1) * (X_{best} - Mutual\_Vector * BF_2) \tag{12.13}$$

$$Mutual\_Vector = \frac{X_i + X_j}{2} \tag{12.14}$$

where X_i is an organism of the ith member of the ecosystem and X_j is randomly selected from the ecosystem to interact with X_i. *rand(0,1)* denotes a vector of random numbers. BF_1 and BF_2 denote the benefit factors and is kept either 1 or 2. *Mutual_Vector* represents the mutual relationship between the organisms X_i and X_j.

12.4.2 Commensalism phase

Commensalism exists in nature between individuals of two species, where one species gathers its food or other benefits from the other without harming or benefitting the latter. The remora fish, for an instance, always remains adhered to a shark and devours the left-over food of a shark without harming or benefitting it. In this way there exists a commensalism relationship between the shark and remora fish. Similar to the mutualism phase, X_j is randomly chosen from the ecosystem to interact with X_i and a new organism X_{inew} can be formed using (10).

$$X_{inew} = X_i + rand(-1,1) * (X_{best} - X_j) \tag{12.15}$$

where $(X_{best} - X_j)$ portrays the beneficial advantage provided by X_j to help X_i increase its survival advantage in the ecosystem to the highest degree X_{best} in the current organism.

12.4.3 Parasitism phase

Parasitism is that relationship between two organisms in the ecosystem where one benefits at the cost of some harm caused to the other. The organism that benefits is called the 'parasite' and the one that faces the harm is called the 'host'. An example can be taken of the deer tick, which attaches itself to the host to suck its blood and thus gets benefitted. Not only this, the deer tick is also the carrier of Lyme disease, which causes joint damage and kidney problems and subsequently the animal suffers from a lack of blood.

In SOS, X_j is selected arbitrarily and designated as the 'host'. *Parasite_Vector* is an artificial organism conceived in the ecosystem. If the objective function evaluated of the *Parasite_Vector* is better than X_j, the latter will be replaced. And if the objective function evaluated of X_j is better, it will be assumed to possess immunity and the *Parasite_Vector* will be eliminated from the ecosystem.

The pseudo-code for the SOS algorithm is detailed below:

Initialize ecosystem E for n organisms for a specified ecosize and specify maximum number of fitness evaluations (maxFE)
Calculate fitness of the organisms

Identify the best organism and name it as X_{best}
while *FE < maxFE*
for *i = 1: ecosize*
Mutation
Select an organism X_j *randomly from the ecosystem such that* $X_j \neq X_i$
Determine Mutual_Vector and benefit factors as follows:

$$Mutual\_Vector = \frac{X_i + X_j}{2}$$

BF1 = round(1 + rand)
BF2 = round(1 + rand)
Modify organisms X_i *and* X_j *based on their mutual relationship as follows:*

$$X_{inew} = X_i + (X_{best} - Mutual\_Vector * BF_1) * rand(0,1)$$

$$X_{jnew} = X_j + (X_{best} - Mutual\_Vector * BF_2) * rand(0,1)$$

Calculate fitness of the modified organisms, $f_i(X_{i_{new}})$ *and* $f_j(X_{j_{new}})$
FE = FE + 2
if $f_i(X_{inew}) < f_i(X_i)$
Set $f_i(X_i) = f_i(X_{inew})$ *and* $X_i = X_{inew}$
end
if $f_j(X_{jnew}) < f_j(X_j)$
Set $f_j(X_j) = f_j(X_{jnew})$ *and* $X_j = X_{jnew}$
end
Commensalism
Select an organism X_j *randomly from the ecosystem such that* $X_j \neq X_i$
Modify organisms X_i *and* X_j *based on commensalism as follows:*

$$X_{inew} = X_i + (X_{best} - X_j) * rand(-1,1)$$

Calculate fitness value of modified organism, $f_i(X_{inew})$
FE = FE + 1
if $f_i(X_{inew}) < f_i(X_i)$
Set $f_i(X_i) = fi(Xinew)$ *and* $X_i = X_{inew}$
end
Parasitism
Select an organism X_j *randomly from the ecosystem such that* $X_j \neq X_i$
Create Parasite_Vector from organism X_i
FE = FE + 1
Calculate fitness of new organism, f_i*(Parasite_Vector)*
if f_i*(Parasite_Vector)* $< f_i(X_i)$
Set $f_i(X_i) = f_i$*(Parasite_Vector) and* X_i *= Parasite_Vector*
end
end
end

12.5 The whale optimization algorithm

The humpback whale is known for its most diverse hunting repertoire for capturing prey in what is known as a bubble-net feeding strategy. Motivated by this technique, this algorithm was initially put forward by Mirjalili and Lewis [28] in 2016. The main features and procedure of WOA are described in the following subsection.

12.5.1 Features

Whales are highly intelligent marine mammals of the open ocean; known to cooperate, scheme, engage in complex behavior, and use sonar for various vocalizations. Being descendants of land-living mammals, they have to breathe from the surface of the oceans. Their brains must be constantly alert to breathe, hence they can only shut down half of their senses while sleeping. Whales have double the number of spindle cells, compared to human beings and this is the major reason for their increased smartness. Humpback whales (*Megaptera movaeangliae*) are one of the biggest species of baleen whales and are energetic predators, employing a diverse hunting technique called the bubble-net feeding method.

12.5.2 Methodology

The humpback whale applies the bubble-net feeding strategy to search and attack its prey.

This strategy employs a definite pattern of encircling the prey. The mathematical design of the conduct of bubble-net feeding is discussed below:

12.5.2.1 *Search for the prey (exploration phase)*

In this phase, the location of a certain search agent is updated according to the search agent chosen randomly, instead of capturing the best search agent. This characteristic can be mathematically expressed as follows:

$$\vec{D} = \left|\vec{C} \cdot \vec{X}_{rand} - \vec{X}\right| \tag{12.16}$$

$$\vec{X}(iter+1) = \vec{X}_{rand} - \vec{A} \cdot \vec{D} \tag{12.17}$$

where $\vec{X}_{rand}$ is the Position vector of whale taken randomly from current population.

12.5.2.2 *Encircling prey*

Humpback whales have the natural aptitude of recognizing the position of the herd of prey and encircling and confining it to hunt. This behaviour of encircling is expressed by the following equations:

$$\vec{D} = \left|\vec{C} \times \vec{X_P}(iter) - \vec{X}(iter)\right| \tag{12.18}$$

$$\vec{X}(iter+1) = \vec{X_P}(iter) - \vec{A} \times \vec{D} \tag{12.19}$$

where X and X_P indicates the position of the whale and prey respectively. *'iter'* indicates the current iteration and A and C are coefficient vectors which are calculated using (12.20) and (12.21).

$$\vec{A} = 2\vec{a} \times \vec{r_1} - \vec{a} \tag{12.20}$$

$$\vec{C} = 2 \times \vec{r_2}, \qquad r_1, r_2 \in [0,1] \tag{12.21}$$

Vector $\vec{a}$ decreases linearly from 2 to 0 through the iteration course for both the exploration and exploitation phases.

12.5.2.3 Bubble-net attacking method (exploitation phase)

Two approaches are considered for the bubble-net feeding technique of the humpback whales which are elaborated below:

12.5.2.4 Shrinking-encircling mechanism

This state is accomplished by bringing down the value of *'a'* in the (12.20). Hence the variation range of $\vec{A}$ is also diminished by $\vec{a}$ and we can say that $\vec{A} \in [-a, a]$.

12.5.2.5 Spiral-updating positions

Humpback whales enclose their prey by winding them into a tight circle. This method is obtained by determining the distance between the humpback whale and the position of the encircled prey. To imitate the helix-shaped movement of the whales, a spiral equation has been created which is as follows:

$$\vec{X}(iter+1) = \vec{D} \cdot e^{bl} \cdot Cos(2\prod l) + \vec{X}(iter) \tag{12.22}$$

where $\vec{D} = \left|\vec{X_P}(iter) - \vec{X}(iter)\right|$ signifies the distance between ith whale to its prey (best solution), b is the constant for defining the shape of the logarithmic spiral and $l \in [-1, 1]$.

The humpback whale, while swimming around its prey, employs the technique of constantly shrinking in a circular motion as well as a spiral-shaped path at the same time. Due to this characteristic, it can be assumed that there is a 50% probability of choosing between the shrinking encircling mechanism and the spiral model to update the position of the whale during optimization. This behaviour of the whale can be mathematically formulated as:

$$\vec{X}(iter+1) = \begin{cases} \vec{X_P}(iter) - \vec{A} \cdot \vec{D} & if \quad p < 0.5 \\ \vec{D} \cdot e^{bl} \cdot Cos(2\prod l) + \vec{X_P}(iter) & if \quad p \geq 0.5 \end{cases} \tag{12.23}$$

p varies randomly between [0,1].

WOA is able to choose between either a spiral or a circular movement of the whale depending on the value of *'p'*. Finally, the WOA comes to an end by satisfying all the termination conditions which were given initially. The pseudo-code of the WOA [28] is described stepwise below:

Initialize the whales population X_i $(i = 1, 2\ldots n)$
Calculate the MG generation cost associated with each whale
X^* = *the best whale (whale with minimum generation cost)*
while *(t < iter_max)*
for *each whale*
Update a, A, C, l, and p
if1 *(P < .5)*
if2 *(|A| < 1)*
Update the position of the current whale by Eq. (12.18)
else if2 *(|A| ≥ 1)*
Select a random whale (X_{rand})
Update the position of the current whale by Eq. (12.19)
end if2
else if1 *(P ≥ .5)*
Update the position of the current search by the Eq. (12.22)
end if1
end for
Check if any whale goes beyond the search space and amend it
Calculate the fitness of each whale
Update X^* *if there is a further minimized generation cost*
iter = iter +1
end while
return X^*

12.6 Results and discussion

12.6.1 Description of the system

A strong and powerful WOA is used to evaluate its performance in the optimal sizing of a MG and minimizing its operating cost. The MGs considered in this work are all LV, grid-connected types. MG test system 1 and MG test system 3 are both grid-connected MGs comprising DG sources like MT, FC, PV system, and windfarm along with a battery storage device. MG test system 2 also includes the same DG sources as 1 and 3 connected to the grid, but does not include a battery storage system. The complete system data which includes the maximum and minimum capacity of the DGs, their fuel cost coefficients, load demand profile for 24 hours and the RES output for those seasons is gathered from Ref. [29] for microgrid test system 1, Ref. [30] for microgrid test system 2, and Ref. [31] for microgrid test system 3. The MATLAB R2013a platform is used to code and execute the algorithm on a desktop computer with 2.53GHz core I3 processor and 2GB RAM. The program is run for 20 trials considering various population sizes and 1500 iterations. The weightage factor and crossover probability constant is tuned to 0.7 and 0.2 for the DE algorithm respectively. The benefit factor for SOS is considered as 2 (Figs. 12.1–12.5).

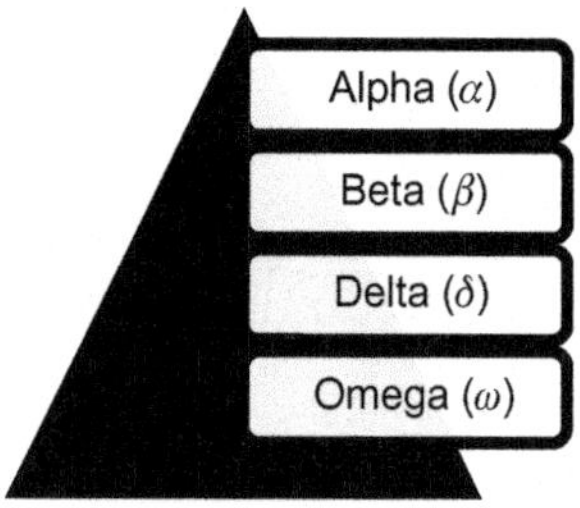

Alpha (α)
Beta (β)
Delta (δ)
Omega (ω)

FIGURE 12.1 Hierarchy of dominance of grey wolves in their packs.

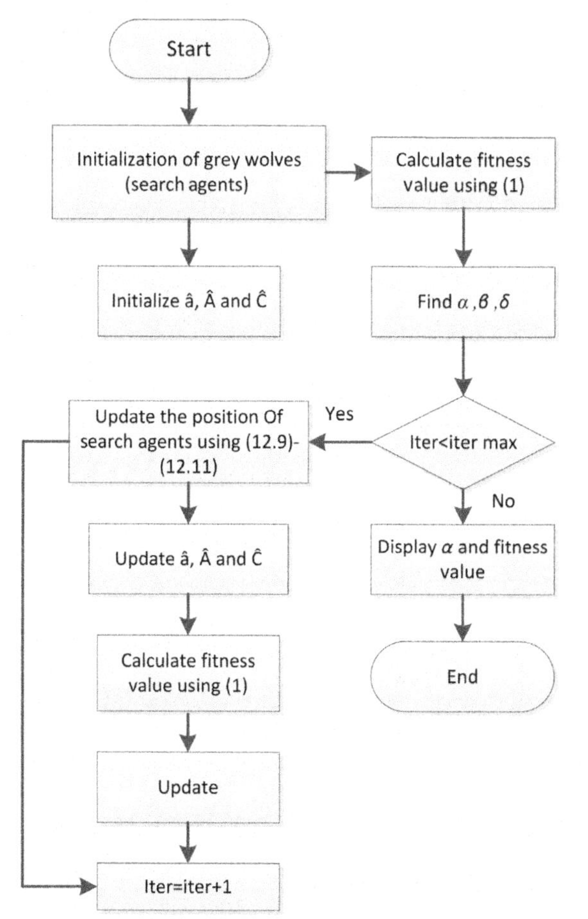

FIGURE 12.2 Flowchart of the grey wolf optimization (GWO) algorithm.

FIGURE 12.3 Honey bee and flower.

FIGURE 12.4 Remora fish and shark.

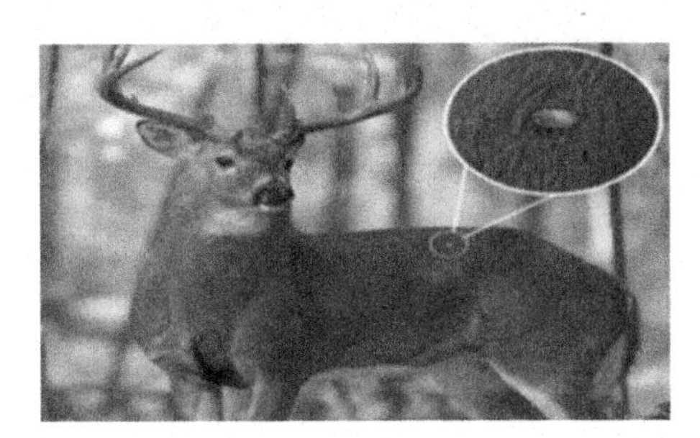

FIGURE 12.5 Deer tick feeding on the blood of the host.

TABLE 12.1 Cost comparative analysis of microgrid test system 1.

Method	Best value ($)	Average value ($)	Worst value ($)	SD	Time (s)	No of hits	*P*-value
Case 1							
MRC-GA [29]	219.05	–	–	–	–	–	–
DE	213.0476	214.7023	216.3571	2.34	86	18	NA
SOS	209.0525	419.3847	210.3322	0.90	92	19	NA
GWO	208.3861	208.6376	208.8892	0.35	80	19	NA
WOA	**207.7140**	207.7652	208.7388	0.23	78	19	1.1933e − 05
Case 2							
MRC-GA [29]	251.03	–	–	–	–	-	NA
DE	228.8410	229.7694	230.6978	1.31	83	19	NA
SOS	225.2363	225.7682	226.3001	0.75	88	18	NA
GWO	224.7334	224.8683	225.0033	0.19	78	19	NA
WOA	**222.9614**	222.9956	223.6471	0.15	75	19	1.1933e − 05

Bold indicates best results

12.6.2 Comparative analysis

Table 12.1 shows the cost-comparative analysis of the MG for both the cases involved in microgrid test system 1. It can be seen that for the cases, the WOA yielded a much better and minimized result of 207.7140€ for case 1 and 222.9614€ for case 2 than DE, SOS, GWO, and the algorithm available in the literature. For MG test system 2, it is observed

TABLE 12.2 Cost comparative analysis of microgrid test system 2.

Method	Best value (€)	Average value (€)	Worst value (€)	S.D	Time (s)	No. of hits	*P*-value
AIMD [30]	170.7947	–	–	–	–	–	–
MAIMD [30]	165.2139	–	–	–	–	–	–
DE	156.3803	157.007	157.6337	0.88	77	17	**NA**
SOS	154.7692	155.3847	156.0002	0.87	80	19	**NA**
GWO	153.5054	153.7361	153.9669	0.33	70	19	**NA**
WOA	**152.9667**	153.0130	153.8941	0.21	67	19	1.1933e – 05

TABLE 12.3 Cost comparative analysis of microgrid test system 3.

Method	Best value (€ĉ)	Average value (€ĉ)	Worst value (€ĉ)	S.D	Time (s)	No. of hits	*P*-value
MSFLA [31]	405.14	–	405.16	–	–	–	–
DE	309.5538	310.3772	311.2007	1.16	65	18	**NA**
SOS	282.8308	283.195	283.5603	0.51	76	16	**NA**
GWO	270.3249	270.581	270.8372	0.36	63	18	**NA**
WOA	**267.3079**	267.3476	268.1024	0.18	60	19	1.1933e – 05

from Table 12.2 that the WOA algorithm minimized the MG cost to as low as 152.9667€ compared to 153.5054€ by GWO 156.3803€ by DE, 154.7692€ by SOS, and other values available in the literature. Similarly for MG test system 3, WOA also proved to be a far more beneficial and efficient algorithm as it yielded a much reduced price of 267.3079€ĉ compared to 405€ĉ available in literature and other prices by DE, SOS, and GWO as listed in Table 12.3. Figs. 12.6–12.9 shows the convergence curves when the optimization of all the three cases was performed using DE, SOS, GWO, and WOA. The early convergence of WOA can be realized clearly from these figures. Figs. 12.10–12.13 shows the hourly output of the DG sources when the MG cost was minimized by the WOA. The participation of RES and utility at peak load and high market prices can be analyzed from these figures.

12.6.3 Solution quality

Finally, Tables 12.1–12.3 lists the statistical analysis of the proposed algorithm, which was done performing a Wilcoxon Signed Rank Test. This test was used for one sample data set, which was received as the outcome of the mentioned algorithm. It is a pairwise test done to find substantial variances in the behavior of two diverse algorithms. Any given algorithm may be considered robust if it is able to prove its statistical worth.

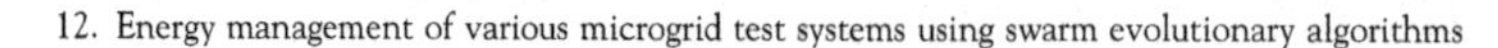

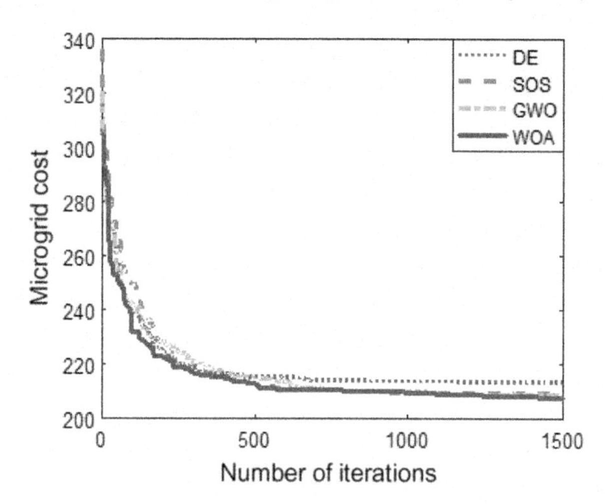

FIGURE 12.6 Convergence curve of minimal micro grid cost for test sys 1_case 1.

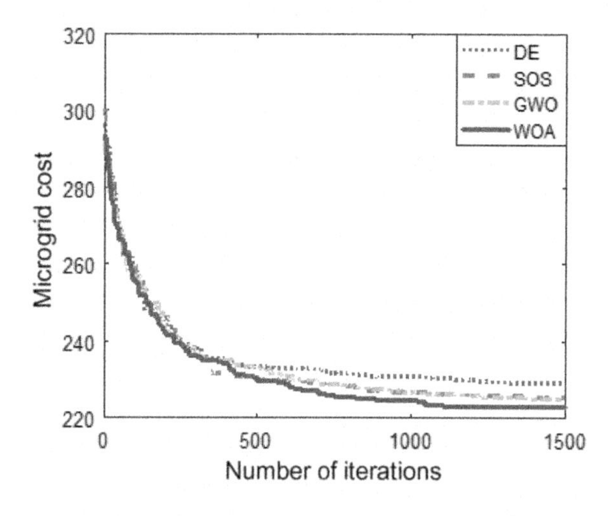

FIGURE 12.7 Convergence curve of minimal micro grid cost for test sys 1_case 2.

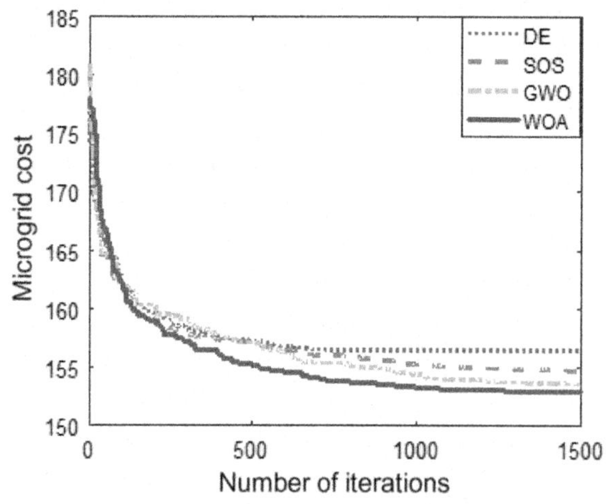

FIGURE 12.8 Convergence curve of minimal micro grid cost for test sys 2.

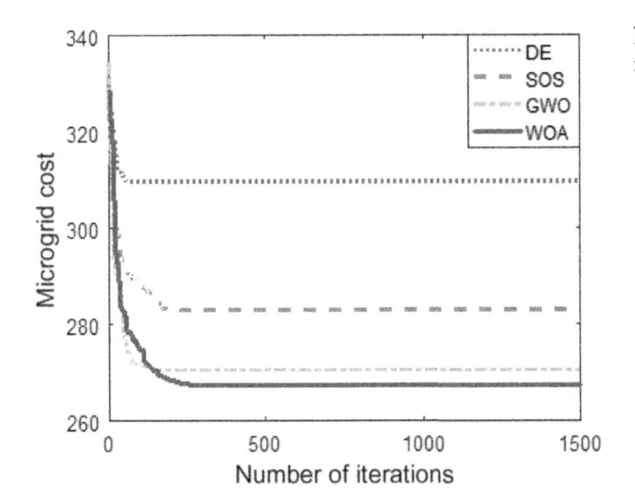

FIGURE 12.9 Convergence curve of minimal micro grid cost for test sys 3.

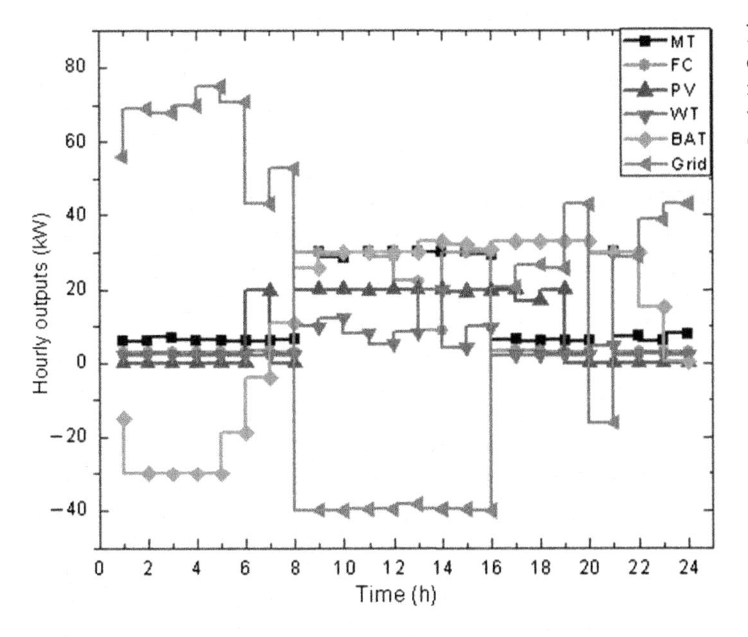

FIGURE 12.10 Hourly outputs of distributed generations (DGs) for microgrid (MG) test sys 1_case 1 using whale optimization algorithm (WOA).

For this purpose, it has to provide sufficient evidence against a null hypothesis. The *P*-value (probability value) which comes out to be less than .05 achieved by employing this test gives clear proof against the proposed null hypothesis. The *P*-values received from this test for all the cases with their minimum, maximum, average values, and standard deviation are also listed in the tables. From these tables, it was observed that the *P*-value in every case was much lower than the desired value of .05 thereby establishing the statistical significance of the results.

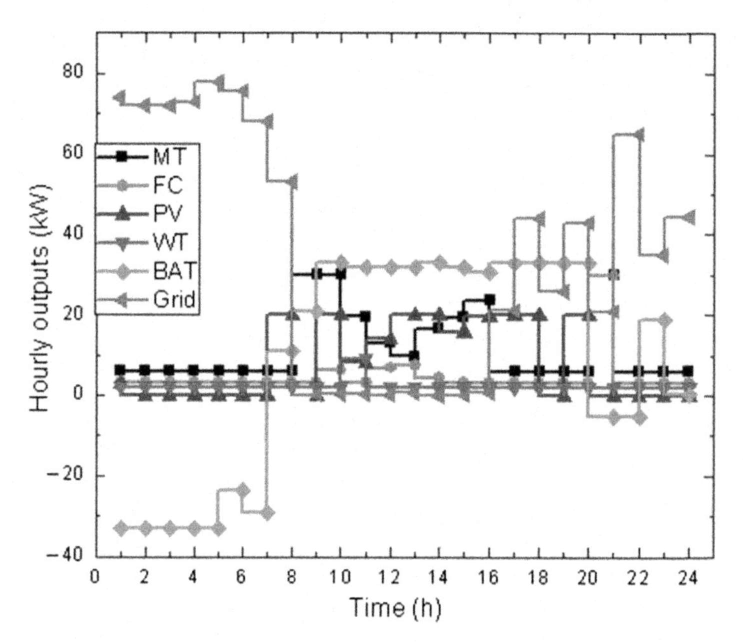

FIGURE 12.11 Hourly outputs of distributed generation (DGs) for microgrid (MG) test sys 1_case 2 using whale optimization algorithm (WOA).

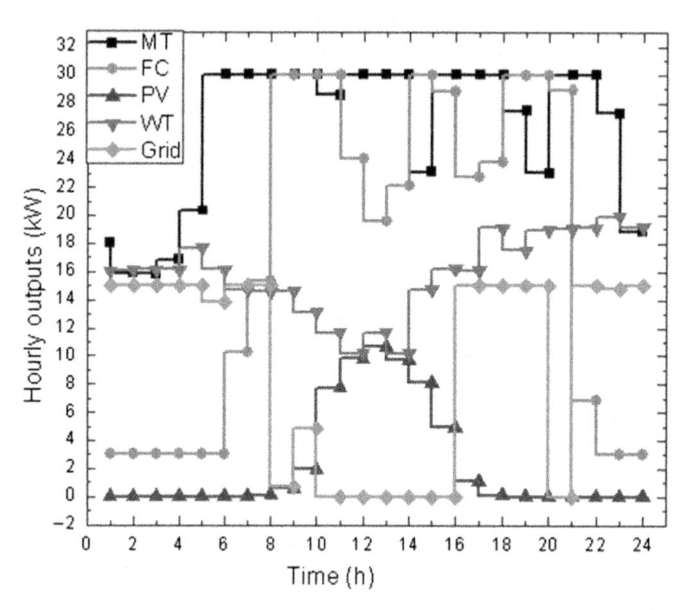

FIGURE 12.12 Hourly outputs of distributed generation (DGs) for microgrid (MG) test sys 2 using whale optimization algorithm (WOA).

12.6.4 Robustness

Initialization of evolutionary algorithms is always done randomly which is why multiple trial runs are needed to arrive at a decision regarding robustness of the same. WOA was evaluated for 20 trial runs for all cases. The number of times it hit the minimum

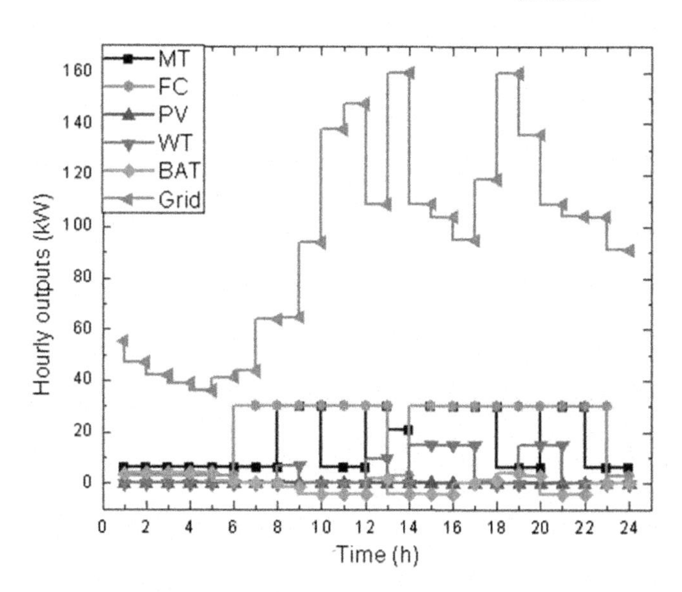

FIGURE 12.13 Hourly outputs of distributed generation (DGs) for microgrid (MG) test sys 3 using whale optimization algorithm (WOA).

solution is shown in Tables 12.1 −12.3. It can be seen that the lowest number of times it hit the minimum solution was 16 whereas the highest number was 19. The average success rate came out to be 95% which is highly appreciable.

12.7 Conclusion

A WOA method was used in this chapter for optimal sizing of three typical LV grid-connected MGs, which included DG sources such as storage devices, PVs, WTs, MTs, and FCs. Three cases were studied for the optimal sizing of DG sources so that the MG functions in the most efficient as well as economical way by satisfying the load demands and constraints with varying loads and RES outputs. The results obtained are then compared to a few other optimization algorithms found in the literature – DE, SOS, and GWO – and it was seen that WOA outperformed them all. Due to its better computational speed and exploitability, WOA can therefore be considered one of the strongest optimization tools to solve various power system and MG problems.

References

[1] M. Dali, J. Belhadj, X. Roboam, Hybrid solar–wind system with battery storage operating in grid-connected and standalone mode: control and energy management–experimental investigation, Energy 35 (6) (2010) 2587–2595.

[2] E. Pouresmaeil, D. Montesinos-Miracle, O. Gomis-Bellmunt, J. Bergas-Jané, A multi-objective control strategy for grid connection of DG (distributed generation) resources, Energy 35 (12) (2010) 5022–5030.

[3] A. Derghal, N. Golea, N. Essounbouli, A fuzzy interval optimization-based approach to optimal generation scheduling in uncertainty environment, J. Renew. Sustain. Energy 8 (6) (2016) 065501.

[4] A.A. Moghaddam, A. Seifi, T. Niknam, M.R.A. Pahlavani, Multi-objective operation management of a renewable MG (micro-grid) with back-up micro-turbine/fuel cell/battery hybrid power source, Energy 36 (11) (2011) 6490–6507.
[5] J. Radosavljević, M. Jevtić, D. Klimenta, Energy and operation management of a microgrid using particle swarm optimization, Eng. Optim. 48 (5) (2016) 811–830.
[6] Z. Wang, Q. Zhu, M. Huang, B. Yang, Optimization of economic/environmental operation management for microgrids by using hybrid fireworks algorithm, Int. Trans. Electr. Energy Syst. 27 (12) (2017).
[7] M.J. Kasaei, Optimal integrated scheduling of distributed energy resources in power systems by virtual power plant, ISA Trans. (2017). Available from: https://doi.org/10.1016/j.isatra.2017.12.003.
[8] C.A.C. Coello, A comprehensive survey of evolutionary-based multiobjective optimization techniques, Knowl. Inf. Syst. 1 (3) (1999) 269–308.
[9] G.C. Liao, A novel evolutionary algorithm for dynamic economic dispatch with energy saving and emission reduction in power system integrated wind power, Energy 36 (2) (2011) 1018–1029.
[10] P.H. Chen, H.C. Chang, Large-scale economic dispatch by genetic algorithm, IEEE Trans. Power Syst. 10 (4) (1995) 1919–1926.
[11] J.B. Park, K.S. Lee, J.R. Shin, K.Y. Lee, A particle swarm optimization for economic dispatch with nonsmooth cost function, IEEE Trans. Power Syst. 20 (1) (2005) 34–42.
[12] J. Sun, W. Fang, D. Wang, W. Xua, Solving the economic dispatch problem with a modified quantum-behaved particle swarm optimization method, Energy Convers. Manag 50 (12) (2009) 2967–2975.
[13] W.M. Lin, F.S. Cheng, M.T. Tsay, Nonconvex economic dispatch by integrated artificial intelligence, IEEE Trans. Power Syst. 16 (2) (2001) 307–311.
[14] A. Derghal, N. Golea, N. Essounbouli, A fuzzy interval optimization-based approach to optimal generation scheduling in uncertainty environment, J. Renew. Sustain. Energy 8 (2016) 65501.
[15] T. Khatib, A. Mohamed, K. Sopian, A review of photovoltaic systems size optimization techniques, Renew. Sustain. Energy Rev. 22 (2013) 454–465.
[16] A.K. Daud, M.S. Ismail, Design of isolated hybrid systems minimizing costs and pollutant emissions, Renew. Energy 44 (2012) 215–224.
[17] J.C. Hernández, A. Medina, F. Jurado, Optimal allocation and sizing for profitability and voltage enhancement of PV systems on feeders, Renew. Energy 32 (10) (2007) 1768–1789.
[18] J. Radosavljević, M. Jevtić, D. Klimenta, Energy and operation management of a microgrid using particle swarm optimization, Eng. Optim. 273 (2015) 1–20.
[19] A.K. Basu, A. Bhattacharya, S. Chowdhury, Planned scheduling for economic power sharing in a CHP-based micro-grid, IEEE Trans. Power Syst. 27 (1) (2012) 30–38.
[20] T. Niknam, F. Golestaneh, A. Malekpour, Probabilistic energy and operation management of a microgrid containing wind/photovoltaic/fuel cell generation and energy storage devices based on point estimate method and self-adaptive gravitational search algorithm, Energy 43 (1) (2012) 427–437.
[21] A.A. Moghaddam, A. Seifi, T. Niknam, M.R. AlizadehPahlavani, Multi-objective operation management of a renewable MG (micro-grid) with back-up micro-turbine/fuel cell/battery hybrid power source, Energy 36 (11) (2011) 6490–6507.
[22] A.G. Anastasiadis, S.A. Konstantinopoulos, G.P. Kondylis, G.A. Vokas, P. Papageorgas, Effect of fuel cell units in economic and environmental dispatch of a microgrid with penetration of photovoltaic and micro turbine units, Int. J. Hydrog. Energy (2016) 6–13.
[23] G. Mohy-ud-din, Hybrid dynamic economic emission dispatch of thermal, wind, and photovoltaic power using the hybrid backtracking search algorithm with sequential quadratic programming, J. Renew. Sustain. Energy 9 (2017) 15502.
[24] M.J. Kasaei, Energy and operational management of virtual power plant using imperialist competitive algorithm, Int. Trans. Electr. Energy Syst. 28 (11) (2018) e2617.
[25] M.J. Kasaei, M. Gandomkar, J. Nikoukar, Optimal management of renewable energy sources by virtual power plant, Renew. Energy 114 (2017) 1180–1188.
[26] S. Mirjalili, S.M. Mirjalili, A. Lewis, Grey wolf optimizer, Adv. Eng. Softw. 69 (2014) 46–61.
[27] M.-Y. Cheng, D. Prayogo, Symbiotic organisms search: a new metaheuristic optimization algorithm, Comput. Struct. 139 (2014) 98–112.
[28] S. Mirjalili, A. Lewis, The whale optimization algorithm, Adv. Eng. Softw. 95 (2016) 51–67.

[29] C. Chen, et al., Smart energy management system for optimal microgrid economic operation, IET Renew. Power Gen. 5 (3) (2011) 258–267.

[30] K.P. Kumar, B. Saravanan, K.S. Swarup, A two stage increase–decrease algorithm to optimize distributed generation in a virtual power plant, Energy Procedia 90 (2016) 276–282.

[31] M. Haghshenas, H. Falaghi, Environmental/economic operation management of a renewable microgrid with wind/PV/FC/MT and battery energy storage based on MSFLA, J. Electr. Syst. 12 (1) (2016) 85–101.

CHAPTER

13

Development of the synchronverter for green energy integration

S. Kumaravel, Vinu Thomas, Tumati Vijay Kumar and S. Ashok

Department of Electrical Engineering, National Institute of Calicut, Kattangal, Kerala, India

13.1 Introduction

There has been an increasingly focus on renewable energy sources throughout the world due to the many socio-economic issues related to conventional energy sources. Furthermore, the growing demand for electrical power has led to the integration of a large number of distributed energy sources into the power system. Among the various distributed energy sources, solar, and wind energy show the greatest promise of all sustainable sources of renewable energy. A power electronic converter is always mandatory to integrate such sources to enable them to provide their output power to the grid or local loads. But, these renewable sources are unpredictable, highly intermittent, site-specific, responsive to climatic circumstances, and in diluted form. A storage device such as a battery is always added to these energy sources to provide a reliable source of power to the load. These renewable sources are relatively small compared to conventional generators. A renewable energy source, a storage device, and local loads comprise a small grid called a microgrid (MG). The structure of a typical MG is shown in Fig. 13.1. There are three modes of operation of the MG: (i) grid-connected mode, (ii) isolated mode, and (iii) islanded mode. The MG is tied to the utility grid in grid-connected mode. In isolated mode, the MG is never connected to the utility grid. In islanded mode, the MG link has been disconnected from the main grid because of some disturbances or some special requirements.

The stability of a power system is secured primarily by the large rotational inertia of synchronous machines. Ensuring a reliable, efficient, and secure power system is a real

Distributed Energy Resources in Microgrids
DOI: https://doi.org/10.1016/B978-0-12-817774-7.00013-2

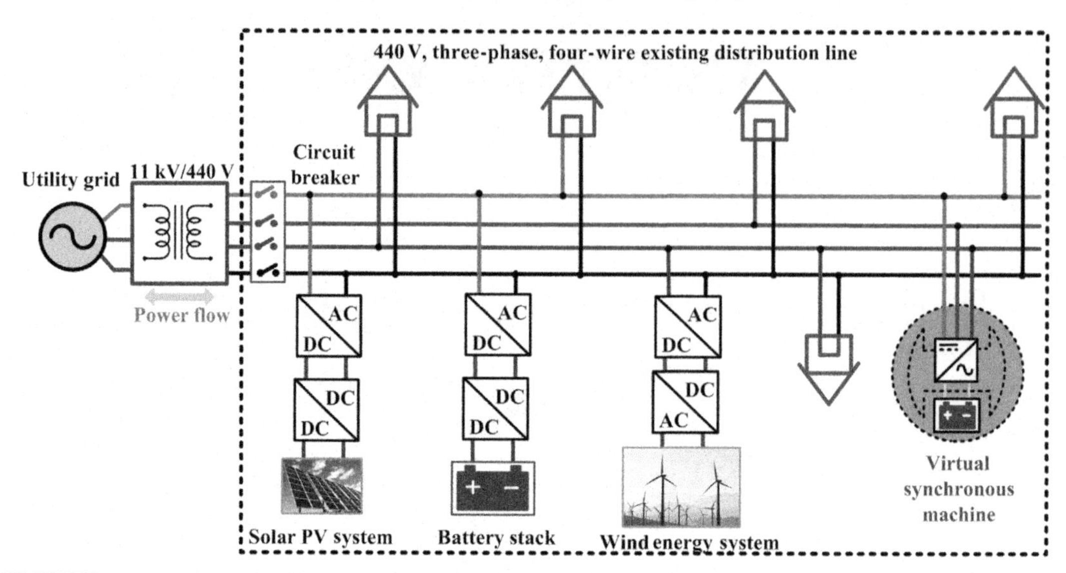

FIGURE 13.1 Structure of a typical microgrid (MG) with distributed generators and a synchronverter.

challenge at present because of the penetration of a number of nonconventional energy sources. Power electronic converters are used for the interconnection of nonconventional energy sources such as photovoltaic (PV) panels, batteries, etc., with the utility grid. These power converters isolate the inertia present in some of the nonconventional energy sources. However, sources like PVs, fuel cell, batteries, etc. do not have any mechanical inertia. Increased penetration of nonconventional energy sources, which have low inertia, results in a more vulnerable grid.

A variety of literature has reported the stability of an inverter interfaced with distributed generators (DGs). Research has focused mainly on controlling methods, the compensation of reactive power, and the shedding of loads connected to the MG. In a weak system, a large value of droop-gain provides better load sharing. However, if the value goes beyond the limit it might lead to instability of the system. Therefore properly tuning the value of droop gain helps to maintain stability [1]. Improvement of transient stability is provided by introducing power derivative integral terms to traditional droop characteristics. The negative impact of providing a large droop gain for proper load sharing has been eliminated in Ref. [2] using a supplementary control loop around the traditional droop control mechanism. In Ref. [3], a method is proposed to provide decentralized communication between inverters which maintains the power sharing of each of the DGs of the system and helps in active and reactive power regulation and also improves system stability. The concept of providing an arctan function for the droop control mechanism to improve the small-signal stability of the system is given in Ref. [4].

The lack of inertia in inverter-interfaced DGs is one of the important issues. The impedance value is implemented virtually using some adaptive droop mechanism to enhance transient stability of the MG. In Ref. [5], control methods are proposed to help improve the voltage stability using reactive sources. The combination of integral control and droop mechanism provides a smooth transition of the MG operation from grid-connected mode to islanded mode [6]. Providing virtual impedance helps to improve constant power stability in AC MGs [7].

A synchronverter imitates the behavior of a synchronous machine and thus provides a control strategy for the interconnection of renewable sources with the grid [8]. A synchronverter has a storage unit and a voltage source converter [9]. The converter is controlled using a control strategy which is developed based on the characteristics of a synchronous machine. The synchronverter is interconnected with the AC distribution grid to control real and reactive power [10].

13.2 Modeling of a synchronverter

The model of synchronous machines is reported in this section. Various assumptions, such as balanced sinusoidal steady state currents/voltages, are considered during the modeling of synchronous machines to simplify the analysis. All inductance of stator are considered as constant based on the round rotor machine. The eddy currents, magnetic saturation effects of the iron core, and damper windings present in the rotor, etc., are not included in the model. The modeling of a synchronverter is done in two parts that is, (i) electrical part and (ii) mechanical part.

13.2.1 Electrical part of a synchronverter

The simplified model of a three-phase round-rotor synchronous machine is shown in Fig. 13.2. The three identical windings of stator and field are located in slots with a uniform air gap. The windings of the stator can be observed as concentrated coils with self and mutual inductances as shown in Fig. 13.1. The field windings can also be observed as a concentrated coil with self-inductance. The mutual inductance of the field coil with respect to three stator coils is expressed bellow.

$$M_{af} = M_f \cos(\theta); M_{bf} = M_f \cos\left(\theta - \frac{2\pi}{3}\right); M_{cf} = M_f \cos\left(\theta - \frac{4\pi}{3}\right) \tag{13.1}$$

The flux linkages of the rotor and stator windings can be written as follows:

$$\begin{gathered} \Phi_f = M_{af} i_a + M_{bf} i_b + M_{cf} i_c + L_f i_f \\ \Phi_a = L i_a - M i_b - M i_c + M_{af} i_f; \; \Phi_b = L i_b - M i_a - M i_c + M_{bf} i_f \\ \Phi_c = L i_c - M i_a - M i_b + M_{cf} i_f \end{gathered} \tag{13.2}$$

where i_f is the excitation current of rotor; i_a, i_b, and i_c are the phase currents of stator. The stator flux linkages can be rewritten as

$$\Phi = L_s i + M_f i_f \widetilde{\cos\theta} \tag{13.3}$$

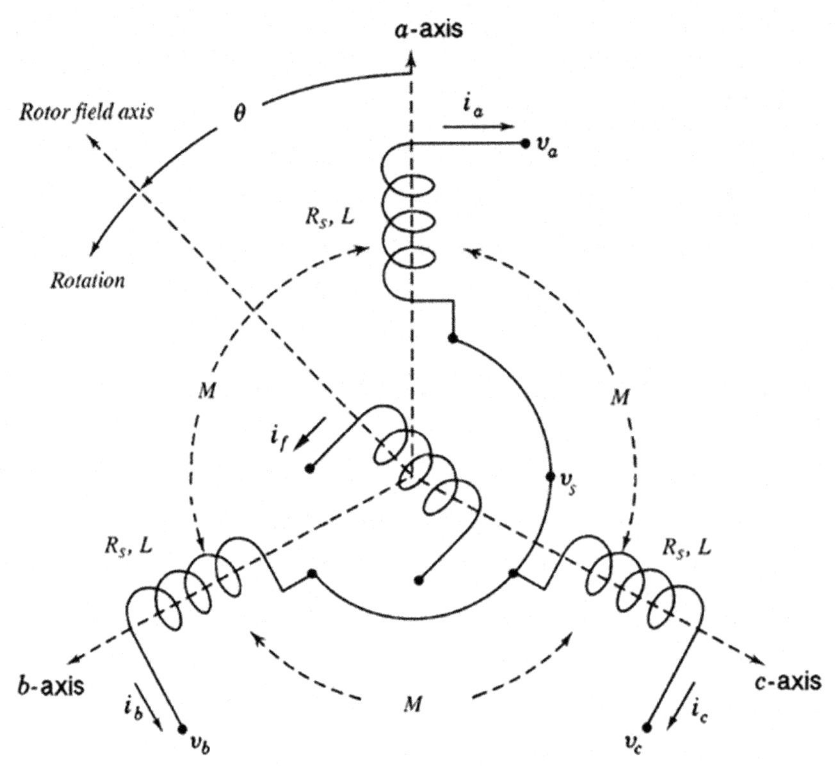

FIGURE 13.2 A simplified model of a three-phase synchronous machine [11].

where $\Phi = \begin{bmatrix} \Phi_a \\ \Phi_b \\ \Phi_c \end{bmatrix}$, $i = \begin{bmatrix} i_a \\ i_b \\ i_c \end{bmatrix}$, $\widetilde{cos}\theta = \begin{bmatrix} \cos\theta \\ \cos\left(\theta - \frac{2\pi}{3}\right) \\ \cos\left(\theta - \frac{4\pi}{3}\right) \end{bmatrix}$ and $\widetilde{sin}\theta = \begin{bmatrix} \sin\theta \\ \sin\left(\theta - \frac{2\pi}{3}\right) \\ \sin\left(\theta - \frac{4\pi}{3}\right) \end{bmatrix}$

Here, $L_s = L + M$. Similarly, field flux linkages are expressed as follows.

$$\Phi_f = L_f i_f + M_f \langle i, \widetilde{cos}\theta \rangle \tag{13.4}$$

where $\langle .,. \rangle$ represents the conventional inner product. For example,

$$\langle i, \widetilde{cos}\,\theta \rangle = i_a \cos\theta + i_b \cos\left(\theta - \frac{2\pi}{3}\right) + i_c \cos\left(\theta - \frac{4\pi}{3}\right)$$

The terminal voltages based on the stator resistance R_s are expressed as:

$$v = -R_s i - \frac{d\Phi}{dt} = -R_s i - L_s \frac{di}{dt} + e \tag{13.5}$$

where $v = [v_a\, v_b\, v_c]^T$ and $e = [e_a\, e_b\, e_c]^T$. The back emf, e, is expressed as:

$$e = M_f i_f \dot{\theta}\, \widetilde{sin}\, \theta - M_f \frac{di_f}{dt}\, \widetilde{cos}\theta \tag{13.6}$$

The terminal voltage of the field winding with resistance R_f is given below:

$$v_f = -R_f i_f - \frac{d\Phi_f}{dt} \tag{13.7}$$

13.2.2 Mechanical part of a synchronverter

The rate of change of speed of the synchronous machine can be expressed based on the swing equation as follows:

$$\ddot{\theta} = \dot{\omega} = \left(T_m - T_e - D_p \Delta\dot{\theta}\right)\frac{1}{J} \tag{13.8}$$

where ω is the angular speed of the synchronous machine, T_m is the mechanical torque, T_e is electromagnetic torque, D_p is the damping factor and J is the moment of inertia. The expression of energy stored in the magnetic field can be given as,

$$\begin{aligned} E &= \frac{1}{2}\langle i, \Phi\rangle + \frac{1}{2} i_f \Phi_f = \frac{1}{2}\left\langle i, L_s i + M_f i_f\, \widetilde{cos}\, \theta\right\rangle + \frac{1}{2} i_f (L_f i_f + M_f \left\langle i, \widetilde{cos}\, \theta\right\rangle \\ E &= \frac{1}{2}\langle i, L_s i\rangle + M_f i_f \left\langle i, \widetilde{cos}\, \theta\right\rangle + \frac{1}{2} L_f i_f^2. \end{aligned} \tag{13.9}$$

From simple energy considerations, the electromagnetic torque is expressed as:

$$T_e = \frac{\partial E}{\partial \theta}|_{\Phi, \Phi_f\ constant}. \tag{13.10}$$

However, the torque equal to the change in energy with respect to the current is as follows:

$$T_e = -\frac{\partial E}{\partial \theta}|_{i, i_f\ constant} \tag{13.11}$$

The final expression of the electromagnetic torque is expressed as:

$$T_e = -M_f i_f \left\langle i, \frac{\partial\, \widetilde{cos}\, \theta}{\partial x}\right\rangle = M_f i_f \left\langle i, \widetilde{sin}\, \theta\right\rangle. \tag{13.12}$$

The generated real and reactive power of the synchronous machine are given below:

$$P = \dot{\theta} M_f i_f \left\langle i, \widetilde{sin}\, \theta\right\rangle \tag{13.13}$$

$$Q = -\dot{\theta} M_f i_f \left\langle i, \widetilde{cos}\theta\right\rangle \tag{13.14}$$

13.3 Implementation of a synchronverter

The synchronverter consist of an electric part and a controller part. The electrical part includes a voltage source, a three-phase voltage-source converter, and an inductor-capacitor (LC) filter as shown in Fig. 13.3. There are two control approaches possible to control the power part: (i) a conventional phase locked loop (PLL) based approach and (ii) a virtual synchronous machine (VSM) approach. The controller receives the output current that is, i_a, i_b, and i_c and the terminal voltages that is, v_a, v_b, and v_c are sensed at the point of common coupling (PCC) as feedback. It produces the pulse width modulation (PWM) pulses for the three-phase voltage-source converter as shown in Fig. 13.3. The control part of the VSM is developed based on the mathematical model of synchronous machines as the VSM discussed in the above section.

The electrical torque of the VSM is calculated using Eq. (13.12) and the angular speed of the rotor of the VSM is calculated using Eq. (13.8) as shown in the control part of Fig. 13.3. The mechanical friction is represented in the VSM as a parameter called the damping factor. The output power of the synchronverter is controlled based on the mechanical torque reference and damping torque. When the net torque [RHS of Eq. (13.8)] is negative, the rotor speed of the VSM accelerates. Suddenly, the damping torque balances the torque

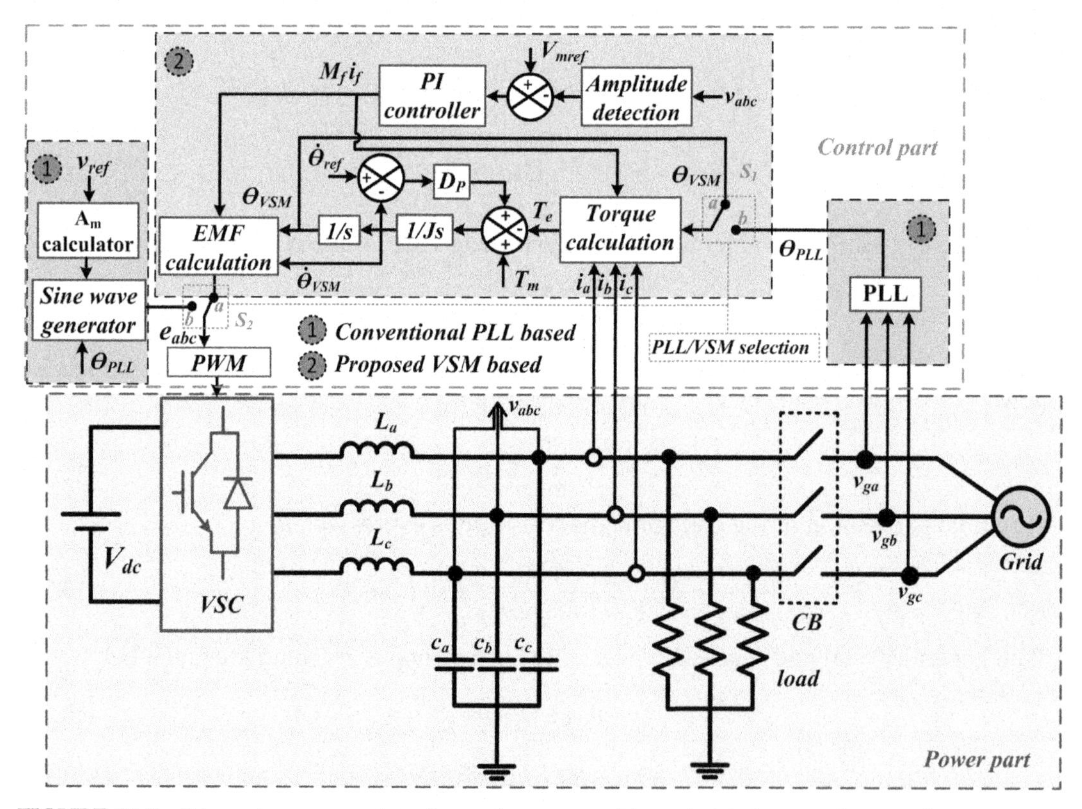

FIGURE 13.3 Schematic representation of a synchronverter with an electrical part and a control part.

equation and maintains the speed of the VSM at the rated value. The same corrective action happens when the net torque is positive. In the synchronverter, characteristics of frequency-droop are implemented using the damping factor, D_p, which is calculated from the angular frequency reference $\dot{\theta}_{ref}$ and angular speed of VSM $\dot{\theta}_{VSM}$. The value of D_p is given below for a frequency droop factor of 0.5%

$$D_p = \frac{-\Delta T}{\Delta \omega} \tag{13.15}$$

The value of moment of inertia J is calculated using the following Eq. (13.16):

$$J = D_p t_f \tag{13.16}$$

where t_f is the active power regulation loop-time constant. Similarly, the voltage regulation of synchronverter is achieved by controlling the reactive power. The term voltage droop coefficient, D_q, is used to achieve the voltage regulation. The amplitude of the actual voltage is calculated using Eq. (13.17). The value of D_q is then calculated using the error difference term of both voltage and reactive power as given in Eq. (13.18).

$$v_a v_b + v_b v_c + v_c v_a = -\frac{3}{4} V_m^2 \tag{13.17}$$

$$D_q = -\frac{\Delta Q}{\Delta v} \tag{13.18}$$

Setting the values of the moment of inertia and the damping coefficient gives additional freedom to the control in VSM when compared to the method. The complete control part of the synchronverter is shown in Fig. 13.3.

13.4 The simulation of a synchronverter

The performance of the synchronverter is verified through a simulation study. A simulation platform of the synchronverter is developed on the MATLAB/Simulink platform. A 400 V direct current (DC) voltage source as an input power source, a two-level voltage source converter, and an LC filter are used to realize the synchronverter. In the autonomous mode, a resistive load of 3 kW is connected at the output of the synchronverter to analyze the performance of the synchronverter. For the grid-connected mode of operation, the synchronverter is connected with the grid using a circuit breaker (*CB*). The three-phase output current and voltage of the inverter after filtering are sensed and fed back to the controller. Table 13.1 shows the values of various parameters of the electrical and control parts of the synchronverter used in the simulation.

To operate the synchronverter in autonomous mode, the *CB* is kept in an off position and the switches S_1, and S_2 are kept in position *a* as shown in Fig. 13.3. The mechanical torque of the VSM is set to zero. The three-phase output voltage of the synchronverter is sensed and used to calculate the voltage amplitude. This amplitude is compared with the reference voltage and an error signal is produced. Using this error signal, the PI controller produces the value of $M_f\, i_f$. The electromagnetic torque is calculated using Eq. (13.12).

TABLE 13.1 Parameters of synchronverter.

Parameters	Values	Parameters	Values
Grid nominal voltage (L-L) V_{rms}	200 V	Nominal grid frequency (f_n)	50 Hz
Rated power	3000 W	DC-link voltage (V_{dc})	400 V
Filter inductance (L_f)	20 mH	Filter resistance (R_f)	0.135 Ω
Filter capacitance (C_f)	40 μF	Voltage amplitude (A_m)	0.817
Switching frequency (f_s)	10 kHz	Damping factor (D_p)	6.085
Moment of inertia (J)	0.0122	Gain of PLL (K_p, K_i, K_d)	180,3200,1
Proportional constant (K_p)	3.18×10^{-3}	Integral constant (K_i)	4×10^{-5}
Grid resistance (R_g)	0.135 Ω	Grid inductance (L_g)	20 mH

Angular speed of the VSM is calculated using the electromagnetic torque and the damping factor as given in Eq. (13.8). By integrating the angular speed of the VSM further, the value of θ_{VSM} is calculated. The modulating waveform is generated finally using the value of M_f i_f and θ_{VSM}. This modulation signal is then compared to the triangular carrier signal of 10 kHz frequency and PWM pulses are produced to drive the voltage source converter of the synchronverter.

A detailed simulation study is carried out to check the performance of the synchronverter for the specification given in Table 13.1. A DC voltage of 400 V is considered to produce an output voltage of 200 V Line to Line (L-L) at the PCC. The steady state waveforms, such as the electromagnetic torque, angular speed, voltage at PCC output current after LC filter, and the active and reactive power of the synchronverter, are observed from the simulation as shown in Fig. 13.4.

Fig. 13.4 shows the simulation results of the synchronverter in autonomous mode when the load of 1.5 kW is connected at the output side of the voltage source converter. The steady state waveform of the angular speed and the electrical torque of the synchronverter is shown Fig. 13.4A. A small ripple content can be observed from the electrical torque. The line-to-line voltage and corresponding output current of the synchronverter is shown in Fig. 13.4B. Since the load connected is resistive, the output current is in phase with the output voltage. The active power and reactive power are plotted in Fig. 13.4C. The active power output is close to 1500 W due to the connected load and the reactive power is zero since the connected load is purely resistive. Fig. 13.4D–F shows the results when the load is changed from 1.5 kW to 3 kW. During the transient period, there is a surge in electrical torque and finally it settles down to the value corresponding to 3 kW. The output voltage and output current waveforms are shown in Fig. 13.4E which indicates the increase in current when the load is increased. Fig. 13.4F shows the variation in active power when the load is changed from 1.5 to 3 kW. The reactive power is found to remain at zero since the load is resistive.

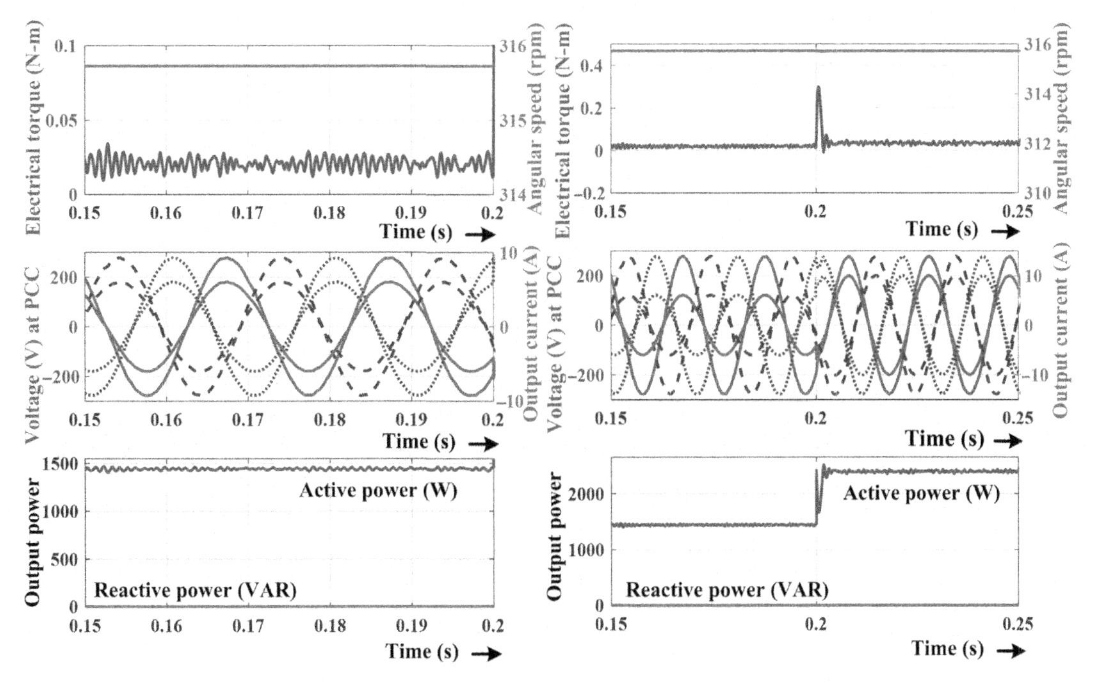

FIGURE 13.4 Simulation result of the synchronverter in autonomous mode.

13.5 The hardware implementation of the synchronverter

The simulation results obtained are verified using an experimental setup of the laboratory-scale model of a synchronverter. The block diagram for the experimental setup is shown in Fig. 13.5 and a corresponding laboratory set-up is shown Fig. 13.6. A 400 V lead acid battery bank acts as the power source for the three-phase inverter. The data acquisition interface OP 8660 is used for sensing the inverter output voltage and current. The data acquisition interface can sense up to 15 A peak current and 600 V peak voltage, which are converted to the voltage range suitable for the real-time simulator. OP 8660 scales the current of 15 A to 10 V and the voltage of 600 V is scaled to 10 V. The sensed output voltage and currents are given to the real time simulator OPAL-RT OP 5700. The synchronverter control circuit and PWM switching blocks are implemented in OPAL-RT OP 5700. A PWM switching frequency of 10 kHz is used for the inverter operation. A resistive load of 0.75 kW is connected to the output of the inverter.

The synchronization of the inverter with the grid is done by closing the *CB* which is actuated using a relay mechanism which monitors the synchronverter output voltage and the grid voltage. The hardware results are displayed on a digital storage oscilloscope (DSO) after acquiring the physical signals using the data acquisition interface. The real power, reactive power, and angular frequency are converted into analogue signals and connected to DSO through the real-time simulator.

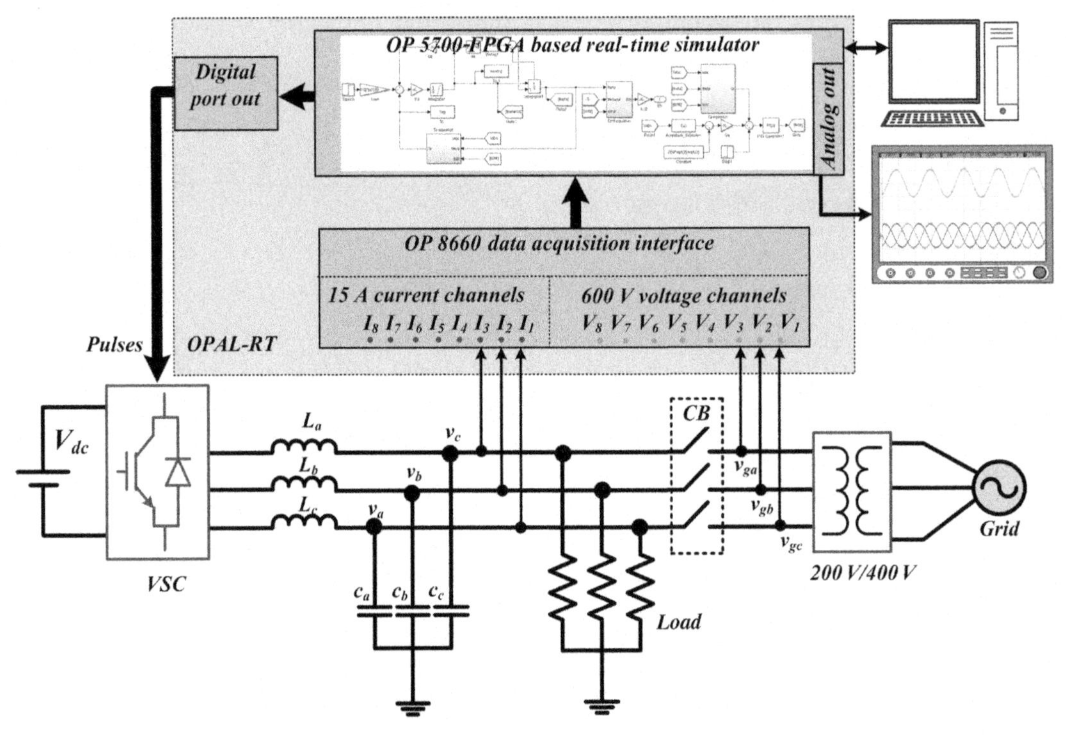

FIGURE 13.5 A block diagram of an experimental setup.

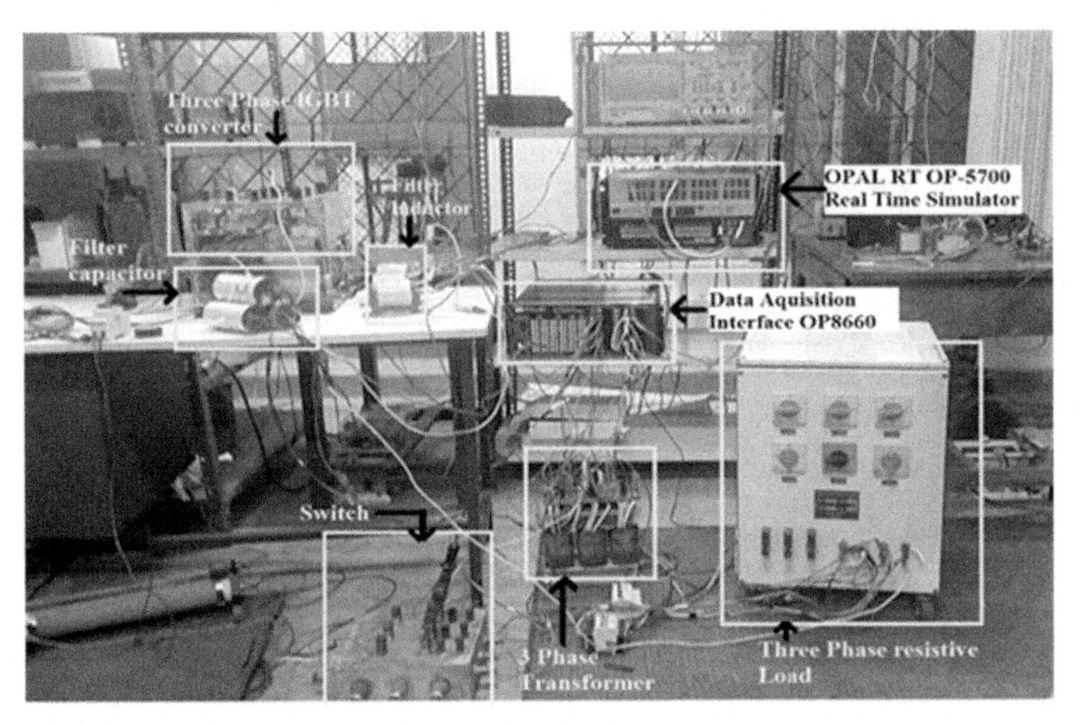

FIGURE 13.6 An experimental set-up of a synchronverter.

The three-phase output voltage (v_a, v_b, v_c) and the current (i_a) of a synchronverter during the autonomous mode of operation are shown in Fig. 13.7A. The output voltage waveforms are sinusoidal in shape with a total harmonic distortion (THD) content less than 3% and the output current is in phase with the output voltage due to the resistive load of 0.75 kW. The output current is sinusoidal in shape with a THD content less than 5%. The real power output and the angular speed of the synchronverter during autonomous mode are shown in Fig. 13.7B. Since the connected load is maintained at 0.75 kW, the real power output and the angular speed of the synchronverter remain at a steady value. The output voltage and current of phase a of the synchronverter is also shown in Fig. 13.7B. It can be seen clearly that the output current (i_a) is in phase with the output voltage. The magnitudes of the voltage and current are scaled down to fit into the same oscilloscope screen. Fig. 13.7C and D shows the hardware results for the grid-connected mode of operation of the synchronverter. The gate pulses are given to the synchronverter at t_1 and it is operated in autonomous mode and then at the instant, t_2, the CB is closed and the synchronverter put into grid-connected mode. When the synchronverter is changed from autonomous mode to grid-connected mode, there is a slight change in the angular speed of the

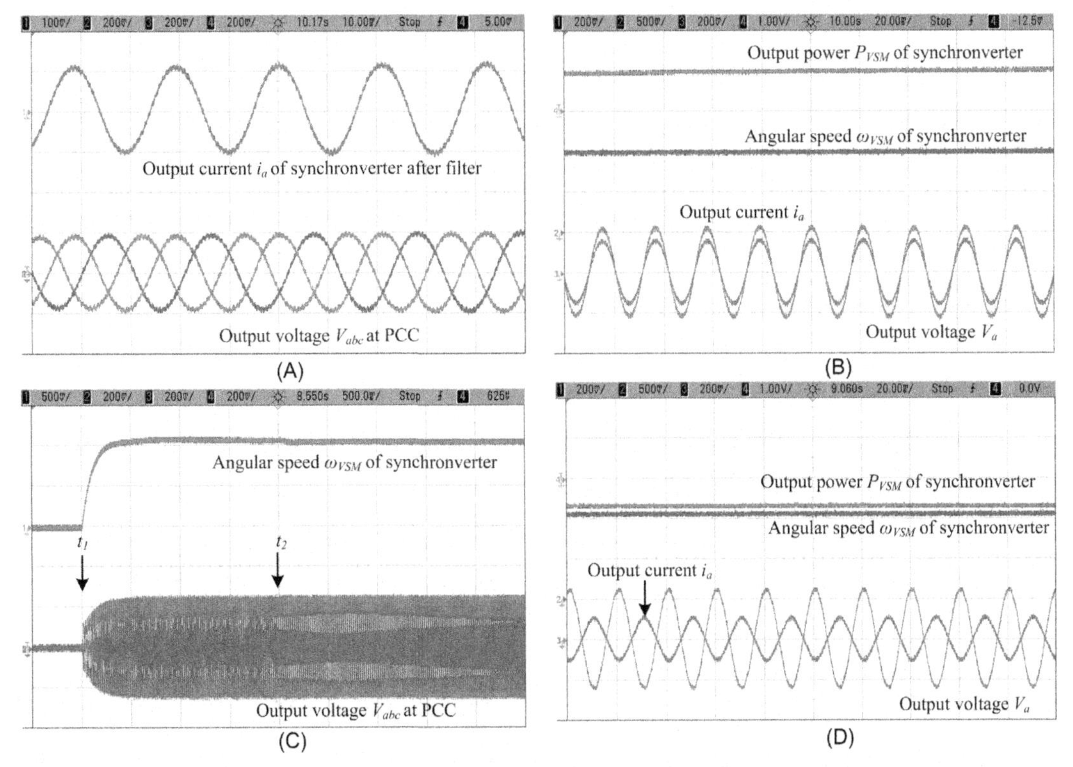

FIGURE 13.7 Hardware results of the synchronverter (i) autonomous mode (A) steady state waveforms of phase a current and three-phase voltages at PCC, (B) voltage and current of phase a, angular speed and active power; (ii) grid-connected mode (C) three-phase voltages at PCC and angular speed, (D) voltage and current of phase a, angular speed, and active power.

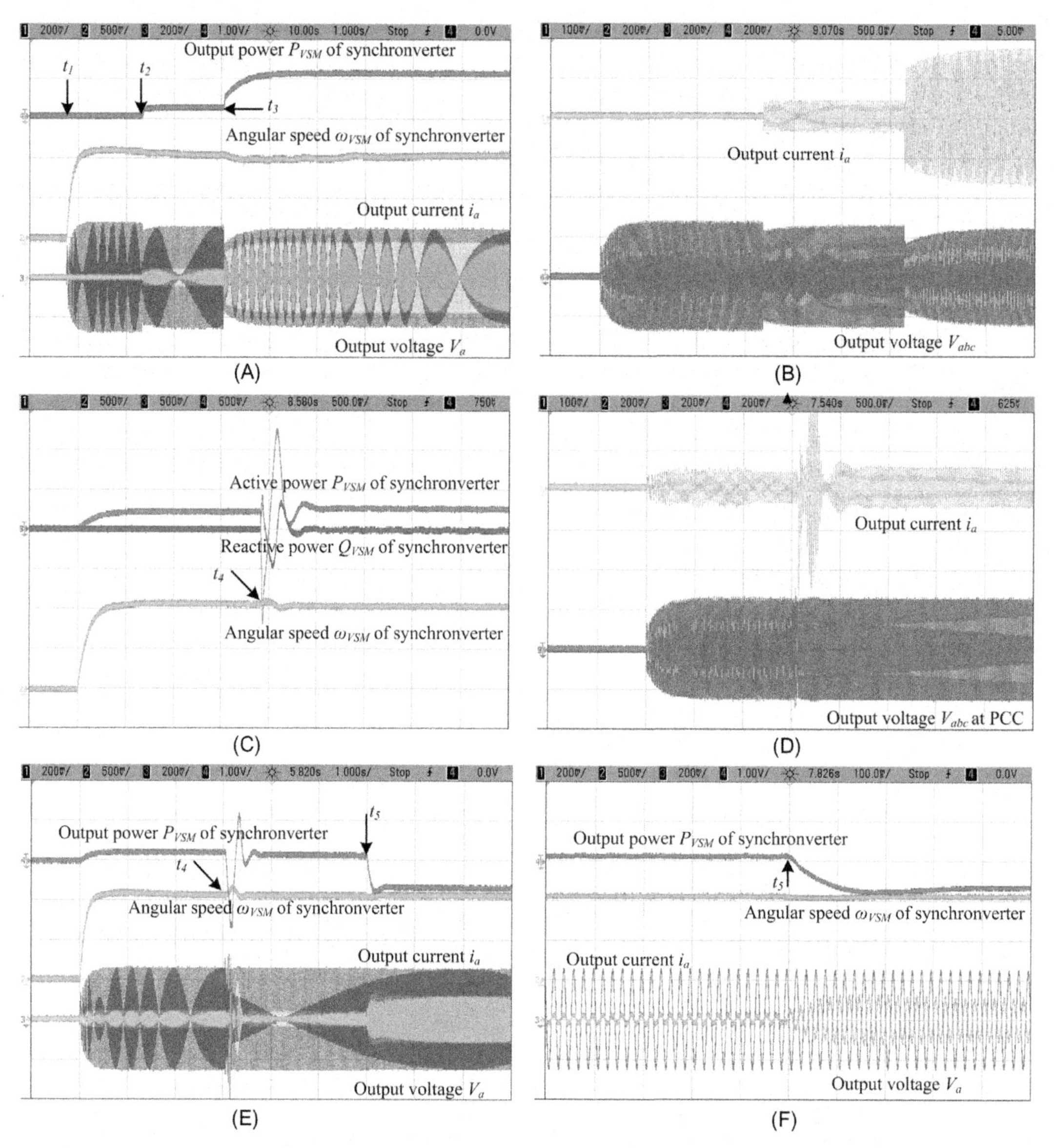

FIGURE 13.8 Performance of the synchronverter in a laboratory environment: in autonomous mode (A) voltage and current of phase *a*, angular speed; and active power (B) phase *a* current and three phase voltages at PCC; autonomous to grid-connected mode (C) active and reactive power and angular speed (D) phase *a*; current and three-phase voltages at PCC (E) active power, angular speed and voltage and current of phase *a*; grid-connected mode (F) voltage and current of phase *a*, angular speed and active power.

synchronverter as shown in Fig. 13.7C. But the output voltage of the synchronverter does not change much change during the mode change. The steady state waveform of the output power, angular speed, and voltage and current of phase *a* of the synchronverter is shown in Fig. 13.7D when the power is drawn by the synchronverter from the grid. During this power transfer, the output current in phase a (i_a) is 180° out of phase with the output voltage.

13.6 Performance evaluation of the synchronverter

In this section, a detailed performance evaluation of the synchronverter in both grid-connected and autonomous modes is carried out. Fig. 13.8A and B shows the performance of the synchronverter in autonomous modes. The synchronverter is turned on at the instant t_1 in no-load (the value of P_{VSM} is zero) and the angular speed reaches the steady state from zero as shown in Fig. 13.8A. At the instant t_2, the synchronverter is loaded by a 200 W load and Fig. 13.8A confirms the change in phase a current and output power according to the load change. At the instant t_3, a load of 1 kW is added and the real power and current increases accordingly. Throughout the operation, there is a slight change in the angular speed of the synchronverter as shown in Fig. 13.8A. The closer view of the three-phase output voltage and phase a current for the above operation is shown in Fig. 13.8B.

Fig. 13.8C shows the operation of the synchronverter in autonomous mode till the instant t_4 and in grid connected mode. When the synchronverter is changed from the autonomous mode to grid connected mode, there is a slight oscillation occur in real power and reactive power and it comes back to the steady state after some time as shown in Fig. 13.8C. Similar oscillation occurs in the output current of the synchronverter as shown in Fig. 13.8D. The performance of the synchronverter is verified to prove the capability of bidirectional power transfer with the grid. From the instant t_4 to t_5, the synchronverter injects power into the grid as shown in Fig. 13.8E. At the instant t_5, the active power reference of the synchronverter, P_{VSMref}, is changed to absorb 500 W power from the grid. The output waveforms shown in Fig. 13.8E and F confirm that the above operation is successfully performed by the synchronverter. The output power of the synchronverter becomes negative after the instant t_5 as shown in Fig. 13.8F and, accordingly, the output current of the synchronverter also changes: the output current is 180° out of phase with the output voltage as shown in Fig. 13.8F.

13.7 Conclusion

Integrating inertialess nonconventional energy sources into the modern power system creates a vulnerable situation. Operating a storage device and a voltage source converter with an appropriate control technique such as a synchronverter was discussed in this paper to support the integration of such inertialess nonconventional energy sources. A VSM control technique was developed based on the basic equation of the conventional synchronous machine. The operation of the synchronverter has been validated using the simulation which was developed in the MATLAB/Simulink platform, and the results of the synchronverter in autonomous mode were presented in this chapter. A three-phase 200 V rms (L-L) prototype of the synchronverter was fabricated and an experimental setup was assembled in the laboratory environment. The VSM control technique was successfully implemented using OPAL-RT. The three-phase voltage and current signals were sensed from the power hardware and fed into OPAL-RT. The control signals were derived from OPAL-RT to drive the six switches of the VSC of the synchronverter The performance of the synchronverter is verified in both autonomous and grid-connected mode through experimentation. The experimental results confirm the satisfactory performance of the synchronverter.

References

[1] M.M. Krishnan, R. Ramaprabha, Design and analysis of grid connected photovoltaic inverter under normal and fault conditions, in: 3rd International Conference on Electrical Energy Systems (ICEES), Chennai, 2016, pp. 272–275.

[2] Q.C. Zhong, G.C. Konstantopoulos, B. Ren, M. Krstic, Improved synchronverters with bounded frequency and voltage for smart grid integration, IEEE Trans. Smart Grid 9 (2) (2018) 786–796.

[3] R.V. Ferreira, S.M. Silva, D.I. Brandao, H.M.A. Antunes, Single-phase synchronverter for residential PV power systems, in: 2016 17th International Conference on Harmonics and Quality of Power (ICHQP), Belo Horizonte, pp. 861–866.

[4] R. Aouini, B. Marinescu, K. Ben Kilani, M. Elleuch, Synchronverter-based emulation and control of HVDC transmission, IEEE Trans. Power Syst. 31 (1) (2016) 278–286.

[5] S. Dong, Y. Chi, Y. Li, Active voltage feedback control for hybrid multiterminal HVDC system adopting improved synchronverters, IEEE Trans. Power Deliv. 31 (2) (2016) 445–455.

[6] C. Li, J. Xu, C. Zhao, A coherency-based equivalence method for MMC inverters using virtual synchronous generator control, IEEE Trans. Power Deliv. 31 (3) (2016) 1369–1378.

[7] Q.C. Zhong, Z. Ma, P.-L. Nguyen, PWM-controlled rectifiers without the need of an extra synchronisation unit, in: IECON2012 – 38th Annual Conference on IEEE Industrial Electronics Society, Montreal, QC, pp. 691–695.

[8] P.-L. Nguyen, Q.C. Zhong, F. Blaabjerg, J.M. Guerrero, Synchronverter-based operation of STATCOM to mimic synchronous condensers. in: 2012 7th IEEE Conference on Industrial Electronics and Applications (ICIEA), Singapore, pp. 942–947.

[9] B.S. Rigby, N.S. Chonco, R.G. Harley, Analysis of a power oscillation damping scheme using a voltage-source inverter, IEEE Trans. Ind. Appl. 38 (4) (2002) 1105–1113.

[10] Q.-C. Zhong, T. Hornik, Synchronverters: grid-friendly inverters that mimic synchronous generators, Control of Power Inverters in Renewable Energy and Smart Grid Integration, 1, Wiley-IEEE Press, 2012, p. 400.

[11] J.J. Grainger, W.D. Stevenson, Power Systems Analysis, McGraw-Hill, New York, 1994.

Further reading

Q.C. Zhong, G. Weiss, Synchronverters: inverters that mimic synchronous generators, IEEE Trans. Ind. Electron. 58 (4) (2011) 1259–1267.

Q.C. Zhong, G. Weiss, Static synchronous generators for distributed generation and renewable energy, in: Proceedings of IEEE PES PSCE, pp. 1–6, March 2009.

CHAPTER

14

Power converter solutions and controls for green energy

Vijay K. Sood and Haytham Abdelgawad
University of Ontario Institute of Technology (UOIT), Oshawa, ON, Canada

14.1 Introduction

The increase in photovoltaic (PV) installed capacity in recent years has sparked a continuous evolution of the PV power conversion stage [1,2]. The PV converter industry has evolved rapidly from infancy within the last two decades. One of the drivers behind this progress has been the PV converter market's stringent specifications, such as high efficiency (above 98%), long warranty periods (to get closer to PV module warranties of 25 years), high-power quality, transformerless operation, leakage current minimization (which imposes restrictions on the topology or modulation), and special control requirements such as the maximum power point tracking (MPPT). Another driver behind this development is the fact that, for a long time, the power converter represented only a small fraction of the cost of the whole PV system due to high PV module prices, allowing PV inverter manufacturers room for developing higher performance and more sophisticated topologies. The development of new PV converter topologies also motivated manufacturers to develop proprietary technology to differentiate themselves from their competitors and achieve a competitive advantage in the growing PV converter market [3].

14.2 Literature review

Ref. [4] presents an overview of grid-connected PV systems and then compares their advantages and disadvantages.

Ref. [5] introduces a different inverter topology that uses a block of energy storage in a series-connected path with the line interface block. This design facilitates an independent control over the capacitor voltage and soft switching for all semiconductor devices. It increases the converter complexity compared to traditional designs, but it provides control

Distributed Energy Resources in Microgrids
DOI: https://doi.org/10.1016/B978-0-12-817774-7.00014-4

for an energy storage voltage and reduced ripple by using electrolytic or film-type capacitors. This topology also provides a facility for reactive power transfer and maintains high efficiency.

Ref. [6] discusses different topologies based on cascaded H-bridge multilevel inverters. In one topology, they describe a multilevel inverter, with two PV arrays for each phase. In a second topology, they introduce a transformer and decrease the number of PV arrays to one for all three phases. In a third topology, the number of switches is much reduced compared to the other above-described topologies and uses the same number of PV arrays for all the three phases. They simulate these topologies with an resistance - inductance (RL) load.

Ref. [7] presents a grid-tied PV inverter with simulation and experimental tests for a reference-voltage-fixation method for direct current-alternating current (DC-AC) inversion, switching techniques with a peripheral interface controller (PIC) microcontroller, insulated gate bipolar transistor (IGBT) gate drive circuit operation with proper filtering, and finally power delivery to the grid with proper isolation. The achievements of the practical testing of a grid-tied inverter lies in successful dc-ac conversion, along with the capability of matching inverter output voltage and frequency with continuous fluctuating grid voltage and frequency.

Ref. [8] presents a new current source converter topology for 1-phase PV application. The main principle for this proposed topology is that instantaneous power transfer across the switching bridge is maintained constant. With the help of this topology, the low-frequency ripple common to single-phase inverters is eliminated or reduced enough to reduce the size of passive components to achieve necessary stiffness. With the low current ripple, MPPT performance is readily achieved. They verify the modulation and control methods using a detailed Saber model.

Ref. [9] focuses on inverter topologies for single-phase grid-connected PV modules. They describe some of the standards for PV and grid application like DC current injection into grid, power quality, and islanding operation detection. They classify the inverter topologies based on a number of power processing stages, the power decoupling type between the PV modules and grid, the types of grid interfaces, transformers, and type of interconnection between stages.

Ref. [10] describes a novel flyback-type single-phase inverter circuit, which is suitable for an AC-module system. They remove the problem of low power from a PV array due to partial shading on a few modules with the use of an AC-module strategy. They describe the use of a small-rating film capacitor instead of electrolytic capacitors, which is necessary for decoupling and with the help of film capacitors, the life of the inverter is increased.

Ref. [11] describes a single-phase PV transformerless inverter. They propose a new high-efficiency topology that generates limited varying common-mode voltage; this new topology uses a bipolar pulse width modulation (PWM) full bridge. They verify this topology with a 5-kW prototype.

Ref. [12] describes a transformerless inverter for a grid-connected PV system. They design an inverter with a very high rate of reliability and efficiency. The proposed inverter utilizes two different AC-coupled inductors for positive and negative half-grid cycles, and super junction metal oxide–semiconductor field effect transistor (MOSFETs) to achieve high efficiency.

Ref. [13] describes a transformerless inverter for a grid-connected PV system. They propose two stepdown converters where each converter modulates a half wave of output current. They describe transformerless topologies like H5 and highly efficient and reliable inverter concept (HERIC), which gives a very high level of efficiency for a low-power grid-tied system.

Ref. [14] describes a multilevel inverter, which offers a high-power capability with lower harmonic and commutation losses. The proposed multilevel inverter structure consists of two basic parts. The parts are classified as Level and H-Bridge Modules. They note that this multilevel inverter is good for unequal DC sources. They introduce a super imposed carrier pulse width modulation (SICPWM) technique for harmonic reduction.

Ref. [15] proposes a full-bridge series-resonant buck-boost inverter. This inverter includes a full-bridge topology and an inductance - capacitance (LC) resonant tank without auxiliary switches. The proposed inverter provides the main switch for turn-on at zero current switching (ZCS) by a resonant tank. This inverter has a very high efficiency.

Ref. [16] presents three-phase, two-stage, grid-connected module-integrated converters (MICs). Generally, for MICs single-stage conversion is required but here they propose zero-voltage switching (ZVS) in two-stage operations for the grid-connected PV system. In the first stage, they interface a high-efficiency full-bridge LLC resonant DC-DC converter which is interfaced to the PV array that produces a DC-link voltage. In the second stage, they consider a three-phase DC to AC inverter circuit, which employs an easy soft switching method without any auxiliary components. This inverter reduces the per watt cost and improves reliability.

Ref. [17] presents a single-phase, grid-connected PV system with a power quality conditioner. They use an incremental conductance method for MPPT and a shunt controller for voltage support.

14.3 Usual requirements for photovoltaic converters

In the past, PV modules were very expensive and the efficiency was very low; PV-power integration into the distribution grid was not obvious. In addition, the safety requirements imposed by electric companies and governments were lacking. Today, PV installations are a relatively substantial part of the electrical market, and as PV becomes a more relevant actor in power systems, the corresponding requirements and regulations for the safe and reliable use of PV systems are being standardized.

In general, two groups of requirements have to be met when a PV installation is considered. These two groups are the performance requirements and the legal regulations that the PV installations have to meet.

14.3.1 Performance requirements of photovoltaic converters

14.3.1.1 Efficiency

The losses of PV inverters have reduced in time, and efficiency figures are possible above 97% (see, for instance, Sunny Boy 5000TL by system, mess and anlagentechnik - german solar energy supplier (SMA) for domestic applications below 5.25 kW) and even more for central inverters (see, for instance, SunnyCentral 760CP XT by SMA, a central inverter with nominal power up to 850 kW with 98% of efficiency) [18]. The PV inverter efficiency for state-of-the-art branded products stands around 98%. However, it is notable that the efficiency is expected to improve further when silicon carbide (SiC) and gallium nitride (GaN) devices become the basic power semiconductors for PV inverters in the next decade [19].

14.3.1.2 Power density

Since power density is important for domestic and commercial applications (below 20 kW), several such solutions have been presented recently (e.g., the ABB PVS 300 inverter which is based on a neutral point clamped (NPC) topology) [20].

14.3.1.3 Installation and manufacturing cost

The installation and manufacturing costs of an inverter are important factors in selecting an appropriate inverter. The manufacturing cost is a trade-off between the power quality and the performance capabilities of inverter. However, the installation cost may vary greatly from one country or region to another as the land, labor, and other local factors may have a great influence on the total cost.

14.3.1.4 Minimization of leakage current

The leakage current appears because of the high-stray capacitance between the PV cells and the grounded metallic frame of each module, and the high-frequency (HF) harmonics caused by the modulation of the power converter. Galvanic isolation can help to interrupt the leakage path, but the use of a transformer presents drawbacks such as higher cost and additional losses, leading in general to a reduction in efficiency. Nevertheless, the transformer is mandatory in some countries because of local regulations. If the transformer is not mandatory, as a second solution, several power converter topologies have been designed specifically to minimize the effect of the HF harmonics on the leakage currents [21].

14.3.2 Legal requirements of photovoltaic converters

14.3.2.1 Galvanic isolation

One important requirement for PV systems is galvanic isolation for safety reasons. This feature is required only by some national codes, such as RD-1699/2011, which applies to the connection of a PV systems at the low-voltage (LV) distribution grid in Spain. This requirement means that PV topologies are not standardized and they have to be designed specifically to fulfill this galvanic isolation requirement, which is usually achieved by using a transformer (either high or low frequencies).

14.3.2.2 Antiislanding detection

The islanding phenomenon for grid-connected PV systems occurs when the PV inverter does not disconnect after the grid has tripped and continues to provide power to the local load [22].

In the conventional case of a residential electrical system, cosupplied by a rooftop PV system, the grid disconnection can appear as a result of a local equipment failure detected by the ground fault protection or of an intentional disconnection of the line for servicing. In both situations, if the PV inverter does not disconnect, some hazardous situations can occur, such as

- Retripping the line with an out-of-phase closure, damaging some equipment, and
- A safety hazard for utility line workers who assume that the lines are de-energized.

To avoid these serious situations, safety measures and detection methods called antiislanding requirements have been required in standards. In IEEE 1574, it is defined that after an unintentional islanding where the PV system continues to energize a portion of the power system (island) through the point of common coupling, the PV system shall detect the islanding and stop to energize the area within 2 seconds [23].

14.4 Photovoltaic system configurations

Grid-connected PV power-generation systems can be found in different sizes and power levels for different needs and applications, ranging from a single PV module from around 200 W to more than a million modules for PV plants over 100 MW [24]. Therefore the generic PV energy conversion systems' structure (Fig. 14.1), can vary significantly from one plant to another [25]. For simplicity, grid-connected PV systems are subdivided depending on their power rating: small scale from a few watts to a few tens of kilowatts, medium scale from a few tens of kilowatts to a few hundred kilowatts, and large scale from a few hundreds of kilowatts to several hundreds of megawatts. In addition, PV systems can be classified further depending on the PV module arrangement: a single module, a string of modules, and multiple strings and arrays (parallel-connected strings) [26]. The PV module arrangement also gives the inverter configuration its name: AC-module inverter, string inverter, multistring inverter, or central inverter, as shown in Fig. 14.2 and Table 14.1.

The AC-module configuration uses a dedicated grid-tied inverter for each PV module of the system [27]. Therefore this configuration is also known as a module-integrated inverter and microinverter because of the small size and low-power rating of the

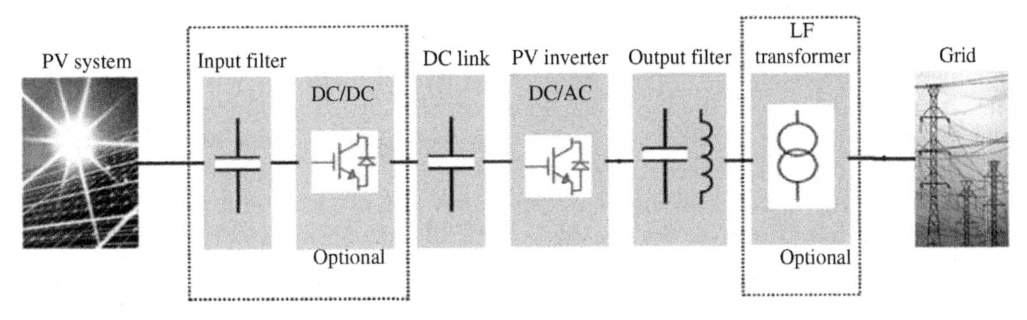

FIGURE 14.1 The generic structure of a grid-connected photovoltaic (PV) system [25].

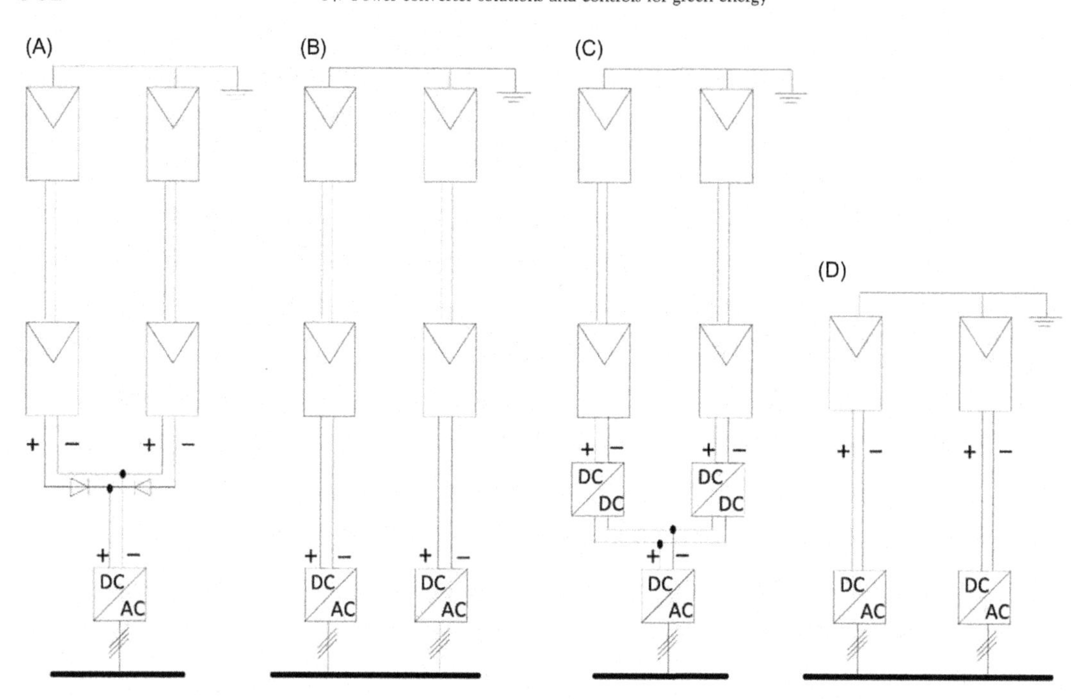

FIGURE 14.2 Photovoltaic (PV) system configurations: (A) central, (B) string, (C) multistring, and (D) module integrated.

converter. The LV rating of PV modules (generally around 30 V) requires voltage elevation for grid connection. Therefore AC-module inverters are only found with an additional DC-DC stage, usually with an HF transformer to provide galvanic isolation and elevate the voltage. Because of the additional DC-DC stage and HF isolation, this is the configuration with the lowest power converter efficiency, which is compensated somewhat by the highest MPPT accuracy due to the dedicated converter. This configuration is useful for places with lots of partial shading, complex roof structures, small systems, or combinations of different roof orientations. The small size of the converter allows a very compact enclosure design that can be attached to the back of each PV module, hence the name module-integrated inverter. Because of their LV operation, MOSFET devices are most commonly found in these topologies.

String inverters interface a single PV string to the grid [28]. They can be subdivided into single- and two-stage conversion topologies, depending on the addition (or not) of a DC-DC stage used to adapt the DC voltage output from the PV string to the DC side voltage of the grid inverter. In addition, the DC-DC stage decouples the MPPT control from the grid-side control (active and reactive power) by enabling a fixed voltage at the inverter's DC side. Furthermore, grid inverters can be found with or without galvanic isolation. Isolation can be introduced at the grid side with LF transformers or within the DC-DC stage with a HF transformer.

The different combinations between single- or two-stage, with transformer or transformerless string inverters, has led to a wide range of different configurations (Fig. 14.3).

TABLE 14.1 Grid-connected photovoltaic (PV) energy conversion systems configurations overview.

	Small scale	Medium scale		Large scale
	AC module	**String**	**Multistring**	**Central**
Power range	<350 W	<10 kW	<500 kW	<850 kW (<1.6 MW for dual)
Devices	MOSFET	MOSFET & IGBT	MOSFET & IGBT	IGBT
MPPT efficiency	Highest (one module—one MPPT)	Good (one large string—one MPPT)	High (one small string—one MPPT)	Good (one array—one MPPT)
Converter efficiency	Lowest (up to 96.5%)	High (up to 97.8%)	High (up to 98%)	Highest (up to 98.6%)
Features *Positive*	*Flexible/modular*	*Good MPPT efficiency*	*Flexible/modular*	*Simple structure*
	Highest MPPT efficiency	*Reduced DC wiring*	*High MPPT efficiency*	*Highest converter efficiency*
	Easy installation	*Transformerless (very common)*	*Low cost for multiple string system*	*Reliable*
Negative	Higher losses Higher cost per watt	High component count	Two stage is mandatory	Needs blocking diodes (for array)
	Two stage is mandatory	One string, one inverter		Poor MPPT performance Not flexible
Examples	Power One Aurora MICRO-0.3-I and Siemens SMIINV215R60	Danfoss DLX 4.6 and ABB PVS 300	SMA SB5000TL and SATCON Solstice	SMA MV Power Platform and 1.6 Siemens SINVERT PVS630

Compared to AC-module inverters, the string inverter has a less accurate MPPT of the PV systems and, under partial shading, would reduce the energy yield. However, for a PV system of the same power rating, the string inverter has a lower cost per watt and is more efficient. The string inverter is very popular for small-to-medium-scale PV systems, particularly for residential rooftop PV plants.

To add more flexibility to the string inverter and improve the MPPT performance of the PV system, the multistring concept was developed [30]. The strings are divided into smaller pieces (fewer modules in series) and connected through independent MPPT DC-DC converters to the grid-tied inverter. The DC-DC stage also boosts the voltage of the smaller strings. The additional DC-DC stages are a cost-effective solution compared to having several string inverters. As can be seen in Fig. 14.3, multistring inverters can also be found with or without isolation. Since they reduce partial shading and mismatching, they are suitable not only for rooftop PV systems but also for medium- and large-scale plants.

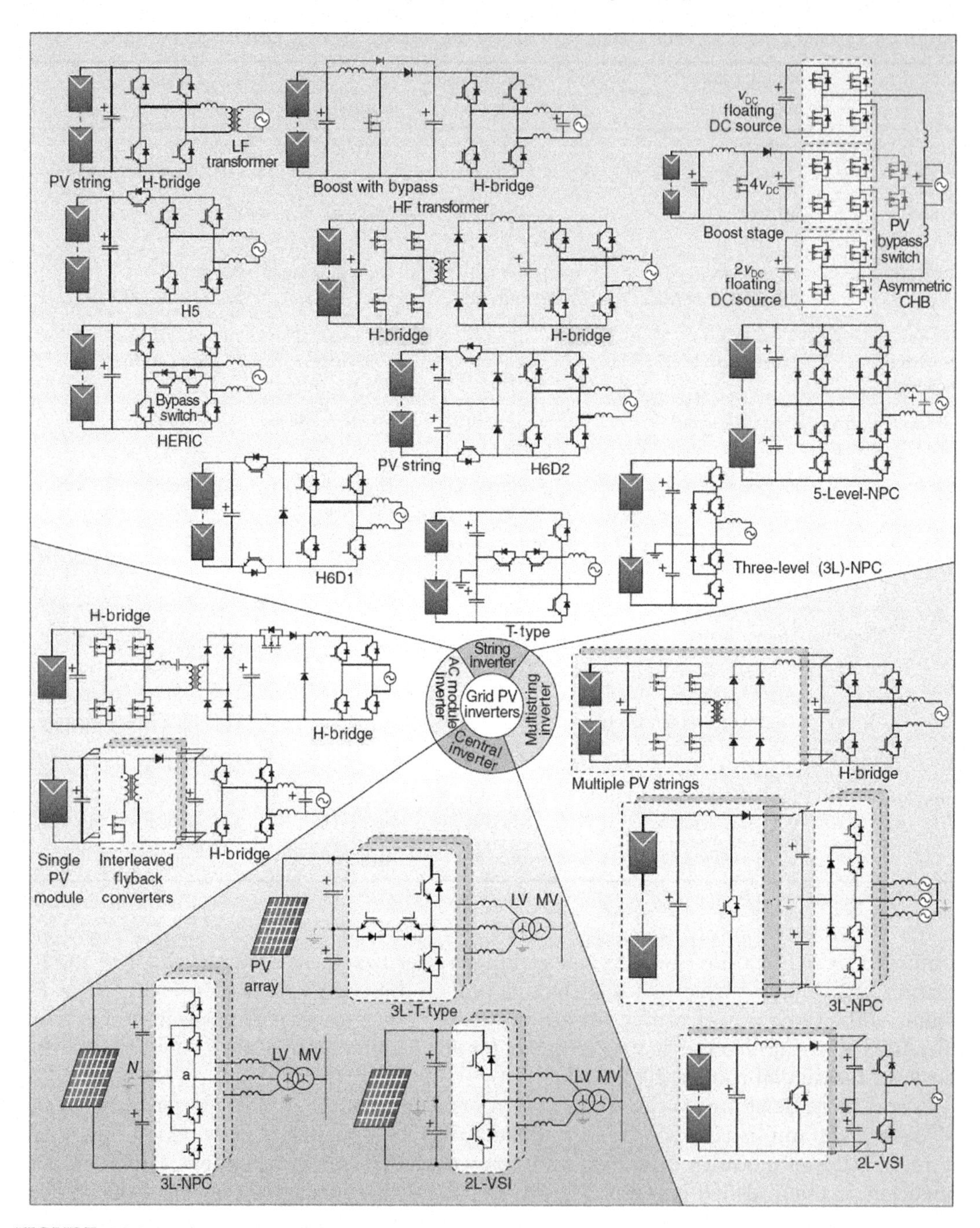

FIGURE 14.3 Industrial photovoltaic (PV) inverter topologies for central, string, multistring, and AC-module configurations. *MV*, Medium voltage; *2L-VSI*, two-level voltage source inverter [29].

Finally, the central inverter interfaces a whole PV array to the grid through a single inverter [2]. The array is composed of parallel-connected strings. A blocking diode in series to each string is necessary to prevent them from acting as load when partial shading or mismatch occurs. Because the whole array is connected to a single inverter, this configuration can only provide a single MPPT operation, leading to the lowest MPPT efficiency of all configurations. Nevertheless, it provides a simple structure, reliable, and efficient converter, making it one of the most common solutions for large-scale PV plants. Since they operate at a LV (11,000 V), the limit of IGBT technology enables converters of up to 850 kW. To increase the power rating, some manufacturers commercialize two central inverters connected through a 12-pulse transformer, with a rating up to 1.6 MW. Nevertheless, currently very large PV plants can reach several hundreds of megawatts. Therefore several hundreds of dual central inverters are needed in large PV farms.

14.5 Industrial photovoltaic inverters

The evolution in power converter technology for PV applications, driven by the growth in the PV installed capacity and the search for the ultimate PV inverter, has led to the existence of a wide variety of power converter topologies used in practice. Fig. 14.3 shows several industrial PV inverter topologies for central, string, multistring, and AC-module configurations, which will be analyzed in this section. Table 14.2 summarizes some of the characteristics of some commercial power converter topologies for these PV inverter applications.

14.5.1 String inverter topologies

The most common string inverter topology is the full- or H-bridge inverter. Several modified and enhanced versions have found their way into the market [31]. The H-bridge with a grid-side LF transformer features a simple power circuit, galvanic isolation, and voltage elevation provided by the transformer, which enables a larger range of input voltages. This converter can be controlled with three-level carrier-based PWM techniques since the common-mode voltages cannot generate a leakage current due to isolation. The bypass switching state (zero-voltage level) prevents a reactive current flow between the filter inductor and the DC-link capacitor. Nevertheless, the bulky transformer has several disadvantages, making this topology less popular.

The transformerless H-bridge, also known as an H4 inverter (shown in a two-stage configuration with a boost DC-DC stage), gets rid of the LF transformer by splitting the grid inductor into the phase and neutral wires of the systems and using a bipolar PWM (two-level) to solve the issues of the switched common-mode voltage and leakage currents and by using a boost stage for a wider input voltage range. The downside is that the two-level modulation reduces the power quality at the grid connection and lowers the efficiency since there is a reactive current flow between the passive elements of the circuit at zero voltage through the freewheeling diodes as the DC-link capacitor is not isolated from the grid at any time. To overcome the problem of the reactive current transfer between the

TABLE 14.2 Summary of characteristics and examples of industrial photovoltaic (PV) inverter topologies.

Topology	2L-VSI	HERIC	3L-NPC	H-NPC	1:2:4-CHB	H5	HFH-BRIDGEDC-DC	HF FLYBACKDC-DC
Pros	Simple	No freewheeling	Constant CM	Low THD	High-power	No freewheeling	Small	Small
	Robust	current losses	voltage	Transformerless	quality	current losses	Compact	Compact
	Large capacity	Transformerless	Low THD		Transformerless	Transformerless	Easy installation	Easy installation
Cons	Higher THD	Bidirectional	HF isolation	No. 5 level waveform	Complex module	Special PWM	High input- output voltage ratio	High input-output voltage ratio
	Large transformer	bypass switch	High number of devices	High number of devices	Complex control	modulation	Soft switching	Less efficient HF transformer concept
	Poor MPPT							
Configuration	Central	String	String	String	String	Multistring	AC module	AC module
Input voltage	550–850 V	900 V	600 V	900 V	380 V	750 V	60 V	45 V
Rated AC power	1.5 MW	4.8 kW	4.8 kW	8 kW	4 kW	5250 W	200 and 300 W	190–260 W
Grid connection	Three phase	Single phase	Single phase	Single phase	Single phase	Single phase	Single phase	Single, three phase
Efficiency	98.5%	97.8%	97.3%	97%	97.5%	97%	96.5%	96.3%
Isolation	LF transformer	Transformerless	HF transformer	Transformerless	Transformerless	Transformerless	HF transformer	HF transformer
Number of independent MPPT	Two arrays	One string	One string	One string	One string	Two strings	One module	One module
Brand/model	Satcon Prism	Sunways NT 5000	Danfoss DLX	ABB PVS 300 TL 8000	Mitsubishi PV-PN40G	SMA Sunny Boy	Power One	Siemens Microinverter System
	Platform Equinox		4.6			5000TL	Aurora MICRO-0.3-I	

grid filter and the DC-link capacitor in transformerless H-bridge string inverters during freewheeling, several proprietary solutions have been introduced by manufacturers [31–33].

The H-bridge with the HF isolated DC-DC stage is composed of a MOSFET full-bridge inverter, an HF transformer, and a diode full-bridge rectifier. This approach greatly reduces the size of the converter, improving the power density compared to LF transformer-based topologies. However, the additional converter stages introduce higher losses.

The H5 string inverter by SMA adds an additional switch between the DC-link and the H-bridge inverter to open the current path between both passive components, increasing the efficiency and reducing the leakage current.

The HERIC, introduced by Sunways, uses instead a bidirectional switch that bypasses the whole H-bridge inverter, separating the grid filter from the converter during freewheeling.

The H6 topology, introduced by Ingeteam [34], adds an additional switch in the negative DC bar to the H5 topology. Two versions were introduced: one with a diode connected in parallel to the DC side of the H-bridge of the H6 topology, called the H6D1, and the H6D2, which adds two auxiliary freewheeling diodes instead of one. Both allow freewheeling without interaction between passive components while enabling a unipolar output compared to the H5. The difference between the H6D1 and the H6D2 is that in the former, the additional switches block the total DC voltage, while in the latter, they only block half.

The three-level neutral point clamped inverter (3L-NPC) also has several modified and enhanced versions for PV string inverters [35]. The advantage of the 3L-NPC over the H-bridge is that it provides a three-level output without a switched common-mode voltage since the neutral of the grid is grounded to the same potential as the midpoint of the DC link. This enables transformerless operation without the problem of the leakage currents and modulation methods that do not use the potential of the converter. The main drawback compared to the H-bridge is that it requires a total DC link of double the voltage to connect to the same grid. Hence, more modules need to be connected in series or an additional boost stage is required.

A full-bridge of two 3L-NPC legs was introduced by ABB, resulting in the 5L-NPC inverter [36]. As with the H-bridge, this converter also requires a symmetrical grid filter distributed between the grid phase and neutral wires. A special modulation technique can achieve a line frequency common-mode voltage; hence, no leakage currents are generated while enabling a transformerless operation.

The T-type or three-level transistor-clamped string inverter was introduced by Conergy. The converter can clamp the phase of the grid directly to the neutral to generate the zero-voltage level using a bidirectional power switch. For the same reason as the 3L-NPC, it can operate transformerless. The main difference is that it does not require the two additional diodes of the 3L-NPC. The bidirectional switches block each half of the voltage blocked by the phase-leg switches.

The asymmetric cascaded H-bridge was introduced by Mitsubishi [37] and features three series-connected H-bridge cells operating with unequal DC voltage ratios (1:2:4). The PV system is connected through a boost DC-DC stage to only one of the H-bridge cells, which is the only one processing active power to the grid. The other two cells use floating

DC links for power quality improvement through the generation of 13 voltage levels. This enables a reduction of the switching frequency without compromising the power quality. The topology requires a bidirectional bypass switch connected to the large cell to reduce the changing potential between the PV system and the ground to reduce the possibility of leakage currents and enable a transformerless operation.

14.5.2 Multistring topologies

The main difference between the multistring and string configuration is that multistring is exclusively a two-stage system composed by more than one DC-DC stage [30]. Hence, all inverter topologies in the "string inverter topologies" section could be used in a multistring configuration. As with string inverters, the same combinations of isolated and transformerless configurations apply with or without symmetric grid filters.

One of the first multistring inverters introduced in practice was the half-bridge inverter with boost converters in the DC-DC stage by SMA [30]. Other topologies that have followed include the H-bridge, the H5, the three-phase two-level voltage source inverter (2L-VSI), the 3L-NPC, and the three-phase three-level T-type converter (3L-T) [31]. Fig. 14.3 shows some examples of practical multistring configurations. The most common DC-DC stages used for multistring configurations are the boost converter and the HF isolated DC-DC switch-mode converter based on an H-bridge, HF transformer, and diode rectifier.

14.5.3 Central topologies

Central inverter configurations are mainly used to interface large PV systems to the grid. The most common inverter topology found in practice is the 2L-VSI, composed of three half-bridge phase legs connected to a single DC link. The inverter operates below 1000 V at the DC side (typically between 500 and 800 V), limited by the PV module's insulation, which prevents larger strings. Grid connection is done through a LF transformer to elevate the voltage already within the collector of the power plant to reduce losses. More recently, the three-phase 3L-NPC and the three-phase 3L-T converter have been used also for this configuration (Fig. 14.3). The characteristics, advantages, and disadvantages analyzed for the single-phase versions of these topologies for PV string systems also hold for the central inverter version.

14.5.4 Alternating current module topologies

A commercial AC-module topology is the interleaved flyback converter (Fig. 14.3), developed by Enphase Energy [38]. The converter performs MPPT and voltage elevation and provides galvanic isolation while the H-bridge inverter controls the DC-link voltage, grid synchronization, and active/reactive power control. Several flyback converters are connected in parallel, which enables a higher switching frequency, resulting in a further reduction of the HF transformer and, hence, a very compact inverter. It also allows for a reduction in the current ripple both at the input and output of the DC-DC stage due to the phase-shifted carrier modulation, extending the life span of the capacitors.

Another commercial AC-module-integrated converter (Fig. 14.3) includes a resonant H-bridge stage with an HF isolation transformer and a diode bridge rectifier as a DC-DC converter instead of the flyback developed by Enecsys [39]. The H-bridge DC-DC stage has better power conversion properties compared to the flyback.

14.6 Control techniques for grid-connected solar photovoltaic inverters

The control of a grid-connected PV system can be divided into two important parts:

1. An MPPT controller to extract the maximum power from the PV modules, and
2. An inverter controller, which ensures the control of active/reactive power fed to the grid; the control of DC-link voltage; high quality of the injected power and grid synchronization.

14.6.1 Maximum power point tracking controller

The basic principle of the MPPT algorithm depends on the exploitation of voltage and current variations caused by pulsations of instantaneous power. Analyzing these variations allows us to obtain the power gradient and evaluate if the solar PV system operates close to the maximum power point [40]. The maximum power delivered by the solar PV array is given by the relation

$$P_{\max} = V_{\text{mpp}} I_{\text{mpp}}$$

where V_{mpp} and I_{mpp} are respectively the optimal operating voltage and current of PV array at the condition of maximum power output.

The solar cell exhibits nonlinear $V-I$ characteristics; therefore, an MPPT controller must track the maximum power and match the current environmental changes [41]. The MPPT is achieved by using a DC-DC converter between the PV array and inverter. From the measured voltage and current, the MPPT algorithm generates the optimal duty ratio (D) in order to maintain the electrical quantities at values corresponding to the maximum power point [42]. The most widely used MPPT techniques include perturbation and observation (P&O), incremental conductance, open circuit voltage, short circuit current, fuzzy logic, and neural network-based methods.

Ref. [43] proposes a hybrid MPPT method which combines the P&O and particle swarm optimization (PSO) methods with the advantage that search space for the PSO is reduced, and hence, the time required for convergence can be greatly improved. In Ref. [44], the authors propose a single-stage three-phase PV system that features an enhanced MPPT capability, and an improved energy yield under partial shading conditions. An MPPT method based on controlling an AC/DC converter connected at the PV array output, such that it behaves as a constant input power load is proposed in Ref. [45]. A one dimensional Newton–Raphson method based calculation for evaluation of MPP of PV array is proposed in Ref. [46]. Chaos search theory [47], self-synchronization error dynamics formulation [41], and MPPT methods based on ripple correlation control [40] have been reported. A comparison of multiple MPPT techniques is presented in Ref. [48]. A performance comparison of various MPPT techniques applied to a single-phase, single-stage, grid-connected PV system are presented in Ref. [49]. A comparative study of MPPT techniques for PV systems available until January 2012, is presented in Ref. [50].

14.6.2 Inverter controller

The control strategy applied to the inverter consists mainly of two cascaded loops. Usually, there is a fast internal current loop, which regulates the grid current, and an external voltage loop, which controls the DC-link voltage. The current loop is responsible for power quality issues and current protection; thus, harmonic compensation and dynamics are the important properties of the current controller. The DC-link voltage controller is designed for balancing the power flow in the system. Usually, the design of this external controller is aimed at the optimal regulation and stability of systems having slow dynamics. This voltage loop is designed for a stability time higher than the internal current loop by 5–20 times. The internal and external loops can be considered decoupled; therefore, the transfer function of the current control loop is not considered when the voltage controller is designed [51–59].

In some works, the control of the inverter connected to the grid is based on a DC-link voltage loop cascaded with an inner power loop instead of a current one. In this way, the current injected into the grid is indirectly controlled.

14.6.2.1 Control structure for three-phase inverter connected to the grid

To study stationary and dynamic regimes in three-phase systems, the application of "vector control" (Park vector) is used for the analysis and control of DC-AC converters, enabling abstraction of differential equations that govern the behavior of the three-phase system in independent rotating frames.

14.6.2.1.1 *dq*-control

The concept of decoupled active/reactive power control of three-phase inverter is realized in the synchronous reference frame by using the *abc-dq* transformation for converting the grid current and voltages. In this way, the AC current is decoupled into active and reactive power components, I_d and I_q, respectively. These current components are then regulated in order to eliminate the error between the reference and measured values of the active and reactive powers. In most cases, the active power current component, I_d, is regulated through a DC-link voltage control aimed at balancing the active power flow in the system [60–62]. As shown in Fig. 14.4, the power control loop is followed by a current control system. By comparing the reference and measured currents, the current controller should generate the proper switching states for the inverter to eliminate the current error and produce the desired AC current waveform [63,64]. In cases where the reactive power has to be controlled, a reactive power reference must be imposed on the system. Linear PI controller is an established reference-tracking technique associated with the *dq* control structure due their satisfactory combinational performance. Eq. (14.1) states the transfer function on the *dq* coordinate structure.

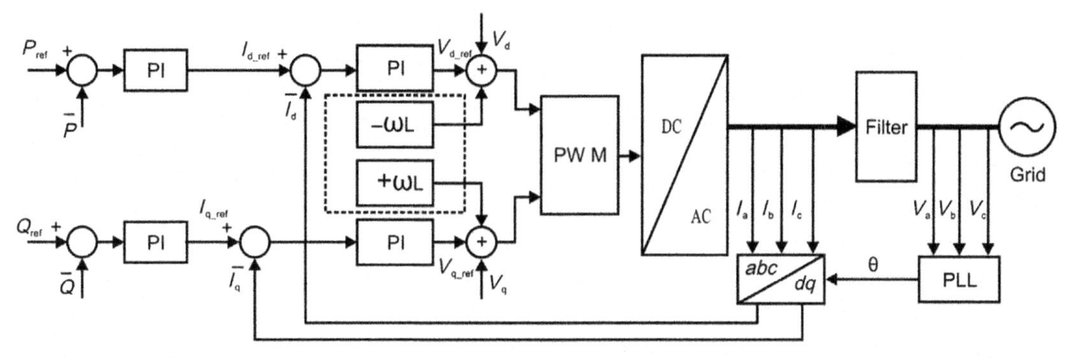

FIGURE 14.4 General structure for *dq* control strategy.

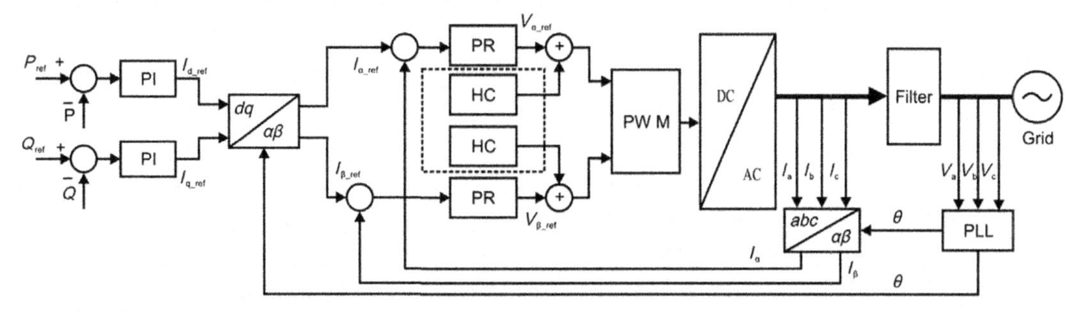

FIGURE 14.5 General structure for $\alpha\beta$ control strategy.

$$G_{\mathrm{PI}}^{dq}(s) = \begin{pmatrix} K_{\mathrm{P}} + \dfrac{K_{\mathrm{I}}}{s} & 0 \\ 0 & K_{\mathrm{P}} + \dfrac{K_{\mathrm{I}}}{s} \end{pmatrix} \tag{14.1}$$

where K_{P} is the proportional gain and K_{I} is the integral gain of the controller.

For improving the performance of the PI controller in such a structure (Fig. 14.4), cross-coupling terms and voltage feed forward are usually used [57–59]. In any case, with all these improvements, the compensation capability of the low-order harmonics in the case of PI controllers is very poor. Refs. [65–68] propose the use of proportional resonant + harmonic compensator (PR + HC) controller to improve the system's dynamic response and harmonic distortion, eliminate steady-state error and prevent the use of the feed-forward. The phase-locked loop (PLL) technique [55,56] is usually used in extracting the phase angle of the grid voltages in the case of PV systems.

14.6.2.1.2 $\alpha\beta$-control

In this case, the grid currents are transformed into a stationary reference frame using the $abc \rightarrow \alpha\beta$ module [59–69], as shown in Fig. 14.5. The *abc* control is to have an individual controller for each grid current; characteristic to this controller is the fact that it achieves a very high gain around the resonance frequency, thus being capable of eliminating the steady-state error between the controlled signal and its reference. High-dynamic characteristics of the proportional resonant (PR) controller have been reported in different works, and which are gaining common popularity in the current control for networked systems, are an alternative solution for performance under the proportional integral (PI) controller. The basic operation of the controller (PR) is based on the introduction of an infinite gain at the resonant frequency to eliminate the steady-state error at this frequency between the control signal and the reference. It does not require the use of feed forward [67–70]. The transfer matrix of the PR controller in the stationary reference frame is given by:

$$G_{\mathrm{PR}}^{\alpha\beta}(s) = \begin{pmatrix} K_{\mathrm{P}} + \dfrac{K_{\mathrm{I}}}{s^2 + \omega^2} & 0 \\ 0 & K_{\mathrm{P}} + \dfrac{K_{\mathrm{I}}}{s^2 + \omega^2} \end{pmatrix} \tag{14.2}$$

14.6.2.1.3 *abc*-control

As mentioned in Ref. [59], in *abc* control an individual controller for each grid current is used. However, in any case, having three independent controllers is possible by having extra considerations in the controller design. *abc* control is a structure where nonlinear controllers like hysteresis or dead beat are preferred due to their high dynamics. The performance of these controllers is proportional to the sampling frequency; hence, the employment of digital signal processors (DSPs) or field-programmable gate array (FPGA) is an advantage for such an implementation. A possible implementation of *abc* control is depicted in Fig. 14.6 [59], where the output of a DC-link voltage controller sets the active current reference. Using the phase angle of the grid voltages provided by a PLL system, the three current references are created. Each of them is compared with the corresponding measured current, and the error goes into the controller. If hysteresis or dead beat controllers are employed in the current loop, the modulator is not necessary. The output of these controllers is the switching states for the switches in the power converter. In the case that three PI or PR controllers are used, the modulator is necessary to create the duty cycles for the PWM pattern.

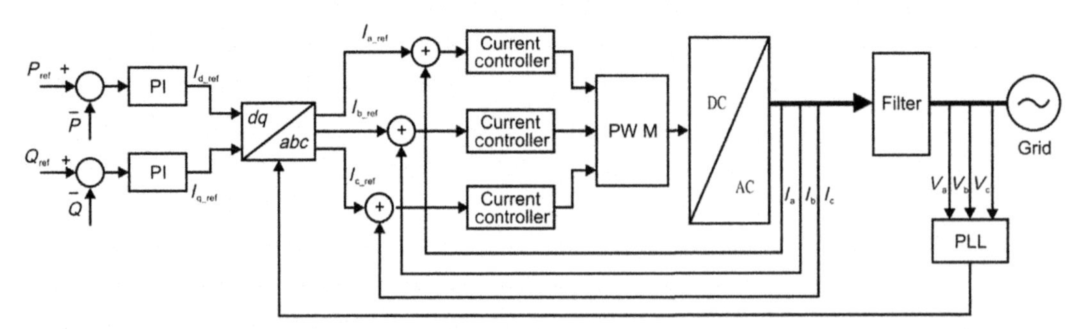

FIGURE 14.6 General structure for *abc* control strategy [59].

TABLE 14.3 Advantages and disadvantages of control structures for three-phase inverters.

Control strategies	Associated controller type	Advantages	Disadvantages
dq-control	PI	• Filtering and controlling can be achieved more easily • Simplicity	• Very poor compensation capability of the low-order harmonics • The steady-state error is not eliminated
$\alpha\beta$-control	PR	• Very high gain around the resonance frequency is achieved • The steady-state error is eliminated • Highly dynamic	• No full control of power factor (PF) Complex • Hardware circuit
abc-control	PI		• The transfer function is complex
	PR	• The transfer function is simple	• More complex than hysteresis and dead beat
	Hysteresis	• High dynamic • Rapid development	• High complexity of the control for current regulation
	Dead beat	• Simple control for current regulation • Highly dynamic • Rapid development	• Implementation in high-frequency micro-controller

The PI controller is widely used in conjunction with the *dq* control, but its implementation in the *abc* frame is also possible, as described in Ref. [58].

The implementation of a PR controller in *abc* control is simple since the controller is already in a stationary frame and the implementation of three controllers is possible as expressed in Eq. (14.3)

$$G_{PR}^{abc}(s) = \begin{pmatrix} K_P + \dfrac{K_I}{s^2 + \omega^2} & 0 & 0 \\ 0 & K_P + \dfrac{K_I}{s^2 + \omega^2} & 0 \\ 0 & 0 & K_P + \dfrac{K_I}{s^2 + \omega^2} \end{pmatrix} \tag{14.3}$$

Table 14.3 summarizes the pros and cons of control structures in three-phase inverters.

14.6.2.1.4 Single-phase inverters

The control structures for single-phase grid-connected inverters fall into three categories:

1. Control structure for single-phase inverter with DC-DC converter,
2. Control structure for single-phase inverter without DC-DC converter, and
3. Control structure based on power control shifting phase (PCSP).

14.6.2.1.5 Control structure for single-phase with DC-DC converter

The control structure for the single-phase with a DC-DC converter, proposed in Refs. [55–65], is shown in Fig. 14.7. The most common control structure for the DC-AC grid converter is a current-controlled H-bridge PWM inverter having low-pass output filters. Typically, L filters are used but the new trend is to use inductance - capacitance - inductance (LCL) filters that have a higher order, which leads to more compact designs:

- Control of instantaneous current values
- Current is injected in phase with the grid voltage (i.e., PF = 1)
- Use PLL for synchronization of the current I_{grid} and V_{grid}.

In order to control the output DC voltage to a desired value, an inverter control system which can adjust the duty cycle automatically is needed (Fig. 14.7). This controller has two control loops: the internal current control loop and the external DC-bus voltage control loop.

- The internal control loop is used to control the instantaneous values of AC current in order to generate a sinusoidal current in phase with the grid voltage. The reference current I_{ref}, is generated from a PLL sinusoidal signal reference which synchronizes the output inverter current with grid voltage [55]. The current amplitude is regulated from the external voltage loop.
- The external loop ensures the regulation of DC-bus voltage V_{DC}. It is necessary to limit the V_{DC} voltage, however, the control of V_{DC} guarantees the regulation of power injected into the grid.

In Fig. 14.8 [51], a control structure for topology with a DC-DC converter and L filter is presented. In this case, the reference current I_{ref}, is generated from the sinusoidal signal reference determined from a grid voltage sample. This structure is associated with PI controllers. To improve the performance of the PI controller in such a current-control structure and to cancel the voltage ripples of the PV generator, due to variations in the instantaneous power flow through the PV system which depend on the change of atmospheric conditions (mainly the irradiance and temperature), a faster response of the boost control loop, the inverter, and the DC bus capacitor is required. On the other hand, the output voltage (the main voltage) represents an external disturbance of considerable magnitude at

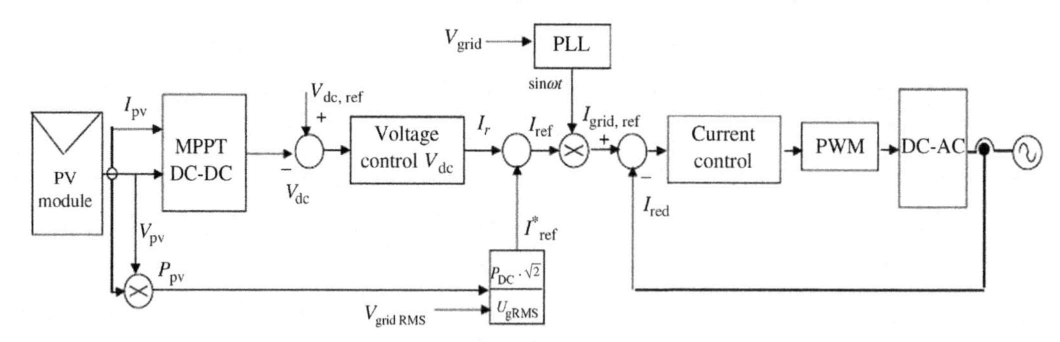

FIGURE 14.7 Control structure topology for single-phase with direct current (DC-DC) converter.

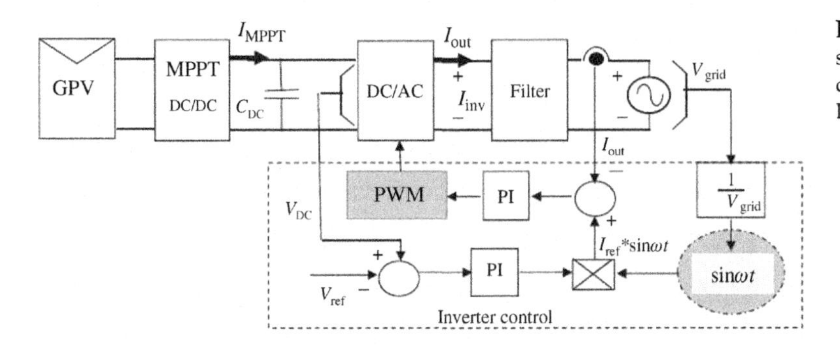

FIGURE 14.8 Control structure with direct current-direct current (DC-DC) converter and L filter.

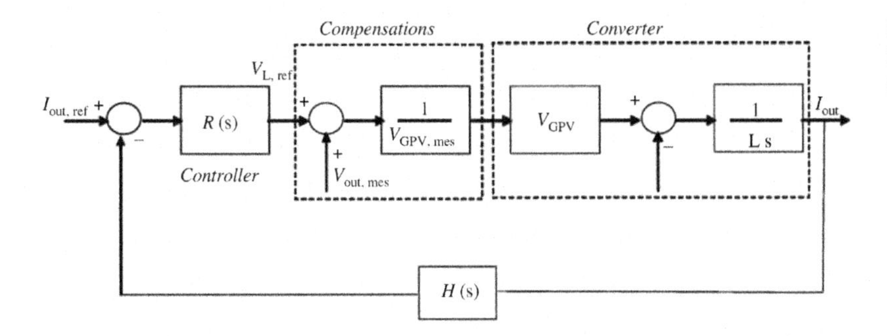

FIGURE 14.9 Control loop structure of the inverter output current.

60 Hz for the system. There exists a compensation of these effects at the output of the PI controller so as to calculate directly the reference voltage for the inductance [51]. Fig. 14.9 shows the control loop of the inverter output current.

The inverter output current expression is given by:

$$I_{out}(s) = \frac{d \times V_{GPV}(s) - V_{out}(s)}{Ls} \tag{14.4}$$

The feed-forward technique [51] is based on including new terms to control variables, in this case the duty cycle (d), in order to eliminate the dependence related to perturbations of the control system.

To compensate the effect of output voltage, the average and filtered output voltage values, called $V_{out,mes}$, are used in Fig. 14.9. However, to compensate the voltage V_{GPV}, it is necessary to use, the measured value before filtering. In this case, it is necessary to calculate the duty cycle (d) as follows:

$$d = \frac{V_{L,ref} + V_{out,mes}}{V_{GPV,mes}} \tag{14.5}$$

as K_{sv} is the same step of the measured circuits. So,

$$d = \frac{V_{L,ref} + K_{sv}V_{out}}{K_{sv}V_{GPV}} \tag{14.6}$$

Hence, the inductance voltage V_L can be deduced as:

$$V_L = dV_{GPV} - V_{out} = \frac{V_{L,ref}}{K_{sv}} \tag{14.7}$$

The advantage of this control structure is the control of the instantaneous power injected into the grid from the solar module and the synchronization of the current signal with the grid voltage (i.e., to keep the voltage and current in phase) which guarantees unity PF and improves the MPPT dynamic. The disadvantage is the noise in the inverter output current signal due to the use of the grid signal sample for generating and synchronizing the reference current with the grid signal.

Refs. [55,56] propose a control structure for topologies with DC-DC converter and LCL filter as shown in Fig. 14.10. This structure has the following characteristics:

- Typical structure for powers up to 5 kW max.
- PI or proportional resonant (PR) controller for current control.
- PWM control, hysteresis, or predictive.
- PI controller for voltage control.
- Optional transformer.

The principal elements of this control structure are the control algorithm based on the PLL, the MPPT and the control of the provided power, and the injected current into the grid.

14.6.2.1.6 Control structure for single-phase system without direct current-direct current (DC-DC) converter

Fig. 14.11 shows the control structure for a single-phase system without a DC-DC converter [65]. The same control loops are used: the internal current one and the external voltage loop [67]. The difference with respect to the control structure for topologies with a DC-DC converter is that the DC-AC inverter determines the maximum power point.

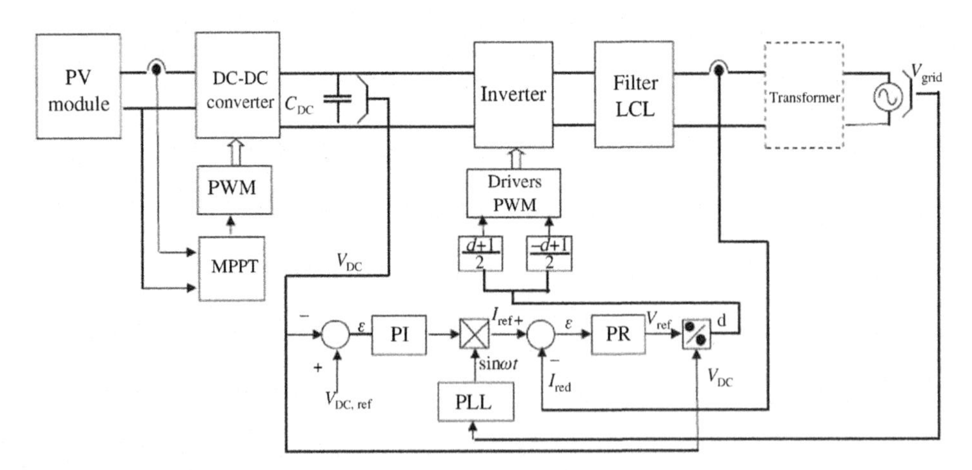

FIGURE 14.10 Control structure with direct current-direct current (DC-DC) converter and LCL filter.

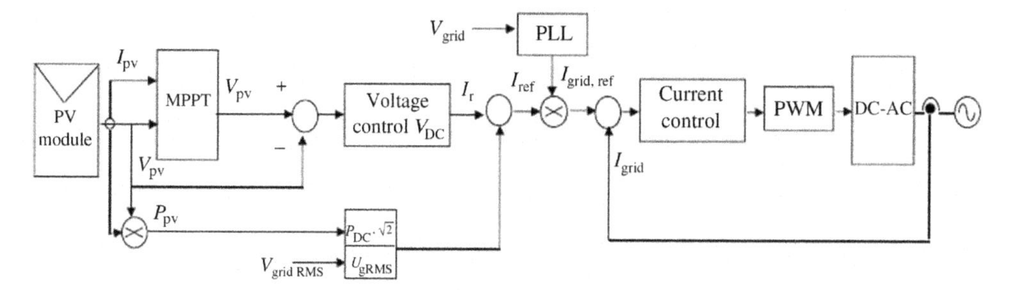

FIGURE 14.11 Control structure for single phase without direct current-direct current (DC-DC) converter.

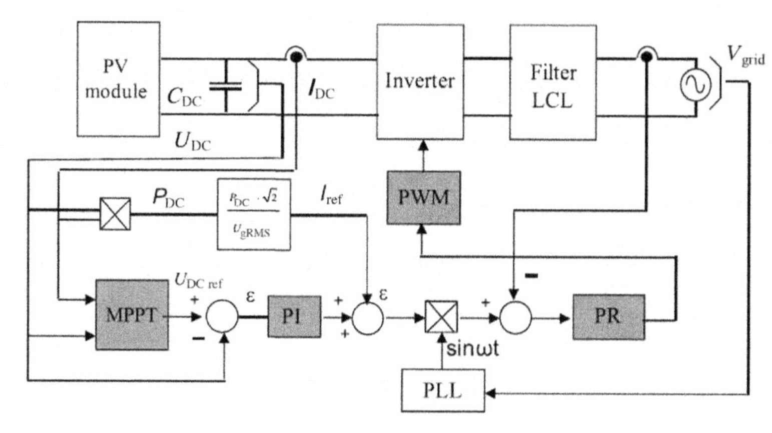

FIGURE 14.12 Injected power control structure.

Another control structure is proposed in Ref. [55]; here the power control is based on the current control injected into the grid. The power control structure for the PV system connected to the grid is in the range of 1–5 kW. The full-bridge inverter connected to the grid across the LCL filter is shown in Fig. 14.12. This power control structure is divided principally on the synchronize algorithm based on the PLL, a MPPT, the input power control of the continuous side, and the injected current control into the grid.

- *PLL*: used for the synchronization of the inverter output current with the grid voltage, the PF equal to the unity, also allowing for the generation of the sinusoidal and clean reference signal [56,71–73].
- *Input power control*: in this case, the control strategy for the power configuration of the PV system uses a feed-forward and does not include the DC-DC converter and is presented in Fig. 14.13. The amplitude value of the reference current is calculated from the solar modules power P_{pv} and the RMS voltage grid ($V_{grid,RMS}$), adding the controller value (I_r) of the output continued DC bus (V_{dc}). The result is expressed with the amplitude reference (I_{ref}) as shown in Fig. 14.13 [55]. The use of the feed-forward improves the dynamic response of the PV system. The DC bus voltage controller ensures a fast PV system response to the input power change [51,55].

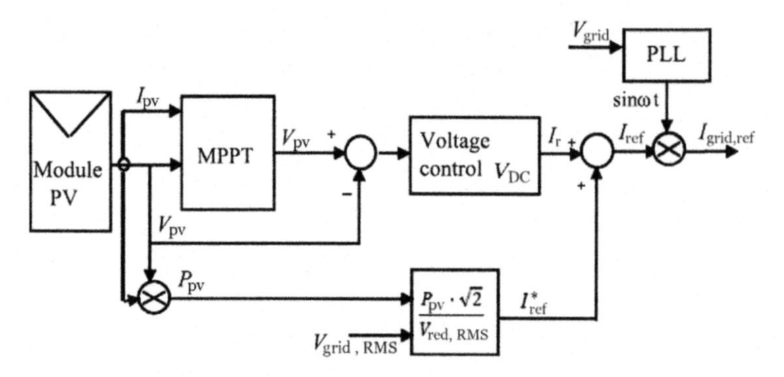

FIGURE 14.13 Control structure of input power (solar panel power).

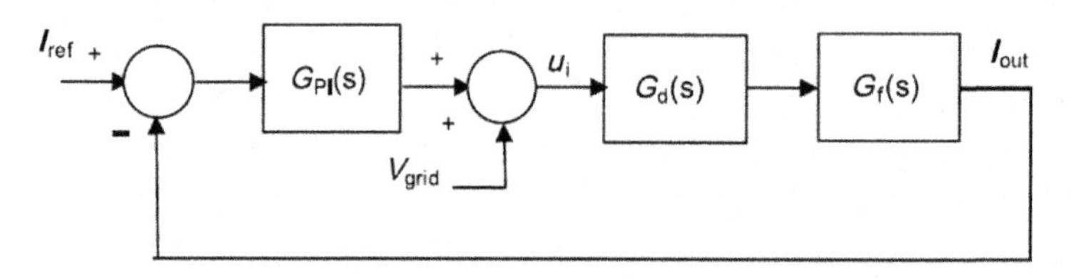

FIGURE 14.14 Inverter current loop with proportional integral (PI) controller.

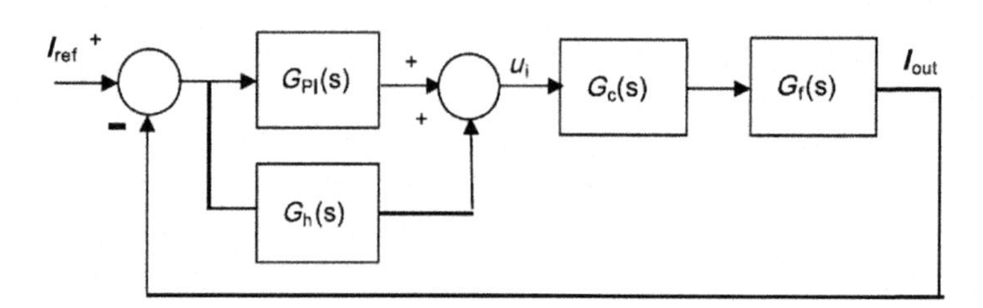

FIGURE 14.15 Inverter current loop with proportional controller (PR) controller.

- *Current control*: the PI controller is used with the feed-forward technique of the grid voltage as shown in Fig. 14.14. The transfer function of the PI controller, $G_{PI}(s)$ is defined as:

$$G_{PI}(s) = K_P + \frac{K_I}{s} \tag{14.8}$$

As mentioned previously, the feed-forward technique improves the dynamic response. This guarantees the stability of perturbations in the system introduced by the feedback voltage [55,56,65,66] and proposes an alternative solution for the poor performances of the PI controller, which includes the use of the proportional resonant controller PR.

The current loop of the PV inverter with the PR controller is presented in Fig. 14.15. The transfer function of the PR current controller is defined in Refs. [57,58,66,74] as:

$$G_c(s) = K_p + K_I \frac{s}{s^2 + \omega_0^2} \tag{14.9}$$

and the transfer function $G_h(s)$ of the harmonic compensator (HC) is defined in Ref. [65] as:

$$G_h(s) = \sum_{h=3,5,7} K_{Ih} \frac{s}{s^2 + (\omega_0 h)^2} \tag{14.10}$$

The HC is designed to compensate for the third, fifth, and seventh harmonics as they are the most predominant ones in the current spectrum.

In this case, it is shown that the use of the PR + HC controller gives a better dynamic response of the system, very low harmonic distortion (0.5%) and eliminates the error in the steady state without using the feed-forward technique. Adding the HC to the PR controller makes the system more reliable with better elimination of harmonics.

14.6.2.1.7 Control structure based on power control shifting phase

Fig. 14.16 shows the control structure for a PWM single-phase inverter connected to the grid as proposed in Refs. [75,76]. The PV system consists of a PV generator (PVG), an MPPT block, and a PWM single-phase inverter (DC-AC).

The DC-DC converter is employed to boost the PV array voltage to an appropriate level based on the magnitude of utility voltage, while the controller of the DC-DC converter is designed to operate at the maximum power point (MPP) to increase the economic feasibility of the PV system. Several algorithms can be used in order to implement the MPPT [77,78]: P&O, incremental conductance, parasitic capacitance, constant voltage etc. For the MPPT controller, the P&O method is adopted owing to its simple structure and the fact that it requires fewer measured parameters. This strategy is implemented to operate under rapidly changing solar radiation in a power-PV-grid-connected system, using only one variable: PV output current.

The control loop for the PWM inverter is assured by the output current control, the DC bus control and synchronizing to the grid, to inject power into the grid at all times. In this case, the voltage at the point of common coupling (PCC – the point where the load would be connected in parallel to the two sources), is not considered. The inverter is decoupled from the grid. The output voltage of the PWM inverter is already set by the utility PV

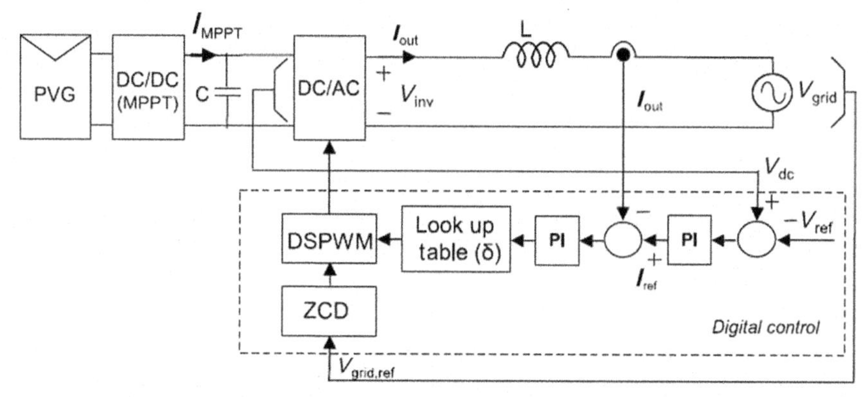

FIGURE 14.16 Control structure based on the shifting phase for a single-phase inverter connected to the grid.

TABLE 14.4 Advantages and disadvantages of control structures for single-phase inverters.

Topologies	Advantage	Disadvantages
Single-phase inverter with DC-DC converter	• Instantaneous current control • Fast dynamic	• No full control of PF • Complex hardware circuit
Single-phase inverter without DC-DC converter	• Instantaneous current control • Simplicity of the conversion system • Fast dynamic	• No full control of PF • Complex hardware circuit
Single-phase inverter with PCSP	• Simplicity • Less circuitry • Fewer resources • Reactive power controlled	• No full control of current • No fast dynamics

modules. Therefore the inverter is current-controlled to ensure only power injection into the grid.

Power control is obtained by means of the inverter output voltage shifting phase, PCSF. Fig. 14.16 represents a controller with two control loops: an inner one, that allows control of the inverter output current and an outer one to control the DC bus voltage (V_{DC}).

The reference of the output current (I_{ref}) depends on the DC bus voltage (V_{DC}) and its reference (V_{ref}). A low-pass filter is incorporated in order to ensure that high-frequency switching noise present in the measured inverter output current signal does not pass through to the PI controller.

The control structure is associated with PI controllers, since they demonstrate satisfactory behavior when regulating DC variables.

In this case the output current I_{out} is not controlled by varying the amplitude modulation index m_a, since it is considered constant, but by the phase shifting of the inverter output voltage with respect to the grid voltage. The adequate value of the phase shifting is obtained by taking into the account the zero-crossing detector (ZCD) of the reference grid voltage ($V_{grid,ref}$). The digital sinusoidal pulse width modulation (DSPWM) block generates the driving signals for the PWM inverter according to the switching pattern, with the corresponding phase shifting, in order to satisfy the current reference, I_{ref}. So, the PF is indirectly controlled. As a result, a certain amount of reactive power can be generated.

The main advantage of this control strategy is its simplicity with respect to the computational requirements of the control circuit and hardware implementation. On the other hand, it allows reconfiguring the control in a fast and simple way in case that not only an active power needs to be injected but also a reactive one.

Table 14.4 resumes the advantages and inconveniences of each control structure for single-phase topologies.

14.6.2.2 Reactive power requirements

PF control and reactive power regulation are the most important issues in connecting PV arrays to the grid. The grid-connected inverter must be controlled in such a way that not only does it inject a current with low total harmonic distortion (THD), but it also allows controlling the injected reactive power into the grid selecting a proper PF according to the grid demands. Thus, the most efficient systems are those that allow varying the power injected into the grid, both active and reactive, depending on the power grid's needs [79–81].

Some solutions are proposed in Refs. [82–85] to obtain high-reliability inverters and many control techniques of grid-connected PV inverters. Multiple closed-loop control structures for grid current and DC-link voltages are described in Refs. [86–90]. Other control structures consisting of classical PI and/or bang-bang controllers are given in Refs. [91,92]. Other authors propose the use of PLL control of the grid current in Refs. [93–95]. An extended direct power control (EDPC), based on geometrical considerations about inverter voltage vectors and their influences on active and reactive power change, is proposed in Refs. [96]. The input/output feedback linearization control (FLC) technique, widely applied to electrical motor control [97] and PWM rectifier control [98], has been applied to PV inverters by Ref. [99], which gives a complex model of the inverter, including switching functions.

In Ref. [100], the PF of a grid-connected PV inverter is controlled using the input/output FLC technique. This technique transforms the nonlinear state model of the inverter in the *dq* reference frame into two equivalent linear subsystems, in order to separately control the grid PF and the DC#l-ink voltage of the inverter. This method allows control of both PF and DC-link voltage using the same control algorithm. Also, in this control method, the MPP control is moved towards the DC-AC converter, hence, there is no need to use a DC-DC converter, which increases the simplicity of the conversion system. Compared to other control methods, in Refs. [75,76], the grid PF is controlled using a previously calculated and tabulated PWM and acting on the phase shift between grid voltage and inverter output voltage as a control parameter. The proposed control strategy is capable of controlling, not only the current injected into the grid, but also the PF, with a minimum number of DSPWM patterns. Varying the PF, within a certain range, the injected reactive power (inductive or capacitive) can be changed and controlled dynamically, in order to obtain the high reliability of the inverter. This method breaks the limitations of existing grid-connected system where the inverter topology is designed to supply only active power to the grid without injecting a reactive power.

14.7 Conclusion

A comprehensive literature review of grid-connected PV systems is carried out. This chapter presents a detailed survey of grid-connected PV systems used to integrate solar energy into the utility grid. Topologies of single-phase grid-connected inverters have been analyzed critically and a comparative study of these topologies is presented. The three-phase grid-connected inverters have also been outlined. The control techniques for the single and three-phase grid-connected inverters are critically reviewed and presented.

According to the developed review, it can be concluded that the PV market has experienced exponential growth in the last decade, becoming an important alternative and a clean energy source in many countries. Along with the decrease in price and the increase in efficiency of the PV modules, PV converter topologies have been changing continuously, following more demanding requirements and standards. These regulations are being adapted to a new power system scenario where renewable energy sources are an important part of the energy mix. Today, and meeting these legal requirements, PV converter topologies deal with issues such as high efficiency, high-power density, grid-code compliance, reliability, long warranties, and economic costs.

The efficiency characteristic of parallel inverters with a common DC bus is deliberated along with the optimal operation strategy. An inverter system performance ratio (ISPR) is proposed as an overall index of life-time energy conversion efficiency. It shows that the configuration with a common DC bus is a potential solution to reducing the energy cost of PV power-generation systems.

A good number of PV converter topologies can be found on the market for string, multistring, central, and AC-module PV applications. Among these converter topologies, it can be affirmed that one of the most important appearances is that of the multilevel converter, mainly the NPC, the T-type, and the H-bridge, not only for high-power applications, but also for residential applications in the kilowatt and LV range.

In the near future, it is expected that a completely new family of PV converters will be developed based on SiC power semiconductors (some commercial PV converters exist, but only with SiC diodes.) These new SiC-based PV converters and the next-generation GaN PV converters will reduce the compromise between performance and efficiency, enabling the next generation of grid-connected PV systems.

Some implementation structures for three-phase inverters, like dq, $\alpha\beta$, and abc control, were reported. The PI controller is widely used in conjunction with the dq control. The implementation of PR controller in $\alpha\beta$ is commonly used. In the abc control, nonlinear controllers like hysteresis or dead beat are preferred due to their high dynamics.

A discussion of the different single-phase inverters controllers and their ability to compensate for low-order harmonics presented in the grid was given. The resonant proportional controller + harmonic compensator (PR + HC) controller gives a better dynamic response of the system, very low harmonic distortion and eliminates error in the steady state without using the feed-forward voltage. Adding the HC to the PR makes the system more reliable with better elimination of harmonics.

Power factor control and reactive power regulation are known as the most important issues in connecting PV arrays to the grid. Control based on the shifting phase for grid-connected PV inverters allows the control in a fast and simple way in case not only an active power needs to be injected but also a reactive one.

Abbreviations

CSI current source inverter
DSP digital signal processor
DSPWM digital sinusoidal pulse width modulation
EDPC extended direct power control
FLC feedback linearization control
FPGA field-programmable gate array
GaN gallium nitride
HC hysteresis controller

hc	harmonic compensator
HERIC	highly efficient and reliable inverter concept
HF	high frequency
IEEE	institute of electrical and electronics engineers
IGBT	insulated gate bipolar junction transistor
LF	low frequency
LV	low voltage
MICs	module-integrated converters
MOSFET	metal oxide–semiconductor field effect transistor
MPP	maximum power point
MPPT	maximum power point tracking
MV	medium voltage
NPC	neutral point clamped
OCC	one cycle control
PCC	point of common coupling
PCSP	power control shifting phase
PI	proportional integral controller
PIC	peripheral interface controller
PLL	phase-locked loop
PR	proportional resonant controller
PV	photovoltaic
PVG	photovoltaic generator
PWM	pulse width modulation
SiC	silicon carbide
SICPWM	super imposed carrier pulse width modulation
SPWM	sinusoidal pulse width modulation
THD	total harmonic distortion
ZCD	zero-crossing detector
ZCS	zero current switching
ZVS	zero voltage switching

References

[1] R. Teodorescu, M. Liserre, P. Rodríguez, Grid Converters for Photovoltaic and Wind Power Systems, IEEE Press/Wiley, Piscataway, NJ, 2011.

[2] J.M. Carrasco, L.G. Franquelo, J.T. Bialasiewicz, E. Galvan, R.C. PortilloGuisado, M.A.M. Prats, et al., Power-electronic systems for the grid integration of renewable energy sources: a survey, IEEE Trans. Ind. Electron. 53 (4) (2006) 1002–1016.

[3] I.E. Agency, PVPS Report Snapshot of Global PV 1992–2013, 2014, pp. 1–16.

[4] H. AbdEl-Gawad, V.K. Sood, Overview of connection topologies for grid-connected PV systems, in: 2014 IEEE 27th Canadian Conference on Electrical and Computer Engineering (CCECE), 2014, pp. 1–8.

[5] B.J. Pierquet, D.J. Perreault, A Single-phase photovoltaic inverter topology with a series-connected energy buffer, IEEE Trans. Power Electron. 28 (10) (2013) 4603–4611.

[6] K.N.V. Prasad, N. Chellamma, S.S. Dash, A.M. Krishna, Y.S.A. Kumar, Comparison of photo voltaic array based different topologies of cascaded H-bridge multilevel inverter, in: International Conference on Sustainable Energy and Intelligent Systems (SEISCON 2011), 2011, pp. 110–115.

[7] F.B. Zia, K.M. Salim, N.B. Yousuf, R. Haider, M.R. Alam, Design and implementation of a single phase grid tie photo voltaic inverter, in: 2nd International Conference on the Developments in Renewable Energy Technology (ICDRET 2012), 2012, pp. 1–4.

[8] C.R. Bush, B. Wang, A single-phase current source solar inverter with reduced-size DC link, in: IEEE Energy Conversion Congress and Exposition, 2009, pp. 54–59.

[9] S.B. Kjaer, J.K. Pedersen, F. Blaabjerg, Power inverter topologies for photovoltaic modules—a review, in: Conference Record of the 2002 IEEE Industry Applications Conference. 37th IAS Annual Meeting (Cat. No. 02CH37344), vol. 2, 2002, pp. 782–788.

[10] T. Shimizu, K. Wada, N. Nakamura, Flyback-type single-phase utility interactive inverter with power pulsation decoupling on the DC input for an ACphotovoltaic module system, IEEE Trans. Power Electron. 21 (5) (2006) 1264–1272.
[11] R. Gonzalez, J. Lopez, P. Sanchis, L. Marroyo, Transformerless inverter for single-phase photovoltaic systems, IEEE Trans. Power Electron. 22 (2) (2007) 693–697.
[12] B. Gu, J. Dominic, J.-S. Lai, C.-L. Chen, T. LaBella, B. Chen, High reliability and efficiency single-phase transformerless inverter for grid-connected photovoltaic systems, IEEE Trans. Power Electron 28 (5) (2013) 2235–2245.
[13] S.V. Araujo, P. Zacharias, R. Mallwitz, Highly efficient single-phase transformerless inverters for grid-connected photovoltaic systems, IEEE Trans. Ind. Electron. 57 (9) (2010) 3118–3128.
[14] J. Lakwal, M. Dubey, Modeling and simulation of a novel multilevel inverter for PV systems using unequal DC sources, in: 2014 IEEE International Conference on Advanced Communications, Control and Computing Technologies, 2014, pp. 296–300.
[15] C.-M. Wang, A novel single-stage full-bridge buck-boost inverter, IEEE Trans. Power Electron. 19 (1) (2004) 150–159.
[16] L. Chen, A. Amirahmadi, Q. Zhang, N. Kutkut, I. Batarseh, Design and implementation of three-phase two-stage grid-connected module integrated converter, IEEE Trans. Power Electron 29 (8) (2014) 3881–3892.
[17] R.A. Mastromauro, M. Liserre, T. Kerekes, A. Dell'Aquila, A single-phase voltage-controlled grid-connected photovoltaic system with power quality conditioner functionality, IEEE Trans. Ind. Electron. 56 (11) (2009) 4436–4444.
[18] Products—Overview | SMA solar (Online). Available from: <https://www.sma.de/en/products/overview.html> (accessed 09.10.18).
[19] ABB, (Online). Available from: <http://www.abb.com/product/seitp322/df4308429896b1a3c1257c8a0054355c.aspx>, 2013.
[20] L. Feitknecht, 100-days electricity from the sun. 2013, (Online). Available from: <https://www.damproject.org/book/773680936/download-100-days-luc-feitknecht.pdf> (accessed: 22.12.15).
[21] T. Kerekes, R. Teodorescu, M. Liserre, Common mode voltage in case of transformerless PV inverters connected to the grid, in: 2008 IEEE International Symposium on Industrial Electronics, 2008, pp. 2390–2395.
[22] A. Yafaoui, B. Wu, S. Kouro, Improved active frequency drift anti-islanding detection method for grid connected photovoltaic systems, IEEE Trans. Power Electron. 27 (5) (2012) 2367–2375.
[23] IEEE 1547-2003—IEEE Standard for Interconnecting Distributed Resources with Electric Power Systems, (Online). Available from: <https://standards.ieee.org/standard/1547-2003.html> (accessed 09.10.18).
[24] First solar Sarnia PV power plant, Ontario, Canada, (Online). Available from: <http://www.solaripedia.com/13/303/3431/sarnia_solar_farm_photovoltaics.html> (accessed 19.12.13).
[25] S. Kouro, B. Wu, H. Abu-Rub, F. Blaabjerg, Photovoltaic energy conversion systems, Power Electronics for Renewable Energy Systems, Transportation and Industrial Applications, John Wiley & Sons, Ltd, Chichester, 2014, pp. 160–198.
[26] F. Blaabjerg, Z. Chen, S.B. Kjaer, Power electronics as efficient interface in dispersed power generation systems, IEEE Trans. Power Electron. 19 (5) (2004) 1184–1194.
[27] R. Carbone, A. Tomaselli, Recent advances on AC PV-modules for grid-connected PV plants, in: Proc. Int. Conf. Clean Electrical Power (ICCEP 2011), 2011, pp. 124–129.
[28] M. Meinhardt, G. Cramer, Multi-string-converter: the next step in evolution of string converter technology, in: 9th European Power Electronics and Applications Conf. (EPE 2001), 2001, pp. 1–9.
[29] S. Kouro, J.I. Leon, D. Vinnikov, L.G. Franquelo, Grid-connected photovoltaic systems: an overview of recent research and emerging PV converter technology, IEEE Ind. Electron. Mag 9 (1) (2015) 47–61.
[30] A.M. Pavan, S. Castellan, S. Quaia, S. Roitti, G. Sulligoi, Power electronic conditioning systems for industrial photovoltaic fields: centralized or string inverters? in: 2007 International Conference on Clean Electrical Power, 2007, pp. 208–214.
[31] B. Burger, D. Kranzer, Extreme high efficiency PV-power converters, in: 2009 13th European Conference on Power Electronics and Applications, 2009, pp. 1–13.
[32] J.R. Dreher, F. Marangoni, L. Schuch, M.L. da S. Martins, L. Della Flora, Comparison of H-bridge single-phase transformerless PV string inverters, in: 2012 10th IEEE/IAS International Conference on Industry Applications, 2012, pp. 1–8.

[33] S.B. Kjaer, J.K. Pedersen, F. Blaabjerg, A review of single-phase grid-connected inverters for photovoltaic modules, IEEE Trans. Ind. Appl. 41 (5) (2005) 1292–1306.
[34] R. Gonzalez, J. Coloma, L. Marroyo, J. Lopez, P. Sanchis, Single-phase inverter circuit for conditioning and converting DC electrical energy into AC electrical energy, Eur. Patent EP 2 053 730 A1, 2009.
[35] S. Saridakis, E. Koutroulis, F. Blaabjerg, Optimal design of modern transformerless PV inverter topologies, IEEE Trans. Energy Convers. 28 (2) (2013) 394–404.
[36] D.W. Karraker, K.P. Gokhale, M.T. Jussila, Inverter for solar cell array, U.S. patent 2011/0299312 A1, 2011.
[37] T. Urakabe, K. Fujiwara, T. Kawakami, N. Nishio, High efficiency power conditioner for photovoltaic power generation system, in: The 2010 International Power Electronics Conference (ECCE ASIA), 2010, pp. 3236–3240.
[38] M. Fornage, Method and apparatus for converting direct current to alternating current, U.S. patent 7,796,412B2, 2010.
[39] P. Garrity, Solar PV power conditioning unit, U.S. patent 8,391,031B2, 2013.
[40] D. Casadei, G. Grandi, C. Rossi, Single-phase single-stage photovoltaic generation system based on a ripple correlation control maximum power point tracking, IEEE Trans. Energy Convers. 21 (2) (2006) 562–568.
[41] C.-L. Kuo, C.-H. Lin, H.-T. Yau, J.-L. Chen, Using self-synchronization error dynamics formulation based controller for maximum photovoltaic power tracking in micro-grid systems, IEEE J. Emerg. Sel. Top. Circuits Syst. 3 (3) (2013) 459–467.
[42] B. Bendib, H. Belmili, F. Krim, A survey of the most used MPPT methods: conventional and advanced algorithms applied for photovoltaic systems, Renew. Sustain. Energy Rev. 45 (2015) 637–648.
[43] K.L. Lian, J.H. Jhang, I.S. Tian, A maximum power point tracking method based on perturb-and-observe combined with particle swarm optimization, IEEE J. Photovoltaics 4 (2) (2014) 626–633.
[44] H. Ghoddami, A. Yazdani, A single-stage three-phase photovoltaic system with enhanced maximum power point tracking capability and increased power rating, IEEE Trans. Power Deliv. 26 (2) (2011) 1017–1029.
[45] E. Koutroulis, F. Blaabjerg, A new technique for tracking the global maximum power point of PV arrays operating under partial-shading conditions, IEEE J. Photovolt. 2 (2) (2012) 184–190.
[46] M. Uoya, H. Koizumi, A calculation method of photovoltaic array's operating point for MPPT evaluation based on one-dimensional Newton–Raphson method, IEEE Trans. Ind. Appl. 51 (1) (2015) 567–575.
[47] L. Zhou, Y. Chen, K. Guo, F. Jia, New approach for MPPT control of photovoltaic system with mutative-scale dual-carrier chaotic search, IEEE Trans. Power Electron. 26 (4) (2011) 1038–1048.
[48] E. Romero-Cadaval, G. Spagnuolo, L.G. Franquelo, C.A. Ramos-Paja, T. Suntio, W.M. Xiao, Grid-connected photovoltaic generation plants: components and operation, IEEE Ind. Electron. Mag 7 (3) (2013) 6–20.
[49] S. Jain, V. Agarwal, Comparison of the performance of maximum power point tracking schemes applied to single-stage grid-connected photovoltaic systems, IET Electr. Power Appl. 1 (5) (2007) 753.
[50] B. Subudhi, R. Pradhan, A comparative study on maximum power point tracking techniques for photovoltaic power systems, IEEE Trans. Sustain. Energy 4 (1) (2013) 89–98.
[51] A. Barrado Bautista, A. Lázaro Blanco, Power Electronics Problems, Pearson Prentice Hall, 2007.
[52] E.J. Bueno Peña, Optimization of the behavior of a three-level NPC converter connected to the electricity grid (Ph.D. thesis), University of Alcalá, 2005.
[53] S. Buso, L. Malesani, P. Mattavelli, A. Member, Comparison of current control techniques for active filter applications, IEEE Trans. Ind. Electron 45 (5) (1998) 722–729.
[54] M. Ciobotaru, R. Teodorescu, F. Blaabjerg, Improved PLL structures for single-phase grid inverters, in: Proc. Power Electron. Intell. Control Energy Conserv. Conf., February, 2005, pp. 1–6.
[55] M.P. Kazmierkowski, L. Malesani, Current control techniques for three-phase voltage-source PWM converters: a survey, IEEE Trans. Ind. Electron. 45 (5) (1998) 691–703.
[56] M. Ciobotaru, R. Teodorescu, F. Blaabjerg, Control of single-stage single-phase PV inverter, in: 2005 European Conference on Power Electronics and Applications, 2005, p. 10
[57] S. Fukuda, T. Yoda, A novel current-tracking method for active filters based on a sinusoidal internal model [for PWM invertors], IEEE Trans. Ind. Appl. 37 (3) (2001) 888–895.
[58] D.N. Zmood, D.G. Holmes, Stationary frame current regulation of PWM inverters with zero steady-state error, IEEE Trans. Power Electron. 18 (3) (2003) 814–822.

[59] F. Blaabjerg, R. Teodorescu, M. Liserre, A.V. Timbus, Overview of control and grid synchronization for distributed power generation systems, IEEE Trans. Ind. Electron. 53 (5) (2006) 1398–1409.
[60] I. Agirman, V. Blasko, A novel control method of a VCS without AC line voltage sensors, IEEE Trans. Ind. Appl. 39 (2) (2003) 519–524.
[61] H. Zhu, B. Arnet, L. Haines, E. Shaffer, J.-S. Lai, Grid synchronization control without AC voltage sensors, in: Eighteenth Annual IEEE Applied Power Electronics Conference and Exposition, 2003 (APEC '03), vol. 1, pp. 172–178.
[62] L. Zhang, K. Sun, Y. Xing, L. Feng, H. Ge, A modular grid-connected photovoltaic generation system based on DC bus, IEEE Trans. Power Electron. 26 (2) (2011) 523–531.
[63] Z. Zeng, H. Yang, R. Zhao, C. Cheng, Topologies and control strategies of multi-functional grid-connected inverters for power quality enhancement: a comprehensive review, Renew. Sustain. Energy Rev. 24 (2013) 223–270.
[64] M. Monfared, S. Golestan, Control strategies for single-phase grid integration of small-scale renewable energy sources: a review, Renew. Sustain. Energy Rev. 16 (7) (2012) 4982–4993.
[65] R. Teodorescu, F. Blaabjerg, Overview of renewable energy system, in: ECPE Seminar Renewable Energy, ISET, Kassel, Germany, 2006.
[66] R. Teodorescu, F. Blaabjerg, U. Borup, M. Liserre, A new control structure for grid-connected LCL PV inverters with zero steady-state error and selective harmonic compensation, in: Nineteenth Annual IEEE Applied Power Electronics Conference and Exposition, 2004 (APEC '04), vol. 1, pp. 580–586.
[67] R. Teodorescu, F. Blaabjerg, M. Liserre, P.C. Loh, Proportional-resonant controllers and filters for grid-connected voltage-source converters, IEE Proc. Electr. Power Appl. 153 (5) (2006) 750.
[68] J.-Y. Chen, C.-H. Hung, J. Gilmore, J. Roesch, W. Zhu, LCOE reduction for megawatts PV system using efficient 500 kW transformerless inverter, in: 2010 IEEE Energy Conversion Congress and Exposition, 2010, pp. 392–397.
[69] J. Svensson, Synchronisation methods for grid-connected voltage source converters, IEE Proc. Gen. Transm. Distrib. 148 (3) (2001) 229.
[70] Y.C. Qin, N. Mohan, R. West, R. Bonn, Status and needs of power electronics for photovoltaic inverters, Technical Report, DE2002-800985; SAND2002-1535, 2002, (Online). Available from: <http://adsabs.harvard.edu/abs/2002STIN...0301912Q> (accessed 09.10.15).
[71] L. Hassaine, Implementation of a digital control of active and reactive power for investors. application to photovoltaic systems connected to the network, Int. J. Mechatron. Electr. Comput. Technol. 4 (12) (2010) 857–885.
[72] A. Pregelj, M. Begovic, A. Rohatgi, Impact of inverter configuration on PV system reliability and energy production, in: Conference Record of the Twenty-Ninth IEEE Photovoltaic Specialists Conference, 2002, pp. 1388–1391.
[73] S.-J. Lee, H. Kim, S.-K. Sul, F. Blaabjerg, A novel control algorithm for static series compensators by use of PQR instantaneous power theory, IEEE Trans. Power Electron. 19 (3) (2004) 814–827.
[74] G.-C. Hsieh, J.C. Hung, Phase-locked loop techniques. A survey, IEEE Trans. Ind. Electron. 43 (6) (1996) 609–615.
[75] A. Timbus, M. Liserre, R. Teodorescu, F. Blaabjerg, Synchronization methods for three phase distributed power generation systems. An overview and evaluation, in: IEEE 36th Conference on Power Electronics Specialists, 2005, pp. 2474–2481.
[76] L. Hassaine, E. Olias, J. Quintero, M. Haddadi, Digital power factor control and reactive power regulation for grid-connected photovoltaic inverter, Renew. Energy 34 (1) (2009) 315–321.
[77] V. Salas, E. Olías, A. Lázaro, A. Barrado, New algorithm using only one variable measurement applied to a maximum power point tracker, Sol. Energy Mater. Sol. Cells 87 (1–4) (2005) 675–684.
[78] M.A. Eltawil, Z. Zhao, MPPT techniques for photovoltaic applications, Renew. Sustain. Energy Rev. 25 (2013) 793–813.
[79] G. Tsengenes, G. Adamidis, Investigation of the behavior of a three phase grid-connected photovoltaic system to control active and reactive power, Electr. Power Syst. Res. 81 (1) (2011) 177–184.
[80] K. Turitsyn, P. Sulc, S. Backhaus, M. Chertkov, Options for control of reactive power by distributed photovoltaic generators, Proc. IEEE 99 (6) (2011) 1063–1073.

[81] M. Oshiro, K. Tanaka, T. Senjyu, S. Toma, A. Yona, A.Y. Saber, et al., Optimal voltage control in distribution systems using PV generators, Int. J. Electr. Power Energy Syst 33 (3) (2011) 485–492.

[82] European Photovoltaic Industry Association (EPIA), Annual Report 2008, (Online). Available from: <https://www.google.com/url?sa = t&rct = j&q = &esrc = s&source = web&cd = 1&cad = rja&uact = 8&ved = 2ahUKEwjGlI7ti_bhAhURvlkKHc4-AmwQFjAAegQIBBAC&url = http%3A%2F%2Fwww.iea-pvps.org%2Findex.php%3Fid%3D6%26eID%3Ddam_frontend_push%26docID%3D39&usg = AOvVaw1OE244V89RaOHt0l4Q_7BT> (accessed: 03.12.14).

[83] S.B. Kjær, F. Blaabjerg, A novel single-stage inverter for the AC-module with reduced low-frequency ripple penetration, in: Proceedings of EPE'2003, Toulouse, France, 2–4 September 2003, EPE Association, 2003, p. 10.

[84] M.S. ElNozahy, M.M.A. Salama, Technical impacts of grid-connected photovoltaic systems on electrical networks—a review, J. Renew. Sustain. Energy 5 (3) (2013) 032702.

[85] A. Cagnano, F. Torelli, F. Alfonzetti, E. De Tuglie, Can PV plants provide a reactive power ancillary service? A treat offered by an on-line controller, Renew. Energy 36 (3) (2011) 1047–1052.

[86] R. Teodorescu, F. Blaabjerg, Flexible control of small wind turbines with grid failure detection operating in stand-alone and grid-connected mode, IEEE Trans. Power Electron. 19 (5) (2004) 1323–1332.

[87] E. Twining, D.G. Holmes, Grid current regulation of a three-phase voltage source inverter with an LCL input filter, IEEE Trans. Power Electron. 18 (3) (2003) 888–895.

[88] M.P. Kaźmierkowski, R. (Ramu) Krishnan, F. Blaabjerg, Control in Power Electronics: Selected Problems., Academic Press, 2002.

[89] M.G. Molina, P.E. Mercado, Modeling and control of grid-connected photovoltaic energy conversion system used as a dispersed generator, in: 2008 IEEE/PES Transmission and Distribution Conference and Exposition: Latin America, 2008, pp. 1–8.

[90] R. Gonzalez, E. Gubia, J. Lopez, L. Marroyo, Transformerless single-phase multilevel-based photovoltaic inverter, IEEE Trans. Ind. Electron. 55 (7) (2008) 2694–2702.

[91] N.A. Rahim, J. Selvaraj, C. Krismadinata, Five-level inverter with dual reference modulation technique for grid-connected PV system, Renew. Energy 35 (3) (2010) 712–720.

[92] S. Jain, V. Agarwal, New current control based MPPT technique for single stage grid connected PV systems, Energy Convers. Manag. 48 (2) (2007) 625–644.

[93] N. Hamrouni, M. Jraidi, A. Chérif, New control strategy for 2-stage grid-connected photovoltaic power system, Renew. Energy 33 (10) (2008) 2212–2221.

[94] T. Kerekes, R. Teodorescu, M. Liserre, C. Klumpner, M. Sumner, Evaluation of three-phase transformerless photovoltaic inverter topologies, IEEE Trans. Power Electron. 24 (9) (2009) 2202–2211.

[95] J.-M. Kwon, K.-H. Nam, B.-H. Kwon, Photovoltaic power conditioning system with line connection, IEEE Trans. Ind. Electron. 53 (4) (2006) 1048–1054.

[96] J. Alonso-Martínez, J. Eloy-García, S. Arnaltes, Direct power control of grid connected PV systems with three level NPC inverter, Sol. Energy 84 (7) (2010) 1175–1186.

[97] S. Mehta, J. Chiasson, Nonlinear control of a series DC motor: theory and experiment, IEEE Trans. Ind. Electron. 45 (1) (1998) 134–141.

[98] L. Yacoubi, K. Al-Haddad, F. Fnaiech, L.-A. Dessaint, A DSP-based implementation of a new nonlinear control for a three-phase neutral point clamped boost rectifier prototype, IEEE Trans. Ind. Electron. 52 (1) (2005) 197–205.

[99] A. Ovono Zue, A. Chandra, State feedback linearization control of a grid connected photovoltaic interface with MPPT, in: 2009 IEEE Electrical Power & Energy Conference (EPEC), 2009, pp. 1–6.

[100] D. Lalili, A. Mellit, N. Lourci, B. Medjahed, E.M. Berkouk, Input output feedback linearization control and variable step size MPPT algorithm of a grid-connected photovoltaic inverter, Renew. Energy 36 (12) (2011) 3282–3291.

CHAPTER

15

Safety and reliability evaluation for electric vehicles in modern power system networks

Foad H. Gandoman[1,2], Joeri Van Mierlo[1,2], Abdollah Ahmadi[3], Shady H.E. Abdel Aleem[4] and Kalpana Chauhan[5]

[1]Research Group MOBI—Mobility, Logistics, and Automotive Technology Research Centre, Vrije Universiteit Brussel, Brussels, Belgium [2]Flanders Make, Heverlee, Belgium [3]The Australian Energy Research Institute and the School of Electrical Engineering and Telecommunications, the University of New South Wales, Sydney, NSW, Australia [4]Department of Mathematical, Physical and Engineering Sciences, 15th of May Higher Institute of Engineering, Helwan, Cairo, Egypt [5]Department of Electrical and Electronics Engineering, Galgotias College of Engineering and Technology, Greater Noida, India

15.1 Introduction

With the changes that have taken place in the structure of electrical distribution networks, various concepts too have changed. Distributed generation (DG) sources, such as solar photovoltaic (PV) cells and wind turbines, have played a significant role in these changes. Also, to tackle the problem of increasing global pollution, original equipment manufacturers (OEMs) have introduced hybrid vehicles and full electric vehicles (EVs) [1,2]. Thus, there has been growing enthusiasm for conventional combustion vehicles to be replaced with EVs in order to reduce the impact of these problems. In this regard, new issues and challenges have arisen in the field of modern electrical networks, such as power quality evaluation, reliability, safety, and monitoring.

Reliability and safety evaluation in modern electrical networks can also be studied in internal DG systems or larger dimensions (smart grids or microgrids). Therefore, reliability

Distributed Energy Resources in Microgrids
DOI: https://doi.org/10.1016/B978-0-12-817774-7.00015-6

assessments of the electrical components for EVs, such as the battery pack, power electronic converters, and electric motor are vital in order to develop approaches that will perform well during the lifetime of the system [3,4].

Researchers have been exploring new methods for evaluating the reliability of EVs' performance in modern power systems. Research has also addressed issues of reliability and safety in the smart grid [5,6] and the internal electric system of EVs [7,8]. Vehicle-to-grid (V2G) and grid-to-vehicle (G2V) strategies are often cited as promising to improve the reliability of electric power grids. However, the impact of the degradation of the EVs' electrical components has not been investigated in detail. This chapter aims to explain the concepts of reliability and safety evaluation for EVs' electrical components.

This chapter studies and discusses the concept of reliability and safety assessment for the main electrical components of EVs through the following issues:

- Assessment of the reliability of techniques in modern electrical networks.
- Study of the different types of failures in electrical components of EVs.
- The role of the reliability and safety concepts in the present and future EVs' applications.

The chapter is organized as follows: Section 15.2 presents issues related to safety and reliability in the modern electrical network by focusing on EVs. The EV trend is described in Section 15.3, while an evaluation of the reliability and safety of electrical components for EVs is explained in Section 15.4. Section 15.5 explains the reliability evaluation and challenges facing the EV sector and, finally, conclusions are drawn in Section 15.6.

15.2 Safety and reliability in the modern electrical network by focusing on electric vehicles

Typically, reliability issues fall within two particular classes, (1): power quality disturbance, and (2) short-term shutdown and long-term shutdown (blackouts). Also, There are three main issues which are used to assess reliability in electrical systems [5]:

1. The number of components (consumers) influenced,
2. The recurrence whenever blackouts or voltage disturbances occur, and
3. The duration of failures.

Correspondingly, the reliability of electrical components is guaged in terms of how often possible reliability issues occur (failures and duration) to determine how many customers or components are influenced. Fig. 15.1. shows the area related to reliability performance in a power system.

The influenced zones appear in Fig. 15.2, which indicates three particular, various hierarchical levels I (HL I, II and III) and the sufficiency appraisal which can be conducted at each of these levels [6]. At HL I, the sufficiency of the creating system to load necessities is inspected. Both generation and transmission facilities are considered in HL II. This action is sometimes introduced as a composite system adequacy assessment. The HL III adequacy evaluation includes all these practical zones in order to assess customer load point indices.

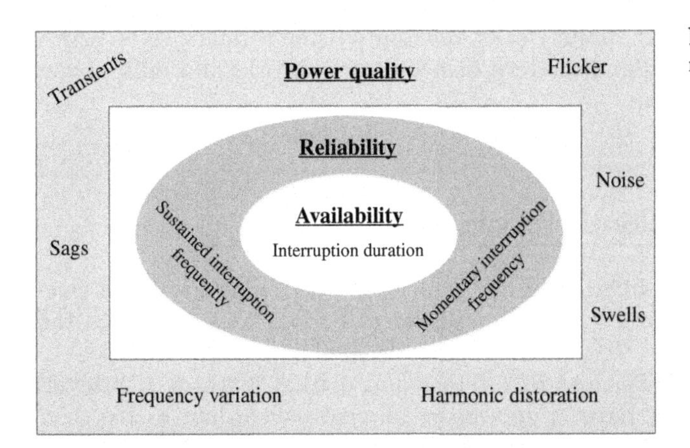

FIGURE 15.1 Reliability area performance in power system.

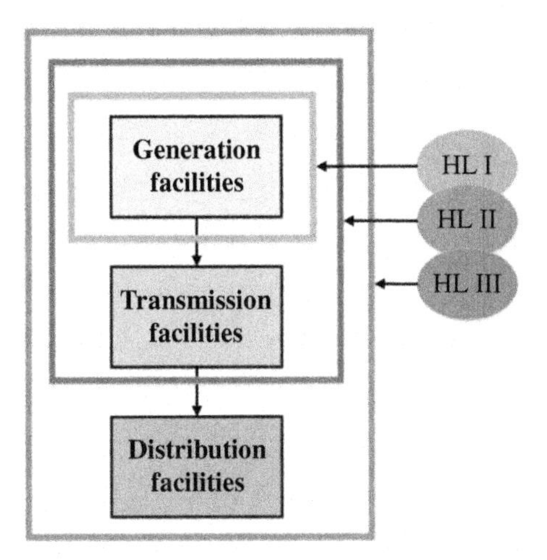

FIGURE 15.2 Hierarchical levels of electrical system reliability.

15.3 Reliability assessment in modern electrical networks

Considering the use of smart grids and increasing the number of devices used in networks, reliability assessment becomes even more relevant in such networks. In the past, the assessment of reliability in traditional networks was previously considered by investigating the uncertainty of power plants. In today's systems, the complexity of the reliability assessment is increased with the sensitivity of electrical devices and the possibility of generating electricity by using DG sources in the distribution network [7].

The aims of the modern electrical network are enhanced customer service, and improved reliability, monitoring, and control of the distribution system. Thus, in order to increase the reliability of modern electrical networks, the reliability of distributed dispersed sources of sensitive consumer and modern consumers (such as EVs) are of particular importance. In modern electrical networks, reliability is the primary layer which plays a vital role between a smart grid and modern distribution networks and bulk power systems.

15.4 Electric vehicle trend

There are three main types of EVs: hybrid electric vehicles (HEVs), plug-in hybrid electric vehicles (PHEVs) and battery electric vehicles (BEVs) [8,9]. Fig. 15.3 illustrates the structure of EVs.

Fig. 15.4 shows an overview of PHEVs and BEVs stock and market shares in the developed world where these technologies have largely been adopted. According to Fig. 15.1 [10–12], the energy density of batteries has increased, and the prices of EVs have dropped since 2010. By 2040, it is predicted that 20 million EVs will be on the road in China, and while in the United States and Europe (United Kingdom, Germany, and France) these

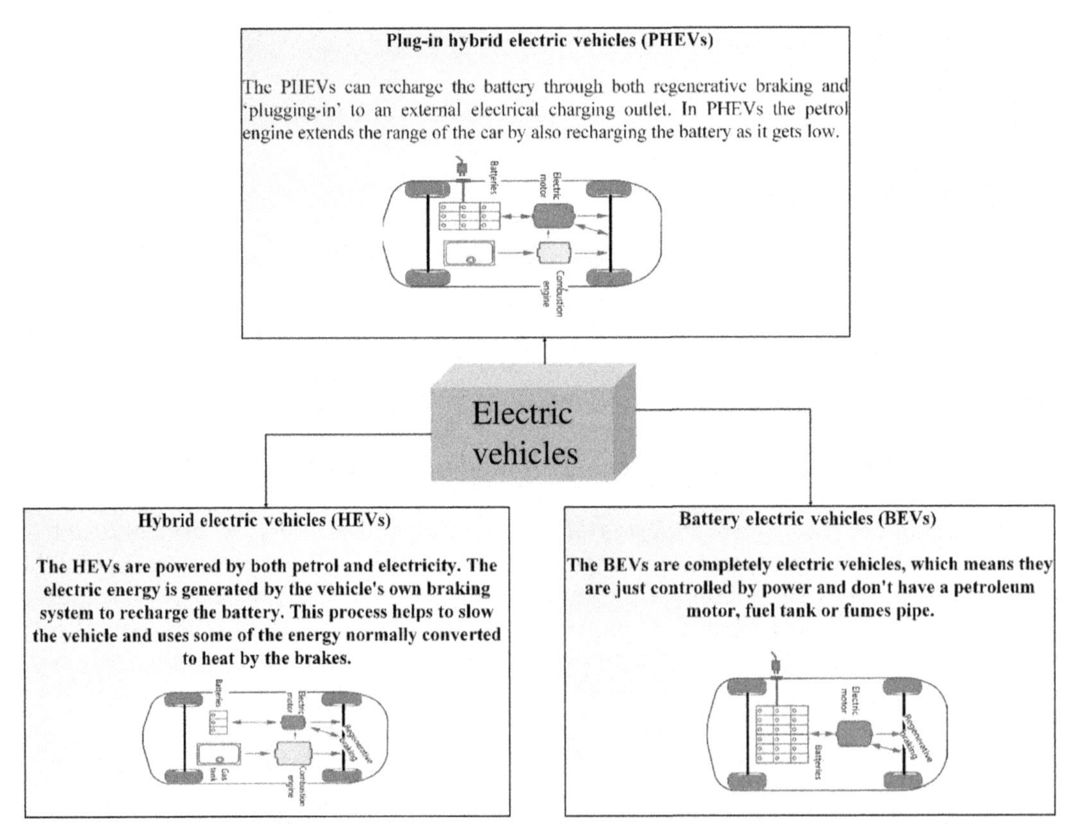

FIGURE 15.3 Electric vehicles (EVs) structure.

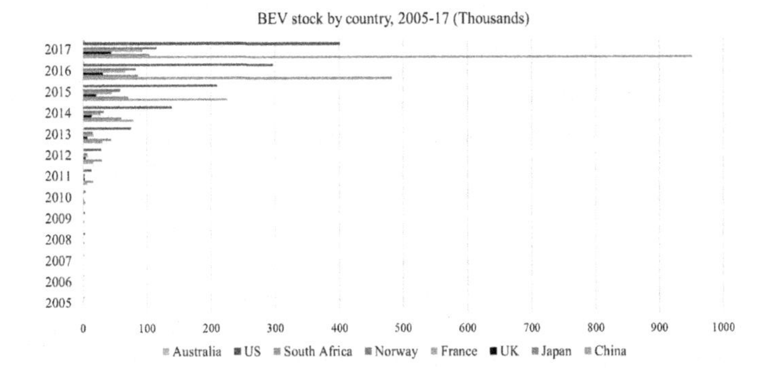

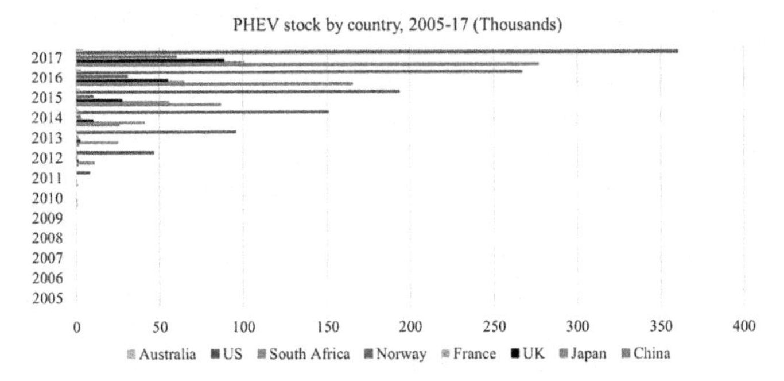

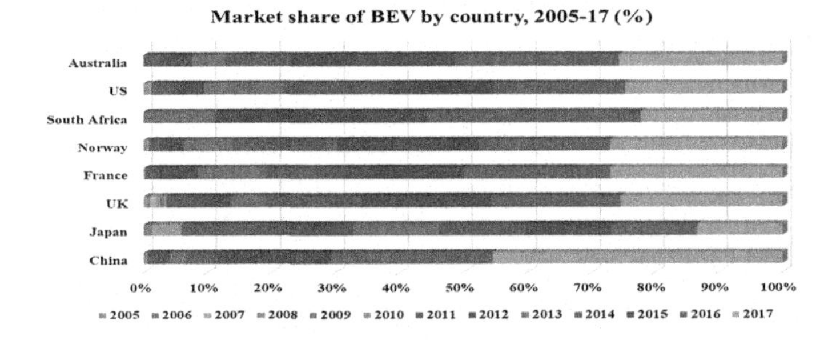

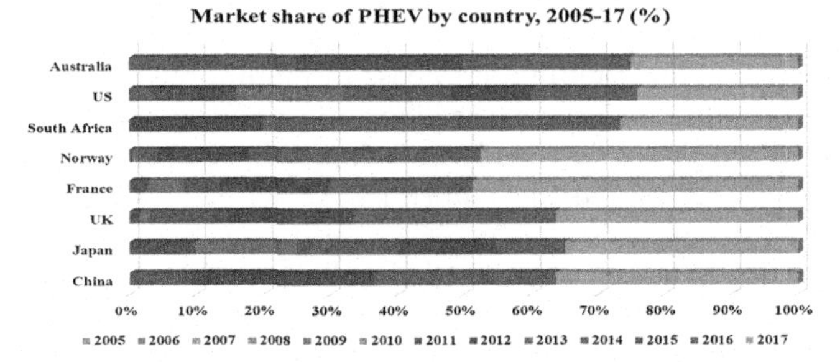

FIGURE 15.4 Overview of battery electric vehicles (BEVs) and plug-in hybrid electric vehicles (PHEVs) stock and market shares in the world's developed countries.

rates are expected to be as high as 10 million and more than 10 million, respectively. In total the number of EV sales in China, Germany, and the United States could be more than 50 million by 2040 [12].

15.5 Reliability and safety evacuation of electrical components for electric vehicles

The main electrical components of EV are electric motors, power electronic converters, and battery packs. In this section, the causes and consequences of the various failures that occur in the electrical parts of EVs will be studied.

15.5.1 Lithium-ion battery pack

The important aspects of reliability and safety concepts in the lithium-ion (Li-ion) battery pack are (1) chemical issues, (2) thermal issues, (3) mechanical issues, and (4) electrical issues.

15.5.1.1 Chemical issues

Failures from a chemical perspective can occur during the production and the lifetime process of the Li-ion battery. Fig. 15.5 shows the experimental results (voltage and capacity) of the lifetime of NMC Li-ion cell after 25×100 cycle in 25°C with state of charge (SoC) 50% and depth of discharge (DoD) 90%.

Many physical and chemical processes contribute to battery degradation. These mechanisms depend on many factors such as the battery chemistry construction, its method of operating, and so on. The complexity and interdependence of the processes involved make them very difficult to model, thus predicting future degradation is very challenging for Li-ion battery failures. Fig. 15.6 shows the main chemical failures in the different parts of the Li-ion battery cell.

- *Solid Electrolyte Interphase (SEI) formation growing*:
  SEI had a mosaic structure and is composed of different degradation products, such as lithium oxide and lithium fluoride. SEI serves as a protection on the graphite electrode, which enables the lithium-ion to pass through it but not the electrons, avoiding further degradation of the electrolyte and providing good efficiency of the battery. Thus, if the growth of this inside layer increases too much, it impedes the battery and ultimately reduces its lifespan [13].
- *Lithium plating*:
  If the charging current is very high in a Li-ion battery, the transport rate of $Li+$ to the negative graphite electrode exceeds the rate that the $Li+$ can be intercalated into the graphite/anode. Under this condition, $Li+$ may be deposited as metallic Li, which can degrade the battery's life [14]. Additionally, the formation of lithium dendrites during lithium plating can result in a short circuit, causing instant battery failure, and tearing of the separator [15]. Fig. 15.7 shows lithium plating formation during the lifetime of a Li-ion cell.

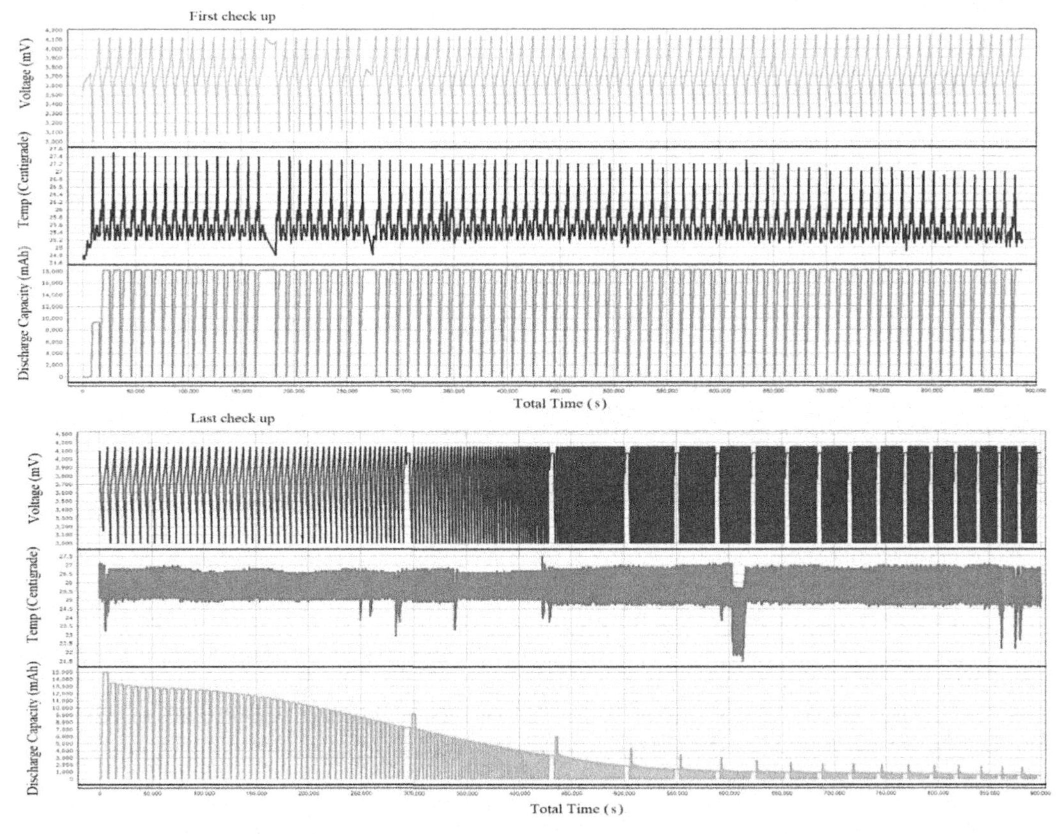

FIGURE 15.5 The experimental results (voltage and capacity and temperature) of the lifetime of a Li-ion cell after 25 × 100 cycle in 25°C with state of charge (SoC) 50% and depth of discharge (DoD) 90%.

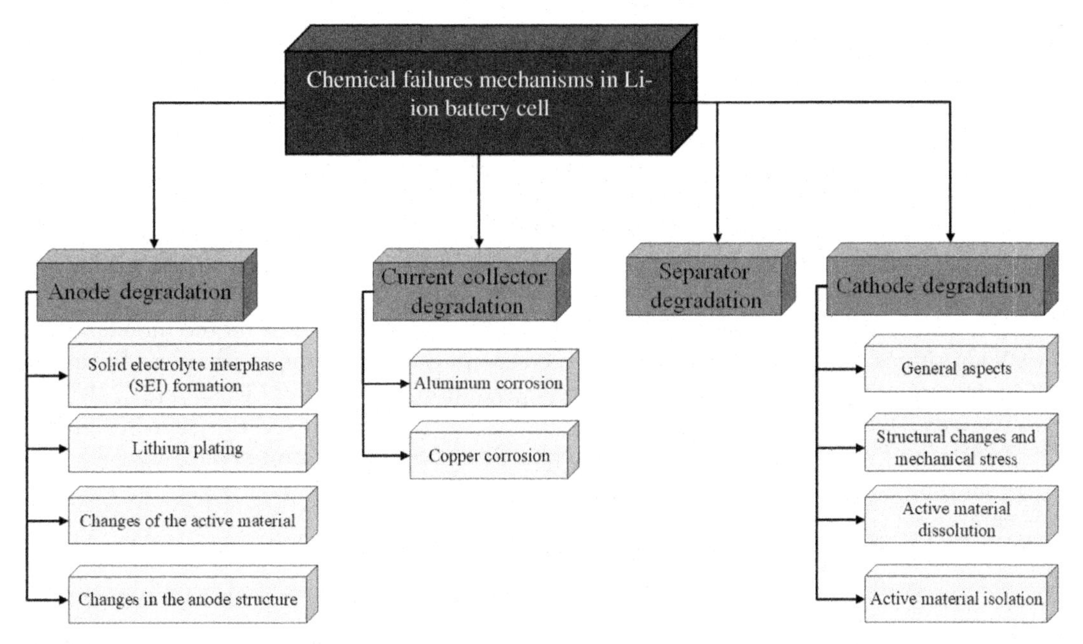

FIGURE 15.6 The classification of chemical failures mechanisms in a Li-ion battery cell.

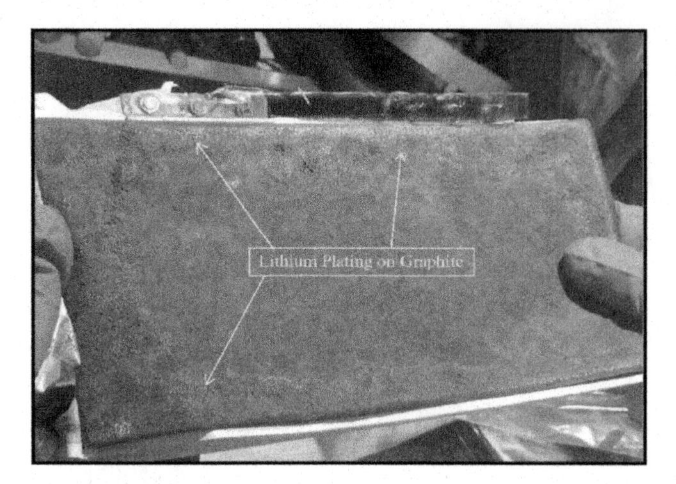

FIGURE 15.7 Lithium plating on graphite (anode).

- *Active material changing*:
  Generally, aging effects occur inside the anode, which is an active material as a result of the insertion and extraction of lithium ions (less than 10%) [16]. This action may lead to mechanical stresses and result in structural damages and cracking of carbon. Also, carbon particle cracking and carbon exfoliation caused by the cointercalation of solvents, electrolyte reduction within carbon, and gas evolution within carbon will contribute to the accelerated electrode degradation [17].
- *Changes in the anode structure*:
  Contact loss regarding electrical and mechanical issues in the anode side of Li-ion cells, such as carbon particles, carbon and the current collector, carbon and binder, and the current collector and binder, results in higher cell impedance [18]. Moreover, the contact loss allows electrolytes to penetrate the bulk of the electrode, which changes the porosity features and improves anode performance. Thus, the penetration of electrolytes to anode formation can cause internal cell pressure which contributes to the degradation of the mechanical electrode properties during the lifetime (long run) of a Li-ion battery cell [19].
- *Structural changes and mechanical stress*:
  There are different issues, which may cause failures from the cathode side of the Li-ion battery cell. The structural changes in the Li-ion cell has a greater effect on the cathode more than on the anode [20]. In another words, during the lithiation process in the crystal lattice distortion, some of the cathode oxides undergo a phase change.
- *Active material dissolution*:
  The outcome of the negative behavior in the Li-ion battery based on the manganese cathode side is a loss of capacity and an increase in the rate of the inside impedance. The process of active material dissolution degradation is shown in Fig. 15.8 [21,22].
- *Active material isolation*:
  Isolation of the active particles is known as one of the major causes of Li-ion degradation from the cathode side [23]. There are three disadvantage regarding such isolation [24] (1) crack formation in the active materials and probably fragmentation, (2) the breaking of the binder, and (3) binder adhesion property deterioration.

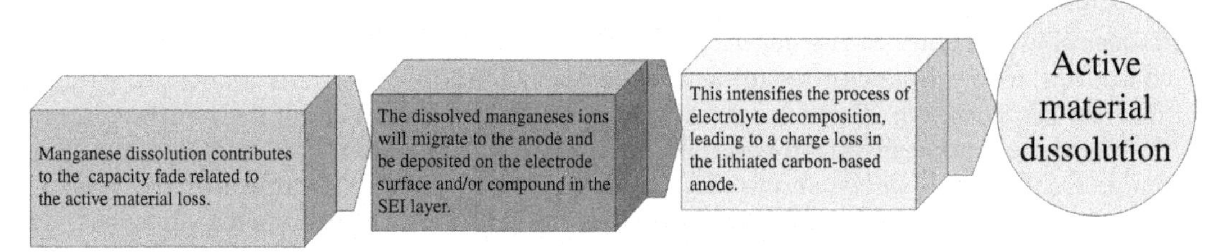

FIGURE 15.8 The process of active material dissolution degradation.

- *Separator failures*:

  Separator failures can lead to a reduction of power in a Li-ion battery. The principle reason regarding separator failures are the lithium dendrite growth caused by the separator pores, an attack through the electrolyte, the blocking of passageways in the separator during cycling (shutdown separators), and structural degradation arising from high temperatures or a high cycle number [25,26].
- *Current collector failures*:

  The corrosion of an aluminum current collector in Li-ion batteries is another possible factor that can affect the long-term performance and safety of Li-ion batteries [27,28]. The potential window of copper as a current collector is around 1 V. If the rate of the voltage on the anode side rises from that point, copper will oxidize and dissolve within the electrolyte. The dissolved copper during the recharging process contributes by making a plate from the anode side. The plate via oxidized copper plays a capacity fade role and also contributes to its catastrophic failure by causing an internal short circuit in the Li-ion battery [29,30].

15.5.1.2 Thermal issues

The main factors which contribute to the thermal runaway in Li-ion battery packs are poor design and integration, poor safety monitoring, weak manufacturing, and poor handling (packing conditions) [31,32]. Moreover, the factor which most commonly causes degradation based on the temperature of the Li-ion battery is a hot temperature. The most important degradation based on the low-temperature effect on the Li-ion battery is the formation of lithium grains, intercalation gradients (with cycling), and lithium plating. The primary failures in a Li-ion battery from a thermal point of view are:

1. *Intercalation gradients (with cycling)*: Fast-charging the Li-ion cells at a low temperature induces a mechanical strain on the graphite. Crack formation, fissures, and splits in the graphite particles are some of the results of this mechanical strain [33–35].
2. *Electrolyte decomposition*: Thermal impact is a factor which influences electrolyte decomposition. Common solvents such as ethylene carbonate and ethyl methyl carbonate or various mixtures of both, result in reduced onset temperatures and increased heat release with increased concentrations of $LiPF_6$ [36].
3. *Oxidation of electrolytes*: Temperature plays an important role in electrolyte oxidation. Electrolyte oxidation on the surface of the cathode is a typical problem for lithium-ion batteries under normal operating conditions, and especially in high-temperature applications of the battery [37].

4. *Formation of lithium grains*: The formation of lithium grains is one of the most common undesirable reactions on the anode side of a Li-ion battery in low-temperature conditions, as it influences the aging of the battery [38,39].
5. *Electrode decomposition*: There are many items that increase the rate of the temperature of the electrode (anode side) and cause of electrode decomposition. Critical parameters for the thermal stability of the anode side are the graphite material's properties, the rate of temperature of Li + in the graphite structure, the stability of the SEI layer, and electrolyte composition [40–42].

15.5.1.3 Mechanical issues

A study of the failure of Li-ion batteries from a mechanical viewpoint can be divided into two main categories: (1) external causes and (2) internal causes. The most important external mechanical cause of failure in a Li-ion battery is a crash. Also, the internal mechanical degradation in Li-ion cells is cyclic mechanical stress and a rupture of the cell casing.

- *Rupture of the cell casing*: The failure is a result of a slow deterioration of the electrolyte each time it undergoes heat-cycling, which can result in a swelling of the cell and eventually a rupture of the cell casing [43,44].
- *Cyclic mechanical stress*: Due to the charging and discharging process, the movement of the lithium ions in and out of the crystal structure of the electrodes causes stresses, which can lead to cracking of the electrode particles. The normal reaction is increased internal impedance in the cell, and a failure of the anode SEI layer which can lead to overheating and cell failure [45,46].
- *Crash*: One of the mechanical failures in EVs is caused by crashes. Fig. 15.9 shows the most important issues, from a safety perspective, which may happen after a crash in the Li-ion battery pack.

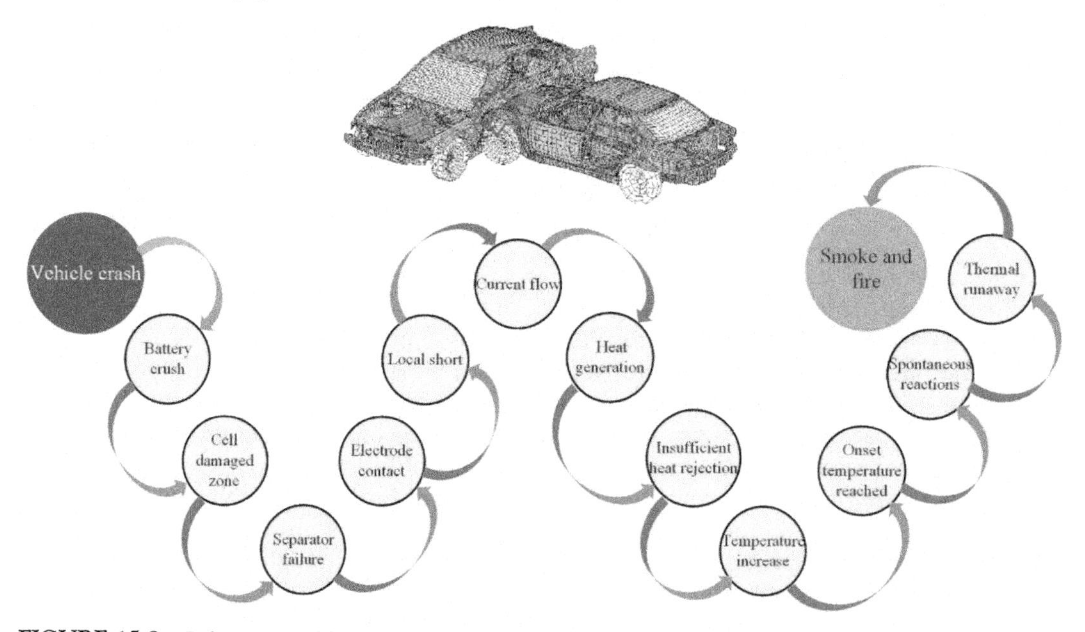

FIGURE 15.9 Failures caused by crash in the Li-ion battery pack.

15.5.1.4 Electrical issues

Several electrical factors can result in Li-ion battery failures. The common factors are discussed below:

- *Overcharge*: During an overcharge more Li + than is necessary moves from the cathode to the anode side in the Li-ion battery cell, with this unwanted reaction changing the atomic structure of the cathode side. Besides, during normal discharging after overcharge, plating production is inevitable. These chemical reactions result from an increased rate in the temperature and voltage of the battery during the overcharge which affects the lifetime and fade capacity of the battery [47].
- *Overvoltage*: The activity of the electrolytes during an overvoltage condition is mostly defeated. If the rate of the voltage is more than the maximum voltage of the cells, this changes the atomic structure of the electrolyte and the Li-ion passing action is limited (fade capacity) [48].
- *Low voltage*: Discharging a Li-ion battery at low voltage is another factor causing degradation of the electrolyte and the cathode side [49]. The Li-ion battery is based on the active production material (especially on the cathode side) which should be discharged under safe voltage.
- *Short circuit*: When a short circuit occurs in the Li-ion cell, the inside temperature increases, facilitating the decomposition of lithium salt [50]. This chemical reaction results in cracks forming on the surface of the graphite.

15.5.2 Power electronic converter

In the charging of EVs, different components and possible faults need to be considered regarding reliability and safety issues [51].

Failure in one of the system's components might cause an entire system failure, which may lead to reduced reliability and safety of the charger system. Factors used to evaluate the reliability of the converter in the chargers are its sensitivity, the safety factor, capacitor, and temperature. Table 15.1 illustrates the main power of electronic devices' system-level reliability.

15.5.3 Electric motor

Fig. 15.10 shows the distribution factors related to an electric motor's failures [53,54]. A reliability evaluation of an electric induction motor can be difficult due to the contribution of many factors. Electric motor failures can be a result of one of the following five categories [53–55]:

- Mechanical and dynamic stress,
- Electromagnetic stress,
- Thermal stress,
- Environmental stress, and
- Other stresses (the errors that can result either from the design or the installation procedures).

TABLE 15.1 Reliability level in the power electronics system [52].

Items	Major influencing factors	Failure-rate formulas from MIL-HDBK-217
Diode	✓ Temperature ✓ Rated voltage ✓ Reverse Blocking	$\lambda_P = \lambda_b\,\pi_T\,\pi_S\,\pi_C\,\pi_Q\,\pi_E$, $\lambda_b = 0.003$ (base failure rate) $\pi_T = \text{Exp}\left(-3091\left(\frac{1}{T_J+273} - \frac{1}{298}\right)\right)$ (temperature factor), T_J is the junction temperature $\pi_S = 0.054$ for $V_S \le 0.3$, $\pi_S = V_S^{2.43}$ for $0.3 < V_S \le 1$ (electrical stress factor), V_S is the voltage stress ratio which is equal to the ratio of applied blocking voltage to the rated blocking voltage in the diode $\pi_C = 1$ (contact construction factor), $\pi_Q = 5.5$ (quality factor), $\pi_E = 1$ (environment factor)
MOSFET	✓ Temperature ✓ Power dissipation ✓ Voltage breakdown	$\lambda_P = \lambda_b\,\pi_T\,\pi_A\,\pi_Q\,\pi_E$, $\lambda_b = 0.012$ (base failure rate) $\pi_T = \text{Exp}\left(-1925\left(\frac{1}{T_J+273} - \frac{1}{298}\right)\right)$ (temperature factor), T_J is the junction temperature $\pi_A = 10$ for $Pr \ge 250$ W (application factor) $\pi_Q = 5.5$ (quality factor), $\pi_E = 1$ (environment factor)
Capacitor	✓ Temperature ✓ Voltage ✓ Type	$\lambda_P = \lambda_b\,\pi_T\,\pi_A\,\pi_Q\,\pi_E$, $\lambda_b = 0.0028\left[\left(\frac{S}{0.55}\right)^3 + 1\right]\exp\left(4.09\left(\frac{T+273}{358}\right)^{5.9}\right)$ (base failure rate), T is the ambient temperature, S is the ratio of operating to the rated voltage in the capacitor $\pi_{CV} = 0.32C^{0.19}$ (capacitor factor), $\pi_Q = 10$ (quality factor), $\pi_E = 1$ (environment factor)
Inductor	✓ Temperature ✓ Current ✓ Voltage ✓ Insulation	$\lambda_P = \lambda_b\,\pi_T\,\pi_A\,\pi_Q\,\pi_E$, $\lambda_b = 0.00003$ (base failure rate for inductors), $\lambda_b = 0.019$ (base failure rate for transformers) $\pi_T = \text{Exp}\left(\frac{-0.11}{8.617\times10^{-5}}\left(\frac{1}{T_J+273} - \frac{1}{298}\right)\right)$ (temperature factor), T is the ambient temperature, $\pi_Q = 30$ (quality factor), $\pi_E = 1$ (environment factor)

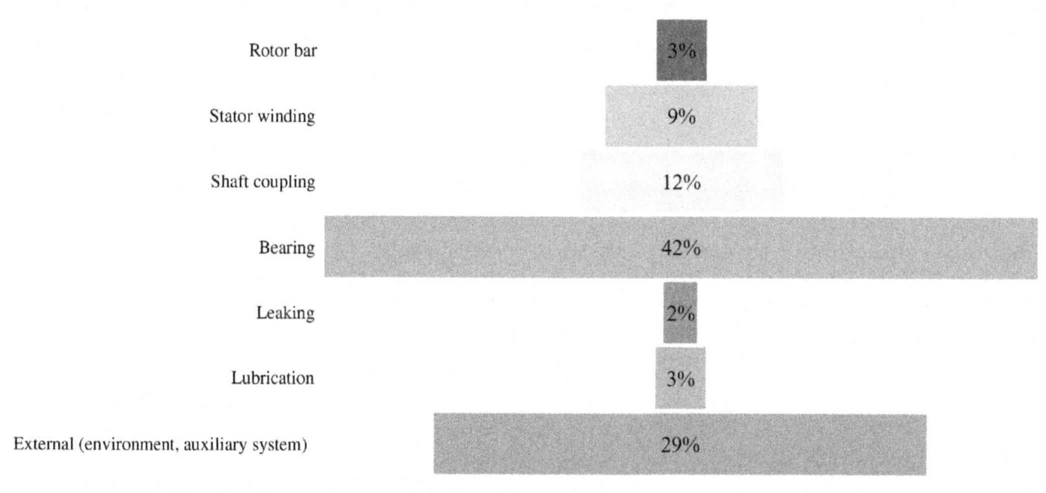

FIGURE 15.10 The distribution factors relating to failures of electric motors.

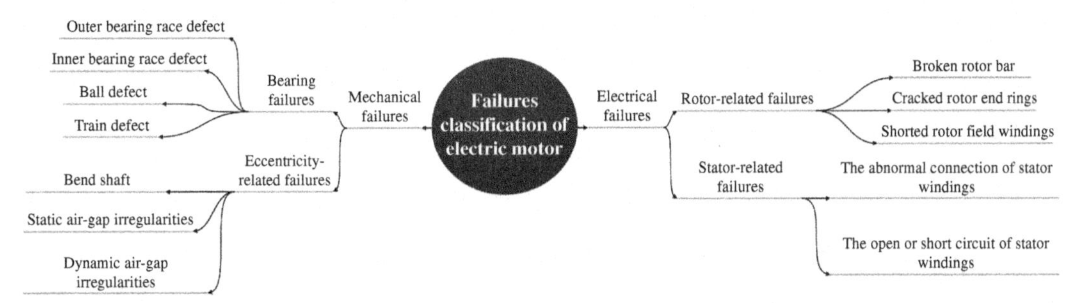

FIGURE 15.11 Classification of failures of electric motors.

The classification of the primary electrical and mechanical failures for electric motors are shown in Fig. 15.3 (Fig. 15.11).

15.6 Electric vehicle's reliability evaluation and challenges

EVs' various components can result in potential new dangers, such as short circuit, fire, loss of traction control, and electric shock in relation to overall safety [56]. Additionally, to enhance the safety and reliability assessment for EVs, it can be helpful to introduce new methods. Therefore, the following aspects need to be considered when designing a reliable EV:

- *Functional safety analysis*: Functional safety requirements need to be considered during the production of the components [57].
- *Design for safety*: The focus of the design phase should be based on a safety test of the system.
- *Safety aspects during the vehicle lifetime*: Customers should be informed on how to keep the EV in a healthy condition [58].
- *Failure modes and effects analysis (FMEA) improvement*: The FMEA results will be applied to perform test procedures for thorough gap analysis and analysis of existing standards [59].
- *Reliability methods*: Using new flexible methodologies to evaluate EVs' reliability will be appropriate for OEMs and researchers [56,60].

The construction of EVs and their software developments depend upon a centralized topology of scalable electronic control units to carry out all hardware-independent software functions. Moreover, actuators and sensors will be smart elements which communicate over standardized data interfaces with the electrical/electronic construction. Therefore, new technology has to be extended for extra-functional properties from a software perspective in the EVs [59,60]. Also, the EV communication network must include very high availability and reliability requirements. Bus protocols candidates, time-triggered systems, and integration are challengeable bus protocols for EVs' internal vehicle networks [60].

In addition, the design of a successful software system requires a general understanding of the overall system structure of EVs, which involves hardware and software components and their complex communication architecture. Therefore, this combination of system knowledge along with the highly specialized expertise in the field is a specific challenge in the embedded system design for the modern automotive industry [60].

15.7 Conclusion

This chapter presents the concept of a reliability and safety assessment of the main electrical components of EVs. Considering the developments in recent decades in the field of EVs, reliability and safety are critical parameters for assessing the lifetime of EVs' components regarding the planning, service, and maintenance. Essential items in EVs, which should be evaluated from a safety and reliability perspective, include the battery pack, power electronic components, and the electric motor. Thermal, electrical, chemical, and mechanical issues are essential factors in failure analysis in the EVs. The major challenge in the EV industry is the battery pack system, which must be reliable and safe with a long lifetime for use in the new generations of EVs.

Acknowledgment

We acknowledge the support of our research team from Flanders Make.

References

[1] S. Meliopoulos, P. Leader, Power System level impacts of plug-in hybrid vehicles power system level impacts of plug-in hybrid vehicles, in: IEEE Energy 2030 Conference, 2008.
[2] B.W. Lane, J. Dumortier, S. Carley, S. Siddiki, K. Clark-Sutton, J.D. Graham, All plug-in electric vehicles are not the same: predictors of preference for a plug-in hybrid versus a battery-electric vehicle, Transp. Res. Part D Transp. Environ. 65 (2018) 1–13.
[3] B. Wang, G. Tian, Y. Liang, T. Qiang, Reliability modeling and evaluation of electric vehicle motor by using fault tree and extended stochastic petri nets, J. Appl. Math 2014 (2014). Article ID 638013.
[4] J. Drobnik, P. Jain, Electric and hybrid vehicle power electronics efficiency, testing and reliability, in: World Electric Vehicle Symposium and Exhibition (EVS27), 2013, pp. 1–12.
[5] R. Allan, Power system reliability assessment—a conceptual and historical review, Reliab. Eng. Syst. Saf. 46 (1) (1994) 3–13.
[6] Roy Billinton, Quantitative Reliability Assessment of Electricity Supply, Wiley, 2014.
[7] A. Escalera, B. Hayes, M. Prodanović, A survey of reliability assessment techniques for modern distribution networks, Renew. Sustain. Energy Rev. 91 (2018) 344–357.
[8] C.C. Chan, An overview of electric vehicle technology, Proc. IEEE 81 (9) (1993) 1202–1213.
[9] A. Poullikkas, Sustainable options for electric vehicle technologies, Renew. Sustain. Energy Rev. 41 (2015) 1277–1287.
[10] International Energy Agency (IEA), Global EV Outlook 2018, 2018.
[11] International Energy Agency (IEA), Global EV Outlook 2017: two million and counting, IEA Publications, 2017, p. 66.
[12] F.H. Gandoman, et al., Status and future perspectives of reliability assessment for electric vehicles, Reliab. Eng. Syst. Saf. 183 (2019) 1–16.
[13] S.C. Nagpure, B. Bhushan, S.S. Babu, Multi-scale characterization studies of aged Li-ion large format cells for improved performance: an overview, J. Electrochem. Soc. 160 (11) (2013) A2111–A2154.
[14] N. Harting, N. Wolff, U. Krewer, Identification of lithium plating in lithium-ion batteries using nonlinear frequency response analysis (NFRA), Electrochim. Acta 281 (2018) 378–385.

[15] X.-G. Yang, Y. Leng, G. Zhang, S. Ge, C.-Y. Wang, Modeling of lithium plating induced aging of lithium-ion batteries: transition from linear to nonlinear aging, J. Power Sources 360 (2017) 28–40.

[16] N.D. Williard, Degradation analysis and health monitering of lithium ion batteries, Current (2011). Available from: https://drum.lib.umd.edu/handle/1903/12381.

[17] A. Barré, B. Deguilhem, S. Grolleau, M. Gérard, F. Suard, D. Riu, A review on lithium-ion battery ageing mechanisms and estimations for automotive applications, J. Power Sources 241 (2013) 680–689.

[18] L. Suo, et al., How solid-electrolyte interphase forms in aqueous electrolytes, J. Am. Chem. Soc. 139 (51) (2017) 18670–18680.

[19] S. Bertolini, P.B. Balbuena, Buildup of the solid electrolyte interphase on lithium-metal anodes: reactive molecular dynamics study, J. Phys. Chem. C 122 (2018). p. acs.jpcc.8b03046.

[20] R. Hausbrand, et al., Fundamental degradation mechanisms of layered oxide Li-ion battery cathode materials: methodology, insights and novel approaches, Mater. Sci. Eng. B 192 (2015) 3–25.

[21] Y. Liu, Z.Y. Wen, X.Y. Wang, A. Hirano, N. Imanishi, Y. Takeda, Electrochemical behaviors of Si/C composite synthesized from F-containing precursors, J. Power Sources 189 (1) (2009) 733–737.

[22] G.P. Nayaka, K.V. Pai, G. Santhosh, J. Manjanna, Dissolution of cathode active material of spent Li-ion batteries using tartaric acid and ascorbic acid mixture to recover Co, Hydrometallurgy 161 (2016) 54–57.

[23] M. Wohlfahrt-Mehrens, C. Vogler, J. Garche, Aging mechanisms of lithium cathode materials, J. Power Sources 127 (1–2) (2004) 58–64.

[24] D.E. Demirocak, B. Bhushan, Probing the aging effects on nanomechanical properties of a LiFePO4 cathode in a large format prismatic cell, J. Power Sources 280 (2015) 256–262.

[25] S. Kalnaus, Y. Wang, J.A. Turner, Mechanical behavior and failure mechanisms of Li-ion battery separators, J. Power Sources 348 (2017) 255–263.

[26] R. Kostecki, L. Norin, X. Song, F. McLarnon, Diagnostic studies of polyolefin separators in high-power Li-ion cells, J. Electrochem. Soc. 151 (4) (2004) A522.

[27] A. Tang, G. Hu, M. Liu, Mechanical degradation of electrode materials within single particle model in Li-ion batteries for electric vehicles, J. Math. Chem. 55 (10) (2017) 1903–1915.

[28] J. Wen, Y. Yu, C. Chen, A review on lithium-ion batteries safety issues: existing problems and possible solutions, Mater. Express 2 (3) (2012) 197–212.

[29] J. Vetter, et al., Ageing mechanisms in lithium-ion batteries, J. Power Sources 147 (1–2) (Sep. 2005) 269–281.

[30] J. Cannarella, C.B. Arnold, The effects of defects on localized plating in lithium-ion batteries, J. Electrochem. Soc. 162 (7) (2015) A1365–A1373.

[31] N.E. Galushkin, N.N. Yazvinskaya, D.N. Galushkin, The mechanism of thermal runaway in alkaline batteries, J. Electrochem. Soc. 162 (4) (2015) A749–A753.

[32] J. Zhang, L. Zhang, F. Sun, Z. Wang, An overview on thermal safety issues of lithium-ion batteries for electric vehicle application, IEEE Access 6 (2018) 23848–23863.

[33] C. Yuqin, L. Hong, W. Lie, L. Tianhong, Irreversible capacity loss of graphite electrode in lithium-ion batteries, J. Power Sources 68 (2) (1997) 187–190.

[34] B.P. Matadi, et al., Irreversible capacity loss of Li-ion batteries cycled at low temperature due to an untypical layer hindering li diffusion into graphite electrode, J. Electrochem. Soc. 164 (12) (2017) A2374–A2389.

[35] J. Shim, K.A. Striebel, The dependence of natural graphite anode performance on electrode density, J. Power Sources 130 (1–2) (2004) 247–253.

[36] G.G. Botte, R.E. White, Z. Zhang, Thermal stability of LiPF6–EC:EMC electrolyte for lithium ion batteries, J. Power Sources 97–98 (2001) 570–575.

[37] L. Yang, B.L. Lucht, Inhibition of electrolyte oxidation in lithium ion batteries with electrolyte additives, Electrochem. Solid State Lett. 12 (12) (2009) A229.

[38] F. Shi, et al., Strong texturing of lithium metal in batteries, Proc. Natl. Acad. Sci. U.S.A. 114 (46) (2017) 12138–12143.

[39] B.L. Mehdi, et al., The impact of Li grain size on coulombic efficiency in Li batteries, Sci. Rep. 6 (2016) 1–8.

[40] E.P. Roth, D.H. Doughty, J. Franklin, DSC investigation of exothermic reactions occurring at elevated temperatures in lithium-ion anodes containing PVDF-based binders, J. Power Sources 134 (2) (2004) 222–234.

[41] M. Herstedt, H. Rensmo, H. Siegbahn, K. Edström, Electrolyte additives for enhanced thermal stability of the graphite anode interface in a Li-ion battery, Electrochim. Acta 49 (14) (2004) 2351–2359.

[42] Y.S. Park, H.J. Bang, S.M. Oh, Y.K. Sun, S.M. Lee, Effect of carbon coating on thermal stability of natural graphite spheres used as anode materials in lithium-ion batteries, J. Power Sources 190 (2) (2009) 553–557.
[43] D. Gonzalez-Rodriguez, et al., Mechanical criterion for the rupture of a cell membrane under compression, Biophys. J. 111 (12) (2016) 2711–2721.
[44] D.P. Finegan, et al., Identifying the cause of rupture of Li-ion batteries during thermal runaway, Adv. Sci. 5 (1) (2018).
[45] A. Mukhopadhyay, B.W. Sheldon, Deformation and stress in electrode materials for Li-ion batteries, Prog. Mater. Sci. 63 (2014) 58–116. no. February.
[46] G. Bucci, T. Swamy, S. Bishop, B.W. Sheldon, Y.-M. Chiang, W.C. Carter, The effect of stress on battery-electrode capacity, J. Electrochem. Soc. 164 (4) (2017) A645–A654.
[47] Q. Yuan, F. Zhao, W. Wang, Y. Zhao, Z. Liang, D. Yan, Overcharge failure investigation of lithium-ion batteries, Electrochim. Acta 178 (2015) 682–688.
[48] J. Hong, Z. Wang, P. Liu, Voltage fault precaution and safety management of lithium-ion batteries based on entropy for electric vehicles, Energy Procedia 104 (2016) 44–49.
[49] P. Prochazka, D. Cervinka, J. Martis, R. Cipin, P. Vorel, Li-ion battery deep discharge degradation, ECS Trans. 74 (1) (2016) 31–36.
[50] R. Zhao, J. Liu, J. Gu, Simulation and experimental study on lithium ion battery short circuit, Appl. Energy 173 (2016) 29–39.
[51] M. Ghavami, C. Singh, Reliability evaluation of electric vehicle charging systems including the impact of repair, in: 2017 IEEE Ind. Appl. Soc. Annu. Meet., 2017, pp. 1–9.
[52] Y. Song, B. Wang, Survey on reliability of power electronic systems, IEEE Trans. Power Electron. 28 (1) (2013) 591–604.
[53] M. AlMuhaini, F.S. Al Badawi, Reliability modelling and assessment of electric motor driven systems in hydrocarbon industries, IET Electr. Power Appl. 9 (9) (2015) 605–611.
[54] A. Ristow, M. Begović, A. Pregelj, A. Rohatgi, Development of a methodology for improving photovoltaic inverter reliability, IEEE Trans. Ind. Electron. 55 (7) (2008) 2581–2592.
[55] E. Frederick Ojiemhende, A. Titus Olugbenga, S. Emmanuel Oludare, K. Olufemi Oluseye, A. Nafiu Sidiq, A comparative overview of electronic devices reliability prediction methods-applications and trends, Majlesi J. Telecom. Dev. 5 (4) (2016) 129–137.
[56] D. Huitink, Thermomechanical reliability challenges and goals and design for reliability methodologies for electric vehicle systems, in: International Technical Conference and Exhibition on Packaging and Integration of Electronic and Photonic Microsystems, 2017, pp. 1–10.
[57] Y. Li, X. Wang, Functional safety analysis for the design of VCU used in pure electric vehicles, in: Proc. 2017 4th Int. Conf. Inf. Sci. Control Eng. ICISCE 2017, 2017, pp. 1004–1008.
[58] S. Davidov, M. Pantoš, Planning of electric vehicle infrastructure based on charging reliability and quality of service, Energy 118 (2017) 1156–1167.
[59] C. Lin, P. Li, C. Li, W. Chiang, Investigation on IGBT failure effects of EV/HEV inverter using fault insertion HiL testing, in: EVS28 Int. Electr. Veh. Symp. Exhib., 2015, pp. 1–9.
[60] S. Collong, R. Kouta, Fault tree analysis of proton exchange membrane fuel cell system safety, Int. J. Hydrogen Energy 40 (25) (2015) 8248–8260.

Further reading

G. Georgakos, U. Schlichtmann, R. Schneider, S. Chakraborty, Reliability challenges for electric vehicles, in: Proc. 50th Annu. Des. Autom. Conf. (DAC '13), 2013, p. 1.

CHAPTER

16

Load forecasting using multiple linear regression with different calendars

Hannah Nano[1], *Savanna New*[1], *Achraf Cohen*[2] *and Bhuvaneswari Ramachandran*[1]

[1]Department of Electrical and Computer Engineering, Hal Marcus College of Science and Engineering, University of West Florida, Pensacola, FL, United States [2]Department of Mathematics and Statistics, Hal Marcus College of Science and Engineering, University of West Florida, Pensacola, FL, United States

16.1 Introduction

The production of electricity differs from other manufactured products as electric energy must be generated as soon as the demand is needed. To deal with this, commercial electric power companies provide their customers with a stable flow of energy, which is assisted with electric power load forecasting. With accurate forecasting, power companies can be more efficient in power supply, maintenance costs, etc. As an important part of the energy management system, load forecasting has been a hot topic of research in power systems [1–3]. With the rapid development of the electric power industry and market-oriented reforms in recent years, the importance of load forecasting, especially short-term load forecasting, is becoming increasingly imperative. There are many methods of short-term load forecasting [4–11], with commonly used prediction methods being the time series method, similar day method, neural network method, regression analysis, and time series analysis. In addition, there are several emerging techniques, such as gradient boosting machines and random forests, [12–16]. Exploring the various variables for load forecasting models has also been attempted. Typical load forecasting research articles have used the load or log-transformed load as the dependent variable [17,18]. Weather and calendar variables were very frequently used [19–21].

Consumption is different from demand: demand is the rate of using electricity and consumption is the actual load that the user consumes in a given time span. However, both have a direct correlation to seasonal patterns and hours of the day. For instance, in a cold

Distributed Energy Resources in Microgrids
DOI: https://doi.org/10.1016/B978-0-12-817774-7.00016-8

winter month and hot summer month the demand for electricity will be higher in comparison to the cool, neutral temperatures of spring and fall. This is the same idea when comparing hours in a day, that is, daytime usage is higher than the hours of the day when the household or business is sleeping or closed. Because of this trend and correlation, load forecasting often takes into consideration calendar/time variables, that is, month of the year, day of the week, and hour of the day.

The close tracking of the system load by the system generation at all times is a basic requirement in the operation of power systems. For economically efficient operation and for effective control, this must be accomplished over a broad spectrum of time intervals. In the range of seconds, when load variations are small and random, the automatic generation control (AGC) function ensures that the online generation matches the load [22]. For a timescale of minutes, when larger load variations are possible, the economic dispatch function is used to ensure that the load matching is economically allocated among the committed generation sources. For periods of hours and days, still wider variations in the load occur, and meeting the load over such time frames entails the start-up or shutdown of entire generating units or the interchange of power with neighboring systems. This is determined by several generation control functions such as hydro scheduling, unit commitment, hydro-thermal coordination, and interchange evaluation. Over the time range of weeks, when very wide swings in the load are present, functions such as fuel, hydro, and maintenance scheduling are performed to ensure that the load can be met economically with the installed resource mix. In addition, to ensure the secure operation of the power system, at some future time requires the study of its behavior under a variety of postulated contingency conditions by the offline network analysis functions. In a real-time environment, state estimators are used to validate telemetered measurements from which the estimated values of the voltage magnitude and angle at each bus are determined. These values may be used to compute estimates for the instantaneous load. Procedures for very-short-term load prediction are embedded in the AGC and economic dispatch functions with lead times of the order of seconds and minutes, respectively. The load information for the hydro scheduling, unit commitment, hydro-thermal coordination, and the interchange evaluation functions is obtained from the short-term load forecasting system. The fuel and hydro allocation and maintenance scheduling functions require load forecasts for periods longer than one week. These load predictions are obtained from operational planning forecasting systems with lead times as long as one to two years.

Smoothing techniques are used to reduce random fluctuations in time series data. They provide a clearer view of the true underlying behavior of the series. The moving average (MA) approach ranks among the most popular techniques for the preprocessing of time series. The simple MA, exponential MA, and weighted MA methods are widely used. The exponential smoothing method is also a straightforward time-series prediction method which also has the characteristics of simple calculation and high precision. Recently researchers [15] used an exponential smoothing grey prediction model combining an exponential smoothing model and the characteristics of short-term load demand to carry out short-term load forecasting using the historical data of load demand. Their simulated results showed that the method has a satisfactory prediction effect on the short-term load demand. Authors have developed load-forecasting models to predict hourly load demand for various seasons of the year [23]. They have explored various artificial intelligence techniques for short-term load forecasting [24] and have proposed and compared several univariate approaches for short-term load forecasting based on neural networks [25]. In this research, the authors have used the MA smoothing method and the exponential MA smoothing method along with multiple linear regression to train, test, and compare the versatility and efficacy of both methods.

A typical electricity demand series presents multiple seasonal patterns. When modeling the annual seasonality, the Gregorian calendar is usually used in the load-forecasting models. It dissects the days of the year into 12 months based on the moon's orbit around the earth. However, the season marked by changes in the weather are a result of the yearly orbit of the earth around the sun. Therefore the widely used Gregorian calendar, which is based on the moon's orbit around the earth, for load forecasting may not be an accurate indicator for the change of the season. Alternatively, authors in [26] came up with the idea, for load-forecasting purposes, of using the 24 solar terms that originated in China. The 24-solar-term calendar is used to guide agricultural activities and was developed based on the sun's position in the zodiac. Is the Indian calendar better than the Gregorian calendar for load-forecasting purposes? Hence, in our research, we want to use the Indian calendar to categorize days of the year. The approach was evaluated by applying the methodology to two datasets: one representing six states in the United States called International Standards Organization—New England (ISO-NE) dataset and the other a North American Utility (NAU) dataset. A forecast-evaluation setting, namely cross-validation is used to evaluate the performance of the proposed methodology. The results obtained are compared with those obtained from the Gregorian calendar.

The remaining of this chapter is categorized as follows: Section 16.2 presents the material and methodology. Section 16.3 describes and discusses the results, and Section 16.4 shows the conclusion and future research.

16.2 Material and methods

The accuracy of load forecasting is measured by an index called the mean absolute percentage error (MAPE). In general, load forecasts with approximately 10% MAPE can be obtained easily; however, the cost of this error is significantly high and research efforts to help reduce it by a few percentage points would be considered a significant contribution [4]. If the accuracy of short-term load forecasts can be improved, it will result in significant financial savings for utilities and cogenerators [27,28]. Hence, in this research, the authors have used a multiple linear regression (MLR) model to forecast the load demand in a smart grid.

16.2.1 Test systems datasets

In this chapter, the authors considered two sets of data to demonstrate the applicability and performance of the proposed methodology. The first set of data used is seven years of users' data published by ISO-NE from 2011 to 2017. ISO-NE serves multiple regions, however, to simplify our study we conducted the study based on the total system (ISO-NE). ISO-NE serves the six states in the north-eastern United States including Connecticut, Massachusetts, Maine, New Hampshire, Rhode Island, and Vermont. The load data supplied is adjusted for daylight saving time. Fig. 16.1 shows the monthly load and temperature

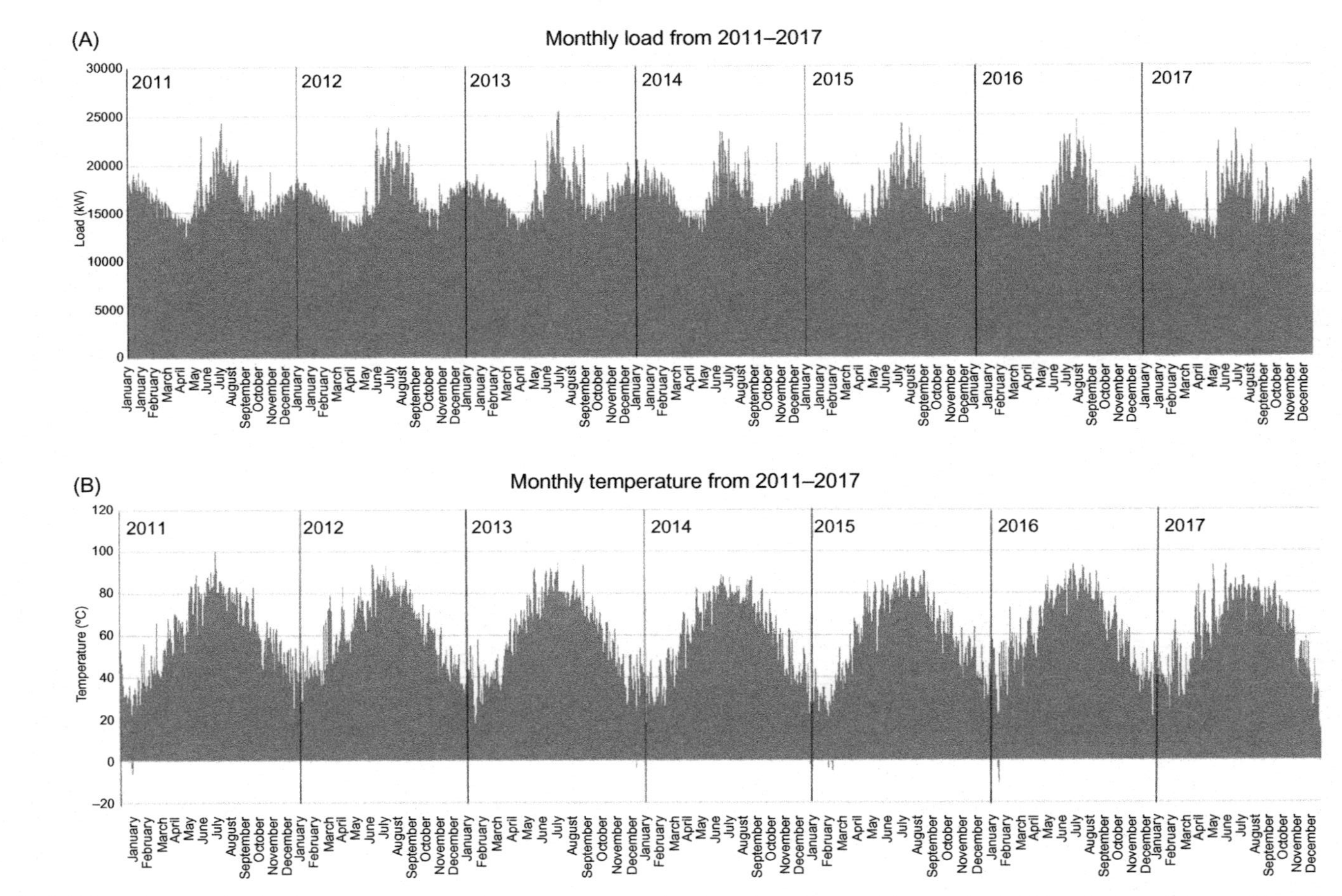

FIGURE 16.1 Monthly load and temperature of the International Standards Organization-New England (ISO-NE) dataset from 2011 to 2017.

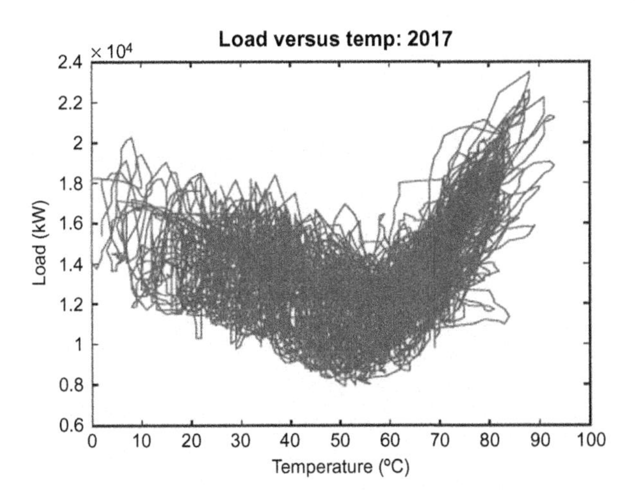

FIGURE 16.2 Load versus temperature for International Standards Organization-New England (ISO-NE) dataset for 2017 year.

profiles for the ISO-NE dataset. Fig. 16.2 shows the relationship between the load and temperature for each month from 2011 until 2017. Other years have the same shape of relationship.

The second set of data corresponds to the NAU dataset from 1985 to 1991. Fig. 16.3 shows the monthly load and temperature profiles for the NAU dataset. Fig. 16.4 shows the relationship between the load and temperature for each month from 1985 until 1991. Other years have similar shape of relationship.

16.2.2 Multiple linear regression model

Multiple linear regression is a technique that is most commonly used when several variables are known, and a response is predicted. This method is used to explain a relationship between one continuous dependent variable and at least one other variable. The vanilla model is a linear regression model that was proposed in order to predict the load. This model is given as follows:

$$Y_{\text{Load}} = \beta_0 + \beta_1 \text{Trend} + \beta_2 M + \beta_3 W + \beta_4 H + \beta_5 WH + \beta_6 T + \beta_6 T + \beta_7 T^2 + \beta_8 T^3 + \beta_9 TM + \beta_{10} T^2 M + \beta_{11} T^3 M + \beta_{12} TH + \beta_{13} T^2 H + \beta_{14} T^3 M \tag{16.1}$$

where Y_{Load} is the given load and β_i are the regression coefficients; trend represents the chronological trend; M, W, and H are variables representing month of the year, day of the week and hour of the day, respectively; T is the dry bulb temperature. The multiplications between the variables represent the interactions. The variables M, W, and H are coded as string variables in the code program.

We used this model to predict the load with different calendars (Gregorian and Indian), which means the "week" and the "month" variables will change in Eq. (16.1). The trend component in the regression model is estimated using two different methods, that are MA and exponential weighted moving average (EWMA) with $\lambda = 0.01$.

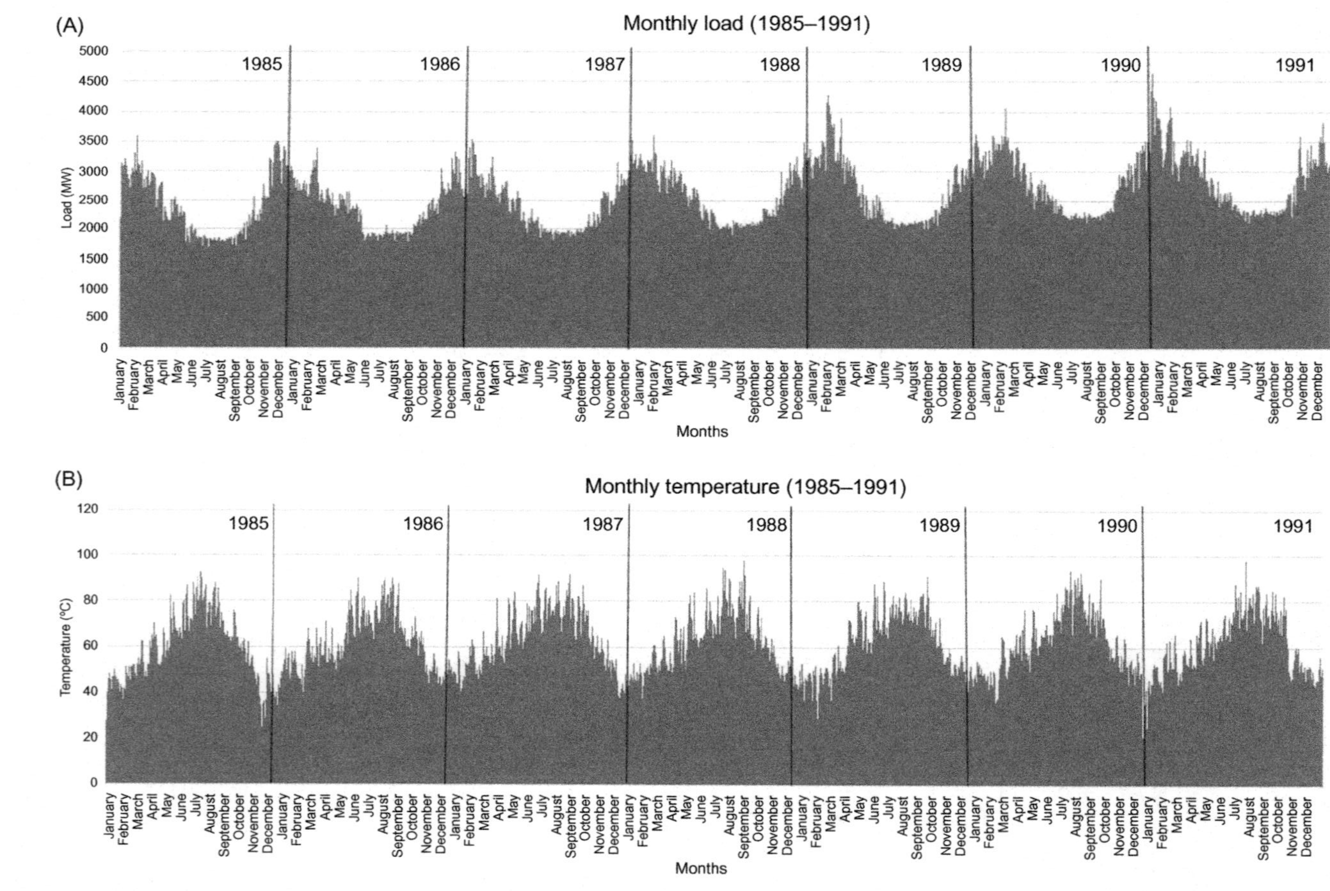

FIGURE 16.3 Monthly load and temperature of the North American Utility (NAU) dataset from 1985 to 1991.

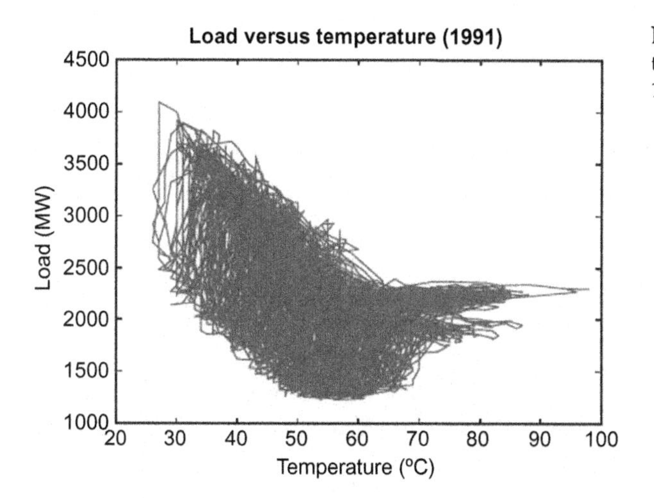

FIGURE 16.4 Load versus temperature for the North American Utility (NAU) dataset for 1991 year.

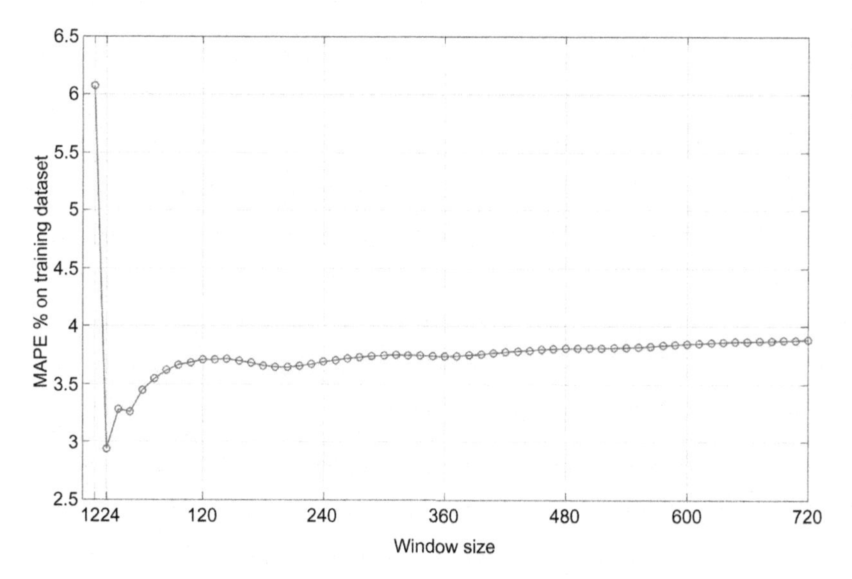

FIGURE 16.5 Variation of mean absolute percentage error (MAPE) with window size.

16.2.2.1 Moving average

When using MA smoother, a moving window running through the data and arithmetic mean is computed. In order to determine what window size to use, we did a prior study to determine the window size that could provide better results in terms of MAPE. Fig. 16.5 shows that the best window size is 24 hours in terms of MAPE. Therefore 24 hours will be the window size for the MA further in the manuscript.

16.2.2.2 Exponential weighted moving average

When using the exponential smoother, the most recent observations are given less weight than the past observations, as follows:

$$Y_t = \lambda X_t + (1 - \lambda)\, X_{t-1} \tag{16.2}$$

where $Y_0 = X_0$ and $\lambda = 0.01$.

16.2.2.3 K-fold cross-validation

Cross-validation is one of the most widely used techniques when assessing the generalization capabilities of a predictive model. In this study we use the cross-validation technique. It first separates the data into K equal, or partially equal, parts. $K - 1$ is used as training data and, based on those results, one of the K segments is used as validation data. This process is repeated until all the K values are found. In this study, we used one year of data as testing data and the remaining years as training data. For example, in order to predict the load for 2013, the training data is the data from 2011 to 2016 (excluding the year 2013), then the estimate is found and compared to the actual load information.

To compare calendars and methods the MAPE must be compared. MAPE is used to measure error. The lower the error the better the result. MAPE is found by:

$$\text{MAPE} = \frac{100}{N} \sum_{i=1}^{N} \left| \frac{A_i - P_i}{A_i} \right| \tag{16.3}$$

where N is the number of observations; A_i is the actual load data; and P_i is the predicted load data. MAPE is found for each year and then an average is found.

16.3 Results

Both test systems are tested with two calendars viz., Gregorian and Indian calendars. The user-load data and temperatures were supplied for the total years for each test system (ISO-NE and NAU). The cross-validation was used for evaluating the performance of the two calendars setup. Table 16.1 provides the starting dates of each month in the Indian calendar. It is well known that India has 29 states, seven union territories and 22 official languages. Due to the diversity of cultures and traditions followed throughout the country, India has several calendars based on their region of usage. Out of them, the authors have considered the Tamil calendar for comparison against the Gregorian calendar for short-term load forecasting. The number of days in a month varies between 29 and 32. The Tamil New Year follows the spring equinox and generally falls on April 14 of the Gregorian year every year. The following list compiles the 12 months of the Tamil (Indian) calendar.

The Tamil calendar is a sidereal Hindu calendar used in Tamil Nadu, India. Sidereal time is a time-keeping system that astronomers use to keep track of the direction in which to point their telescopes to view a given star in the night sky. Briefly, sidereal time is a "time scale that is based on the Earth's rate of rotation measured relative to the fixed

TABLE 16.1 Starting dates of Indian (Tamil) calendar.

Number	Month name	Gregorian calendar equivalent	Seasons
1	Chithirai	April 14–May 13	Spring
2	Vaikasi	May 14–June 14	Spring
3	Aani	June 15–July 15	Summer
4	Aadi	July 16–August 16	Summer
5	Aavani	August 17–September 16	Monsoon
6	Purattasi	September 17–October 16	Monsoon
7	Aippasi	October 17– November 15	Autumn
8	Karthikai	November 16–December 15	Autumn
9	Margazhi	December 16–January 13	Winter
10	Thai	January 14–February 12	Winter
11	Maasi	February 13–March 13	Prevernal
12	Panguni	March 14–April 13	Prevernal

stars" rather than the Sun. Thus, the Tamil calendar is primarily based on two large planets, viz. Jupiter and Saturn and their unique property of revolution around Sun. The 60-year Tamil calendar is based on this fact and is derived from following calculations: The traditional Tamil year starts on April 14. The Tamil calendar follows the vernal equinox (equinox is the day on which the sun's axis is not tilted; vernal is autumn in Latin). Tropical vernal equinox falls around March 22 and adding 23 degrees of trepidation or oscillation to it, we get the Hindu sidereal (Sun's transition into Aries). Hence, the Tamil calendar begins on the same date in April which is observed by most traditional calendars of the rest of India. Five revolutions of Jupiter around the Sun take 60 years and two revolutions of Saturn around the Sun takes the same time. The relative positions of Jupiter and Saturn in one year will be repeated once every 60 years. The 60-year cycle was essentially conceived for predicting the climate of a year, as the relative position of the two major planets, Jupiter and Saturn, is recognized for its impact on climate. Twelve Tamil months are named after the star on which the full moon day occurs. Tamil months begin on the full moon day of every month.

Tables 16.2–16.5 represent the results for case 1 (ISO-NE). Table 16.2 lists all MAPE test (in %) values for each validation year along with the average using exponential MA smoothing (considering trend for the previous year only) and Table 16.3 shows all the MAPE (in %) with trend taken as average of all other years. The Gregorian calendar did better when compared to the Indian calendar. On average, the Gregorian calendar performed better six out of seven times. Tables 16.4 and 16.5 show the percentage results of MAPE for ISO-NE test system, but now using cross-validation with the MA smoothing approach. The results from all these implementations underline the fact that the conventional Gregorian calendar of representing seasonal variations in electric loads yields better

TABLE 16.2 Mean absolute percentage error (MAPE) values using exponential weighted average (EWMA) ($\lambda = 0.01$) as trend and using the previous year trend as the testing year trend: International Standards Organization-New England (ISO-NE) dataset.

Calendar	2012	2013	2014	2015	2016	2017
Gregorian	3.79	3.96	3.68	4.00	4.06	6.04
Indian	6.06	4.65	3.72	4.31	4.70	9.67

TABLE 16.3 Mean absolute percentage error (MAPE) values using exponential weighted average (EWMA) ($\lambda = 0.01$) as trend and using the average of the all other years trends as the testing year trend: International Standards Organization-New England (ISO-NE) dataset

Calendar	2011	2012	2013	2014	2015	2016	2017
Gregorian	3.85	3.77	3.97	3.84	4.13	3.89	6.17
Indian	7.91	4.30	5.01	4.47	4.06	5.17	9.47

TABLE 16.4 Mean absolute percentage error (MAPE) values using moving average (MA) as trend and using the average of the all other years trends as the testing year trend: International Standards Organization-New England (ISO-NE) dataset.

Calendar	2012	2013	2014	2015	2016	2017
Gregorian	5.18	5.86	5.18	5.69	5.79	7.93
Indian	9.48	6.08	6.21	6.78	5.92	12.5

TABLE 16.5 Mean absolute percentage error (MAPE) values using moving average (MA) as trend and using the average of the all other years trends as the testing year trend: International Standards Organization-New England (ISO-NE) dataset

Calendar	2011	2012	2013	2014	2015	2016	2017
Gregorian	4.44	4.84	5.18	5.02	5.49	5.04	7.27
Indian	10.10	5.52	6.61	6.09	5.50	6.52	11.53

results than the Indian calendar. The results of the second test system using the NAU dataset are shown in Tables 16.6–16.9.

16.4 Discussion

The application of the MLR model to short-term load forecasting was demonstrated with two real datasets and their results were shown in the previous section. From the

TABLE 16.6 Mean absolute percentage error (MAPE) values using exponential weighted average (EWMA) ($\lambda = 0.01$) as trend and using the previous year trend as the testing year trend: North American Utility (NAU) dataset.

Calendar	1986	1987	1988	1989	1990	1991
Gregorian	7.18	4.46	5.28	5.90	8.13	7.93
Indian	6.07	5.21	5.40	5.54	8.84	6.49

TABLE 16.7 Mean absolute percentage error (MAPE) values using exponential weighted average (EWMA) ($\lambda = 0.01$) as trend and using the average of the all other years trends as the testing year trend: North American Utility (NAU) dataset.

Calendar	1985	1986	1987	1988	1989	1990	1991
Gregorian	14.85	11.57	7.68	4.28	5.86	10.27	11.62
Indian	15.79	11.33	7.47	4.59	5.75	10.79	10.40

TABLE 16.8 Mean absolute percentage error (MAPE) values using moving average (MA) as trend and using the previous year trend as the testing year trend: North American Utility (NAU) dataset.

Calendar	1986	1987	1988	1989	1990	1991
Gregorian	7.99	6.75	6.86	7.76	8.46	7.94
Indian	7.77	7.86	7.32	6.83	9.07	8.09

TABLE 16.9 Mean absolute percentage error (MAPE) values using moving average (MA) as trend and using the average of the all other years trends as the testing year trend: North American Utility (NAU) dataset.

Calendar	1985	1986	1987	1988	1989	1990	1991
Gregorian	13.82	12.26	8.79	5.72	7.16	10.4	11.57
Indian	15.75	11.63	7.98	6.24	6.56	11.06	10.1

results, it is evident that the Gregorian calendar resulted in a better forecast of the load due to its seasonal patterns and the Indian calendar resulted in higher errors (in most of the cases, it rarely gave lower MAPE than the Gregorian calendar) when tested on two American regional/utility datasets. One reason that could be attributed to the inaccuracy of Indian calendar is the test data itself, both of which are in North America and since the load and temperature variations in North America do not correspond to the climate pattern in India as shown in Table 16.1, it results in higher error. To rectify this problem and also to investigate the truth in this reasoning, the authors will test the Indian calendar for load forecasting with Indian regional/utility data and verify the hypothesis. Following a similar analysis framework, research can be conducted for other climatic regions to obtain a more comprehensive understanding on using the regional calendars for load forecasting. Also with the integration of renewable energy sources into the smart grid and with the increasing penetration of electric vehicles (EVs), more practical and real-life case studies are needed to examine closely the variation in factors, such as forecast horizons, seasonal weather variations, user patterns, and charging/discharging pattern of batteries/EVs.

Other studies can be done with other calendars, other forecast horizons, and/or other forecasting techniques. Hybrid methods using artificial neural networks, deep neural networks, support vector machines, wavelet transforms, and other genetic and swarm-optimization methods are being examined by different researchers in combination with learning algorithms to speed up the approach and also to result in a more accurate forecast. These techniques can be tested on the different regional calendars to see if the methods can be universally applied or if the choice of regional calendars do influence regional data. The goal of this research was to test the feasibility and applicability of the Indian calendar to forecast load using two test cases. Another issue that the researchers ran into was, for some instances, the load read 0 in our user data ISO-NE, which interfered with the whole program, and resulted in MAPE becoming infinite. To prevent infinite error, those data points must be ignored or estimated. Though load forecasting is an offline load prediction tool, the time it took for our proposed method to converge was less than a minute.

16.5 Conclusions

This research used the Indian (Tamil) calendar along with the Gregorian calendar for forecasting electric load. In an Indian calendar, the beginning of the year starts in April and it is based on planetary alignment. The proposed approach was tested on two test systems to study how effective and accurate the forecast is. The results for K-fold cross validation is shown. Two different smoothing techniques were used to demonstrate the feasibility and accuracy of Gregorian calendar compared against Indian calendar. Another point worth noting is that while the paper mentioned getting poor results using the Gregorian calendar, more studies can attempt changing all the variables to strings. Changing the month of the year, day of week, and hour of day to strings helps create a much more precise regression model and therefore better results.

References

[1] P. Ju, W. Jiang, X. Zhao, J. Wang, S. Zhang, Y. Liu, Ninety-six points short-term load forecasting—theory & applications, Autom. Electr. Power Syst. 25 (22) (2001) 32–36.

[2] X.-X. Zhang, Q. Zhou, H.-J. Ren, C.-X. Sun, Q.-Y. Cheng, Input parameters selection in short-term load forecasting model based on incremental reduction algorithm, Autom. Electr. Power Syst. 13 (2005) 008.

[3] Y. Dong, Z. Jianguo, Practicable method for load forecasting of distribution network, Autom. Electr. Power Syst. 25 (22) (2001). I–4.

[4] H.S. Hippert, C.E. Pedreira, R.C. Souza, Neural networks for short-term load forecasting: a review and evaluation, IEEE Trans. power syst. 16 (1) (2001) 44–55.
[5] T. Hong, S. Fan, Probabilistic electric load forecasting: a tutorial review, Int. J. Forecast. 32 (3) (2016) 914–938.
[6] K. Lee, Y. Cha, J. Park, Short-term load forecasting using an artificial neural network, IEEE Trans. Power Syst. 7 (1) (1992) 124–132.
[7] D.C. Park, M. El-Sharkawi, R. Marks, L. Atlas, M. Damborg, Electric load forecasting using an artificial neural network, IEEE Trans. Power Syst. 6 (2) (1991) 442–449.
[8] P. Wang, B. Liu, T. Hong, Electric load forecasting with recency effect: a big data approach, Int. J. Forecast. 32 (3) (2016) 585–597.
[9] R. Weron, Modeling and Forecasting Electricity Loads and Prices: A Statistical Approach, 403, John Wiley & Sons, 2007.
[10] J. Yu, B.-B. Lee, D. Park, Real-time cooling load forecasting using a hierarchical multi-class SVDD, Multimed. Tools Appl. 71 (1) (2014) 293–307.
[11] S. Cheng, A new approach to load forecasting based on similar day, Proc. Jiangsu Electr. Eng. Assoc. 18 (1999) 28–32.
[12] T. Hong, P. Pinson, S. Fan, Global energy forecasting competition 2012, Int. J. Forecast. 30 (2014) 357–363.
[13] T. Hong, P. Pinson, S. Fan, H. Zareipour, A. Troccoli, R.J. Hyndman, Probabilistic energy forecasting: Global energy forecasting competition 2014 and beyond, Int. J. Forecast. 32 (3) (2016) 896–913.
[14] R. Ramanathan, R. Engle, C.W. Granger, F. Vahid-Araghi, C. Brace, Short-run forecasts of electricity loads and peaks, Int. J. Forecast. 13 (2) (1997) 161–174.
[15] J. Xie, B. Liu, X. Lyu, T. Hong, D. Basterfield, Combining load forecasts from independent experts, in: North American Power Symposium (NAPS), 2015, pp. 1–5. IEEE.
[16] B.-J. Chen, M.-W. Chang, et al., Load forecasting using support vector machines: a study on eunite competition 2001, IEEE Trans. Power Syst. 19 (4) (2004) 1821–1830.
[17] T. Hong, J. Wilson, J. Xie, Long term probabilistic load forecasting and normalization with hourly information, IEEE Trans. Smart Grid 5 (1) (2014) 456–462.
[18] S. Fan, R.J. Hyndman, The price elasticity of electricity demand in south australia, Energy Policy 39 (6) (2011) 3709–3719.
[19] J.R. Lloyd, Gefcom2012 hierarchical load forecasting: gradient boosting machines and gaussian processes, Int. J. Forecast. 30 (2) (2014) 369–374.
[20] N. Charlton, C. Singleton, A refined parametric model for short term load forecasting, Int. J. Forecast. 30 (2) (2014) 364–368.
[21] Y. Goude, R. Nedellec, N. Kong, Local short and middle term electricity load forecasting with semi-parametric additive models, IEEE Trans. Smart Grid 5 (1) (2014) 440–446.
[22] G. Gross, F.D. Galiana, Short-term load forecasting, Proc. IEEE 75 (12) (1987) 1558–1573.
[23] J. Mi, L. Fan, X. Duan, Y. Qiu, Short-term power load forecasting method based on improved exponential smoothing grey model, Math. Probl. Eng. 2018 (2018).
[24] M.Q. Raza, A. Khosravi, A review on artificial intelligence based load demand forecasting techniques for smart grid and buildings, Renew. Sustain. Energy Rev. 50 (2015) 1352–1372.
[25] G. Dudek, Neural networks for pattern-based short-term load forecasting: a comparative study, Neurocomputing 205 (2016) 64–74.
[26] X. Jingrui, H. Tao, Load forecasting using 24 solar terms, J. Mod. Power Syst. Clean Energy 6 (2) (2018) 208–214.
[27] H.K. Alfares, M. Nazeeruddin, Electric load forecasting: literature survey and classification of methods, Int.J. Syst. Sci. 33 (1) (2002) 23–34.
[28] G.-C. Liao, T.-P. Tsao, Application of fuzzy neural networks and artificial intelligence for load forecasting, Electr. Power Syst. Res. 70 (3) (2004) 237–244.

CHAPTER

17

Unintentional islanding detection

M. Suman[1] and M. Venkata Kirthiga[2]

[1]National Institute of Technology, Tiruchirappalli, India [2]Department of Electrical and Electronics Engineering, National Institute of Technology, Tiruchirappalli, India

17.1 Introduction

The unintentional islanding formation (UIF) in microgrids (MGs) is supposed to be detected, and the distributed generators (DGs) are denied from energizing the loads owing to certain reasons viz.,

1. *Health hazards to working personnel*
   If a portion of the system remains energized, this can cause health hazards to working or maintenance personnel. If the circuit breaker A (CB A) in Fig. 17.1 is opened due to any fault in the distribution system, and subsequently, the maintenance crew is dispersed to repair the faulty feeder in the islanded system, it can result in the electrocution of linemen since some part of the system remains energized.
2. *Unsynchronized reclosing causes excessive transients*
   If the CB B in Fig. 17.1 is opened due to some reasons consequently the voltage magnitude, phase and frequency of the two electrical regions change. Upon unsynchronized reclosing of the breaker, excessive transients which can damage the electrical equipment in the utility grid are developed and also pose a great threat to the MG indulging the DGs.
3. *Ineffective grounding*
   If the CB A in Fig. 17.1 is opened due to some reasons, consequently the effective grounding is lost in the islanded region. If a ground fault occurs at this operating area in the islanded region, it results in a voltage rise of more than 30% which may lead to insulation failure.
4. *Uncoordinated protection*
   In general, distribution systems are radial in nature, and hence overcurrent relays are used for protection. However, on an islanded system, the short circuit current capacity of the inverter on which the DGs are based becomes too small and might not be sufficient enough to energize the overcurrent relays. Hence, the relays may be

Distributed Energy Resources in Microgrids
DOI: https://doi.org/10.1016/B978-0-12-817774-7.00017-X

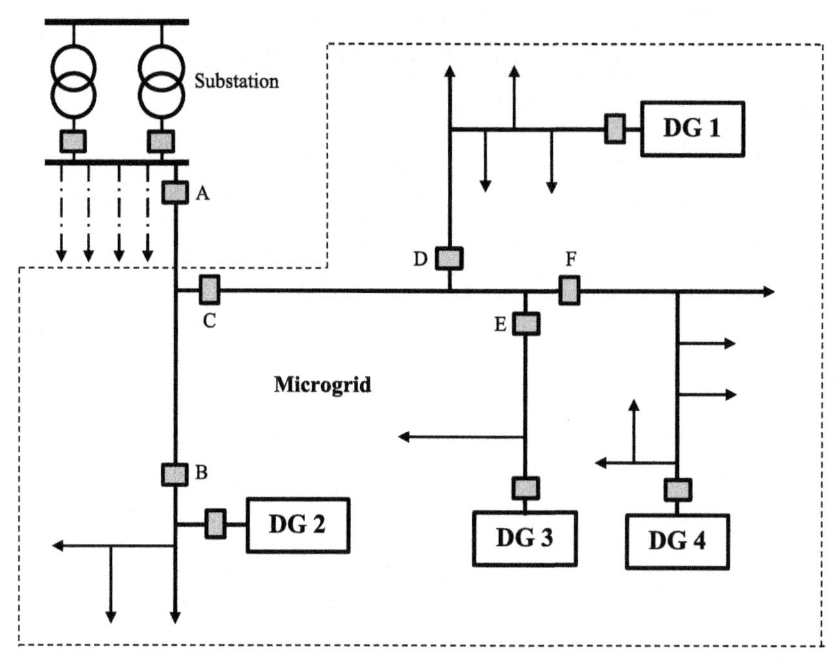

FIGURE 17.1 Schematic diagram of sample test system.

insensitive to faults. In the presence of multiple DGs, the change in short circuit current from the DGs leads to uncoordinated operation of the relays.

5. *Loss of control over voltage and frequency*

Considering breaker F is opened, DG 4 is islanded from the grid with loads. Prior to isolation, the DG is operated in power quality control mode and the voltage and frequency are controlled by the grid. However, during the islanded condition, the control over the voltage and frequency is lost. Load-switching events in the islanded region, causes voltage and frequency to vary abruptly, which eventually damages the electrical equipment in the islanded system.

Owing to the aforementioned whys and wherefores, it is obligatory to detect the UIF. The techniques which detect the UIF are called islanding detection techniques (IDTs).

17.2 Performance evaluation indices of islanding detection techniques

The performance of IDTs is evaluated using certain criteria such as false detection ratio, nondetection zone, feasibility in multiple DGs scenario, islanding detection time, effect on power quality, and implementation cost. Even though a plethora of researchers have proposed numerous techniques, it is hard to find even a single IDT which satisfies all the criteria of detecting the islanding formation.

17.2.1 False detection ratio

False detection is the term for the mal-detection of grid-connected mode of operation as an islanded condition by the IDTs. Many transient conditions like load switching, shut down of sources and fault condition etc., would result in the measured value exceeding the prespecified threshold leading to fault detection of UIF, even though in grid-connected conditions. The false detection ratio (F_{dr}) is expressed as follows,

$$F_{dr} = \frac{N_{False}}{N_{False} + N_{True}} \times 100 \quad (17.1)$$

where N_{False} is the no. of times the IDT wrongly detects the grid-connected mode of operation as an islanded condition and N_{True} is the number of of times the IDT detects the UIF correctly. The percentage of F_{dr} should be as small as possible.

17.2.2 Nondetection zone

The nondetection zone (NDZ) is the operating zone within which the IDTs are insensitive to detecting the island formation. NDZ is evaluated based on either power mismatch or load parameter space.

17.2.2.1 Power mismatch space

In a grid-connected mode of operation, the voltage magnitude and frequency are dictated by the grid's operating conditions. Upon island formation, the voltage magnitude and frequency are dependent on the power mismatch between the generation and demand in the islanded system. For lower active and passive power mismatches, the voltage magnitude and frequency variation on an islanded condition are insignificant. Hence, many IDTs are insensitive to detecting the island formation. The NDZ in power mismatch space is expressed as follows [1],

$$\left(\frac{V}{V_{max}}\right)^2 - 1 \leq \frac{\Delta P}{P_G} \leq \left(\frac{V}{V_{min}}\right)^2 - 1 \quad (17.2)$$

$$Q_f\left(1 - \left(\frac{f}{f_{min}}\right)^2\right) \leq \frac{\Delta Q}{P_G} \leq Q_f\left(1 - \left(\frac{f}{f_{max}}\right)^2\right) \quad (17.3)$$

where Q_f is the quality factor, V_{min} and V_{max} are the minimum and maximum voltages permissible limits in MG, P_G is the active power generation by the DGs in the islanded system, f_{min} and f_{max} are the minimum and maximum frequencies permissible limits in the MG, and ΔP and ΔQ are the active power and reactive power differences between generation and demand.

17.2.2.2 Load mismatch space

The normalized capacitance (C_{norm}) versus load inductance (L_{load}) variations, for a chosen resistance value are used to define the NDZ of the IDTs in the load mismatch space

(LPS). This method is suitable for frequency-drifting-based IDTs [2]. The C_{norm} is expressed as follows:

$$C_{norm} = \frac{C}{C_{res}} = \omega_0^2 LC \quad (17.4)$$

The NDZ is also described in an alternative LPS, that depends on Q_f versus f_{res} axes. [3]. The system frequency is considered as the resonant frequency (f_{res}) of the load. Unlike the former method, for various resistance values of the load, this method does not plot different curves to investigate NDZ, which is a notable advantage. For specified load values of L and C, the quality factor increases with an increase in resistance.

17.2.3 Feasibility in multiple distributed generators scenario

While choosing an IDT for islanding detection, it is necessary to verify the IDT for a scenario involving multiple DGs. The efficacy of the IDT need not be the same for the single and multiple DG scenario. In active IDTs, the effect caused by various DGs may cancel out one another or reduce the cumulative effect in a multiple DGs scenario due to nonsynchronized disturbance injection among various DGs.

17.2.4 Islanding detection time

Islanding detection time (T_{IDTE}) is defined as the time taken by the IDT to detect UIF from the instant when a MG disconnects from the utility grid (T_{GD}). The islanding detection time (IDTE) is expressed by the following equation,

$$T_{IDTE} = T_{IDT} \sim T_{GD} \quad (17.5)$$

where T_{IDT} is the time instant of detecting the UIF.

17.2.5 Effect on power quality

Few active techniques inject disturbance or add feedback to the control parameters of the inverter to reduce the nondetection zone and detect islanding faster at the cost of power quality degradation. Thus, IDT, which has less or a minimal effect on power quality, is desired.

17.3 Islanding detection islanding detection techniques

The different types of IDTs [4–6] are depicted in Fig. 17.2.

17.3.1 Islanding detection time of various international standards

The minimum islanding detection time and quality factors of the load suggested by various international standards are tabulated in Table 17.1.

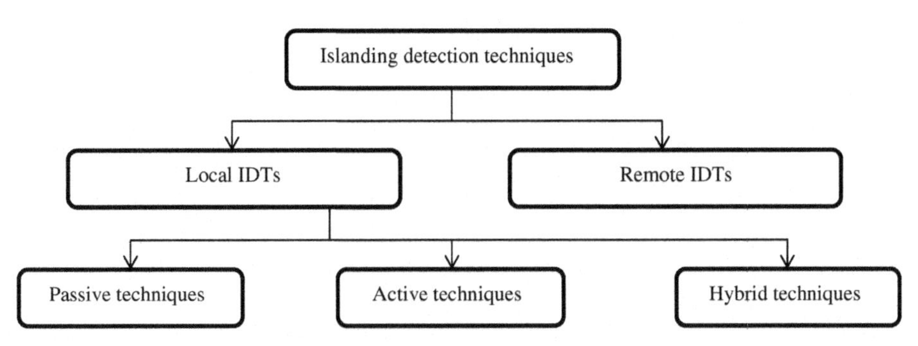

FIGURE 17.2 Types of islanding detection techniques (IDTs).

TABLE 17.1 Minimum islanding detection time and quality factors suggested by international standards.

Standard	Quality factor	Island detection time (s)
IEEE 1547 [37]	1	2
IEC 62116 [38]	1	2
Korean standard	1	0.5
UL 1741 [39]	≤1.8	2
VDE 0126-1-1 [40]	2	0.2
IEEE 929-2000 [41]	2.5	2

17.4 Local islanding detection techniques

Local techniques are confined to the parameters available at the point of connection or at the output of the DG terminals. The schematic diagram of a typical test system is shown in Fig. 17.3. The DG can be inertial (rotating machines) or noninertial (inverter based).

17.4.1 Passive techniques

In passive techniques, the parameters such as voltage magnitude, phase angle, frequency, and total harmonic distortion (THD) at the output terminal of the DGs or at the point of connection are measured and analyzed to detect UIF. The basic flowchart of passive techniques is depicted in Fig. 17.4.

If the measured value of the parameters such as voltage magnitude, frequency, harmonic distortions etc. exceeds the threshold continuously for more than the intentional time delay, then the islanding detection signal (IDS) is set to logical 'HIGH', indicating the islanded condition, else IDS remains "LOW." The intentional time delay assists the system not to mal-detect the transient conditions as islanded conditions. The intentional time delay (t_{in}) helps not to mal-detect the nonislanded as islanded condition.

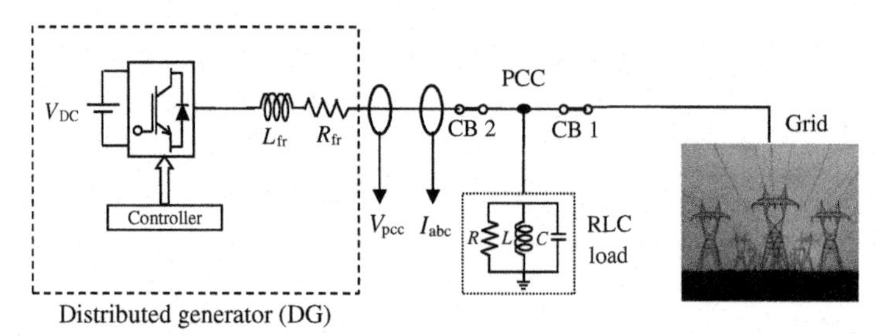

FIGURE 17.3 Schematic diagram of a typical test system.
V_{pcc} is the three phase instantaneous voltage at point of common coupling or point of connection or at the output terminal of the DG; I_{abc} is the three phase instantaneous output current of the DG; L_{fr} is the filter inductance; and R_{fr} is the resistance of the filter inductor.

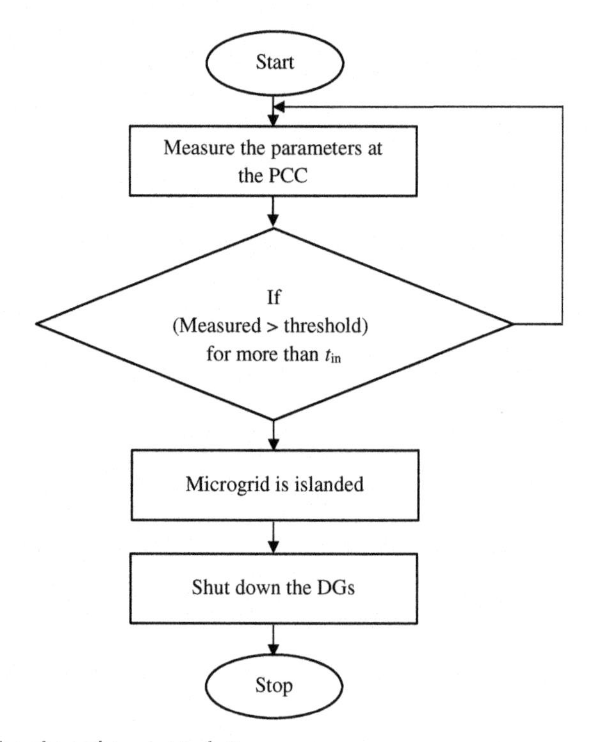

FIGURE 17.4 Basic flowchart of passive techniques.

17.4.1.1 Over/under voltage and frequency

The basic passive IDTs depend on over/under voltage (OUV) and over/under frequency (OUF). Whenever there is a distinct power mismatch between generation and demand (real power or reactive power), there will be a remarkable change in voltage or frequency. Hence, the island formation is detected easily by using OUV and OUF. IEA PVPS T5-09 [7]. However, on a less power mismatched condition or perfectly matched power condition, the voltage or frequency variations from nominal are insignificant.

So, the OUV and OUF become incapable of detecting island formation. The changes in real and reactive power are given by the following expressions,

$$\Delta P = P_{Demand} - P_{Generation} \tag{17.6}$$

$$\Delta Q = Q_{Demand} - Q_{Generation} \tag{17.7}$$

The change in voltage magnitude in post-islanding conditions is given by the following expression

$$V_{islanded} = V_{pre\text{-}islanded} + \Delta V \tag{17.8}$$

where, $V_{islanded}$, and $V_{pre\text{-}islanded}$ are pre and post-islanding voltage at point of common coupling (PCC); ΔV is based on the active power mismatch between the generation and demand. ΔV is positive if $P_{generation} > P_{demand}$ and ΔV is negative if $P_{generation} < P_{demand}$. Similarly, the frequency of the voltage signal in a post-islanding condition is based on the reactive power mismatch between the generation and demand.

$$f_{islanded} = f_{pre\text{-}islanded} + \Delta f \tag{17.9}$$

When there is a considerable mismatch in active and reactive power, there will be a substantial change in voltage magnitude and frequency respectively on post-islanding conditions. On measuring the voltage magnitude and frequency and comparing these with the predefined thresholds/operating ranges, the island formation is detected. There need not be any physical OUV/OUF relays, since the controller itself is incorporated with OUV/OUF functions in order to detect island formation as well as to protect the DG from OUV and OUF. Whenever the voltage magnitude or frequency exceeds their operating limits, the DG needs to take the necessary actions.

If $P_{generation} = P_{demand}$, and $Q_{generation} = Q_{demand}$, then the changes in voltage magnitude and frequency of the voltage do not change significantly in a post-islanding condition. Nonetheless, these techniques suffer from the nondetection zone problem during a close or less power mismatch between generation and demand. Later on, several passive IDTs were developed to detect island formation based on analyzing the parameters such as voltage magnitude, frequency, voltage phase angle, impedance, and THD, etc.

17.4.1.2 *Rate of change of frequency*

When the MG is isolated, the power mismatch conditions cause a notable variation in frequency in the isolated system. Upon evaluating the rate of change of frequency (ROCOF) for one or more cycles, the island formation is detected easily owing to the remarkable variation in ROCOF on an islanded system. Freitas et al. [8] have evaluated ROCOF shown in (17.10) which exceeds the predefined threshold for more than the deliberate time delay, and the island detection signal is set to logical HIGH which indicates the island formation. Thereafter, the DGs in the isolated system are turned OFF within four or five cycles. For small- and medium-rated DGs, the turn OFF time is between 0.3 and 0.7 seconds for a frequency variation of 0.3 Hz/s. The step-by-step procedure of islanding detection is shown as a block diagram in Fig. 17.5.

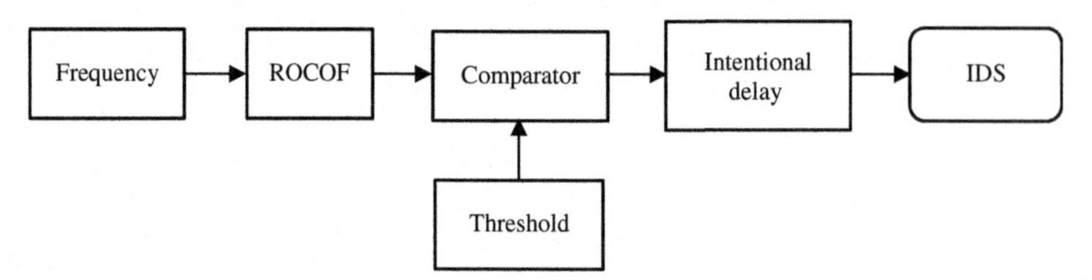

FIGURE 17.5 Block diagram of passive islanding detection technique (IDT) based on rate of change of frequency (ROCOF).

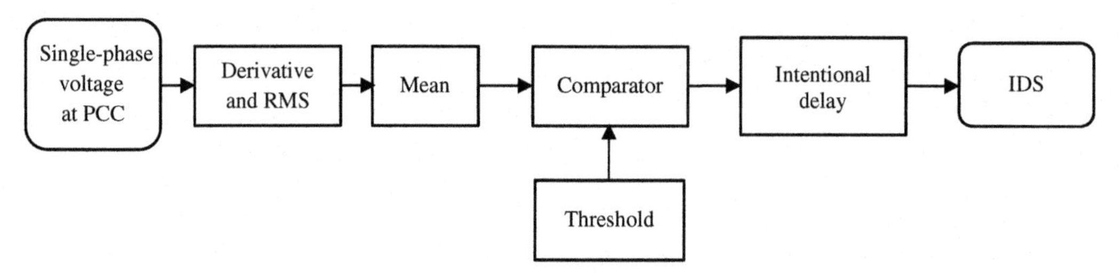

FIGURE 17.6 Block diagram of passive islanding detection technique (IDT) based on voltage ripples.

$$\frac{df}{dt} = \frac{\Delta P \times f}{2 \times H \times G} \tag{17.10}$$

where ΔP is the variation in output power; f is the system frequency; H is the inertia constant of the generator; G is the rated capacity of the generator; and IDS is the islanding detection signal.

17.4.1.3 *Impact of islanding on point of common coupling voltage*

Typically, the voltage magnitude at PCC remains constant at the grid-connected mode of operation. On the contrary, during the islanded mode of operation, the voltage magnitude experiences abrupt and continued variations. Hence, the voltage ripples in the root mean square (RMS) voltages at PCC are determined and compared with a predefined threshold. When the computed value exceeds the threshold for more than the time delay, then the IDS is set as "1," indicating the islanded condition; otherwise it is a grid-connected system [9]. The step-by-step procedure of islanding detection is shown as a block diagram in Fig. 17.6.

In general, a phase locked loop (PLL) is used with all the inverter-based DGs and phase angle is obtained from the PLL. The energy of the rate of change of the phase angle is observed and used as a detection index. Whenever, the observed value exceeds the defined threshold for more than the intentional time delay, it is considered an islanded condition. This technique has a negligible nondetection zone [10].

17.4.1.4 *Combination of one or more passive methods*

The accuracy of detecting island formation can be improved by using multivariable detection indices. The islanding detection time and nondetection zone are reduced by using combinations of change in active power, reactive power, frequency, and voltage phase angle [11].

17.4.1.5 *Harmonic components*

In general, harmonic components in voltage signals are undesirable from a power system point of view. However, they can act as an effective islanding detection index too when appropriately handled [12]. The harmonic distortions in voltage waveform are negligible in the grid-connected mode of operation since the grid acts as a stiff voltage source. Upon island formation, the voltage distortions vary considerably owing to the current harmonics generated by the inverter. Subsequently, if the total harmonic distortion of the voltage waveform exceeds a certain threshold, it is considered to be an islanded condition. So, the PCC voltage signals are measured and the voltage magnitudes of specific harmonic components such as third, fifth, and seventh are evaluated and compared with their predefined threshold to detect and distinguish between the islanding and nonislanding conditions.

17.4.1.6 *Phase jump*

In the-grid-connected mode of operation, the output current of the inverter is synchronized with the voltage at the point of connection, generally attained through PLL. However, during the islanded condition, there will be an abrupt jump in the voltage phase angle which results in a large phase angle difference between the voltage at the point of connection and the inverter output current signals. Hence, the abrupt phase jump of voltage at the point of connection as well as the phase error between the voltage at the point of connection and the current signals are used as a detection index. When the phase error exceeds the prespecified voltage for a certain period of time, then it is identified as an islanded condition [2].

17.4.1.7 *Impedance change*

The impedance offered by the islanded system is remarkably greater than the grid-connected system. The alteration in impedance due to the shift from grid-connected to islanded mode of operation is used to detect the island formation. The computed impedance is compared with that of the predefined threshold to discriminate the islanding from the nonislanding conditions [13,14].

17.4.1.8 *Oscillators*

The measured frequency is fed as input to the duffing oscillator, and the change from "chaotic state" to "great periodic state" and vice versa of the duffing oscillator is used to detect the island formation [15]. Similarly, the apparent change between chaotic and regular motions in a forced Helmholtz oscillator is used to detect the island formation [16]. Even though these techniques detect UIF effectively, the IDTEs of these techniques are higher than a few standards.

17.4.1.9 *Exploration of passive techniques*

Passive techniques do not affect the power quality in both modes of operation (grid-connected and islanded), do not destabilize the system on post islanding, are suitable for both single and multiple DG systems, are applicable to both inertial and noninertial DGs, and are easy and cheaper to implement. Most of the passive techniques suffer from large nondetection zones and decision on a threshold value also becomes a tedious task.

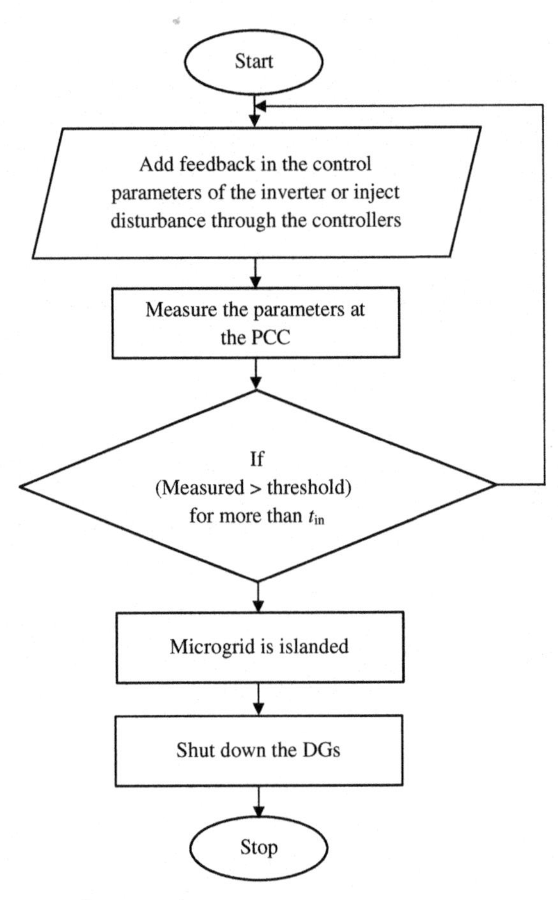

FIGURE 17.7 Basic flowchart of active techniques.

17.4.2 Active techniques

Active techniques add positive or negative feedback to the control parameters of the converter or inject disturbance deliberately through the DG(s) into the system and, subsequently measure the parameters at the output terminal of the DGs or at the point of connection and are analyzed to detect UIF. Adding feedback to the control parameters or injecting disturbance into the system make the parameters vary notably in an islanded condition. Subsequently, the parameters are analyzed to detect and distinguish between the UIF and nonislanded conditions.

The basic flowchart of active techniques is shown in Fig. 17.7.

The earlier works, which have proposed active IDTs based on adding positive or negative feedback to the control parameters, are discussed and categorized according to the detection indices.

17.4.2.1 Active frequency drift

In this technique, the inverter injects distorted current into the PCC and slightly deviates with respect to the PCC voltage. Upon island formation, the frequency of voltage signal at the PCC drifts up or down which causes the frequency to violate, upper or lower, operating limits. The nondetection zone in active frequency drift (AFD) is reduced by implementing positive feedback. Ropp et al. [17]. The chopping fraction (*cf*) is used as a detection index. The value of *cf* given in (17.11) is less in the grid-connected mode of operation whereas it increases in the islanded condition. Once the *cf* value exceeds the prespecified threshold, it is considered to be an islanded condition.

$$cf = 2\frac{t_{\mathrm{d}}}{T_{\mathrm{Vgrid}}} \tag{17.11}$$

where t_{d} is the dead time or zero time and T_{Vgrid} is the time period of voltage waveform at PCC.

17.4.2.2 Sandia frequency drift

This technique is an extension of AFD which adds positive feedback to the frequency signal obtained from the PLL. In the grid-connected mode of operation, the positive feedback does not affect the frequency. However, once the MG is islanded, the positive feedback causes the frequency to exceed the operating limit. Eventually, the UIF is detected. Wang et al. [18]. The equation of the *cf* with feedback gain is expressed in the following equation,

$$cf = cf_0 + K(f_{\mathrm{pcc}} - f) \tag{17.12}$$

where cf_0 is the cf in absence of any error; K is the gain; f_{pcc} is the frequency of the voltage signal at PCC; and f is the nominal frequency.

17.4.2.3 Slip mode frequency shift

The variation in frequency of the voltage signal at PCC is used as feedback to the phase angle controller of the inverter. During the presence of grid, the feedback to the phase angle controller does not have any impact on the output frequency of the inverter since the grid acts a stiff source. Since the feedback depends on the variation of frequency at PCC, when the MG is disconnected, the frequency keeps on varying and exceeding the operating limits or threshold. Therefore the island formation is detected [19].

17.4.2.4 Voltage phase angle

In this technique, the variation of voltage phase angle is applied as a positive feedback to the control parameters of the inverter to detect the island formation [20]. When the MG is islanded, the real and reactive power references are varied according to the change in phase angle as expressed in (17.13)–(17.15). The voltage and frequency are intended to operate outside the operating limits during post-islanding conditions, and on measuring voltage and frequency, the island formation is detected. The system performance in the grid-connected mode of operation remains unaffected by positive feedback.

$$P = k \times \sin(\delta) \tag{17.13}$$

$$Q = k \times \cos(\delta) \tag{17.14}$$

$$k = \frac{V_{\mathrm{DG}} \times V_{\mathrm{PCC}}}{X_f} \tag{17.15}$$

17.4.2.5 *Correlation between* P–V *and* Q–f

The correlation between real power and voltage, and reactive power and frequency droop are used to detect the island formation, which are expressed in (17.16) and (17.17). Zeineldin and Kirtley [21,22] The MG system operates well during the grid-connected mode of operation. However, it loses its stability during the post-islanding condition, since the variation of voltage and frequency of the voltage signal at PCC are utilized to vary the active power and reactive power references. Just by implementing the OUV/OUF protection methods, the island formation is effectively detected.

$$P_{\mathrm{ref}} = P_0 + 2P_0(V - V_0) \tag{17.16}$$

$$Q_{\mathrm{ref}} = -2P_{\mathrm{ref}}\frac{Q_{\mathrm{f}}}{f_0}(f - f_0) \tag{17.17}$$

17.4.2.6 *Signal injection*

The islanding detection based on disturbance injection through the converters is a suitable solution for detecting the island formation without destabilizing the system with a zero nondetection zone. During the grid-connected mode of operation, the disturbance signal flows into the grid owing to the low impedance offered by the grid as shown in Fig. 17.8 and the fundamental current flows into the load or grid depending upon the necessity.

Nonetheless, on the islanded mode of operation, the disturbance signal is forced to flow into the load due to the absence in the grid of a low-impedance path as shown in Fig. 17.9, which makes the parameters such as voltage and frequency vary significantly.

- *Low-frequency signal injection*

  Injecting a low frequency (LF) signal through the *d*-axis and *q*-axis controller makes a notable variation in the voltage magnitude or frequency of the voltage signal, respectively, at PCC during the post-islanding condition [23]. When the disturbance is injected through the *d*-axis controller, the UIF is detected effectively even with a smaller change in PCC voltage using a hybrid analyzing technique, which is based on the mean

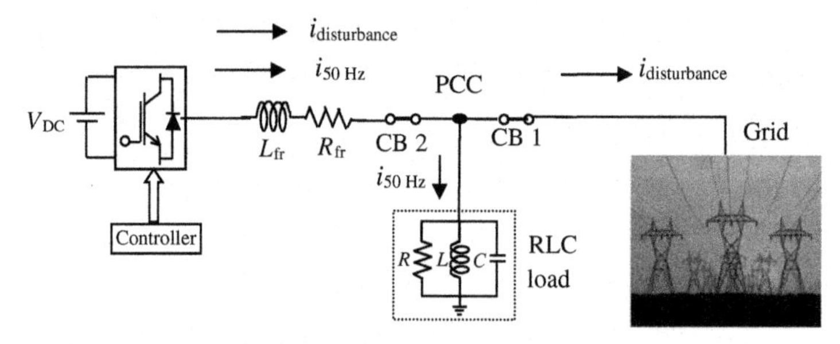

FIGURE 17.8 Effect of disturbance injection in grid-connected mode of operation.

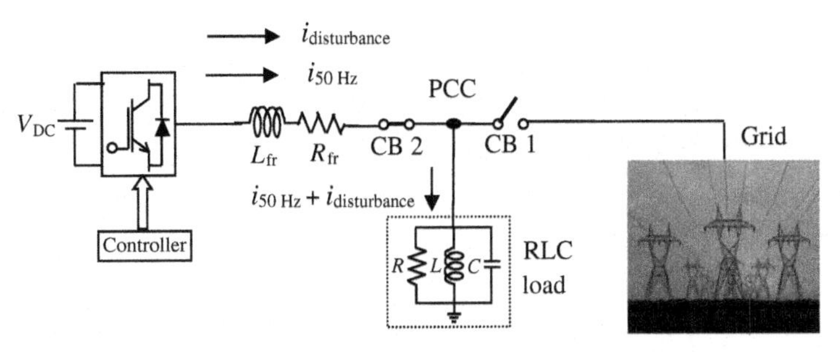

FIGURE 17.9 Effect of disturbance injection in the islanded mode of operation.

of absolute d-axis voltage variation and the mean of absolute rate of change of the d-axis voltage variation [24]. The d-axis voltage variation (v_d) at PCC on the post-islanding condition follows the following expression,

$$v_d = V_{\omega_0} + V_{\omega_1}(\sin(\omega_d t + \phi_1)) + V_{\omega_2}(\sin(\omega_d t + \phi_2)) \tag{17.18}$$

$$\text{where}\phi_1 = -\tan^{-1}\left(R\left(\omega_i C - \frac{1}{\omega_i L}\right)\right), i = 1,2 \tag{17.19}$$

$$V_{\omega_0} = \frac{i_d}{\left(\left(\frac{1}{R}\right)^2 + \left(\omega_0 C - \frac{1}{\omega_0 L}\right)^2\right)^{\frac{1}{2}}} \tag{17.20}$$

$$V_{\omega_2} = \frac{i_{dr}}{2\left(\left(\frac{1}{R}\right)^2 + \left(\omega_i C - \frac{1}{\omega_i L}\right)^2\right)^{\frac{1}{2}}}, i = 1,2 \tag{17.21}$$

When the disturbance is injected through the q-axis controller, the UIF is identified effectively upon analyzing the frequency of the voltage signal using the rate of change of frequency. The island detection time and accuracy can be varied according to the analyzing methodology [25,26]. This technique detects the UIF with a zero nondetection zone. However, applying this technique to multiple DG scenarios is a challenging task owing to the possibilities of disturbance signal interactions of various DGs.

- *Negative sequence current injection*

  Injecting a negative sequence current through the converter causes a substantial change in negative sequence voltage on an islanded system [27]. The island formation is detected when a negative sequence voltage or a negative sequence impedance exceed their thresholds. Even though there is a negative sequence current injection, these techniques do not destabilize the system during the post-islanding condition, unlike the techniques based on adding feedback to the control parameters of the inverter.

- *High-frequency signal injection (HFSI)*

  In this method of HFSI, the DG injects an HF voltage/current signal into the system, and on measuring and analyzing the voltage or impedance at PCC, the UIF formation is detected. The variations of voltage or impedance due to the injected HF signal, are insignificant in the grid-connected mode of operation. On the contrary, the variations are significant on islanded condition. On evaluating voltage or impedance, the islanded condition is detected easily [28,29]. Though HFSI helps to detect the island formation effectively, the major issues with the HFSI are the facts that the injected HF signal should not be removed by the active power filter connected at the same PCC and it is difficult to eliminate the transient harmonics using digital filters.

17.4.2.7 Exploration on active techniques

Merits

- They possess zero or less nondetection zone.
- Detection time is shorter.

Demerits

- Most of the techniques are not applicable to both inertial and noninertial DGs.
- Most of the active techniques destabilize the system on post-islanding condition.
- There is a slight or significant degradation of power quality during the grid-connected mode of operation.
- Most of the techniques are not suitable for a multiple DG scenario.

17.4.3 Hybrid techniques

Hybrid techniques use passive techniques permanently and implement active techniques depending on the necessity. The basic flowchart of hybrid techniques is shown in Fig. 17.10.

17.4.3.1 Average rate of change of voltage and real power shift

This technique detects island formation based on the average rate of change of voltage (ARCV) for higher power mismatches. At lower power mismatches, ARCV is unable to differentiate the islanding condition owing to insufficient voltage variation and subsequently the active technique based on real power shift is introduced. Due to the real power shift, the voltage signal violates the operating limits, and then ARCV easily distinguishes the island formation at this condition [30].

17.4.3.2 Decision tree and positive feedback

A decision tree (DT) classifier as a passive technique and positive feedback signals as an active technique are implemented to detect island formation [31]. The DT classifier is used to distinguish the islanding from the nonislanding events during higher power mismatches. On the other hand, if the power mismatch is less, then the positive feedback schemes are encompassed based on the variation of voltage and frequency. Due to the positive feedback, the voltage and frequency deviate notably, and the DT detects the island formation.

17.4.3.3 Rate of change of frequency and Sandia frequency shift

The Sandia frequency shift (SFS) technique devises the frequency to operate outside the operating range on post-islanding, but it is augmented into the control parameters of the inverter only if the ROCOF is unable to distinguish the islanding from the nonislanding condition [32].

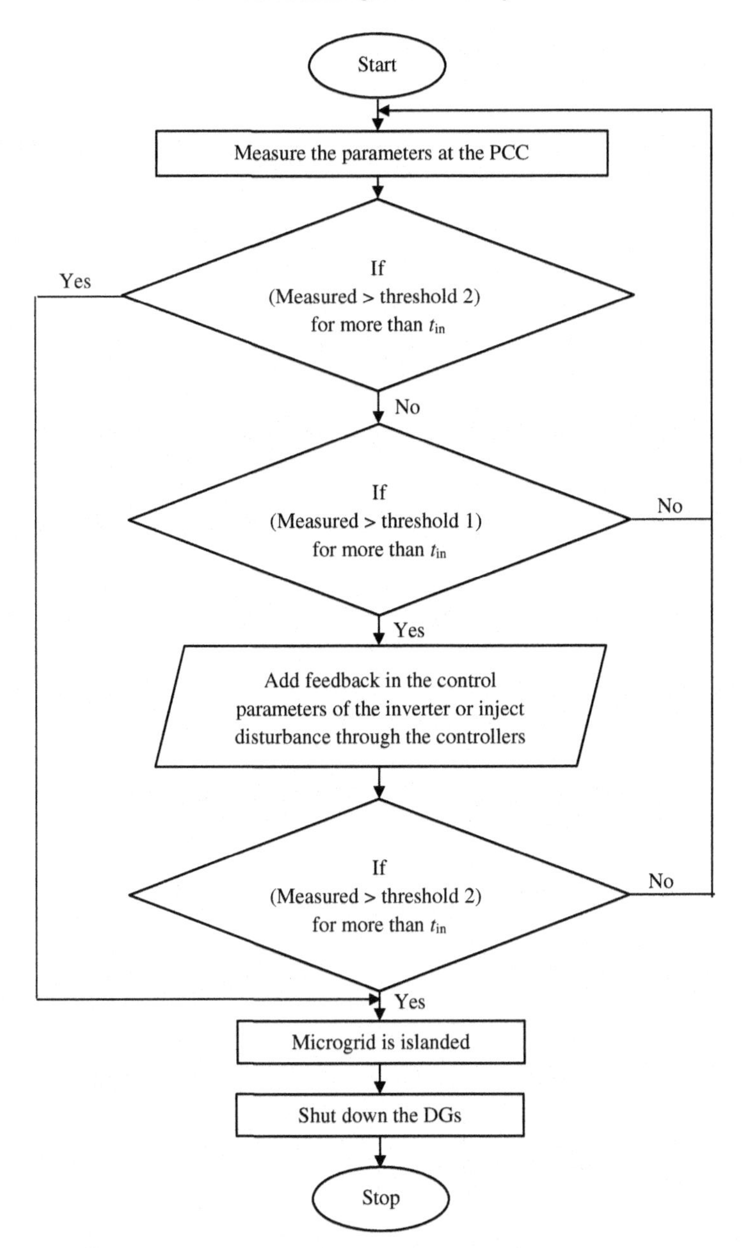

FIGURE 17.10 Basic flowchart of hybrid techniques.

17.4.3.4 Wavelet energy entropy and positive feedback

In the first level, wavelet energy entropy is used as an indicator to inspect the possibility of islanding and initiates the active technique in the second level upon the incapable identification of island formation. Shrivastava et al. [33]. Active frequency drift is used as an active technique which accelerates the frequency to move outside the operating range on post islanding by giving positive feedback to the control parameters.

17.4.3.5 *Exploration on hybrid techniques*

Merits

- They possess zero or less nondetection zone.
- Detection time is higher than passive and active techniques.
- Power quality degradation during the grid-connected mode of operation is reduced notably compared to active techniques.

Demerits

- Most of the techniques are not applicable to both inertial and noninertial DGs.
- Most of the active techniques destabilize the system during the post islanding condition.
- Most of the techniques are not suitable for a multiple-DG scenario.

17.5 Remote islanding detection techniques

Remote IDTs use wired or wireless communication between the substation/breaker and the DG to detect islanding formation.

17.5.1 Power line carrier communication

Signal is broadcast along the distribution feeder (power line) from the substation to the DG site. If the signal is not established at the DG site then it is concluded as an islanded condition. The transmitter (T) and receiver (R) are implemented at the sending (substation) and receiving end (DG site) respectively as shown in Fig. 17.11. The signal sent over the power line has to flow through the CB, switches or any other circuit-disconnecting component existing between the substation and DG site. Hence, the disconnection between the substation and DG site is detected automatically by this method, known as power line carrier communication (PLCC). This method does not need any telecom lines and the fact that it is independent from the system configuration is an added advantage [34].

17.5.2 Signal produced by disconnect

Even though this technique is similar to the PLCC method, the power line is not used as a mode to transfer the signals in this approach. The switch or CB status is continuously transferred to the DG site through dedicated telephone lines, microwave, and radio wave

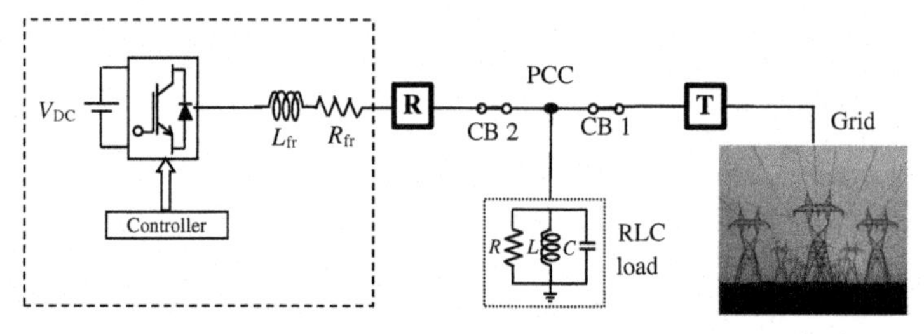

FIGURE 17.11 Schematic diagram of a power line carrier communication (PLCC) system.

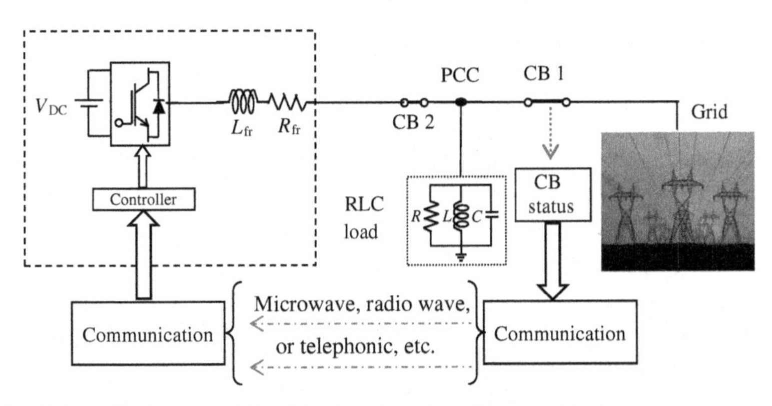

FIGURE 17.12 Schematic diagram of the signal produced by disconnect technique.

communication etc., as shown in Fig. 17.12. This method is expensive owing to the necessity of having dedicated communication lines between the DG site and the circuit-disconnecting components. IEA PVPS T5-09 [7].

17.5.3 Transfer trip scheme

The status of the CB, switch, and recloser are monitored continually since they are the means of circuit disconnection. If the MG is islanded due to the opening of the breaker, recloser, or switch, the status of the disconnecting components is divulged to the centralized controller. The centralized controller sends the signal to trip the DGs. Walling [35]. However, to provide decentralized control, the state of the disconnecting components can be sent to each DG separately. For better coordination and control of the DGs, a supervisory control and data acquisition (SCADA) system is used along with the transfer trip scheme. The SCADA system encompasses a wide communication system for monitoring and control purposes. Upon disconnection from the grid, several alarm signals are activated for disconnecting the DGs.

17.5.4 Phasor measurement units

The parameters measured in different phasor measurement units are sent to a centralized controller to detect the island formation. Skok et al. [36]. The synchronized data sent from the various PMUs are compared with the default value. Iif the measured value exceeds the default value, it is identified as the sign of island formation. Hence, the DGs should be turned OFF based on the indication from the centralized controller.

17.5.5 Exploration of remote techniques

Remote techniques are suitable for environments with both single and multiple DGs; have zero nondetection zone; are applicable to both inertial and noninertial DGs; have fewer detection times; do not degrade the power quality; and do not destabilize the system on post-islanding condition.

Irrespective of numerous advantages, the initial investment cost is too high to make remote techniques applicable for small MGs.

17.6 Results and discussion

According to several international standards, DGs should be turned OFF within the prescribed time to safeguard working personnel and electrical equipment during an UIF. The test system of a pre-islanded condition is shown in Fig. 17.13. It is simulated for perfectly matched power conditions, in which the power generation in the islanded system is exactly equal to the demand. In order to mimic UIF, the DG and RLC load are isolated from the grid at 1 second by opening the circuit breaker (CB 1) and it is shown in Fig. 17.14. The voltage and frequency responses are depicted in Figs. 17.15 and 17.16 respectively for pre- and post-islanded conditions.

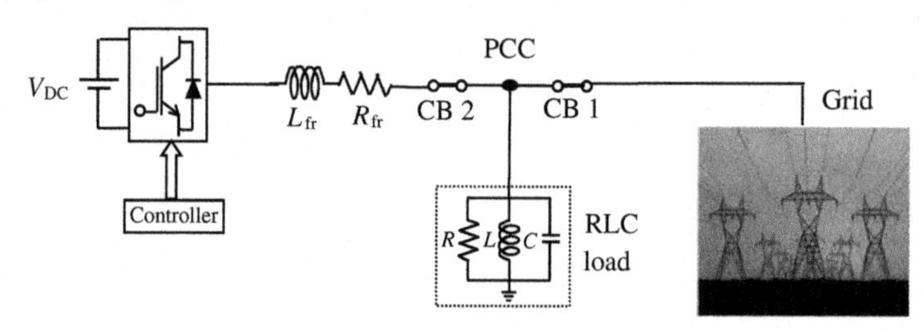

FIGURE 17.13 Test system of a preislanded condition.

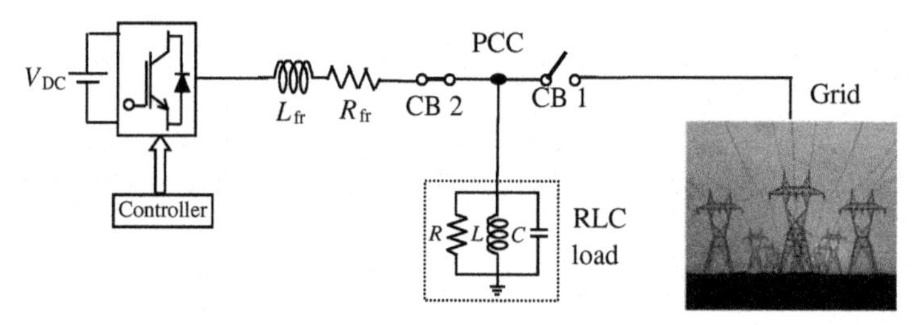

FIGURE 17.14 Test system of a postislanded condition.

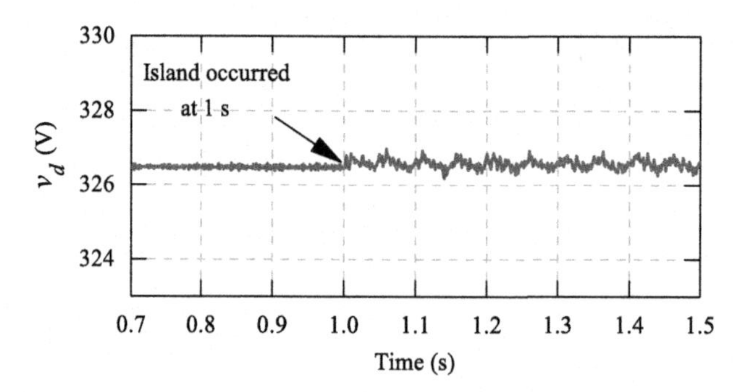

FIGURE 17.15 *d*-axis voltage response of pre- and past-islanded conditions.

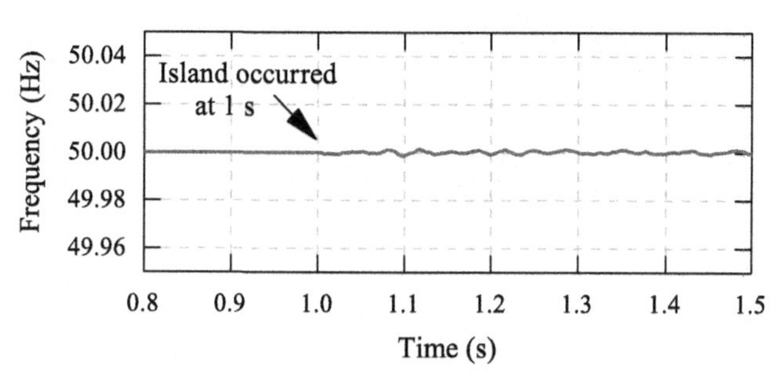

FIGURE 17.16 Frequency response of pre- and post-islanded conditions.

17.6.1 *d*-axis voltage response of pre- and post-islanding for perfectly matched power conditions

The *d*-axis voltage variation of pre- and post-islanding for a perfectly matched power condition is shown in Fig. 17.15.

17.6.2 Frequency response of pre- and post-islanding for perfectly matched power conditions.

The frequency variation of pre- and post-islanding for a perfectly matched power condition is shown in Fig. 17.16.

It is clear from Figs. 17.15 and 17.16 that there is no notable variation in voltage and frequency between pre- and post-islanding condition for perfectly matched power conditions. Due to the insignificant variation, most of the passive techniques fail to detect the island formation. To overcome this drawback, active techniques are used. Active techniques based on adding positive or negative feedback to the control parameters make the voltage or frequency drift outside the higher or lower operating limit on post-islanding condition [17–22]. However, these techniques destabilize the system on post-islanding condition. Active techniques without destabilizing the system on post-islanding condition are proposed [5,24–26] based on disturbance injection, which make the voltage or frequency to vary depending on the type of injection. The island formation is detected easily on analyzing the voltage or frequency or some other parameters.

17.7 Comparisons between various islanding detection techniques

The performance indices of different IDTs are compared and tabulated in Table 17.2.

17.8 Discussion on future scope

- The number of classifier-based passive techniques is increasing steadily. However, advanced processing capability computers are mandatory for such techniques and the feasibility on the real-time scenario needs to be verified.

TABLE 17.2 Comparison of performance indices of different islanding detection techniques (IDTs).

Technique	Nondetection zone (NDZ)	Islanding detection time (IDTE)	Feasibility in multiple distributed generators (DGs) scenario	Implementation cost	Effect on power quality
Passive	Large	Less	Preferred	Cheap	Nil
Active	Small/zero	Less	Not preferred	Cheap	Highly degrades
Hybrid	Small/zero	More	Not preferred	Cheap	Slightly degrades
Remote	Zero	More	Highly preferred	Too costly	Nil

- The active technique suitable for both inertial and noninertial DGs is still a wide-open area for research.
- The hybrid techniques suitable for both inertial and noninertial DGs and multiple DG environments can be developed because the power quality degradation during the grid-connected mode of operation is eliminated.
- In future, the implementation of smart technologies including smart meters and phasor measurement units in the distribution system can be used as a reliable and effective remote islanding detection method without any additional cost, which reduces the cost of remote methods drastically.

17.9 Conclusion

The detection of unintentional island formation is critical to safeguard working personnel and electrical equipment. The most crucial condition in detecting the UIF is the perfectly matched power condition, where the power generation in the islanded system is equal to the demand. At this operating condition, there is no significant change in frequency or voltage on the post-islanding condition. Several IDTs have been developed so far to detect the UIF and also to distinguish between islanded and nonislanded conditions. However, there is hardly any technique that works effectively for all operating conditions. An outperforming IDT is required to work efficiently to detect the islanded condition faster and also to discriminate accurately between nonislanded and islanded conditions.

References

[1] Z. Ye, A. Kolwalkar, Y. Zhang, P. Du, R. Walling, Evaluation of anti islanding schemes based on nondetection zone concept, IEEE Trans. Power Electron. 19 (5) (2004) 1171–1176.
[2] M.E. Ropp, M. Begovic, A. Rohatgi, G.A. Kern, R.H. Bonn, S. Gonzalez, Determining the relative effectiveness of islanding detection methods using phase criteria and nondetection zones, IEEE Trans. Energy Convers. 15 (3) (2000) 290–296.
[3] L.A.C. Lopes, Huili Sun, Performance assessment of active frequency drifting islanding detection methods, in IEEE Transactions on Energy Conversion. 21 (1) (2006) 171–180.

[4] C. Li, C. Cao, Y. Cao, Y. Kuang, L. Zeng, B. Fang, A review of islanding detection methods for microgrid, Renew. Sustain. Energy Rev. 35 (2014) 211–220.

[5] S. Murugesan, V. Murali, Hybrid analysing technique based active islanding detection for multiple DGs, in: IEEE Transactions on Industrial Informatics. 15 (3) (2019) 1311–1320.

[6] S. Dutta, P.K. Sadhu, M.J.B. Reddy, D.K. Mohanty, Shifting of research trends in islanding detection method—a comprehensive survey, Prot. Control Mod. Power Syst. (3) (2018).

[7] IEA PVPS T5-09, Evaluation of islanding detection methods for photovoltaic utility interactive power systems, in: International Energy Agency Implementing agreement on Photovoltaic Power Systems, U.S.A., 2002.

[8] W. Freitas, W. Xu, C.M. Affonso, Z. Huang, Comparative analysis between ROCOF and vector surge relays for distributed generation applications, IEEE Trans. Power Deliv. 20 (2) (2005) 1315–1324.

[9] B. Guha, R.J. Haddad, Y. Kalaani, Voltage ripple-based passive islanding detection technique for grid-connected photovoltaic inverters, IEEE Power Energy Technol. Syst. J. 3 (4) (2016) 143–154.

[10] H. Samet, F. Hashemi, T. Ghanbari, Islanding detection method for inverter-based distributed generation with negligible non-detection zone using energy of rate of change of voltage phase angle, IET Gener. Transm. Distrib. 9 (15) (2015) 2337–2350.

[11] R. Sirjani, C.F. Okwose, Combining two techniques to develop a novel islanding detection method for distributed generation units, Measurement 81 (2016) 66–79.

[12] J. Merino, P. Mendoza-Araya, G. Venkataramanan, M. Baysal, Islanding detection in microgrids using harmonic signatures, IEEE Trans. Power Deliv. 30 (5) (2015) 2102–2109.

[13] N. Liu, A. Aljankawey, C. Diduch, L. Chang, J. Su, Passive islanding detection approach based on tracking the frequency-dependent impedance change, IEEE Trans.Power Deliv. 30 (6) (2015) 2570–2580.

[14] N. Liu, C. Diduch, L. Chang, J. Su, A reference impedance-based passive islanding detection method for inverter-based distributed generation system, IEEE J. Emerg. Sel. Top. Power Electron. 3 (4) (2015) 1205–1217.

[15] H. Vahedi, G.B. Gharehpetian, M. Karrari, Application of duffing oscillators for passive islanding detection of inverter-based distributed generation units, IEEE Trans. Power Deliv. 27 (4) (2012) 1973–1983.

[16] M. Bakhshi, R. Noroozian, G.B. Gharehpetian, Novel islanding detection method for multiple DGS based on forced Helmholtz oscillator, IEEE Trans. Smart Grid (2018) (in press).

[17] M.E. Ropp, M. Begovic, A. Rohatgi, Analysis and performance assessment of the active frequency drift method of islanding prevention, IEEE Trans. Energy Convers. 14 (3) (1999) 810–816.

[18] X. Wang, W. Freitas, W. Xu, V. Dinavahi, Impact of DG interface controls on the Sandia frequency shift antiislanding method, IEEE Trans. Energy Convers. 22 (3) (2007) 792–794.

[19] F. Liu, Y. Kang, Y. Zhang, S. Duan, X. Lin, Improved SMS islanding detection method for grid-connected converters, IET renew. Power Gener. 4 (1) (2010) 36–42.

[20] H. Pourbabak, A. Kazemi, A new technique for islanding detection using voltage phase angle of inverter-based DGS, Int. J. Electr. Power Energy Syst. 57 (2014) 198–205.

[21] H.H. Zeineldin, A QF DROOP curve for facilitating islanding detection of inverter based distributed generation, IEEE Trans. Power Electron. 24 (3) (2009) 665–673.

[22] H.H. Zeineldin, J.L. Kirtley, A simple technique for islanding detection with negligible non detection zone, IEEE Trans. Power Deliv. 24 (2) (2009) 779–786.

[23] G. Hernandez-Gonzalez, R. Iravani, Current injection for active islanding detection of electronically-interfaced distributed resources, IEEE Trans. Power Deliv. 21 (3) (2006) 1698–1705.

[24] S. Murugesan, V. Murali, S.A. Daniel, Hybrid analyzing technique for active islanding detection based on d-axis current injection, IEEE Syst. J. 12 (4) (2018) 3608–3617.

[25] S. Murugesan, V. Murali, Q-axis current perturbation based active islanding detection for converter interfaced distributed generators, Turk. J. Electr. Eng. Comput. Sci. 26 (5) (2018) 2633–2647.

[26] A. Emadi, H. Afrakhte, A reference current perturbation method for islanding detection of a multi-inverter system, Electr. Power Syst. Res. 132 (2016) 47–55.

[27] H. Karimi, A. Yazdani, R. Iravani, Negative-sequence current injection for fast islanding detection of a distributed resource unit, IEEE Trans. Power Electron. 23 (1) (2008) 298–307.

[28] K. Jia, H. Wei, T. Bi, D.W.P. Thomas, M. Sumner, An islanding detection method for multi-DG systems based on high-frequency impedance estimation, IEEE Trans. Sustain. Energy 8 (1) (2017) 74–83.

[29] D. Reigosa, F. Briz, C. Blanco, P. Garca, J.M. Guerrero, Active islanding detection for multiple parallel-connected inverter-based distributed generators using high-frequency signal injection, IEEE Trans. Power Electron. 29 (3) (2014) 1192–1199.
[30] P. Mahat, Z. Chen, B. Bak-Jensen, A hybrid islanding detection technique using average rate of voltage change and real power shift, IEEE Trans. Power Deliv. 24 (2) (2009) 764–771.
[31] B. Zhou, C. Cao, C. Li, Y. Cao, C. Chen, Y. Li, et al., Hybrid islanding detection method based on decision tree and positive feedback for distributed generations, IET Gener.Transm. Distrib. 9 (14) (2015) 1819–1825.
[32] M. Khodaparastan, H. Vahedi, F. Khazaeli, H. Oraee, A novel hybrid islanding detection method for inverter-based dgs using SFS and ROCOF, IEEE Trans. Power Deliv. 32 (5) (2017) 2162–2170.
[33] S. Shrivastava, S. Jain, R.K. Nema, V. Chaurasia, Two level islanding detection method for distributed generators in distribution networks, Int. J. Electr. Power Energy Syst. 87 (2017) 222–231.
[34] W. Xu, G. Zhang, C. Li, W. Wang, G. Wang, J. Kliber, A power line signaling based technique for anti-islanding protection of distributed generators—Part I: scheme and analysis, IEEE Trans. Power Deliv. 22 (3) (2007) 1758–1766.
[35] R.A. Walling, Application of direct transfer trip for prevention of DG islanding, In: Proceedings of IEEE Power and Energy Society General Meeting, 2011.
[36] S. Skok, K. Frlan, K. Ugarkovic, Detection and protection of distributed generation from island operation by using pmus, Energy Procedia 141 (2017) 438–442. Power and Energy Systems Engineering.
[37] IEEE Std 1547.4, IEEE guide for design, operation, and integration of distributed resource island systems with electric power systems. Standard, 2011.
[38] IEC 62116, Testing procedure of islanding prevention measures for grid connected photovoltaic power generation systems. Standard, 2008.
[39] UL 1741, Ul standard for safety for inverters, converters, controllers and interconnection system equipment for use with distributed energy resources. Standard, 2010.
[40] DIN-VDE, Automatic disconnection device between a generator and the low voltage grid. Standard, 2005.
[41] IEEE Std 929, IEEE recommended practice for utility interface of photovoltaic (PV) systems. Standard, 2000.

CHAPTER

18

An analysis of the current- and voltage current–based characteristics' impact on relay coordination for an inverter-faced distributed generation connected network

Krutika R. Solanki[1], *Vipul N. Rajput*[1], *Kartik S. Pandya*[2] *and Rajeev Kumar Chauhan*[3]

[1]Electrical Engineering Department, Dr. Jivraj Mehta Institute of Technology, Mogar, India [2]Electrical Engineering Department, Charotar University of Science and Technology, Changa, India [3]Department of Electronics and Instrumentation Engineering, Galgotias Educational Institutions, Greater Noida, India

18.1 Introduction

An assembly of distributed generations (DGs) to the distribution power system includes several benefits such as clean energy, high consistency, and improved quality of power [1]. These DGs can be either nonrenewable types (fuel cell, gas turbine, reciprocating engines, etc.) or renewable types (wind turbine, photovoltaic generation, etc.) [2]. Furthermore, the presence of DGs into the distribution networks brings some challenges to the protection systems due to low fault current level and bidirectional power flow. The inverter-faced distributed generation (IDG) contributes a small amount of fault current (1–2 pu) compared to DGs with rotating machines (5 pu) [3]. This small fault current contribution by IDGs causes the mis-coordination of protective relays. Therefore the protection of an IDG connected network is still a very critical issue.

Distributed Energy Resources in Microgrids
DOI: https://doi.org/10.1016/B978-0-12-817774-7.00018-1

To tackle this issue, several studies have been presented in the literature. As converters used in IDG are a virtuous source of harmonics, the harmonic content-based protection is presented in Ref. [4]. In this study, the total harmonic distortion (THD) is constantly supervised and, if the THD level surpasses a defined value, the local circuit breaker will be tripped. However, this protection scheme does not give satisfactory operation in case of networks with several dynamic loads [5] and against the high impedance fault (HIF) [6]. In the past decade, adaptive protection schemes have become one of the popular protection methods as they readjust the settings of the relays according to the circumstances of the power systems [2,7]. This protection scheme is mainly classified as adaptive overcurrent [8], adaptive differential [9], and adaptive protection with symmetrical components [10]. The main issues associated with these protection schemes include the cost of communication infrastructure, the inability to give protection for looped microgrids, and the fact that they are affected by unbalanced loads and transients [6]. Dewadasa et al. [11] have presented the distance protection for the IDG connected network, but fault resistance can cause an error in impedance measurement [2]. In Ref. [12], the authors have presented voltage-based protection for IDG connected networks. This scheme is able to protect against internal as well as external faults to any zone, but the symmetrical faults, HIFs, and single-phase tripping are not extensively considered [13].

Furthermore, in the literature, various intermediate devices are also utilized to change the fault current level supplied by the DGs such as fault current limiters (FCLs), energy storage devices (ESDs), and static switches. By using an FCL, the fault current supplied by the DGs is reduced to restore the original settings of protective relays [14,15]. Nonetheless, determining the accurate impedance value for FCLs is very difficult in case of higher penetration of DGs. On the other hand, ESDs are used to increase the fault current supplied by the DGs, so protective relays can be operated conventionally [16]. Recently, Samet et al. [17] have suggested using static switches for bypassing the converters in case of a fault, which ultimately maximizes the short-circuit current contribution. Conversely, the use of external devices requires timely maintenance and also increases overall investments. Furthermore, researchers have also suggested some advanced protection schemes for DG connected networks, including differential protection based on time-frequency transform [18], wavelet transformed-based protection [19], traveling wave-based protection [20], and off-nominal frequency injection-based differential frequency protection [21]. However, these protection schemes provide good response against different types of faults, but training for pattern recognition based on real cases, suitable filters to isolate noise in voltage and current signals, and low bandwidth communication systems are required [6].

One of the critical challenges in the protection of an IDG connected network is that the fault current supplied by the IDG is too low to be sensed by the plain overcurrent relays. Therefore the use of conventional overcurrent protection will affect the coordination of relays. On the other hand, according to IEEE Standards 1547, the installations of DGs should not disturb the coordination of the protection system [1]. To address this issue, the modified voltage-based overcurrent characteristic is utilized to develop reliable protection systems for IDG connected networks. In addition, an optimum coordination overcurrent relay is formulated and solved by considering both conventional and modified characteristics of overcurrent relays for an IDG connected 19-bus system. Also, the firefly algorithm (FA) is utilized to resolve the problem of optimum relay coordination for IDG connected networks, and the efficacy of the obtained results is evaluated by comparative analysis.

18.2 Problem formulation for optimum relay coordination

The main objective of optimum relay coordination is to reduce the tripping time of relays by finding the optimum settings of relays, namely, plug setting and time multiplier setting. At the same time, all the constraints associated with coordination and boundary criteria should be satisfied.

18.2.1 Objective function formulation

The objective function (OF) of the optimum coordination of relays is to reduce the sum of the operating time of relays for maximum fault current. It can be defined as follows [15,22,23].

$$\min Z = \sum_{i=1}^{m} T_i \tag{18.1}$$

where Z represents the OF, T_i is the operational time of the relay, and m represents relay numbers.

18.2.2 Constraints formulation

The essential constraints also need to be satisfied, while achieving the goal of minimizing the operating time of relays. These constraints are described in the following subsections.

1. *Relay characteristic*: The inverse definite minimum time characteristic of overcurrent relays is widely used in protection systems. It is expressed by Eq. (18.2) [22].

$$T = \frac{0.14}{I_f/(\text{CT} \times \text{PS})^{0.02} - 1} \times \text{TMS} \tag{18.2}$$

where T denotes the functioning time of the relay, I_f represents the short-circuit current, and TMS and PS are time multiplier setting and plug setting of the relay, respectively.

In Ref. [24], the modified characteristic of the overcurrent relay is presented which considers voltage and current for calculating the tripping time. This characteristic is expressed as:

$$T = \frac{0.14}{I_f/(\text{CT} \times \text{PS})^{0.02} - 1} \times \text{TMS} \times \left(\frac{1}{e^{1-V}}\right)^{\alpha} \tag{18.3}$$

where V is the per unit phase voltage magnitude measured by relay, α is the constant parameter and considered continuously from 0 to 2. This characteristic incorporates the conventional overcurrent characteristic as shown in Eq. (18.2) plus the term associated with fault voltage magnitude. The exponential function of the fault voltage magnitude part can be used to confirm that the tripping time will not become zero for zero magnitude of fault voltage. By setting the value of α as zero, the conventional characteristic of overcurrent relay is obtained. Another feature of this characteristic is that the effect of reduction of voltage on tripping time can be controlled by using the additional parameter of α. Because the tripping time of the relay depends on both current and voltage for calculating the tripping time, this modified characteristic can be implemented to sense the low fault current supplied by IDG.

2. *Coordination criteria*: The coordination criteria can be defined as the required time period between the tripping of primary and backup (P/B) relays for maintaining the selectivity. Essentially, the primary relay should trip first in a faulty situation. If it does not operate, a corresponding backup relay or relays should operate after the coordination time interval (CTI). Normally, the CTI is preferred in the range of 0.3–0.5 seconds for electromechanical relays and 0.1–0.2 seconds for numerical relays [25]. The CTI is mainly required to prevent mis-coordination. It can be defined as:

$$T_{BR} - T_{PR} \geq \text{CTI} \tag{18.4}$$

where T_{BR} and T_{PR} are the tripping time in order of backup and primary relays.

3. *Constraints of time multiplier settings (TMS)*: TMS can be used to control the operational time of the relay. It can be expressed as:

$$\text{TMS}_{\text{min}} \leq \text{TMS} \leq \text{TMS}_{\text{max}} \tag{18.5}$$

where TMS_{min} and TMS_{max} are considered as the upper and lower bounds of TMS of the relay, respectively. The restrictions on TMS values are usually provided by the manufacturer [26].

4. *Constraints of PS*: The bounds on PS provide precautions to the operation of relay in such a way that the relay does not operate in a little overload situation, and similarly it can easily detect the lowest fault current. The boundaries of PS are calculated by the following equations [25]:

$$\text{PS}_{\text{min}} = \max\left(\text{TS}_{\text{min}}, \frac{1.25 \times I_{\text{L,max}}}{\text{CT}}\right) \tag{18.6}$$

$$\text{PS}_{\text{max}} = \min\left(\text{TS}_{\text{max}}, \frac{2 \times I_{\text{f,min}}}{3 \times \text{CT}}\right) \tag{18.7}$$

where PS_{min} and PS_{max} are the calculated lowest and highest bounds of the PS of relay, respectively, $I_{\text{L,max}}$ is the full load current, and $I_{\text{f,min}}$ is the minimum fault current of the relay. The TS_{min} and TS_{max} are the lowest and highest tap settings of the relay, respectively, which is given by the manufacturer, and CT is the current transformer ratio of the relay.

5. *Constraints of operational time of relay*: The operation of relays is restricted to certain minimum and maximum time for maintaining security of the protection system. The bounds on operational time can be expressed as:

$$T_{\text{min}} \leq T \leq T_{\text{max}} \tag{18.8}$$

where T_{min} and T_{max} are the lower and higher operating times of relay, respectively [26].

18.3 Application of the firefly algorithm for the optimum coordination of relays

The FA is a nature-inspired algorithm introduced by Yang [27] and derived from the flashing-light characteristics of fireflies. However, two fundamental functions of the flashing-light are to attract mating partners (communication), and to attract potential prey. In addition, flashing may also serve as a protective warning mechanism. A group of fireflies has the characteristic of moving toward brighter and higher attractive places by the intensity of light which can be considered as the fitness function of the optimization problem [27–29].

Basically, FA uses three ideal rules presented as:

1. Each firefly is presumed to be unisex so that the attraction of fireflies to one another is considered to be irrespective of their sex.
2. The brightness and attractiveness are proportionate to one another, and also both parameters reduce if their distance increases. Therefore the less bright firefly will step in the direction of a brighter one.
3. The brightness capability of fireflies can be found in the view of the fitness function. In the minimization problem, the least fitness function value closely relates to the firefly with maximum light intensity.

1. *Attractiveness*:

   When the distance from the origin rises, the lighting intensity and so attraction is reduced. Both effects of inverse-square rule as well as absorption are estimated by using the Gaussian law as follows:

$$I_r = I_0 \times e^{(-\gamma \times r^2)} \tag{18.9}$$

   where r represents the distance among any two fireflies. I_r presents lighting intensity which varies with distance r, whereas I_0 presents the original intensity of light. γ stands for light absorption constant which maintains the reduction in intensity of light.

   The actual formulation of the attractiveness function of the firefly can be defined as:

$$\beta_r = \beta_0 \times e^{(-\gamma \times r^m)}, \quad m \geq 1 \tag{18.10}$$

   where β_0 notation represents the initial attraction at r equal to 0.

2. *Distance*:

   The distance between any two fireflies p and q at X_p and X_q respectively is calculated as:

$$r_{p,q} = ||X_p - X_q|| = \sqrt{\sum_{i=1}^{j} \left(x_{p,i} - x_{q,i}\right)^2} \tag{18.11}$$

   where $r_{p,q}$ represents the distance among fireflies p and q, $x_{p,i}$ is considered as ith section of the spatial coordinate X_p, and jth stands for numbers of dimension.

3. *Movement*:

   Eq. (18.12) shows the movement of firefly p toward the brighter firefly q.

$$X_p = X_p + \left(\beta_0 \times e^{(-\gamma \times r^m)}\right)\left(X_q - X_p\right) + \left(\alpha \times \varepsilon_p\right) \tag{18.12}$$

where x_p represents the current situation or solution of the firefly and ε_p shows a random value generated by uniformly distribution in [0,1]. Eq. (18.12) consists of three terms, in which first term represents the current place of firefly p. The middle term demonstrates the movement of a less bright firefly toward a brighter firefly. Finally, the third term shows the randomized steps of a firefly in the range of [0,1].

The chief advantage of the FA is that it is able to solve nonlinear and multiobjective problems efficiently. It does not use any velocity problem. It provides a solution with a high convergence rate and does not need a good starting solution to initiate the iteration process [30]. Fig. 18.1 illustrates the FA application for solving the optimum coordination problem of relays.

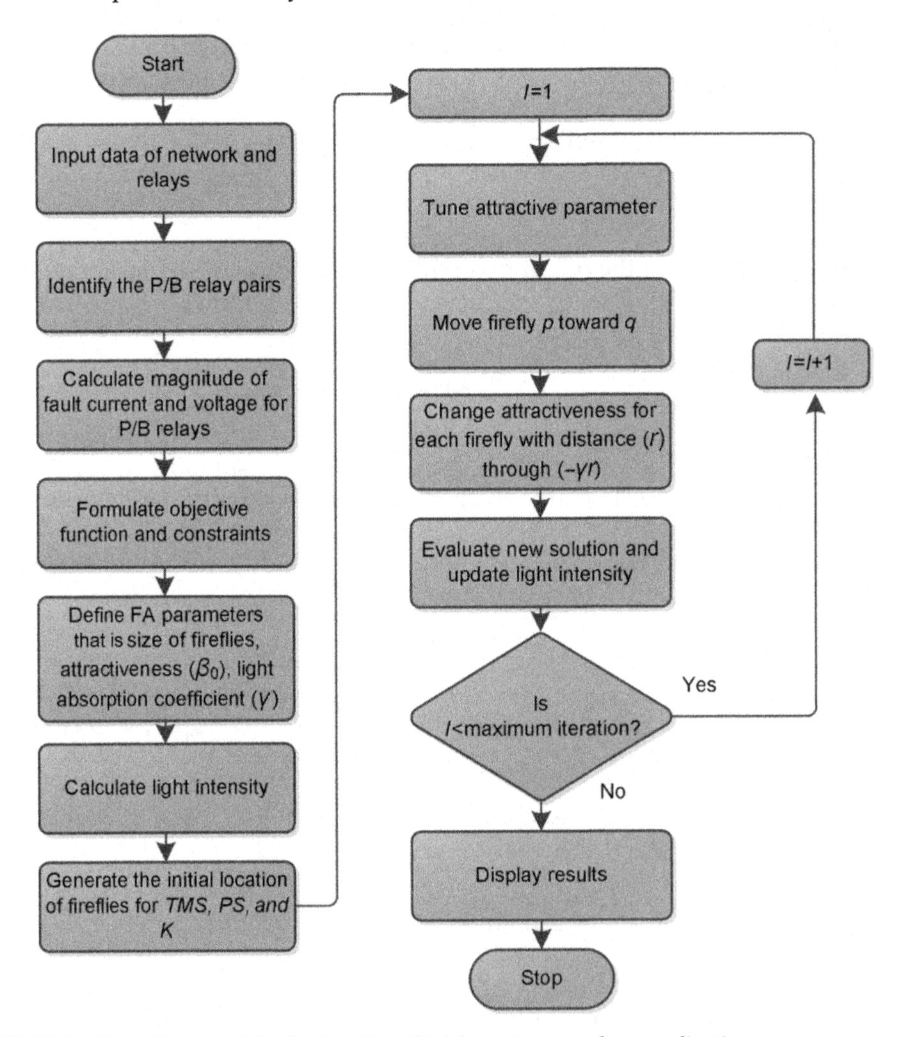

FIGURE 18.1 Flow diagram of firefly algorithm (FA) for optimum relay coordination.

18.4 Test results

This segment describes the implementation of the FA for solving the relay coordination problem on an IDG connected 19-bus test system. A photovoltaic distributed generator (PVDG) has been considered as an IDG. Also, the relay coordination problem has been solved by considering conventional and modified voltage-based overcurrent characteristics. In addition, the comparative analysis is performed to show the viability of a voltage-based protection scheme.

18.4.1 System information and implementation

Fig. 18.2 presents the single-line diagram of three PVDGs connected 19-bus system. These PVDGs are connected at buses 8, 4, and 11 through 22.9/0.4 kV transformers. The rating of transformers is considered equal to those of the DGs [31]. Each PVDG has a rating of 2 MVA. Detailed information of the system can be found in Ref. [32]. The CT ratios associated with their relays are presented in Table 18.1, whereas Table 18.2 provides the short-circuit results of a near-end 3-Φ fault. The TS_{min} and TS_{max} are considered in order of 0.5 and 2.5, respectively. Table 18.3 shows the lower and upper limits of PS for each relay, which are premeditated using Eqs. (18.6) and (18.7). The limits of TMS are taken in a continuous fashion from 0.1 to 1.1. The minimum limits of both CTI and operating time of relays are presumed as 0.2 seconds. The FA parameters are selected as: firefly numbers (n) = 25, attractiveness = 0.2, light absorption coefficient = 1, and randomization = 0.5.

18.4.2 Result and discussion

In case of only the current-based characteristic, TMS and PS are the only variables to be optimized. On the other hand, three variables, namely, PS, TMS, and α, are optimized while considering voltage-based modified characteristic. By executing the FA, the proposed results of relay settings are presented in Table 18.4 for both conventional and improved characteristics. The operating time of primary relays (T_{PR}) and backup relays (T_{BR}), as well as CTI between P/B relay pairs are calculated using these relay settings, which are tabulated in Table 18.5. Also, the total operating time of primary relays and backup relays, along with total CTI between them, are presented in the last row of Table 18.5.

As observed from Table 18.5, the total operating time of primary relays is obtained as 10.2958 seconds by considering the conventional characteristic given by Eq. (18.2), whereas it is 5.8818 seconds for a voltage-based characteristic given by Eq. (18.3). It is noted that the obtained sum of the operational time of all primary relays for a voltage-based characteristic is about 57% less than a conventional characteristic. Furthermore, the sum totals of the operational time of backup relays are 14.1102 and 10.9909 seconds for conventional and modified characteristics respectively. Again, the sum of the operational time of backup relays considering the modified characteristic is about 78% less than a conventional characteristic. In the same way, the obtained total CTI value for the new characteristic is 78% of those obtained using a conventional characteristic. In addition, while

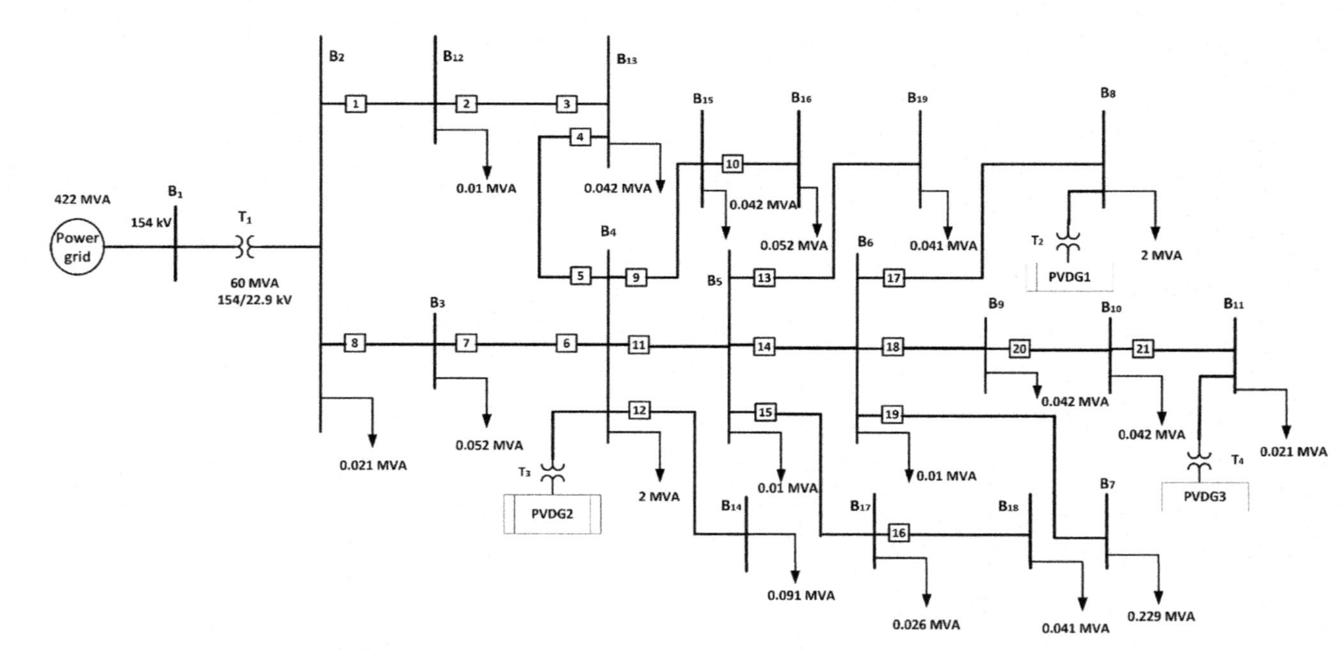

FIGURE 18.2 Single-line diagram of a 19-bus test system.

TABLE 18.1 Current transformer (CT) ratio of relay.

Relay no.	CT ratio
6,7,8,11,14,18,20,21	320/1
1,2,3,4,5,17	240/1
9,10,12,13,15,16,19	160/1

TABLE 18.2 Short-circuit results.

PR	Fault current (A)	Fault voltage (kV)	BR	Fault current (A)	Fault voltage (kV)	PR	Fault current (A)	Fault voltage (kV)	BR	Fault current (A)	Fault voltage (kV)
1	4760	0	–	–	–	11	1790	0	4	452	8.2100
2	952	0	1	952	17.140	12	2050	0	7	1030	0.1400
3	645	0	5	645	11.620	12	2050	0	4	452	8.2100
4	951	0	2	951	0.1100	13	1020	0	11	665	14.510
5	1610	0	7	1030	0.1400	14	679	0	11	665	14.510
6	1030	0	4	452	8.2100	15	1020	0	11	665	14.510
7	1040	0	8	1040	16.310	16	489	0	15	489	12.000
8	4600	0	–	–	–	17	717	0	14	628	1.7200
9	2500	0	7	1030	0.1400	18	933	0	14	628	1.7200
9	2500	0	4	452	8.2100	19	963	0	14	628	1.7200
10	966	0	9	966	12.210	20	935	0	18	935	0.0700
11	1790	0	7	1030	0.1400	21	901	0	20	896	0.9700

TABLE 18.3 Calculated limits of plug settings (PS) for relays.

Relay no.	PS_{min}	PS_{max}
1	0.5	2.5000
2	0.5	2.5000
3	0.5	1.7832
4	0.5	1.2554
5	0.5	1.7776
6	0.5	2.1248
7	0.5	2.1456
8	0.5	2.1456

(Continued)

TABLE 18.3 (Continued)

Relay no.	PS_{min}	PS_{max}
9	0.5	2.5
10	0.5	2.5000
11	0.5	2.5000
12	0.5	2.0685
13	0.5	1.0374
14	0.5	2.5000
15	0.5	2.0290
16	0.5	1.7123
17	0.5	1.4561
18	0.5	2.5000
19	0.5	2.5000
20	0.5	2.4886
21	0.5	1.8623

discussing the operation of individual relays, primary relays are operated near to minimum prescribed time (0.2 seconds) in case of the modified characteristic as seen in Table 18.5.

The average values of the operational time of individual primary and backup relays as well as average CTI values between each P/B relay pair are illustrated in Fig. 18.3. It has been perceived from Fig. 18.3 that the average value of the operational time of primary relay and backup relay in case of conventional characteristic is in order of 0.4903 and 0.6719 seconds, whereas these are 0.2801 and 0.5324 seconds in case of modified characteristic. Also, the average CTI value is obtained as 0.8424 and 0.6553 seconds for conventional and modified characteristic respectively. These results imply that the modified characteristic can improve the tripping time of relays and coordination between P/B relay pairs.

The low level of fault magnitude and responsibility of providing backup protection to more than one relay affect the tripping time of the relay, which ultimately deteriorates the coordination of relays. In the presented work, this problem has been experienced with the relay numbers 14 and 11. Relay 14 senses very low fault current, such as 679 A as primary relay and 628 A as backup relay. Also, this relay provides backup protection to relays 17, 18, and 19. Due to this low fault current and responsibility of the backup protection to three relays, it takes an operational time of 1.0191 seconds for the fault in its zone to be sensed in case of the conventional characteristic. Also, it receipts 1.0784 seconds while functioning as backup protection. The voltage-based modified characteristic is able to solve this problem by involving fault voltage magnitude for calculating the operating

TABLE 18.4 Relay settings obtained using the firefly algorithm (FA).

	Conventional characteristic		Improved characteristic		
Relay no.	TMS	PS	TMS	PS	K
1	0.1398	2.2500	0.2056	1.1032	0.8835
2	0.1133	1.1628	0.9031	0.5873	1.9998
3	0.1000	0.5175	0.1561	0.9199	1.6179
4	0.1000	0.8345	0.2239	1.0268	1.7328
5	0.1416	0.6478	0.1956	1.1991	1.3634
6	0.1000	0.6512	0.1633	0.9845	1.5283
7	0.1227	1.1077	0.2897	1.6391	1.9602
8	0.1001	1.9109	0.4377	1.2423	1.8092
9	0.2591	0.5002	0.4202	1.8712	1.8960
10	0.1002	0.5228	0.1626	1.0255	1.1217
11	0.1000	1.7531	0.2484	1.3229	1.7822
12	0.1000	0.7485	0.4665	1.3614	1.8434
13	0.1001	0.9151	0.4870	0.5727	1.9310
14	0.2124	0.5038	0.4097	0.8757	1.6699
15	0.1004	0.9210	0.4121	0.7621	1.7450
16	0.1000	0.5029	0.1938	0.9726	1.7121
17	0.1000	0.5054	0.1247	1.2197	1.5747
18	0.1125	1.1992	0.6380	0.7030	1.7233
19	0.1001	0.5951	0.3001	0.7066	1.5621
20	0.1000	1.0569	0.5143	0.7288	1.9701
21	0.1000	0.5138	0.3577	0.5250	1.9920
$\sum_{i=1}^{21} T_i$ (s)	10.2958		5.8818		

time. Therefore relay 14 trips within 0.6048 and 0.7525 seconds for the fault in its zone and its primary relay protection zone respectively. Further, relay 11 senses the fault current magnitude of 1790 A for the fault in its zone, and 665 A when a fault is in its primary relay protection zone. Similar to relay 14, relay 11 has to provide the backup to three relays, that is, relays 13, 14, and 15. So, it is obvious that this noticeable difference between short-circuit current experienced by the relay and the duty to provide the backup protection for

TABLE 18.5 Operating time of primary and backup relays, and coordination time interval (CTI) of primary and backup (P/B) relay pair.

P/B relay pair		Conventional characteristic			Improved characteristic		
PR	BR	T_{PR} (s)	T_{BR} (s)	CTI (s)	T_{PR} (s)	T_{BR} (s)	CTI (s)
1	–	0.4400	–	–	0.2000	–	–
2	1	0.6384	1.7195	1.0811	0.4395	0.8896	0.4502
3	5	0.4180	0.6908	0.2728	0.2000	0.8684	0.6684
4	2	0.4424	0.6423	0.2000	0.2024	0.4455	0.2430
5	7	0.4143	0.7963	0.3820	0.2000	0.4255	0.2255
6	4	0.4311	0.8530	0.4218	0.2069	0.8451	0.6382
7	8	0.7891	1.3374	0.5483	0.4144	1.8940	1.4795
8	–	0.3404	–	–	0.2000	–	–
9	7	0.5090	0.7963	0.2873	0.2037	0.4255	0.2217
9	4	0.5090	0.8530	0.3440	0.2037	0.8451	0.6413
10	9	0.2797	0.7102	0.4305	0.2053	1.0239	0.8186
11	7	0.5963	0.7963	0.2000	0.2000	0.4255	0.2255
11	4	0.5963	0.8530	0.2567	0.2000	0.8451	0.6451
12	7	0.2396	0.7963	0.5568	0.2254	0.4255	0.2000
12	4	0.2396	0.8530	0.6134	0.2254	0.8451	0.6196
13	11	0.3542	4.1083	3.7541	0.2002	1.9946	1.7943
14	11	1.0191	4.1083	3.0893	0.6048	1.9946	1.3898
15	11	0.3563	4.1083	3.7520	0.2322	1.9946	1.7624
16	15	0.3809	0.5810	0.2000	0.2114	0.8953	0.6839
17	14	0.3870	1.0784	0.6914	0.2000	0.7525	0.5525
18	14	0.8783	1.0784	0.2001	0.5524	0.7525	0.2001
19	14	0.2957	1.0784	0.7826	0.2013	0.7525	0.5512
20	18	0.6814	0.8815	0.2001	0.3565	0.5566	0.2000
21	20	0.4046	0.7115	0.3070	0.2000	0.3999	0.2000
Sum (s)		10.2958	14.1102	18.5319	5.8818	10.9909	14.4161

more relays are the reasons for the greater operating times of backup relays. Therefore relay 11 with a conventional characteristic takes 4.1083 seconds while performing as backup protection. On the other hand, it takes 1.9946 second with the modified characteristic which is significantly lower than the conventional characteristic. Thus it is verified that the voltage-based modified characteristic performs better than the conventional characteristic for a low magnitude of fault current.

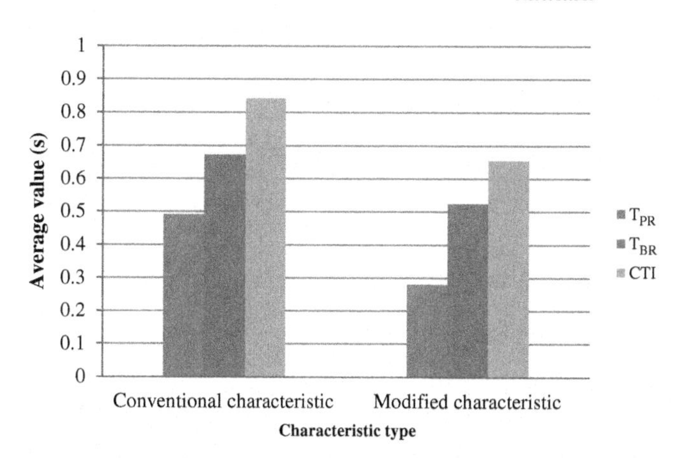

FIGURE 18.3 Average value of T_{PR}, T_{BR}, and coordination time interval (CTI).

18.5 Conclusion

In this study, the FA method is applied productively to resolve the optimum relay coordination problem. The proposed problem is implemented on PVDGs connected 19-bus system taking into consideration the conventional as well as modified characteristics of overcurrent rely. In addition, the obtained outcomes using the voltage-based characteristic are compared with the outcomes found using the conventional characteristic. The comparative analysis exhibits that the operational time of relays as well as coordination between relays have been improved while considering the modified characteristic as compared to the conventional characteristic. It is also discussed that the operational time of relays and coordination of P/B relay pairs are affected when the relay senses a low fault current and works as backup for more than one relay. In this situation, the consideration of conventional characteristic causes extensive operational time of relays. This problem is solved satisfactorily using the voltage-based characteristic.

References

[1] N. Nimpitiwan, G.T. Heydt, R. Ayyanar, S. Suryanarayanan, Fault current contribution from synchronous machine and inverter based distributed generators, IEEE Trans. Power Deliv. 22 (2007) 634–641.

[2] S. Mirsaeidi, D.M. Said, M.W. Mustafa, M.H. Habibuddin, K. Ghaffari, An analytical literature review of the available techniques for the protection of micro-grids, Int. J. Electr. Power Energy Syst. 58 (2014) 300–306.

[3] S.P. Pokharel, S.M. Brahma, S.J. Ranade, Modeling and simulation of three phase inverter for fault study of microgrids, in: 2012 North American Power Symposium (NAPS), 2012, pp. 1–6.

[4] H. Al-Nasseri, M.A. Redfern, Harmonics content based protection scheme for micro-grids dominated by solid state converters, in: 12th International Middle-East Power System Conference, 2008, pp. 50–56.

[5] A.R. Haron, A. Mohamed, H. Shareef, A review on protection schemes and coordination techniques in microgrid system, J. Appl. Sci. 12 (2012) 101–112.

[6] S. Mirsaeidi, X. Dong, S. Shi, D. Tzelepis, Challenges, advances and future directions in protection of hybrid AC/DC microgrids, IET Renew. Power Gener. 11 (2017) 1495–1502.

[7] B. Fani, M. Dadkhah, A. Karami-Horestani, Adaptive protection coordination scheme against the staircase fault current waveforms in PV-dominated distribution systems, IET Gener., Transm. Distrib. 12 (2018) 2065–2071.

[8] A.H. Etemadi, R. Iravani, Overcurrent and overload protection of directly voltage-controlled distributed resources in a microgrid, IEEE Trans. Ind. Electron. 60 (2013) 5629–5638.
[9] E. Sortomme, S.S. Venkata, J. Mitra, Microgrid protection using communication-assisted digital relays, IEEE Trans. Power Deliv. 25 (2010) 2789–2796.
[10] S. Mirsaeidi, D. Mat Said, M.W. Mustafa, M. Hafiz Habibuddin, A protection strategy for micro-grids based on positive-sequence component, IET Renew. Power Gener. 9 (2015) 600–609.
[11] M. Dewadasa, R. Majumder, A. Ghosh, et al., Control and protection of a microgrid with converter interfaced micro sources, in: IEEE International Conference on Power Systems, 2009. (ICPS'09), 2009, pp. 1–6.
[12] H. Al-Nasseri, M. Redfern, F. Li, A voltage based protection for micro-grids containing power electronic converters, in: Power Engineering Society General Meeting, 2006, p. 7. Available from: https://doi.org/10.1109/PES.2006.1709423.
[13] S.A. Gopalan, V. Sreeram, H.H.C. Iu, A review of coordination strategies and protection schemes for microgrids, Renew. Sustain. Energy Rev. 32 (2014) 222–228.
[14] W. Rebizant, K. Solak, B. Brusilowicz, G. Benysek, A. Kempski, J. Rusiński, Coordination of overcurrent protection relays in networks with superconducting fault current limiters, Int. J. Electr. Power Energy Syst. 95 (2018) 307–314.
[15] W. El-Khattam, T.S. Sidhu, Restoration of directional overcurrent relay coordination in distributed generation systems utilizing fault current limiter, IEEE Trans. Power Deliv. 23 (2008) 576–585.
[16] F. Van Overbeeke, Fault current source to ensure the fault level in inverter-dominated networks, in: 20th International Conference and Exhibition Electricity Distribution—Part 1, 2009, pp. 1–4.
[17] H. Samet, E. Azhdari, T. Ghanbari, Comprehensive study on different possible operations of multiple grid connected microgrids, IEEE Trans. Smart Grid 9 (2018) 1434–1441.
[18] S. Kar, S.R. Samantaray, Time-frequency transform-based differential scheme for microgrid protection, IET Gener. Transm. Distrib. 8 (2014) 310–320.
[19] D.P. Mishra, S.R. Samantaray, G. Joos, A combined wavelet and data-mining based intelligent protection scheme for microgrid, IEEE Trans. Smart Grid 7 (2016) 2295–2304.
[20] X. Li, A. Dyśko, G.M. Burt, Traveling wave-based protection scheme for inverter-dominated microgrid using mathematical morphology, IEEE Trans. Smart Grid 5 (2014) 2211–2218.
[21] A. Soleimanisardoo, H.K. Karegar, H. Zeineldin, Differential frequency protection scheme based on off-nominal frequency injections for inverter-based islanded microgrids, IEEE Trans. Smart Grid 10 (2) (2019) pp. 2107–2114. Available from: https://doi.org/10.1109/TSG.2017.2788851.
[22] P.P. Bedekar, S.R. Bhide, Optimum coordination of directional overcurrent relays using the hybrid GA-NLP approach, IEEE Trans. Power Deliv. 26 (2011) 109–119.
[23] A.S. Noghabi, J. Sadeh, H.R. Mashhadi, Considering different network topologies in optimal overcurrent relay coordination using a hybrid GA, IEEE Trans. Power Deliv. 24 (2009) 1857–1863.
[24] K.A. Saleh, H.H. Zeineldin, A. Al-Hinai, E.F. El-Saadany, Optimal coordination of directional overcurrent relays using a new time–current–voltage characteristic, IEEE Trans. Power Deliv. 30 (2015) 537–544.
[25] V.N. Rajput, F. Adelnia, K.S. Pandya, Optimal coordination of directional overcurrent relays using improved mathematical formulation, IET Gener., Transm. Distrib. 12 (2018) 2086–2094.
[26] F.A. Albasri, A.R. Alroomi, J.H. Talaq, Optimal coordination of directional overcurrent relays using biogeography-based optimization algorithms, IEEE Trans. Power Deliv. 30 (2015) 1810–1820.
[27] X.-S. Yang, Firefly algorithms for multimodal optimization, Stochastic Algorithms: Foundations Applications, Springer, 2009, pp. 169–178.
[28] X.-S. Yang, S.S.S. Hosseini, A.H. Gandomi, Firefly algorithm for solving non-convex economic dispatch problems with valve loading effect, Appl. Soft Comput. 12 (2012) 1180–1186.
[29] X.-S. Yang, X. He, Firefly algorithm: recent advances and applications, Int. J. Swarm Intell. 1 (2013) 36–50.
[30] A. Tjahjono, D.O. Anggriawan, A.K. Faizin, A. Priyadi, M. Pujiantara, T. Taufik, et al., Adaptive modified firefly algorithm for optimal coordination of overcurrent relays, IET Gener., Transm. Distrib. 11 (2017) 2575–2585.
[31] W.K. Najy, H.H. Zeineldin, W.L. Woon, Optimal protection coordination for microgrids with grid-connected and islanded capability, IEEE Trans. Ind. Electron. 60 (2013) 1668–1677.
[32] H.-J. Lee, G. Son, J.-W. Park, Study on wind-turbine generator system sizing considering voltage regulation and overcurrent relay coordination, IEEE Trans. Power Syst. 26 (2011) 1283–1293.

CHAPTER

19

On the topology for a smart direct current microgrid for a cluster of zero-net energy buildings

Francisco Gonzalez-Longatt[1], *Francisco Sanchez*[1] *and Sri Niwas Singh*[2]

[1]Wolfson School of Mechanical, Electrical and Manufacturing Engineering, Centre for Renewable Energy Systems Technology – CREST. Loughborough University, Loughborough, United Kingdom [2]Department of Electrical Engineering, Indian Institute of Technology Kanpur, Kanpur, India

19.1 Introduction

The challenges related to the forthcoming energy infrastructure are anticipated to increase and could potentially constitute a threat to universal energy security and economic prosperity. Among those challenges are [1–3]: (1) environmental climate change, (2) scheme security, and (3) economic globalization. The pressure on energy infrastructure is increasing since resources across the world are becoming rare and the necessity for sustainable development is increasingly important [4,5]. As part of a global strategy on climate change, administrations worldwide are initiating labors towards becoming maintainable and decarbonized economies. The growth to a decarbonized economy includes at least three key characteristics [2]: (1) improving energy efficiency measures, (2) fostering renewable energy competencies, and (3) handling adaptation requirements appearing due to climate change. One of the main features and enablers of the transition to a decarbonized economy is energy efficiency. The EU has set a target to reduce its energy consumption by least 30% by 2050 compared with the 2005 baseline.

Distributed Energy Resources in Microgrids
DOI: https://doi.org/10.1016/B978-0-12-817774-7.00019-3

A decarbonized culture comprises individuals living and working in low-energy, low-emission structures, together with smart heating and cooling systems, as well as additional low-energy and extremely high-efficiency energy industries such as the transport system.

A tremendous potential to improve the energy usage of both individual and industry clients exists. Families and industries would benefit from a more protected and well-organized energy system together with at least a couple of essential profits: a reduction in energy costs and a simulatneous move towards a more maintainable society. There are numerous enterprises worldwide which are focusing on advancing very low or zero energy technologies [6–11].

The EU is currently imposing the *Nearly Zero-Energy Buildings* (nZEB) initiative, corresponding to Article 9 of the Directive 2010/31/EU [8], which states: "Member States shall ensure that by 31 December 2020 all new buildings are nearly zero-energy buildings; and after 31 December 2018, new buildings occupied and owned by public authorities are nearly zero-energy buildings" [12]. Likewise, the aforementioned directive defines a nZEB as "a building that has a very high energy performance. The nearly zero or very low amount of energy required should be covered to a very significant extent by energy from renewable sources, including energy from renewable sources produced on-site or nearby" [12]. The United Kingdom was among the original countries worldwide to create a national legal framework for addressing climate change [13] and to commit to the research and development of the nZEB concept. The Climate Change Act of 2008 [14], creates a long-term legal basis for dealing with climate change. It puts forward a goal of decreasing the global UK carbon balance for all greenhouse-contributing gases by as much as 80% related to mid-1990s levels, by 2050, with a decrease of at least 40% by the year 2020.

In December 2006, the country's administration pledged that from that year onwards, all new houses would be carbon neutral, and presented the Code for Sustainable Homes [15,16] (CSH), that still serves as a basis to compare the environmental qualities of new households. The pledge was ratified in the "Building a Greener Future: Policy Statement" [17] in 2007, which projected an advanced narrowing of the building guidelines to attain the 2016 goal, initially by 25% in 2010 and subsequently by 45% in 2013. Nevertheless, the CSM was developed into a distinct program in March 2015 and the UK administration vowed instead to combine the home guidelines and benchmarks, including the CSM standards, into "Part L of the Building Regulations" [18].

Numerous universities around the world and in the United Kingdom in particular have established widespread investigations in nZEBs, taking into account different methods such as eco-architecture and market interaction, among others. However, there is little research in the development of intelligent DCMGs with more than one terminal that are capable of allowing the advance of independent zero-net energy constructions. The main reasons for this is the challenge related to the operation of a MG totally disconnected from the traditional alternating current (AC) power system and ensuring a zero-net emission using 100% renewable resources.

This chapter starts by introducing the most relevant concepts of a smart multi-terminal DCMG for a cluster of *Zero-Net Energy Building* (ZNEB) as well as the strategy required to permit a close-to-zero energy objective over the envisaged period. The rest of the chapter is devoted to outlining the topology optimization problem, in which the design requirements are employed as restrictions.

19.2 Multi-terminal direct current microgrid for a cluster zero-net energy buildings

Progressing a smart multi-terminal direct current microgrid (MTDCMG) for self-sufficient zero-net energy structures is a difficult challenge because there are numerous separate notions linked together and their interaction must allow the overall benefit to be maximized.

The smart MTDCMG notion is a crucial component for creating a cluster of buildings with an annual carbon emission coefficient of zero as well as with zero-net energy consumption, known as ZNEB.

A ZNEB is utterly autonomous from the electricity grid (off-grid), and consequently labors exclusively with self-generated and spread renewable energy sources (RESs). It facilitates the goal of 100% sustainable production as well as no carbon emissions, while simultaneously takes into account the modern use of electricity for additional services (such as water, heating, communications, nourishment, etc.). A ZNEB can be worked autonomously from infrastructural support facilities as an independent building, thereby constituting an outstanding answer for the electrification of rustic zones in unindustrialized countries or a substitute for developed zones everywhere else. There are numerous trials to overcome in the realization of the aforementioned integrated smart MTDCMG [19], including:

1. Combined design and procedures planning, bearing in mind nondispatchable RESwith uncorrelated demand-primary sources and controlled distributed energy storage systems.
2. Live energy equilibrium with 100% of nondispatchable RESs (wind and solar power).
3. Regaining the energy consumption of electricity storage schemes after an important eventuality.
4. The danger of blackout and massive outages throughout opposing climate conditions.
5. Other developing trials connected to technology, business models, or legal frameworks.

This chapter presents key implicit ideas related to the progress of smart MTDCMGs in order to enable a successful, 100% autonomous ZNEB. Likewise, it encompasses thoughts on the viability and influence of discrete notions, as well as on the entire communication between ideas.

19.2.1 Smart grid components

The term "smart grid" is not novel; it first made appearance in the piece "Towards a Smart Grid" from Amin and Wollenberg [20]. There are a vast quantity of smart grid explanations worldwide [21,22], however, a universal definition does not exist.

The Department of Energy and Climate Change (DECC) and the Office of Gas and Electricity Markets (Ofgem), in the United Kingdom, study the smart grid as a mechanism for allowing innovative energy sources as well as new methods of consumption [23]. The smart or intelligent electric grid is envisioned to advance and enable an effective and opportune shift to a nonpolluting economy that will aid the United Kingdom to develop its carbon diminution goals as well as to ensure its energy security while diminishing charges to customers.

In this work, a specific notion of the smart grid is supposed, grounded on the most accepted comprehension of the essential traits of a smart grid [19], that includes it having:

1. a self-healing capability to amend difficulties at initial phases (resilience).
2. improved stages of connections among agents, such as markets or consumers.
3. optimized action and control, safeguarding the optimal usage of resources.
4. a very predictive process to avoid dangers.
5. spread resources and statistics to combine all critical data.
6. boosted cyber-physical safety from risks originating from each threat.

Therefore throughout this chapter, the term smart grid is employed to characterize a cyber-physical organization with three crucial constituents: immense distribution in ICT and optimum placement of smart systems. Initially, the notion seems achievable, nonetheless it includes some rough involvement with other notions linked to the progress of an intelligent MTDCMG idea for an autonomous ZNEB.

19.2.2 Autonomous system component

An autonomous or independent scheme is planned to work self-sufficiently, and as a consequence the system is isolated or off-grid. Typical desirable ancillary services include an electricity grid, public water organizations, a gas grid, storm-water drainage, sewage-treatment plants, and communication enterprises in a few cases as well as in public or communal streets. Removing the requirement on the electricity system is comparatively modest, but this implies numerous consequences on the short- as well as on the long-term protected action of the scheme. It is important to point out that, as some academics working with biomass and/or biofuels have documented, the likelihood of the systems working entirely off-grid while simultaneously growing all necessary food is a significantly more difficult as well as a laborious endeavor [19].

This section explains why the linking of the standard AC-power grid is not taken into account, and, as a consequence, the smart MTDCMG is completely autonomous. There are two primary conditions for going completely autonomous [24]:

1. Long-term energy security: Ensure that the energy source over a long time period is sufficient to provide the minimum energy demand (or at least that it is able to provide further electricity to the MG that is deprived of a grid connection (e.g., a generator serving as backup, demand management, etc.).
2. Enough flexibility: In order to have acceptable flexibility, it is implied that the MG (e.g., load modulation, generation, as well as storage assets) can at any time match the electrical consumption with the production.

Dr. Damien Ernst from Liege University anticipated four conceivable possibilities that allow for a completely off-grid future [25]:

1. Oversizing of the supply of energy: Photovoltaic (PV) fixing could be dimensioned for generating in periods with less solar irradiation, a value of energy that is equal at least to the energy used up throughout these stages. That alternative allows for a practical reduction in storage charges.
2. Storage devices based on hydrogen that rely on electrolysis to yield hydrogen as well as fuel-cells that produce electrical energy from hydrogen are significantly cheaper than batteries in levelling out longer-term variations. Certainly, the cost of a hydrogen tank increases gradually with its volume.
3. Reduced interseasonal fluctuations: The closer you are to the equator, the smaller the interseasonal variation of PV energy generation. For instance, in Germany PV panels generate three times the amount of energy in the five or six clearest months in year than throughout the remainder of the year altogether; however, this aspect drops to 1.8 in southern Portugal. An improved primary energy evaluation and estimate constitutes then a critical aspect in decreasing the variations and diminishing the magnitude as well as the cost of energy storage.
4. Outsourcing of the mobile usage: Throughout the winter months, autonomous MG holders could charge their electric vehicle at the office to decrease usage or yet inject energy at evening periods from a car to the MG in what is known as vehicle-to-home (V2H).

A statement by Tesla Motors regarding the company's novel battery system, the Tesla Powerwall [26], with 10 kWh of storage and with a marketing price of ~ £2,250 (US $3,400), seems like a gamechanger on the electric-grid-development paradigm.

The Powerwall comprises the following components: a lithium-ion battery that is able to recharge, a liquid-thermal supervision scheme, a battery supervision scheme and an intelligent DC-to-DC device for monitoring power flows. Therefore this is a device that is 100% DC, as a DC-to-AC converter is not incorporated. The system looks almost ready to be part of a smart MTDCMG and it constitutes a credible answer to going 100% autonomous.

19.3 Multiterminal system direct current

In this chapter, a notion of a multiterminal DC (MTDC) scheme is employed to characterize an electrical grid in which three or more converters are connected to one another through a transmission system at a unique DC voltage value.

The MTDC serves as the fundamental element of the entire smart MTDCMG idea, and at least three dissimilar categories of DC/DC power converters are taken into consideration [19]:

1. Traditional diode rectifiers,
2. Boost converters, and
3. Bidirectional converters.

A couple of voltage values are included in the structural preparation of the MTDCMG. A high voltage (HV) is used for high-power demand loads, connecting high-power generation and high-density energy storage as well as for allowing a power exchange among associates of the MTDCMG. On the other hand, a low voltage (LV) value is used only for low-power demands (see Fig. 19.1)

19.3.1 Microgrid

The traditional idea of a MG entails a set of organized distributed energy resources that are able to provide adequate and nonstop energy to an important portion of the inner demand. An AC MG has autonomous controls, and deliberate islanding occurs with negligible service disruption (constituting a near-continuous changeover from grid to isolated operation).

The intelligent MTDCMG notion is grounded on the interface between numerous organized nanogrids (NGs). The NG is a very tiny MG, classically servicing just a building or a single-demand sink. Fig. 19.1 shows a schematic depiction of an NG, in which some electric sections are encompassed: nondispatchable power production (PV and eolic), local loads (flexible loads), as well as limited electrical energy storage (EES). The NG is linked to the rest of the NGs by means of an HV DC distribution system as shown in Fig. 19.2. The MTDC constitutes the high-power stage, in which all the energy interactions among NGs enables the smart MTDCMG notion. Regional EES is situated inside of each NG, and,

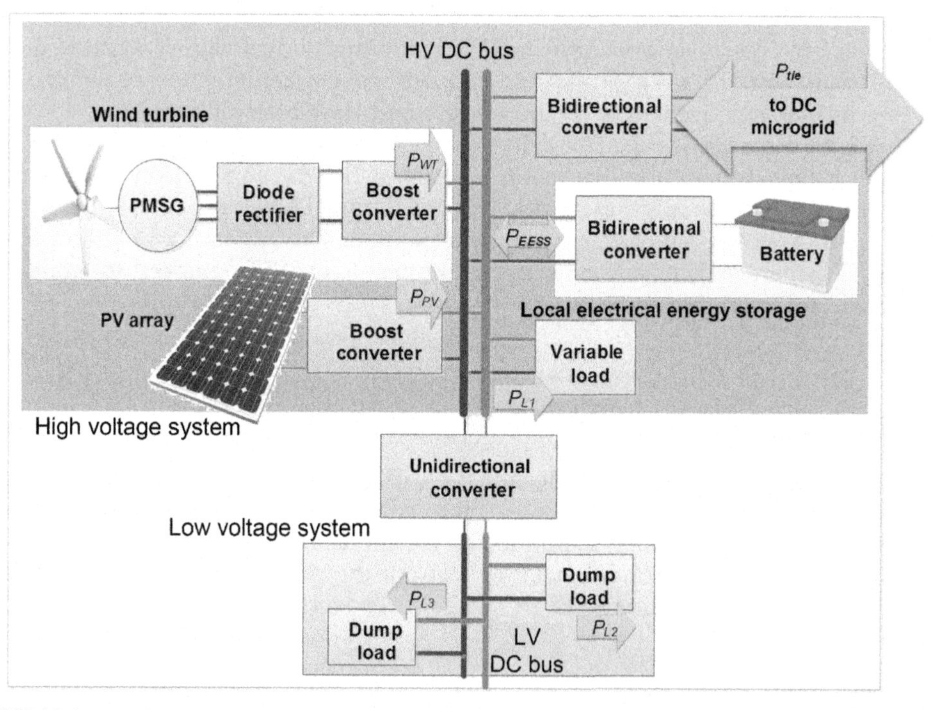

FIGURE 19.1 A schematic diagram of a nanogrid representing a house inside the smart MTDCMG concept. Power losses are negligible. *MTDCMG*, Multi-terminal direct current microgrid.

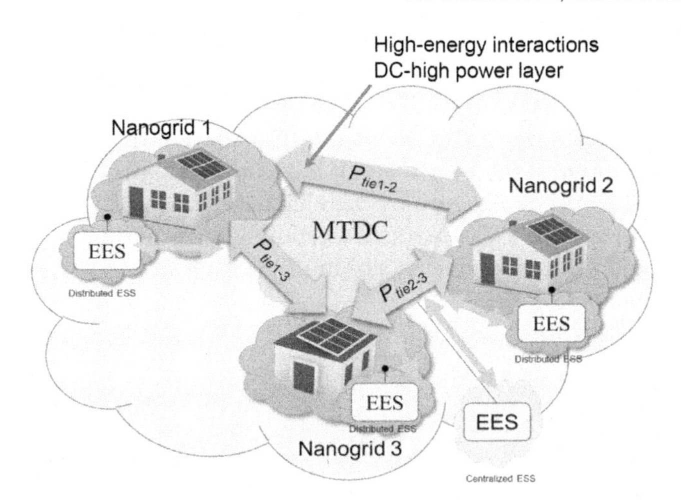

FIGURE 19.2 An illustrative diagram of a DC multiterminal MG representing the key elements of the smart MTDCMG concept and the energy interactions.

to enhance the energy security, a centralized EES or communal EES is incorporated. Power and energy relations among generation-demand-storage are among the critical elements of the smart MTDCMG concept; also, it is necessary to maximize the use of nondispatchable renewable generation, enabling the aspects of a zero-net energy system.

19.3.2 Zero-net energy building

A common definition of a ZNEB does not exist and this matter has been extensively debated in academic circles [27]. There are numerous elements tangled in the general ZNEB description: indicator of the equilibrium, balancing phase, type of energy used that is encompassed in the equilibrium, type of energy stability, accepted renewable energy (RE) supply selections, linking to the energy organization, energy efficacy necessities, the inside climate, as well as the building-grid interaction for the case of a grid-connected ZNEB. Four of the most popular notions on the ZNEB are shown below [19]:

1. *Site energy net-zero*: A site ZNEB generates, as a minimum, at least as much energy as it employs yearly when accounted in the site.

$$\sum_{i=1}^{year} E_p(t_i) = \sum_{k=1}^{year} E_c(t_k) \tag{19.1}$$

where $E_p(t_i)$ and $E_c(t_i)$ are produced and consumed energy at time t_i.

2. Energy costs net-zero: The budget that the utility repays to the structure proprietor for the grid-exported energy is at least equivalent to the quantity the proprietor repays the utility for the energy facilities.

$$\sum_{i=1}^{\text{year}} T_{\text{export}}(t_i) = \sum_{k=1}^{\text{year}} T_{\text{import}}(t_k) \tag{19.2}$$

where $T_{\text{explort}}(t_i)$ and $T_{\text{import}}(t_i)$ are the total amount of money paid by exported and imported energy at time t_i.

3. Energy emissions net-zero: A net-zero emission structure generates, as a minimum, as much pollution-free sustainable energy as it uses from emission-generating sources of energy.
4. Source energy net-zero: A net-zero source energy ZNEB generates, as a minimum, at least as much energy as it employs througout the year when it is accounted source-wide. This concept differs from the site energy net-zero by the characterization of a site ZNEB, instead of using a source, for example, PV or wind. To estimate the building's entire source energy, both imports and outputs of energy are multiplied by the necessary site-to-source conversion factors.

$$\sum_{i=1}^{\text{year}} E_{\text{p}}(t_i) = \sum_{k=1}^{\text{year}} E_{\text{c}}(t_k) \tag{19.3}$$

where $E_{\text{p}}(t_i)$ and $E_{\text{c}}(t_i)$ are produced and consumed energy at time t_i.

In this chapter, the definitions associated with grid possibilities are rejected as the autonomous concept is presumed as in the previous chapter. Also, the most basic ZNEB model is employed in this chapter, which studies a structure that produces as much power as it consumes. This notion provides an essential condition for the smart MTDCMG notion.

19.4 Model architecture of nanogrid: direct current house

Energy usage inside a DC house is the cause of the electricity consumption in an NG, and the energy consumption pattern is very reliant on the regime and activities of the residents. Therefore the apt recreation of the electrical demand inside an NG requires a model capable of reflecting the relationship between the occupant's activities as well as their associated use of electrical devices. In this chapter, a detailed model of domestic electrical consumption is employed. This model is founded upon a mixture of forms of active occupancy (e.g., when individuals are at home or not, wide-awake or sleeping, etc.), as well as everyday movement profiles that define how people devote their time to acomplishing specific actions.

The unique concept and progress of a fine-grained model for internal electric use was introduced initially by I Richardson et al. [28], and there have been numerous improvements over the years [29–31]. The creation and corroboration of the model used in this chapter is presented in an earlier published work [28,29,32].

In this chapter, the high-resolution representation introduced on [31] was heavily altered and enhanced to include the effect of wind power production as well as battery energy storage (a topic that it out of the scope of this chapter). The sketch assembly of the NG model is shown in Fig. 19.3.

The outdoor solar irradiance, the outdoor ambient temperature, as well as the wind speed are employed as model inputs. The central element of this system contains the replication of the main kinds of internal electrical appliances, including those used for heating and cooking.

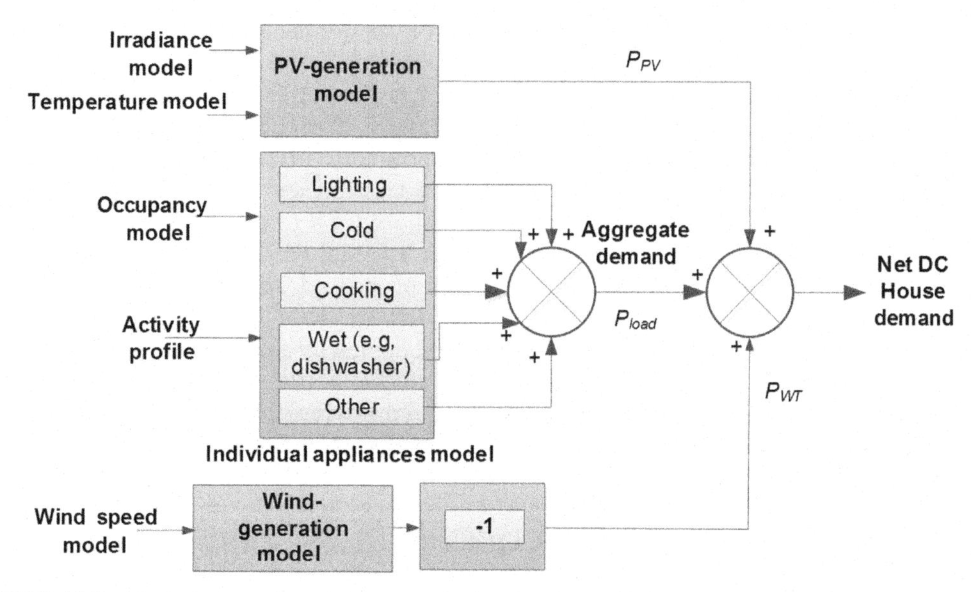

FIGURE 19.3 Block diagram illustrating the structure of the high-resolution model of NG used in this chapter. *NG*, Nanogrid.

The main components of this model are the active occupancy sub-model. The active occupancy is used to describe when one person or more are inside the house and involved in any activity involving the use of electricity. As a consequence this model defines the total electricity use. It introduces the social conduct relating to the local energy load. The subsequent subsection outlines the key components of the representation.

19.4.1 Model defining the domestic occupancy

The model of active domestic occupancy is used to define the number of dwellers within the DC house in a particular moment of the day. This constitutes a core element of the modeling because it has been identified to be a key defining element of the electricity consumption. In this chapter, the authors decided to use the model presented by Richardson et al. [28].

The active occupancy is represented by a whole number that changes throughout the day in a pseudo-aleatory way, reflecting the natural behavior of individuals going about their daily lives [29].

The active occupancy of a specific dwelling at a time t is a random process which can be represented by a random variable that undergoes shifts from one state $X^{(t)}$ to another, $X^{(t+1)}$ on a space-state. As a consequence, a first-order Markov-Chain can be used to represent the occupancy model and create synthetic occupancy data based on the following equation:

$$X^{(t+1)} = \mathbf{P}X^{(t)} \tag{19.4}$$

where $X^{(t)}$ represents the occupancy at the time a specific time t, and $\mathbf{P}$ is the transition probability matrix (TPM). The TPM is used to describe the transitions from one occupancy state to other and each of its entries is represented by a nonnegative real number indicating a probability. The occupancy model needs to derive a set of TPM from the basis data that specifies the likelihood of a change from one stage of active occupancy to an alternative, at each individual time-step during the day [32].

A comprehensive survey regarding the topic of how individuals spend their time in the United Kingdom is presented in the UK 2000 Time Use Survey (TUS) [33]. It is based on several thousands of one-day journal diaries recorded with a resolution of 10 minutes. For simplicity, the TPM used in this chapter is based on the real-world data obtained from the TUS [28,33]. This data and the Markov-Chain have been used in several publications with outstanding results [34,35].

The active domestic occupancy modeling process begins by establishing the initial state, for example, at 04:00am, then the Markov-Chain subsequently is established initially by taking an arbitrary figure for each time-step and employing it in combination with the adequate TPM (considering time of the day as well as the day of the week and with the total number of residents as well) to establish the next state, $X^{(t+1)}$. The authors of this chapter have implemented the active domestic occupancy model using MATLAB in order to combine the data into a larger simulation platform.

19.4.2 Individual device model

The model for an individual device is employed to calculate the power demand of the main variants of internal electrical devices based on a simulated activity during the day; in this chapter, individual appliances include those used for (1) heating, (2) cooking, and (3) cleaning. The representation consists of specific devices characterized by their power consumption grounded upon activity sketches. It is created from time series statistics that represent how individuals spend their time. Lastly, the representation adds the power curve from every piece of equipment at a given instant, delivering the combined demand of the DC establishment at a particular moment of the day.

Implementation of the aforementioned model for individual devices is publically available in the form of a Microsoft Excel spreadsheet at [36], which contains activity profiles and other pertinent information. In this chapter, the model of individual appliances has been implemented in MATLAB.

19.4.3 Photovoltaic generation model

The PV generation and production representation is built by two constituents [36]: (1) the dynamic irradiance model and (2) the PV generation model.

19.4.3.1 Dynamic irradiance model

The level of open-air solar irradiation is computed as the product of two key features (see Fig. 19.4):

1. The surface-wide irradiance model during "clear-sky" circumstances. The pure sky irradiation, at an exact position (latitude and longitude), is estimated by the design of

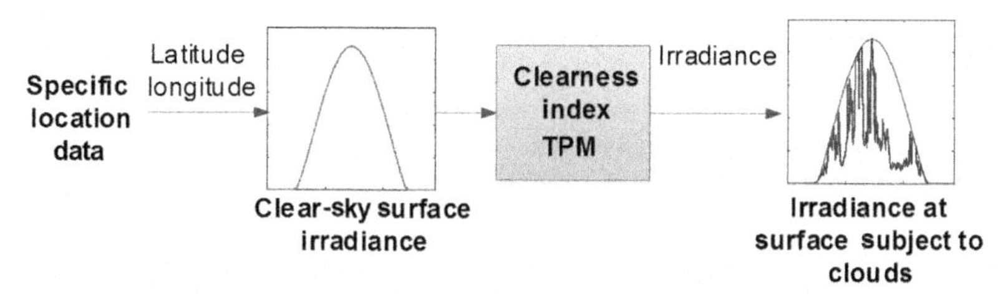

FIGURE 19.4 Block diagram demonstrating the structure of the irradiance model.

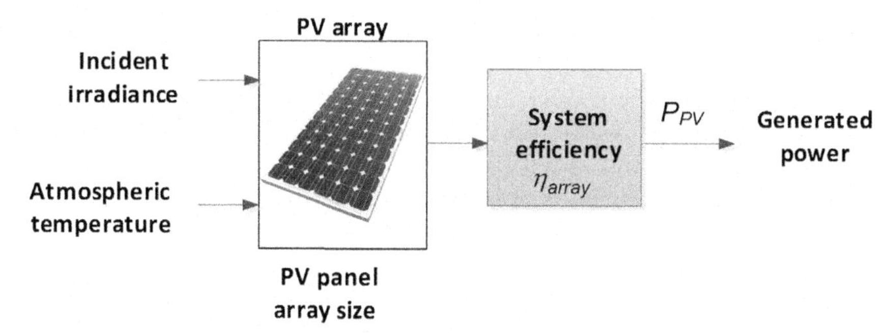

FIGURE 19.5 Block diagram demonstrating the structure of the photovoltaic (PV) production model.

the solar location. The Haurwitz clear sky paradigm is also used to calculate global horizontal irradiance (GHI). This model [37] has been implemented for this specific chapter by the authors.

2. Model of irradiance attenuation. Solar irradiance is attenuated because of changing climatic conditions; it is especially influenced by relevant conditions such as clouds passing above the PV installation throughout the day. The thorough model of the covering of clouds moving across the sky by means of a detailed atmospheric model is a very intricate and complicated process. However, a widely accepted and used approach uses the data of the attenuation formed by clouds from synthetic data created by some approximations. A simple mechanism to approximate the model of irradiance attenuation is by using a clearness index. A first-order Markov-Chain, a proven technique, is used to create the synthetic data of the clearness index [38–40]. The TPM employed in this chapter is grounded on [31].

19.4.3.2 Model of photovoltaic production

A basic energy transformation archetype is utilized for the PV system. The example uses the incoming iradiation at the PV surface as the input and requires the external area of the piece arrangement as well as a general overall efficiency to compute the produced power (see Fig. 19.5). This method has been employed successfully in academic works with excellent outcomes [41,42]. The output power (P_{PV}) produced by a PV scheme is simply approximated by the following equation [43]:

$$P_{PV} = A_{PV,array}\eta_{PV}I \tag{19.5}$$

where $A_{PV,array}$ represents the total area of the PV array, I is the pertaining solar insolation, and η_{PV} the system transformation efficacy of the PV system. The solar irradiation on the surface of the array is computed by using the irradiance model described in the previous subsection. The overall system effectiveness η_{PV} is composed chiefly of the product of two aspects [43] ($\eta_{PV} = \eta_{module} \times \eta_{inverter}$): the solar efficiency of the solar module (η_{module}) and its dependence on temperature, and the DC–AC conversion effectiveness ($\eta_{inverter}$) which includes cable losses. For simplicity, this chapter considers only the DC–AC conversion efficacy ($\eta_{inverter} = 85\%$) and solar module efficacy ($\eta_{module} = 15\%$).

19.4.4 Wind energy conversion model

A full, detailed wind energy conversion model is not required in this chapter. Instead, the wind turbine is simulated considering a straightforward energy conversion model (see Fig. 19.6). The input is the wind speed from the wind turbine and the output is the produced electrical power (P_{WT}), which is calculated as the product between a global system efficiency ($P_{WT} = \eta_{WT} P_{wind}$) and the power available from the incident wind on the turbine blade (P_{wind}):

$$P_{wind} = \frac{1}{2} C_p \rho \pi r^2 V_w^3 \tag{19.6}$$

where, C_p is the power-coefficient of a wind turbine, ρ corresponds to the air density (kg/m^3) at the wind turbine location, r is the length of the turbine blade (m), and V_w corresponds to the wind speed (m/s).

19.4.5 Smart operation multi-terminal direct current microgrid

The smart MTDCMG is a core on the development of a cluster of buildings in order to reach the zero-net energy utilization and zero-carbon emanation annual targets. The DC-high power layer is intended to interconnect several NGs, allowing energy interaction among all parties connected to the MTDCMG (see Fig. 19.2).

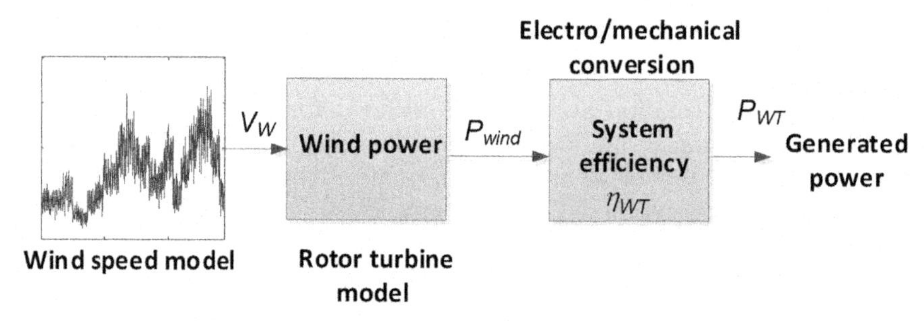

FIGURE 19.6 Block diagram demonstrating the structure of the wind turbine power production model.

The control of NGs is a topic well developed in the literature [44,45], covering several angles of the operation: normal/faults [46], dynamic/steady-state [47], etc. Also, the DC distribution grid has been well-covered in recent literature [48–50]. However, the optimal operation of integrated NGs into an MDTCMG has not been explored extensively in the literature as is presented in this chapter.

In this chapter, the smart operation of the MTDCMC refers to the optimal steady-state considering the uncertainties related to the primary energy source (wind and sun) and the human behavior on the domestic electric demand. The topology of the MTDCMG provides the opportunity to increase the overall operational efficacy of the system by reducing its losses. However, when deciding on the optimal topology of a complex planning problem, it is a nondeterministic polynomial-time complex problem which involves a relationship between many variables, and finding the solution can be a computational overburden when uncertainties are included. This section defines the optimization problem related to the smart steady-state operation of an MTDCMG.

19.4.6 Power flow of multi-terminal direct current microgrid

Considering an MTDCMG which consists of n_{dc} DC nodes, every node is considered by the node voltage ($U_{dc,i}$), and the node ($P_{dc,i}$) power injection (positive for a generation, and negative for demand). The current injected into the MTDCMG is written using the matrix form as [51]:

$$\mathbf{I_{dc}} = \mathbf{Y_{dc}} \mathbf{U_{dc}} \tag{19.7}$$

where the $\mathbf{I_{dc}} = [I_{dc,1}, I_{dc,2}, \ldots, I_{dc,ndc}]^T$ represents the DC current vector and each entry corresponds to each node current injection ($I_{dc,i}$), $\mathbf{U_{dc}} = [U_{dc,1}, U_{dc,2}, \ldots, U_{dc,ndc}]^T$ corresponds to the DC voltage vector in which each entry corresponds to each node voltage ($U_{dc,i}$); and finally, $\mathbf{Y_{dc}}$ represents the DC-node admittance matrix.

The current injections $\mathbf{I_{dc}}$ are related to DC voltage and the power injection $\mathbf{P_{dc}}$ in the following equation [52]:

$$\mathbf{P_{dc}} = K_{conv} \mathbf{U_{dc}} \otimes (\mathbf{Y_{dc}} \mathbf{U_{dc}}) \tag{19.8}$$

where $\mathbf{P_{dc}} = [P_{dc,1}, P_{dc,2}, \ldots, P_{dc,ndc}]^T$ is related to the load flow towards the DC grid by way of the DC terminal and the symbol $\otimes$ means element-wise (point to point) matrix multiplication operator, and $K_{conv} = 1$ for a single-pole device whereas $K_{conv} = 2$ for a two-pole device [53]. The power-voltage current shown on (19.7) defines the power balance of the MTDCMG, and the solution of that set of nonlinear Eq. (19.8) provides the steady-state values of the DC-high power layer voltages and its power flow.

19.4.7 Optimal power flow of multi-terminal direct current microgrid

The optimal power-flow (OPF) constitutes a much widely used tool to optimize the steady-state control and performance of a traditional AC-power grid. The OPF calculation aims to find a group of values for the system variables that optimizes one or more of the functionalities of the system [54], for example, operational limits, system active and reactive losses, the overall cost of generation, or system security. The steady-state behavior of an MTDCMF system can be described easily by the following matrix equation [5,25]:

$$\mathbf{g}_{\mathbf{NL}}(\mathbf{x}i,\mathbf{y}i) = \mathbf{0} \tag{19.9}$$

where $\mathbf{g}_{\mathbf{NL}}$ is the group of nonlinear arithmetical equations (g_{NLi}, $i = 1\ldots\, n$) used to establish the power equilibrium at grid nodes as pointed in (19.5), **x** corresponds to the *state vector*, and **y** corresponds to the *independent variable vector*. The state vector includes the state-variables that describe the MTDCMG state. Voltages on DC buses are reliant on variables or autonomous variables liable on the voltage controller employed at the terminal converter. The swing node and other node types provide the acknowledged or independent variables which are included in vector **y**.

OPF is expressed mathematically as an overall constrained optimization question in which a set of restrictions are taken into account. One of the most straightforward and general OPF formulations is actually based on a problem of minimization with no inequality restrictions as [55]:

$$\min fun(\mathbf{x},\mathbf{y}) \tag{19.10}$$

Subject to:

$$h(\mathbf{x},\mathbf{y}) = \mathbf{0} \tag{19.11}$$

where *fun*(**x**,**i**) is the function to be optimized.

19.4.8 Objective function definition

The problem of optimizing the execution of an MTDCMG requires the definition of the viewpoint in which the system actions will be optimized. Classically, the goal is to minimize the global generating costs.

The vast majority of the available OPF procedures try to optimize only one goal. Nevertheless, various additional objective functions are possible [56]: minimize variations in the controls, minimize overall grid wasted energy, maximize reliability, etc. After a comprehensive literature evaluation, there is a relatively large quantity of published papers that tackle the OPF multiobjective question [56], and the preferred joint aims can contain the cost of generating the power, environmental variables and also security of supply. One of the objective functions of the OPF that is most widely used minimizes the system overall losses as in Refs. [54,57]. A power administration scheme is envisaged to be a constituent of the smart MTDC: it regulates and displays the action of the MTDC relating to one or some of the DC houses and therefore minimizing the system overall losses is envisaged to be among the set of goals throughout the standard as well as during the long-term operation [55].

In this chapter, the overall system losses are situated on the DC-high power layer and it corresponds to the Joule losses or "ohmic waste" in the cables. Under the former supposition, the total overall system losses in an MTDC are simply inscribed as [55]:

$$h(\mathbf{x},\mathbf{y}) = P_{\text{losses}} = \sum_{i=1}^{n_{\text{dc}}} P_{\text{dc},i} \tag{19.12}$$

where $P_{\text{dc},i}$ represents the power injection or demand (<0) at the ith node, and it is an element in $\mathbf{P_{dc}} = [P_{\text{dc},1}, P_{\text{dc},2}, \ldots, P_{\text{dc},n}]^{\text{T}}$ which is calculated regarding the nodal voltages ($U_{\text{dc},i}$) using (19.8).

19.4.9 Constraint definition

The OPF problem in an MTDC constitutes a precise optimization question, that is usually termed restriction optimization. Here, the objective function, $h(\mathbf{X},\mathbf{Y})$, is optimized regarding a number of variables and crucially, with limitations on the aforementioned variables. The restraints split the probing space into two fields: the achievable field where the restrictions are fulfilled, and the unachievable field in which as a minimum one of the restrictions is unfulfilled. Generally, the OPF question can contain numerous particular arrangements for restrictions [55]: nonlinear restrictions, linear inequality restrictions, bound restrictions, and linear-equality restrictions. A broad explanation of the description of the restrictions employed throughout this chapter is offered in the following subsections.

19.4.9.1 Bound set constraints

The limit in the constituents of the solution vector $\mathbf{X}$ are defined by inferior ($\mathbf{X_{min}}$) and superior ($\mathbf{x_{max}}$) bounds. In this chapter the Bound constraints are explicitly inscribed like the following inequality function:

$$\mathbf{x_{min}} < \mathbf{x} < \mathbf{x_{max}} \tag{19.13}$$

DC-to-DC devices are used to regulate the DC voltage inside a MTDC. The aforementioned power converters typically use insulated gate bipolar transistors (IGBTs) as commutation or switching mechanisms which are very sensitive and possess a meager capacity to handle voltage variations. A DC overvoltage, which might stress the switching mechanisms, as well as extremely low undervoltages could produce damaging overcurrent on the IGBT. Consequently, there exist limits in the steady-state voltage ranges at the converter/changing stations. In this chapter, the ith node DC voltage at the transformation stations ($U_{\text{dc},i}$) is written with box restrictions based on the operational confines:

$$U_{\text{dc,MIN}} < U_{\text{dc},i} < U_{\text{dc,MAX}} \tag{19.14}$$

where $U_{\text{dc,MIN}}$ and $U_{\text{dc,MAX}}$ represent both the minimum as well as the maximum permitted voltage potential. The use of box restrictions allows for technical and also, fundamentally, operational limits to be met. Also, adding this type of constraint provides a mathematical advantage because it allows a reduction in the searching space and provides a faster and more reliable solution.

19.4.9.2 Nonlinear equality constraints

Nonlinear equality restrictions are typicaly in the form $\mathbf{g_{NL}}(\mathbf{x},\mathbf{y}) = \mathbf{0}$, where $\mathbf{g_{NL}}$ is a restriction vector, with one constituent for every restriction. The precise devising for the OPF comprises a set of nonlinear equivalence restrictions. The restrictions characterize the power-flow equations at each node. In the vast majority of real problems, the function lowest in value is obtained on the frontier flanked by the achievable and the unachievable areas.

19.4.9.3 Linear inequality constraints

Linear inequalities restrictions are of the formula:

$$\mathbf{A_i x} < \mathbf{B_i} \tag{19.15}$$

where $\mathbf{A_i}$ is an n-by-m matrix (a_{jk}), that characterizes m restrictions for an n-dimensional vector $\mathbf{x}$. $\mathbf{B_i}$ is m-dimensional. The inequality restrictions prescribe the limit for the mechanisms of the solution $\mathbf{X}$ in most optimization problems. A severe existing inadequacy exists on the DC-to-DC converters employed in the MTDCMG schemes. The switching mechanisms of the power converters, generally IGBTs, have the minimum overcurrent capacity. The converter controller makes sure that the maximum current in the valves is not exceeded. Linear inequality restrictions are employed in the OPF formulation of MTDCMG to signify the extreme current boundary in electronic devices [55]:

$$\mathbf{I_{conv}} < \mathbf{I_{conv}^{max}} \tag{19.16}$$

where $\mathbf{I_{conv}^{max}}$ characterizes a vector covering the extreme charging current allowed in each device. Using node examination, the node current is converted into a group comprising linear inequality restrictions as follows:

$$\mathbf{I_{conv}} = \mathbf{Y_{dc}U_{dc}} < \mathbf{I_{conv}^{max}} \tag{19.17}$$

where $\mathbf{A_i} = \mathbf{Y_{dc}}$,

$\mathbf{X} = \mathbf{U_{dc}}$ and $\mathbf{B_i}$ as is defined previously.

19.4.9.4 Linear-equality constraints

Linear-equality restrictions are typically depicted by [55]:

$$\mathbf{Ax} = \mathbf{B} \tag{19.18}$$

in which $\mathbf{A}$ is an n-by-m matrix, representing m restrictions for an n-dimensional vector $\mathbf{x}$. $\mathbf{B}$ is m-dimensional.

19.5 Simulation and results

This subsection uses simulation results over an illustrative test case to evaluate the optimal topology of a smart MTDCMG for a cluster of houses. This section begins with a description of the test system used on the proposed assessment, then OPF is used to obtain the total system loses and to determine the optimal topology for the proposed test system.

A group of three ZNEBs is employed for demonstrative purposes: DC-house 1, DC-house 2, and DC-house 3. Each house is represented by an NG as depicted on Fig. 19.1. Internal details of the DC connectivity and the DC−DC converters are not included in this chapter.

The fine-grained model of an NG is employed to model each DC House. This model includes up to 28 appliances within each dwelling, classified into: cold, consumer electronics + information and communication technologies (ICT), cooking, wet and water heating, and electric space heating. To include multiple proprietorship, devices are listed explicitly: for instance, three of the 28 devices are TV sets. In the NG model, a single house could, thus, have up to three TV sets. A register of the devices characterized in this test system is shown in Table 19.1.

The number of residents is assumed to be different for each DC house, and an occupancy model has been used to create active occupancy pattern throughout the day. Fig. 19.7 displays the replicated one-minute resolution, active occupancy form for the three DC houses: the dark-grey shade embodies weekdays (wd) and light-grey color weekends (we). Every horizontal line specifies the number of people active in the dwelling at that instant. From this data, typical human behavior during the 24 hours can be noted: there is a minimal motion throughout the evening (00:00am−07:00am). During weekdays there is slight activity (at the dwelling) throughout the day and most activity occurs during the night. However, the pattern changes slightly with more activity during the day in the case of weekends. Also, it can be observed from the simulated active occupancy, that a maximum of occupancy is reached during nights as expected.

Fig. 19.8 shows the simulated aggregate demand of the three DC houses being studied (as before the dark-grey color represents weekdays, and the light-grey color weekends). The load demand is minimum throughout the evening period, when members of the households are sleeping, and only the fridge/freezer is actually consuming power. A rise in power demand is evident as soon as the inhabitants become dynamic in the morning. There are spikes or peaks in consumption; they could be produced by using short-time, large consumption devices, like for example a microwave oven or the kettle, etc. An important rise in the starting demand is realized during the teatime period because of the use of the washing-machine.

The 24-hour pattern of three global solar irradiances are replicated by using the dynamic irradiance model presented in the previous subsection; results are shown in Fig. 19.9. The replicated total irradiance, which considers cloud attenuation, is used to calculate the PV power production (W) at each house. The total electricity generated by each PV array on the day is 5.89 KhW, 5.25 kWh, 4.17 kWh for DC-house 1, DC-house 2 and DC-house 3 respectively.

A MATLAB program [58] was used to generate a synthetic wind speed time series at 10 m height. The wind speed simulation program uses the Von Kármán wind turbulence model considering average wind speed and roughness length. Fig. 19.10 shows the simulated wind speed at the location of DC-house 1. A typical low-speed rooftop wind turbine is used for illustrative purposes. The characteristic of the 2000-Watt wind turbine generator is presented to the wind-turbine representation introduced in the preceding chapter, and the simulated output power of the rooftop wind turbine(s) installed at DC-house 1 is shown in Fig. 19.11.

TABLE 19.1 Details of appliance model configuration for each DC house.

Appliance category	Appliance type	DC-house 1	DC-house 2	DC-house 3
Cold	Fridge/freezer	Yes	Yes	No
	Refrigerator	Yes	No	No
	Upright freezer	Yes	No	No
Consumer electronics + ICT	Answering machine	Yes	Yes	Yes
	Cassette/CD player	Yes	Yes	Yes
	Clock	Yes	No	Yes
	Wireless telephone	Yes	Yes	Yes
	Hi-fi audio	Yes	Yes	Yes
	Iron	Yes	Yes	Yes
	Vacuum cleaner	Yes	Yes	Yes
	Computer	Yes	Yes	Yes
	Printer machine	No	Yes	Yes
	TV-1	Yes	Yes	Yes
	TV-2	Yes	Yes	Yes
	TV-3	No	Yes	No
	DVD	Yes	Yes	Yes
	TV decoder box	Yes	Yes	Yes
Cooking	Hob	Yes	Yes	No
	Oven	Yes	No	Yes
	Microwave	Yes	Yes	Yes
	Kettle	Yes	Yes	Yes
	Cooking apparatus	Yes	Yes	Yes
Wet	Dishwasher	No	No	Yes
	Dryer machine	Yes	Yes	Yes
	Washing (o) machine	Yes	Yes	Yes
Water heating	Electric shower	Yes	No	Yes
Electric space heating	Storage heaters	Yes	No	Yes
	Other electric space heating	No	Yes	No

A DC-high power layer should interconnect a cluster of DC Houses in order to establish a smart MTDCMG. The network topology defines the connection properties of the MTDCMG. A graph can be used to create a circuit model, defining the connectivity between the DC houses. In this chapter, a connected graph is used to define the topology

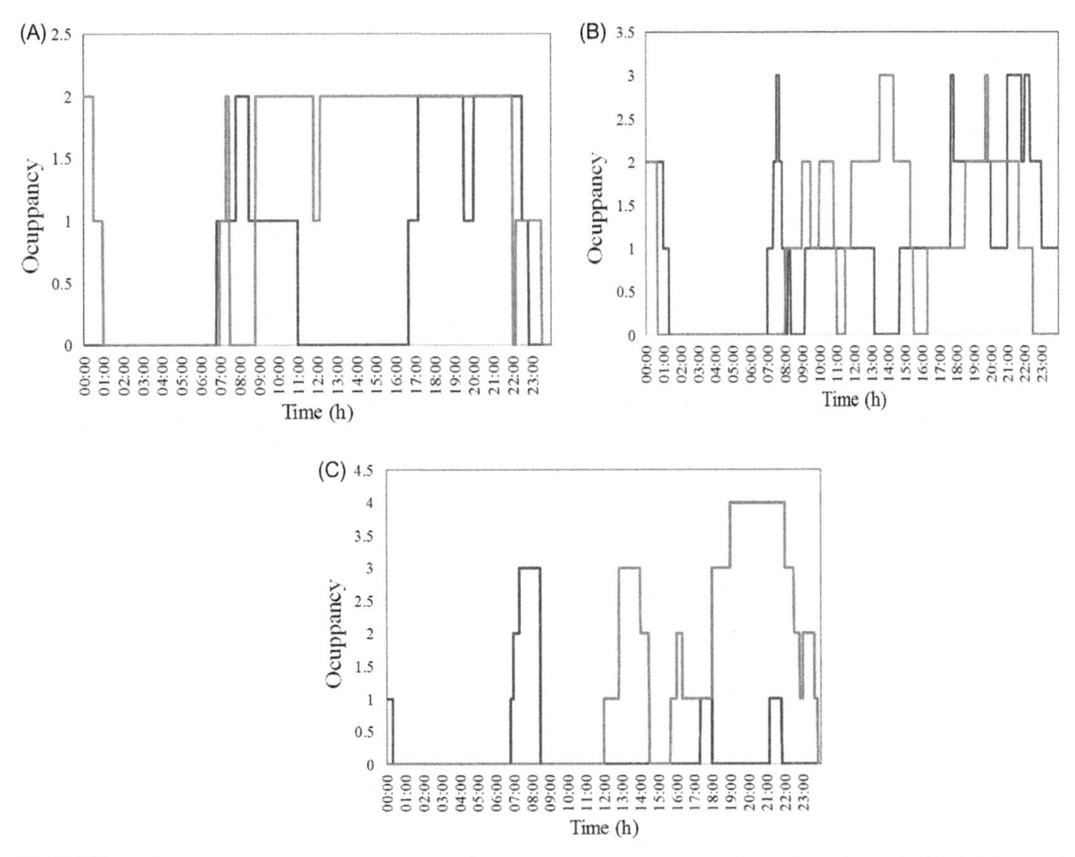

FIGURE 19.7 Simulated active occupancy pattern. (A) DC-house 1: Number of residents: 2, (B) DC-house 2: Number of residents: 3, (C) DC-house 3: Number of residents: 4.

of the MTDCMG. A graph is realized when a path exists between every pair of vertices. Therefore it transpires that in a connected graph, there are no unapproachable vertices, and, as a consequence, there is inherent redundancy in this connection which allows an increase in supply security [59]. A complete graph of n_{dc} nodes has $n_{\mathrm{dc}}(n_{\mathrm{dc}}-1)/2$ branches. For the particular case of three DC houses, $n_{\mathrm{dc}}=3$, and the complete graph represents a standard delta connection. An alternative graph representation can be created adding one extra node and connecting a terminal node (DC house) to it; the created complete bipartite graph is named star. Star is the only connected graph in which at most one node has a degree greater than one. Considering the previous explanation there are only two possible ways to connect three homes: Case 1: Delta linking and Case 2: Star linking (as shown Fig. 19.12). The delta connection offers the advantage of redundancy: if one branch is out of service for any reason, the other two branches can supply power to the other nodes (if appropriate sizing is considered). On the other hand, the star connection offers the advantage of adding an extra node, where a centralized energy storage system can be easily installed; however, if one branch is out of service the corresponding DC

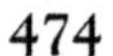

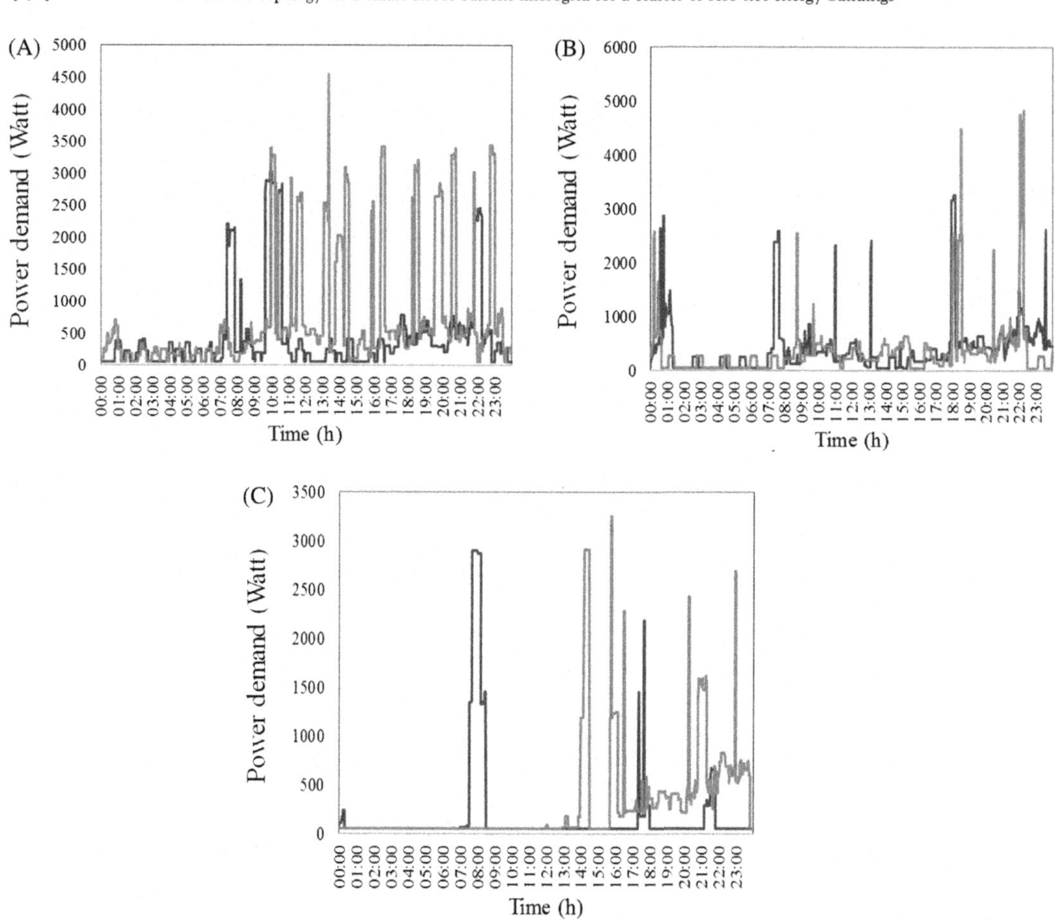

FIGURE 19.8 Simulated electricity demand pattern. Values presented in parenthesis represent average daily demand in kW of weekday and weekend (wd/we). (A) DC-house 1: (436.94/711.60), (B) DC-house 2: (416.24/365.44), (C) DC-house 3: (165.19/290.32).

house should operate in islanding mode. Case I and II are depicted in Fig. 19.12 and represent extremely rudimentary topologies. Nevertheless, it is explanatory enough for the optimum topology assessment intended in this chapter. The previous discussion can be extended easily to a larger number of DC houses. However, this chapter is limited to two cases for illustrative purposes.

Bipolar DC systems are employed typically within the MTDCMG, and the established cable-resistive properties are typical of medium voltage DC wires that are widely accessible in the marketplace. Resistances vales were assumed to be alike among Case 1 and Case 2.

A Matlab - R2016a (version 8.3.0 64-bits) package was advanced to compute the ideal long-term process of the MTDCMG. The MATLAB implementation includes two sub-programs: (1) the model of the DC house which is implemented as a set of functions based on the modeling described in the preceding section, and (2) in which the optimization

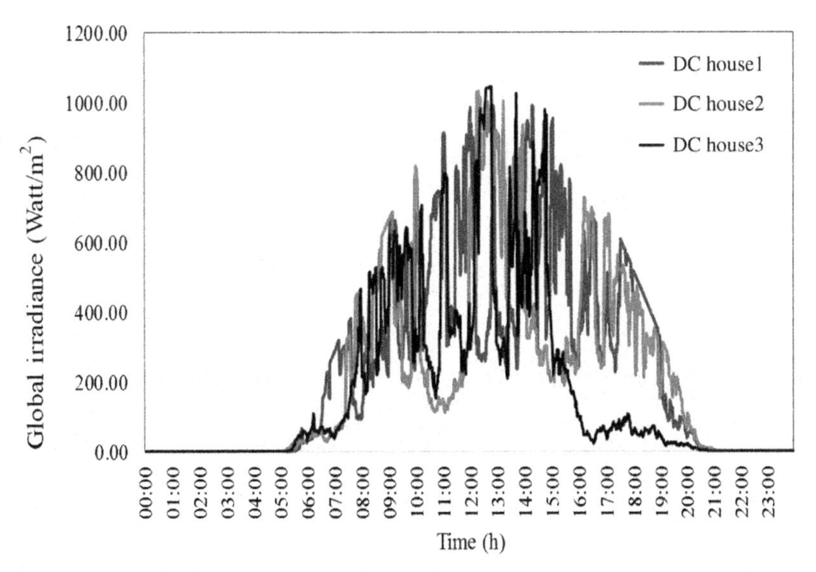

FIGURE 19.9 Simulated global irradiance with attenuation due to clouds. Latitude = 52.80, longitude = −1.20, panel area $A_c = 10\ m^2$.

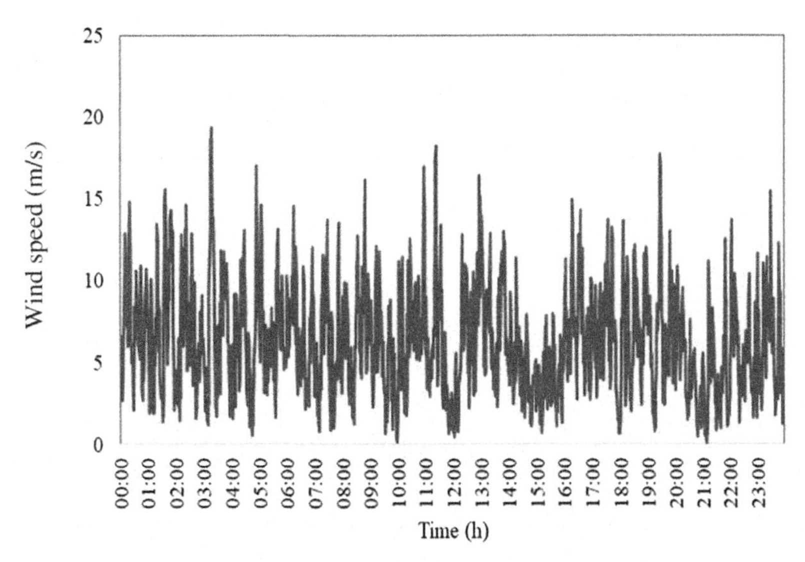

FIGURE 19.10 Synthetic wind speed time series at DC-house 1. ave = 6.29 m/s at 10 m, $z_0 = 0.01$.

problem is solved for the high-resolution model of the group of DC Houses. The losses minimization solution is calculated using an interior point algorithm for the solution of the optimization problem.

All simulations are completed by means of a PC based on Intel, Core™ i7 CPU 3.2 GHz, 32 GB RAM with Windows 10 64-bits OS.

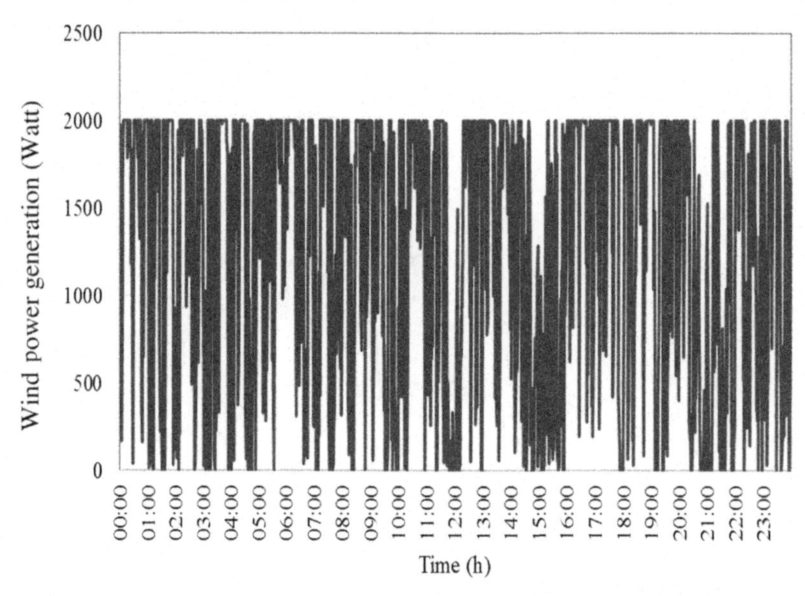

FIGURE 19.11 Simulated wind power production at DC-house 1. Average power = 1298.08 kW, electricity generated on the day = 30.95 kWh.

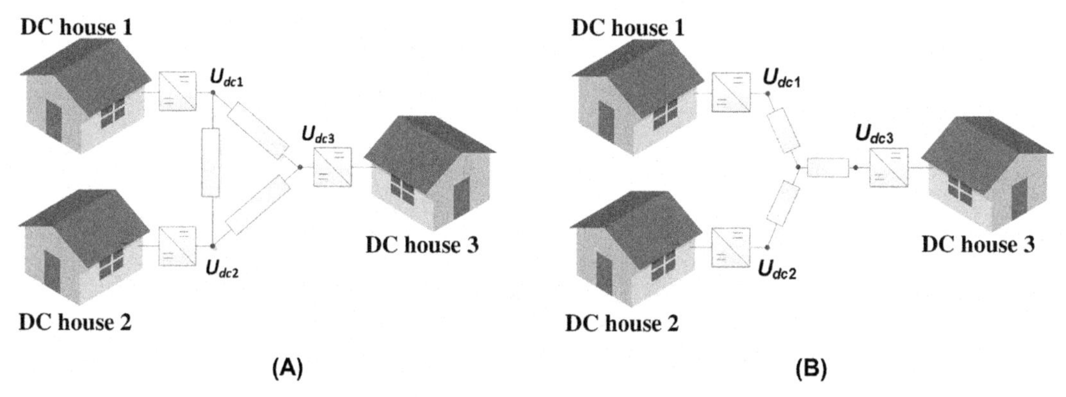

FIGURE 19.12 Simulated topologies scenarios of high-power layer -MTDC. The resistances used in the DC power cables are $R_{13} = 0.045$, $R_{23} = 0.052$, and $R_{12} = 0.073$ per unit. Asymmetrical configuration is selected. (A) Case I: Delta connection (B) Case II: Star connection.

Fig. 19.12 shows the replication outcomes of the DC voltage in all the terminal nodal points (U_{dci}) involved in the high-power level. The voltage shapes display a significant requirement on the generation as well as on the demand equilibrium in relation to the MTDC. DC-house 1 has fitted a wind turbine and a battery energy storage (BES), and, as a consequence, the smallest voltage deviations are sensibly situated at U_{dc1}. Replication outcomes reveal a minor voltage profile in the MTDC throughout weekends and holidays, which seems to be the effect of the greater power consumption during this time period as a result of the highly active residence. Simulation outcomes for Case 2 show a smaller voltage deviation compared to Case 1 Fig. 19.13.

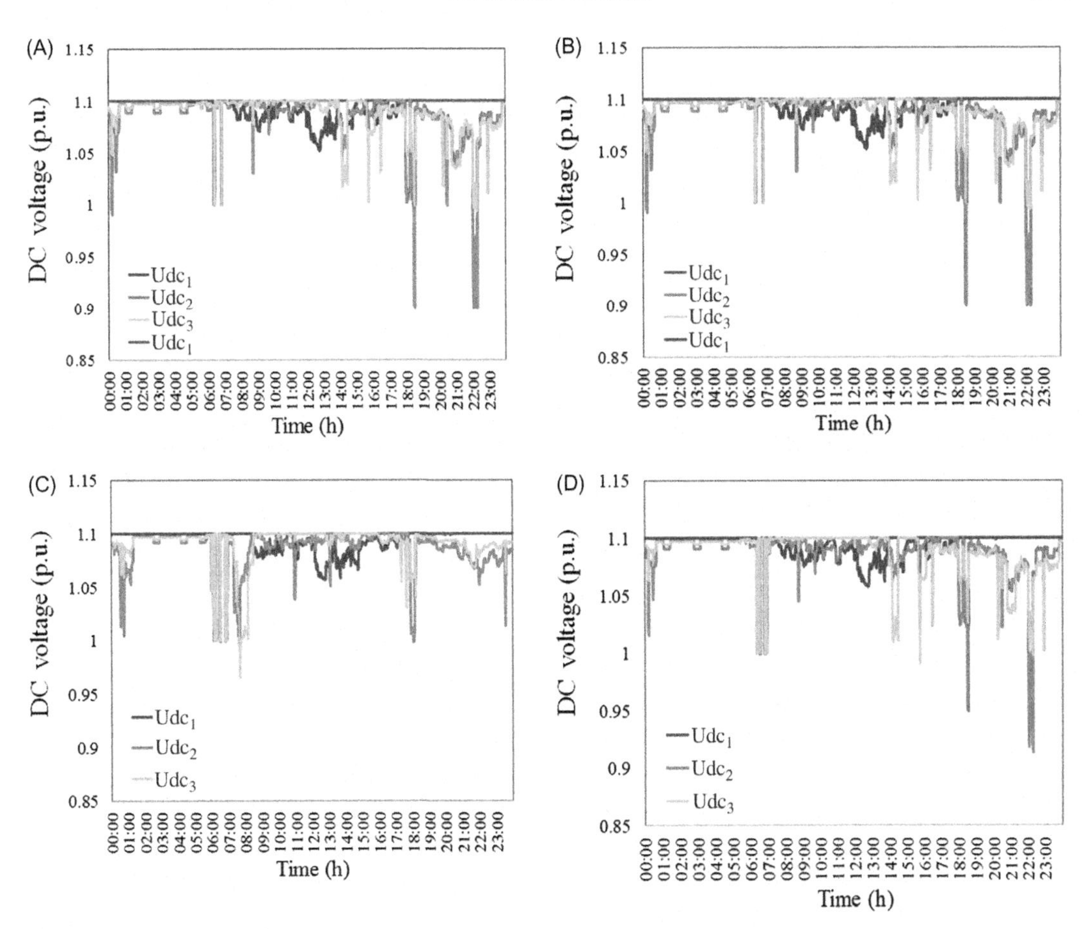

FIGURE 19.13 Simulated results: direct current (DC) voltages (pu). (A) Case I: Weekdays (wd), (B) Case I: Weekend (we), (C) Case II: Weekdays (wd), (D) Case II: Weekend (we).

Fig. 19.14 displays the total power losses in the MTDCMG, considering a one-minute graining, using the demand profiles for a typical day of the week and weekends/holidays and considering two different topological formations (Case 1 and Case 2). As anticipated, extreme power losses occur over the weekend and during holidays; this is because of the high active occupancy of the home by the residents, which increases the demand. Simulation outcomes also reveal more significant losses exclusively in Case II. There is a clear correlation between human behavior (active occupancy) and the electricity demand affecting the total power losses

Regarding, average losses during the 24-hour period, Case I exhibits lower losses compared to Case II. Lastly, based on simulation outcomes, the Delta (Case I) topology exhibits the best presentation regarding the power losses under the measured situations of uncertainties.

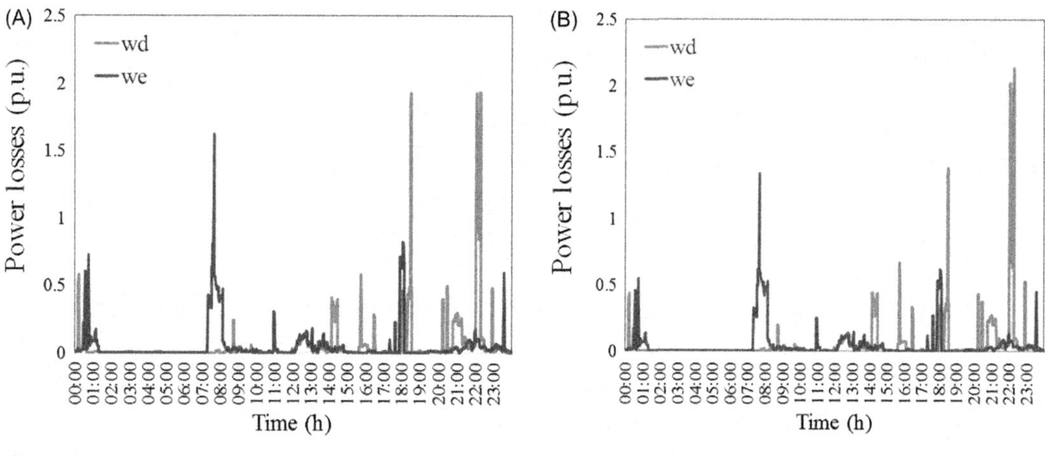

FIGURE 19.14 Simulated results: power losses (pu). (A) Case I, (B) Case II.

19.6 Conclusions

The main aspects regarding the shift to a decarbonized economy system are renewable energy, energy-efficient devices, and economic development. Progress of an intelligent multiterminal, DCMG for autonomous and ZNEB, that permits the development of the profits of numerous previously identified ideas into a original idea permits the evolution to a decarbonized economy.

The intelligent MTDCMG notion therefore is a crucial element to create a cluster of buildings with negligible net energy consumption and nil carbon emanations per annum. However, also, further development is required in relation to other aspects: (1) combined plan and procedures scheduling considering nondispatchable renewable energy systems without a correlated primary basis, as well as constrained distributed energy storage systems, (2) instantaneous energy equilibrium, bearing in mind the use of 100% nondispatchable renewable energy systems (wind power or solar), (3) the recovery of energy usage of ESS after a significant eventuality, (4) reducing the hazard of power failures or massive brownouts throughout hostile meteorological conditions, and (5) the development of additional trials connected to the technology and/or legal frameworks.

The form of use of electrical energy in a specific domestic ZNEB is extremely dependent on the actions of the dwellers and their related usage of electricity. A fine-grained model of local electrical energy use is introduced in this chapter simultaneously with the amalgamation of the active occupancy relevant patterns (e.g., when people are at their house and consuming energy), and day-to-day activity patterns that typify how individuals fill their hours carrying out specific actions. The fine-grained energy consumption representation is employed simultaneously with numerous random situations, such as weather and human behavior, to develop some realistic situations of electrical consumption at every ZNEB and then resolve the appropriate constrained optimization problem of a system topology for the group of interrelating ZNEBs. This chapter has established the proper execution for the anticipated method by means of an illuminating example, namely a group of three DC houses.

References

[1] B. Johansson, Security aspects of future renewable energy systems–a short overview, Energy 61 (2013) 598–605.

[2] F. Gonzalez-Longatt, Frequency control and inertial response schemes for the future power networks. Green Energy and Technology in: J. Hossain, A. Mahmud (Eds.), Large Scale Renew. Power Gener., Springer, Singapore, 2014, pp. 193–231.

[3] F. Gonzalez-Longatt, Impact of emulated inertia from wind power on under-frequency protection schemes of future power systems (in English) J. Mod. Power Syst. Clean Energy (2015) 1–8. /08/12 2015.

[4] F.M. Gonzalez-Longatt, Activation schemes of synthetic inertia controller for full converter wind turbine generators, in: PowerTech. 2015 IEEE Eindhoven, 2015, pp. 1–5.

[5] F. Gonzalez-Longatt, B.S. Rajpurohit, S.N. Singh, Optimal structure of a Smart DC micro-grid for a cluster of zero net energy buildings, in: 2016 IEEE International Energy Conference (ENERGYCON), 2016, pp. 1–7.

[6] COOPERaTE, COOPERaTE: control and optimization for energy positive neighbourhoods, 2013. Available from: <http://www.cooperate-fp7.eu/index.php/home.html>.

[7] DIMMER, DIMMER: District Information Modeling and Management for Energy Reduction, 2013. Available from: <http://dimmer.polito.it/>.

[8] EPBD, Towards 2020 – Nearly Zero Energy Buildings, 2014. Available from: <http://www.epbd-ca.eu/themes/nearly-zero-energy>.

[9] MESMERISE-CCS, MESMERISE-CCS: Multi-Scale Energy Systems Modelling Encompassing Renewable, Intermittent, Stored Energy and Carbon Capture and Storage, 2014. Available from: <https://ukccsrc.ac.uk/resources/ccs-projects-directory/multi-scale-energy-systems-modelling-encompassing-renewable>.

[10] SHINE-ZC, SHINE-ZC: Sustainable Housing Innovation Network of Excellence - Zero Carbon, 2011. Available from: <http://gtr.rcuk.ac.uk/project/DB712AB2-B58A-4947-AB0C-C890F2BB1BCB>.

[11] ZenN, ZenN: Zero Energy Building Renovation, 2013. Available from: <http://www.zenn-fp7.eu/>.

[12] EU. Directive 2010/31/EU of the European Parliament and of the Council of 19 May 2010 on the energy performance of building, 2010. Available from: <http://eur-lex.europa.eu/legal-content/EN/TXT/HTML/?uri = CELEX:32010L0031&from = EN>.

[13] J.B. Cullingworth, V. Nadin, Town and Country Planning in the UK, 14th ed, Routledge, Taylor & Francis Group, London; New York, 2006, pp. xxxiii, 588 p.

[14] U. S. L. Database, Climate Change Act 2008, 2008. Available from: <http://www.legislation.gov.uk/ukpga/2008/27>.

[15] (2006). Code for Sustainable Homes. Available from: <http://www.planningportal.gov.uk/buildingregulations/greenerbuildings/sustainablehomes>.

[16] DCLG, Policy Paper: 2010 to 2015 Government Policy: Building Regulation, 2015. Available from: <https://www.gov.uk/government/publications/2010-to-2015-government-policy-building-regulation/2010-to-2015-government-policy-building-regulation#appendix-5-technical-housing-standards-review>.

[17] DCLG, Building a Greener Future: policy statement, 2007. Available from: <https://www.rbkc.gov.uk/PDF/80%20Building%20a%20Greener%20Future%20Policy%20Statement%20July%202007.pdf>.

[18] HMG, Part L of the Building Regulations—2013 Edition Available from: <https://www.gov.uk/government/uploads/system/uploads/attachment_data/file/441415/BR_PDF_AD_L1A_2013.pdf>.

[19] F. Gonzalez-Longatt, B.S. Rajpurohit, S.N.S., Smart Multi-terminal DC micro-grids for autonomous zero-net energy buildings: implicit concepts, in: Presented at the EEE PES Innovative Smart Grid Technologies 2015 Asia, ISGT Asia 2015, Bangkok, Thailand, 4–6 November 2015.

[20] S.M. Amin, B.F. Wollenberg, Toward a smart grid: power delivery for the 21st century, IEEE Power Energy Mag. 3 (5) (2005) 34–41.

[21] DOE, Smart Grid | Department of Energy, 2013. Available from: <http://energy.gov/oe/services/technology-development/smart-grid>.

[22] EU, Smart Grids European Technology Platform, 2015. Available from: <http://www.smartgrids.eu/>.

[23] DECC, Smart Grid Vision and Routemap, 2014.

[24] D. Ernst, Microgrids and their destructuring effects on the electrical industry, 2014. Available from: <http://orbi.ulg.ac.be/bitstream/2268/173600/1/ernst-microgrids.pdf>.

[25] V. Francois-Lavet, R. Fonteneau, D. Ernst, Using approximate dynamic programming for estimating the revenues of a hydrogen-based high-capacity storage device, in: 2014 IEEE Symposium on Adaptive Dynamic Programming and Reinforcement Learning (ADPRL), 2014, pp. 1–8.

[26] TeslaMotors, Energy Storage for a Sustainable Home, 2015. Available from: <http://www.teslamotors.com/powerwall>.

[27] A.J. Marszal, et al., Zero energy building – a review of definitions and calculation methodologies, Energy Build. 43 (4) (2011) 971–979.

[28] I. Richardson, M. Thomson, D. Infield, A high-resolution domestic building occupancy model for energy demand simulations, Energy Build. 40 (2008) 1560–1566.

[29] I. Richardson, M. Thomson, D. Infield, C. Clifford, Domestic electricity use: a high-resolution energy demand model, Energy Build. 42 (10) (2010) 1878–1887.

[30] I. Richardson, M. Thomson, D. Infield, A. Delahunty, Domestic lighting: a high-resolution energy demand model, Energy Build. 41 (7) (2009) 781–789.

[31] I. Richardson, M. Thomson, Integrated simulation of photovoltaic micro-generation and domestic electricity demand: a one-minute resolution open-source model, Proc. Inst. Mech. Eng. Part A J. Power Energy 127 (1) (2012) 73–81.

[32] I. Richardson, Integrated high-resolution modelling of domestic electricity demand and low voltage electricity distribution networks (PhD Theses), Electronic, Electrical and Systems Engineering, Loughborough University, Loughborough, UK, 2011.

[33] Ipsos-RSL, Ipsos-RSL and Office for National Statistics: The United Kingdom 2000 Time Use Survey, Colchester, Essex, September 2003.

[34] A. Navarro-Espinosa, L.F. Ochoa, Probabilistic impact assessment of low carbon technologies in LV distribution systems, Power Syst. IEEE Trans. no. 99 (2015) 1–12. vol. PP.

[35] E. McKenna, M. Krawczynski, M. Thomson, Four-state domestic building occupancy model for energy demand simulations, Energy Build. 96 (2015) 30–39.

[36] I. Richardson, M. Thomson. (2011). Integrated simulation of photovoltaic micro-generation and domestic electricity demand: a one-minute resolution open source model. Available from: <https://dspace.lboro.ac.uk/dspace-jspui/handle/2134/8774>.

[37] B. Haurwitz, Insolation in relation to cloudiness and cloud density, J. Meteorol. 2 (3) (1945) 154–166. /09/01 1945.

[38] B.O. Ngoko, H. Sugihara, T. Funaki, Synthetic generation of high temporal resolution solar radiation data using Markov models, Sol. Energy 103 (2014) 160–170.

[39] M. Hofmann, S. Riechelmann, C. Crisosto, R. Mubarak, G. Seckmeyer, Improved synthesis of global irradiance with one-minute resolution for PV system simulations, Int. J. Photoenergy 2014 (2014) 10. Art. no. 808509.

[40] W. Tushar, H. Shisheng, Y. Chau, J.A. Zhang, and D.B. Smith, Synthetic generation of solar states for smart grid: a multiple segment Markov chain approach, in: Innovative Smart Grid Technologies Conference Europe (ISGT-Europe), 2014 IEEE PES, 2014, pp. 1–6.

[41] M.A. Eltawil, Z. Zhao, Grid-connected photovoltaic power systems: technical and potential problems—a review, Renew. Sustain. Energy Rev. 14 (1) (2010) 112–129.

[42] S. Conti, S. Raiti, Probabilistic load flow using Monte Carlo techniques for distribution networks with photovoltaic generators, Sol. Energy 81 (12) (2007) 1473–1481.

[43] J.V. Paatero, P.D. Lund, Effects of large-scale photovoltaic power integration on electricity distribution networks, Renew. Energy 32 (2) (2007) 216–234.

[44] D. Dong, I. Cvetkovic, D. Boroyevich, W. Zhang, R. Wang, P. Mattavelli, Grid-interface bidirectional converter for residential DC distribution systems—part one: high-density two-stage topology, IEEE Trans. Power Electr. 28 (4) (2013) 1655–1666.

[45] A. Werth, N. Kitamura, K. Tanaka, Conceptual study for open energy systems: distributed energy network using interconnected DC nanogrids, IEEE Trans. Smart Grid 6 (4) (2015) 1621–1630.

[46] J. Schonberger, R. Duke, S.D. Round, DC-bus signaling: a distributed control strategy for a hybrid renewable nanogrid, IEEE Trans. Ind. Electr. 53 (5) (2006) 1453–1460.

[47] S.I. Ganesan, D. Pattabiraman, R.K. Govindarajan, M. Rajan, C. Nagamani, Control scheme for a bidirectional converter in a self-sustaining low-voltage DC nanogrid, IEEE Trans. Ind. Electr. 62 (10) (2015) 6317–6326.

[48] D. Sciano *et al.*, Evaluation of DC-links on dense-load urban distribution networks, IEEE Trans. Power Delivery. 99, 1-1, 2016.

[49] K.T. Tan, B. Sivaneasan, X.Y. Peng, P.L. So, Control and operation of a DC grid-based wind power generation system in a microgrid, IEEE Trans. Energy Convers. 99 (2015) 1–10.

[50] D. Dong, F. Luo, X. Zhang, D. Boroyevich, P. Mattavelli, Grid-interface bidirectional converter for residential DC distribution systems—2014; Part 2: AC and DC interface design with passive components minimization, IEEE Trans. Power Electr. 28 (4) (2013) 1667–1679.

[51] F. Gonzalez-Longatt, J.M. Roldan, C.A. Charalambous, Solution of ac/dc power flow on a multiterminal HVDC system: Illustrative case supergrid phase I, in: 47th International Universities Power Engineering Conference (UPEC 2012), 2012, pp. 1–7.

[52] T.M. Haileselassie, K. Uhlen, Impact of DC line voltage drops on power flow of MTDC using Droop control, IEEE Trans. Power Syst. 27 (3) (2012) 1441–1449.

[53] F. Gonzalez-Longatt, J. Roldan, C.A. Charalambous, Power flow solution on multi-terminal HVDC systems: supergid case, in: Presented at the International Conference on Renewable Energies and Power Quality (ICREPQ'12), Santiago de Compostela, Spain, 28–30 March 2012.

[54] R.T. Pinto, P. Bauer, S.F. Rodrigues, E.J. Wiggelinkhuizen, J. Pierik, B. Ferreira, A novel distributed direct-voltage control strategy for grid integration of offshore wind energy systems through MTDC Network, IEEE Trans. Ind. Electr. 60 (6) (2013) 2429–2441.

[55] F. Gonzalez-Longatt, Optimal steady-state operation of a MTDC system based on DC-independent system operator objectives, in: 11th IET International Conference on AC and DC Power Transmission, 2015, pp. 1–7.

[56] S.A. Soliman, A.-A.H. Mantawy, *Modern Optimization Techniques with Applications in Electric Power Systems* (Energy systems), Springer, New York, 2012, pp. xvii, 414 p.

[57] M. Aragüés-Peñalba, A. Egea-Àlvarez, O. Gomis-Bellmunt, A. Sumper, Optimum voltage control for loss minimization in HVDC multi-terminal transmission systems for large offshore wind farms, Electr. Power Syst. Res. 89 (0) (2012) 54–63.

[58] F.M. González-Longatt, et al., Modelación y simulación de la velocidad de viento por medio de una formulación estocástica (in Spanish), Revista Ingeniería UC 14 (3) (2007) 7–15.

[59] E. Kranakis, MITACS (Network), *Advances in Network Analysis and Its Applications* (Mathematics in Industry, No. 18), Springer, Heidelberg, 2013, pp. xvi, 409 pages.

CHAPTER

20

Energy-management solutions for microgrids

Mostafa H. Mostafa[1], *Shady H.E. Abdel Aleem*[2], *Samia Gharib Ali*[3] *and Almoataz Y. Abdelaziz*[4]

[1]Department of Electrical Power and Machines, International Academy of Engineering and Media Science, Cairo, Egypt [2]Department of Mathematical, Physical and Engineering Sciences, 15th of May Higher Institute of Engineering, Helwan, Cairo, Egypt [3]Department of Electrical Power and Machines, Kafrelsheikh University, Cairo, Egypt [4]Faculty of Engineering & Technology, Future University in Egypt, Cairo, Egypt

20.1 Introduction

Globally, the use of microgrids (MGs) is growing exponentially with time due to their intelligent characteristics and significant impacts on the efficiency and reliability of grid-provided power [1,2]. The United States' Department of Energy defines an MG as a collection of loads and distributed energy resources (DERs) that are interconnected together under predefined standards to form a single controllable structure that can be worked while linked with the main grid or isolated from it [3]. The authors in Ref. [4] introduce an MG as a collection of DERs that can supply energy consistently and adequately to a large load. In addition, the MG must have autonomous control and isolated mode must occur with the least interruption. The authors in Ref. [5] classify the MG as a local grid that includes distributed generation, power electronics converters, ESSs, and telecommunication. Telecommunication is important for the MG to work in isolated mode or grid-connected mode. The authors in Ref. [6] expand the definition of an MG to include the necessary control and protective devices. The above discussion shows that MG consist of DERs, ESSs, dispatchable load, protection devices, and a telecommunication system working together to generate heat and power for a local area in isolated mode or parallel to the grid [3–6].

Distributed generation (DG) can be described as a slight scale generation element based on burning, such as reciprocating engines and turbines, or nonburning such as fuel cells, photovoltaic (PV) panels, wind turbines (WTs), etc. These are situated near the end clients

Distributed Energy Resources in Microgrids
DOI: https://doi.org/10.1016/B978-0-12-817774-7.00020-X

and are characterized as renewable or cogeneration sources [7,8]. DG is known as electricity generation by facilities that can be connected to the power system at any location [9].

RESs and ESSs are widely used around the world to lessen the pollution which is produced by traditional power plants and reduce purchasing energy costs [10,11]. ESSs help the renewable DG to be integrated with the MG with a high penetration level [12]. Currently, renewable energy and ESSs are widely integrated into MGs to achieve these objectives. Also, ESSs are receiving more interest with regard to improving their efficiency, density, cost, and lifespan. ESSs are important in the operation of MGs because they provide many benefits for MGs. ESSs can be utilized to store power during off-peak times and to provide recharge during peak time. As the cost of kWh in the off-peak is less than the cost of kWh in peak times, this price difference is beneficial to the ESS proprietor [13,14]. Other benefits include reliability and power quality development [13]. RESs are affected by ecological factors. Thus, ESSs are utilized to store extra power to preserve the frequency and voltage system and to supply power to the load at storage state [15]. Due to the importance of ESSs, their location and sizing must be selected very carefully to reduce the capital costs of ESS and confirm the correct operation of the MG. ESSs are used to deal with the intermittent renewable energy sources (RES) efficiently, enhance MG reliability and power quality, smooth the uncertainty while handling loads variation, alter peak load, develop efficient control system, and regulate frequency and voltage system [10–16]. The integration of ESSs at non-optimal sizes leads to various problems such as high cost, power losses, frequency problems, and bigger ESS capacity. Therefore the capacity and location of ESSs must be selected very carefully to ensure the correct operation of the MG.

20.2 Statistical analysis of related energy management for microgrids

Global energy demand has been growing exponentially. Traditional energy resources (e.g., coal, oil, and gas) have contributed significantly to global electricity generation. However, they have also contributed to an increase in pollution in the world and a deterioration of human health [2]. Thus, RESs have been employed in recent years to lessen the pollution produced by traditional energy resources [3]. Fig. 20.1 shows that global renewable power capacity reached 2195 GW in 2017 [17]. In 2017, the global deployment of RESs, without hydropower, reached 1081 GW, with China being the top country to deploy RESs followed by the United States and Germany, as shown in Fig. 20.2 [17].

The statistical analysis presents data about documents in the field of electrical energy storage, renewable energy, and MGs which have been published until 2017. The topics have been searched using the Web of Science Database. Fig. 20.3 shows the number of publications for six topics over 17 years (2000–2017). It shows that the number of publications related ESS and MG has been increasing.

20.3 Basic concepts of energy management for microgrids

Worldwide, electric power plants are predicted to more intelligent in the near future. Thus, the intelligent MG, which is able to work in isolated mode or grid-connected mode, has begun to attract considerable attention because it increases system reliability and

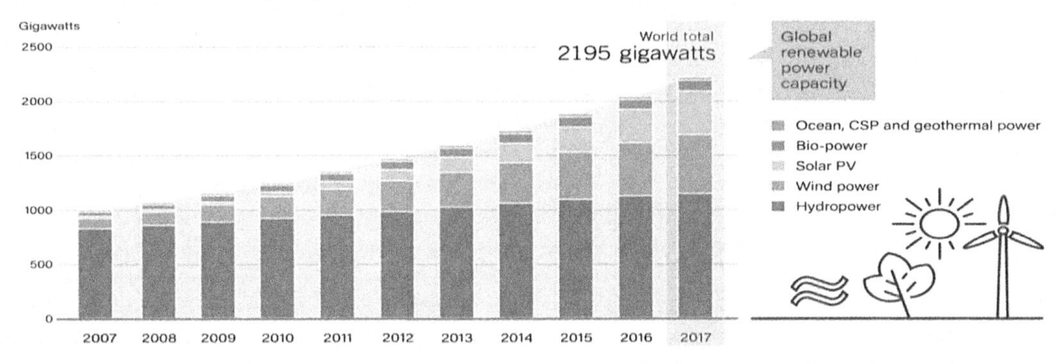

FIGURE 20.1 Global renewable power capacity in the world [17].

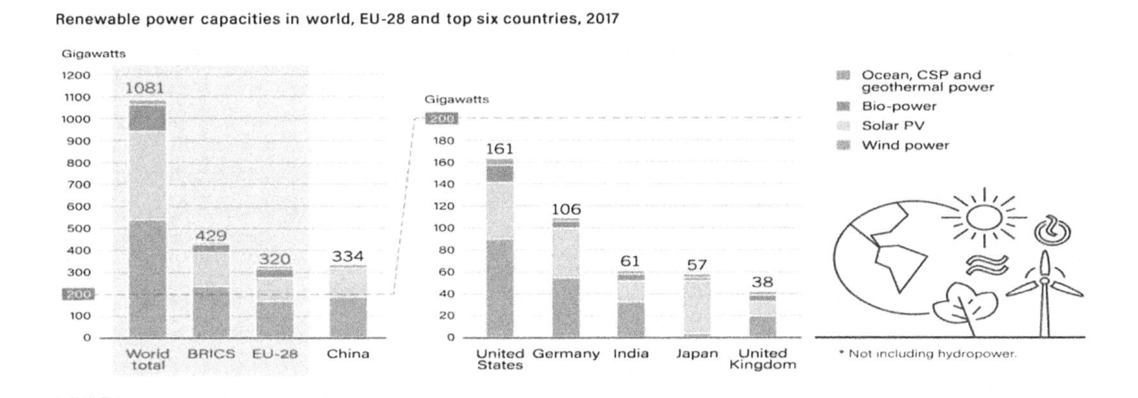

FIGURE 20.2 Global deployment of renewable energy systems (RESs) [17].

flexibility, decreases the investment of transmission lines, and increases the renewable energy deployment. However, MGs face many challenges in reaching their objectives, such as system stability, energy management, and power quality [18]. The concept of energy management for MGs is shown in Fig. 20.4. It consists of controllable DG (e.g., fuel cells and cogeneration or combined heat and power), noncontrollable and limited DG (e.g., PV panels and WTs), controllable load, a bidirectional energy storage system, and bidirectional utility. The system has two modes: isolated mode and grid-connected mode. An MG manager should be responsible for the energy dispatch of DG and supply power for the customer depending on predictions related to renewable energy generation. However, due to the uncertainty of RES, the MG may not be able to supply the customer alone and may have to buy energy from the utility grid [19].

Energy management for MGs is not limited to supply loads, but also about providing an optimal operation for MGs to reach their minimum total operation cost (TOC) or

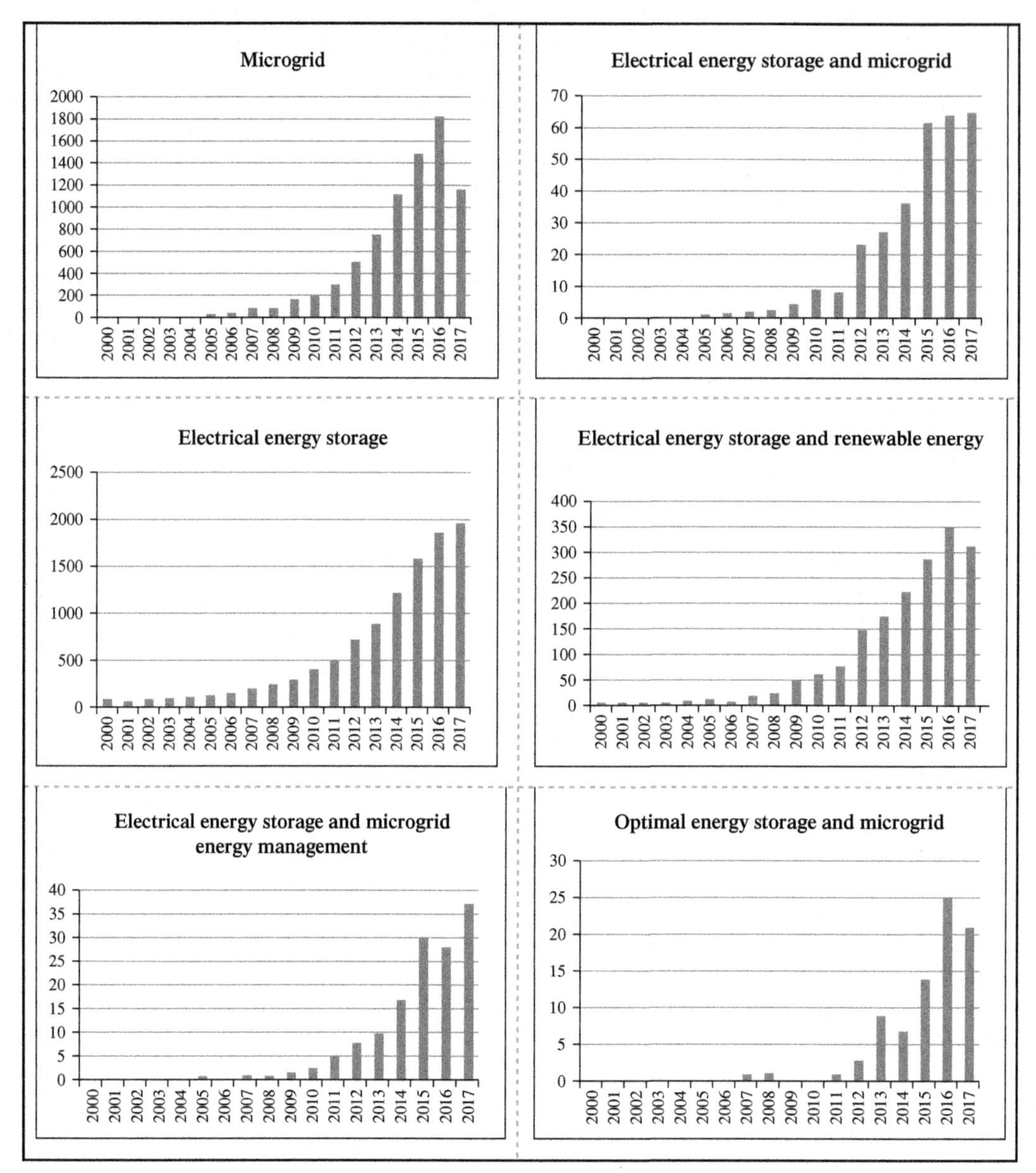

FIGURE 20.3 A summarized statistical analysis of the trend of research associated with energy storage system (ESS) and microgrids (MGs).

minimizing the total daily operation cost. MG management also needs to consider keeping power losses and carbon dioxide emissions to a minimum, maximizing the output power of renewable energy, ensuring maximum efficiency, and smoothing the influence of power exchanges at the point of common coupling (PCC). In addition, numerous constraints should be considered, such as ensuring a power balance between generation and load, as well as awareness of the output power limit of DG, the power capacity limit of the

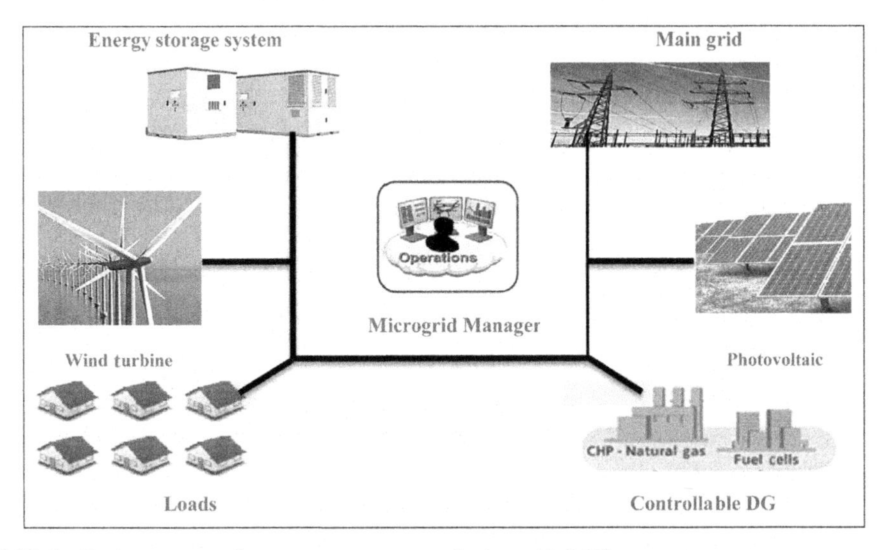

FIGURE 20.4 Basic concepts of energy management of microgrid (MG).

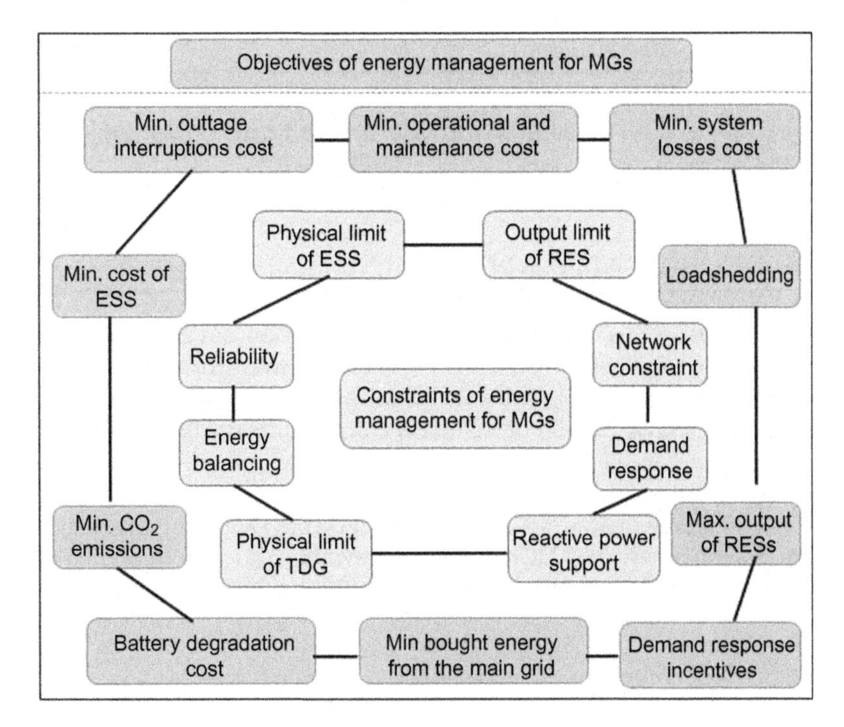

FIGURE 20.5 Objectives and constraints of energy management for microgrids (MGs).

transmission line, the start-up/shut down limit of the DG, the charging and discharging limit of ESS, the state of charge (SOC) limit of ESS, voltage magnitude and angle limits, and spinning reserve limit. Therefore energy management for MG faces a complex multi-objective optimization issue [20]. Fig. 20.5 shows the multiple objectives that need to be

TABLE 20.1 Examples of microgrids in the world.

Site	Microgrid
Europe	Model City of Manheim, Germany; Cell Controller Project, Denmark; CRES-Gaidouromantra, Kythnos, Greece; Liander's Holiday Park at Bronsbergen, Zutphen, The Netherlands; RSE-DER test facility, Italy; TECNALIA-DER test facility, Bilbao, Spain; PIME'S project, Dale, Norway; Szentendre, Hungary; Salburua, Spain; La Graciosa Island microgrid, Spain; Optimagrid, Spain; iSare project, Guipúzcoa, Spain
Australia	CSIRO, Roltnest Island, Coral Bay, King Island, Denhem, Esperence, Hopetoun, Kings Canyon, and Brcmer Bay
Africa	Diakha Madina, Senegal
Asia	Rural PV hybrid microgrid, West Bank; Hangzhou Dianzi University, China; NbDO microgrid, Aichi, Kyotang, Elaciiinohc, Japan; NEDO Tohoku Fukushi University, Sendai, Japan; Shimi/u Corp. microgrid; Tokyo Gas microgrid, Aiclii Institute of Technology microgrid, Japan; INER microgrid, Taiwan
South America	Chile; OHagUe's microgrid, Chile; Huatacondo microgrid, Chile, Robinson Crusoe Island.
North America	Fort Zed, Fort Collins, Colorado; University of San Diego, California; Santa Rita Jail, Santa Rita, California; Perfect Power, Chicago, Illinois; BCiT microgrid, Vancouver, BC, Canada; Balls Gap Station, Milton, West Virginia

met by an energy-management solution for MGs and the many constraints that should be considered in the solution.

20.4 Microgrid configuration and design

MGs consist of DERs, ESSs, dispatchable loads, protection devices, and telecommunication systems working together to generate heat and power for a local area in isolated mode or parallel to the grid [3–6]. Table 20.1 shows examples of general MGs around the world [21]. Table 20.2 shows details of the main components of the simulation MGs which are studied in this chapter. The types of DERs, which are integrated in the studied MGs, vary considerably. The diversity in the type of DG indicates that numerous DGs are considered. Most of the studies using MGs include renewable DG and nonrenewable DG, and differ in the types of renewable DG and nonrenewable DG. [15,26,27,37,38,40,43] study MGs that include renewable DG only.

20.5 Energy storage system in the microgrid

Table 20.3 shows an overview of the type of ESS, as well as technical and economic information on such a system, which is studied in this chapter. The types of ESS, which are integrated in the MGs, vary considerably. The selection of an ESS type and size depends on the objective function of the study because ESSs have significant impacts on

TABLE 20.2 Main component of the simulation microgrid (MG).

References	Renewable					NonrRenewable			Mode of operation of microgrid	Simulated location	Year of publication
	PV	WT	Hydro generator	Tidal power	Fuel cell	Microturbine	Diesel generator	Gas turbine			
[22]	20 kW	—	—	—	30 and 20 kW	30 kW	—	—	Grid-connected Mode	Generic	2011
[23]	1 MW	—	—	—	—	—	—	Two 5 MW each & Two 3 MW each	Grid-connected mode	Generic	2012
[24]	500 kW	1000 kW	1000 kW	—	—	—	500 k kW	500 kW	Grid-connected mode	Generic	2013
[25]	25 kW	15 kW	—	—	30 kW	30 kW	—	—	Grid-connected mode	Test MG model	2013
[26]	780 kW	29 and 23 kW	—	—	—	—	—	—	Grid-connected mode	Modified IEEE-37 bus system	2014
[13]	—	20 kW	—	—	50 kW	30 kW	Two diesel generators of 50 kW each	—	Isolated and grid-connected mode	Generic	2015
[27]	4 MW	8 MW	—	—	—	—	—	—	Isolated mode	North China	2015
[28]	25 kW	15 kW	0	—	30 kW	30 kW	—	—	Grid-connected mode	Typical MG system	2015
[15–37]	3 MW	—	1.2 and 2 MW	—	—	—	—	—	Isolated mode	Thailand	2016
[38]	0.78 MW	—	—	—	—	—	—	—	Grid-connected mode	Modified IEEE-33 bus system	2016
[39]	—	3 MW	—	—	—	—	Four diesel generators (3.5, 3, 3, 4.1) MW	—	Grid-connected mode	Generic	2016

(*Continued*)

TABLE 20.2 (Continued)

References	Renewable					NonrRenewable			Mode of operation of microgrid	Simulated location	Year of publication
	PV	WT	Hydro generator	Tidal power	Fuel cell	Microturbine	Diesel generator	Gas turbine			
[40]	1 MW	1 MW	—	—	—	—	—	—	Grid-connected mode	Modified IEEE-14 bus system	2016
[41]	320 kW	850 kW	—	150 kW	80 kW	150 kW	120 kW	—	Grid-connected mode	Zhanruo Island	2016
[42]	79.2 kWp	750 kW	—	—	—	50 kW	—	220 kW	Isolated Mode	China	2017
[43]	57 MW	187 MW	—	—	—	—	—	—	Grid-connected mode	Dammam City	2017
[44]	—	3 × 15 MW	—	—	—	2 × 3 1 × 3.5 1 × 4.1 MW	—	—	Grid-connected mode	Modified 33-bus system	2017

TABLE 20.3 Technical and economic information of energy storage system (ESS).

References	Type of ESS		Output power of ESS	Energy capacity of ESS	Power cost of ESS	Energy cost of ESS	C_{fm} $/kW/year	C_{vm} $/kWh/year	Lifetime years	Efficiency	Depth of charge	Mode operation of microgrid
[13]	Deep Cycle flooded lead acid battery		30 kW	300 kWh	$210/kW	$150/kWh	—	—	A weighted ampere hour (Ah) aging model is employed	—	—	Isolated Mode
[13]	Deep Cycle flooded lead acid battery		50 kW	400 kWh	$210/kW	$150/kWh	—	—	A weighted Ah aging model is employed	—	—	Grid-connected mode
[15]	polysulphide bromide		3.2405 MW	16.2025 MWh	$150/kW	$65/kWh	$9	0	15	65%		Isolated Mode
[15]	Vanadium redox (VR)		3.2405 MW	16.2025 MWh	$426/kW	$100/kWh	$9	0	15	70%		Isolated Mode
[22]	Lead acid		30 Kw	300 kWh	0	$100/kWh	$0.02		5	85%	50–70	Isolated Mode
[22]	Vanadium redox (VR)		40 Kw	400 kWh	$426/Kw	$100/kWh	0		15	65%–75 %	95	Isolated Mode
[23]	Not documented		3.6 MW	18 MWh	—	—	—	—	—	—	—	Grid Connected
[27]	Hybrid energy storage system (HESS)	Battery	1.5 MW	7807.8 MWh	—	670/kWh	—	—	—	70	80	Isolated Mode
		UC	2 MW	1985.1 MWh	—	4000/kWh	—	—	—	98	98	
[28]	Li-ion		30 kW	500 kWh	—	$465/kWh	—	$15	3	—	90	Grid Connected
[37]	Lead-acid BESS		1.3124 MW	2.6248 MWh	$206/kW	$200/kWh	$20	0	65	70%	30	Isolated Mode
[37]	Vanadium redox (VR)		1.3124 MW	2.6248 MWh	$426/kW	$100/kWh	$9	0	15	70%	75	Isolated Mode

(Continued)

TABLE 20.3 (Continued)

References	Type of ESS		Output power of ESS	Energy capacity of ESS	Power cost of ESS	Energy cost of ESS	C_{fm} \$/kW/year	C_{vm} \$/kWh/year	Lifetime years	Efficiency	Depth of charge	Mode operation of microgrid
[39]	Not documented		3.466 MW	24.773 MWh	—	—	—	—	—	—	—	Grid Connected
[42]	Hybrid ESS	Lion	—	36 kWh	—	—	—	—	13.7	—	—	Isolated Mode
		lead acid	—	77 kWh	—	—	—	—	2.74	—	—	
		super capacitor	—	19 kWh	—	—	—	—	20	—	—	
[43]	Sodium sulfur		20 MW	63 MWh	$450/ kW	$450/ kWh	$10	0	10	0	0	Grid Connected
[44]	Intensium flex high-energy lithium-ion		16 kW	—	$3000/ kW	$1000/ kWh	—	—	20	—	70–90	Isolated Mode

system stability, security, power quality, peak load serving, and load management [29]. A minimum SOC is necessary to extend the battery life [15]. Pumped hydro storage (PHS) and compressed air energy storage (CAES) both have a large power output, greater than 50 MW, and energy storage capacities, greater than 100 MWh. Batteries are categorized into 1–50 MW power rating and 5–100 MWh energy capacities, which can be known as medium-power and energy ESSs. Hydrogen energy storage (H_2S) and thermal energy storage (TES) are small power and energy ESSs. Flywheel energy storage (FES), superconductor magnetic energy storage (SMES), and double layer capacitor storage (DLCS) can have large power or large energy capacity but not both [41]. Large-scale ESSs can be used as power plants, while the medium and small ones are useful for remote communities and electric vehicle (EV) and satellite applications [22].

Fig. 20.6 shows that there is a relationship between the power and energy capacity of various ESS technologies and their discharge time. FES, DLCS, and SMES are functionally most appropriated to power quality, bridging power, and smoothing the intermittent supply of RESs as their discharge time is small (less than 1 hour) [30]. The energy management, smoothing the intermittency of RESs, and peak-shifting objective-based targets are more adapted to the small CAES, lead acid (LA), lithium-ion (Li-ion), nickel–cadmium (Ni–Cd), polysulphide bromide battery (PSB), and zinc bromide (ZnBr) due to their medium discharge time (between 1 and 10 hours) [25]. Back-up generation and seasonal ESS objective-based targets are more adapted to PHS, large CAES, solar fuel, fuel cell, TES, and vanadium redox battery (VRB) due to their large discharge time (over 10 hours) [30].

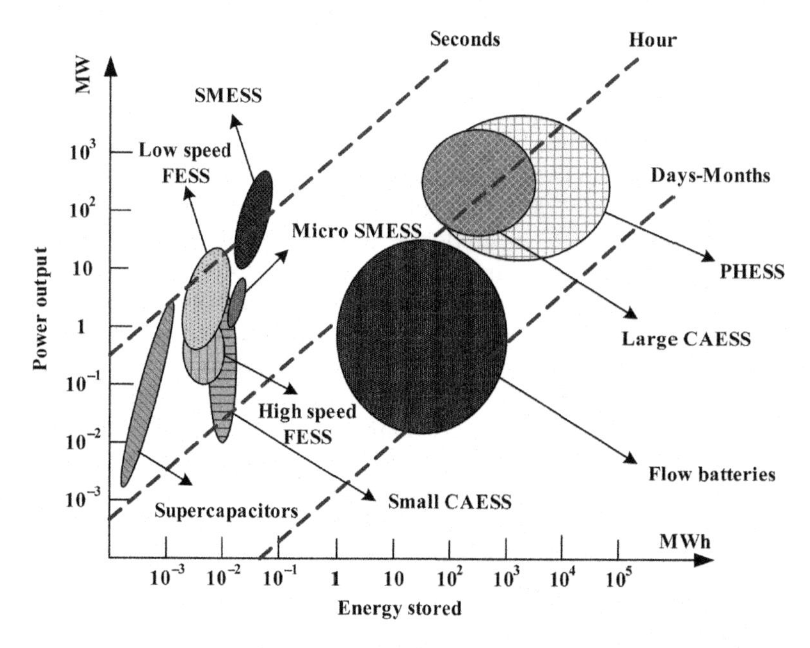

FIGURE 20.6 Energy storage system (ESS) technologies power and energy outputs [31].

20.6 Energy management solutions for microgrids

The energy-management solutions for MGs can be viewed as an optimization issue [32]. The TOC of the MG can be decreased by growing demand control, installing various renewable resources, and trading electricity in the electricity market [33]. Advanced energy management techniques and smart communication techniques are used in MGs to increase the size of renewable energy resources (RERs) and reduce the daily operating costs [34]. The variation of power at the PCC can be supported by organizing and optimizing the penetration level of the different energy sources and loads (energy management) [35,36]. Table 20.4 provides the objective function, constraint, simulation, and optimization program, and mode of the MG which are utilized in each reviewed research.

In Ref. [24], the authors analyze the generation and storage capacity energy of the MG. The presented method is divided into three steps. The first step discusses the RES commitment, the second step discusses the ESS problems, and the third step discusses the operation of thermal units to achieve minimized operation costs and CO_2 emissions by maximizing the energy which is obtained by the RESs. The results show that, during some periods, generation is greater than load (the ESS is charged at these periods) and at other periods generation is lower than load (ESS is discharged at these periods). In Ref. [25], the harmony search algorithm (HSA) is used to get an optimal operation of MG to minimize the daily operation cost of the MG. The results show that, the daily operation cost is reduced when considering ESS in the operation of the MG. In Ref. [45], improved particle swarm optimization (PSO) is used to get an optimal operation of the MG to maximize the output of renewable energy (PV and hydropower) in MG to enhance the stability and reliability of MG and efficiency of DG in all conditions. The uncertainty of RESs, the life of ESSs, and characteristics of each DG are considered in this study. The results show that the proposed method is appropriate for MGs that include hydropower generation. In Ref. [41], PSO is used to get the optimal real-time capacity of DGs and ESS for MGs to decrease power losses and improve economy. Real-time optimization methods are better than the traditional methods as they obviate the forecast errors. It is known that the price of electricity changes based on the time of use (TOU). TOU in this study is sorted into three types: high price at peak time, moderate price at moderate time, and low price at valley time. This methodology comprises three steps. Firstly, it collects the information of the MG, which includes the forecasting output of renewable energy, SOC of the battery, the maintenance cost of the DG and the battery, and electricity price. Secondly, the constraints of the operation of MG are taken into account to enhance the performance of the MG. Thirdly, the optimal capacity of the DG and ESS are obtained to achieve minimum cost and reduce power losses. Fig. 20.7 shows the influence that the optimal capacity of the ESS has on the TOC of the MG. It shows that there is one point which has a minimum total cost.

In Ref. [40], the authors present an optimization strategy for the MG energy scheduling that takes into account the storage energy supplied by EVs. They present a stochastic model for scheduling the energy of MG in grid-connected mode. This method depends on the Benders decomposition algorithm (BDA) to obtain the MG energy scheduling. Two approaches are presented in this study for scheduling the MG energy. The change of total

TABLE 20.4 Objective function and constraints of microgrid (MG) energy management.

References	Objective function	Constraint	Simulation and optimization program	Mode of microgrid	Year of publication
[24]	• Minimize operation cost and CO_2 emissions	1. Output power limit 2. Power-balanced limit 3. Start-up/shut-down limit	FICO expresses programming optimization	Grid-connected mode	2013
[25]	• Minimize total daily operating cost	1. Output power limit 2. Power-balanced limit 3. SOC limit 4. Charge and discharge power limit	HSA	Grid-connected mode	2013
[45]	• Maximize output power of renewable energy	1. Power-balanced limit 2. PV output limit 3. Hydropower balance limit 4. ESS limit	Improved PSO	Grid-connected mode	2015
[41]	• Improve the economy and decrease power losses	1. Output power limit 2. Power-balanced limit 3. Ramp of DG limit 4. Power transmission limit 5. The SOC limit 6. ESS limit	PSO	Grid-connected mode	2016
[40]	• Minimize the forecasted total operation of MG	1. Power balanced between generation and load 2. Capacity of parking limit 3. Departing electric vehicles (EVs) equal arriving EVs 4. Energy injected and absorbed from the battery limit 5. Apparent power limit 6. The voltage magnitude and angle limits	Benders decomposition method	Grid-connected mode	2016
[46]	• Minimize total daily operating cost	1. Power-balanced limit 2. ESS limit	GAMS	Grid-connected mode	2016

(*Continued*)

TABLE 20.4 (Continued)

[47]	• Minimize the bought energy and maximize the employment of the bought energy	—	Linear programing + stochastic linear programming	Grid-connected mode	2017
[48]	• Minimize the daily operation of MG and smooth the influence of power exchanges at the PCC	1. Power limit 2. Cooling load limit 3. Building thermal balanced equation 4. Temperature limit 5. Power purchase limit	Optimal dispatch program (Matlab)	Grid-connected mode	2017
[49]	• Minimize the forecasted MG operational costs	1. Power balanced limit 2. Output Power limit 3. Ramp up/down of microturbine limit 4. Start-up/shut down limit 5. Spinning reserve limit 6. Charging and discharging power limit 7. Stored energy in ESS limit	GAMS	Grid-connected mode	2017

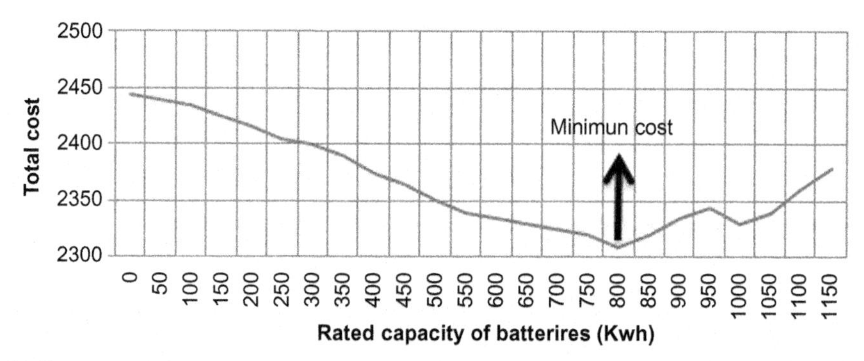

FIGURE 20.7 The influence optimal capacity of ESS on the operational cost of the microgrid (MG).

demand, the obtainable EVs, and the uncertainty of renewable resources are considered in this strategy. The EVs are utilized to interchange the electricity in the electricity market and help to reduce the economic cost of the TOC of system [50]. The position of the EVs is called a parking facility. A parking facility, which is near a mall and office building, can provide the energy to the MG [51]. EVs can be used to store the energy during low price (off-peak time) and inject the stored energy into the grid when the price of electricity is high. The results show that, the TOC with the parking facility is lower than the TOC without the parking facility, so TOC is reduced due to the use of the parking facility. Also, this work discusses the influence of the parking facility on the buying and selling of energy in the day-ahead market. Without the parking facility, its predominant energy is bought. The study debates the influence of uncertainty of total load and number of EVs on the TOC individually and integrally. The TOC is reduced when uncertainty of total load and number of EVs is considered in the computation. This method uses various values for battery wear cost ($/MWh) to show the influence of battery wear cost on the rate that the EV's consume or inject energy and the cost saving. When the wear cost of battery gets lower, the cost saving will increase. However, the average charging/discharging time of EVs' batteries is high. Thus lowering the value of battery wear cost can decrease the EV owners' motivation for participating in the vehicle to grid program. In Ref. [47], the authors introduce an energy management strategy to minimize the purchased energy. This strategy is based on optimizing the energy which is produced from the energy resources. Surplus energy is stored in the ESS and EVs, and used by the MG as needed. This strategy involves two-stage energy scheduling. The first stage uses linear programming (LP) to assume the exhaustion energy, size of the RES, and the energy cost with a minimum of purchased energy used and a maximum of stored energy in ESSs and EVs. The second stage uses stochastic linear programming (SLP) to evaluate the impact of uncertainty of exhaustion energy and RES, and energy cost. The energy system, which is used to verify this strategy, is included off-the-grid and the number of MGs as shown in Fig. 20.8. The selling of energy isn't considered in this strategy.

In Ref. [48], the authors present an energy scheduling strategy in an office building to decrease operating costs and variation of power at PCC. The MG, built comprising PV, an electric chiller, and an EV, is used to verify this method as shown in Fig. 20.9. In approach

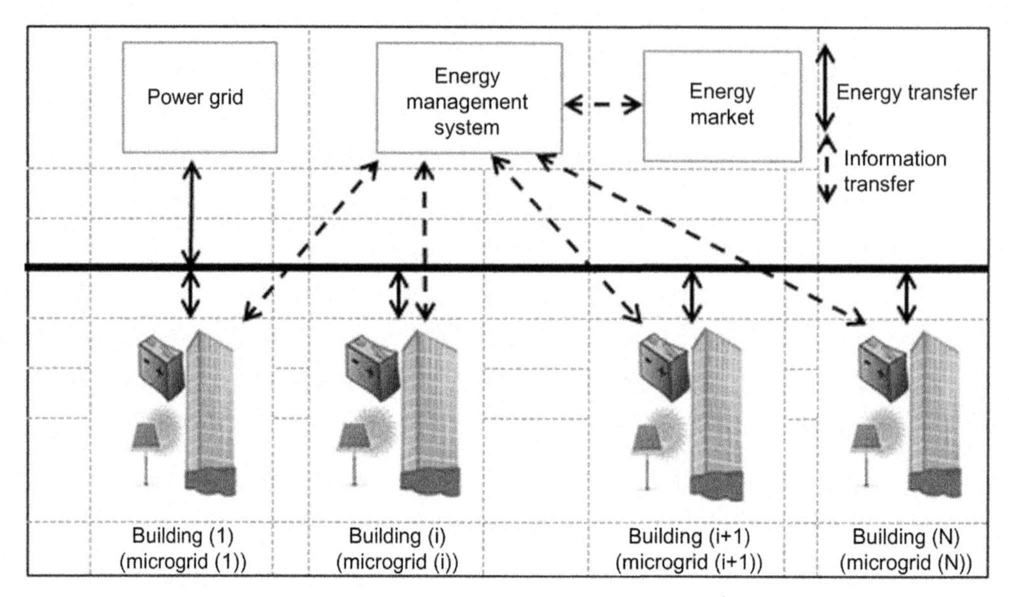

FIGURE 20.8 Simplified energy systems.

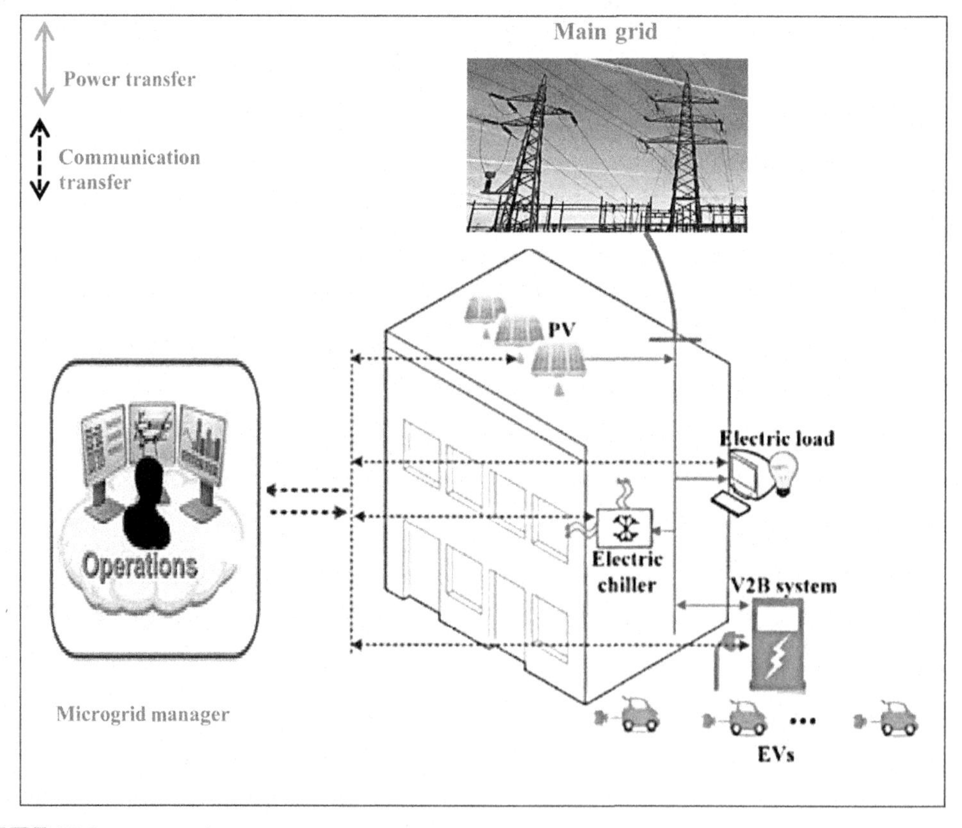

FIGURE 20.9 Microgrid energy management system.

one, an optimal dispatch program is used to get the day-ahead dispatch by using the hourly electricity demand, temperature and radiation values, and the prediction size of RESs for minimizing the daily operation costs. In approach two, a two-layer method is used to streamline the variation of power at the PCC by organizing the ESS and EVs. In this study, a control method is presented to dispatch the output of EVs. In Ref. [49], a stochastic method is utilized to represent the change of load, the output of RES, and the electricity price. An optimal stochastic operation of the MG in grid-connected mode is discussed by considering ESS and a demand response program (DRP) to decrease the forecasted operation cost. It also takes into account the TOU used to shift the load from peak periods to off-peak periods [52]. TOU is utilized in this study to evaluate the influence it has on the operation cost. The results show that, the TOC of MG is the least when considering ESS and DRP in the operation [46,52].

20.7 Energy storage system optimization for microgrid

Due to the important role of ESSs, their location and sizing must be selected very carefully to ensure the correct operation of the MG. The integration of an ESS of nonoptimum size can result in an increase in costs, power losses, frequency problems, and bigger ESS capacity [15]. For these reasons, the optimal location and sizing (power and energy capacity) of ESSs are the main objectives for MGs. Table 20.5 offers a general review of the optimization objective function, the constraints and obligations of the optimization method, the optimization and simulation programs used in each search, and the year of publication. Table 20.5 shows the similarities and differences between each paper. It shows that the target of most research is related to the total operating costs of the MG. Table 20.6 presents the total operating cost of an MG.

The issue of ESS optimization for an MG has been the focus of various optimization programs in recent years. In Ref. [22], the matrix real coded genetic algorithm (MRCGA) based on a net present value (NPV) is presented as a means of obtaining the optimal size and economic operation of an ESS in the MG to improve economic performance of MG. This study presented the optimal number and type energy storage units in addition to their optimal operation values. This method seeks to maximize the power injected from the MG to the grid. The ESS injects power when electricity price is high and absorbs power when the electricity price is low, according to the hourly electricity cost in the market. The exchange of power in this study doesn't exceed 30 kW. The result shows that the total NPV is maximized, when the optimal size of the ESS is obtained. Also, the VRB is more cost effective than the lead-acid battery (LAB). In Ref. [23], the optimal size of an ESS in grid-connected mode is introduced to reinforce the MG reliability and minimize the capital cost of the ESS and therefore the operating costs of the MG by using mixed integer programming (MIP). In this work, MG reliability is calculated by using a practical stochastic process (Monte Carlo simulation), while loss of load expectation (LOLE) is used to investigate the reliability of the MG, and the impact of transmission lines on MG reliability is neglected. Fig. 20.10 shows that when the capital cost of an ESS increases, the operating cost of the MG decreases, but the total cost increases. So, an optimal size of an ESS is needed to obtain the minimum total cost. Three cases are discussed in this work to show

TABLE 20.5 Objective function and constraints of optimal size of energy storage system (ESS) for microgrid (MG).

References	Objective function	Constraint	Optimization and simulation program	Target	Year of publication
[22]	• Maximum net present value of MG $MaxOC^{NPV} = OC_{WO} - OC_{W}$ $OC_{WO} = C_{Cap}^{DG} + C_{RC}^{DG} + C_{OM}^{DG} + C_{FC}^{DG} + C_{SC}^{DG} - C_{E}^{DG}$ $OC_{W} = C_{Cap}^{DG} + C_{RC}^{DG} + C_{OM}^{DG} + C_{FC*}^{DG} + C_{SC*}^{DG} - C_{R}^{DG} + C_{Cap}^{BAT} + C_{OM}^{BAT} + C_{RC}^{BAT} - C_{R}^{BAT}$	1. Power balancing 2. Policy of power exchange at inverter capacity shortage	GA	Optimal size	2011
[23]	• Minimum capital cost of ESS and minimum operating cost of MG $Min\ OC + C_{cap}$ $C_{cap} = C_{p}.P_{cbat} + C_{E}.E_{cbat}$ $OC = \sum_{s=1}^{M_s} P_s \left(\sum_{p=1}^{M_p} MD(p) \sum_{h=1}^{M_h} \sum_{i=1}^{N_i} \left[F_i\left(P_{ihp}\right) \right] \right) + \sum_{s=1}^{M_s} P_s$ $\sum_{p=1}^{M_p} MD(p) \left(\sum_{h=1}^{M_h} \sum_{i=1}^{N_i} \left[SUC_{iph} + SDC_{ihp} \right] \right)$ $+ \sum_{s=1}^{M_s} P_s \left(\sum_{p=1}^{M_p} MD(p) \sum_{h=1}^{M_h} \left[EP_{hp} P_{grid,hp} \right] \right.$	1. Generation limit $\sum_{i=1}^{N} P_i^t + P_{wt}^t + P_{Grid}^t + P_{bat}^t = (1 - LSI) P_l^t$ 2. ESS limit $-P_{cbat} \le P_{bat} \le DP_{cbat}$ 3. Reliability limit $LL \le LL_{Target}$	MIP	Optimal size	2012
[26]	• Minimize the power bought from the main grid	—	HSA	Optimal location	2014
[28]	• Minimize the total operating cost of the MG $Min(F) = C_{grid} + C_{DG} + SUC_{DG} + SDC_{DG} + OMC_{DG} + TCPD$	1. Power limit $P_{MT}^t + P_{FC}^t + P_{PV}^t + P_{wt}^t + P_{Grid}^t + P_{bat}^t = P_l^t$ 2. Output power limit $P_{MT}^{min} \le P_{MT} \le P_{MT}^{max}$ $P_{PV}^{min} \le P_{PV} \le P_{PV}^{max}$ $P_{WT}^{min} \le P_{WT} \le P_{WT}^{max}$ $P_{FC}^{min} \le P_{FC} \le P_{FC}^{max}$	GWO	Optimal size	2015

		3. Grid limit $P_{grid}^{min} \leq P_{grid} \leq P_{grid}^{max}$ 4. ESS output limit $P_{Bat}^{min} \leq P_{Bat} \leq P_{Bat}^{max}$ 5. Reserve limit $P_{MT}^{t} + P_{FC}^{t} + P_{Grid}^{t} + P_{bat}^{t} \geq R^{t} + P_{l}^{t}$			
[27]	• Minimize daily cost of HESS $Min\ f_1 = \frac{1}{365}(C_{Cap} + C_O + C_M)$ $C_{Cap} = E_{bat}\ C_{EBat} f_{PBat} + E_{uc}\ C_{Euc} f_{Puc}$ $C_O = E_{bat}\ C_{EBat} f_{OBat} + E_{uc}\ C_{Euc} f_{Ouc}$ $C_M = E_{bat}\ C_{EBat} f_{MBat} + E_{uc}\ C_{Euc} f_{Muc}$	1. Constraints about energy storage a. storage limit $SOC_{Min,Bat} \leq SOC_{t,Bat} \leq SOC_{Max,Bat}$ $SOC_{Min,uc} \leq SOC_{t,uc} \leq SOC_{Max,uc}$ b. Max power limitation $P_{cbat,max} = -min\{P_{bat}, \frac{(SOC_{bat,max} - SOC_{bat,t-1}).E_{bat}}{\eta_{cbat}.\Delta t}\}$ $P_{dbat,max} = min\{P_{bat}, \eta_{dbat} * \frac{(SOC_{bat,t-1} - SOC_{bat,min}).E_{bat}}{\Delta t}\}$ $P_{cuc,max} = -min\{P_{uc}, \frac{(SOC_{uc,max} - SOC_{uc,t-1}).E_{uc}}{\eta_{cuc}.\Delta t}\}$ $P_{duc,max} = min\{P_{uc}, \eta_{duc} * \frac{(SOC_{bat,max} - SOC_{bat,t-1}).E_{bat}}{\Delta t}\}$ 2. Constraints about operation of MG a. Power Limit $P_{PV}^{t} + P_{wt}^{t} + P_{uc}^{t} + P_{bat}^{t} + P_{sh}^{t} = P_{l}^{t} + P_{waste}^{t}$ b. LPSP & SPSP Limit $LPSP = \frac{\sum_{t=1}^{24} P_{sh}}{\sum_{t=1}^{24} P_{l}} \leq LPSP_{max}$ $SPSP = \frac{\sum_{t=1}^{24} P_{waste}}{\left(\sum_{t=1}^{24} P_{wt} + \sum_{t=1}^{24} P_{PV}\right)} \leq SPSP_{max}$	QPSO	Optimal size	2015

(Continued)

TABLE 20.5 (Continued)

[13]	• Minimize schedule cost of MG and minimize total cost of ESS $Min(MOC_D) = MSC_D + \text{TCPD}$ $\text{TCPD} = \frac{1}{365}\left(\frac{r.(1+r)^{lt}}{(1+r)^{Lt}-1}.C_{cap}\right)$ $C_{cap} = C_p.P_{cbat} + C_E.E_{cbat}$ $SC_D.= \sum_{t=1}^{T}(\sum_{i=1}^{N}(F_i(P_i^t).U_i^t + OMC_i(P_i^t)$ $+ SUC_i(1-U_i^{t-1}).U_i^t)$ $+ EP^t.P_{Grid}^t)$ $F_i(P_i^t) = a_i.\ (P_i^t)^2 + b_i.P_i^t + C_i$ $OMC_i(P_i^t) = K_{OMC_i}.P_i^t$	1. Power limit $\sum_{i=1}^{N} P_i^t + P_{wt}^t + P_{Grid}^t + P_{bat}^t = P_l^t$ 2. Reserve limit $\sum_{i=1}^{N}(P_i^{max}.U_i^t - P_i^t) + BSR^t$ $+ (P_{Grid}^{max} - P_{Grid}^t) \geq R^t$ $BSR^t = P_{dmax}^t - P_{batt}^t$ if $E^{t-1} \geq E_{min}^t$ Otherwise equal zero 3. Generation limit $P_i^{min}.U_i^t \leq P_i^t \leq P_i^{max}U_i^t$ 4. Grid limit $-P_{grid}^{max}. \leq P_{Grid}^t \leq P_{grid}^{max}$ 5. Time limit $U_i^t = 1$ if $T_i^{on} < MUT_i$ $U_i^t = 0$ if $T_i^{off} < MDT_i$ $U_i^t = 0$ or 1 otherwise	GA based on fuzzy system	Optimal size	2015
[15]	• Minimize size of ESS and minimize total cost of ESS Minimize $F_1 = Min\ (P_{cbat})$ Minimize $F_2 = Min\ (C_{cap} + C_{OM})$ $C_{cap} = C_p.P_{cbat} + C_E.E_{cbat}$ $C_{OM} = C_{fm}.P_{cbat} + C_{vm}.E_{cbat}$	1. Output power of ESS Limit $P_{Bat}^{min} \leq P_{Bat} \leq P_{Bat}^{max}$ 2. Energy capacity of ESS limit $E_{Bat}^{min} \leq E_{Bat} \leq E_{Bat}^{max}$ 3. Frequency limit $F_{min} \leq \text{F} \leq F_{max}$	PSO	Optimal size	2016

[38]	• Minimize power losses of MG $Min \sum_{(i,j)\in E} r_{ij} I_{ij}^2$	1. Current limit $0 \le I_{ij}^2 \le (I_{ij}^{max})^2$ 2. Voltage limit $V_i^{min} \le V_i \le V_i^{max}$ 3. Reactive power limit $Q_{RPC}^{min} \le Q_{RPC} \le Q_{RPC}^{max}$	Monte Carlo	Optimal size & location	2016
[37]	• Minimize size of ESS and minimize capital and the operating and maintenance cost of ESS Minimize $F_1 = Min\ (P_{cbat})$ Minimize $F_2 = Min\ (C_{cap})$ Minimize $F_3 = Min\ (C_{OM})$ $C_{capital} = C_p.P_{cbat} + C_E.E_{cbat}$ $C_{OM} = C_{fm}.P_{cbat} + C_{vm}.E_{cbat}$	1. Output power of ESS Limit $P_{Bat}^{min} \le P_{Bat} \le P_{Bat}^{max}$ 2. Energy capacity of ESS limit $E_{Bat}^{min} \le E_{Bat} \le E_{Bat}^{max}$ 3. Frequency limit $F_{min} \le F \le F_{max}$ 4. Cost limit $C_{cap} \le C_{cap}^{Max}$ $C_{OM} \le C_{OM}^{Max}$	PSO	Optimal size	2016
[39]	• Minimum capital cost of ESS and minimum operating cost of MG $Min\ OC + C_{cap}$ $C_{cap} = C_p.P_{cbat} + C_E.E_{cbat}$ $OC = \sum_{s=1}^{M_s} P_s(\sum_{p=1}^{M_p} MD(p)(\sum_{h=1}^{M_h}\sum_{i=1}^{N_i}[F_i(P_{ihp})])$ $+ \sum_{s=1}^{M_s} P_s(\sum_{p=1}^{M_p} MD(p)(\sum_{h=1}^{M_h}\sum_{i=1}^{N_i}[SUC_{iph} + SDC_{ihp}])$ $+ \sum_{s=1}^{M_s} P_s(\sum_{p=1}^{M_p} MD(p)(\sum_{h=1}^{M_h}[EP_{hp}P_{grid,hp}]$ $\sum_{s=1}^{M_s} P_s(\sum_{p=1}^{M_p} MD(p)(\sum_{h=1}^{M_h}(P_{L,hp}LSI_{hp}\ LL)))$	1. Generation limit $P_i^{min}.U_i^t \le P_i^t \le P_i^{max} U_i^t$ $P_{i,hp} + P_{i,(h-1)p} \le \text{RU}\ (1-\alpha_{i,hp})$ $+ P_i^{min}\alpha_{i,hp}$ $P_{i,(1-h)p} - P_{i,hp} \le \text{RD}\ (1-\beta_{i,hp})$ $+ P_i^{min}\beta_{i,hp}$ $\sum_{h'}^{DT}(1 - U_{i,(h-h')p}) \ge DT_i\alpha_{i,hp}$ $\sum_{h'}^{UP} U_{i,(h-h')p}) \ge UP_i\beta_{i,hp}$ $\sum_{i=1}^{N} P_i^t + P_{wt}^t + P_{Grid}^t + P_{bat}^t$ $= (1-LSI)P_l^t$ 2. ESS limit $0 \le\ P_{bat} \le\ P_{c,max}$ $0 \le\ P_{bat} \le\ P_{d,max}$	GAMS	Optimal size	2016

(*Continued*)

TABLE 20.5 (Continued)

[42]	• Minimize the lifecycle cost of MG $Min\ C_T = C_{cap} + C_{OM} + C_r + C_V + C_{ESS}$ $C_{cap} = \sum_{i=1}^{m}\sum_{K=0}^{K_i} N_i.C_i.\frac{1}{(1+r)^{K.L_i}}$ $C_{om} = \sum_{i=1}^{m}\sum_{j=1}^{j_{max}} K_{om,i}.\frac{1}{(1+r)^j}$ $C_V = \sum_{i=1}^{m}\sum_{j=1}^{j_{max}} N_{p,i}C_{p,i}.\frac{1}{(1+r)^j}$ $C_r = \sum_{i=1}^{m}\sum_{K=1}^{K_i} C_{r,i}\frac{1}{(1+r)^{K.L_i}}$ $C_l = \sum_{i=1}^{m}\sum_{j=1}^{j_{max}} E_{l,i}C_{l,i}.\frac{1}{(1+r)^j}$	1. Output power limit $0 \le P_{pv} \le P_{pv,peak}$ $0 \le P_{WT} \le P_r$ $0 \le P_{WT} \le P_r$ $P_{gt,mix} \le P_{gt} \le P_{gt,max}$ 2. ESS limit $P_{c,min} \le P_{bat} \le P_{c,max}$ $P_{d,min} \le P_{bat} \le P_{d,max}$ 3. ESS energy limit $E_{cbat,min} \le E_{cbat} \le E_{cbat,max}$ 4. Time interrupt limit $T_i < T_{i,max}$	PSO	Optimal size	2017
[32]	• Minimum MG operation cost • $C_T(\text{DODR}, E_{cbat}) = C_{LTD}(\text{DODR}, E_{cbat}) + C_{ENS}(\text{DODR}, E_{cbat})$	1. Power balance limit $P_{bat} = (P_{PV} - P_{PVC}) - (P_L - P_{LC})$ 2. ESS output limit $0 \le P_{Bat} \le P_{Bat}^{max}$ 3. DODR limit $DODR_{LOW} \le \text{DODR} \le DODR_{UP}$	MATLAB simulation	Optimal size	2017
[44]	• Minimum total operation cost $Min\ OC + C_{cap}$ $C_{cap} = C_p.P_{cbat} + C_E.E_{cbat}$ $OC = \sum_{s=1}^{M_s} P_s(\sum_{p=1}^{M_p} MD(p)(\sum_{h=1}^{M_h}\sum_{i=1}^{N_i}[C_{DG}(P_{ihp})])$ $+ \sum_{s=1}^{M_s} P_s(\sum_{p=1}^{M_p} MD(p)(\sum_{h=1}^{M_h}\sum_{i=1}^{N_i}[SUC_{iph} + SDC_{ihp}])$ $+ \sum_{s=1}^{M_s} P_s(\sum_{p=1}^{M_p} MD(p)\ (\sum_{h=1}^{M_h}[EP_{hp}P_{grid,hp}]$	1. Power balanced limit $\sum_{i=1}^{N} P_i^t + P_{wt}^t + P_{Grid}^t + P_{bat}^t - P_{DRP}^t = (1 - LSI)P_l^t$ 2. Voltage limit $V_i^{min} \le V_i \le V_i^{max}$ 3. Main grid limit $P_{grid}^{min} \le P_{grid} \le P_{grid}^{max}$ 4. Output power limit $P_i^{min}.U_i^t \le P_i^t \le P_i^{max}U_i^t$	General algebraic modeling system (GAMS)	Optimal location and size	2017

- Minimum loss of load expectation

$$\text{Min LOLE} = \sum_{s=1}^{M_s} P_s \left(\sum_{h=1}^{M_h} \sum_{i=1}^{N_i} \left[w_{ih}^{s} \right] \right)$$

5. Ramp-up limit

$$P_{i,hp} + P_{i,(h-1)p} \leq \text{RU} \\ (1 - \alpha_{i,hp}) + P_i^{min} \alpha_{i,hp}$$

6. Ramp-down limit

$$P_{i,(1-h)p} - P_{i,hp} \leq \text{RD} \\ (1 - \beta_{i,hp}) + P_i^{min} \beta_{i,hp}$$

7. Minimum down time of DG

$$\sum_{h'}^{DT} (1 - U_{i,(h-h')p}) \geq DT_i \alpha_{i,hp}$$

8. Minimum up time of DG

$$\sum_{h'}^{UP} U_{i,(h-h')p}) \geq UP_i \beta_{i,hp}$$

TABLE 20.6 The consideration of total operation cost of an microgrid (MG).

References	Capital cost of ESS	Operating and maintenance cost of energy storage system (ESS)	Replacement cost	Environment compensation	Energy shortage cost	Fuel cost	Operating & maintenance cost of Distributed generation (DG)	Start-up cost	Shut-down cost	Power bought from the main grid
[13]	✓	×	×	×	×	✓	✓	✓	×	✓
[15]	✓	✓	×	×	×	×	×	×	×	×
[42]	✓	✓	✓	✓	✓	×	×	×	×	×
[37]	✓	✓	×	×	×	×	×	×	×	×
[22]	✓	✓	✓	×	×	✓	✓	✓	×	✓
[23]	✓	✓	×	×	×	✓	✓	✓	✓	✓
[26]	×	×	×	×	×	×	×	×	×	✓
[27]	✓	✓	×	×	×	×	×	×	×	×
[39]	✓	✓	×	×	×	✓	✓	✓	✓	✓
[43]	✓	✓	✓	×	×	×	×	×	×	✓
[28]	✓	✓	×	×	×	✓	✓	✓	✓	✓
[53]	✓	✓	✓	×	✓	✓	✓	×	×	×
[44]	✓	×	×	×	×	✓	×	✓	×	✓

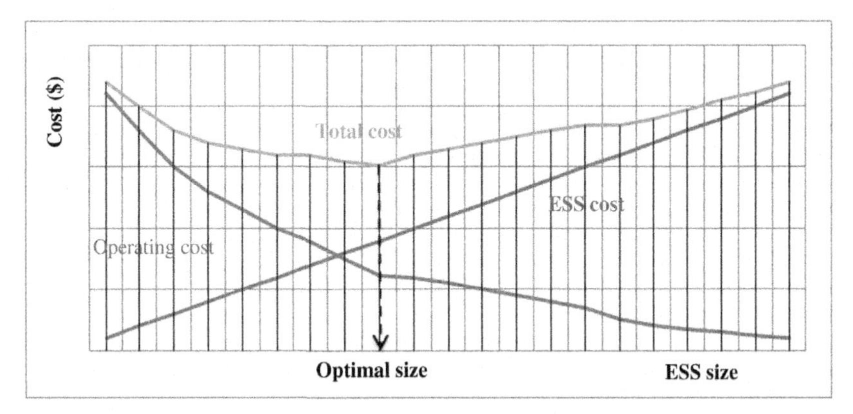

FIGURE 20.10 Relation between capital cost of energy storage system (ESS), operation cost of a microgrid, and total cost.

the influence of the optimal size of an ESS on total cost and reliability of MG. The result shows that an ESS of optimal size enhances the MG's reliability and decreases the total cost of the MG.

In Ref. [26], a methodology is used to obtain the optimal location of ESSs and DERs in an MG to minimize the bought energy from the main grid. The 37-bus IEEE standard MG which includes WTs, PV panels, and 4 ESSs is used to test this method in grid-connected mode. In this study, it is supposed that power generation and power absorption is fixed through a one-hour and the energy is stored evenly in all ESSs. The results show that the bought energy from the main grid is reduced when ESSs are located optimally, so the operation cost of the MG is reduced. In Ref. [28], gray wolf optimization (GWO) is used to obtain an optimum capacity of ESSs and DGs to minimize the total operating cost of the MG. To verify the result of this method, the results, which are obtained by GWO in all cases, are compared with other optimizations (PSO, genetic algorithm (GA), transition state (TS), improved bat algorithm (IBA), teaching learning based optimisation (TLBO), bat algorithm (BA), biogeography based optimisation (BBO), and differential evolution (DE)). The result shows that GWO is the best optimization technique for the operation of an MG without ESS, with ESS without initial charge, and with ESS with initial charge. In Ref. [27], the authors present a new quantum-behaved PSO (QPSO) to obtain the optimal size of the hybrid energy storage system (HESS) in order to achieve the minimum daily cost of the HESS. This study doesn't utilize a single ESS but a combination of batteries and an ultra-capacitor. The ultra-capacitor is characterized by a high-speed response and high-power density, which treats the repeated vibration of power. Batteries are used to treat the simple vibration in power. The results show that an ultra-capacitor can improve the operation of the MG and prevent the repeated charge/discharge of batteries. By comparing the result of the proposed QPSO and the result of the conventional PSO, it is found that an optimal solution is obtained after 47 iterations by using QPSO and after 73 iterations by using conventional PSO. So, the proposed QPSO represents a high convergence rate compared to conventional PSO. Also, the obtained results by QPSO are better than those obtained by the conventional PSO. The results show that the proposed QPSO results in savings on the daily cost of HESS. In Ref. [13], the GA technique is proposed for sizing the ESSs in MGs to minimize the operating cost of the MG. A suitably sized ESS is required for the safe operation of the MG. The optimal energy and optimal power capacity of the ESS are obtained by using this technique, but the location of the ESS isn't considered in this technique. EMS is optimized to get the best operation because the output power of the ESS is controlled by EMS. The advantage of this technique uses an aging-based model to predict the life time of the ESS. A battery management system (BMS) is utilized to stop the battery from working in an environment where there is an oversupply of current, over/under voltage, or over/under temperature. The BMS is also used to compute the SOC and report it to the smart energy manager. So, a BMS is utilized in this study to guarantee the safe operation of the ESS. The results show that the operation mode of the MG is influenced by the optimal size of ESS. Also, the integration of the optimal size of the ESS decreases the operating costs for isolated-mode MGs and also, for grid-connected-mode MGs.

In Ref. [15], PSO which depends on frequency control is introduced for sizing the ESS in the MG to minimize the total ESS cost and prevent instability and voltage collapse in the MG when there is a loss of grid-provided power. In grid-connected mode, the grid is responsible for regulating the stability of the system, but in isolated mode the ESS is utilized to regulate system stability. When AC MG frequency is less than 50 or 60 Hz, ESS is discharged to increase the system frequency. Also, when MG frequency is greater than 50 or 60 Hz, ESS is used to reduce the system frequency. So, ESS is utilized to support the MG's stability. Thus, an isolated mode of operation of the MG must be implemented to test this method. A three-phase fault occurs in the study in which the MG is isolated from the grid. In this method, it is assumed that the time which needs to clear the three-phase fault, return the system to operating mode and connect it to the grid is 1–2 hours and the ESS can supply the load of an MG for about 5 hours. Moreover, determining the optimal size depends on PSO, analytic methods, and conventional methods which are compared in Ref. [15]. The results show that PSO determined the best system frequency and optimal capacity of the ESS. As a result, it was determined that integrating polysulfide-bromine energy storage systems is cheaper than integrating a vanadium redox energy storage system (VRESS). The influence of ESS and a capacitor bank (reactive power compensator) on power losses and voltage profile is presented in Ref. [38]. In Ref. [38], the optimal location of RPC and ESS, which are presented to minimize the power losses and improve the voltage profile of the MG in grid-connected mode, is obtained as an optimal power-flow method. Uncertainty of PV installation is considered by using the Monte Carlo technique to obtain the best result. The effect of penetration level of RPCs and ESS is discussed. The results show that the power losses of the system decrease from 174.9 to 158.9 kW (9.32%) when RPC is integrated into the system. Also, the power losses of the system decrease from 174.9 to 154.9 kW (11.44%) when an ESS is integrated into the system. As a result, the power losses of the system decrease from 174.9 to 137.2 kW (21.55%), when both RPS and ESS are integrated into the system. In Ref. [41], PSO dependent on the dynamic demand response (DDR) is presented in order to obtain the optimum sizing of BESS to preserve the MG's stability, develop the voltage collapse, and minimize the total cost of MG operation at the isolated MG operation. DDR is discussed in this study as a time domain reacting to load cooperation or frequency load shedding. DDR can be utilized to increase MG stability, reliability, and flexibility. In this work, the influence of BESS costs with redox flow BESS and lead acid BESS is studied. Peak load can be reduced and the power output of an MG can be optimized by using PSO with DDR. BESS decreases frequency and voltage defections due to an outage of utility. DDR is modulated in the load due to the operation of the MG. In Ref. [31], two-stage optimization is suggested to get the optimal capacity of PV, WT, and ESS to minimize the total cost and CO_2 emissions in grid-connected mode. In the first stage, the optimal capacity of PV and WT is obtained, while in the second stage, the optimal capacity of the ESS is obtained. The reliability, power failure of PV and WT, use factor of battery, and total cost are considered in this study. The results show that CO_2 emissions decrease by 70% when compared with an MG without renewable energy and ESS. Also, this solution is the best for reducing total cost and CO_2 emissions together. In Ref. [42], PSO is utilized to get the optimal sizing of distributed generators and ESS when considering load classification, the uncertainty of renewable energy, and characteristics of ESS to minimize total cost. ESS, which is used in this study, consists of two layers: HESS with three storage devices (super capacitor, Li-ion battery, and LAB) to obtain the best performance at the most economic cost. The super capacitor (high-power density), which is used to supply the important load, can be used to supply power in the event of an unexpected power failure or short-term power variation. The storage batteries (high-energy density), which are used to supply normal load, are used for discharge at the long time. To verify the optimization technique in this work, an MG model with a 20-year lifetime is utilized. This MG consists of a wind turbine, PV, combined heat and power generation (CHP), ESS, and an EV. In this study, three cases are discussed to show the influence of load classification (important load or normal load) and demand side management (DSM) on the optimal size of DG and ESS and therefore the total cost. The first case obtains the optimal sizing of DG and ESS without load classification and DSM, while the second case obtains the optimal sizing of DG and ESS with load classification and without DSM. The third case obtains the optimal sizing of DG and ESS with load classification and DSM. The optimal size of DG in the first case is the same as the optimal size of DG in the second case because the daily load in the first case is the same as the daily load in the second case. But, the optimal size of the HESS in the second case differs from the first case due to load classification, which also results in a reduction of the total operating cost and HESS cost. In the third case, the EVs were involved dispatching of the MG. The daily load differs from the first and second cases due to the application of DSM. In Ref. [53], an optimal depth of discharge (DOD) and size of ESS are presented by a Matlab simulation to enhance the economy. The output change of energy sources and load varying are taken into account in this algorithm. The influence of battery lifetime and energy shortage in the TOC of the MG are considered in this optimization. The range that constrains the battery operation between minimum and maximum DOD is called depth of discharge range (DODR) [47]. DODR is used to avoid overcharge and overdischarge. Many factors affect battery life, such as temperature, DOD, and discharge rate [48,49]. The results show that the DODR and size of the battery affect the total operating cost of the MG. In Ref. [44], the optimal capacity and location of ESSs are presented as a means of minimizing the TOC and enhancing the reliability of an MG. A DRP is utilized to achieve the target of this study and to shift the load from peak periods to off-peak periods to make the load curve flat. Two cases are introduced, and the results are compared to evaluate the effect of DRP on the total operating cost and reliability of MG. The results of this study show that the total operating cost of the MG is reduced when considering DRP.

20.8 Conclusions and future trends

ESSs have considerable influences on grid stability, security, power quality, peak load serving, control, total operation cost, power losses, and load change. This chapter presents and reviews papers concerning the impact of ESS on MG operation. Both the technical and economic advantages of various ESSs have been reviewed. The details of the objective function and constraints in each reviewed paper are discussed and compared to other ones. In the selection of ESS type and capacity, various elements should be considered concurrently, including objective function, constraints, lifetime, capital cost, load change, and depth of charge. In general, the total operational cost of the MG is reduced when integrated with an ESS. Future trends need to consider what influences the optimal number, location, and size of ESS on the MG operation. Additional factors that should also be considered are load management, number of charging/discharging cycles, SOC, and voltage and frequency regulation.

Abbreviations

ESS energy storage system
MG microgrid
PV photovoltaic
WT wind turbine
SMES superconductor magnetic energy storage
FES flywheel energy storage
PHS pumped hydro storage
H_2S hydrogen energy storage
TES thermal energy storage
PBESS polysulfide-bromine energy storage system
LA lead acid
Li-ion lithium-ion
Ah ampere hour
TOC total operating cost of MG
CO_2 carbon dioxide
HSA harmony search algorithm
TOU time of use
BDA benders decomposition algorithm
GA genetic algorithm
MIP mixed-integer programming
GWO gray wolf optimization
HESS hybrid energy storage system
LPSP loss of power supply probability
LOLE loss of load expectation
DG distributed generation
DER distributed energy resources
RES renewable energy source
DLCS double layer capacitor storage
CAES compressed air energy storage
NaS sodium-sulfur
ZnBr zinc bromide
PSB polysulfide bromide battery

VRB vanadium redox
VRESS vanadium redox energy storage system
Ni–Cd nickel-cadmium battery
UC ultra-capacitor
PCC point of common coupling
EV electric vehicle
SOC state of charge
PSO particle swarm optimization
GAMS general algebraic modeling system
NPV net present value
BAT battery
OC operating cost
QPSO quantum-behaved particle swarm optimization
SPSP surplus of power supply probability
DODR depth of discharge range
TDG raditional distributed generation

Nomenclature

C_{vm} variable cost of ESS
OC_{WO} operating cost without ESS
C_{cap} capital cost
C_{SC} start-up cost
C_R earning cost
C_p cost of ESS per kWh
P_{cbat} power capacity of ESS
$s = 1, 2, 3, \ldots M$ number of scenarios
$h = 1, 2, 3, \ldots M$ number of hours
P_s probability of scenarios
F_i fuel cost of unit i
SDC_{iph} shutdown cost of unit i, at hour h, at period p
P_{grid} power imported from/exported to the grid
P_L^t load demand during time t
P_{wt}^t output power of wind turbine during period t
P_c charge power of ESS
LL lost load
OMC_i operating and maintenance cost
C_{grid} cost of power imported from/exported to the grid
P_{MT}^t output power of micro-turbine at time t
R^t spinning reserve
C_O operating cost
f_O The operation coefficient
f_M the maintenance coefficient
f_P the depreciation coefficient
P_{uc}^t output power of ultra-capacitor at time t
P_{waste} power waste
MSC_D the mean daily schedule cost
L_i lifetime of unit
a_i, b_i, c_i, **and** K_{OM} Parameter cost of generator source
MUT minimum up-time
F nominal frequency
r_{ij} resistance of branch ij
Q_{RPC} shunt capacitor

RU	ramping-up limit
RD	ramping-down limit
DT	down time of unit I
C_l	energy shortage cost
C_t	total cost
N_i	number of units of DG or ESS
$N_{p,i}$	emission amount of each pollutant gas
$C_{p,i}$	treatment cost
$C_{p,i}$	treatment cost
P_r	rated output power
P^t_{DRP}	active load with DRP
w^s_{th}	binary factor for load curtailment
P_{LC}	load curtailment power
T^{off}_i	off-time of generator *i*
C_{fm}	fixed cost of ESS
OC_W	operating cost with ESS
C_{RC}	replacement cost
C_{OM}	operating and maintenance cost
C_{FC}	fuel cost
C_E	cost of ESS per kWh
E_{cbat}	energy capacity of ESS
$p = 1, 2, 3, \ldots M$	number of periods
$i = 1, 2, 3, \ldots N$	number of units
MD(p)	number of days in a period
SUC_{iph}	shutdown cost of unit *i*, at hour *h*, at period *p*
EP_{ph}	electricity price at hour h, at period p
P^t_i	output power of unit i at time *t*
P^t_{bat}	output power of ESS during period *t*
LSI_{hp}	load shedding at hour *h*, at period *p*
D	depth of charge
TCPD	total cost per day of energy storage system
C_{DG}	total cost of DG
P_{pv}	output power of PV
P^t_{FC}	output power of fuel cell at time *t*
C_M	maintenance cost
E_{uc}	energy of ultra-capacitor
P_d	discharge power of ESS
η_d	efficiency of discharge
η_c	efficiency of charge
P_{sh}	power shortage
MOC_D	mean daily operating cost of MG
r	interest rate
U^t_i	state of unit *i* at time *t*
BSR^t	reserve provided by the battery
MDT	MINIMUM down time
I_{ij}	current of branch *ij*
V_i	voltage magnitude at bus *i*
β_i	shut-down of unit *i*
α_i	start-up of unit *i*
UP	uptime of unit *i*
C_V	environmental compensation cost
C_r	recycling cost
C_i	cost of unit *i*

K	replacement times of unit
j_{Max}	service life time of microgrid
E_l	energy shortage
P_{PVC}	PV generation curtailment power
P_{gt}	output power of gas turbine
T_i	interrupt time to load
C_{LTD}	lifetime degradation cost
C_{ENS}	energy not supply cost
T_i^{on}	on time of generator i

References

[1] M. Elsied, A. Oukaour, T. Youssef, H. Gualous, O. Mohammed, An advanced real time energy management system for microgrids, Energy 114 (2016) 742–752.

[2] T.R. Nudell, M. Brignone, M. Robba, B. Michela, T.R. Andrea, A. Anuradha, A dynamic market mechanism for combined heat and power microgrid, in: Energy Management, IFAC Congress, Toulouse, France, (1), 2017, pp. 10033–10039.

[3] S. Parhizi, H. Lotfi, A. Khodaei, S. Bahramirad, State of the art in research on microgrids: a review, IEEE Trans. 3 (2015) 890–925.

[4] D. Magdefrau, T. Taufik, M. Poshtan, M. Muscarella, Analysis and review of DC microgrid implementations, in: IEEE International Seminar on Application for Technology of Information and Communication (ISemantic), Semarang, Indonesia, 2016, pp. 241–246.

[5] Q. Shafiee, J.C. Vasquez, J.M. Guerrero, Distributed secondary control for islanded microgrids—a networked control systems approach, in: 38th Annual Conference on IEEE Industrial Electronics Society, 2012, pp. 5637–5642.

[6] X. Zhao-Xia, F. Hong-Wei, Impacts of P–f and Q–V droop control on microgrids transient stability, in: International Conference on Applied Physics and Industrial Engineering, vol. 24, 2012, pp. 276–282.

[7] J. Momoh, G.D. Boswell, Improving power grid efficiency using distributed generation, in: Proceedings of the 7th International Conference on Power System Operation and Planning, January 2007, pp. 11–17.

[8] T. Ackerman, G. Anderson, L. Soder, Distributed generation: a definition, Electr. Power Syst. Res. 57 (2001) 195–204.

[9] H. Zareipour, K. Bhattacharya, C.A. Canizares, Distributed generation: current status and challenges, in: Annual North American Power Symposium (NAPS), Chicago, May 2004, pp. 1–8.

[10] Q. Tabart, I. Vechiu, A. Etxeberria, S. Bacha, Hybrid energy storage system microgrids integration for power quality improvement using four-leg three-level NPC inverter and second-order sliding mode control, IEEE Trans. Ind. Electron. 65 (2018) 424–435.

[11] K. Rahbar, J. Xu, R. Zhang, Real-time energy storage management for renewable integration in microgrid: an off-line optimization approach, IEEE Trans. Smart Grid 6 (1) (2015) 124–134.

[12] D. Parraa, M. Swierczynski, D.I. Stroe, S.A. Norman, A. Abdon, J. Worlitschek, et al., An interdisciplinary review of energy storage for communities: challenges and perspectives, Renew. Sustain. Energy Rev. 79 (2017) 730–749.

[13] Y. Jiang-Feng, W. Wei, P. Yong-Gang, A.S. Jacob, R. Banerjee, P.C. Ghosh, et al., A method for optimal sizing energy storage systems for microgrids, Renew. Energy 212 (2015) 539–549.

[14] A.S.A. Awad, T.H.M. EL-Fouly, M.M.A. Salama, Optimal ESS allocation for load management application, IEEE Trans. 30 (2015) 327–336.

[15] T. Kerdphol, K. Fuji, Y. Mitani, M. Watanabe, Y. Qudaih, Optimization of a battery energy storage system using particle swarm optimization for stand-alone microgrids, Int. J. Electr. Power Energy Syst. 81 (2016) 32–39.

[16] A. Joseph, M. Shahidehpour, Battery storage systems in electric power systems, in: IEEE Power Engineering Society General Meeting Montreal, QC, Canada, 2006.

[17] REN21. Renewables 2018-global status report, Paris, REN21 Secretariate; 2018.

[18] M.F. Zia, E. Elbouchikhi, M. Benbouzid, Microgrids energy management systems: a critical review on methods, solutions, and prospects, Appl. Energy 222 (2018) 1033–1055.

[19] Siemens, How Microgrids Can Achieve Maximum Return on Investment (ROI): The Role of the Advanced Microgrid Controller, Energy Efficiency Markets, LLC, 2016.
[20] C.M. Colson, M.H. Nehrir, A review of challenges to real-time power management of microgrids, in: IEEE Power and Energy Society General Meeting, Calgary, AB, Canada, 2009.
[21] C. Abbey, D. Cornforth, N. Hatziargyriou, K. Hirose, A. Kwasinski, E. Kyriakides, et al., Powering through the storm, IEEE Power Energy Mag. 12 (3) (2014) 67–76.
[22] C. Chen, S. Duan, T. Cai, B. Liu, G. Hu, Optimal allocation and economic analysis of energy storage system in microgrids, IEEE Trans. Power Electron. 26 (10) (2011) 2762–2773.
[23] S. Bahramirad, W. Reder, A. Khodaei, Reliability-constrained optimal sizing of energy storage system in a microgrid, IEEE Trans. Smart Grid 3 (4) (2012) 2056–2062.
[24] I. Strnad, D. Skrlec, An approach to the optimal operation of the microgrid with renewable energy sources and energy storage systems, Eurocon 2013 (July) (2013) 1135–1140.
[25] B. Jeddi, V. Vahidinasab, Optimal operation strategy of distributed generators in a microgrid including energy storage devices, in: Smart Grid Conference (SGC), 2013, pp. 41–46.
[26] N.A. Ashtiani, M. Gholami, G.B. Gharehpetian, Optimal allocation of energy storage systems in connected microgrid to minimize the energy cost, in: 19th Conf. Electr. Power Distrib. Networks (EPDC 2014), 2014, pp. 25–28.
[27] X. Xie, H. Wang, S. Tian, Y. Liu, Optimal capacity configuration of hybrid energy storage for an isolated microgrid based on QPSO algorithm, in: Proc. 5th IEEE Int. Conf. Electr. Util. Deregulation, Restruct. Power Technol. (DRPT 2015), 2016, pp. 2094–2099.
[28] S. Bhattacharjee, A. Bhattacharya, S. Sharma, Grey wolf optimisation for optimal sizing of battery energy storage device to minimise operation cost of microgrid, IET Gener. Transm. Distrib 10 (3) (2016) 625–637.
[29] J. Eyer, G. Corey, S.N. Laboratories, Energy Storage for the Electricity Grid: Benefits and Market Potential Assessment Guide: a Study for the DOE Energy Storage Systems Program, Sandia National Laboratories, 2010.
[30] X. Luo, J. Wang, M. Dooner, J. Clarke, Overview of current development in electrical energy storage technologies and the application potential in power system operation, Appl. Energy 137 (2015) 511–536.
[31] A. Kumar, K. Priyalakshmi, S. Rangnekar, An overview of energy storage and its importance in Indian renewable energy sector Part I—technologies and comparison, J. Energy Storage 13 (2017) 10–23.
[32] P. Stluka, D. Godbole, T. Samad, Energy management for buildings and microgrids, in: 50th IEEE Conference on Decision and Control and European Control Conference (CDC-ECC), 2011, pp. 5150–5157.
[33] Z. Huang, T. Zhu, Y. Gu, D. Irwin, A. Mishra, P. Shenoy, Minimizing electricity costs by sharing energy in sustainable microgrids, in: Proceedings of the 1st ACM Conference on Embedded Systems for Energy-Efficient Buildings, 2014.
[34] D. Zhang, N. Shah, L.G. Papageorgiou, Efficient energy consumption and operation management in a smart building with microgrid, Energy Convers. Manage. 74 (2013) 209–222.
[35] H. Yang, H. Pan, F. Luo, J. Qiu, Y. Deng, M. Lai, et al., Operational planning of electric vehicles for balancing wind power and load fluctuations in a microgrid, IEEE Trans. Sustain. Energy 8 (2) (2016) 592–604.
[36] X. Xu, X. Jin, H. Jia, X. Yu, K. Li, Hierarchical management for integrated community energy systems, Appl. Energy 160 (2015) 231–243.
[37] T. Kerdphol, Y. Qudaih, Y. Mitani, Optimal battery energy storage system using PSO considering dynamic demand response for microgrids, Int. J. Electr. Power Energy Syst. 83 (2016) 58–66.
[38] S. Liu, F. Liu, T. Ding, Z. Bie, Optimal allocation of reactive power compensators and energy storages in microgrids considering uncertainty of photovoltaics, Energy Procedia 103 (2016) 165–170.
[39] A. Karimi, A stochastic approach to optimal sizing of energy storage systems in a microgrid, in: 2016 Smart Grids Conference (SGC), 2016, pp. 20–21.
[40] E. Mortaz, J. Valenzuela, Microgrid energy scheduling using storage from electric vehicles, Electr. Power Syst. Res. 143 (2017) 554–562.
[41] Y.J. Feng, W. Wei, P.Y. Gang, A real-time optimal energy dispatch for microgrid including battery energy storage, in: Ski. 10th Int. Conf. Software, Knowledge, Inf. Manag. Appl., 2017, pp. 314–318.
[42] Z. Liu, Y. Chen, R. Zhuo, H. Jia, Energy storage capacity optimization for autonomy microgrid considering CHP and EV scheduling, Appl. Energy 210 (2018) 1113–1125.
[43] U. Akram, M. Khalid, S. Shafiq, Optimal sizing of a wind/solar/battery hybrid grid-connected microgrid system, IET Renew. Power Gener. 12 (1) (2018) 72–80.

[44] S. Nojavan, M. Majidi, N.N. Esfetanaj, An efficient cost-reliability optimization model for optimal siting and sizing of energy storage system in a microgrid in the presence of responsible load management, Energy 139 (2017) 89–97.
[45] C. Cai, S. Cheng, B. Jiang, W. Dai, M. Wu, J. Zhang, Optimal Operation of microgrid composed of small hydropower and photovoltaic generation with energy storage based on multiple scenarios technique, in: Proc. 5th IEEE Int. Conf. Electr. Util. Deregulation, Restruct. Power Technol. (DRPT 2015), 2016, pp. 2185–2190.
[46] C.A. Correa, G. Marulanda, A. Garces, Optimal microgrid management in the Colombian Energy Market with demand response and energy storage, in: IEEE Power and Energy Society General Meeting (PESGM), 2016, pp. 1–5.
[47] M.P. Fanti, A.M. Mangini, M. Roccotelli, W. Ukovich, District microgrid management integrated with renewable energy sources, energy storage systems and electric vehicles, IFAC 50 (2017) 10015–10020.
[48] X. Jin, J. Wu, Y. Mu, M. Wang, X. Xu, H. Jia, Hierarchical microgrid energy management in an office building, Appl. Energy 208 (2017) 480–494.
[49] H. Farham, H. Alipour, Effects of demand response program and energy storage system on optimal stochastic short-term generation scheduling of grid-connected microgrid, in: Electr. Power Distrib. Networks Conf. (EPDC 2017), 2017, pp. 80–88.
[50] J.A. Lopes, F.J. Soares, P.M. Almeida, Integration of electric vehicles in the electric power system, Proceedings of the IEEE 99 (1) (2011) 168–183.
[51] H. Khodr, J. Martinez-Crespo, M. Matos, J. Pereira, Distribution systems reconfiguration based on OPF using benders decomposition, IEEE Trans. 24 (2009) 2166–2176.
[52] S. Nojavan, H. Qesmati, K. Zare, H. Seyyedi, Large consumer electricity acquisition considering time-of-use rates demand response programs, Arab.J. Sci. Eng. 39 (12) (2014) 8913–8923.
[53] L. Setyawan, J. Xiao, P. Wang, Optimal depth-of-discharge range and capacity settings for battery energy storage in microgrid operation, in: Asian Conference on Energy, Power and Transportation Electrification (ACEPT), 2017, pp. 1–7.

Further reading

Z. Bo, Z. Xuesong, C. Jian, W. Caisheng, G. Li, Operation optimization of standalone microgrids considering lifetime characteristics of battery energy storage system, IEEE Trans. 4 (2013) 934–943.
Q. Hao, Z. Jianhui, L. Jih-Sheng, Y. Wensong, A high-efficiency grid-tie battery energy storage system, IEEE Trans. 26 (2011) 886–896.
J.D. Dogger, B. Roossien, F.D.J. Nieuwenhout, Characterization of Li-ion batteries for intelligent management of distributed gridconnected storage, IEEE Trans. 26 (2011) 256–263.

Index

Note: Page numbers followed by "*f*" and "*t*" refer to figures and tables, respectively.

E

N

O

T

U

V

W

X

Z

Printed in the United States
By Bookmasters